KNOTS
IN HELLAS '98

Proceedings of the International Conference on
KNOT THEORY AND ITS RAMIFICATIONS

K&E Series on Knots and Everything — Vol. 24

KNOTS IN HELLAS '98

Proceedings of the International Conference on KNOT THEORY AND ITS RAMIFICATIONS

European Cultural Centre of Delphi Greece, 7–15 August 1998

Editors

C. McA. Gordon *(University of Texas, USA)*

V. F. R. Jones *(University of California, Berkeley, USA)*

L. H. Kauffman *(University of Illinois, USA)*

S. Lambropoulou *(University of Göttingen, Germany)*

J. H. Przytycki *(George Washington University, USA)*

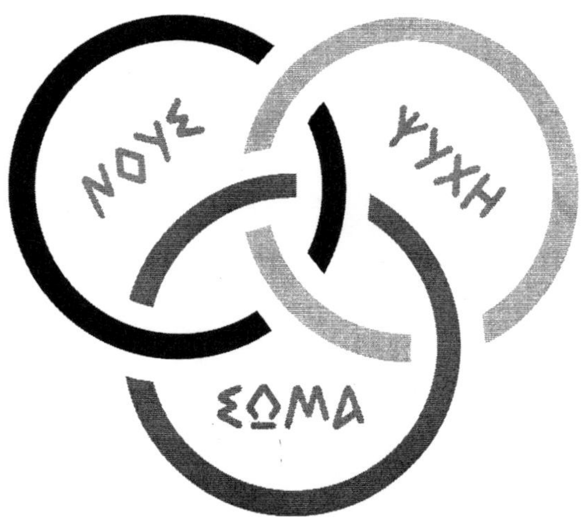

World Scientific
Singapore • New Jersey • London • Hong Kong

Published by

World Scientific Publishing Co. Pte. Ltd.

P O Box 128, Farrer Road, Singapore 912805

USA office: Suite 1B, 1060 Main Street, River Edge, NJ 07661

UK office: 57 Shelton Street, Covent Garden, London WC2H 9HE

British Library Cataloguing-in-Publication Data
A catalogue record for this book is available from the British Library.

KNOTS IN HELLAS '98
Proceedings of the International Conference on Knot Theory and Its Ramifications

ISBN 981-02-4340-5

Printed in Singapore.

Preface

Knot theory is the area of topology that deals with the study and classification of knots, and it is very closely related to the construction and classification of 3-manifolds. Since the ground-breaking discovery of the Jones polynomial, unexpected connections with operator algebras and Hecke algebras, statistical mechanics and quantum field theory have been found, new intriguing invariants of 3-manifolds have been constructed and the long-standing conjectures of Tait have been solved. Meanwhile, there there has been considerable progress in our understanding of the connection between knots and 3-manifolds via Dehn surgery, leading in particular to the solution of Tietze's knot complement problem. By now knot theory is an extremely rich subject, having contact with many branches of mathematics, whilst finding unforeseeable applications in physics (electrical networks, fluid mechanics, energy of knots), in biology (DNA topology, DNA knots and links), in chemistry (synthetic molecules), etc.

The *International Conference in Knot Theory and its Ramifications; Knots in Hellas '98* was held at the European Cultural Centre of Delphi (ECCD), 7–15 August 1998, under the auspices of the Greek Ministry of Education. It was the first conference on the subject to be held in Greece and it was organised in collaboration with the mathematics department of the University of Athens. Its aim was to bring together researchers in all branches of the subject, to provide a forum for the exchange of ideas between experts on the many different aspects of knot theory and to provide an overview of the current state of this rapidly developing field.

There were four series of lectures, which formed the core of the conference, given by: V.F.R. Jones on *planar algebras*, J. Birman on *braids and knots*, L. Kauffman on *virtual knots and finite-type invariants*, and C.McA. Gordon on *surgery on knots*. The themes of the special sessions were: quantum and finite-type invariants of knots and 3-manifolds, braids, surgery on knots, knots and algebra, algebraic topology based on knots, higher dimensional knots and manifolds, and TQFT and invariants of 3-manifolds. In addition, a whole day was devoted to the applications of knot theory to other scientific areas, and an open problem session was held on the last day of the conference.

Cultural events included the presentation of a historical medal by the mayor of Delphi to V.F.R. Jones, a talk on *the mathematical basis of Plato's dialectics* by S. Negrepontis, and a visit to the ancient site of the temple of Apollo. The logo of the conference consists of the Borromean rings, connecting indivisibly the three ancient greek ideals: mind (ΝΟΥΣ), soul (ΨΥΧΗ) and body (ΣΩΜΑ).

More than 100 participants were registered, representing 31 different countries from all over the world: Algeria, Australia, Belgium, Belarus, Bulgaria, Canada, China, Denmark, France, Germany, Greece, Italy, Israel, Japan, South Korea, Mexico, the Netherlands, Poland, Portugal, Rumania, Russia, Saudi Arabia, Spain, Switzerland, Tunisia, Turkey, U.K., U.S.A., Vietnam, Yugoslavia (Serbia), New Zealand. We were particularly pleased at the large participation of people from under-represented countries and groups. For example, there were about 80 graduate students and young researchers and about 40 women researchers.

The conference was financed by the TMR and INCO Euroconferences Programme (Contract No. ERBFMMACT980409), the Greek Ministry of Education, the Greek Ministry of Culture, the University of Athens and the Mathematics Department of the University of the Aegean. We express our gratitude to all these organisations.

We would like to thank: Stylianos Negrepontis of the University of Athens for agreeing to be a co-organiser of the conference and for his great help with many important administrative details; Hugh Morton for compiling the list of open problems, arising out of the problem session, which appears in this volume; Charilaos Aneziris for creating and updating the homepage of the conference, and Kleopatra Kyparissidou for her invaluable help with the running of the conference. Finally, we thank the stuff of the ECCD for their administrative help and for providing a congenial and pleasant atmosphere, and all the participants for making the conference so stimulating and successful.

This volume is a collection of 34 refereed research papers, which are mostly based on lectures given at the conference. In addition to the present proceedings, two special issues of the Journal of Knot Theory and its Ramifications have been dedicated to the conference, containing more papers by participants. We thank all the authors for their many interesting and varied submissions.

February 2000

The editors

C.McA. Gordon
V.F.R. Jones
L.H. Kauffman
S. Lambropoulou
J.H. Przytycki

Contents

TQFT INVARIANTS AT INFINITY FOR THE WHITEHEAD MANIFOLD

H. ABCHIR
Université Hassan II, Ecole Supérieure de Technologie
Route d'El Jadida Km 7, B.P. 8012, 20100 Casablanca, Maroc
abchir@mbox.azure.net

ABSTRACT

We use the Topological Quantum Field Theory derived from the skein theory of the Kauffman bracket to compute TQFT-invariants at infinity for Whitehead's manifold.

Keywords: Whitehead manifold, TQFT, 3-valent graphs, Temperley-Lieb algebra.

Introduction

The Whitehead manifold [10] is an interesting example of an open orientable 3-manifold which is contractible but not homeomorphic to $\mathbb{R}^3$. Recently, L. Funar [4, 5] has observed that topological quantum field theories (TQFT) [9] can be used to define invariants "at infinity" for open 3-manifolds. He considered the Witten-Reshetikhin-Turaev TQFT associated with $SU(2)$, but was able to compute the invariant for the Whitehead manifold only at levels $k = 2$ and $k = 3$. For $k = 3$, the invariant is non-trivial, which reproves the fact that the Whitehead manifold is not homeomorphic to $\mathbb{R}^3$.
The purpose of this paper is to show how to use the techniques of skein theory and the trivalent calculus [7] to compute many more invariants at infinity for Whitehead-like manifolds. Our method gives rise to nice and simple formulas which makes the calculations easier. It allows us to compute the invariant for all the V_p-theories [2] when p is odd or $p \equiv 2 \bmod 4$. Furthermore, it gives an upper bound if $p \equiv 0 \bmod 4$. We will see that the invariant distinguishes the Whitehead manifold from $\mathbb{R}^3$ in most cases.
If $p = 2k + 4$ is even, the TQFT-functor V_p corresponds to the Witten-Reshetikhin-Turaev TQFT for $SU(2)$ at level k. For $p = 8$ and $p = 10$, we find again Funar's result.

We thank the Referee for pointing out to us Hoste and Przytycki's paper [6]. It is shown there that the generic Kauffman bracket skein module of many Whitehead manifolds is not finitely generated, thereby also distinguishing those Whitehead manifolds from $\mathbb{R}^3$. It would be interesting to find a direct link between their result and our computation.

1 Statement of the main result

1.1 Invariants at infinity

In this paper we mean by *int* and *cl* the interior and the closure respectively. Let M be an open 3-manifold. We choose an ascending sequence of compact submanifolds M_i satisfying

$$\forall i,\ M_i \subset int(M_{i+1}),\ \text{and}\ M = \bigcup_i M_i$$

Then, for each positive integer i, the manifold $C_i = cl(M_{i+1} - M_i)$ is an oriented cobordism from ∂M_i to ∂M_{i+1}.
Funar [4, 5] has defined TQFT-invariants at infinity of M, as follows. A TQFT V is a functor (satisfying certain axioms) from a $2+1$-dimensional cobordism category to the category of (complex) vector spaces. The functor V assigns to the cobordism C_i a linear map denoted by $Z_{C_i} : V(\partial M_i) \rightarrow V(\partial M_{i+1})$. We obtain the directed system $(V(\partial M_i), Z_{C_i})$ of vector spaces. We define $V^\infty(M)$ to be simply the inductive limit of that system. It is a topological invariant at infinity associated to the TQFT V and to the open manifold M [4, 5]. To see that one can take an other exhausting sequence N_i [3] and then show that both inductive sequences have the same limit.

Example. If $M = \mathbb{R}^3$ we can write

$$\mathbb{R}^3 = B^3 \cup S^2 \times [1,2] \cup S^2 \times [2,3] \cup \cdots$$

where $S^2 \times [i, i+1]$ is the identity cobordism between two copies of the 2-sphere. It follows from the TQFT axioms that the limit of the associated directed system is $V(S^2) = \mathbb{C}$, because $Z_{S^2 \times [i,i+1]}$ is the identity map. Therefore $V^\infty(\mathbb{R}^3) = \mathbb{C}$.

1.2 Whitehead manifolds

Let K be a framed knot in the solid torus $S^1 \times D^2$. Let $f : S^1 \times D^2 \rightarrow S^1 \times D^2$ be the embedding defined by K. We denote by C the cobordism $C = cl(S^1 \times D^2 \setminus f(S^1 \times D^2))$. Note that C is a cobordism from $S^1 \times S^1$ to itself. We define the manifold W_f to be the following infinite union

$$W_f = D^2 \times S^1 \bigcup_{S^1 \times S^1} C \bigcup_{S^1 \times S^1} C \bigcup_{S^1 \times S^1} C \ldots$$

Then W_f is an open orientable 3-manifold.
If the induced map $f_* : \pi_1(S^1 \times D^2) \to \pi_1(S^1 \times D^2)$ is zero, then the manifold W_f is simply connected and contractible. In this case, we call W_f a *Whitehead manifold.*

1.3 Main theorem

Let K be the oriented knot in $S^1 \times D^2$ with the blackboard framing as in the picture below.

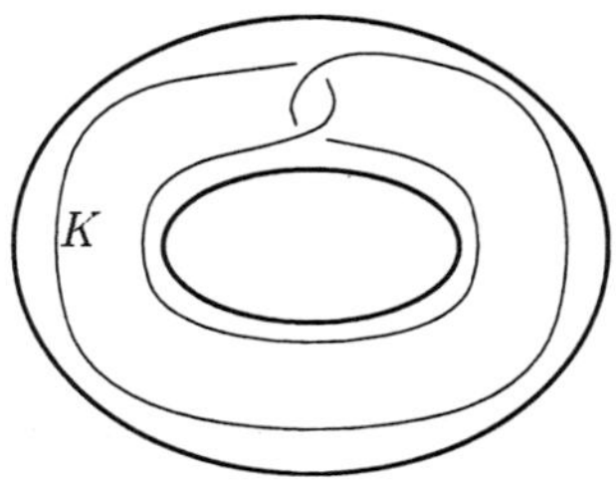

Figure 1: The knot K in the solid torus

Let $f : S^1 \times D^2 \to S^1 \times D^2$ be the embedding defined by K, and let W_f be the associated manifold defined as above. It is easy to see that f is null-homotopic, so that W_f is a Whitehead manifold.

Theorem 1.1 *Let $p \geq 3$ be an integer, and let V_p be the TQFT-functor constructed in [2].*
If $p = 2n + 1$ is odd, $rank(V_p^\infty(W_f)) = n$.
If $p = 2k + 4$ is even, then
- *if k is odd, $rank(V_p^\infty(W_f)) = 1 + [k/2]$,*
- *if k is even , $rank(V_p^\infty(W_f)) \leq 1 + [k/4]$.*

Example.
If $p = 5$ then $V_5^\infty(W_f) \simeq \mathbb{C}^2$ and so W_f is not trivial.

Remark 1.2 If $p = 2k + 4$ is even, the TQFT-functor V_p corresponds to the Witten-Reshetikhin-Turaev TQFT for $SU(2)$ at level k. Using that approach, Funar [4, 5] had obtained the invariant at infinity of the Whitehead manifold for levels $k = 2$ and $k = 3$.

Remark 1.3 One can check that the theorem holds also for the Whitehead manifolds obtained if we change the framing of K, or replace K by its mirror image.

2 Reformulation of the main theorem

2.1 About the TQFT V_p

Let $C_2^{p_1}$ be the cobordism category of smooth closed 2-manifolds with p_1-structure and containing a banded link. The bordisms are thus compact smooth oriented 3-manifolds with p_1-structure containing a banded link (see [1, 2]). (Actually, p_1-structures will not be important for this paper. See remark 2.2 below.) Let p be an integer, $p \geq 3$. The TQFT V_p is a functor from $C_2^{p_1}$ to the category of complex vector spaces satisfying some properties (see [2]), and which assigns to an object Σ of $C_2^{p_1}$ a vector space $V_p(\Sigma)$ and to a cobordism $(M; L)$ a linear map denoted by $Z_{p,(M;L)}$.
One has $V_p(\emptyset) = \mathbb{C}$. If $\partial M = \Sigma$, and L is a banded link in M, then $Z_{p,(M;L)}(1)$ is a vector in $V_p(\Sigma)$ which is denoted by $Z_p(M; L)$. If Σ is connected, then for any fixed M with $\partial M = \Sigma$, the vector space $V_p(\Sigma)$ is generated by the vectors $Z_p(M; L)$.
The assignement $L \mapsto Z_p(M; L)$ satisfies the *Kauffman bracket skein relations* (see the picture below), with Kauffman's skein variable A equal to a fixed $2p$-th root of unity in $\mathbb{C}$.

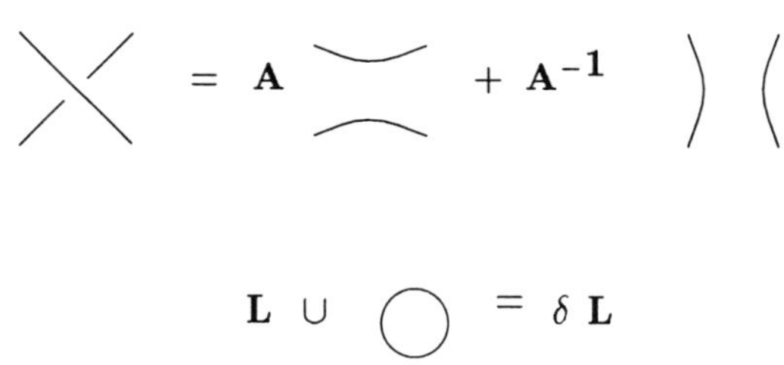

where $\delta = -A^2 - A^{-2}$.
Set $k = n - 1$ if $p = 2n + 2$ is even and $k = 2n - 1$ if $p = 2n + 1$ is odd. We can also consider colored banded links. Here, a color is just an integer j with $0 \leq j \leq k$. Actually, coloring a component of L by the color j means replacing that component with j parallel strands and inserting a Jones-Wenzl idempotent f_j (see below for more details).
The space $V_p(S^1 \times S^1)$ has a basis given by coloring the standardly framed core of the solid torus $D^2 \times S^1$. The color i goes from 0 to k if p is even and from 0 to $k - 1$ and is even if p is odd (see proposition 4.10 in [2]). Using the Kauffman skein algebra $\mathcal{K}(D^2 \times S^1) \simeq \mathbb{C}[z]$ and the Chebyschev polynomials e_i, ($e_0 = 1$, $e_1 = z$ and $ze_i = e_{i+1} + e_{i-1}$), there is an alternative description of that basis given by the vectors $Z_p(D^2 \times S^1; e_i)$ which we denote also, for convenience, by e_i. (See [1, 2] for more details.)

Remark 2.1 If $p = 2n + 1$ is odd, we have the relation $e_{n+i} = e_{n-1-i}$ [1]. It follows that $\{e_0, e_1, e_2, \ldots, e_{n-1}\}$ is also a basis of $V_p(S^1 \times S^1)$.

2.2 New statement

We consider the Whitehead manifold $M = W_f$, where f is given by the framed knot K in the figure above. Then M can be described as the ascending union of compact oriented manifolds M_i, where M_i is just the union of $D^2 \times S^1$ with i copies of C. The cobordism C induces the endomorphism $Z_C : V_p(S^1 \times S^1) \to V_p(S^1 \times S^1)$, which we denote by h. Thus $V_p^\infty(W_f)$ is the limit of the directed system

$$V_p(S^1 \times S^1) \xrightarrow{h} V_p(S^1 \times S^1) \xrightarrow{h} V_p(S^1 \times S^1) \xrightarrow{h} \ldots$$

We will use the above described basis (e_i) of the space $V_p(S^1 \times S^1)$ to compute the operator h, the iterate maps h^r, and so the invariant $V_p^\infty(W_f)$. In fact we note that $h(e_i)$ is the element of $V_p(S^1 \times S^1)$ represented by the framed knot K colored by i, and it suffices to express this in terms of the basis elements e_j.

Note that h here is just a linear map, which has no reason to be a ring homomorphism. (Actually, it will follow from proposition 3.4 that h is *not* a ring homomorphism.)

Remark 2.2 Here we have ignored the choice of p_1-structure on the cobordism, but $h = Z_C$ is independent of this choice up to multiplication by a nonzero scalar. Therefore this choice is irrelevant for our purposes.

Then our main theorem can be stated more precisely as follows

Theorem 2.3 *If $p = 2n + 1$ is odd, then h is an isomorphism, and hence $rank(V_p^\infty(W_f)) = rank(V_p(S^1 \times S^1)) = n$.*
If $p = 2k + 4$ is even (and hence $rank(V_p(S^1 \times S^1)) = k$), then $rank(h) = 1 + [k/2]$. Furthermore,
- *if k is odd, then $Im(h^r) = Im(h)$ for all $r \geq 1$, and hence $rank(V_p^\infty(W_f)) = 1 + [k/2]$.*
- *if k is even, then $rank(h^2) = 1 + [k/4]$, and hence $rank(V_p^\infty(W_f)) \leq 1 + [k/4]$.*

3 Proof

3.1 3-valent graphs

At first we recall some needed ingredients to be found in the paper [7].
Recall that a color is an integer j, $0 \leq j \leq k$. We make the convention that writing i beneath a knot or tangle diagram means that this component has to be replaced by i parallel ones and inserting one little box into each component

which represents an idempotent f_i of the Temperley-Lieb algebra T_i. Recall that A is a $2p$-th root of the unity.
If i is a color, there is a particular element f_i in T_i called Jones-Wenzl idempotent and represented by a box as follows

$f_i =$ 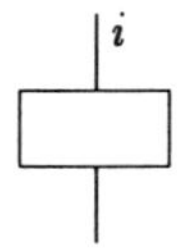

The defining properties of f_i are

(i) $f_i x = x f_i = \varepsilon(x) f_i$, for each $x \in T_i$ and where $\varepsilon(x)$ is the coefficient of 1_i in the decomposition of x in the standard basis of T_i.

(ii) $f_i^2 = f_i$.

Note that the relations (i) and (ii) imply the following graphical identities

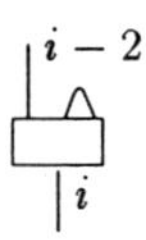

$= 0$ (1) and 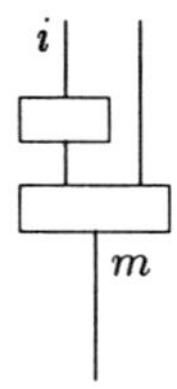

$=$ 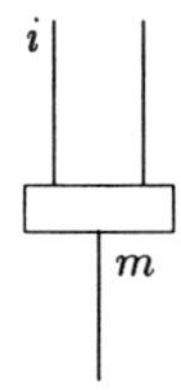

(2)

We shall use the notations,

$$[n] = \frac{A^{2n} - A^{-2n}}{A^2 - A^{-2}},$$

the quantum factorial $[n]! = [1][2]\cdots[n]$, with $[0]! = 1$, and

$$\langle n \rangle = (-1)^n [n+1].$$

A triple (a, b, c) of colors is called p-admissible if $(a+b+c)$ is even, $|\, a - b \,| \leq c \leq a + b$ and $a + b + c \leq 2k$.
Given a p-admissible triple (a, b, c), a trivalent vertex is defined by

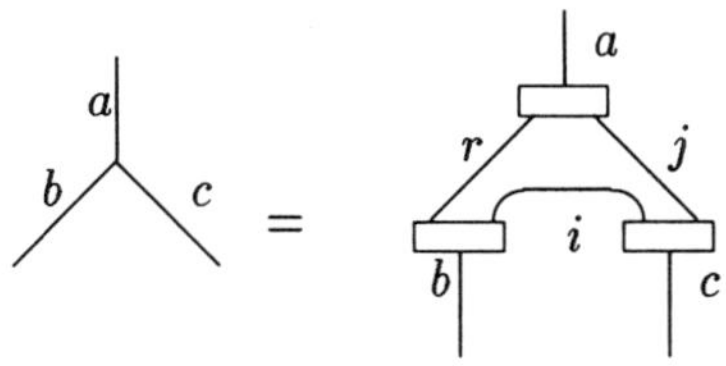

The numbers $i = (b+c-a)/2,\ j = (a+c-b)/2,\ r = (a+b-c)/2$ are called the internal colors. G. Masbaum and P. Vogel had determined the trihedron coefficient, the tetrahedron coefficient and the half-twist coefficient which are respectively

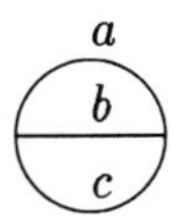

$= \langle a, b, c \rangle$,

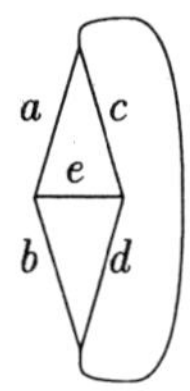

$f = \left\langle \begin{matrix} a & b & e \\ d & c & f \end{matrix} \right\rangle$

and

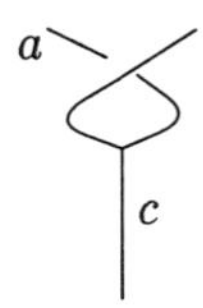

$= \delta(c; a, b)$

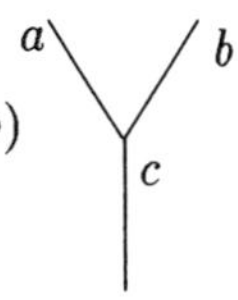

Knowing these 3 coefficients allows one to evaluate every colored link diagram [7].

Remark 3.1 For later use, we note from [7, p. 380] that

$$\delta(c; a, b)^2 = \frac{\mu_c}{\mu_a \mu_b},$$

where $\mu_i = (-1)^i A^{i^2+2i}$ is the twist coefficient.

3.2 Intermediary results

We start this section by recalling that, as mentionned in the remark 2.1, if $p = 2n+1$ is odd, then $\{e_0, e_1, e_2, \cdots, e_{n-1}\}$ is a basis of $V_p(S^1 \times S^1)$.

Proposition 3.2 *If h is the endomorphism of $V_p(S^1 \times S^1)$ induced by the cobordism C, then*

$$h(e_i) = \sum_{\substack{l=0 \\ l \text{ even}}}^{s} \alpha_l^i e_l$$

where

$$\alpha_l^i = \sum_{\substack{m=0 \\ m \text{ even}}}^{s} \delta(m;i,i)^2 \frac{\langle m \rangle}{\langle m,i,i \rangle} \frac{\left\langle \begin{matrix} l & i & i \\ m & i & i \end{matrix} \right\rangle}{\langle l,i,i \rangle}$$

where $s = inf(2i, (2(k-i))$ if $p = 2k+4$ is even and $s = 2i$ if $p = 2n+1$ is odd with $0 \leq i \leq n-1$.

Proof. Remember that we need to show that

$Z_p($ 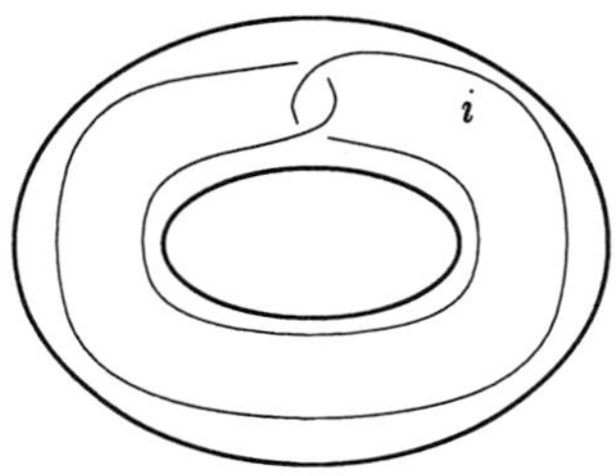

$) = \sum_j \alpha_j^i \, Z_p($ 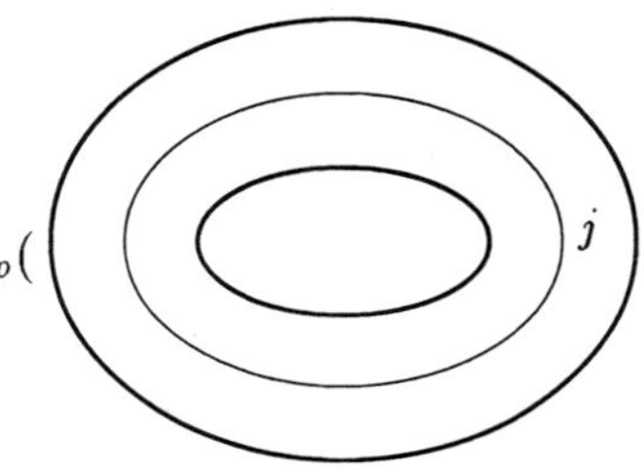

$)$

In other words we need to express K colored by i (that is, K cabled with e_i) in terms of the e_j.

In [7] we have the relation 

$= \sum_m \delta(m;i,i) \frac{\langle m \rangle}{\langle m,i,i \rangle}$ 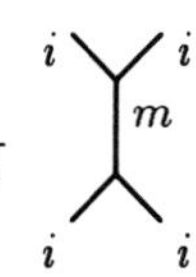

where (m,i,i) is a p-admissible triple. By the multiplication of tangles we get

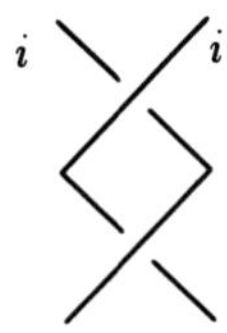

$$= \sum_m \sum_h \delta(m;i,i)\delta(h;i,i)\frac{\langle m\rangle\langle h\rangle}{\langle m,i,i\rangle\langle h,i,i\rangle}$$

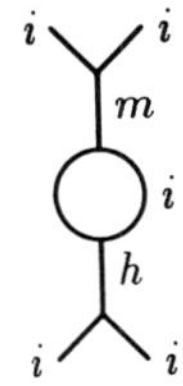

Hence by using the following relations

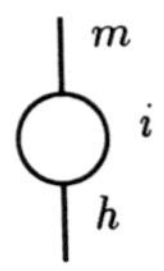

$$= \delta_h^m \frac{\langle i,i,h\rangle}{\langle h\rangle}$$

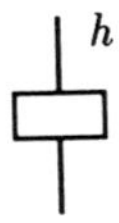

$$= \frac{\langle i,i,m\rangle}{\langle m\rangle}$$

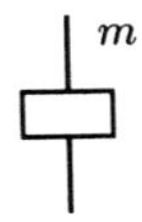

the twist writes

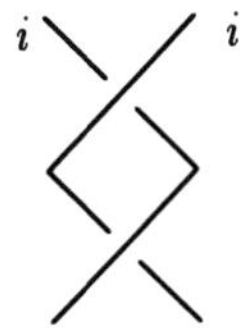

$$= \delta^2(m;i,i)\frac{\langle m\rangle}{\langle m,i,i\rangle}$$

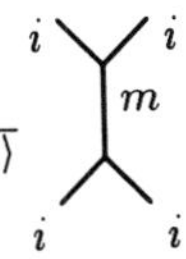

well we know also that

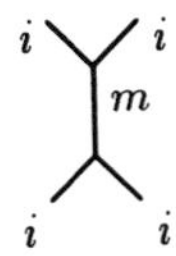

$$= \sum_l \left\{ \begin{array}{ccc} i & i & l \\ i & i & m \end{array} \right\}$$

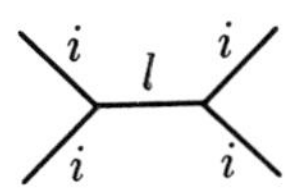

where $\left\{ \begin{array}{ccc} i & i & l \\ i & i & m \end{array} \right\}$ is the 6-j-symbol (see[7]). One has

$$\left\{ \begin{array}{ccc} i & i & l \\ i & i & m \end{array} \right\} = \frac{\langle l\rangle}{\langle l,i,i\rangle} \frac{\left\langle \begin{array}{ccc} l & i & i \\ m & i & i \end{array} \right\rangle}{\langle l,i,i\rangle}$$

Finally the identity

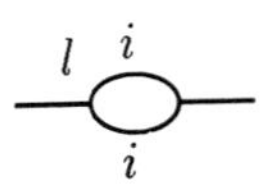

$$= \frac{\langle l,i,i\rangle}{\langle l\rangle}$$

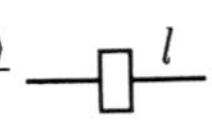

leads to the expected expression.

The ranges of summation are determined by the conditions of p-admissibility on the triples (m,i,i) and (l,i,i) which are the following:
m and l are always even, but $0 \leq m,l \leq inf(2i,(2k-i))$ if $p = 2k+4$ is even and $0 \leq m,l \leq 2i$ if $p = 2n+1$ is odd (in this case remember that we assume $0 \leq i \leq n-1$). □

Proposition 3.3 *If p is even and $p = 2k+4$ then for each color i, $0 \le i \le [\frac{k}{2}]$*

$$h(e_{k-i}) = h(e_i)$$

Proof.
If $p = 2k+4$, let i be an integer such that $0 \le i \le [\frac{k}{2}]$. By proposition 3.2,

$$h(e_i) = \sum_{\substack{l=0 \\ l \text{ even}}}^{2i} \alpha_l^i e_l$$

For the remainder of the proof we will denote by $\bar{j}$ the integer $k-j$, for $0 \le j \le [\frac{k}{2}]$.
Let

$$\beta_{l,m,i} = \delta(m;i,i)^2 \frac{\langle m \rangle}{\langle m,i,i \rangle} \frac{\left\langle \begin{matrix} l & i & i \\ m & i & i \end{matrix} \right\rangle}{\langle l,i,i \rangle} \tag{1}$$

then

$$\alpha_l^i = \sum_{\substack{m=0 \\ m \text{ even}}}^{2i} \beta_{l,m,i}$$

To establish the equality $h(e_{k-i}) = h(e_i)$ we will show that $\beta_{l,m,i} = \beta_{l,m,\bar{i}}$ by comparing their factors.
At first we notice that $\delta^2(m;i,i) = (-1)^k \delta^2(m;\bar{i},\bar{i})$. This can be checked using remark 3.1 (recall $A^{2k+4} = -1$).
Next we assert that there is relation between the two tetrahedron coefficients

$$\left\langle \begin{matrix} l & i & i \\ m & i & i \end{matrix} \right\rangle \quad \textit{and} \quad \left\langle \begin{matrix} l & \bar{i} & \bar{i} \\ m & \bar{i} & \bar{i} \end{matrix} \right\rangle$$

Indeed, if we burst the four vertices of the first of them by the rule

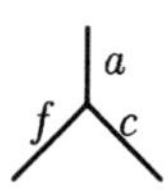

$$= \frac{\langle a\, f\, c \rangle}{\left\langle \begin{matrix} a & b & e \\ d & c & f \end{matrix} \right\rangle}$$

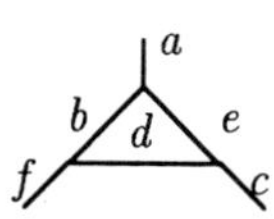

we obtain that

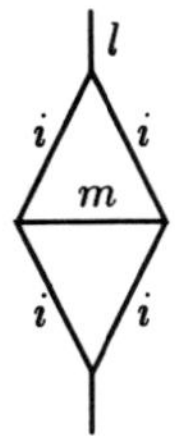

$$= \frac{\langle l,i,i\rangle^2\langle m,i,i\rangle^2}{\left\langle \begin{matrix} l & \bar{i} & \bar{i} \\ k & i & i\end{matrix}\right\rangle^2 \left\langle \begin{matrix} m & \bar{i} & \bar{i} \\ k & i & i\end{matrix}\right\rangle^2}$$

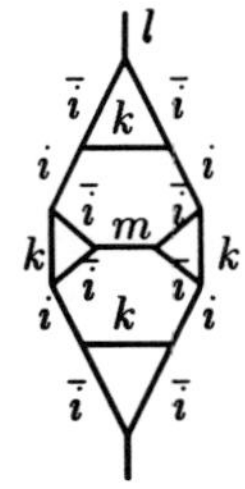

Next, it is not very hard to see that

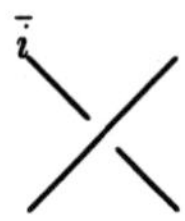

$$= \delta(i;\bar{i},k)\frac{\langle i\rangle}{\langle \bar{i},k,i\rangle}$$

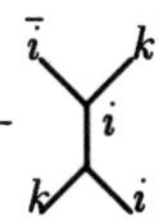

and

$$= \overline{\delta(i;k,\bar{i})}\frac{\langle i\rangle}{\langle k,\bar{i},i\rangle}$$

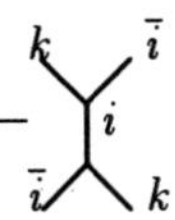

which allows us to write that

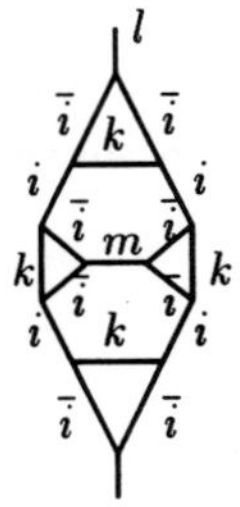

$$= \frac{\langle k,\bar{i},i\rangle^4}{\delta^2(i;\bar{i},k)\overline{\delta^2(i;k,\bar{i})}\langle i\rangle^4}$$

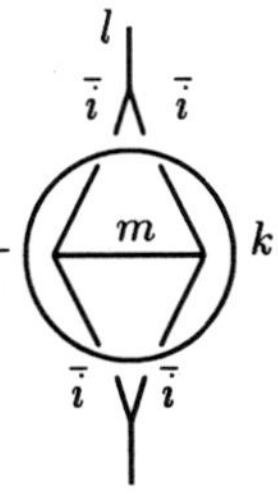

$$= \frac{\langle k, \bar{i}, i\rangle^4}{\delta^2(i;\bar{i},k)\delta^2(i;k,\bar{i})\langle i\rangle^4}\langle k\rangle$$ 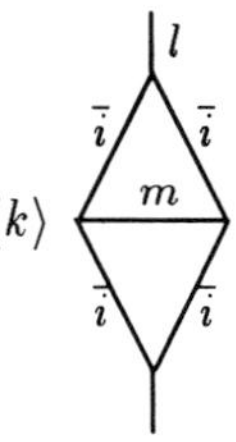 $$= (-1)^k \frac{1}{\langle i\rangle^4}$$ 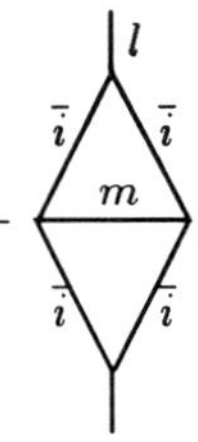

(We have used remark 3.1 and that $\langle k, \bar{i}, i\rangle = \langle k\rangle = (-1)^k$.)
Then we get the following relation between the considered tetrahedron coefficients

$$\left\langle \begin{array}{ccc} l & i & i \\ m & i & i \end{array} \right\rangle = \frac{\langle l,i,i\rangle^2\langle m,i,i\rangle^2}{\left\langle \begin{array}{ccc} l & \bar{i} & \bar{i} \\ k & i & i \end{array} \right\rangle^2 \left\langle \begin{array}{ccc} m & \bar{i} & \bar{i} \\ k & i & i \end{array} \right\rangle^2}(-1)^k \frac{1}{\langle i\rangle^4}\left\langle \begin{array}{ccc} l & \bar{i} & \bar{i} \\ m & \bar{i} & \bar{i} \end{array} \right\rangle$$

From theorem 1 in [7] we deduce the following equalities

$$\langle m,i,i\rangle = (-1)^{i+\frac{m}{2}}([i+\frac{m}{2}+1]![\frac{m}{2}]!^2[i-\frac{m}{2}]!)/([i]!^2[m]!)$$

and

$$\langle l,i,i\rangle = (-1)^{i+\frac{l}{2}}([i+\frac{l}{2}+1]![\frac{l}{2}]!^2[i-\frac{l}{2}]!)/([i]!^2[l]!)$$

Moreover theorem 2 in [7] and the conditions of p-admissibility give the identities:

$$\left\langle \begin{array}{ccc} l & \bar{i} & \bar{i} \\ k & i & i \end{array} \right\rangle = (-1)^k \frac{[\frac{l}{2}]!^2}{[l]!} \quad \textit{and} \quad \left\langle \begin{array}{ccc} m & \bar{i} & \bar{i} \\ k & i & i \end{array} \right\rangle = (-1)^k \frac{[\frac{m}{2}]!^2}{[m]!}$$

By inserting the equalities established above in the expression of $\beta_{l,m,i}$ we obtain that

$$\beta_{l,m,i} = \delta^2(m;\bar{i},\bar{i})\langle m\rangle \frac{\langle m,i,i\rangle\langle l,i,i\rangle[l]!^2[m]!^2}{\frac{l}{2}!^4[\frac{m}{2}]!^4\langle i\rangle^4}\left\langle \begin{array}{ccc} l & \bar{i} & \bar{i} \\ m & \bar{i} & \bar{i} \end{array} \right\rangle$$

Now for each integer j, $0 \le j \le [k/2]$, one can easily check the following formulas $\bar{j} = [j+2]$, $[j]![\bar{j}]! = \frac{[k+1]!}{[j+1]}$ and $\bar{j}+1 = [j+1]$. By using them we show that

$$\frac{\langle m,i,i\rangle\langle l,i,i\rangle[l]!^2[m]!^2}{\frac{l}{2}!^4[\frac{m}{2}]!^4\langle i\rangle^4} = 1/(\langle l,\bar{i},\bar{i}\rangle\langle m,\bar{i},\bar{i}\rangle)$$

Hence $\beta_{l,m,i} = \beta_{l,m,\bar{i}}$ and the proposition follows. □

Proposition 3.4 *(Compare to lemma 14 in [6]) For each integer i, $0 \le i \le [\frac{k}{2}]$ if $p = 2k+4$ and $0 \le i \le n-1$ if $p = 2n+1$, the coordinate α^i_{2i} of the vector $h(e_i)$ is given by*

$$\alpha^i_{2i} = \prod_{j=1}^{i}(1 - A^{-4j})$$

hence, $\alpha^i_{2i} \neq 0$.

Proof. Let T_{2i} be the Temperley-Lieb algebra on $2i$ strings. Let $\mathcal{B}$ be the Jones-Kauffman module of the solid torus $S^1 \times D^2$. There is a "trace map" $tr : T_{2i} \to \mathcal{B}$ which maps a tangle $x \in T_{2i}$ to the closure $\widehat{x}$ of x obtained by identifying the upper and the lower edges of the square. If x_i is the following colored tangle

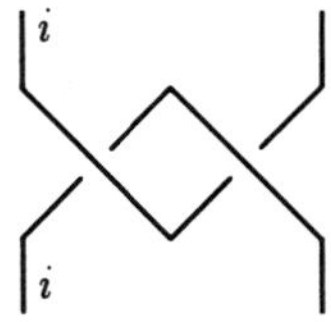

Then $h(e_i) = tr(x_i)$.
Let f_{2i} be the Jones-Wenzl idempotent of T_{2i}. We know that for each $x \in T_{2i}$ we have the relation

$$x f_{2i} = \varepsilon(x) f_{2i}$$

where $\varepsilon(x)$ is the coefficient of the vector 1_{2i} in the decomposition of x in the standard basis of T_{2i}.
The main remark here is that the wanted coefficient α_{2i}^i is exactly $\varepsilon(x_i)$ which we will get by calculating the product $x_i f_{2i} = f_{2i} x_i f_{2i}$ corresponding to the following graph

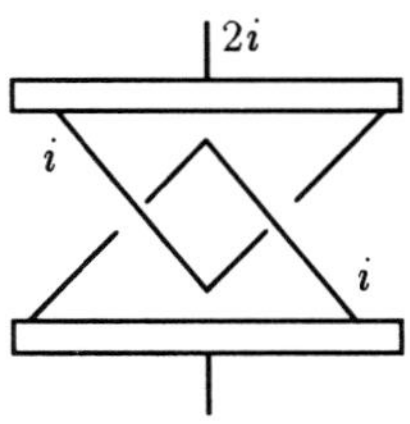

We will show that $x_i f_{2i} = \prod_{j=1}^{i}(1 - A^{-4j}) f_{2i}$ by induction on the index i.
If $i = 0$ it follows from Proposition 3.3 that $\alpha_0^0 = 1$.
If $i = 1$, then the Kauffman relations and the defining properties of the idempotent f_2 immediately give $\alpha_2^1 = 1 - A^{-4}$.
Now assume inductively that $\alpha_{2(i-1)}^{i-1} = (1 - A^{-4(i-1)}) \ldots (1 - A^{-4})$.
At first we consider the following diagram

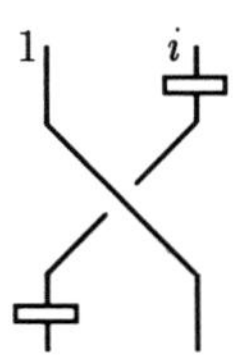

It includes i crossings. If we apply the Kauffmann relations at each of them, then the following equation

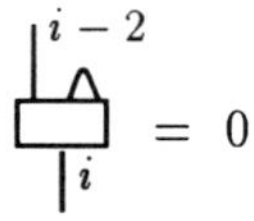

(1)

leads to the equality

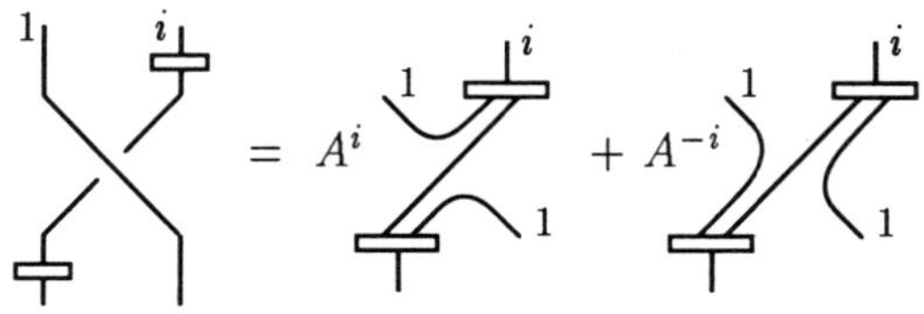

(There are $i-1$ strands connecting the two boxes in each term on the right hand side of this equation. We make the convention that we may omit writing the number of strings if it can easily computed from the rest of the diagram.) Now we note that

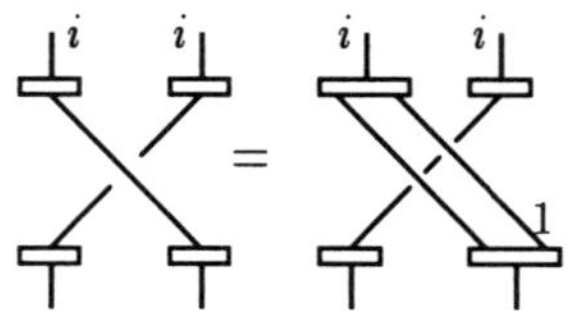

so we can apply the last but one equality and then we obtain

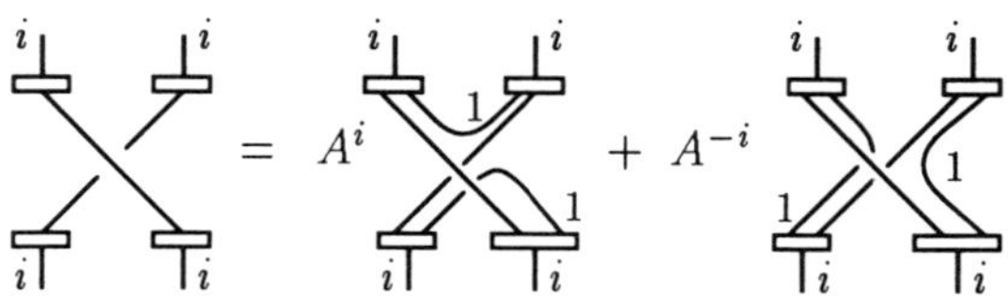

Now by applying the relation

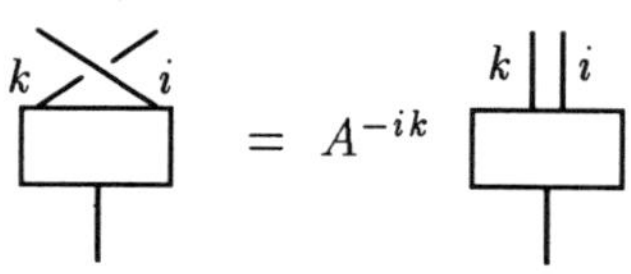

(3)

we obtain that

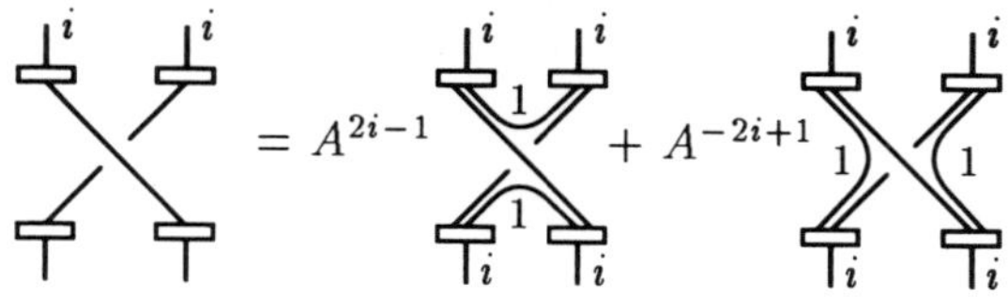

If we glue the remaining crossing in the full twist with the needed boxes inserted, then the last identity becomes

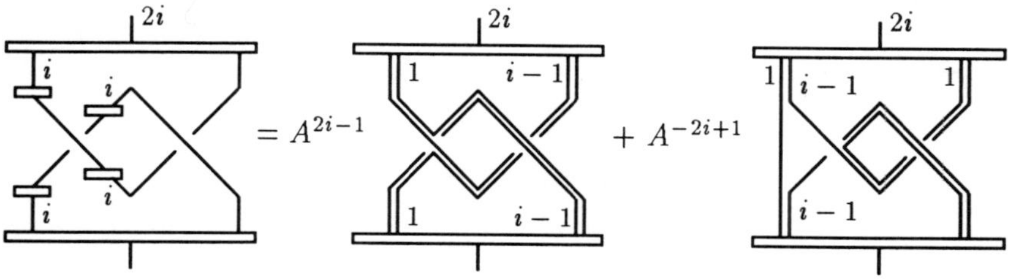

Finally if we use the equations (2) (see page 6), (1) (see page 13) and (3) (see page 14) we obtain that $f_{2i}x_if_{2i}$ writes

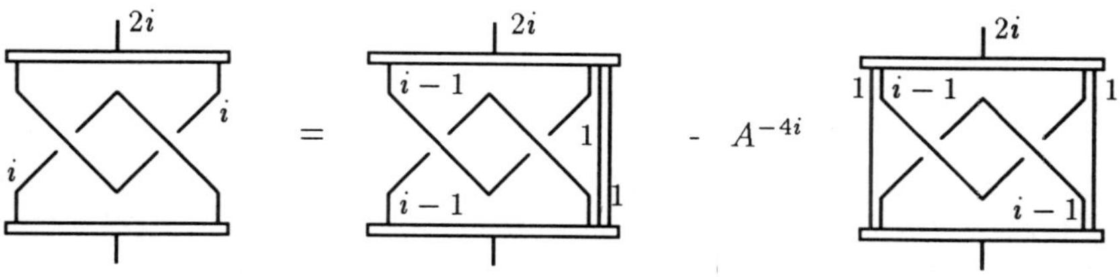

The result for α_{2i}^i now follows from the induction hypothesis.
The coefficient α_{2i}^i does not vanish because the Kauffman bracket is evaluated at the $2p$-th root of the unity. □

3.3 Proof of the main theorem

<u>**Case 1**</u> : If $p = 2n + 1$ is odd, we have seen that

$$h(e_i) = \sum_{\substack{l=0 \\ l \text{ even}}}^{2i} \alpha_l^i e_l$$

for each i, $0 \leq i \leq n-1$, so we get the array

$$\begin{array}{lclclclcl} h(e_0) & = & e_0 & & & & & & \\ h(e_1) & = & \alpha_0^1 e_0 & + & \alpha_2^1 e_2 & & & & \\ \vdots & & \vdots & & \vdots & & & & \\ h(e_{n-1}) & = & \alpha_0^{n-1} e_0 & + & \alpha_2^{n-1} e_2 & + & \cdots & + & \alpha_{2n-2}^{n-1} e_{2n-2} \end{array}$$

By the proposition 3.4, each α_{2i}^i is non zero and hence each e_{2i} is in the image of h. Therefore h is onto and so it is an isomorphism. □

<u>**Case 2**</u> : If $p = 2k + 4$ is even, we have for $0 \leq i \leq [k/2]$

$$h(e_i) = h(e_{k-i}) = \sum_{\substack{l=0 \\ l \; even}}^{2i} \alpha_l^i e_l$$

with $\alpha_{2i}^i \neq 0$. Therefore $Im(h) = Span\{e_{2i} \mid 0 \leq i \leq [k/2]\}$ and then $rank(h) = 1 + [k/2]$. Since $h(e_i) = h(e_{k-i})$, we have that $Span\{e_i - e_{k-i} \mid 0 \leq i \leq [k/2]\}$ is contained in $Ker\, h$. Counting dimensions, we see that those two subspaces are equal.
Furthermore one has

$$Im(h^2) \approx Im(h)/(Im(h) \cap Ker\, h)$$

• **If k is odd**, $Im(h) \cap Ker\, h = 0$, hence $Im(h^2) = Im(h)$ and then $V_p^\infty(W_f) = Im(h^\infty) = Im(h)$.
• **If k is even**, $Im(h) \cap Ker\, h = Span\{e_{2i} - e_{k-2i} \mid 0 \leq i \leq [k/2]\}$. Counting dimensions it follows that $rank(h^2) = 1 + [k/4]$. □

I wish to thank G. Masbaum who proposed me this problem, for his suggestions and for many discussions we had on this subject. I am grateful to P. Vogel for his helpful comments. This work started when G. Masbaum visited the university of Casablanca in Morocco.

References

[1] C. Blanchet, N. Habegger, G. Masbaum and P. Vogel, *Three-Manifold Invariants Derived From The Kauffman Bracket*, Topology 31, No. 4 (1992) 685-699.

[2] C. Blanchet, N. Habegger, G. Masbaum and P. Vogel, *Topological Quantum Field Theories Derived From The Kauffman Bracket*, Topology 34, No. 4 (1995), 883-927.

[3] E. M. Brown, *Unknotted solid tori and genus one Whitehead manifolds*, Trans. of the Ameriacan mathematical society, Vol 333, number 2 (1992).

[4] L. Funar, *TQFT and Whitehead's manifold*, J. Knot Theory and its Ramifications 6, No. 1 (1997), 13-30.

[5] L. Funar, *TQFT for general Lie algebras and applications to open 3-manifolds*, J. Math. Sciences, Tokyo Univ., 7 (1997), 121-181.

[6] J. Hoste and J. H. Przytycki, *The* $(2, \infty)$*-skein module of Whitehead manifolds*, J. Knot Theory and its Ramifications 4, No. 3 (1995), 411-427.

[7] G. Masbaum, P. Vogel, 3*-valent graphs and the Kauffman bracket*, Pacific J. Math., Vol. 164, No. 2 (1994), 361-381.

[8] D. Rolfsen, *Knots and Links*, Publish or Perish (1976), Inc, Washington.

[9] V. Turaev, *Quantum invariants and 3-manifolds*, de Gruyter Studies in Math., Vol. 18 (1994), Berlin, New-York.

[10] J. H. C. Whitehead, *A certain open 3-manifolds whose group is unity*, Quart. Journal Math. 6 (1935), 268-279.

GRIDS AS UNKNOTTING OPERATIONS

NIKOS ASKITAS

ABSTRACT. We show that certain "grid" local moves are unknotting operations. For that we use and extend ideas involved in [A2]. We find basic estimates for the knot invariants these unknotting operations define working as in [A2].

1. Introduction

In [A2] we defined what we called "multi-# local moves" and we showed them to be unknotting operations. Among others we proved that certain grid moves ($2 \times n$ for all n) are also unknotting operations. Let us for the convenience of the reader recall the definition of these grid moves.

Definition 1.1. *A grid local move of type $(d_1, r_1) \times (d_2, r_2)$ will be a local move as in the picture below (and its inverse). In the case $r_1 = r_2 = 0$ the move is simply called a grid move of type $d_1 \times d_2$.*

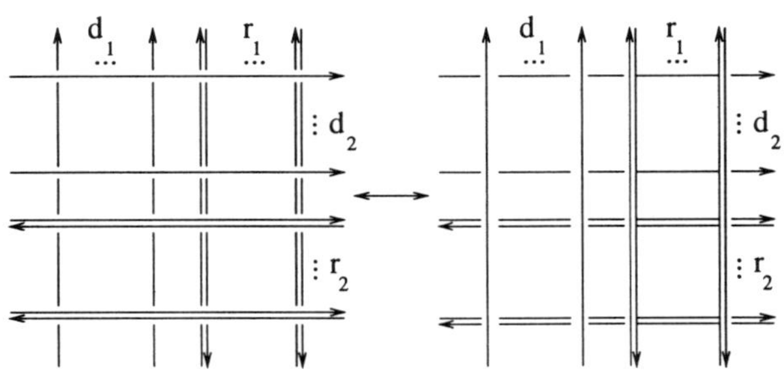

FIGURE 1

Here we will show that for $d = 3, 4$, a $d \times m$ grid is an unknotting operation for all $m \geq 1$, (modulo the fact that a $4 \times 2m$ grid has to preserve the Arf invariant). As in [A2] we will draw our proofs so, in order to simplify them, and (unfortunately) at the expense of making it a bit more painful to read them we need to develop some notation first. Specifically we will be dealing with bundles of strings so we need to abbreviate the way we draw them. The figure below contains an explanation of our notation. It basically makes a conventions on how to deal with crossings when a single drawn strand represents more than one strings. A string with a little vertical

Key words and phrases. 4-manifolds, knots, unknotting numbers, immersions of 2-spheres in 4-manifolds.

Acknowledgements: The author was partially supported by EUTMR Fellowship No ERBFMBICT982870. He would like to thank Prof. G. Stamou for his help during his year at the AUTH and the people at the MPI for their hospitality during the same period.

line decorated with a pair of integers (p, n) will stand for a bundle of p strings lying on a band which is twisted n half times.

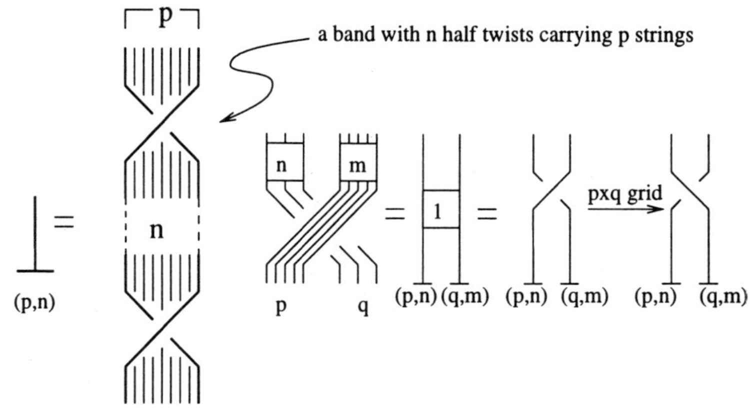

FIGURE 2

The figure below is drawn already in the notation just introduced. We will refer to it as the **key sequence**. It is a commutative diagram of local moves and isotopies: The arrow (0) is the composition of the arrow (1) (identity using notation), followed by the arrow (2) (an isotopy), followed by the arrow (3) (a grid local move of type $q \times (p+is)$ followed by an isotopy) finally followed by the arrow (4) (a grid local move of type $q \times (tl+sk)$). The key sequence therefore provides us with the arrow (0) which is achieved via (besides isotopies) two grid local moves of type: $q \times (p+is)$ and $q \times (tl+sk)$. The parameters involved are integeres p, q, t, s, i, k, l subject only to conditions: $p, q \geq 1$, $t, s \geq 1$ and $t+s=q$, $i \geq 1$ and $k, l \geq 0$. They can otherwise be chosen arbitrarily.

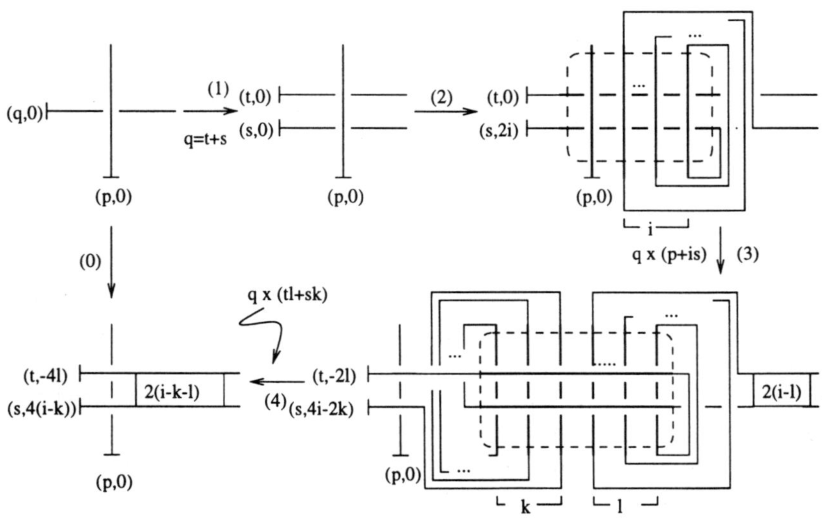

FIGURE 3

The main theorem of this paper is then the following.

Theorem 1.1. *A grid local move of type $a \times b$ is an unknotting operation for $a = 3, 4$ and any $b \geq 1$ modulo the fact that a move of type $4 \times 2b$ has to preserve the Arf invariant.*

The author believes that all grid local moves are unknotting operations. The paper is organized as follows. In Section 2 we prove Theorem 1.1. The proof consists of applying our key sequence to special situations. In Section 3 we discuss the metrics on the space of knots and knot invariants resulting from these unknotting operations and write down some basic estimates.

2. Grid Local Moves

In this section we want to prove our Theorem 1.1. The proof in the case $a = 3$ is immediate. For $b = 1$ it is trivial and for $b = 2$ it was proven in [A2]. The proof for $b \geq 3$ is an immediate consequence of:

Proposition 2.1. *A grid local move of type $3 \times b$ is locally equivalent to the ordinary crossing change (a 1×1 grid).*

Proof. We only need to show that a grid move of type $3 \times b$ can express a grid move of type 1×1 (i.e., the ordinary crossing change). To see this we just plug the key sequence, for $p = 2$, $s = 1$, $t = 2$, $l = 2$ and $k = i - 2$, into the dotted box of Figure 4 below. We take the reader through the arrows. The arrow (1) is simply an isotopy. In the arrow (2) we see the starting position of the arrow (0) of the key sequence (for p=2 and q=3) so that we can apply the key sequence (two $3 \times (i + 2)$ grids for any $i \geq 2$).Arrow (3) is an isotopy. That and the mirror image of the same figures show that a $3 \times b$ grid (for $b \geq 4$) induces a $\pm$-full twist on two strings.

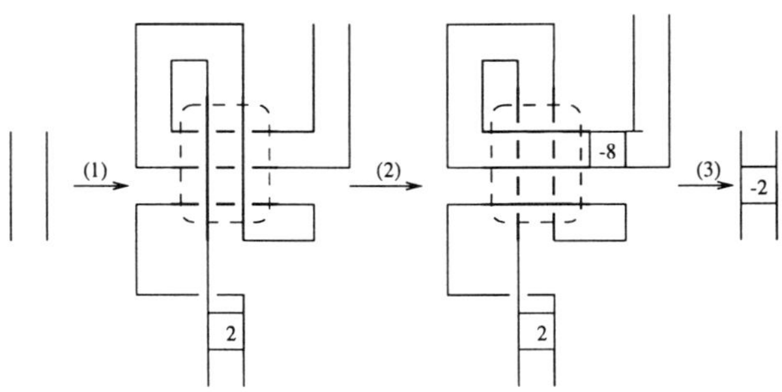

FIGURE 4

To express a 1×1 grid in terms of 3×2 and 3×3 grids just notice that a $a \times bc$ grid is nothing but a sequence of c grids of type $a \times b$. This finishes the proof of the proposition.

To prove Theorem 1.1 in the case $a = 4$ we will have to work somewhat harder. We will first use our key sequence to get various Lemmas which will then be assembled into a proof. The upshot of the following series of Lemmas will be to show that a grid move of type $4 \times b$ reduces every knot to a connected sum of trefoils. From that point on we will use the ideas and tricks developed here to update and modify certain Lemmas in [A2]. The latter ones will then allow us to finish the proof. The proofs of the Lemmas to follow will be identical to that of Proposition 2.1.

Lemma 2.1. *Two* $4 \times m$ *grids (any* $m \geq 6$*) have the effect in* (A) *in Figure 5 below:*

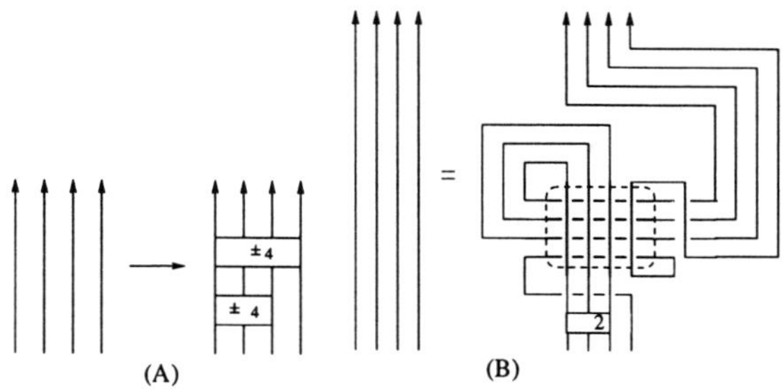

FIGURE 5

Proof. Apply the key sequence for $p = q = 4$, $t = 3$, $s = 1$, $l = 2$ and $k = i - 2$ in (B) of Figure 5 (as after an isotopy we see the starting position of the arrow (0) of the key sequence for $p = q = 4$).

Lemma 2.2. *Two* $4 \times m$ *grids (any* $m \geq 6$*) have the effect in* (A) *in Figure 6 below:*

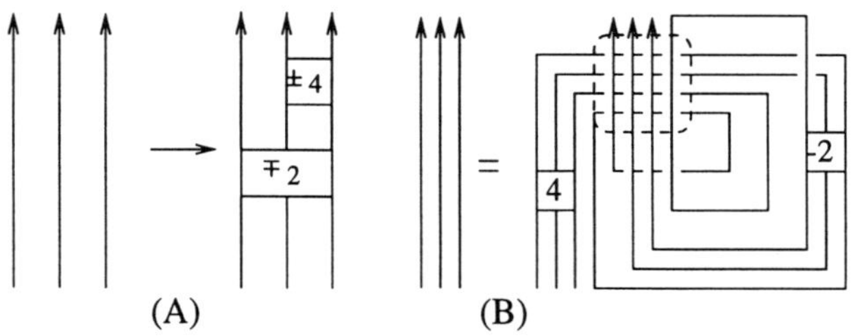

FIGURE 6

Proof. Apply the key sequence for $p = q = 4$, $t = 3$, $s = 1$, $l = 2$ and $k = i - 2$ in (B) of Figure 6.

Lemma 2.3. *Two* $4 \times 2m$ *grids (any* $m \geq 1$*) have the effect in* (A) *in Figure 7 below:*

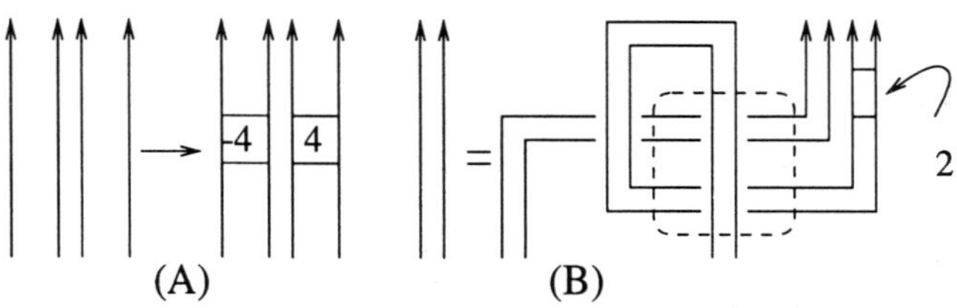

FIGURE 7

Proof. Apply the key sequence for $p = 2, q = 4$, $t = 2$, $s = 2$, $l = 1$ and $k = i$ in (B) of Figure 7 above.

Now we are ready to start putting together the proof of Theorem 1.1 for $a = 4$. Look at the figure below. Arrow (1) is simply an isotopy. Arrow (2) is a $4 \times (2i+2)$ (for any $i \geq 1$) grid followed by an isotopy. Arrow (3) is i times application of Lemma 2.1 (i.e., a number of $4 \times m$ grids (any $m \geq 6$)). Arrow (4) is $2i$ times application of Lemma 2.2 (i.e., a number of $4 \times m$ grids (any $m \geq 6$)). Arrow (5) is an isotopy and finally the arrow (6) is once application of Lemma 2.3 (i.e., a number of $4 \times 2n$ grids (any $n \geq 1$)).

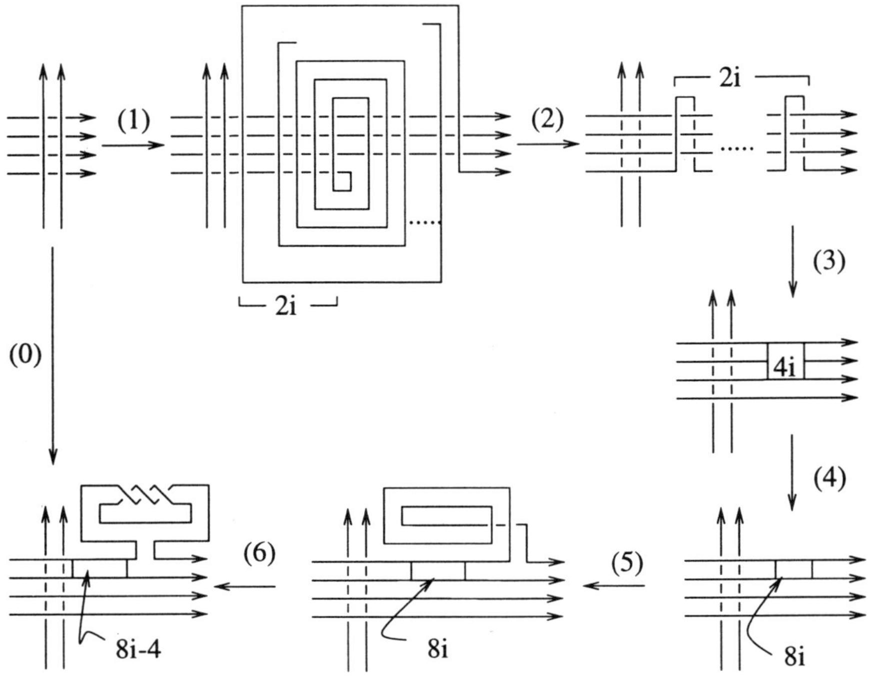

FIGURE 8

Since any knot can be reduced to the unknot or the trefoil via 2×4 grids (cf. [A2]) this figure shows that by applying $4 \times 2b$ grids (for any $b \geq 3$) any knot can be reduced to a connected sum of trefoil knots. The number of connected summands is then odd or even according as the Arf invariant of the knot is 1 or 0. Hence any grid local move of the type $4 \times b$, for any $b \geq 3$, achieves the same since an $a \times bc$ grid is locally nothing but c successive $a \times b$ grids. Notice that in the figure above we achieve a - to + 2×4 via any $4 \times b$, (for any $b \geq 3$) grid modulo introducing an even number of $K(3,2)$ summands. The mirror image of the figure shows that a + to - 2×4 grid can be achieved via $4 \times b$ grids modulo introducing an even number of $K(-3,2)$ connect summands. Now we need to further show that any $4 \times 2b$ grid can reduce the number of summands to one or zero (according to the Arf invariant) and that any grid of the form $4 \times b$, (for any b odd), can unknot the trefoil as well. For this we need to modify some of our Lemmas in [A2]. The idea is basically that the central Lemma there, namely Lemma 2.6, can be restated for $4 \times b$ grids. We will refer to that proof and show how to modify it.

Lemma 2.4. *We can pass from* (a) *to any of* (b) *or* (c) *in the figure below via a number of* $4 \times b$ *grid local moves for any* $b \geq 3$.

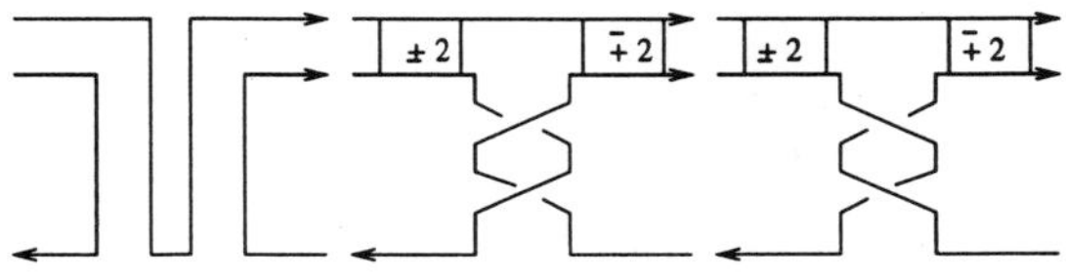

FIGURE 9

Proof. Lemma 2.6 of [A2] is (stated for and) proven by applying grids of the form $2 \times n$, for any $n \geq 2$. Moreover two such are required of opposite direction (- to + or + to - grid-crossing change). But our figure above shows that modulo connect summing with an even number of trefoil knots a 2×4 grid can be achieved by $4 \times b$ grids, for any $b \geq 3$. Since we apply two 2×4 grids we have an even number of $K(3,2)$ and an even number of $K(-3,2)$ summands. Lemma 2.3 then eliminates summands of the form $K(3,2)\#K(-3,2)$. QED.

Now Lemma 2.8 of [A2] holds for $4 \times b$ grids.

Lemma 2.5. *We can pass from* $K(3,2)$ *to* $K(-3,2)$ *via a sequence of* $4 \times b$ *grids, for any* $b \geq 3$.

Putting together Lemmas 2.3, 2.4 and 2.5 we now have proven:

Lemma 2.6. *Any knot* K *can be reduced to either the unknot (if* $Arf(K) = 0$*) or the trefoil (if* $Arf(K) = 1$*) via* $4 \times b$ *grids, for any* $b \geq 3$.

For grids of the form $4 \times b$, b even this cannot be improved. It remains to show that for grids of the form $4 \times b$, for any b one can unknot the trefoil. Notice that so far we have proven that any knot can be reduced to a connected sum of a number of trefoils of the same sign and that this number can be changed by any even number. It suffices therefore to show that a grid of the form $4 \times b$, (for any odd b), can change the parity of the number of summands. This will be achieved with the help of:

Lemma 2.7. *A sequence of* $4 \times b$ *grids, for any odd* b*, achieves* $\pm(8k+2)$ *half twists, some* k *on two parallel like oriented strings.*

Proof. The proof is contained in the figure below. We take the reader through the arrows. We will show that the arrow (0) can be achieved via grid moves of the desired kind. This will finish the proof. Arrow (1) is just an isotopy. Arrow (2) is a $4 \times (2i+3)$ grid followed by an isotopy. Arrow (3) is i applications of Lemma 2.1 (i.e., grids of type $4 \times m$, $m \geq 6$). Arrow (4) is just an isotopy and finally the arrow (5) is $2i+3$ applications of Lemma 2.2 (i.e., $4 \times m$ grids, $m \geq 6$). So we can achieve this result by any $4 \times b$ grid with $b \geq 5$ and any odd b.

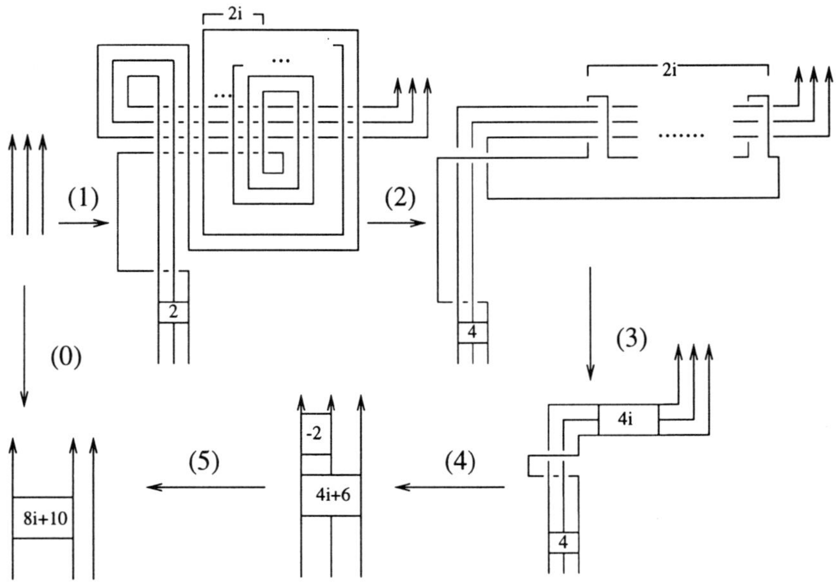

FIGURE 10

In the figure below we show how Lemma 2.7 can be used to change the parity of the number of $K(3,2)$ summands. We describe the arrows. The arrow (1) is an application of Lemma 2.7 (i.e., any $4 \times b$ grid with $b \geq 5$ odd. Arrow (2) is Lemma 2.3 (i.e., any $4 \times 2b$ grid). Arrow (3) finally is repeated applications of the arrow (2). The final result is a connected sum of an odd number of trefoils.

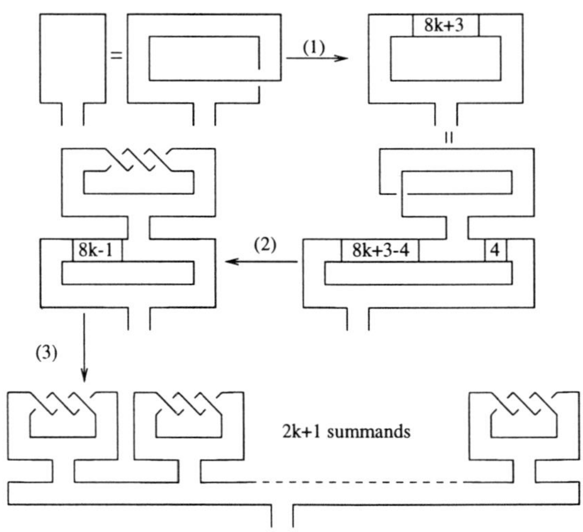

FIGURE 11

This finishes the proof of Theorem 1.1.

3. Metrics, unknotting numbers and estimates

As in [A2] a grid unknotting operation defines a metric on the space of all knots and relevant "unknotting sets". We briefly recall their definition and we proceed to give some basic estimates. Before we do this let us discuss notation. In [A2] we talked about multi-# moves and grid local moves and used the symbol # decorated with a certain vector carrying the parameters. We introduced $\natural$ for a grid move. It is now clear that we should continue to use one symbol decorated with the parameters. We will use # for all of them and we will decorated as follows: $\#_{(d_1,r_1)\times(d_2\times r_2)}$ or $\#_{d_1\times d_2}$ when $r_i = 0$. We can now define the relevant unknotting sets:

Definition 3.1. *By $\mathcal{U}^{\#}_{d_1\times d_2}(p,n)$ we will mean the set of all knots which can be unknotted by p + to - $d_1 \times d_2$-grid crossing changes and n - to + $d_1 \times d_2$-grid crossing changes. The relevant unknotting number $u^{\#}_{d_1\times d_2}(K)$ is then nothing but:*

$$\min\{p+n\colon K \in \mathcal{U}^{\#}_{d_1\times d_2}(p,n)\}$$

As in [A2] one could more generally define slicing numbers and sets as well as (when d_1 or d_2=3 or 4) metrics $d^{\#}_{d_1\times d_2}$ on the space $\mathcal{K}$ of all knots. Let us agree also on notation for some other standard moves. The relevant entities for one full twist one d strings will be decorated with d. So the classical unknotting number for example will be u^2 or simply u. We now would like to begin writing some estimates about these unknotting numbers. We prepare the ground for $u^{\#}_{3\times d}$. Our first Lemma is an analogue of Lemma 2.3 of [A2] and its proof is similar to the one for the latter.

Lemma 3.1. *Suppose that $K \in \mathcal{U}^{\#}_{(d_1,r_1)\times(d_2,r_2)}(p,n)$. Then there is a sphere embedded in $M^4_{p+n} = \#_{p+n}S^2 \times S^2$ which is smooth everywhere except at one point where it is the cone on K and represents on homology $\oplus_p(d_1,d_2) \oplus_n (d_1 - d_2)$ in the natural basis of $H_2(M^4_{p+n})$.*

Now we are ready to write down our basic estimate for $u^{\#}_{3\times d}$.

Theorem 3.1. *Let $K \in \mathcal{K}$ be any knot. If $d \geq 4$ then $u^{\#}_{3\times d}(K) \leq 2u^2(K)$. Moreover if $d \equiv 0 \bmod 3$ then $(8\frac{d}{3}+2)u^{\#}_{3\times d}(K) \geq |\sigma_{\frac{1}{3}}(K)|$.*

Proof. To prove the first inequality notice that the proof of our Proposition 2.1 implies that $\mathcal{U}^2(p,n) \subset \mathcal{U}^{\#}_{3\times(2+i)}(p+n,p+n)$ for any $i \geq 2$. To prove the second inequality use our Lemma 3.1 and P. Gilmer's thesis as in [A2] to get that if $K \in \mathcal{U}^{\#}_{3\times d}(p,n)$ then $2(p+n) \geq |8\frac{d}{3}(p-n) + \sigma_{\frac{1}{3}}(K)$. This finishes the proof of this Lemma.

Now we want to estimate $u^{\#}_{4\times d}$. As always the estimates will fall out of Lemma 3.1 and P. Gilmer's theorem. In this way we can get estimates involving σ_x for $x = \frac{1}{2}, \frac{1}{4}$ as follows:

Theorem 3.2. *Let $K \in \mathcal{K}$ be any knot. Then:*

$$d \equiv 0 \bmod 4 \implies (6d+2)u^{\#}_{4\times d}(K) \geq |\sigma_{\frac{1}{4}}(K)|,$$

$$d \equiv 0 \bmod 2 \implies (4d+2)u^{\#}_{4\times d}(K) \geq |\sigma_{\frac{1}{2}}(K)|, \quad \text{and}$$

$$d \equiv 1 \bmod 2 \implies (4d+6)u^{\#}_{4\times d}(K) \geq |\sigma_{\frac{1}{2}}(K)|$$

Proof. They are all proved combining Lemma 3.1 and P. Gilmer's theorem. We only need to make a remark in the third case when d is odd. In that case if $K \in \mathcal{U}^{\#}_{4\times d}(p,n)$ then there is an embedded sphere in $\#_p(S^2 \times S^2 \#_4 CP^2)\#_n(S^2 \times S^2 \#_4 \overline{CP}^2)$ which is smooth everywhere except at a point where it is a cone on K and on homology it represents $\oplus_p(4,(d-1)) \oplus_4 (2)) \oplus_p (-4,(d-1)) \oplus_4 (2))$ in the obvious standard basis. Then applying Gilmer's inequality we get respectively in these three cases:

$$2(p+n) \geq |6d(p-n) + \sigma_{\frac{1}{4}}(K)|$$
$$2(p+n) \geq |4d(p-n)\sigma_{\frac{1}{2}}(K)|,$$
$$6(p+n) \geq |4d(p-n)\sigma_{\frac{1}{2}}(K)|.$$

This finishes the proof.

We should remind the reader that in case d is even a $4 \times d$ grid unknots only Arf invariant zero knots. Estimates on unknotting number one knots in these unknotting operations can be found using minimal representing genus information which comes from gauge theory.

References

[A1] N. Askitas, *Embeddings of 2-spheres in 4-manifolds,* Manuscripta Math. 89,, 35-47, 1996

[A2] N. A. Askitas, *Multi-# unknotting operations: A new family of local moves on a knot diagram and related invariants of knots,* J. of Knot Theory Ramifications, Vol 7, No. 7 857-871, (1998)

[G] P. M. Gilmer, *Configurations of Surfaces in 4-manifolds,* Trans. of AMS, Vol 264, No. 2 353-380, (1981)

[K] R. Kirby, *The topology of 4-manifolds,* LNM , Springer Verlag, (1981)

[M] H. Murakami, *Some Metrics on Classical Knots,* Math. Ann. 270, 35-45, (1985)

(N. Askitas) Mathematics Department, School of Sciences, Aristoteles University of Thessaloniki, GR-54006, Thessaloniki, Greece

E-mail address, N.Askitas: `askitas@ccf.auth.gr`

On Simple Points Of Character Varieties Of 3-Manifolds

S. Boyer [1] and X. Zhang

We begin by recalling a few basic notions from algebraic geometry.

Let X be a complex affine algebraic set and $\mathbb{C}[X]$ its coordinate ring. The *dimension* of X, denoted $dim(X)$, is the Krull dimension of $\mathbb{C}[X]$. For $x \in X$, the *dimension of X at x*, written $dim_x(X)$, is the maximum dimension of an algebraic component of X which passes through x. We use $T_x^{Zar}(X)$ to denote the Zariski tangent space of X at x. A point $x \in X$ is called *simple* if $dim(T_x^{Zar}(X)) = dim_x(X)$. A proof of the following basic result can be found in §II.2 of [Sh].

Proposition 1 *Let X be a complex affine algebraic set and $x \in X$. Then x is a simple point of X if and only if it is contained in a unique algebraic component of X and is a smooth point of that component.* ◊

For any finitely generated group G, $R(G)$ will denote the set of representations of G with values in $SL_2(\mathbb{C})$. The character χ_ρ of $\rho \in R(G)$ is the function $\chi_\rho : G \rightarrow \mathbb{C}$ defined by $\chi_\rho(g) = trace(\rho(g))$. The set of such characters will be written $X(G)$. Both $R(G)$ and $X(G)$ are complex, affine, algebraic sets and the function $t : R(G) \rightarrow X(G)$ which sends a representation to its character is regular [CS].

Though the generic point of either $R(G)$ or $X(G)$ is simple, it is difficult in general to determine whether this is the case for a given point. In this note we show that when G is the fundamental group of a link L in a closed, connected, orientable 3-manifold, then assumptions on the topology of a manifold resulting from a Dehn surgery on L can be used to deduce that certain representations and characters are indeed simple. Results of this nature are of interest to low-dimensional topologists because they establish quantitative links between the topology of the Dehn surgeries on L and Culler-Shalen seminorms [BZ1], [BZ2] or A-polynomials [BZ3].

Recall that $sl(2,\mathbb{C})$ is the set of all 2×2 complex matrices with trace zero. The adjoint homomorphism $Ad : SL(2,\mathbb{C}) \rightarrow Aut(sl(2,\mathbb{C}))$ is given by matrix conjugation, i.e. $Ad(A)(U) = AUA^{-1}$ for $A \in SL(2,\mathbb{C})$ and for $U \in sl(2,\mathbb{C})$.

If G is a group and $\rho : G \rightarrow SL(2,\mathbb{C})$ a representation, then the composition $Ad \circ \rho : \pi \rightarrow Aut(sl(2,\mathbb{C}))$ defines a G-module structure on $sl(2,\mathbb{C})$ which we shall denote by $Ad \circ \rho$. A fundamental observation, due to A. Weil [W1], is that the Zariski tangent space $T_\rho^{Zar}(R(G))$

[1]Partially supported by grants: NSERC OGP 0009446 and FCAR EQ 3518.

of $R(G)$ at a point $\rho \in R(G)$ can be identified with a subspace of the space of 1-cocycles $Z^1(G; Ad\circ\rho)$. Further, the orbit under conjugation of any $\rho \in R(G)$, $O(\rho)$, is a subvariety of $R(G)$ whose Zariski tangent space coincides with the space of 1-coboundaries, $B^1(G; Ad\circ\rho)$, under this identification.

Let $R^{irr}(G)$ be the set of irreducible representations in $R(G)$ and $X^{irr}(G)$ the set of characters of such representation. It is shown in Proposition 3.2 of [Po] that the map $t| : R^{irr}(G) \to X^{irr}(G)$ is an analytic bundle whose fibres are the orbits $O(\rho)$ where $\rho \in R^{irr}(G)$. It follows (see Proposition 3.5 of [Po]) that for $\chi_\rho \in X^{irr}(G)$, $T^{Zar}_{\chi_\rho}(X(G))$ can be identified as a subspace of the 1-dimensional cohomology group $H^1(G; Ad\circ\rho)$ in such a way that the following diagram commutes.

$$\begin{array}{ccccc}
T^{Zar}_{\rho}\mathcal{O}(\rho) & \longrightarrow & T^{Zar}_{\rho}R(G) & \xrightarrow{(dt)_\rho} & T^{Zar}_{\chi_\rho}X(G) \\
\cong\big\downarrow & & \big\downarrow & & \big\downarrow \\
B^1(G; Ad\circ\rho) & \longrightarrow & Z^1(G; Ad\circ\rho) & \longrightarrow & H^1(G; Ad\circ\rho).
\end{array}$$

In particular, if $H^1(G; Ad\circ\rho) = 0$, then χ_ρ is an isolated point of $X(G)$. If ρ is assumed to be *scheme reduced*, i.e. $T^{Zar}_{\chi_\rho}(R(G)) \cong Z^1(G; Ad\circ\rho)$ (see eg. §3.1.3 of [Po]), then the converse is true. We highlight this discussion with a proposition.

Proposition 2 [W1, Po] *If $\chi_\rho \in X(G)$ is the character of an irreducible representation $\rho \in R(G)$, then $T^{Zar}_{\chi_\rho}(X(G))$ can be identified with a subspace of $H^1(G; Ad\circ\rho)$.* ◇

Consider a compact, connected, orientable 3-manifold M whose boundary consists of n tori. Set $R(M) = R(\pi_1(M))$ and $X(M) = X(\pi_1(M))$. Recent research has shown that $R(M)$ and $X(M)$ contain a great deal of topological and geometric information about the underlying manifold M as well as its Dehn fillings. Below we give sufficient conditions for a character in $X(M)$ to be a simple point whose associated algebraic component is n-dimensional. The basic idea is to combine Weil's result with cohomological calculations. We point out that for any space W and representation $\rho : \pi_1(W) \to SL(2, \mathbb{C})$, there is a canonical identification of $H^1(\pi_1(W); sl(2, \mathbb{C})_\rho)$ with $H^1(W; sl(2, \mathbb{C})_\rho)$ (see Theorem IV.3.3 of [Wh]).

We shall suppose in what follows that $\partial M = T_1 \cup \ldots \cup T_n$, where each T_i is a torus. We use $M(r_1, ..., r_n)$ to denote the manifold obtained by Dehn filling M along T_i with slopes r_i, $i = 1, ..., n$. Inclusion induces a surjective homomorphism $\pi_1(M) \to \pi_1(M(r_1, \ldots, r_n))$, and hence injective maps $R(M(r_1, ..., r_n)) \to R(M)$ and $X(M(r_1, ..., r_n)) \to X(M)$.

Theorem 3 *Suppose that $\rho \in R(M(r_1, ..., r_n)) \subset R(M)$ is a representation such that*

(a) ρ is irreducible,

(b) $\rho(\pi_1(T_i))$ is not contained in $\{\pm I\}$ for $i = 1, ..., n$, where I is the identity matrix,

(c) $H^1(M(r_1, ..., r_n); Ad \circ \rho) = 0$.

Then

(1) ρ is a simple point of $R(M)$ and the unique algebraic component of $R(M)$ which contains it is $n+3$-dimensional.

(2) χ_ρ is a simple point of $X(M)$ and the algebraic component of $X(M)$ which contains it is n-dimensional.

(3) there is a complex analytic n-disk D, smoothly embedded in $R(M)$ and containing ρ, such that $t|D$ is an analytic isomorphism onto a neighbourhood of χ_ρ in $X(M)$.

(4) ρ is scheme reduced and there is a commutative diagram

$$\begin{array}{ccccc}
T^{Zar}_\rho O(\rho) & \rightarrowtail & T^{Zar}_\rho R(M) & \xrightarrow{(dt)_\rho} & T^{Zar}_{\chi_\rho} X(M) \\
\cong \downarrow & & \cong \downarrow & & \cong \downarrow \\
B^1(\pi_1(M); Ad\rho) & \rightarrowtail & Z^1(\pi_1(M); Ad\rho) & \twoheadrightarrow & H^1(\pi_1(M); Ad\rho).
\end{array}$$

Proof. Consider the decomposition

$$M(r_1, ..., r_n) = M \cup V$$

where $V = V_1 \cup \ldots \cup V_n$ and V_i is the solid torus attached to T_i with slope r_i. Applying the Mayer-Vietoris sequence for cohomology with coefficients in $sl(2, \mathbb{C})$, twisted by $Ad \circ \rho$, to this decomposition, condition (c) of the hypotheses imlies that $H^1(M; Ad \circ \rho) \oplus H^1(V_1; Ad \circ \rho) \oplus \ldots \oplus H^1(V_n; Ad \circ \rho) \cong H^1(\partial M; Ad \circ \rho)$. Since $\rho(\pi_1(V_i)) \subset \rho(\pi_1(T_i))$ is not contained in $\{\pm I\}$ for $i = 1, ..., n$, we have $\oplus_{i=1}^n H^1(V_i; Ad \circ \rho) \cong \mathbb{C}^n$ and $H^1(\partial M; Ad \circ \rho) \cong \mathbb{C}^{2n}$. Thus $H^1(M; Ad \circ \rho) \cong \mathbb{C}^n$, and therefore the dimension of $X(M)$ at the point χ_ρ is at most n (cf. Proposition 2). On the other hand, by a result of W. Thurston (see Proposition 3.2.1 of [CS]), the dimension of any component of $X(M)$ which contains χ_ρ is at least n. Thus

$$n \leq dim_{\chi_\rho}(X(M)) \leq dim(T^{Zar}_{\chi_\rho} X(M)) \leq dim_{\mathbb{C}}(H^1(M; Ad \circ \rho)) = n$$

and hence χ_ρ is a simple point of $X(M)$. Further, the unique component X_0 of $X(M)$ which contains χ_ρ is n-dimensional.

Next consider $\rho \in R(M)$. By Corollary 1.5.3 of [CS], $t^{-1}(X_0) = R_0$ is an irreducible component of $R(M)$ of dimension $n+3$. Since the image of ρ is non-abelian we have $dim_{\mathbb{C}}(B^1(M; Ad \circ \rho)) = 3$ and therefore $Z^1(M; Ad \circ \rho) \cong \mathbb{C}^{n+3}$. It follows, as in the previous paragraph, that ρ is a simple point of $R(M)$ contained in a component of dimension $n+3$.

Further we deduce that part (4) of the theorem holds. Finally, the conclusions above imply that if $D \subset R(M)$ is an appropriately small, complex analytic n-disk containing ρ and transverse to $\mathcal{O}(\rho)$ there, then $t|D$ is an analytic isomorphism onto a neighbourhood of χ_ρ in $X(M)$. $\diamond$

Remark 4 Condition (c) of the theorem is, in general, difficult to verify, and it is of interest to replace it by topological conditions on the filled manifold $M(r_1, \ldots, r_n)$. This is indeed what we do in the two corollaries below. Another topologically significant scenario arises when $M(r_1, \ldots, r_n)$ is a *small* 3-manifold, i.e. it contains no essential surfaces. It is known that this hypothesis implies that $X(M(r_1, \ldots, r_n))$ is a finite set (see §2 of [CS]), and so in particular $T^{Zar}_{\chi_\rho} X(M(r_1, \ldots, r_n)) = 0$. If we knew that $\rho \in R(M(r_1, \ldots, r_n))$ were scheme reduced in $X(M(r_1, \ldots, r_n))$, then we could conclude that $H^1(M(r_1, ..., r_n); Ad \circ \rho) = 0$. Unfortunately, the problem of determining whether or not a given ρ is scheme reduced appears to be difficult.

Corollary 5 *Suppose that M is the exterior of a knot in the 3-sphere. If $M(r)$ is a hyperbolic 3-manifold and $\rho \in R(M(r)) \subset R(M)$ is a discrete faithful representation of $\pi_1(M(r))$, then ρ is a simple point of $R(M)$ and the unique irreducible component of $R(M)$ which contains ρ is 4-dimensional. Furthermore, χ_ρ is simple in $X(M)$ and the irreducible component of $X(M)$ which contains it is 1-dimensional.*

Remark 6 The assumption that M be the exterior of a knot in the 3-sphere is not crucial. Though its role was to guarantee that condition (b) of Theorem 3 held, this condition is often a consequence of the hypothesis that $M(r)$ be hyperbolic. For instance if M is itself hyperbolic, i.e. the interior of M admits a complete, hyperbolic metric of finite volume, then by work of W. Thurston [Th], the manifolds $M(r)$ are hyperbolic except for at most finitely many slopes r. Furthermore, the core of the filling solid torus is a geodesic in $M(r)$. It follows that if r' is a slope on ∂M which is dual to r, i.e. $\Delta(r, r') = 1$, and $\alpha' \in H_1(\partial M) = \pi_1(\partial M) \subset \pi_1(M)$ is a primitive class which is carried by r', then $\rho(\alpha')$ is a hyperbolic element of $SL(2, \mathbb{C})$, and so in particular is not $\pm I$.

Proof of Corollary 5. By Theorem 3 we only need to verify that such ρ satisfies conditions (a), (b) and (c) of Theorem 3.

If ρ were to be reducible, then without loss of generality, its image would lie in U, the group of upper-triangular matrices in $SL(2, \mathbb{C})$. The non-trivial elements of the kernel of the projection of U onto the diagonal matrices consists of parabolic matrices, and therefore since ρ is a discrete faithful representation of a *closed* hyperbolic manifold, this kernel is trivial. Thus $\pi_1(M(r))$ is abelian, and it follows that up to conjugation, $\rho(\pi_1(M(r)))$ is a discrete group of diagonal matrices. Since $\pi_1(M(r))$ has no non-trivial torsion, the only possibility is for $\pi_1(M(r))$ to be isomorphic to either the trivial group or $\mathbb{Z}$. Since neither of these groups is the fundamental group of a closed hyperbolic 3-manifold, ρ must be irreducible. Thus condition (a) of Theorem 3 holds.

Condition (b) follows from the fact that any representation ρ of the group of a knot in S^3 which takes values in $\{\pm I\}$ on the peripheral subgroup, has image in $\{\pm I\}$.

Finally, condition (c) was proved by Weil in [W 2]. ◇

Next we consider the case where $M(r)$ is a Seifert fibered space. We call a manifold W a *small-Seifert space* if it admits a Seifert structure whose base orbifold $\mathcal{F}$ is a 2-sphere with at most three singular points. Such manifolds have non-cyclic fundamental group if and only if $\mathcal{F}$ has exactly three singular points. If this is the case, then the fundamental group of W has a presentation of the form

$$\pi_1(W) = \{x, y, h \mid x^{\alpha_1} = h^{\beta_1},\ y^{\alpha_2} = h^{\beta_2},\ (xy)^{\alpha_3} = h^{\beta_3},\ xh = hx,\ yh = hy\},$$

where $\alpha_i > 1$ and β_i are integers for $i = 1, 2, 3$ [J]. The central element h generates a cyclic, normal subgroup $< h >$ of $\pi_1(W)$ and the quotient of $\pi_1(W)$ by $< h >$ is the triangle group $\Delta(\alpha_1, \alpha_2, \alpha_3) = < x, y \mid x^{\alpha_1} = y^{\alpha_2} = (xy)^{\alpha_3} = 1 >$.

Proposition 7 *Let W be a small-Seifert fibered space with noncyclic fundamental group and $\rho \in R(\pi_1(W))$ be a non-abelian representation, i.e. $\rho(\pi_1(W))$ is nonabelian. Then $H^1(W; Ad \circ \rho) = 0$.*

Proof. As ρ is non-abelian, the space of 1-coboundaries, $B^1(\pi_1(W); Ad\circ\rho)$, is 3-dimensional. Thus to prove that $H_1(\pi_1(W); Ad \circ \rho) = 0$, it is enough to show that the dimension of $Z^1(\pi_1(W); Ad \circ \rho))$ is at most 3. Since h is central and ρ is non-abelian, we must have $\rho(h) \in \{\pm I\}$. It follows that $\rho(x), \rho(y)$ and $\rho(xy)$ are of finite order greater than 2. After a possible conjugation of ρ, we may assume that

$$\rho(x) = \pm \begin{pmatrix} \lambda & 0 \\ 0 & \lambda^{-1} \end{pmatrix}, \rho(y) = \pm A \begin{pmatrix} \mu & 0 \\ 0 & \mu^{-1} \end{pmatrix} A^{-1}, \text{ and } \rho(xy) = \pm B \begin{pmatrix} \sigma & 0 \\ 0 & \sigma^{-1} \end{pmatrix} B^{-1}$$

where λ, μ and σ are roots of unity of order larger than 2. By our assumptions on ρ, in each of the matrices $A = \begin{pmatrix} a_1 & a_2 \\ a_3 & a_4 \end{pmatrix}$ and $B = \begin{pmatrix} b_1 & b_2 \\ b_3 & b_4 \end{pmatrix}$, at most one entry is zero.

Any 1-cocycle $u \in Z^1(\pi_1(W); Ad \circ \rho)$ satisfies the *cocycle condition*

$$u(vw) = u(v) + (Ad \circ \rho(v))(w)$$

for each $v, w \in \pi_1(W)$, and so is determined by its values at x, y and h. Let

$$u(h) = \begin{pmatrix} h_1 & h_2 \\ h_3 & -h_1 \end{pmatrix}.$$

Since h is central and $\rho(h) = \pm I$, the cocycle condition implies that $u(h) = \rho(w)(u(h))\rho(w)^{-1}$ for each $w \in \pi_1(W)$. Taking $w = x$ shows that $h_2 = h_3 = 0$, and then taking $w = y$ shows $h_1 = 0$, i.e. $u(h) = 0$. It follows that $u(h^n) = 0$ for each $n \in \mathbb{Z}$. The relations from the presentation of $\pi_1(W)$ described above show that u also satisfies:

$$u(x^{\alpha_1}) = u(h^{\beta_1}) = 0, u(y^{\alpha_2}) = u(h^{\beta_2}) = 0, u((xy)^{\alpha_3}) = u(h^{\beta_3}) = 0.$$

Set $u(x) = \begin{pmatrix} x_1 & x_2 \\ x_3 & -x_1 \end{pmatrix}$ and $u(y) = \begin{pmatrix} y_1 & y_2 \\ y_3 & -y_1 \end{pmatrix}$. We'll show that these identities place three independent linear constraints on $x_1, x_2, x_3, y_1, y_2, y_3$. Thus the dimension of $Z^1(\pi_1(W); Ad \circ \rho)$ is at most 3, and we are done.

Using the cocycle condition, the identities yield:

$$(1) \quad u(x) + \rho(x)u(x)\rho(x)^{-1} + ... + \rho(x)^{\alpha_1 - 1}u(x)\rho(x)^{-(\alpha_1 - 1)} = \begin{pmatrix} 0 & 0 \\ 0 & 0 \end{pmatrix}$$

$$(2) \quad u(y) + \rho(y)u(y)\rho(y)^{-1} + ... + \rho(y)^{\alpha_2 - 1}u(y)\rho(y)^{-(\alpha_2 - 1)} = \begin{pmatrix} 0 & 0 \\ 0 & 0 \end{pmatrix}$$

$$(3) \quad u(xy) + \rho(xy)u(xy)\rho(xy)^{-1} + ... + \rho(xy)^{\alpha_3 - 1}u(xy)\rho(xy)^{-(\alpha_3 - 1)} = \begin{pmatrix} 0 & 0 \\ 0 & 0 \end{pmatrix}.$$

From (1), we obtain

$$\begin{pmatrix} \alpha_1 x_1 & (1 + \lambda^2 + ... + \lambda^{(2\alpha_1 - 2)})x_2 \\ (1 + \lambda^{-2} + ... + \lambda^{-(2\alpha_1 - 2)})x_3 & -\alpha_1 x_1 \end{pmatrix} = \begin{pmatrix} 0 & 0 \\ 0 & 0 \end{pmatrix}$$

and therefore

$$(1a) \qquad x_1 = 0.$$

Next set $W = \begin{pmatrix} w_1 & w_2 \\ w_3 & -w_1 \end{pmatrix} = A^{-1}u(y)A$ and $M = \begin{pmatrix} \mu & 0 \\ 0 & \mu^{-1} \end{pmatrix}$. Equation (2) can be written

$$W + MWM^{-1} + ... + M^{\alpha_2 - 1}WM^{-(\alpha_2 - 1)} = \begin{pmatrix} 0 & 0 \\ 0 & 0 \end{pmatrix}$$

which, as above, yields $w_1 = 0$. Hence from

$$W = \begin{pmatrix} a_4 & -a_2 \\ -a_3 & a_1 \end{pmatrix} \begin{pmatrix} y_1 & y_2 \\ y_3 & -y_1 \end{pmatrix} \begin{pmatrix} a_1 & a_2 \\ a_3 & a_4 \end{pmatrix}$$

$$= \begin{pmatrix} a_4y_1 - a_2y_3 & a_4y_2 + a_2y_1 \\ -a_3y_1 + a_1y_3 & -a_3y_2 - a_1y_1 \end{pmatrix} \begin{pmatrix} a_1 & a_2 \\ a_3 & a_4 \end{pmatrix}$$

$$= \begin{pmatrix} a_1a_4y_1 - a_1a_2y_3 + a_3a_4y_2 + a_2a_3y_1 & a_2a_4y_1 - a_2^2y_3 + a_4^2y_2 + a_2a_4y_1 \\ -a_1a_3y_1 + a_1^2y_3 - a_3^2y_2 - a_1a_3y_1 & -a_2a_3y_1 + a_1a_2y_3 - a_3a_4y_2 - a_1a_4y_1 \end{pmatrix}$$

we obtain

$$\text{(2a)} \qquad a_1a_4y_1 - a_1a_2y_3 + a_3a_4y_2 + a_2a_3y_1 = 0.$$

Finally set $V = \begin{pmatrix} v_1 & v_2 \\ v_3 & -v_1 \end{pmatrix} = B^{-1}u(xy)B$ and $S = \begin{pmatrix} \sigma & 0 \\ 0 & \sigma^{-1} \end{pmatrix}$, so that equation (3) becomes

$$V + SVS^{-1} + \ldots + S^{\alpha_3-1}VS^{-(\alpha_3-1)} = \begin{pmatrix} 0 & 0 \\ 0 & 0 \end{pmatrix},$$

which yields $v_1 = 0$. Now

$$\begin{aligned} u(xy) &= u(x) + \rho(x)u(y)\rho(x)^{-1} \\ &= \begin{pmatrix} 0 & x_2 \\ x_3 & 0 \end{pmatrix} + \begin{pmatrix} \lambda & 0 \\ 0 & \lambda^{-1} \end{pmatrix} \begin{pmatrix} y_1 & y_2 \\ y_3 & -y_1 \end{pmatrix} \begin{pmatrix} \lambda & 0 \\ 0 & \lambda^{-1} \end{pmatrix} \\ &= \begin{pmatrix} y_1 & x_2 + \lambda^2 y_2 \\ x_3 + \lambda^{-2}y_3 & -y_1 \end{pmatrix} \end{aligned}$$

so

$$\begin{aligned} V &= B^{-1}u(xy)B \\ &= \begin{pmatrix} b_4 & -b_2 \\ -b_3 & b_1 \end{pmatrix} \begin{pmatrix} y_1 & x_2 + \lambda^2 y_2 \\ x_3 + \lambda^{-2}y_3 & -y_1 \end{pmatrix} \begin{pmatrix} b_1 & b_2 \\ b_3 & b_4 \end{pmatrix} \\ &= \begin{pmatrix} b_4y_1 - b_2(x_3 + \lambda^{-2}y_3) & b_4(x_2 + \lambda^2y_2) + b_2y_1 \\ -b_3y_1 + b_1(x_3 + \lambda^{-2}y_3) & -b_3(x_2 + \lambda^2y_2) - b_1y_1 \end{pmatrix} \begin{pmatrix} b_1 & b_2 \\ b_3 & b_4 \end{pmatrix}, \end{aligned}$$

from which we deduce

$$v_1 = b_1[b_4y_1 - b_2(x_3 + \lambda^{-2}y_3)] + b_3[b_4(x_2 + \lambda^2y_2) + b_2y_1].$$

Thus

$$\text{(3a)} \qquad b_1b_4y_1 - b_1b_2x_3 - b_1b_2\lambda^{-2}y_3 + b_3b_4x_2 + b_3b_4\lambda^2y_2 + b_2b_3y_1 = 0.$$

Since either $a_1a_2 \neq 0$ or $a_3a_4 \neq 0$, and either $b_1b_2 \neq 0$ or $b_3b_4 \neq 0$, relations (1a), (2a) and (3a) show that $Z^1(\pi; Ad \circ \rho)$ is at most 3-dimensional. $\diamondsuit$

Corollary 8 *Suppose that M is a knot exterior in S^3. If $M(r)$ is a small-Seifert fibered 3-manifold and $\rho \in R(M(r))$ is an irreducible representation, then ρ is a simple point of*

R(M) and the unique irreducible component of R(M) which contains ρ is 4-dimensional. Furthermore the point χ_ρ is simple in $X(M)$ and the irreducible component of $X(M)$ which contains χ_ρ is 1-dimensional.

Remark 9 As in Remark 6, the assumption that M be the exterior of a knot in the 3-sphere is a mere convenience employed to guarantee that $\rho(\pi_1(\partial M)) \not\subset \{\pm I\}$. The latter condition can be shown to hold for quite general manifolds M in many topologically significant situations.

References

[BZ1] S. Boyer and X. Zhang, *Finite Dehn surgery on knots*, J. Amer. Math. Soc. **9** (1996), 1005-1050.

[BZ2] ———, *On Culler-Shalen seminorms and Dehn filling*, to appear in Ann. of Math.

[BZ3] ———, *A proof of the finite filling conjecture*, preprint.

[CS1] M. Culler and P. Shalen, *Varieties of group representations and splittings of 3-manifolds*, Ann. of Math. **117** (1983) 109-146.

[J] W. Jaco, *Lectures on three-manifold topology*, CBMS Regional Conf. Ser. Math. **43** (1980).

[Po] J. Porti, *Torsion de Reidemeister pour les variétés hyperboliques*, Mem. Amer. Math. Soc. **612**, 1997.

[Sh] I. Shafarevich, *Basic algebraic geometry*, Die Grundlehren der mathematischen Wissenschaften, Band **213**, Springer-Verlag, New York, 1974.

[Th] W. Thurston, *The geometry and topology of three manifolds*, lecture notes, Princeton, 1979.

[W1] A. Weil, *Remarks on the cohomology of groups*, Ann. Math. **80** (1964) 149-157.

[W2] A. Weil, *On discrete subgroups of Lie groups*, Ann. Math. **75** (1962) 578-602.

[Wh] G. W. Whitehead, *Elements of homotopy theory*, GTM **61**, Springer-Verlag, 1978.

Steven Boyer
Département de mathématiques, UQAM
P. O. Box 8888, Station A
Montreal H3C 3P8
Canada
e-mail: boyer@math.uqam.ca

Xingru Zhang
Department of Mathematics
SUNY Buffalo
Buffalo, N.Y., 14214-3093
U.S.A.
e-mail: xinzhang@math.buffalo.edu

THE 3-MOVE CONJECTURE FOR 5-BRAIDS

QI CHEN

ABSTRACT. Nakanishi conjectured in 1981 that any link is 3-equivalent to a trivial link.
J. Przytycki proved that some classes of links, including 11 crossing links, are 3-equivalent to trivial links. Based on this result we prove that all closed 4-braids are 3-equivalent to trivial links. Then we show that all links with no more than 12 crossings are 3-equivalent to trivial links. For closed 5-braids only 7 of them (up to conjugacy class) have more than 12 crossings. We reduce five of them by hand. The remaining two are 3-equivalent. Hence there is only one unsettled case which is a 5-braid with 20 crossings as shown in Figure 15. We prove also similar results for $t_3, \bar{t}_4$-moves.

1. INTRODUCTION

The idea of representing a link by a braid was first contemplated by Gauss (cf. [10]). The algebraic notion of a braid was introduced by E. Artin (cf. [2]). n strings form an n-braid if they run vertically in $\mathbf{R}^3$ and the starting points are above the ending points. The starting points and the end points are on two levels in $\mathbf{R}^3$ which are called the top level and the bottom level respectively. Figure 1(a) is a diagram of a 4-braid. Two braids are equivalent if they are ambient isotopic keeping the top level and

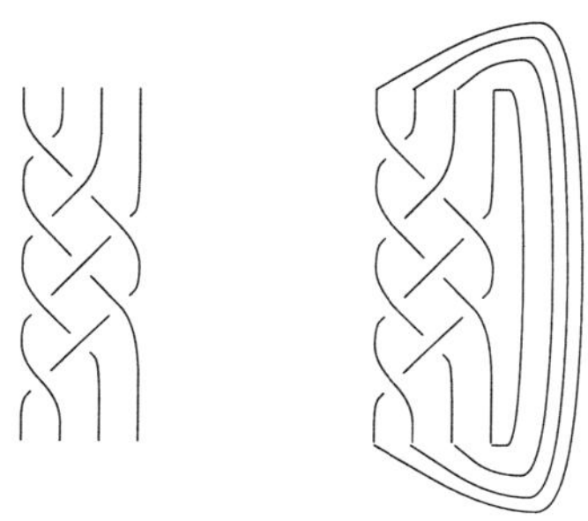

(a) (b)

FIGURE 1. Braid and closed braid.

the bottom level fixed. A braid can be turned into a link by connecting the top and bottom points as illustrated in Figure 1(b). Alexander showed that any link can be obtained in such a way (cf. [1]). If b is a braid then the corresponding closed braid is denoted by L_b. In a braid diagram if the i-th string is over (under) the $i+1$-st string, then this crossing is denoted by σ_i (σ_i^{-1}). Hence, the braid in Figure 1(a) is described as $\sigma_1\sigma_2^{-1}\sigma_1\sigma_3\sigma_2^{-1}\sigma_1\sigma_3\sigma_2^{-1}\sigma_1$. It is clear that every n-braid can be described by a word generated by $\sigma_i, \sigma_i^{-1} (1 \leq i \leq n-1)$. Words describing equivalent braids are considered equal.

Let B_n be the set of words arisen from n-braids in the fashion described above. B_n is clearly a group where multiplication is defined as word juxtaposition. B_n has relations (a) $\sigma_i\sigma_{i+1}\sigma_i = \sigma_{i+1}\sigma_i\sigma_{i+1}$, and (b) $\sigma_j\sigma_k = \sigma_k\sigma_j, |j-k| > 1$, as indicated in Figure 2 where $\approx$ means ambient isotopy.

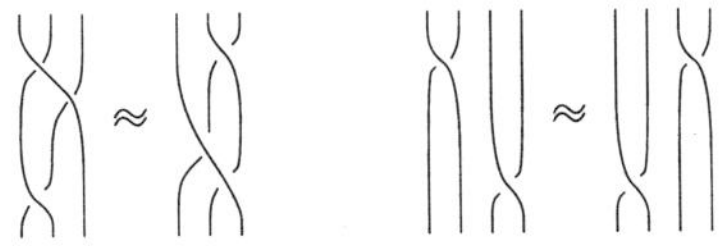

FIGURE 2. $\sigma_1\sigma_2\sigma_1 = \sigma_2\sigma_1\sigma_2, \quad \sigma_1\sigma_3 = \sigma_3\sigma_1$.

E. Artin proved in [2] that these two relations define B_n, i.e.

$$B_n =< \sigma_1, \sigma_2, \dots, \sigma_{n-1} \mid \sigma_i\sigma_{i+1}\sigma_i = \sigma_{i+1}\sigma_i\sigma_{i+1}, \sigma_i\sigma_j = \sigma_j\sigma_i, |i-j| > 1 > .$$

The 3-move is defined as in Figure 3. Two tangles or links are 3-equivalent if one can be transformed into the other by 3-moves and ambient isotopy (cf. [7]).

FIGURE 3. 3-move.

Let

$$C_n = B_n/\sigma_k^3 =< \sigma_1, \sigma_2, \dots, \sigma_{n-1} \mid \sigma_i\sigma_{i+1}\sigma_i = \sigma_{i+1}\sigma_i\sigma_{i+1},$$
$$\sigma_i\sigma_j = \sigma_j\sigma_i, |i-j| > 1, \ \sigma_k^3 = 1, \ k = 1, 2, \dots, n-1 >$$

and

$$p : B_n \to C_n,$$

the canonical projection. Then two braids b_1 and b_2 are 3-equivalent if $p(b_1) = p(b_2)$. Notice that two closed n-braids L_{b_1} and L_{b_2} represent the same link if b_1 and b_2 are conjugate in B_n. Hence, L_{b_1} and L_{b_2} are 3-equivalent if $p(b_1)$ and $p(b_2)$ are conjugate in C_n.

The most important theorem for this paper is the following result of Coxeter [5].

Theorem 1.1. *Let G be a group with the presentation*

$$< \sigma_1, \sigma_2, \dots, \sigma_{n-1} \mid \sigma_i\sigma_{i+1}\sigma_i = \sigma_{i+1}\sigma_i\sigma_{i+1}, \sigma_i\sigma_j = \sigma_j\sigma_i,$$
$$|i-j| > 1, \ \sigma_k^p = 1, \ k = 1, 2, \dots, n-1 >$$

where p is some integer. G is finite if and only if $\frac{1}{p} + \frac{1}{n} > \frac{1}{2}$

Remark 1.2. The 5-braid group under t_3-move is the above group with $n = 5$ and $p = 3$. This group is finite with 155,520 elements and for $n > 5, C_n$ is infinite.

Conjecture 1.3 (1981, Montesinos, Nakanishi). *Any link is 3-equivalent to a trivial link.*

Remark 1.4. Nakanishi first considered the conjecture in 1981. Earlier Montesinos analyzed 3-moves in relation to 3-fold dihedral branch coverings and asked a related but different question (cf. [6]).

We enlarge the set of links for which the conjecture holds.

Theorem 1.5. *Let $c_0 \in C_5, c_0 = (\sigma_1\sigma_2\sigma_3\sigma_4)^{10}$. Then $\forall c \in C_5 \backslash \{c_0^{\pm 1}\}, L_c$ is 3-equivalent to a trivial link. Furthermore L_{c_0} is 3-equivalent to $L_{c_0^{-1}}$.*

Remark 1.6. $(\sigma_1\sigma_2\sigma_3\sigma_4)^5$ generates the center of B_5 (cf. [5]).

In Section 2 we prove all closed 4-braids are 3-equivalent to trivial links. We also prove a similar result for 4-braid tangles. In Section 3 we prove all links with no more than 12 crossings are 3-equivalent to trivial links . Section 4 is devoted to proving Theorem 1.5. In Section 5 we prove similar results for $t_3, \bar{t}_4$-moves.

2. 4-BRAIDS

We start out by considering the following fact.

Theorem 2.1 ([8]; Cor.1.9.). *All links with no more than 11 crossings are 3-equivalent to trivial links.*

Proposition 2.2. *Closed 4-braids are 3-equivalent to trivial links.*

Proof. By using the software package **Groups, Algorithms and Programming** (GAP cf. [12]) we are able to reduce every word in C_4 to be of length no more than 9[1]. Hence, any closed 4-braid is 3-equivalent to a link with no more than 9 crossings. The proposition follows from Theorem 2.1. □

Definition 2.3. *A* 4-braid tangle *is a 2-tangle obtained from a 4-braid by connecting 2 pairs of endings in one of the three ways as illustrated in Figure 4.*

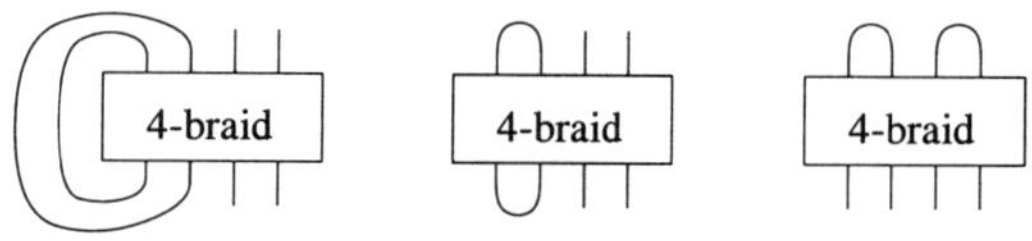

FIGURE 4. 4-braid tangles.

Corollary 2.4. *Any tangle with no more than 10 crossings is 3-equivalent to one of the diagrams in Figure 5, where additional trivial components are allowed in tangles.*

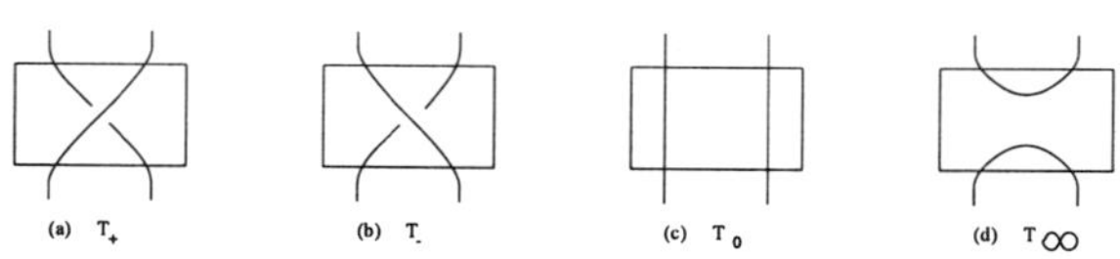

FIGURE 5. Trivial tangles.

Proof. The proof is essentially the same as the proof of Theorem 2.1 (see [8]). □

[1] All but one element $\sigma_1\sigma_2^{-1}\sigma_1\sigma_3\sigma_2^{-1}\sigma_1\sigma_3\sigma_2^{-1}\sigma_1$ (See Figure 1; 9^3_{11} in Rolfsen's notation, cf. [11]) can be reduced to 8 or less crossings.

Corollary 2.5. *Any 4-braid tangle is 3-equivalent to one of the diagrams in Figure 5, where additional trivial components are allowed in tangles.*

Proof. It follows directly from Corollary 2.4 and the proof of Proposition 2.2 where we did not use conjugation. □

Remark 2.6. J. Przytycki conjectured in 1989 that all 2-tangles are 3-equivalent to one of the four 2-tangles in Figure 5, where additional trivial components are allowed in tangles.

3. Links with 12 crossings

Lemma 3.1. *Any link containing the configuration in Figure 6 is 3-equivalent to one with less crossings where inside the circles are single crossings.*

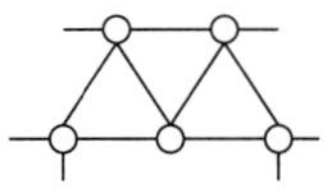

FIGURE 6. Reducible configuration.

This is a simplified version of ([8]; Lemma 2.3).

Proof. Let L be a link containing the configuration in Figure 6. If L contains the configuration as Figure 7(a) then L is 3-equivalent to a link with less crossings as shown in Figure 7 where $\approx$ means ambient isotopy and $\rightarrow$ means that 3-move is applied.

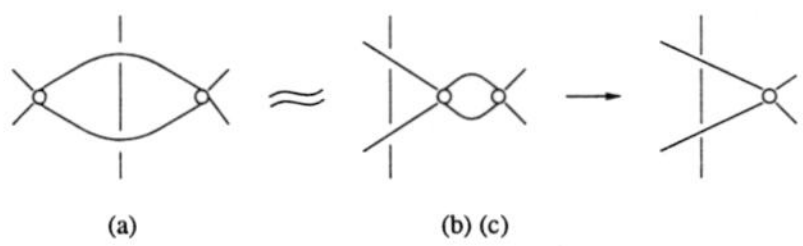

FIGURE 7. Reduction I.

If L does not contain the configuration as Figure 7(a) then L contains the configuration as Figure 8(a) or Figure 9(a), up to mirror image. Once again, L is 3-equivalent to a link with less crossings as indicated in Figure 8 and Figure 9.

FIGURE 8. Reduction II.

Note Figure 9(c) is the mirror image of the configuration Figure 8(a). □

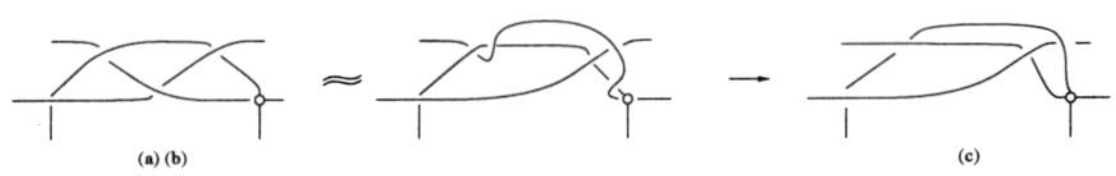

FIGURE 9. Reduction III.

By a *plane arcbody* we mean a pair (R, D) where R is a compact, connected subsurface of a 2-sphere S^2 and D is a part of a link diagram being properly embedded in R. If R is a 2-disk then the arcbody is a tangle. We use L to denote the link diagram of a link L whenever it causes no confusion.

Definition 3.2. *Let L be a link diagram which is neither a connected nor disjoint sum. Let $\{C_i\}$ be a collection of disjoint circles, each C_i cutting L in exactly four points. A* plane arcbody decomposition *of L is the collection of the components of $S^2 - \cup C_i$, where each component is a plane arcbody.*

Proposition 3.3. *Links with no more than 12 crossings are 3-equivalent to trivial links.*

Proof. If L_1 and L_2 are 3-equivalent to trivial links then so are their connected and disjoint sum. Let L be a link with no more than 12 crossings which is neither a connected nor disjoint sum. It suffices to show L is 3-equivalent to a trivial link.

By [4] and [3] L can be decomposed into plane arcbodies in Figure 19 (See Appendix). If the decomposition of L contains only (a)-(c) in Figure 19 then L is clearly 3-equivalent to a trivial link (L is algebraic in Conway's sense). If there are no arcbodies of (s)-(v) in Figure 19 in the decomposition of L then L is 3-equivalent to a trivial link because (d)-(r) in Figure 19 all contain the configuration as Figure 6. Thus L can be reduced to a link with less crossings by Lemma 3.1. By induction L can be reduced to a link whose decomposition contains only (a)-(c) in Figure 19. Hence L is 3-equivalent to a trivial link as claimed.

If the plane arcbodies of (s)-(v) in Figure 19 appear in the decomposition of L, then the decomposition consists of one of the plane arcbodies of (s)-(v) in Figure 19 and (a) and/or (b). Otherwise L has to have more than 12 crossings. In this case L is a 4-braid. (L is a 3-braid if (s) is in the decomposition.) Hence L is 3-equivalent to a trivial link by Proposition 2.2. □

4. 5-BRAID

Now we are ready to prove Theorem 1.5.

Proof of Theorem 1.5. We compute the conjugacy classes of C_5 by using GAP. There are 102 conjugacy classes in C_5. We choose the representatives of these conjugacy classes to have as few crossings as possible. It turns out that only 7 of them have more than 12 crossings. They are c_0, c_0^{-1} and

$$\begin{aligned}
c_1 &= (\sigma_1\sigma_2\sigma_3\sigma_4)^5 = \sigma_1\sigma_2\sigma_3^{-1}\sigma_4\sigma_3^{-1}\sigma_2\sigma_1\sigma_3^{-1}\sigma_2\sigma_3^{-1}\sigma_4\sigma_3^{-1}\sigma_2\sigma_3^{-1},\\
c_2 &= c_1^{-1},\\
c_3 &= \sigma_1\sigma_2\sigma_3^{-1}\sigma_2\sigma_1^{-1}\sigma_2\sigma_3^{-1}\sigma_2\sigma_4^{-1}\sigma_3\sigma_2^{-1}\sigma_1\sigma_2^{-1}\sigma_3\sigma_4^{-1},\\
c_4 &= c_3^{-1},\\
c_5 &= (\sigma_1\sigma_2\sigma_3\sigma_4)^{15} = (\sigma_2^{-1}\sigma_1\sigma_2^{-1}\sigma_3\sigma_4^{-1}\sigma_3)^3.
\end{aligned}$$

Note that if $b_1 = b_2^{-1}$ then L_{b_1} is the mirror image of L_{b_2}. Hence if L_{b_1} is 3-equivalent to a trivial link so is L_{b_2} because the two 3-move operations are mirror image to each other as shown in Figure 3.

Closed c_1, c_3 and c_5 are 3-equivalent to trivial links as shown in Figure 10, Figure 11 and Figure 12[2] respectively. In these graphs $\approx$ and $\rightarrow$ mean, as mentioned before, ambient isotopy and 3-move respectively and $=$ means that we close some string(s) in the braid. The place circled by dotted line is where 3-move is applied.

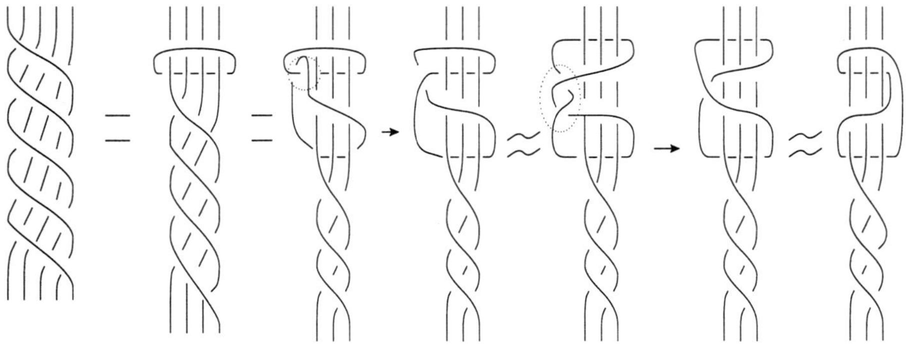

FIGURE 10. Reduction of c_1.

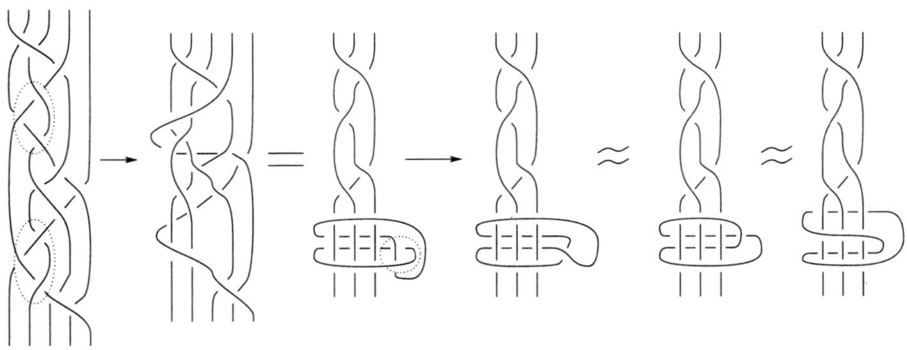

FIGURE 11. Reduction of c_3.

Closed c_0 is 3-equivalent to another closed 5-braid c_0' as shown in Figure 13.

It turns out c_0' is conjugate to c_0^{-1} in C_5. Hence closed c_0 is 3-equivalent to closed c_0^{-1}. Because c_0 is in the center of C_5 (see [5]) it is the only unsettled 5-braid. □

Remark 4.1. The three resulting diagrams in Figure 10, Figure 11 and Figure 12 are 4-braids which are 3-equivalent to trivial links if we close them up by proposition 2.2.

[2]The reduction was made by T. Tsukamoto.

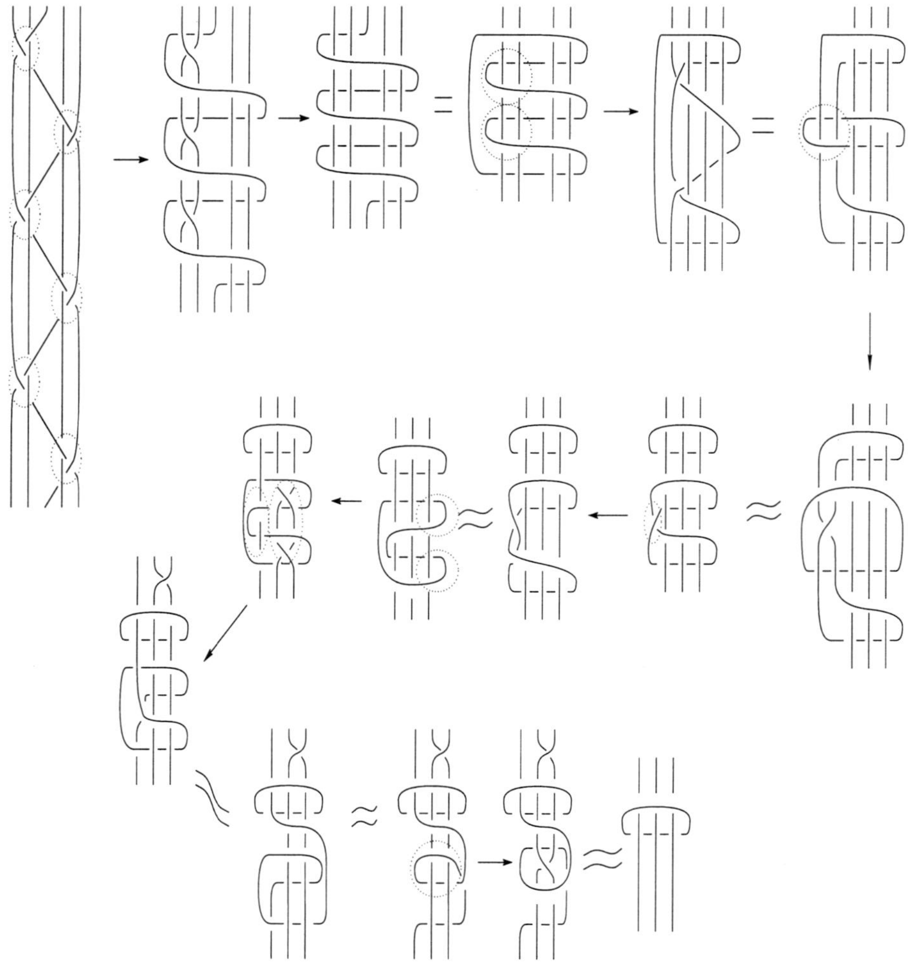

FIGURE 12. Reduction of c_5.

Remark 4.2. In reducing these three closed 5-braids we made use of the following moves, which are consequences of 3-moves and ambient isotopy, in Figure 14 [3].

[3]Observed by T. Tsukamoto.

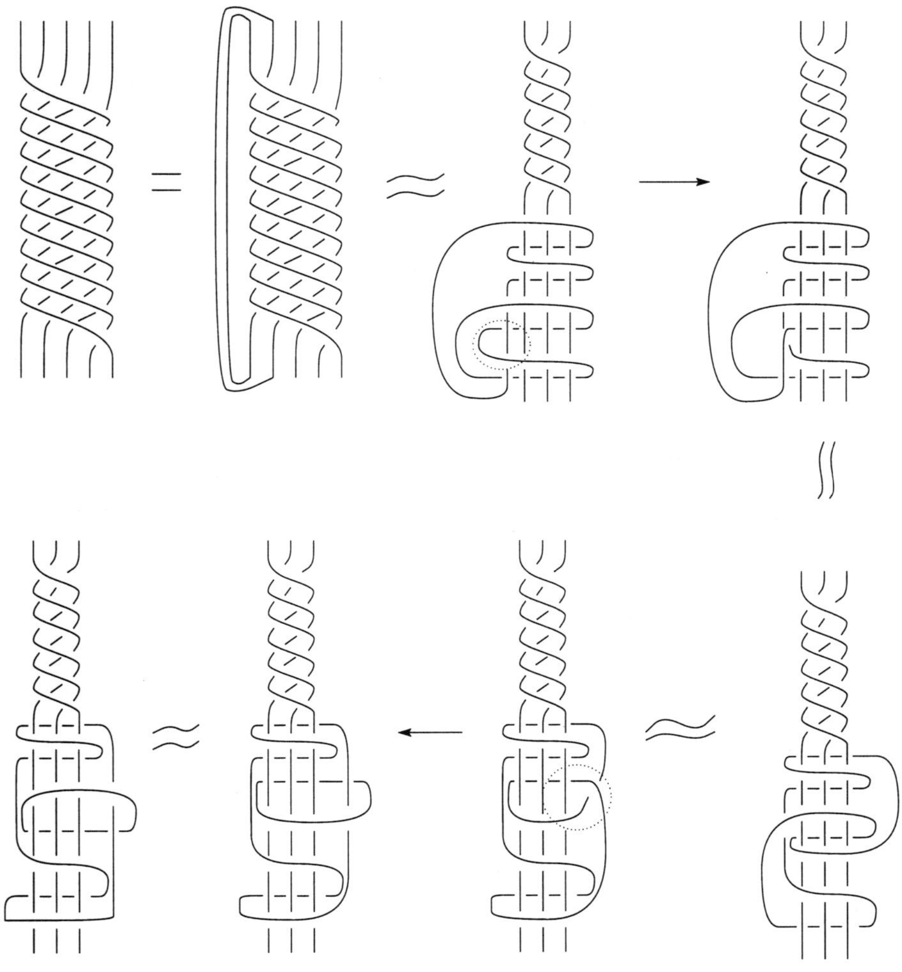

FIGURE 13. Equivalence between c_0 and c_0'.

c_0 is shown in Figure 15. If we can reduce it to a trivial link then all closed 5-braids are 3-equivalent to trivial links. According to Nakanishi (1994) the smallest known potential counterexample to his conjecture is double of Borromean rings, Figure 16, which is a 6-braid with 24 crossings but ours is a 5-braid with 20 crossings.

FIGURE 14. Moves equivalent to 3-move.

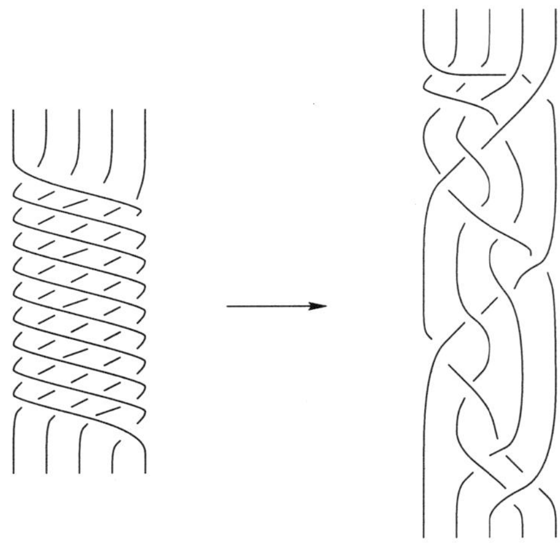

FIGURE 15. Reduction of c_0.

FIGURE 16. Double of Borromean Rings

Remark 4.3. If a link L is 3-equivalent to a trivial link of k components one can find the value k quickly, without a reduction, by analyzing 3-colorings of L. Namely $tri(L) = 3^k$. Here $tri(L)$ is the number of 3-colorings of a diagram of L. We say that a link diagram is 3-colored, or tricolored, if every arc is colored r (red), b (blue) or y (yellow), and at any given crossing either all three colors appear or only one color appears. It was showed that $tri(L)$ is preserved by 3-moves and ambient isotopy (cf. [9]).

5. $t_3, \bar{t}_4$-MOVES

Let t_3-move and $\bar{t}_4$-move defined for oriented links be as in Figure 17 and Figure 18 respectively.

FIGURE 17. t_3-move.

FIGURE 18. $\bar{t}_4$-move.

Two links or tangles are $t_3, \bar{t}_4$-equivalent if one can be transformed into the other by $t_3, \bar{t}_4$-moves and ambient isotopy (cf. [7]). J. Przytycki observed that all links with no more than 11 crossings are $t_3, \bar{t}_4$-equivalent to trivial links (see [8]). Our methods used for 3-equivalence can be applied to prove the following theorem.

Theorem 5.1. (a) *Any oriented link with no more than 12 crossings is $t_3, \bar{t}_4$-equivalent to a trivial link.*

(b) *Any oriented closed 4-braid is $t_3, \bar{t}_4$-equivalent to a trivial link. Furthermore, any oriented 4-braid tangle is $t_3, \bar{t}_4$-equivalent to one of the the diagrams in Figure 5, where additional trivial components are allowed in tangles.*

(c) *Let $c_i, i = 0, \ldots, 5$ be the same as in the proof of Theorem 1.5. Any oriented closed 5-braid other than $c_i, i = 0, \ldots, 5$ is $t_3, \bar{t}_4$-equivalent to a trivial link.*

Sketch of the Proof. The reduction $\sigma_i^3 = 1$ in the braid group is t_3-move and conjugation can be realized as ambient isotopy. Then we follow the method for 3-moves. □

6. ACKNOWLEDGEMENTS

The author wants to thank the Mathematics Department of the George Washington University, especially professor J. Przytycki.

REFERENCES

[1] J. W. Alexander. A lemma on systems of knotted curves. *Proc. Nat. Acad. Sci.*, 9:93–95, 1923.

[2] E. Artin. Theorie der Zöpfe. *abh. Math. Sem. Univ. Hamburg*, 4:47–72, 1926.

[3] A. Caudron. Classification des noeuds et des enlacements. Prepublications, Univ. Paris-Sud, Orsay, 1981.

[4] J. H. Conway. An enumeration of knots and links, and some of their algebraic properties. In *Computational Problems in Abstract Algebra (Proc. Conf., Oxford, 1967)*, pages 329–358. Pergamon, Oxford, 1970.

[5] H. S. M. Coxeter. Factor groups of the braid group. *Proc. Fourth Canadian Math. Congress, Banff*, pages 95–122, 1957.

[6] R. Kirby. Problems in low-dimensional topology. In William H. Kazez, editor, *Geometric topology (Athens, GA, 1993)*, pages 35–473. Amer. Math. Soc., Providence, RI, 1997.

[7] J. H. Przytycki. t_k moves on links. In *Braids (Santa Cruz, CA, 1986)*, pages 615–656. Amer. Math. Soc., Providence, RI, 1988.

[8] J. H. Przytycki. Elementary conjectures in classical knot theory. In *Quantum topology*, pages 292–320. World Sci. Publishing, River Edge, NJ, 1993.

[9] J. H. Przytycki. 3-coloring and other elementary invariants of knots. In Banach Center Publ., Vol.42, *Knot theory (Warsaw, 1995)*, pages 275–295. Polish Acad. Sci., Warsaw, 1998.

[10] J. H. Przytycki. Classical roots of knot theory. *Chaos Solitons Fractals*, 9(4-5):531–545, 1998.

[11] D. Rolfsen. *Knots and links*. Publish or Perish Inc., Houston, TX, 1990. Corrected reprint of the 1976 original.

[12] M. Schönert et al. *GAP - Groups, Algorithms, and Programming*, fifth edition, 1995. web site: http://www-gap.dcs.st-and.ac.uk/ gap/.

Appendix

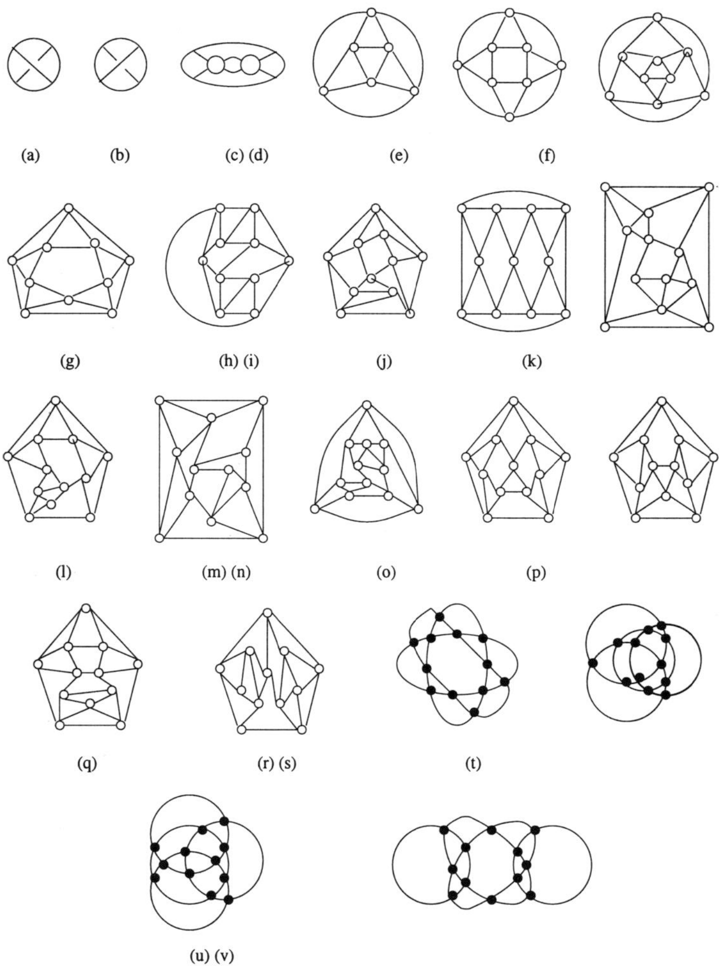

FIGURE 19. Conway polyhedra.

DEPARTMENT OF MATHEMATICS, THE GEORGE WASHINGTON UNIVERSITY
Current address: The Department of Mathematics, SUNY at Buffalo
E-mail address: qichen@acsu.buffalo.edu

A Faithful Representation of the Singular Braid Monoid on Three Strands

Oliver T. Dasbach *†
University of California, Riverside
Department of Mathematics
Riverside, CA 92521 - 0135
USA

Bernd Gemein ‡
Heinrich-Heine-Universität
Mathematisches Institut
Universitätsstr. 1, D-40225 Düsseldorf
Germany

Abstract

We show that the singular braid monoid SB_3 is linear by using a generalization of the classical Burau representation to the singular braid monoid. Furthermore, we use the *HNN*-structure of the group SG_3 in which SB_3 embeds to give a second - group theoretically motivated - solution to the word problem in SB_3. As an application of our approach we compute the homology of SG_3.

1 Preliminaries

While knots are important for our daily life - ties, shoelaces etc. - singular knots are not. However, since the theory of Vassiliev invariants started (see e.g. [BL93]) mathematicians became more and more interested in singular knotted objects, i.e. objects having a finite number of transversal self intersections. One of these objects are singular braids which form the singular braid monoid SB_n, i.e. the monoid generated by the standard generators $\sigma_1, \ldots, \sigma_{n-1}$ of the braid group B_n plus the additional singular generators $\tau_1, \ldots, \tau_{n-1}$, which do not have inverses. A presentation for the singular braid monoid in terms of these generators, build up from the usual choice of a presentation for the braid group is not hard to deduce (cf. [Bir93]).

The theory of Vassiliev invariants suggests a homomorphism from SB_n into the integral group ring $\mathbb{Z}B_n$ of the braid group. This homomorphism is given by mapping a singular generator τ_i to $\sigma_i - \sigma_i^{-1}$ and σ_i to itself (cf. [Bir93]). A famous conjecture of Joan Birman ([Bir93], [FRZ96]) asserts that this homomorphism is injective.

As it appeared to be very useful for knot theorist to have a complete understanding of the braid group B_3 (see e.g. [BM93]) the purpose of this paper is to give some analogous results for the braid monoid SB_3.

By results of Fenn et al. [FKR98] the monoid SB_n embeds in a group SG_n. Using this result we will prove that an extension to SB_n of the Burau representation for B_n that was defined in [Gem97a] is faithful for $n = 3$. Since the Burau representation of B_n is known to be unfaithful for $n \geq 5$ [LP93, Big99] this result cannot carry over at least for $n \geq 5$. (Recently, proofs were announced that the braid groups are linear, though. [Kra99, Big]) So we will give a second solution to the word problem for SB_3 by using the group theoretical structure of SG_3. We will

*supported by the Deutsche Forschungsgemeinschaft (DFG)
†e-mail: *kasten@math.columbia.edu, http://www.math.uni-duesseldorf.de/~kasten*
‡e-mail: *gemein@rz.uni-duesseldorf.de*

extend this idea in a forthcoming paper [DG98] to give a solution to the word problem for all SB_n.

It is worth mentioning that the faithfulness of the singular Burau representation yields an algorithm for the solution of the word problem for SB_3 that has a time complexity which is linear in the length of the word.

This paper was written during a visit of the first author at the Columbia University. He would like to thank Columbia for the warm hospitality and especially Joan Birman for many discussions and fruitful comments on an earlier version of this text. The second author would like to thank Wilhelm Singhof for many useful suggestions and remarks. Furthermore both authors are very grateful to E. Mail for her invaluable help.

2 A new presentation for the singular braid monoid

We recall the well-known presentation for the monoid of singular braids on n strands:

Proposition 2.1 (Baez [Bae92], Birman [Bir93]) *The monoid SB_n is generated by the elements*

$$\sigma_i^{\pm 1},\ i = 1, \ldots, n-1,\ \tau_i,\ i = 1, \ldots, n-1$$

(see Figure 1) satisfying the following relations:

$$\begin{aligned}
\sigma_i \sigma_i^{-1} &= 1 \textit{ for all } i && (1)\\
\sigma_i \sigma_{i+1} \sigma_i = \sigma_{i+1} \sigma_i \sigma_{i+1} \ \textit{ and } \ \sigma_i \sigma_j &= \sigma_j \sigma_i \quad \textit{for } j > i+1 && (2)\\
\tau_{i+1} \sigma_i \sigma_{i+1} &= \sigma_i \sigma_{i+1} \tau_i && (3)\\
\sigma_{i+1} \sigma_i \tau_{i+1} &= \tau_i \sigma_{i+1} \sigma_i && (4)\\
\sigma_i \tau_j &= \tau_j \sigma_i \quad \textit{for } |j-i| \neq 1 && (5)\\
\tau_i \tau_j &= \tau_j \tau_i \quad \textit{for } j > i+1. && (6)
\end{aligned}$$

We use this proposition to derive a new presentation for the singular braid monoid which is more suitable for our purposes:

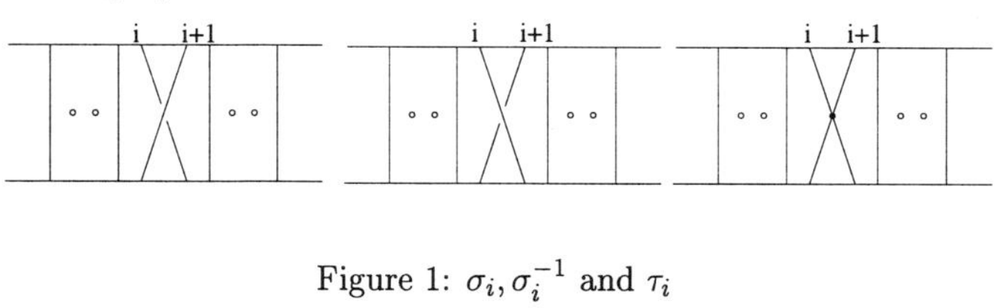

Figure 1: σ_i, σ_i^{-1} and τ_i

Proposition 2.2 *The monoid SB_n is generated by the elements*

$$\sigma_i^{\pm 1},\ i = 1, \ldots, n-1, \ \textit{ and } \tau$$

satisfying the following relations:

$$\begin{aligned}
\sigma_i \sigma_i^{-1} &= 1 \textit{ for all } i && (7)\\
\sigma_i \sigma_{i+1} \sigma_i = \sigma_{i+1} \sigma_i \sigma_{i+1} \ \textit{ and } \ \sigma_i \sigma_j &= \sigma_j \sigma_i \quad \textit{for } j > i+1 && (8)\\
(\sigma_2 \sigma_1)^3 \tau &= \tau (\sigma_2 \sigma_1)^3 && (9)\\
\sigma_i \tau &= \tau \sigma_i \quad \textit{for } i \neq 2 && (10)\\
\sigma_2 \sigma_3 \sigma_1 \sigma_2 \tau \sigma_2 \sigma_3 \sigma_1 \sigma_2 \tau &= \tau \sigma_2 \sigma_3 \sigma_1 \sigma_2 \tau \sigma_2 \sigma_3 \sigma_1 \sigma_2 \textit{ for } n > 3. && (11)
\end{aligned}$$

Proof We will see that the new relations are included in the old ones. So we will concentrate ourselves to show that all other relations can be derived from our relations.

First we note that we have

$$\tau_{i+1} = \sigma_i \sigma_{i+1} \tau_i \sigma_{i+1}^{-1} \sigma_i^{-1}. \tag{12}$$

1. **Relations of the form $\sigma_i \tau_i = \tau_i \sigma_i$**

   We will show that these relations can be deduced from $\sigma_1 \tau_1 = \tau_1 \sigma_1$, Relation (12) and the relations in the braid group (7) - (8). We will proceed by induction on i:

$$\begin{aligned}
\sigma_i \tau_i &= \sigma_i \sigma_{i-1} \sigma_i \tau_{i-1} \sigma_i^{-1} \sigma_{i-1}^{-1} \\
&= \sigma_{i-1} \sigma_i \sigma_{i-1} \tau_{i-1} \sigma_i^{-1} \sigma_{i-1}^{-1} \\
&= \sigma_{i-1} \sigma_i \tau_{i-1} \sigma_{i-1} \sigma_i^{-1} \sigma_{i-1}^{-1} \\
&= \sigma_{i-1} \sigma_i \tau_{i-1} \sigma_i^{-1} \sigma_{i-1}^{-1} \sigma_i \\
&= \tau_i \sigma_i
\end{aligned}$$

2. **Relations of the form $\sigma_i \tau_j = \tau_j \sigma_i$, $|i-j| > 1$**

   We will show that all relations of this form can be derived from the relations $\sigma_i \tau_1 = \tau_1 \sigma_i, \quad i \neq 2$ as well as (12) and the braid group relations (7) - (8).

   Case 1: If $1 < j < i-1$ then we have

$$\begin{aligned}
\sigma_i \tau_j &= \sigma_i \sigma_{j-1} \sigma_j \tau_{j-1} \sigma_j^{-1} \sigma_{j-1}^{-1} \\
&= \sigma_{j-1} \sigma_j \tau_{j-1} \sigma_j^{-1} \sigma_{j-1}^{-1} \sigma_i \\
&= \tau_j \sigma_i
\end{aligned}$$

   where we use (12), (8) and induction on j.

   Case 2: In the braid group with $w := \sigma_{i+1} \sigma_{i+2} \sigma_i \sigma_{i+1}$ it holds:

$$\sigma_i w = w \sigma_{i+2}. \tag{13}$$

   Thus, for $j = i+2$:

$$\begin{aligned}
\sigma_i \tau_{i+2} &= \sigma_i w \tau_i w^{-1} \\
&= w \sigma_{i+2} \tau_i w^{-1} \\
&= w \tau_i \sigma_{i+2} w^{-1} \\
&= w \tau_i w^{-1} \sigma_i \\
&= \tau_{i+2} \sigma_i
\end{aligned}$$

   where we use (12), (13) and Case 1.

   Case 3: If $j \geq i+3$ then we can proceed as in Case 1 to show that

$$\sigma_i \tau_j = \tau_j \sigma_i, \quad i > 1 + j$$

   follows from our relations and Case 2.

3. **Relations of the form** $\tau_i\tau_j = \tau_j\tau_i,\ i < j, j - i > 1$

   Our aim is to reduce these relations to $\tau_1\tau_3 = \tau_3\tau_1$ with the help of (12) and the relation $\sigma_i\tau_j = \tau_j\sigma_i$, for $|j - i| \neq 1$. We use induction on the pair (i, j) that is ordered lexicographically.

   If $i > 1$ then

$$\begin{aligned}\tau_i\tau_j &= \sigma_{i-1}\sigma_i\tau_{i-1}\sigma_i^{-1}\sigma_{i-1}^{-1}\tau_j\\ &= \tau_j\sigma_{i-1}\sigma_i\tau_{i-1}\sigma_i^{-1}\sigma_{i-1}^{-1}\\ &= \tau_j\tau_i.\end{aligned}$$

   In the same way one can deduce $\tau_1\tau_j = \tau_j\tau_1$ for $j > 3$.

4. **Relations of the form** $\sigma_{i+1}\sigma_i\tau_{i+1} = \tau_i\sigma_{i+1}\sigma_i$

   We will show that these relations follow from Relation (12) together with the relations in the braid group and the initial relation:

$$\sigma_2\sigma_1\tau_2 = \tau_1\sigma_2\sigma_1 \tag{14}$$

   We will need the following two relations in the braid group which can be easily tested:

$$\sigma_i^{-1}\sigma_{i+1}^{-1}\sigma_{i-1}^{-1}\sigma_i^{-1} = \sigma_{i-1}^{-1}\sigma_i^{-1}\sigma_{i+1}^{-1}\sigma_{i-1}^{-1}\sigma_i^{-1}\sigma_{i+1} \tag{15}$$

and

$$\sigma_{i-1}^{-1}\sigma_i^{-1}\sigma_{i+1}^{-1}\sigma_{i-1}\sigma_i\sigma_{i+1} = \sigma_i\sigma_{i+1}\sigma_{i-1}^{-1}\sigma_i^{-1} \tag{16}$$

$$\begin{aligned}\sigma_i^{-1}\sigma_{i+1}^{-1}\tau_i\sigma_{i+1}\sigma_i &= \sigma_i^{-1}\sigma_{i+1}^{-1}\sigma_{i-1}^{-1}\sigma_i^{-1}\tau_{i-1}\sigma_i\sigma_{i-1}\sigma_{i+1}\sigma_i\\ &= \sigma_{i-1}^{-1}\sigma_i^{-1}\sigma_{i+1}^{-1}\sigma_{i-1}^{-1}\sigma_i^{-1}\sigma_{i+1}\tau_{i-1}\sigma_{i+1}^{-1}\sigma_i\sigma_{i-1}\sigma_{i+1}\sigma_i\sigma_{i-1}\\ &= \sigma_{i-1}^{-1}\sigma_i^{-1}\sigma_{i+1}^{-1}\sigma_{i-1}^{-1}\sigma_i^{-1}\tau_{i-1}\sigma_i\sigma_{i-1}\sigma_{i+1}\sigma_i\sigma_{i-1}\\ &= \sigma_{i-1}^{-1}\sigma_i^{-1}\sigma_{i+1}^{-1}\sigma_{i-1}\sigma_i\tau_{i-1}\sigma_i^{-1}\sigma_{i-1}^{-1}\sigma_{i+1}\sigma_i\sigma_{i-1}\\ &= \sigma_{i-1}^{-1}\sigma_i^{-1}\sigma_{i+1}^{-1}\sigma_{i-1}\sigma_i\sigma_{i+1}\tau_{i-1}\sigma_{i+1}^{-1}\sigma_i^{-1}\sigma_{i-1}^{-1}\sigma_{i+1}\sigma_i\sigma_{i-1}\\ &= \sigma_i\sigma_{i+1}\sigma_{i-1}^{-1}\sigma_i^{-1}\tau_{i-1}\sigma_i\sigma_{i-1}\sigma_{i+1}^{-1}\sigma_i^{-1}\\ &= \tau_{i+1}\end{aligned}$$

Now we are left with the braid group relations, Relation (10) as well as Relation (12) and

$$\tau_2 = \sigma_1^{-1}\sigma_2^{-1}\tau_1\sigma_2\sigma_1 \tag{17}$$

$$\tau_1\tau_3 = \tau_3\tau_1. \tag{18}$$

Relation (18) is equivalent to Relation (11).
Relation (17) is equivalent to:

$$\begin{aligned}\sigma_2\sigma_1\sigma_1\sigma_2\tau_1 &= \tau_1\sigma_2\sigma_1\sigma_1\sigma_2\\ \Longleftrightarrow \sigma_2\sigma_1\sigma_1\sigma_2\sigma_1^2\tau_1\sigma_1^{-2} &= \tau_1\sigma_2\sigma_1\sigma_1\sigma_2\\ \Longleftrightarrow (\sigma_2\sigma_1)^3\tau_1 &= \tau_1(\sigma_2\sigma_1)^3.\end{aligned}$$

Finally, we can skip the Relations (12) because the τ_j, $j > 1$, only occur in these relations and we set $\tau := \tau_1$. □

Remark 2.3 *As the braid group B_n admits a presentation with two generators σ_1 and $A := \sigma_1 \cdots \sigma_{n-1}$ for every n (cf. Artin's initial paper [Art25]) we can rewrite the presentation for the singular braid monoid in terms of three generators σ_1, A and τ_1. We will omit the details here.*

Corollary 2.4 *The monoid SB_3 is generated by the elements ${\sigma_1}^{\pm 1}, {\sigma_2}^{\pm 1}, \tau_1$ satisfying the following relations:*

1. $\sigma_1 {\sigma_1}^{-1} = \sigma_2 {\sigma_2}^{-1} = 1.$
2. $\sigma_1\sigma_2\sigma_1 = \sigma_2\sigma_1\sigma_2$
3. $\tau_1(\sigma_2\sigma_1)^3 = (\sigma_2\sigma_1)^3\tau_1$
4. $\sigma_1\tau_1 = \tau_1\sigma_1$

The following theorem of Fenn, Keyman and Rourke [FKR98] will make our arguments on SB_n much easier:

Theorem 2.5 *Let SG_n be the group given by the monoid presentation of SB_n considered as a group presentation. Then the natural homomorphism of SB_n into SG_n is an embedding.*

3 A faithful representation of SB_3

One can find a representation of the singular braid groups which is an extension of the famous Burau representation of the braid groups (cf. [Gem97a]). For SB_3 this representation looks like:

Proposition 3.1 *The map β_s given by*

$$\sigma_1 \mapsto \begin{pmatrix} -t & 1 \\ 0 & 1 \end{pmatrix}, \sigma_2 \mapsto \begin{pmatrix} 1 & 0 \\ t & -t \end{pmatrix}, \tau_1 \mapsto \begin{pmatrix} 1-y-ty & y \\ 0 & 1 \end{pmatrix}$$

yields a representation of the singular braid monoid into a matrix ring:

$$\beta_s : SB_3 \to M_2(\mathbb{Z}[t, t^{-1}, y]).$$

We will show that this representation is faithful. For this purpose we need the following easy consequence (cf. [Fin89]) of a theorem of P.M. Cohn [Coh68]:

Theorem 3.2 *Let $d \neq 1, 2, 3, 7, 11$ be square-free, i.e. d is not divisible by the square of an integer, and $\omega := \sqrt{-d}$ if $d \equiv 1$ or 2 modulo 4 or $\omega := (1+\sqrt{-d})/2$ for $d \equiv 3$ modulo 4 and let $\mathcal{O}_d = \mathbb{Z} + \omega\mathbb{Z}$ be the imaginary quadratic integers in $\mathbb{Q}[\omega]$. Furthermore, let A, B and C be the following elements in $PSL_2(\mathcal{O}_d)$:*

$$A := \begin{pmatrix} 1 & 1 \\ 0 & 1 \end{pmatrix}, B := \begin{pmatrix} 0 & -1 \\ 1 & 0 \end{pmatrix} \text{ and } C := \begin{pmatrix} 1 & \omega \\ 0 & 1 \end{pmatrix}.$$

The subgroup $PE_2(\mathcal{O}_d)$ of $PSL_2(\mathcal{O}_d)$ generated by all matrices of the forms

$$\begin{pmatrix} 1 & x \\ 0 & 1 \end{pmatrix} \text{ and } \begin{pmatrix} 1 & 0 \\ y & 1 \end{pmatrix}$$

for x and y in $\mathcal{O}_d$, has the presentation:

$$PE_2(\mathcal{O}_d) = \langle A, B, C \,|\, B^2 = (AB)^3 = [A, C] = 1 \rangle.$$

With $\Sigma_1 := A, T_1 := C$ *and*

$$\Sigma_2 := (ABA)^{-1} = \begin{pmatrix} 1 & 0 \\ -1 & 1 \end{pmatrix}$$

we get - after some easy transformations - the presentation:

$$PE_2(\mathcal{O}_d) = \langle \Sigma_1, \Sigma_2, T_1 \,|\, (\Sigma_1\Sigma_2\Sigma_1)^2 = (\Sigma_1\Sigma_2)^3 = 1,\ \Sigma_1 T_1 = T_1 \Sigma_1 \rangle.$$

Now we have all the tools to show that

Theorem 3.3 *The singular Burau representation* $\beta_s : SB_3 \longrightarrow M_2(\mathbb{Z}[t, t^{-1}, y])$ *is faithful.*

Proof The arguments are essentially the same as in the proof of Magnus and Peluso (cf. [Bir74]) of the faithfulness of the Burau representation for 3-string braids. While in this proof the well-known presentation of the group $PSL_2(\mathbb{Z})$ is used we will make use of the presentation of $PE_2(\mathbb{Z}[\omega])$ of Theorem 3.2 for a suitable ring of integers $\mathbb{Z}[\omega]$.

For reasons of convenience we choose $d = 5$, so $\omega := \sqrt{-5}$.

We will denote the extension of β_s to SG_3 also by β_s. So

$$\beta_s : SG_3 \to M_2(\mathbb{Z}[t, t^{-1}, y, 1/(1 - y - ty)]).$$

The image of β_s in this matrix group is - of course - a quotient of SG_3. Furthermore by setting $t := -1$ and $y := \omega$ it naturally maps onto $PE_2(\mathbb{Z}[\omega])$.

Thus we have a homomorphism from SG_3 onto $PE_2(\mathbb{Z}[\omega])$ given by $\sigma_1 \mapsto \Sigma_1, \sigma_2 \mapsto \Sigma_2$ and $\tau_1 \mapsto T_1$ where Σ_1, Σ_2 and T_1 are as in Theorem 3.2. Moreover by comparing the presentations of SG_3 and $PE_2(\mathbb{Z}[\omega])$ we see that the kernel of this homomorphism is the normal closure of the element $(\sigma_1\sigma_2\sigma_1)^2$, which is - as is easy to see - a central element in SG_3. Hence, the kernel is cyclic.

Thus, the image of SG_3 under β_s is isomorphic to SG_3 modulo a power of $\beta_s((\sigma_1\sigma_2\sigma_1)^2)$. Since $(\sigma_1\sigma_2\sigma_1)^2$ is mapped by β_s to the element

$$\begin{pmatrix} t^3 & 0 \\ 0 & t^3 \end{pmatrix}$$

of infinite order, we see that SG_3 must be isomorphic to its image under β_s.

Because of the Embedding Theorem 2.5 the theorem follows. □

3.1 The Birman-Conjecture for SB_3

Let $\eta : SB_n \longrightarrow \mathbb{Z}B_n$ be the Birman homomorphism which maps $\tau_1 \mapsto \sigma_1 - \sigma_1^{-1}$ and σ_j to itself. In [Gem97b] the following result is proved:

Theorem 3.4 *Let* b *and* b' *be two braids in* SB_n *with* $\eta(b) = \eta(b')$*. Then we have* $\beta_s(b) = \beta_s(b')$*.*

Because the singular Burau representation is faithful for $n = 3$ this means that the Birman homomorphism is also faithful, i.e. the Birman conjecture is valid for $n = 3$. We learned that this result was already obtained by Antal Járai [JJ99].

It might be interesting, however, for the reader to see how the linearity of SB_3 and the injectivity of η relate. Therefore we will give a sketchy and informal proof of Theorem 3.4. For full details the reader is referred to [Gem97b].

For notational reasons we will consider the special case SB_3. However, the argument holds also for higher n. Since the determinant of the Burau matrix of a given braid with m singularities equals $t^k \cdot (1-y-ty)^m$ for some $k \in \mathbb{Z}$, two braids with a different number of singularities cannot map to the same matrix under β_s. The same holds for the Birman homomorphism. Therefore we may restrict ourselves to braids with a fixed number m of singularities. The set of all braids with exactly m singularities will be denoted by $SB_3^{(m)}$ in the sequel.

The main idea of the proof is to imitate the Birman homomorphism on the level of Burau matrices. If we deal with braids having exactly one singularity, this can be done easily: Substituting y by 1 corresponds to a right-handed resolution of the singularity, substituting y by x^{-1} corresponds to a left-handed resolution of the singularity. This induces a well defined homomorphism from the matrix ring $M_2(\mathbb{Z}[t, t^{-1}, y])$ into the group ring $\mathbb{Z}[M_2(\mathbb{Z}[t, t^{-1}])]$ imitating the Birman homomorphism.

If the number of singularities is greater than one, we cannot proceed in the same way. The two indicated substitutions would correspond to a right-handed (resp. left-handed) resolution of all the singularities. Unfortunately, we also have to consider cases where some singularities are resolved in a right-handed way while others are resolved in a left-handed way.

Therefore we change our point of view slightly: First, we number the singularities of our braid b from 1 to m. Afterwards we assign to the j-th singularity of the braid the matrix

$$\begin{pmatrix} 1 - y_j - ty_j & y_j \\ 0 & 1 \end{pmatrix}$$

rather than the usual matrix

$$\begin{pmatrix} 1 - y - ty & y \\ 0 & 1 \end{pmatrix}.$$

In this way we define a modification of the Burau matrix of b. Its entries take values in the polynomial ring $\mathbb{Z}[t, t^{-1}, y_1, \ldots, y_m]$. Of course, the numbering of the singularities was somewhat arbitrary. Therefore we shall regard these modified matrices only up to permutation of the indices of the y_i. (To be more precise, we let the symmetric group Σ_m act on $M_2(\mathbb{Z}[t, t^{-1}, y_1, \ldots, y_m])$ and consider the orbits. By abuse of notation we shall denote the set of these orbits also with $M_2(\mathbb{Z}[t, t^{-1}, y_1, \ldots, y_m])$.) It is obvious that we obtain the Burau representation of b out of its modified Burau matrix by the projection p which sends all the y_i to y.

We have introduced the modified Burau matrix in order to compute the (regular) Burau matrices of all possible resolutions. In fact, define a matrix resolution

$$r : M_2(\mathbb{Z}[t, t^{-1}, y_1, \ldots, y_m]) \to M_2(\mathbb{Z}[t, t^{-1}])$$

as a projection where, in addition, any y_i is mapped either to 1 or to t^{-1}. The index $\mu(r)$ is defined to be the number of y_i which are sent to t^{-1}. Clearly, a given resolution is not well defined on our orbits. However, taking formal sums over all possible resolutions gives a well defined map ρ. So, if M is a modified Burau matrix, then

$$\rho(M) = \bigoplus_r (-1)^{\mu(r)} \cdot r(M).$$

Note that the sum in the formula is a formal sum in the group ring $\mathbb{Z}[M_2(\mathbb{Z}[t,t^{-1}])]$.

Easy calculations show that the application ρ corresponds to the Birman homomorphism η on the level of matrices. In fact, we get the following commutative diagram:

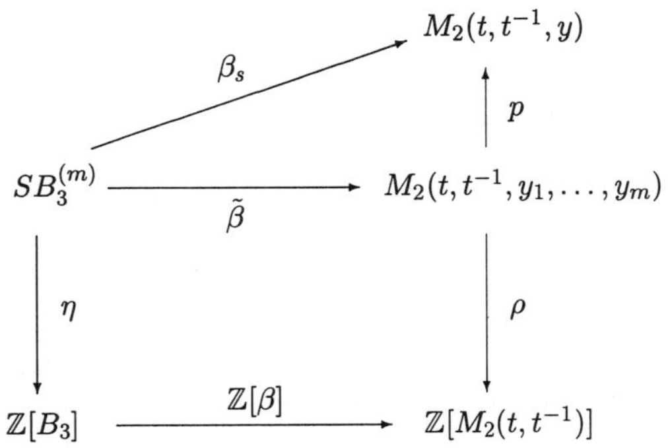

Here $\tilde{\beta}$ denotes the application which maps a braid to its corresponding modified Burau matrix. With $\mathbb{Z}[\beta]$ we denote the extension of the usual (regular) Burau homomorphism to the group rings.

We now claim:

Claim 1 Let $M, M' \in M_2(x^{\pm 1}, y_1, \ldots, y_m)$ be two elements in the image of $\tilde{\beta}$ with $\rho(M) = \rho(M')$. Then we have $p(M) = p(M')$.

Let us assume that the claim is true. then the proof of Theorem 3.4 becomes easy diagram chasing:

Let b, b' be elements of $SB_3^{(m)}$ and suppose that $\eta(b) = \eta(b')$. It follows that $(\mathbb{Z}[\beta]\circ\eta)(b) = (\mathbb{Z}[\beta]\circ\eta)(b')$ and by commutativity of the diagram that $(\rho\circ\tilde{\beta})(b) = (\rho\circ\tilde{\beta})(b')$. Using Claim 1 we get $(p\circ\tilde{\beta})(b) = (p\circ\tilde{\beta})(b')$ and - again by commutativity of the diagram - $\beta(b) = \beta(b')$.

Thus, we only have to show that Claim 1 holds. This is the most technical part of the proof. In fact, we have to figure out in how far the matrices of the formal sum $\rho(M)$ determine the matrix M. This leads to one system of linear equations for each entry of the matrix, which may be solved after having observed the following two facts:

1. If M is in the image of $\tilde{\beta}$, then each y_i cannot appear in the matrix with powers greater than 1.

2. We may use the determinant of the matrices in our formal sum in order to compute the index of the resolution which has produced them. This fact is important when solving the equations.

With these two observations and some tedious computations, we derive that two matrices M and M' are mapped to the same formal sum under ρ if their entries differ by permutations of the indices of the y_i. Hence, they vanish under the projection p.

4 A second solution to the word problem in SB_3

To give a second solution to the word problem for SB_3, i.e. the problem whether two elements in SB_3 are equivalent, we will need Britton's Lemma that can be applied to the group SG_3

Lemma 4.1 (Britton [Bri63]) *Let $H = \langle S | R \rangle$ be a presentation of the group H with a set of generators S and relations R in these generators.*

Furthermore let G be a HNN-extension of H of the following form:

$$G = \langle S, t | R, t^{-1} X_i t = X_i, i \in I \rangle$$

for some index set I, where X_i are words over S.

Let W be a word in the generators of G which involves t.

If $W = 1$ in G then W contains a subword $t^{-1}Ct$ or tCt^{-1} where C is a word in S, and C, regarded as an element of the group H, belongs to the subgroup X of H generated by the X_i.

We will rather solve the word problem for SG_3 than for SB_3. By Corollary 2.4 SG_3 has a presentation as in Britton's Lemma. So to solve the word problem in SG_3 we have to decide whether a given word in the generators of B_3 is element of the subgroup H_3 generated by the elements with which τ_1 commutes: σ_1 and $(\sigma_2\sigma_1)^3$.

This decision problem, called membership problem, would not be hard to solve with the help of the Burau representation of B_3. However we promised to give a puristic proof. So we will choose the approach of Xu [Xu92] for the word problem for B_3 - which was generalized most recently to arbitrary B_n by Birman, Ko and Lee [BKL98] - to solve the membership problem for the subgroup H_3.

We briefly recall this approach using the notation of Birman, Ko and Lee. The first step is to rewrite the presentation of B_3 in terms of the new generators: $a_{21} := \sigma_1, a_{32} := \sigma_2$ and $a_{31} := \sigma_2\sigma_1\sigma_2^{-1}$.

So we get a new presentation

$$B_3 = \langle a_{21}, a_{32}, a_{31} | a_{32}a_{21} = a_{31}a_{32} = a_{21}a_{31} \rangle.$$

Using the element $\delta := a_{32}a_{21}$ one can show now that every element of B_3 can be brought into a unique normal form $\delta^k P$ for some k with P a positive word, i.e. only positive exponents occurs, in the generators a_{21}, a_{32} and a_{31}, such that none of the subwords $a_{32}a_{21}$, $a_{31}a_{32}$ or $a_{21}a_{31}$ appear in P.

Lemma 4.2 *The membership problem for the subgroup H_3 of B_3 generated by the elements σ_1 and $(\sigma_2\sigma_1)^3$ can be solved.*

Proof First of all we see that H_3 is abelian and $(\sigma_2\sigma_1)^3 = \delta^3$. Therefore if we want to bring a word into the normal form we only have to look for σ_1^k for $k \in \mathbb{Z}$. If k is not negative, then the normal form for an element $(\sigma_2\sigma_1)^{3l}\sigma_1^k$ is simply $\delta^{3l}{a_{21}}^k$.

If $k \leq 0$ then the following identities are easy to see:

$$\begin{aligned}
(\sigma_2\sigma_1)^{3l}\sigma_1^{3k} &= \delta^{3l+3k}(a_{31}a_{21}a_{32})^{-k} \\
(\sigma_2\sigma_1)^{3l}\sigma_1^{3k-1} &= \delta^{3l+3k-1}a_{32}(a_{31}a_{21}a_{32})^{-k} \\
(\sigma_2\sigma_1)^{3l}\sigma_1^{3k-2} &= \delta^{3l+3k-2}a_{21}a_{32}(a_{31}a_{21}a_{32})^{-k}.
\end{aligned}$$

Therefore for every word w in the braid group B_3 we can bring it into its unique normal form and compare this form with the normal forms for the elements in H_3. Hence, the membership problem for H_3 is solvable. □

Thus we have proved:

Theorem 4.3 *Let* $w_1 = b_1\tau_1 b_2\tau_1 \cdots \tau_1 b_m$ *and* $w_2 = c_1\tau_1 c_2\tau_1 \cdots \tau_1 c_l$ *be two words in the generators* σ_1, σ_2 *and* τ_1 *in* SB_3*, where the* b_j *and* c_j *are words in* B_3.

Then w_1 *and* w_2 *are equal in* SB_3 *if and only if* $b_m c_l^{-1}$ *is in the subgroup* H_3 *of* B_3 *that is generated by* σ_1 *and* $(\sigma_2\sigma_1)^3$ *and* $b_1\tau_1 b_2\tau_1 \cdots b_{m-1}b_m$ *and* $c_1\tau_1 c_2\tau_1 \cdots c_{l-1}c_l$ *are equal in* SB_3.

This gives a solution to the word problem in SB_3 *because the membership problem for* H *is solvable by Lemma 4.2 and the word problem in* B_3 *is solvable - as mentioned above - as well.*

4.1 The homology of SG_3

We can use the structure of SG_3 as an *HNN*-extension of B_3 to compute the homology of SG_3.

Corollary 4.4

$$H_n(SG_3) = \begin{cases} \mathbb{Z}\times\mathbb{Z} & \textit{for} \quad n = 1,2 \\ \mathbb{Z} & \textit{for} \quad n = 0,3 \\ \{0\} & \textit{else} \end{cases}$$

Proof The group SG_3 is *HNN*-extension of B_3 over a subgroup of B_3 isomorphic to $\mathbb{Z}\times\mathbb{Z}$. Therefore, we have an exact sequence [Bro94]:

$$\ldots \longrightarrow H_n(\mathbb{Z}\times\mathbb{Z}) \longrightarrow H_n(B_3) \longrightarrow H_n(SB_3) \longrightarrow H_{n-1}(\mathbb{Z}\times\mathbb{Z}) \longrightarrow \ldots$$

Since B_3 is isomorphic to the fundamental group of the complement of the trefoil knot, the non-trivial homology groups of B_3 are $H_1(B_3) \cong H_0(B_3) \cong \mathbb{Z}$ and we have for $\mathbb{Z}\times\mathbb{Z}$: $H_2(\mathbb{Z}\times\mathbb{Z}) \cong H_0(\mathbb{Z}\times\mathbb{Z}) \cong \mathbb{Z}$ and $H_1(\mathbb{Z}\times\mathbb{Z}) \cong \mathbb{Z}\times\mathbb{Z}$.

Therefore, the homology groups of SG_3 are trivial for $n \geq 4$ and the third homology group is isomorphic to the second homology group of $\mathbb{Z}\times\mathbb{Z}$.

The abelianization, i.e. $H_1(SG_3)$, of SG_3 is $\mathbb{Z}\times\mathbb{Z}$ and the map $H_1(B_3) \to H_1(SG_3)$ is injective. Hence, $H_2(SG_3) \cong H_1(\mathbb{Z}\times\mathbb{Z}) \cong \mathbb{Z}\times\mathbb{Z}$. □

References

[Art25] E. Artin, *Theorie der Zöpfe*, Abh. Math. Sem. Univ. Hamburg **4** (1925), 47–72.

[Bae92] J. C. Baez, *Link Invariants of Finite Type and Perturbation Theory*, Lett. Math. Phys. **26** (1992), no. 1, 43–51.

[Big] S. Bigelow, *Braid groups are linear*, in preparation.

[Big99] S. Bigelow, *The Burau representation is not faithful for* $n \geq 5$, Geometry and Topology **3** (1999), no. 16, 397 – 404.

[Bir74] J. S. Birman, *Braids, Links and Mapping Class Groups*, Princeton University Press, Princeton, N.J., 1974, Ann. of Math. Studies 82.

[Bir93] J. S. Birman, *New Points of View in Knot Theory*, Bull. Am. Math. Soc. **28** (1993), no. 2, 253–287.

[BKL98] J. S. Birman, K. H. Ko, and S. J. Lee, *A new approach to the word and conjugacy problems in the braid groups*, Adv. Math. **139** (1998), no. 2, 322–353.

[BL93] J. S. Birman and X.-S. Lin, *Knot Polynomials and Vassiliev's Invariants*, Invent. Math. **111** (1993), no. 2, 225–270.

[BM93] J. S. Birman and W.W. Menasco, *Studying links via closed braids III. Classifying links which are closed 3-braids*, Pacific J. Math. **161** (1993), no. 1, 25–113.

[Bri63] J. L. Britton, *The Word Problem*, Ann. of Math. **77** (1963), no. 1, 16–32.

[Bro94] K. S. Brown, *Cohomology of Groups*, Graduate Texts in Mathematics, vol. 87, Springer-Verlag, New York, 1994, Corrected reprint of the 1982 original.

[Coh68] P. M. Cohn, *A presentation of SL_2 for Euclidean quadratic imaginary number fields*, Mathematika **15** (1968), 156–163.

[DG98] O. T. Dasbach and B. Gemein, *The Word Problem for the Singular Braid Monoid*, preprint, available as: math.GT/9809070, August 1998.

[Fin89] B. Fine, *Algebraic theory of the Bianchi groups*, Marcel Dekker Inc., New York, 1989.

[FKR98] R. Fenn, E. Keyman, and C. Rourke, *The singular braid monoid embeds in a group*, J. Knot Theory Ramifications **7** (1998), no. 7, 881–892.

[FRZ96] R. Fenn, D. Rolfsen, and J. Zhu, *Centralisers in the Braid Group and Singular Braid Monoid*, Enseign. Math. (2) **42** (1996), no. 1-2, 75–96.

[Gem97a] B. Gemein, *Polynomials for Singular Knots and Derived Weight Systems*, preprint, Universität Düsseldorf, 1997.

[Gem97b] B. Gemein, *The Burau Representation and the Birman Homomorphism*, preprint, Universität Düsseldorf, 1997.

[JJ99] A. Járai Jr., *On the monoid of singular braids*, Topology Appl. **96** (1999), no. 2, 109–119.

[Kra99] D. Krammer, *The braid group B_4 is linear*, preprint, 1999.

[LP93] D. D. Long and M. Paton, *The Burau representation is not faithful for $n \geq 6$*, Topology **32** (1993), no. 2, 439–447.

[Xu92] P. Xu, *The Genus of Closed* 3*-Braids*, J. Knot Theory and its Ram. **1** (1992), no. 3, 303 - 326.

New knot and link invariants

Thomas Fiedler

Laboratoire E. Picard, Université P. Sabatier,
118, Route de Narbonne, 31062 Toulouse Cedex, France,
e-mail: `fiedler@picard.ups-tlse.fr`

Alexander Stoimenow

Ludwig-Maximilians University Munich, Mathematics
Institute, Theresienstraße 39, 80333 München, Germany,
e-mail: `stoimeno@informatik.hu-berlin.de`,
WWW: `http://www.informatik.hu-berlin.de/~stoimeno`

Current version: September 17, 1999 First version: December 9, 1996

Abstract

We study the new formulas of the first author for the degree-3-Vassiliev invariants for knots in the 3-sphere and solid torus and present some results obtained by them. We show that a knot with Jones polynomial consisting of exactly two monomials must have at least 20 crossings.
Keywords: Vassiliev invariants, Gauß sums, mutation, positive knots, Jones polynomial
AMS subject classification: 57M25

Contents

1 Introduction

Recently, the first author [Fi] introduced some new knot invariants by generalizing the approach of Polyak and Viro [PV] of Gauß sums. Conversely to the Polyak-Viro invariants, which are known to be of finite (Vassiliev) type [BL, BN, BN2, BS, St, Va, Vo], and therefore, at least up to degree ≤ 12 [K] orientation non-sensitive, the main hope of this generalization was to find some new invariants that *do* distinguish knot and link orientation in S^3.

In this mainly expository note we give a definition of these invariants and examples using the computer calculations of the program [St2]. We announce some results, supported by these examples, about these invariants applied to link mutants in S^3. We present some applications of the Gauß sum formulas for the degree-3-Vassiliev invariant to crossing number ineqaulities and positive knots.

A mathematically more detailed description of the first author's approach, including invariance proofs, can be found in his forthcoming monography [Fi2], especially from the (more interesting) point of view of these invariants in the solid torus.

2 Gauß diagrams and Gauß sums

Consider a knot $K : S^1 \hookrightarrow \mathbb{R}^3$ (S^1 and $\mathbb{R}^3$ oriented). Decompose $\mathbb{R}^3 = \mathbb{R}^2 \oplus \mathbb{R}$ so that the projection (henceforth called *knot diagram*) of K into $\mathbb{R}^2$ is generic. To this projection we can assign a *Gauß diagram* (GD), a circle with oriented chords, by connecting points in S^1 mapped to a crossing and orienting the chord from the preimage of the undercrossing to the preimage of the overcrossing. By forgetting the orientation of each chord in a GD we obtain its underlying *chord diagram* (CD) [BN].

Figure 1 shows the knot 6_2 in its standard projection and the corresponding Gauß diagram.

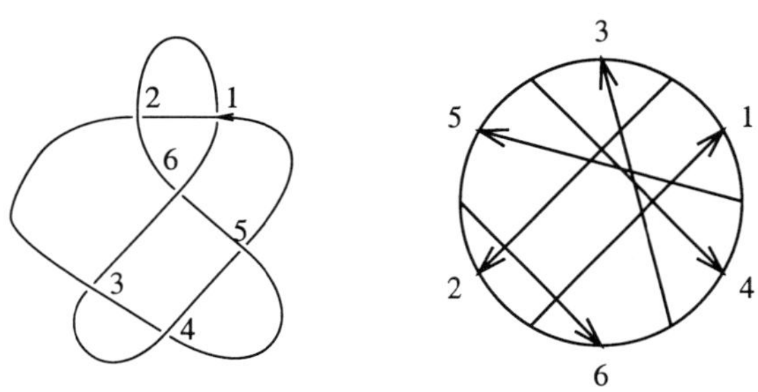

Figure 1: The knot 6_2 and its Gauß diagram.

A Gauß sum of degree k is a term assigned to a knot diagram, which is of the following form

$$\sum_{\substack{\text{ordered choices of } k \text{ crossings of the} \\ \text{knot diagram, satisfying certain condi-} \\ \text{tions}}} \text{function}\big(\text{ data, assigned to the crossings }\big).$$

Each summand we will call *weight* and the function *weight function* (not to be confused with those in [BN]!).

Sums of this type ("small state sums") have been studied for the 1st time for degree ≤ 2 by the first author [Fi3, Fi4] and later by Polyak-Viro [PV]. As data of the crossings they considered the writhe (which is $+1$ on a positive and -1 on a negative crossing) and as weight functions the elementary symmetric polynomials. The aim was to find linear combinations of such terms, invariant under the 3 Reidemeister moves, and therefore giving formulas for knot invariants. All such invariants turned out to be of finite type. This was stated by the authors and proved by Oestlund.

Figure 2: The writhe

Theorem 2.1 (Viro-Polyak-Oestlund) *All invariant Gauß sums of degree k with symmetric polynomials in the writhes of the crossings are Vassiliev invariants (VI) of degree* $\leq k$.

Definition 2.1 *The* winding index *of a plane curve* $C \subset \mathbb{R}^2 = \mathbb{C}$ *around a point* $p \notin C$ *is*

$$w(C,p) := \frac{1}{2\pi i}\oint_C \frac{1}{z-p}\,dz.$$

Pictorially it measures how many times the curve "walks" around p, counting reverse walk negatively.

Definition 2.2 *The* Whitney index $n(C)$ *of a plane curve C is the degree of the map*

$$\frac{C'}{||C'||} : S^1 \longrightarrow S^1.$$

The Whitney index of a knot diagram is the Whitney index of its underlying plane curve.

Definition 2.3 *The* writhe $w(D)$ *of a knot diagram D is the sum of the writhes of all crossings (see figure 2).*

Example 2.1 The standard projection of 6_2 has Whitney index 1 and writhe -2.

In the following we will decisively use the following fact.

Lemma 2.1 *Two plane diagrams of the same knot (i.e., ambient isotopic) with the same n and w are regular isotopic.*

In other words, the ambient isotopy classes of knots decompose into regular isotopy classes parametrized by two numbers n and w. The only restriction to n and w is that their sum is odd.

There are four ways to modify n and w by Reidemeister I moves, as depicted on figure 3 (where '$n\pm\pm$' denotes the operation '$n \to n\pm 1$' and the result of the move on a strand without crossings is displayed).

In the following we will look for regular isotopy invariants of the knot diagram.

Definition 2.4 *A* smoothing *of a crossing is the procedure*

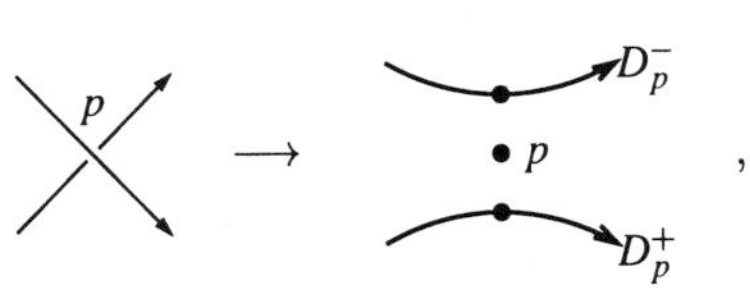

where D_p^+ *denotes the component, where the under- is smoothed to the overcrossing. Note, that apart from the link diagram resulted after this operation, we have the 2 "traces" of p on two of the components of the link diagram, as well as the "trace" of p in its complement. We will sometimes forget about the one or the other and simply omit the corresponding mark in the picture.*

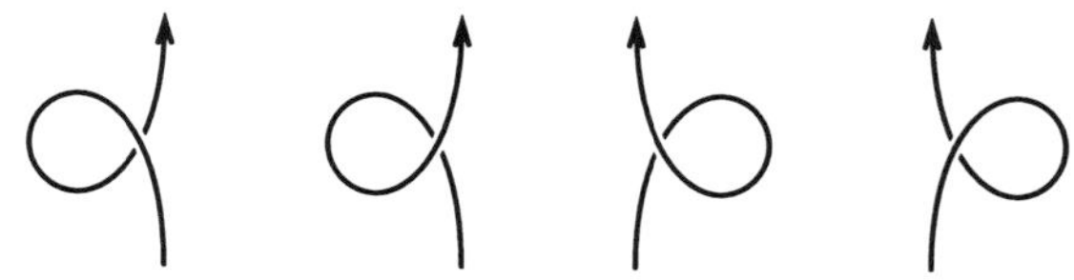

Figure 3: The four ways to modify n and w by Reidemeister I moves.

Definition 2.5 *Smoothing simultaneously 2 crossings p,q in a knot diagram, the resulting link may have 1 or 3 components. In former case call p,q* linked, *else not linked. In terms of the Gauß diagram the chords corresponding to linked crossings intersect.*

Definition 2.6 *Call in a pair p,q of linked crossings p* distinguished, *if D_p^+ contains the* undercrossing *of q, i.e., if the Gauß diagram looks as*

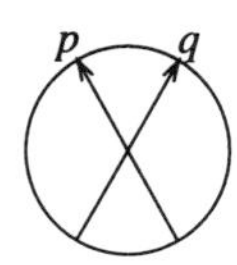

Apart from w, for each crossing p we have four[1] more data:

$$i_p^{\pm} := w(D_p^{\pm}, p), \qquad n_p^{\pm} := n(D_p^{\pm}).$$

(Here in the definition of $i_p^{\pm}$ by p we mean the trace of p in the complement, as described above.) Set

$$i_p := i_p^+ + i_p^-, \qquad n_p := n_p^+ + n_p^-, \qquad \delta_p := i_p^+ - i_p^-.$$

3 The degree-3-Vassiliev invariant and positive knots

3.1 A formula for the degree-3-Vassiliev invariant

Smoothing simultaneously 3 crossings and counting the traces of all crossings on the components of the resulting link, we obtain a partition of 6. The following 5 are possible. Each one corresponds to one of the 5 degree 3 CD's underlying the Gauß diagram (which is not true in higher degree), see figure 4. This partition we will call (rough) configuration.

If p and q are not linked, either $p \in D_q^+$ or $p \in D_q^-$. We will write this sign before q, and analogously the corresponding one before p. The configuration $(4,2)0$ is made of triples of crossings (p,q,r) with (p,r) and (q,r) linked and p and q not linked, $p \in D_q^+$ and $q \in D_p^-$, i.e. "$-p \quad +q$" (so the rough configuration of $(4,2)0$ is $(4,2)$).

To introduce some notation, write a configuration as a bracketed sequence of 6 entries of the form '[sign] letter [digit]'. Walking around the circle, the letter indicates the name of the crossing the chord represents (p,q or r), a basepoint of which we pass, the sign whether we pass the over-(+) or undercrossing (−), and the digit after the first occurence of a letter the type[2] of the crossing the chord represents.

[1] In fact, these are only 3 data, since for all p we have $n_p^+ + n_p^- = n$.

[2] The type of a crossing will be defined in the next section, so at present simply forget about it and about the option to write the digit.

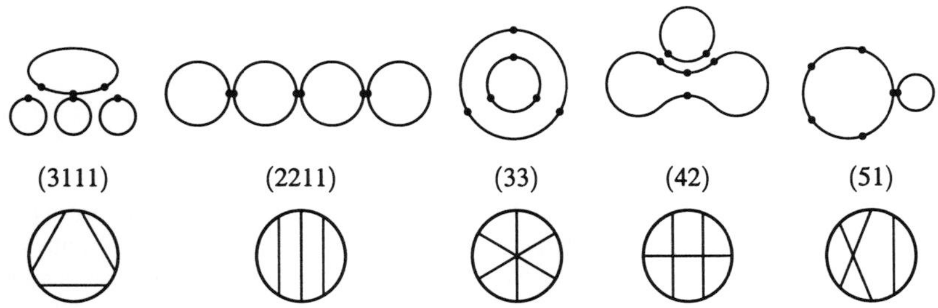

Figure 4: The rough configurations

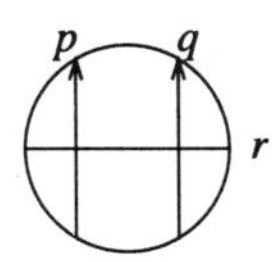

Figure 5: The Gauß diagram of the configuration $(4,2)0$. Both orientations of r are possible.

The sign and the digit are optional. If we omit one of them we do not fix the orientation of the chord or the type of the crossing.

A Gauß sum we will write as “<config>{<weight>}”, e.g.

$$[+q+pr-p-qr]\{w_p+w_q+w_r\} \qquad \text{means} \qquad \sum_{(4,2)0} w_p+w_q+w_r\,.$$

If we omit the weight, we will mean “by default” the product of the writhes of the involved crossings[3]. Everything up to now works as well for 2 crossings instead of 3 with 4 entries within the brackets. So, interpreted as a Gauß sum, e.g.

$$[+q+p-q-p] := \sum_{\substack{p,q \text{ linked} \\ p \text{ distinguished}}} w_p \cdot w_q\,.$$

The summation has to be understood so that we sum over distinct unordered p,q,r but if a configuration has a non-trivial automorphism (i.e., e.g. if p,q,r match, so do q,r,p, as in the case of $(3,3)$), only *one* of the matching permutations is taken.

Here is a formula for the degree-3-Vassiliev invariant (in one of its variations modulo degree-2-Vassiliev invariants).

$$vt_3 = [pqrpqr]+[+q+pr-p-qr]+\frac{1}{2}[pqpq]\{w_p+w_q\}\,. \tag{1}$$

The idea of finding it is (roughly) the following (for an exact proof, see [Fi2]).

We have to check how a linear combination of Gauß sums

$$v = \sum_{\text{conf. } \eta} c_\eta \sum_{\substack{\text{crsg. choices} \\ \text{of conf. } \eta}} \text{function}(\ldots) \tag{2}$$

[3] It will always be clear from the context whether by this notation we mean the Gauß sum or the configuration.

behaves under, say, a type III move.

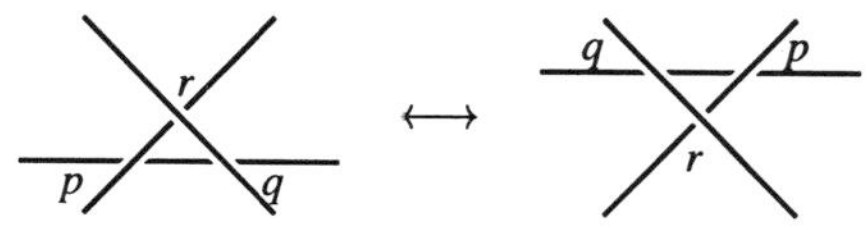

To measure the difference of each Gauß sum on both sides of such a move, we have to care of the choices of 3 crossings which land in different configurations η,μ in both diagrams. For such configurations at least 2 of the chosen crossings must be among p,q,r. In case exactly 2 of them are among p,q,r, we have to deal with an unknown $b_{\eta\mu}$ – the number of choices of the third one (or rather the sum of "function(...)" over such choices). The difference of the Gauß sum on both sides is a certain linear combination of all $b_{\eta\mu}$'s. Each $b_{\eta\mu}$ may appear multiple times for different summands in (2).

There exist linear dependencies between the $b_{\mu\eta}$'s which allow to eliminate some in the sum. For the rest, the difference of v to be 0 (i.e., v to be invariant under Reidemeister III), we have to ensure that all the coefficients of the remaining b's vanish. These are in turn linear combinations of various c_μ's. This way we obtain linear equations for the c_μ's. At this point the system could still be solved by hand, due to the limited number of configurations.

Remark 3.1 As for Vassiliev invariants of the same degree there may be many formulas, differing by scaling and lower degree Vassiliev invariants, it appears desirable to introduce some "standard" set of low-degree Vassiliev invariants, and relate everything to it. vt_3 satisfies the relation

$$vt_3 = 4v_3 = -\frac{1}{3}V^{(2)}(1) - \frac{1}{9}V^{(3)}(1),$$

where v_3 is the formula of [PV]

$$v_3 = \frac{1}{2}[-q+p-r-p+q+r]+[+p-q+r-p+q-r], \tag{3}$$

and V is the Jones polynomial [J].

Here is a consequence we announce of formula (1) (see [St4] for a proof and a sharper result).

Definition 3.1 *Call a knot diagram* positive *if all its crossings have writhe 1.*

Definition 3.2 *Call a knot diagram* reduced *if it does not look as*

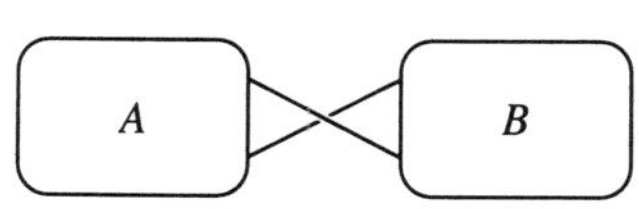

i.e., its corresponding Gauß diagram has no isolated chord.

Theorem 3.1 *The number of crossings of a reduced positive diagram of a knot K is at most $vt_3(K)$.*

This renders it decidable whether for a given knot such a positive diagram exists.

Example 3.1 We have $vt_3(6_2) = 4$, so for our 6_2 there is no such diagram (it would have to have ≤ 4 crossings).

As another application, it follows from (1), that vt_3 is always integral and even more, the identity $vt_3 = 4v_3$ shows, that $vt_3(K)$ is always even for any K. In fact it is even always divisible by 4. This can be shown either via Vassiliev theory arguments or from the other known conditions to the values of the Jones polynomial on knots (summarized by Jones [J2, §12]).

3.2 A lower bound for the crossing number

Here a final consequence of formula (1). Obviously in a diagram of c crossings, we have

$$|vt_3| \leq \binom{c}{2} + \binom{c}{3} \leq \frac{c^3}{6},$$

whence

$$c \geq \sqrt[3]{6|vt_3|} = \sqrt[3]{\left|2V^{(2)}(1) + \frac{2}{3}V^{(3)}(1)\right|}. \tag{4}$$

This way the Jones polynomial can be sometimes more powerful to give lower bounds for the crossing number of a knot than by its span [Ka3, Mu, Th]. In all such cases the knot is non-alternating. A similar lower bound for c was given by Polyak and Viro [PV2] in terms of the degree-2 (Casson) invariant: $c \geq \sqrt{8v_2}$.

There are examples, where the inequality (4) does better than either this of [Ka3, Mu, Th] and [PV2].

Example 3.2 The knot 16_{809894} of Thistlethwaite's tables [HT] (see figure 6) has a non-monic Jones polynomial of span 9, from which one finds $v_2 = 12$, and hence the inequalities of [Ka3, Mu, Th] and [PV2] both give 10 as a lower bound for its crossing number. The diagram of figure 6 has 5 Seifert circles, and the Morton-Williams-Franks inequality [Mo, FW] gives 5 as lower bound for its braid index, hence it indeed has braid index 5 by [Ya]. If 16_{809894} had a representation as a closure of a homogeneous 5 braid, its crossing number 16 would show (see, e. g., corollary 4.1 of [Cr]) that $\max\deg\Delta \geq 1/2(16-5+1) = 6$. However, $\max\deg\Delta = 5$ for this knot, so 16_{809894} is not a closed homogeneous 5 braid, and [Mu2, proposition 7.4] cannot be applied to it. However, $v_3 = 42$, which shows $c \geq 11$, as $24v_3 > 1000$.

In general the combination of the v_2 and span V bounds will perform well, and examples like the one above, at least among not too complicated knots, are difficult to find, but considered separately, one can construct larger classes of knots where (4) gives better results.

For span V an appaling example are torus knots (or a connected sum of such with a knot like 5_2, which leads out of the scope of proposition 7 of [Mu2]) by [J2, proposition 11.9] and the formulas for v_3 on torus knots due to Simon Willerton in [Wi], who used an unpublished (hard) result of Thang Le (see also [St3, §8]).

As for the v_2 bound, one can show from [St8, theorem 3.1] that for all but finitely many positive knots of given genus the v_2 bound for the crossing number will minorate the one of v_3.

16_{809894}

Figure 6: A knot showing that the inequality (4) is sometimes better than the ones of [Ka3, Mu, Th] and [PV2].

The above mentioned conditions of Jones to the value range of the Jones polynomial imply that no other Jones polynomial than the unit one can have only one monomial and that the only polynomials of exactly two monomials, which can occur as Jones polynomials of some knot, are of the form $(a+1)t^{12ay}-at^{12(a+1)y}$ with $a,y\in\mathbb{Z}, a\neq 0,-1$ and $y\neq 0$ (in particular no Jones polynomial of some knot can have span exactly one). The simplest such example is $2t^{12}-t^{24}$. The existence of a knot with such a polynomial is unclear. In contrast to it there is a remarkable result of Traczyk about an infinite number of 2-component links with Jones polynomial with 2 monomials (reproduced by Przytycki in [P3]). Using (4) we can now show that such a knot must have at least 20 crossings. Namely, for a (hypothetic) knot with $V=2t^{12}-t^{24}$ the inequality (4) gives $c\geq\lceil\sqrt[3]{6912}\rceil=20$, whereas the span of V is 12 and $\lceil\sqrt{8v_2}\rceil=10$. By a further bit of argument you can exclude crossing number 20 using that the minimal coefficient is 1 in a positive diagram, a consequence of Kauffman's state model (see [St5]).

As a final application of (1) we mention, that from it it straghtforwardly follows, that the $(2,2n-1)$ torus knot diagram is the unique diagram maximizing v_3 over all diagrams of this (and lower) crossing number. A bit more on argument shows that in fact it maximizes v_3 in crossing number $\leq 2n$.

4 Refined Gauß diagrams

Here we present a generalization of the degree-3-Vassiliev invariant to knots in the solid torus [Go].

Consider a two component link $K\cup T$ in S^3 where T is the trivial knot (unknot). Let K,T,S^3 be oriented. Deform $K\cup T$ in $S^3=\mathbb{R}^3\cup\{\infty\}$ so that $\infty\in T$. This isotopy is unique up to isotopy. Such a link we can represent choosing an appropriate projection $\mathbb{R}^3\to\mathbb{R}^2$ as knot with a point in its complement, on which T projects, assuming the orientation of T to be from the sheet of paper to the reader's eye.

Luckily, lemma 2.1 can be generalized to such diagrams.

Lemma 4.1 *If $K\cup T$ and $K'\cup T'$ isotopic, implying that K and K' isotopic within the solid torus $S^3\backslash T\simeq S^3\backslash T'$, and in the above described projection K and K' have equal n and w, then K and K' are regular isotopic in $S^3\backslash T$.*

The new idea of the first author was to *refine* the crossings in a Gauß diagram, i. e. to assign to each one a *type*. In order to bound the number of configurations and the dimension of the equation system, at present this has been only done for types in $\mathbb{Z}_2$ and $\mathbb{Z}_3$. Here for our introductory purpose we consider $\mathbb{Z}_2$ graduation of the crossings.

Definition 4.1 *The* type *of a crossing p in a refined diagram is* $w(D_p^+,T)$ mod 2.

We will distinguish two cases of refined diagrams according to the parity of $w(K,T)$.

In view of lemma 4.1 the receipt to find invariant Gauß sums remains the same, however, we gain with the type a new useful ingredient.

The price we pay is that the combinatorial trouble arising here is serious – the linear equation systems became too large and had to be solved by computer. We shall later say a word on dimensions. The solutions have been computed by K. Mohnke (even linking number) and myself (odd linking number) using MATHEMATICA™[Wo]. Many of them (and the really interesting ones) are very long and our lack of deeper insight into the solutions, which is a constant negative companion of electronical calculations, prevents us from finding nice expresions for them. Therefore, here we will record just some short (but, unfortunately, less interesting) ones.

We can incorporate the study of knots into this scheme by considering T the meridian of the knot[4] K (of course, we could also try it with $K\cup T$ the split link, but this does not give anything new). T carries a natural orientation from K and S^3. K is isotopic to K' exactly if so are $K\cup T$ and $K'\cup T'$. Since for T the meridian of K we have $w(K,T)=1$, from this point of view the odd linking number invariants are more interesting. However, by applying cabling operations we can also hope to gain some information for knots from the even linking number invariants.

[4]In this case we will omit T and identify the link with the knot.

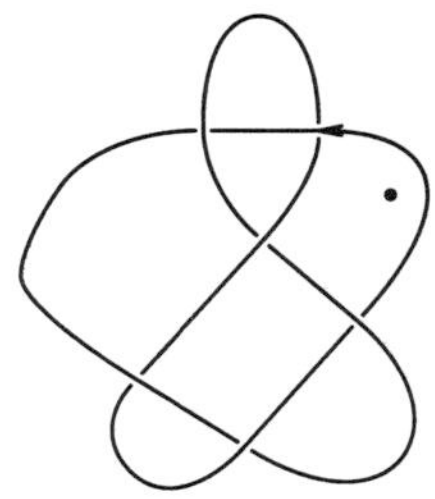

Figure 7: The knot 6_2 with its meridian

As, as we saw, one can treat knots in S^3 as knots with meridian, it was for a brief period tempting to find out what the invariants can do for knots in S^3 this way. In the present note we content ourselves to some failed examples with knots serving to explain the (cabling) idea, which can be more successfully applied to links. We will give more interesting examples in a later note.

Although, after testing the examples given below, finally we found out, that for *knots* this way the solid torus Vassiliev invariants and their cablings don't give more information than the usual degree-3-Vassiliev invariant in S^3, we will give some evidence that (especially at finer crossing type) they may be better than the degree-3-Vassiliev invariants for 2-component links.

4.1 Even linking number case

For the case $w(K,T) \equiv 0 \pmod 2$ the linear equation system was with 50 equations and 48 variables. We obtained 34 invariants. 15 of them are stable and 19 non-stable.

Definition 4.2 *A knot invariant v of regular isotopy is called k-*stable *for some $k \in \mathbb{Z}$, if there exist numbers $n_1, w_1, \dots, n_k, w_k \in \mathbb{N}$ and a function $f_v : \mathbb{Z}^{k+2} \to \mathbb{Z}$, such that for all knots K and all $n, w \in \mathbb{Z}$*

$$v(K,n,w) = f_v(v(K,n_1,w_1), \dots, v(K,n_k,w_k), n, w).$$

Here we use the parametrization of the regular isotopy classes discussed after lemma 2.1 and by "K" we mean the ambient isotopy class of K. "Stable" *means "1-stable".*

In other words, "stable" means that the change of the invariant under change of n or w (Reidemeister I) does not depend on the concrete knot. Roughly, an unstable invariant contains more information than any ambient isotopy invariant derived from it, together with n and w.

Examples of stable invariants are all Vassiliev invariants (where f depends on $v(K,n_1,w_1)$ only) and the Kauffman bracket [Ka] (where f depends on $v(K,n_1,w_1)$ and w).

Here is an example of such an invariant.

$$\begin{aligned} vt_3^{(1)} = & -[+p0+r0-p-r]\{w_r w_p(i_r-i_p)\}+[+p0+q1+r1-q-p-r] \\ & -[+r1+p0+q1-r-q-p]+[+p1-q0-r0+q-p+r] \\ & -[+q0-p1-r0+p-q+r]-[+p1+q0+r0-p-q-r] \\ & -[+r1+p0+q0-r-q-p]-\frac{1}{2}[+p1+r0-p-r]\{w_p\} \\ & +[+p0+q0+r1-p-q-r]+[+p0+q0+r1-q-p-r] \\ & +\frac{1}{2}[+p0+r1-p-r]\{w_p\}. \end{aligned}$$

This invariant is stable, asymmetric (changes the sign under mirroring[5] $K \cup T$) and ambient invariant (that is, invariant under Reidemeister I).

$K \cup T$ is invertible (that is, isotopic to $-(K \cup T) := -K \cup -T$; we reverse the orientation of *all* components!) exactly if $-(K \cup T)$ and $flip(K \cup T) := flip(K) \cup -T$ are regular isotopic (as their n, w are equal) and this is exactly if so are $flip(K) \cup T$ and $-K \cup T$.

Figure 8 shows the 3 operations on a link with $K =$ trefoil.

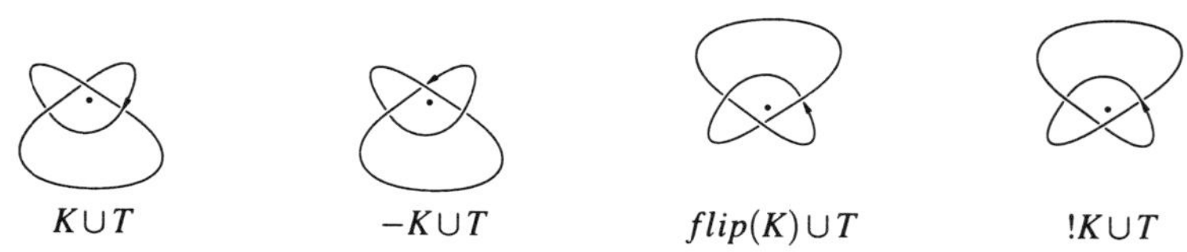

Figure 8: The 3 operations on a link

Another invariant is

$$\begin{aligned} vt_3^{(2)} &= \quad < \text{formula for } vt_3 \text{ only over crossings of type } 0 > \\ &\quad +4[+p1-q0+r0-p+q-r]+[+p1-q0r0+q-pr] \\ &\quad +[+p1-q0r1+q-pr]+[-p1+q0r0-q+pr] \\ &\quad +[-p1+q0r1-q+pr]-[+p1-q1r0+q-pr] \\ &\quad -[+p1-q1r1+q-pr]-[-p1+q1r0-q+pr] \\ &\quad -[-p1+q1r1-q+pr]+[+p0-q0r1+q-pr] \\ &\quad +[-p0+q0r1-q+pr]. \end{aligned}$$

It has the same properties as $vt_3^{(1)}$.

4.2 Odd linking number case

For the case $w(K,T) \equiv 1 \pmod 2$ we had 120 equations and 111 variables. We obtained 33 invariants, 14 stable and 19 non-stable ones. The most complicated ones have 55 terms. Here is a simple one

$$\begin{aligned} vt_3^{(3)} &= -[+p1-q1+r1-p+q-r]-[+p1-q0+r0-p+q-r] \\ &\quad -[+p1-q1-r0+q-p+r]-[+p1-q1-r1+q-p+r] \\ &\quad +[+p0-r0-p+q1+r-q]+[+p0-r1-p+q1+r-q] \\ &\quad +[+p1-q1+r0+q-r-p]+[-q0+r1+p1-p+q-r] \\ &\quad +[-q1+r0+p1-p+q-r]+[+p1-q0+r1+q-r-p] \\ &\quad +[+p1-q1+r1+q-r-p]+[+p1-q0+r0+q-r-p] \\ &\quad +[-q1+r1+p1-p+q-r]+[-q0+r0+p1-p+q-r]. \end{aligned}$$

This invariant is non-stable and asymmetric. One can make the observation that augmenting of n by 2 at fixed w adds one and the same contribution to the value of $vt_3^{(3)}$. We have

$$vt_3^{(3)}(K,n,w) = vt_3^{(3,1)}(K) + vt_3^{(3,2)}(K) \cdot n + vt_3^{(3,3)}(K) \cdot w,$$

where the $vt_3^{(3,i)}$ are ambient isotopy invariants, and therefore $vt_3^{(3)}$ is 3-stable.

[5]Note that mirroring on $K \cup T$ means mirroring in $\mathbb{R}^2$ of the diagram of K with the point T and *not* crossing change, since we need T to point to us.

Remark 4.1 In fact, $vt_3^{(3,2)} = vt_3^{(3,3)}$, so $vt_3^{(3)}$ is 2-stable. This can be deduced from the fact, that we have no $(5,1)$-configuration with isolated chord of type 0. Analogously, if we had no $(5,1)$-configuration with isolated chord of type 1 instead of 0, then we had $vt_3^{(3,2)} = -vt_3^{(3,3)}$.

Remark 4.2 Generally, a Gauß sum invariant induces on an ambient isotopy class a polynomial of n and w, henceforth called (n,w)-polynomial, where w and n appear with exponents ≤ 1 resp. $\leq 1 + \max \deg_n(\text{all weights})$, where $\deg_n$ is the degree with respect to n's of the crossings only.

Remark 4.3 In fact, $vt_3^{(3,2)} = vt_3^{(3,3)}$, so $vt_3^{(3)}$ is 2-stable. This can be deduced from the fact, that we have no $(5,1)$-configuration with isolated chord of type 0. Analogously, if we had no $(5,1)$-configuration with isolated chord of type 1 instead of 0, then we had $vt_3^{(3,2)} = -vt_3^{(3,3)}$.

4.3 Chirality

From the ambient invariance and asymmetry of $vt_3^{(2)}$ it follows that each amphicheral link must have zero invariant. This can be used to detect chirality of links. The following example is due to the first author.

Example 4.1 For the link $K \cup T$ on figure 9 with K the figure 8 knot we have $vt_3^{(2)} = 4$, so it is non-amphicheral. The components themselves are amphicheral, and all other invariants of degree ≤ 3 (lk and the two component link degree 3 Gauß sum of [PV]) are 0 and fail.

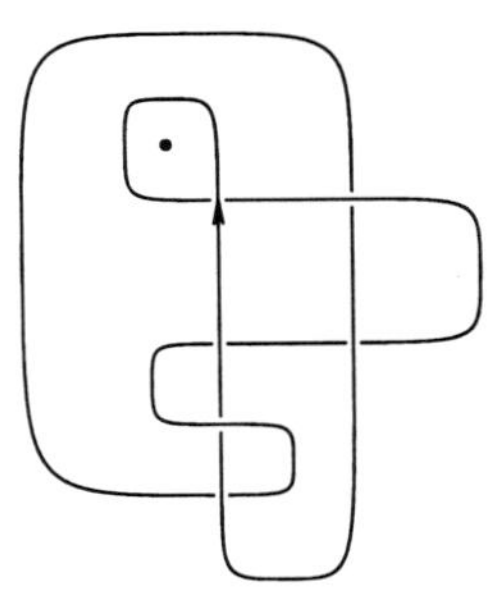

Figure 9: A non-amphicheral link with even linking number.

Two hard examples of chiral knots are [Ka, p.218] 10_{48} (where the HOMFLY polynomial [H] fails to detect chirality) and 9_{42} (where the HOMFLY and Kauffman [Ka2] polynomials fail). After we experimantally observed and the second author subsequently proved more genrally in [St3] that our solid torus invariants will not be able to do it either, there is another idea based on Gauß sums, which can be used at least in the case the knot is alternating.

The flyping theorem of Menasco and Thistlethwaite [MT] shows that in alternating diagrams not only vt_3 is an invariant, but all its three single terms in (1) are! The first and third one are clearly invariants of the intersection graph of the associated Gauß diagram.

The intersection graph of a Gauß diagram is a graph with vertices corresponding to arrows in the Gauß diagram and edges connecting intersecting arrows/vertices. It is an exercise to see that the intersection graph is preserved by flypes, and therefore so are the single terms in (1). (A little bit more argument shows invariance also for the second term in the Polyak-Viro formula (3) and hence also for the other one.)

The third of (1) term is easily computed on 10_{48} and its mirror image to be ± 8 (note, that all three terms are negated by obversion), and its non-vanishing renders it easy to decide about 10_{48}'s chirality. The first term deals (even easier) with the other three alternating troublemakers in Rolfsen's tables [Ro, appendix A] – 10_{91}, 10_{104} (HOMFLY) and 10_{71} (HOMFLY and Kauffman). The argument does not work for the non-alternating 9_{42} and 10_{125} (HOMFLY), but they both have signature ± 2, and so a signature trick works there (see [St6]).

4.4 Skein equivalence and mutation

As Gauß sums change unpredictibly under crossing changes, we could hope that they manage to detect skein equivalent knots. However, checking one such example – the skein equivalent triple 8_8, $!10_{129}$ and 13_{6714} (see figure 10) discovered by Thistlethwaite [LM] (and later in greater variety by Kanenobu [Kn]) – we observed the same disappointment.

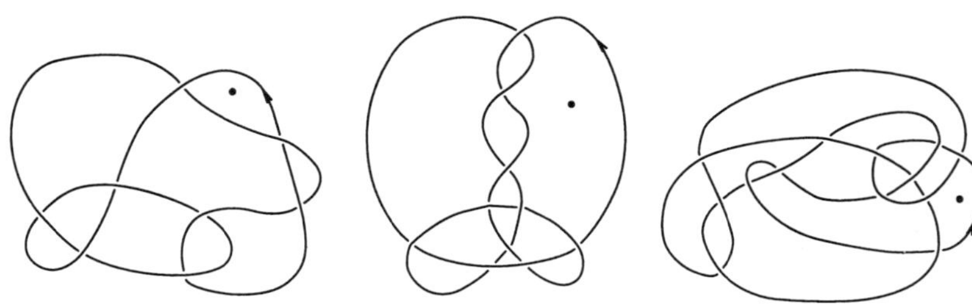

Figure 10: The skein equivalent knots of Thistlethwaite. They form the beginning of a larger series of skein equivalent knots constructed by Kanenobu [Kn].

It is known that mutants always make trouble to distinguish. According to Przytycki, one applicable method is to calculate truncations (in the Alexander variable) of the HOMFLY polynomial of a 3-cable of the link. Nothing simpler works: that the knot polynomials themselves do not detect mutation is an easy consequence of the local diagrammatic relations they satisfy, and subsequently it was shown that all cablings of the Jones polynomial [MT], and the 2-cablings of the HOMFLY [P, P2] and Kauffman [LL] polynomial fail. Such truncations have been shown to have subexponential complexity by Przytycka and Przytycki [PP] and later with sharper estimates by Vertigan (see appendix to [PP2] and [Ki, p. 111]) and were used by Morton and Traczyk and independently by J. Murakami to distinguish, *inter alia*, the Kinoshita-Terasaka and Conway knots (figure 11), the most famous example of mutants.

Again, there is no sense in trying to distinguish mutant knots by applying the solid torus invariants on the meridian of the knots or their cables. However, interestingly, some invariants *do* distinguish (and even without cabling) pairs of solid torus knots which are not (necessarily) mutants in the solid torus but give mutated links when considered with T in S^3. The second author gave such examples in [St7].

5 The computer program

The program [St2] can compute the Fiedler solid torus invariants. It is written in C++ [S] and at present it has ≈ 4000 lines.

It reads in a knot from a projection file produced by J. Weeks' program SnapPea[6] [W], generates the associated extended Gauß diagram (Gauß diagram with all data assigned to each chord) and computes

[6]Unfortunately, the link entry program of SnapPea is not yet available on other systems than Apple™.

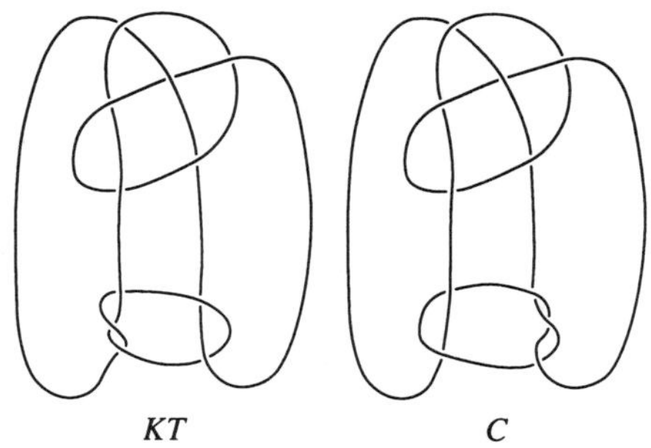

Figure 11: The Kinoshita-Terasaka/Conway mutants

a given Gauß sum on it. It can also create a PostScript (PS) [Ad] image of the knot. The pictures in this paper have been generated this way. It contains detailed comments, so we hope that it will be (re)usable. Below we provide instructions for use and we are always grateful for comments and suggestions about it.

An additional small program is provided to generate input files of a closed braid and its axis (so as to examine the invariants as conjugacy invariants for braids) from a braid word. Therefore, this way it is possible to use the invariants (at least for braids) without SnapPea's graphical interface.

A modification of this program is possible to handle general diagrams by using the extended Dowker notation (see §5.3), but we will develop one only at request.

5.1 How to install and use it

The (`gzipped`) `tar`-archive [St2] should be expanded on a UNIX machine with a command like

```
tar -xzvf gsinv.tgz
```

It contains the following files:

`gsinv.C` – the main program. It is to be compiled using the GNU C++ compiler `g++`, version 2.7.0 ff., with a command like

```
g++ -o gsinv gsinv.C &
```

(Note: due to the large number of complicated expressions for the invariants with $\mathbb{Z}_3$ graduation, compiling will probably last up to an hour, so it is a good idea to do it in the background.)

`comp.h`, `rat.h`, `mydefs.h`, `x.h`, `y.h`, `eveninv.h`, `oddinv.h`, `lk0mod3.h`, `lkZ_0.h` and `lk1mod3.h` – files included into `gsinv.C`.

`8-17-2-1` – the SnapPea projection file of the 2-cable of 8_{17} provided to reproduce the below example.

`mypic.psh` – header file for `dvips` for including the graphics output of the program into a TeX or LaTeX document using the `epsf` package. Include the header file using the command

```
\special{header=mypic.psh}
```

on the first page of the TEX or LATEX document.

`br2proj.C` – This program should be compiled with

```
g++ -o br2proj br2proj.C
```

The program reads in a braid (word) and generates an input file with the link consisting of its closure (which must be a knot) and axis (the point T). The format of this file slightly differs from this of SnapPea, so to make `gsinv` recognize it, envoke it with the additional parameters '`-f 1`'. The braid has to be sent to the program `br2proj` from `stdin`, indicating first the length of the braid (i. e., the number of generators and their inverses counted with multiplicities if appearing in powers) and then the braid word starting with a '{', followed by a comma separated list of non-zero integers and terminated by a '}'. An integer i represents σ_i or σ_{-i}^{-1} accoring to its sign. The number of strands is computed automatically. The braid must close to a knot. The program writes output on `stdout`, so redirect it into a file. See example 5.3.

A call of the program `gsinv` from a UNIX shell should look like

```
gsinv [ -ps ] [ -i <inv#s> ] [ -w <PSwgt> ] [ -m <mode> ] [ -q ]
  [ -K ] [ -mod [ 1 | 2 | 3 ] ] [ -h <homcl> ]  [ -a <arwid> ]
  [ -c <crwid> ] [ -J ] [  [ -f [ 0 | 1 ] ] -l  <input file> |
  <input file> [ <Re(T)> <Im(T)> ] |  [ -f [ 0 | 1 ] ]  -p
  [ -l ]  { <input file>  } ]
```

Here "`[ thing_1 | thing_2 ]`" means *either* `thing_1` *or* `thing_2`, "`[ thing ]`" means at *most* one occurence of `thing` and "`{ thing }`" means at *least* one occurence of `thing`.

Unlike many other programs, if an option of this program requires a parameter, it has to be supplied as the next command line argument, not as the rest of the argument indicating the option. E.g., "`-i6`" is not correct, you have to write "`-i␣6`". Options are processed in order of appearence from left to right on the command line; a later option may override a previous one.

When an option of this program requires a parameter, latter *must* be specified; no defaults exist. However, all option parameters have default settings, which are taken when the option is not specified on the command line.

The program has two modi. In invariant mode (which is the default), each input file is processed by writing the results on `stdout` (unless the `-q` option is specified) and into a log-file whose name is obtained by adding "`.log`" to the name of the input file. In PS mode (activated by the `-ps` option) no output is generated on `stdout` (so the `-q` option is obsolete); instead PS images of the input files are written into corresponding `.eps` files (see example 5.2).

The meaning of the options is the following:

`-ps` sets the PS mode – write PS images instead of computing invariants. The `-ps` option overrides the `-mod` option.

`-m` The next argument should be a number between 1 and 31. Its lowest 5 bits indicate whether an action should be done for $K, -K, flip(K), !K$ and $flip(!K)$. Default setting is 1 for PS (only K) and 15 for invariant computation (first 4).

-K This option indicates in PS mode that the dot for T should not be drawn.

-h The next argument should be a non-zero integer indicating the homology class a to be used in $\mathbb{Z}$-graduation. In $\mathbb{Z}$-graduation, the Gauß sums are computed like in $\mathbb{Z}_3$-graduation with the only difference that a chord labelled by '1' matches (only) a crossing of type a and a chord labelled by '2' matches a crossing of type $-a$, and the linking number must coincide with the one the invariants are designed for in $\mathbb{Z}$ (and not only $\mathbb{Z}_3$). Presently only invariants for zero linking number are available. Default setting for a is 1. This option is obsolete when not specifying '`-mod 1`' (see `-mod` option).

-f Next argument should be either `0` or `1`. 0 (default setting) makes the program expect SnapPea format in the input file(s), 1 sets the format produced by the program `br2proj`. Both inputs formats simultaneously are not allowed. If 1 is set, the possibility to indicate coordinates of T on the command line (see `-p` option) is not allowed.

-w Sets the PS weight to the next argument. This should be a number with $1/2 < \# \leq 1$ indicatng the degree of smoothing the vertices. 1 (default setting) draws sharp vertices, $1/2+\varepsilon$ smooth vertices. A good value is 0.8. A higher value may be useful to avoid intersections of the smoothed curves which can occur at long edges in the diagram polygon. This option automatically sets `-ps`.

-i Only some special invariants with $\mathbb{Z}_2$ graduation given in the next argument should be computed. This option was included at some evolutory stage and is not particularly recommended. The next argument should be something like "`2,5,7-10,4`" (meaning that we are interested in invariants 2,4,5 and 7 to 10 only). Default (without the `-i` option) is all invariants. However, only special numbers in the list should be used. For even linking number they are 1 for $vt_3^{(1)}$, 16 for $vt_3^{(4)}$ and 5 for $vt_3^{(2)}$, for odd linking number – 15 for $vt_3^{(3)}$, 16 for $vt_3^{(4)}$ and 17 for $vt_3^{(5)}$. Additionally for both linking numbers the usual Gauß sum formulas are available (evaluated on K by forgetting T) – invariant number 2 for vt_3 and 3 for v_3. The invariants provided with this option are *not* included in the invariant list computed without the `-i` option, but are linear combinations of invariants in the list. A list of $\mathbb{Z}_2$-invariants available for computation is given in [St7]. The `-i` option overrides the `-mod` option.

-p Indicates that all arguments remaining after parsing in all options should be taken as file names of input files to process (options are all leading arguments starting with a '`-`' and the arguments required to follow some of them). In this case each input file is assumed to be a SnapPea projection file of a knot with a point for T therein (resp. a file of the format rendered by `br2proj`, when '`-f 1`' is specified). The point should be added *after* drawing the knot by a double mouse click. (Note that such a projection is useless for SnapPea, as it is not a "regular" link projection.) If without this option there is only 1 argument remaining after parsing in all options, the program assumes a SnapPea projection file of a knot with a point therein. If (without this option) after parsing in all options there are 3 arguments remaining, the program assumes in the file only a knot and reads in the coordinates of the point T from the last 2 arguments. (If the `-l` option is used, the possibility to give the second and third argument is not allowed.) Another number of remaining arguments than 1 or 3 is incorrect without this option.

-q Quiet mode. Do not print anything on `stdout`, just into the log-files. Obsolete in PS mode.

-J Compute the series of (formally infinite sum) invariants J of degree 2, depending on a parameter a. The program automatically computes the highest absolute value of a type of a crossing and gives the series only foe a between its negated and itself. For each non-zero $a \in \mathbb{Z}$ the invariant J is defined by

$$J(a) := [+pa - q0 - p + q] + \sum_{\substack{m>0,\, n<0,\\ \gcd(m,-n)=a}} [+pm - qn - p + q]\,,$$

where crossing (type) graduation is taken in $\mathbb{Z}$.

-mod The next argument should be either '2' (default setting), '1" or '3' indicating the invariants using which graduation should be computed. '1' means $\mathbb{Z}$-graduation (see the -h option).

-c The next argument should be a number. It explains which part of the diagonal length of the PS image should be taken as crossing size. When passing an undercrossing the PS pen is taken off the sheet at this distance from the crossing and put back at the same distance on the other side.

-a The next argument should be a number. It explains which part of the diagonal length of the PS image should be taken as arrow length. A high value (e.g., 1000) causes the arrow to become invisibly small.

-l This option forces the program to read from each input file a *regular* link projection for SnapPea and to generate a PS image file of the link. It is therefore similar to the -ps -K options, with the difference that no point T is allowed in the input files and the link may have multiple (not only one or two) components. This option is hence unrelated to the main functionality of the program. The -l option sets automatically the -ps option and overrides the -mod option.

Example 5.1 If we write

```
gsinv -m 6 -i 15 8-17-2-1
```

assuming in the file 8-17-2-1 the knot $(8_{17})_{2,1}$ with its own meridian, we obtain the following output

```
file: 8-17-2-1
lk(K,T)%2=1
w=1
n=6

inv #15:
vt33(-k)= 10
vt33(flip(k))= 10
```

The program computes the parity of $lk(K,T)$ and depending on it chooses the right set of invariants. Each invariant is indicated by its name (e.g., vt33 for $vt_3^{(3)}$). So we see that, unfortunately, $vt_3^{(3)}$ does not detect the (possible) non-invertibility of $(8_{17})_{2,1}$ (implying this of 8_{17}).

Example 5.2 If we want to see how $K = (8_{17})_{2,1}$ with its own meridian looks like, type

```
gsinv -ps -m 15 8-17-2-1
```

You obtain 4 PS files: 8-17-2-1.eps with K, 8-17-2-1-.eps with $-K$ and 8-17-2-1_fl.eps and 8-17-2-1_mi.eps with $flip(K)$ and $!K$ respectively. You can include them into a LaTeX file as described above.

Example 5.3 To write the image of the braid $\sigma_1\sigma_2^{-1}\sigma_1\sigma_2^{-3}$ and its axis into the file 6_2.out, type

```
echo "6 {1,-2,1,-2,-2,-2}" | br2proj > 6_2.out
```

and to create an image of it, type

```
gsinv -ps -w 0.8 -f 1 6_2.out
```

Remark 5.1 The polynomial complexity of the program allows it to handle examples of about 80 crossings in a few hours. However, at least at about 100 crossings it exhausts the limit of virtual memory available on an average-sized computer.

5.2 Transforming diagrams

Given a diagram with a point the invariants can be effectively computed. However, ususally a link may be given by a regular (oriented) diagram consisting of components K and T (with T unknotted), so it is worth saying a word on how to transform such a regular diagram into a diagram of the kind we need.

If we do not pose any condition on the (regular) diagram we start with, such a procedure would in particular mean to transform the diagram of T therein into the zero crossing diagram, which is a very subtle question and easily fills several dozens of pages for itself (see [Bi] for an exposition on the problem and some recent results). Hence, we assume that the diagram of T has no self-crossings. E. g., the (closed) mixed braid diagrams of [La] are of this kind.

Given sucgh a diagram of $K \cup T$,

1) use moves of the kind (called in [A, fig. 5.58] "tongue moves")

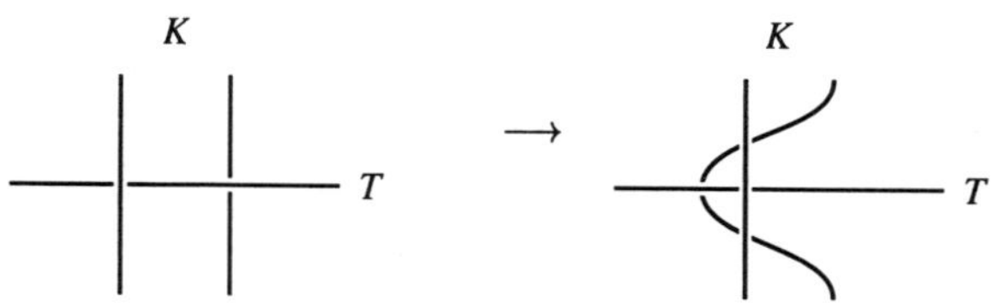

to permute the over- and undercrossings of T with K (at the cost of adding self-cossings of K) so that, running along (the orientation of) T, we have just one sequence of overcrossings and one sequence of undercrossings, i. e., only one *bridge* of T;

2) use moves in the 2-sphere

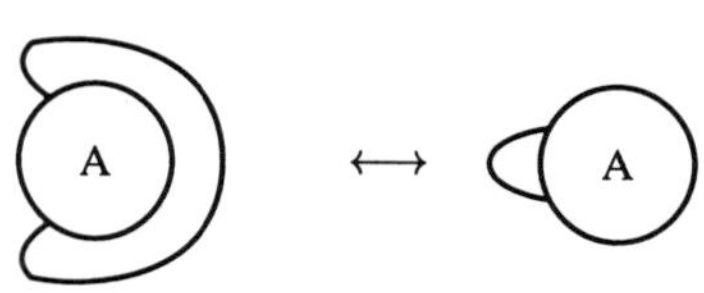

to get the component of the complement of the diagram of K, where the bridge of T ends (that is, the component entered by T by an overcrossing and exited by an undercrossing)

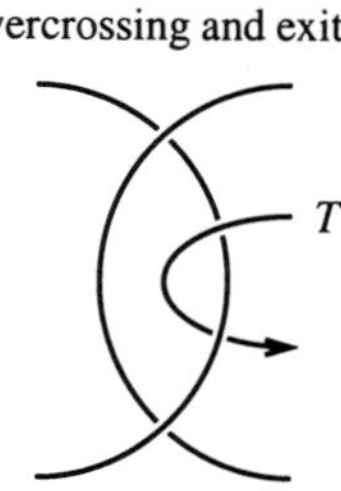

to be the unbounded component in the plane, and finally

3) delete T from the diagram, placing a point into the component of the complement of the diagram of K, where the bridge of T begins (that is, the component entered by T by an undercrossing and exited by an overcrossing).

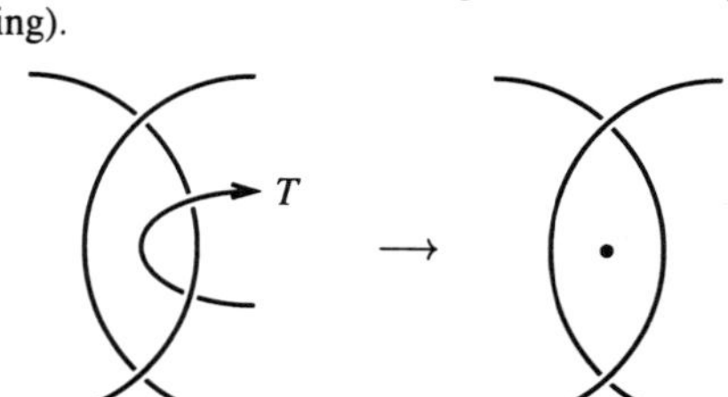

To reverse the procedure (to come from our type of diagram to a regular one) you just need to reverse the last step.

5.3 A notation for diagrams

It is often convenient to represent link diagrams in an alphanumeric format, different from braids. The first systematic way of describing link diagrams in such a way was introduced by Conway [Co]. His notation was developed rather to reflect algebraic properties of the knot (or link) and is hence not particularly convenient for a computer. In this regard much better does the notation of Dowker and Thistlethwaite [DT] for knot diagrams. It has been used to compile by computer prime knot tables up to 16 crossings [HT].

A Dowker notation of an n crossing knot diagram is a permutation (vector) $d = (d_1, \dots, d_n)$ of the numbers $2, 4, \dots, 2n$, with possibly some of them negated. When walking along the line in a knot diagram and numbering the over/under-crossings in this order from 1 to $2n$, the n over- and n under-crossings pair up in n pairs of an over- and under-crossing, such that any pair consists of an even and odd entry. If the i-th entry d_i in a Dowker notation is positive, it indicates that d_i should be the over-crossing in a crossing, where $2i-1$ is the under-crossing, else d_i the under-crossing of the over-crossing $2i-1$. (See, e. g., [A] for more details.) This determines an (oriented) knot diagram up to mirroring of each composite component and up to moves in S^2. There was no convention made by the authors how to fix the mirroring ambiguity, but one way to do so is to take the sign of the left-most entry in the vector d corresponding to a crossing in a given composite component to be equal to the writhe of the crossing it denotes. (For connected diagrams, the program KnotScape [HT] apparently uses exactly this convention, taking the sign of the first entry to be the writhe of the corresponding crossing).

Here we propose how to extend this notation for our purpose (and call it extended Dowker notation).

For a link diagram of our type we must additionally to the Dowker notation of the diagram of K specify (first) the connected component of the complement of the knot diagram of K, where the point T lies (the one, where the bridge of T begins), and (second) the unbounded one (where the bridge ends). We use for this the convention, that for $1 \leq i \leq 2n$ the component described by i (resp. $-i$) is the one to the left (resp. right) of the segment of the line between over/under-crossing i and $i \bmod (2n) + 1$ in orientation direction.

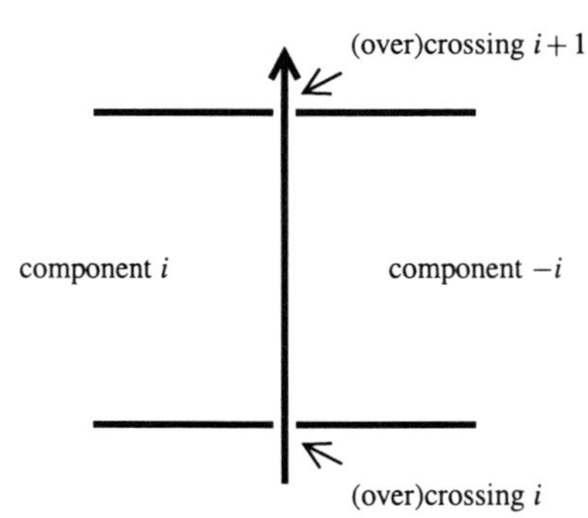

Example 5.4 The diagram of the knot 8_8 and its meridian on figure 10, with starting counting the over/under-crossings at the position of the arrow, could be denoted by 10 8 2 14 4 16 6 2 16 -16 or by 10 8 2 14 4 16 6 2 10 3 (note, that the crossing (1,10) is positive).

Acknowledgement. The second author wishes to thank to the first author for teaching him about Gauß sums and to his former advisor E. Vogt for many helpful discussions. A further thanks goes to J. Weeks at the University of Minnesota, D. Balfanz at Princeton University and R. Lenard at GNU for their help in developing the program, and also to D. Hrencecin at the University of Illinois at Chicago, and to S. Chmutov, S. Duzhin and the referee for their remarks on previous versions of this paper.

The second author would also wish to thank to the organizers of the "Knots in Hellas '98" conference in Delphi, Greece, and among them especially to Jozef Przytycki and Sofia Lambropoulou for the invitation and giving him the possibility to give a talk and for their valuable help and support on both mathematical and non-mathematical matters during the conference.

References

Here are the email addresses of some of the authors of the cited papers. We hope email correspondence could be helpful. Note that email addresses often change, so some may be already out of date. We are grateful for any hints and corrections. A more detailed list of email addresses is maintained by Bar-Natan [BN2].

Dror Bar-Natan	`drorbn@math.huji.ac.il`
Joan S. Birman	`jb@math.columbia.edu`
Louis H. Kauffman	`kauffman@uic.edu`
Xiao-Song Lin	`xl@math.ucr.edu`
Hugh Morton	`su14@liv.ac.uk`
Jozef Przytycki	`przytyck@math.gwu.edu`
Vladimir Turaev	`turaev@math.u-strasbg.fr`
Oleg Viro	`viro@math.uu.se`
Pierre Vogel	`vogel@mathp7.jussieu.fr`
J. Weeks	`weeks@geom.umn.edu`

[A] C. C. Adams, *Das Knotenbuch*, Spektrum Akademischer Verlag, Berlin, 1995 (*The knot book*, W. H. Freeman & Co., New York, 1994).

[Ad] Adobe Systems Inc., *PostScript Language, Tutorial and Cookbook*, Addison-Wesley, 1993.

[BN] D. Bar-Natan, *On the Vassiliev knot invariants*, Topology **34** (1995) 423–472.

[BN2] ——— " ———, *Bibliography of Vassiliev invariants*, available via anonymus file transfer from `ftp.ma.huji.ac.il`, subdirectory `drorbn`.

[BS] ——— " ——— and A. Stoimenow, *The Fundamental Theorem of Vassiliev invariants*, "Geometry and Physics", Lecture Notes in Pure & Appl. Math. **184**, M. Dekker, New York, 1996, 101–134.

[Bi] J. S. Birman, *Recognizing the unknot*, preprint.

[BL] ——— " ——— and X-S. Lin, *Knot polynomials and Vassiliev's invariants*, Invent. Math. **111** (1993) 225–270.

[Co] J. H. Conway, *On enumeration of knots and links*, in "Computational Problems in abstract algebra" (J. Leech, ed.), 329-358. Pergamon Press, 1969.

[Cr] P. R. Cromwell, *Homogeneous links*, J. London Math. Soc. (series 2) **39** (1989), 535–552.

[DT] C. H. Dowker and M. B. Thistlethwaite, *Classification of knot projections*, Topol. Appl. **16** (1983), 19–31.

[Fi] Th. Fiedler, *Talks given in Toulouse & Berlin*, summer 1996.

[Fi2] ——— " ———, *Gauss sum invariants for knots and links*, monography, to appear.

[Fi3] ——— " ———, *A small state sum for knots*, Topology **32 (2)** (1993), 281–294.

[Fi4] ——— " ———, *Die Casson Invariante eines positiven Knotens ist nicht kleiner als sein Geschlecht*, talk given at the knot theory workshop in Siegen, Germany, 1993.

[FW] J. Franks and R. F. Williams, *Braids and the Jones-Conway polynomial*, Trans. Amer. Math. Soc. **303** (1987), 97–108.

[Go] V. Goryunov, *Finite Order Invariants of Framed Knots in a solid torus and in Arnold's J^+-Theory of plane curves*, "Geometry and Physics", Lecture Notes in Pure & Appl. Math. **184**, M. Dekker, New York, 1996, 549–556.

[H] P. Freyd, J. Hoste, W. B. R. Lickorish, K. Millett, A. Ocneanu and D. Yetter, *A new polynomial invariant of knots and links*, Bull. Amer. Math. Soc. **12** (1985), 239–246.

[HT] J. Hoste and M. Thistlethwaite, *KnotScape*, a knot polynomial calculation program, available at `http://www.math.utk.edu/~morwen`.

[J] V. F. R. Jones, *A polynomial invariant of knots and links via von Neumann algebras*, Bull. Amer. Math. Soc. **12** (1985), 103–111.

[J2] —— " ——, *Hecke algebra representations of of braid groups and link polynomials*, Ann. of Math. **126** (1987) 335–388.

[Kn] T. Kanenobu, *Examples of polynomial invariants of knots and links I*, Math. Ann. **275** (1986), 555–572.

[Ka] L. H. Kauffman, *Knots and physics* (second edition), World Scientific, Singapore 1993.

[Ka2] —— " ——, *An invariant of regular isotopy*, Trans. Amer. Math. Soc. **318** (1990), 417–471.

[Ka3] —— " ——, *New invariants in the theory of knots*, Amer. Math. Mon. **3** (1988), 195–242.

[Ki] R. Kirby (ed.), *Problems of low-dimensional topology*, book available on `http://math.berkeley.edu/~kirby`.

[K] J. Kneissler, *The number of primitive Vassiliev invariants up to degree twelve*, University of Bonn preprint, June 1997. See also `q-alg/9706022`.

[La] S. Lambropoulou, *Solid torus links and Hecke algebras of B-type*, Proceedings of the Conference on Quantum Topology (D. N. Yetter, ed.), World Scientific, 1994, 225–245.

[LL] W. B. R. Lickorish and A. S. Lipson, *Polynomials of 2-cable-like links*, Proc. Amer. Math. Soc. **100** (1987), 355–361.

[LM] —— " —— and K. C. Millett, *A polynomial invariant for oriented links*, Topology **26 (1)** (1987), 107–141.

[MT] W. W. Menasco and M. B. Thistlethwaite, *The Tait flyping conjecture*, Bull. Amer. Math. Soc. **25 (2)** (1991), 403–412.

[Mo] H. R. Morton, *Seifert circles and knot polynomials*, Proc. Camb. Phil. Soc. **99** (1986), 107–109.

[MT] —— " —— and P. Traczyk, *The Jones polynomial of satellite links around mutants*, In 'Braids', (Joan S. Birman and Anatoly Libgober, eds.), Contemporary Mathematics **78**, Amer. Math. Soc. (1988), 587–592.

[Mu] K. Murasugi, *Jones polynomial and classical conjectures in knot theory*, Topology **26** (1987), 187–194.

[Mu2] —— " ——, *On the braid index of alternating links*, Trans. Amer. Math. Soc. **326 (1)** (1991), 237–260.

[PV] M. Polyak and O. Viro, *Gauss diagram formulas for Vassiliev invariants*, Int. Math. Res. Notes **11** (1994), 445–454.

[PV2] —— " —— and —— " ——, *On the Casson knot invariant*, preprint.

[PP] T. Przytycka and J. Przytycki, *Subexponentially computable truncations of Jones-type polynomials*, in "Graph Structure Theory", Contemporary Mathematics **147** (1993), 63–108.

[PP2] —— " —— and —— " ——, *Subexponentially computable truncations of Jones-type polynomials*, with appendix on Vertigan's algorithm. Preprint **22**, Odense Univ., 1992.

[P] J. H. Przytycki, *Search for different links with the same Jones' type polynomials: Ideas from graph theory and statistical mechanics*, Panoramas of Mathematics, Banach Center Publications, Vol. **34**, Warszawa 1995.

[P2] —— " ——, *Equivalence of cables of mutants of knots*, Canad. J. Math., **41 3(2)**, 1989, 250–273.

[P3] —— " ——, *Symmetric knots and billiard knots*, in "Ideal Knots", Series on Knots and Everything, Vol. **19** (A. Stasiak, V. Katrich, L. Kauffman, eds.), World Scientific, 1999, 374–414.

[Ro] D. Rolfsen, *Knots and links*, Publish or Perish, 1976.

[S] Bjarne Stroustrup, *The C++ Programming Language*, Addison-Wesley, 1995.

[St] A. Stoimenow, *On Harrison cohomology and a conjecture by Drinfel'd*, diploma thesis, Humboldt University Berlin, 1996.

[St2] —— " ——, *Program for computing Gauß sums*, C++ files available as `gsinv.tgz` on my web page.

[St3] —— " ——, *Gauß sum invariants, Vassiliev invariants and braiding sequences*, to appear in Jour. of Knot Theory and its Ramifications.

[St4] —— " ——, *Positive knots, closed braids and the Jones polynomial*, preprint.

[St5] —— " ——, *On some restrictions to the values of the Jones polynomial*, Humboldt University Berlin preprint, October 1997.

[St6] —— " ——, *Some minimal degree Vassiliev invariants not realizable by the HOMFLY and Kauffman polynomial*, preprint.

[St7] —— " ——, *Mutant links distinguished my degree-3-Gauß sums*, submitted to these Proceedings.

[St8] —— " ——, *Knots of genus one*, preprint.

[Th] M. B. Thistlethwaite, *A spanning tree expansion for the Jones polynomial*, Topology **26** (1987), 297–309.

[Va] V. A. Vassiliev, *Cohomology of knot spaces*, Theory of Singularities and its Applications (Providence) (V. I. Arnold, ed.), Amer. Math. Soc., Providence, 1990.

[Vo] P. Vogel, *Algebraic structures on modules of diagrams*, Université Paris VII preprint, June 1996.

[W] J. Weeks, *Documentation for SnapPea Vers. 2.0 Projection file format*, available from the author of the program.

[Wi] Simon Willerton, *Cabling the Vassiliev invariants*, preprint.

[Wo] S. Wolfram, *Mathematica — a system for doing mathematics by computer*, Addison-Wesley, 1989.

[Ya] S. Yamada, *The minimal number of Seifert circles equals the braid index*, Invent. Math. **88** (1987), 347–356.

GLUCK SURGERY AND FRAMED LINKS IN 4-MANIFOLDS

KAZUO HABIRO,
Graduate School of Mathematical Sciences University of Tokyo 3-8-1 Komaba Meguro-ku Tokyo 153-8914, Japan
habiro@ms.u-tokyo.ac.jp

YOSHIHIKO MARUMOTO
Department of Mathematics Osaka Sangyo University Nakagaito 3-1-1, Daito Osaka 574-8530, Japan
marumoto@las.osaka-sandai.ac.jp

and YUICHI YAMADA
Department of Systems Engineering the University of Electro-Communications 1-5-1 Chofu-gaoka, Chofu Tokyo 182-8585, Japan
yyyamada@e-one.uec.ac.jp

ABSTRACT

Gluck surgery is the operation of cutting out $S^2 \times D^2$, a tubular neighborhood of a 2-knot and pasting it back in a 4-manifold. It may be expected to make a *fake pair* of 4-manifolds, which means a pair that are homotopy equivalent but non-diffeomorphic. We will give an alternative proof of a theorem: Gluck surgery along a banded 2-knot is independent of the bands, which was proved in P. Melvin's thesis and prove that: if a 2-knot is obtained by ribbon moves from another 2-knot, then the Gluck surgeries along them are diffeomorphic. Our method, which we call "*framed links in 4-manifolds*", is a (4,5)-dimensional version of the usual ((3,4)-dimensional) framed links.

Keywords: Gluck surgery, band sum, ribbon knot

1. Introduction and Main Theorem

By the word "2-knot", we mean an embedded 2-sphere in the 4-sphere S^4 or in a general 4-manifold such that the tubular neighborhood of the 2-sphere is diffeomorphic to the product $S^2 \times D^2$; by "2-link", we mean a disjoint union of 2-knots in the 4-manifold.

Gluck surgery is the operation of cutting out a tubular neighborhood of a 2-knot (or a higher dimensional knotted sphere) and pasting it back via a map τ called twisting (see [3]). The twisting map τ is the self-diffeomorphism of $S^2 \times S^1 = \partial(S^2 \times D^2)$ defined by $\tau(z, \alpha) = (\alpha z, \alpha)$, where we identify S^1 with the unit circle in $\mathbf{C}$ and S^2 with the Riemann sphere $\hat{\mathbf{C}} = \mathbf{C} \cup \{\infty\}$. Gluck surgery naturally generalizes to a 2-link $L = L_1 \cup L_2 \cup \cdots \cup L_n$ (each L_i is a connected component) in an arbitrary smooth 4-manifold X. By $\Sigma_X(L)$, or by $\Sigma_X(L_1 \cup L_2 \cup \cdots \cup L_n)$, we mean the 4-manifold obtained by Gluck surgery along each component of L in the 4-manifold X simultaneously.

When L is null-homotopic, Gluck surgery along L does not change the homotopy type of the manifold, but may change the diffeomorphism type. Thus it may be expected that Gluck surgery creates a *fake pair*, which means a pair of manifolds that are homotopy equivalent but non-diffeomorphic. However, to the authors' knowledge, no such fake pair of orientable 4-manifolds obtained by Gluck surgery is known (a non-orientable example is known, see [1],[2]).

Let L_i and L_j be two different components of a 2-link $L = L_1 \cup L_2 \cup \cdots \cup L_n$ in a connected 4-manifold X. Let B^2 denote a 2-disk and I an interval $[-1, 1]$. If an embedded cylinder $b = f(B^2 \times I)$, where $f : B^2 \times I \to X$ is an embedding, satisfies $b \cap L = b \cap (L_i \cup L_j)$, $b \cap L_i = f(B^2 \times \{-1\})$ and $b \cap L_j = f(B^2 \times \{1\})$, b is called a *band* connecting L_i and L_j, and the new link below is called a *banded sum* or a *fusion* of (two components L_i and L_j of) L along the band b:

$$(L \backslash f(B^2 \times \{\pm 1\})) \cup f(\partial B^2 \times I).$$

If the original link is a disjoint union of two 2-knots K_1 and K_2, the band sum is denoted by $K_1 \sharp_b K_2$, see Figure. 1.

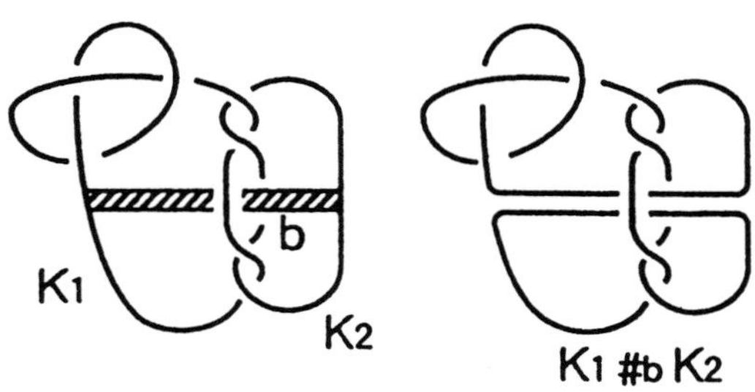

Fig. 1.

If a 2-knot is obtained from an n-component link by taking $n-1$ fusions along bands, the 2-knot is called a *banded knot*. In knot theory, especially in higher dimensional knot theory, *ribbon knots* have been the focus of much research. From the point of view of banded knots, a ribbon 2-knot is defined as a banded knot which is obtained from the *trivial* n-component 2-link by $n-1$ fusions. For definitions, history, and latest results on ribbon 2-knots, see for example [13,14,8,4,6,7,9,10,12].

In [3,Theorem(17.1)] and [14,Theorem(2.3)], it is proved that Gluck surgery along a ribbon 2-knot in the standard 4-sphere S^4 is trivial, i.e., the resulting manifold is diffeomorphic to the original manifold S^4. In the proof there, the symmetry of a ribbon 2-knot along the equator in S^4 played an important role (see [14,p.450]). By a generalization of the method in [14], it is shown that Gluck surgery along a ribbon 2-knot in any 4-manifold is also trivial. Our main theorem includes this fact as a corollary, see Corollary 4.

In this paper, we study Gluck surgeries along 2-links and banded knots in general 4-manifolds by a method, which we call "*framed links in 4-manifolds*". The usual use of framed links (in the theory of 3,4-dimensional manifolds) is to describe a 3- or 4-manifold. A framed link represents a handlebody decomposition of a 4-manifold with boundary. Each component of a framed link in S^3 is the attaching circle of a 4-dimensional 2-handle. By contrast, in our method, a framed link in a 4-manifold X will be defined as a disjoint union of framed circles and a 2-link (2-spheres) in X. Each framed circle is the attaching circle of a 5-dimensional 2-handle, and each 2-sphere is the attaching sphere of a 3-handle. Thus a framed link in X represents a 5-dimensional cobordism obtained by attaching the handles to $X \times \{1\}$ in $X \times [0,1]$. We are concerned with the 4-manifolds that are the boundaries of such cobordisms.

In this paper, we prove the following theorems.

Theorem 1. *Let $L = K_1 \cup K_2$ be a 2-component 2-link in a 4-manifold X and $K_1 \natural_b K_2$ a band sum of the two components along a band b. Then Gluck surgery along $K_1 \natural_b K_2$ in X is diffeomorphic to that along $L = K_1 \cup K_2$, i.e., $\Sigma_X(K_1 \natural_b K_2) \cong \Sigma_X(K_1 \cup K_2)$.*

After writing the first version of this paper, the authors were informed that Theorem 1 is a special case of Lemma 5.1 in P. Melvin's thesis [11] and that our proof is different from his. We will give an alternative proof, which is essentially 5-dimensional rather than 4-dimensional.

In [12], a local move, called a *ribbon move*, on 2-links is defined. On the relation between Gluck surgery and ribbon moves, we will prove the following theorem (see also [5,Corollary 3.5]):

Theorem 2. *Let K, K' be two 2-knots in a 4-manifold X. Suppose that K' is obtained by ribbon moves from K. Then Gluck surgery along K' in X is diffeomorphic to that along K, i.e., $\Sigma_X(K') \cong \Sigma_X(K)$.*

Remark. It is shown by Theorem 2 that Gluck surgery along a 2-knot obtained by band sum of two embedded 2-spheres with *non-zero* opposite self-intersection numbers depends only on the homotopy class of the band. Theorem 2 follows directly from Theorem 1 in the case where the annulus (see Section 4) of the ribbon move connects two components of a 2-link in our sense, because the ribbon move is achieved by changing of the band.

In our proof of Theorem 2, we do not cut nor paste the 2-knot but represent the ribbon move itself by a framed link in the complement of the knot, in a sense. That is an application of framed links in 4-manifolds.

As corollaries we have the following:

Corollary 3. *Let L be an n-component 2-link in a 4-manifold X and K a banded knot obtained by $n-1$ fusions from L. Then Gluck surgery along K in X is diffeomorphic to that along L, i.e.,* $\Sigma_X(K) \cong \Sigma_X(L)$.

Corollary 4. *Gluck surgery along a ribbon 2-knot K in a 4-manifold X is trivial, i.e.,* $\Sigma_X(K) \cong X$, *in other words, this does not create a fake manifold pair.*

Theorem 1 includes the following proposition, which was shown in [6].

Proposition 5. *Let K_1, K_2 be 2-knots in the 4-sphere S^4, Then the following additivity holds:* $\Sigma(K_1 \sharp K_2) \cong \Sigma(K_1) \sharp \Sigma(K_2)$.

In Section 2, we give a definition and some calculus of framed links in 4-manifolds and describe the Gluck surgery $\Sigma_X(K)$ by the method. We prove Theorem 1 in Section 3. In Section 4, we redefine ribbon moves using local coordinates and prove Theorem 2.

2. Framed links in 4-manifolds

In this section, we give a definition and some calculus of the method, which we call "framed links in 4-manifolds". For a given 2-knot K in a 4-manifold X, we define a framed link $\mathcal{L}_{K,1}$ in X and show that the framed link describes $\Sigma_X(K)$, the Gluck surgery along K in X.

Definition. Let X be a connected 4-manifold. A *framed link* $\mathcal{L} = (\mathcal{S}, \mathcal{C})$ in X is the following object:

• $\mathcal{S} = \{S_1, S_2, \cdots, S_m\}$ or ϕ : a 2-link in X, i.e., mutually disjoint 2-spheres in the interior of the manifold X whose normal bundles are trivial,

• $\mathcal{C} = \{(C_1, f_1), (C_2, f_2), \cdots, (C_n, f_n)\}$ or ϕ : where the C_j are mutually disjoint embedded circles which are disjoint from $\cup_i S_i$ in the interior of the manifold X, and f_j is a framing of C_j, i.e., an isotopy class of trivializations of the normal bundle of C_j in X.

On the framings of circles in condition 2 above, in the special case that C_j bounds a 2-disk D_j in X, we call the two choices of the framings the *0-framing* and *1-framing* as follows: Since a normal bundle of D_j in X has a unique trivialization up to isotopy, the restriction of the trivialization to the boundary C_j defines a framing of C_j. This framing is called the 0-framing (with respect to D_j). There is only one other possible choice of framing, since $\pi_1(SO(3)) \cong \mathbf{Z}/2\mathbf{Z}$. The latter framing is called the 1-framing. In the case that the disk D_j is understood from the context, the circle C_j with the 0-framing (resp. 1-framing) is denoted by $(C_j, 0)$ (resp. $(C_j, 1)$), for short.

For a given framed link $\mathcal{L} = (\mathcal{S}, \mathcal{C})$ in a 4-manifold X, as the usual surgery theory and handlebody theory, we construct a new 4-manifold $X(\mathcal{L})$ via a 5-dimensional cobordism as follows:

First attach 5-dimensional 3-handles $H_i^3 = D^3 \times D^2$ to $X \times [0,1]$ along S_i in $X \times \{1\}$ $(1 \leq i \leq m)$. Note that any two trivializations of the normal bundle of S_j are isotopic because $\pi_2(SO(2)) = \{0\}$. Next attach 5-dimensional 2-handles $H_j^2 = D^2 \times D^3$ to $X \times [0,1]$ along C_j in $X \times \{1\}$ by f_j $(1 \leq j \leq n)$, where the attaching part of an r-handle is $\partial D^r \times D^{5-r}$ ($r = 2$ or 3). These handle attachings change $X \times \{1\}$ to a new 4-manifold, which we denote $X(\mathcal{L})$. Note that the process above does not change ∂X $(= \partial X \times \{1\})$.

Next, we study some sufficient conditions for two framed links $\mathcal{L}$ and $\mathcal{L}'$ in a 4-manifold X to ensure that the resulting manifolds $X(\mathcal{L})$ and $X(\mathcal{L}')$ are diffeomorphic. Here we list three basic operations: 2-handle slides, 3-handle slides and 2,3-handle cancellations on framed links in X; these operations do not change the resulting 4-manifold.

(1) 2-handle sliding.
Let $\mathcal{L}' = (\mathcal{S}', \mathcal{C}')$ be obtained from $\mathcal{L} = (\mathcal{S}, \mathcal{C})$ by the following operation:

$$(C_k, f_k), (C_l, f_l) \text{ in } \mathcal{C} \quad \Rightarrow \quad (C_k \sharp_b C_l', f_k + f_l), (C_l, f_l), \text{ in } \mathcal{C}'$$

where C_l' is a disjoint slight push-off of C_l by f_l in X, $\sharp_b$ is a 1-dimensional band sum (see Figure 2) and $f_k + f_l$ is the framing defined by f_k and f_l. All other components and $\mathcal{S}$ are kept unchanged during the operation. We call this operation 2-handle slide. This operation corresponds to sliding of the 2-handle C_k over C_l, and thus does not change the 5-dimensional cobordism itself.

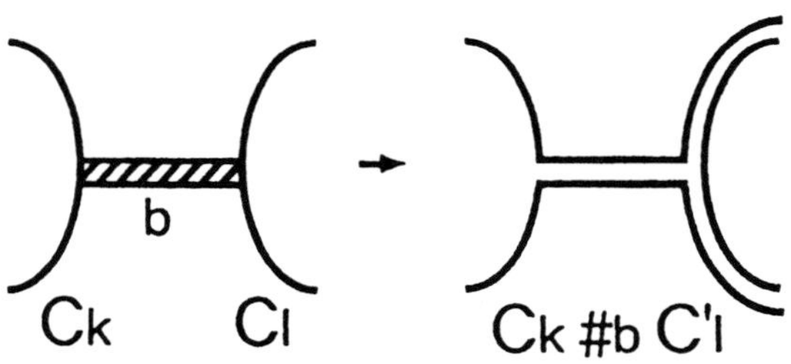

Fig. 2.

(2) 3-handle sliding.
Let $\mathcal{L}' = (\mathcal{S}', \mathcal{C}')$ be obtained from $\mathcal{L} = (\mathcal{S}, \mathcal{C})$ by the following operation:

$$S_k, S_l \text{ in } \mathcal{S} \quad \Rightarrow \quad S_k \sharp_c S_l', S_l \text{ in } \mathcal{S}'$$

where S_l' is a disjoint slight push-off of S_l in X, $\sharp_c$ is a band sum along c (see Figure 3). All other components and $\mathcal{C}$ are kept unchanged during the operation. We call this operation a 3-handle slide. This operation corresponds to sliding of the 3-handle S_k over S_l, and thus does not change the 5-dimensional cobordism.

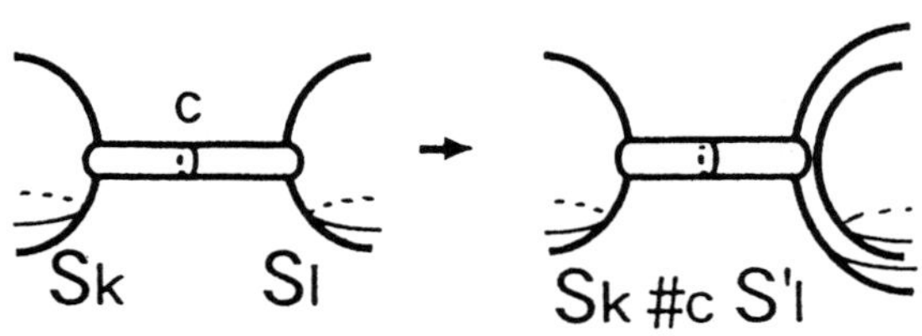

Fig. 3.

(3) 2,3-handle cancelling.
Let $\mathcal{L}' = (\mathcal{S}', \mathcal{C}')$ be obtained from $\mathcal{L} = (\mathcal{S}, \mathcal{C})$ by the following operation:

$$\left\{ \begin{array}{r} S \text{ in } \mathcal{S} \\ (C_S, 0) \text{ in } \mathcal{C} \end{array} \right\} \quad \Rightarrow \quad \text{pairwise cancelled,}$$

where C_S is a meridian circle of the 2-knot S such that C_S bounds a meridian disk disjoint from other parts of the framed link. Note that S is possibly knotted and linked with other components in X. All other components are kept unchanged during the operation. We call this operation a 2,3-handle cancellation, since this operation does not change the resulting 4-manifold, but the 5-dimensional cobordism is not a product (for example, if S is a trivial 2-knot in S^4, the cobordism corresponding to the framed link $(\{S\}, \{(C_S, 0)\})$ is twice punctured $S^2 \times S^3$, which is a cobordism from S^4 to itself). We do not prove that this operation does not change the resulting 4-manifold, as its proof is easier than that of Lemma 6, see the final paragraph of the (conceptional) proof of Lemma 6.

Example $\mathcal{L}_{K,\epsilon}$. For a given 2-knot K in a 4-manifold X, we give an example of a framed link in X, which is denoted by $\mathcal{L}_{K,\epsilon}$ ($\epsilon = 0, 1$). Let K be a 2-knot in X, D_K a meridian disk and C_K the boundary of D_K, the meridian circle of K. We call the framing of C_K induced from D_K "0", and the other framing "1". We define two framed links $\mathcal{L}_{K,0}$ and $\mathcal{L}_{K,1}$ in X by $\mathcal{L}_{K,\epsilon} = (\{K\}, \{(C_K, \epsilon)\})$.

By 2,3-handle cancellation, the resulting manifold $X(\mathcal{L}_{K,0})$ is diffeomorphic to the original manifold X, i.e., $X(\mathcal{L}_{K,0}) \cong X$. Regarding $X(\mathcal{L}_{K,1})$, we have the following;

Lemma 6. *The resulting manifold* $X(\mathcal{L}_{K,1})$ *of the framed link* $\mathcal{L}_{K,1}$ *in* X *is diffeomorphic to the 4-manifold* $\Sigma_X(K)$ *given by Gluck surgery along the 2-knot* K *in* X, *i.e.*, $X(\mathcal{L}_{K,1}) \cong \Sigma_X(K)$.

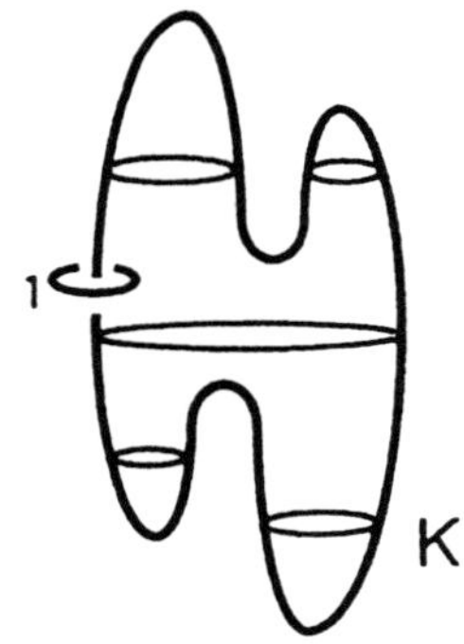

Fig. 4.

We give only a conceptional proof, as this lemma may already be a folklore fact.

Proof. As the first step, we start with the 3-handle correspond to the 2-knot K, the framed link $(\{K\}, \phi)$ in the original 4-manifold X. The resulting manifold $X(\{K\}, \phi)$ of this framed link is obtained from X by a surgery on the 2-knot K, i.e., cutting out $S^2 \times D^2$, the tubular neighborhood of K and pasting $D^3 \times S^1$ along their boundaries.

Next, as the second step, we study the 2-handle correspond to the circle C_K, the framed link $(\phi, \{(C_K, \epsilon)\})$ in the manifold $X(\{K\}, \phi)$ obtained in the last step. We cut out $D^3 \times S^1$, the tubular neighborhood of C_K and past $S^2 \times D^2$ along their boundaries by the framing of C_K. Note that the circle C_K is isotopic to the core $\{o\} \times S^1$ of the surgery, where o is the center of D^3, in $X(\{K\}, \phi)$.

If the framing ϵ is 0 with respect to the meridian disk to K, the second step is just an inverse operation of the first step. Thus the result of the above two steps is cutting out and trivial pasting back of $S^2 \times D^2$. On the other hand, if the framing ϵ is 1, it is cutting out and non-trivial pasting back of $S^2 \times D^2$, which is Gluck surgery along the 2-knot K. □

3. Proof of Theorem 1

In this section, we prove the main theorem by the following three steps:

(i) For given 2-knots K_1, K_2 and a band b connecting them in an arbitrary 4-manifold X (for the definition, see Section 1), we construct a framed link $\mathcal{L}_{K_1,K_2,b}$ in X.

(ii) In Lemma 7, we show that the 4-manifold $X(\mathcal{L}_{K_1,K_2,b})$ is diffeomorphic to $\Sigma(K_1\sharp_b K_2)$.

(iii) In Lemma 8, we show that the 4-manifold $X(\mathcal{L}_{K_1,K_2,b})$ is also diffeomorphic to $\Sigma(K_1 \cup K_2)$.

In this section, we use the notation below about intervals and disks. Here we list some of them. Let $\epsilon > 0$ be a sufficiently small number.

$$I = [-1,1],\ I_\epsilon = [-1-\epsilon, 1+\epsilon],\ I_{2\epsilon} = [-1-2\epsilon, 1+2\epsilon], \quad (I \subset I_\epsilon \subset I_{2\epsilon}).$$
$$J_\epsilon = [-\epsilon, \epsilon],\ J_{2\epsilon} = [-2\epsilon, 2\epsilon], \quad (\{0\} \subset J_\epsilon \subset J_{2\epsilon}).$$
$$B^2 = \text{the unit disk},\ \hat{B^2} = \text{a disk of radius } 1+\epsilon, \quad (B^2 \subset \hat{B^2}).$$

For given 2-knots K_1, K_2 and a band $b = f(B^2 \times I)$ connecting them in a 4-manifold X, we construct a framed link as follows:

First, extending the embedding f over $\hat{B^2} \times I_{2\epsilon} \times J_{2\epsilon}$, we construct an embedding $\hat{f} : \hat{B^2} \times I_{2\epsilon} \times J_{2\epsilon} \to X$ which satisfies $\hat{f}(\hat{B^2} \times I_{2\epsilon} \times J_{2\epsilon}) \cap K_1 = \hat{f}(\hat{B^2} \times \{-1\} \times \{0\})$ and $\hat{f}(\hat{B^2} \times I_{2\epsilon} \times J_{2\epsilon}) \cap K_2 = \hat{f}(\hat{B^2} \times \{1\} \times \{0\})$. In the image of $\hat{f}$, we construct a 2-disk E and a circle C by $E = \hat{f}(\{o\} \times I_\epsilon \times J_\epsilon)$ and $C = \partial E$, where o is the center of $\hat{B^2}$. The disk E defines a framing of C (see Section 2). By "0" we denote this framing of C. For $i = 1$ and 2, let E_i be a meridian 2-disk of the 2-knot K_i disjoint from the band b, and C_i its boundary ∂E_i, a meridian circle. The disk E_i defines a framing of C_i and there is another choice of framing, which is denoted by "1". Define $\mathcal{L}_{K_1,K_2,b}$ to be the framed link in X

$$\mathcal{L}_{K_1,K_2,b} = (\{K_1, K_2\},\ \{(C_1, 1), (C, 0)\}).$$

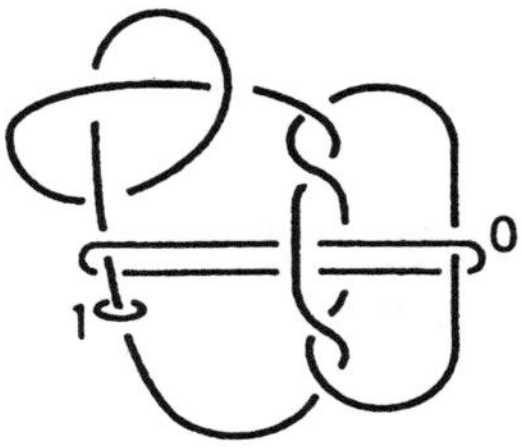

Fig. 5.

In the proof of the following two lemmas, we use the same notation for both a 2-(resp. 3-)handle and the attaching circle (resp. sphere) of the handle.

Lemma 7. *Let K_1, K_2 be two disjoint 2-knots and b a band connecting them in a 4-manifold X and $\mathcal{L}_{K_1,K_2,b}$ the framed link in X. Then the 4-manifold $X(\mathcal{L}_{K_1,K_2,b})$*

is diffeomorphic to the 4-manifold $\Sigma_X(K_1\sharp_b K_2)$ *given by Gluck surgery along the banded sum* $K_1\sharp_b K_2$ *in* X, *i.e.*, $X(\mathcal{L}_{K_1,K_2,b}) \cong \Sigma_X(K_1\sharp_b K_2)$.

Proof. We show that $X(\mathcal{L}_{K_1,K_2,b}) \cong X(\mathcal{L}_{K_1\sharp_b K_2,1})$ by a single handle slide and cancellation. The lemma then follows from Lemma 6.

First, slide the 3-handle K_1 over K_2 along the band b. This sliding changes the attaching sphere K_1 to $K_1\sharp_{b'} K_2'$, where K_2' is a slight disjoint push-off of K_2; b' is a band connecting K_1 and K_2' which is slightly changed on a part of $f(\partial B^2 \times I)$ of the original band b corresponding to the change from K_2 to K_2', see Figure 6. We call the new 3-handle K_{12}.

Fig. 6.

As a result of the handle slide, C becomes a circle isotopic to a meridian circle of K_2, thus the 2-handle C and the 3-handle K_2 are cancelled since the framing of C is 0. The resulting framed link is $(\{K_{12}\}, \{(C_1, 1)\})$, which is isotopic to the framed link $\mathcal{L}_{K_1\sharp_b K_2,1}$ defined in Section 2. □

Lemma 8. *Let* K_1, K_2 *be two disjoint 2-knots and* b *a band connecting them in a 4-manifold* X *and* $\mathcal{L}_{K_1,K_2,b}$ *the framed link in* X. *Then the 4-manifold* $X(\mathcal{L}_{K_1,K_2,b})$ *is diffeomorphic to the 4-manifold* $\Sigma_X(K_1 \cup K_2)$ *given by Gluck surgery along the 2-link* $K_1 \cup K_2$ *in* X, *i.e.*, $X(\mathcal{L}_{K_1,K_2,b}) \cong \Sigma_X(K_1 \cup K_2)$.

Proof. Let $\mathcal{L}_{K_1,1} \cup \mathcal{L}_{K_2,1}$ denote the framed link $(\{K_1, K_2\}, \{(C_1, 1), (C_2, 1)\})$. We show that $X(\mathcal{L}_{K_1,K_2,b}) \cong X(\mathcal{L}_{K_1,1} \cup \mathcal{L}_{K_2,1})$ by a single handle slide, and then the lemma follows from Lemma 6.

We slide C over C_1 and call the new circle C'. The circle C' no longer links K_1, and bounds a 2-disk E' given by slightly changing E near the point $K_1 \cap E_1$. The framing of the circle is changed from 0 to 1 with respect to E', see Figure 7. The circle C' is moved by an isotopy until it becomes disjoint from the meridian disk E_1 of K_1, because circles do not link in a 4-dimensional manifold. Furthermore, C' is moved to a meridian circle of K_2 in the image of $\hat{f}$, the neighborhood of the original band b. Thus the resulting framed link is $(\{K_1, K_2\}, \{(C_1, 1), (C', 1)\})$, which is isotopic to the framed link $\mathcal{L}_{K_1,1} \cup \mathcal{L}_{K_2,1}$. This proves the lemma.
□

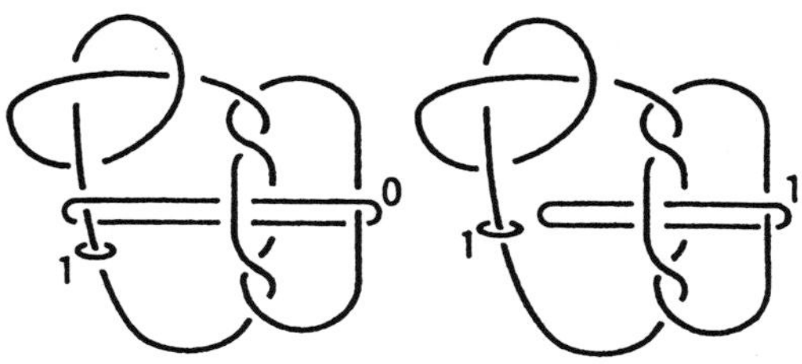

Fig. 7.

4. Ribbon moves and Gluck Surgery

In this section, we redefine a ribbon move on 2-knots in 4-manifolds by explicit coordinates, and prove Theorem 2. This proof is an application of framed links in 4-manifolds. In the proof, the ribbon move itself is represented by a framed link, in a sense.

Let $f : \mathbf{R} \to \mathbf{R}$ be a smooth function obtained from the following map $\hat{f}$ (a semicircle) by smoothing near the non-smooth points (corners) $\{r = \pm 3\}$:

$$\hat{f}(r) = \begin{cases} \sqrt{3^2 - r^2} & |r| < 3, \\ 0 & |r| \geq 3. \end{cases}$$

In the 4-dimensional Euclidean space $\{(x, y, z, t)\}$, we take a 4-ball B_5 of radius 5 centered at the origin and we construct subsets A (an annulus) and $D_\pm$ (2-disks) of B_5 as follows, see Figure 8:

$$\begin{aligned} B_5 &= \{(x, y, z, t)|\ x^2 + y^2 + z^2 + t^2 < 5^2\}, \\ A &= \{(x, y, z, t) \in B_5|\ x^2 + y^2 = 1,\ t = 0\}, \\ D_\pm &= \{(x, y, z, t) \in B_5|\ y = \pm f(\sqrt{x^2 + t^2}),\ z = 0\}. \end{aligned}$$

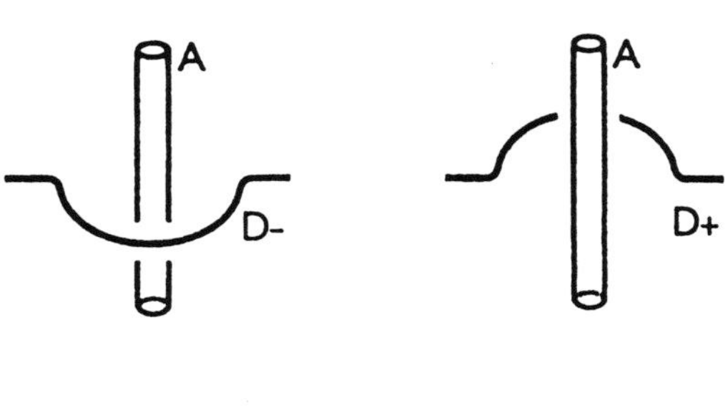

Fig. 8.

In Figures 8,9,10 and 11, we only draw the intersections of the subsets with the 3-dimensional subspace $\{(x,y,z,t)|\, t=0\}$ in B_5.

Let K_- be a 2-knot in a 4-manifold X. We say that a 2-knot K_+ is obtained from K_- by a *ribbon move* (in U) if there exist a local coordinate chart (U,ψ) where $\psi : B_5 \to U$ is a diffeomorphism which satisfies

(i) The map ψ induces a pairwise diffeomorphism
$\psi : (B_5, A \cup D_-) \overset{\cong}{\to} (U, U \cap K_-)$, and
(ii) $(K \cap (X \backslash U)) \cup \psi(A \cup D_+) = K_+$ in X.

Thus the old 2-knot K_- and the new one K_+ differ only by the 2-disks D_- and D_+ in the open 4-ball U in X.

To prove Theorem 2, it is sufficient to show it in the case $K = K_-$ and $K' = K_+$. Thus in what follows, we assume this. For the ribbon move in U from K to K' defined as above, all operations will be done in U. Thus we identify objects in B_5 with their images under the map ψ, and use the coordinate map ψ to define subsets in U.

Before starting the proof, for the given ribbon move in U, we construct a framed link $\mathcal{L}_U$ in U as follows: Let C_K be a meridian circle of D_- in U (thus C_K is a meridian circle of the 2-knot K); S a 2-sphere and C a circle defined below, see Figure 9

$$S = \{(x,y,z,t) \in B_5|\; x^2+y^2+t^2 = (2+\epsilon)^2, z=0\},$$
$$C = \{(x,y,z,t) \in B_5|\; (y+3)^2+z^2 = (1+\epsilon)^2, x=0, t=0\}$$

where ϵ is a sufficiently small positive number. Note that S is the boundary of a normal 3-disk of the center line (core) of the annulus A, and that C links both S and the disk D_-. Because U (or B_5) is contractible, each circle C_K and C bounds a 2-disk in U which is unique up to isotopy, and the 2-disk defines a framing of each circle (see Section 2). Let "0" be this framing and "1" be the other choice of framing. We define a framed link $\mathcal{L}_U$ as $(\{K,S\},\{(C_K,1),(C,0)\})$, see Figure 9.

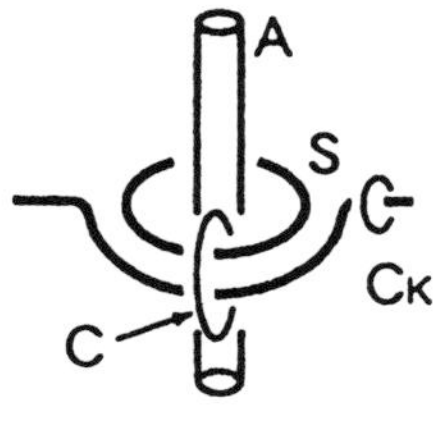

Fig. 9.

Note that a framed link $(\{S\}, \{(C,0)\})$ in X does not change the 4-manifold X, because the pair can be cancelled, but after the process of handle attachings, the 2-knot K in X is changed. Thus the framed link $(\{S\}, \{(C,0)\})$ represents the ribbon move itself.

We prove Theorem 2 in the case $K = K_-$ and $K' = K_+$ by two lemmas:

(i) In Lemma 9, we show that the 4-manifold $X(\mathcal{L}_U)$ is diffeomorphic to $\Sigma_X(K')$.

(ii) In Lemma 10, we show that the 4-manifold $X(\mathcal{L}_U)$ is also diffeomorphic to $\Sigma_X(K)$.

In the proof of the following two lemmas, we use same notation for both a 2-(resp. 3-)handle and the attaching circle (resp. sphere) of the handle.

Lemma 9. *Let $\mathcal{L}_U$ be the framed link in X constructed as above. Then the 4-manifold $X(\mathcal{L}_U)$ is diffeomorphic to $\Sigma_X(K')$ given by Gluck surgery along K' in X, i.e., $X(\mathcal{L}_U) \cong \Sigma_X(K')$.*

Proof. We show that $X(\mathcal{L}_U) \cong X(\mathcal{L}_{K',1})$ by a single handle slide and cancellation. The lemma then follows from Lemma 6.

First, slide the 2-knot (3-handle) K over S, see Figure 10. By the sliding, the 2-knot K is changed to K' in X, and C becomes a meridian circle of S and bounds a meridian disk disjoint from other parts of the framed link. Thus the 2-handle C and the 3-handle S are cancelled. The resulting framed link is $(\{K\}, \{(C_K, 1)\}) = \mathcal{L}_{K,1}$, see Figure 10. This proves the lemma. □

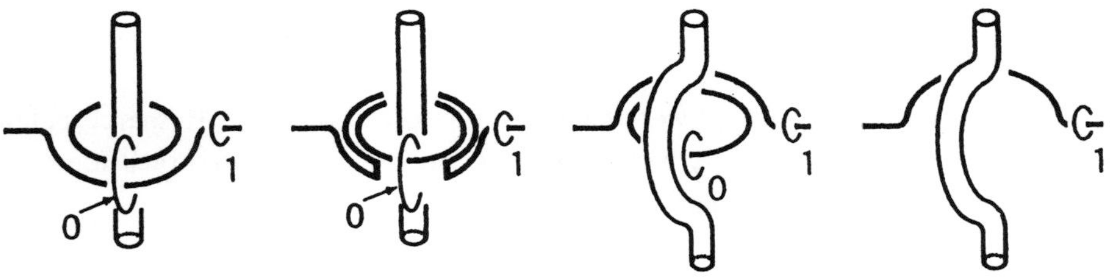

Fig. 10.

Lemma 10. *Let $\mathcal{L}_U$ be the framed link in X constructed as above. Then the 4-manifold $X(\mathcal{L}_U)$ is diffeomorphic to $\Sigma_X(K)$ given by Gluck surgery along K in X, i.e., $X(\mathcal{L}_U) \cong \Sigma_X(K)$.*

Proof. We show that $X(\mathcal{L}_U) \cong X(\mathcal{L}_{K,1})$ by a 2-handle slide and the triviality of Gluck surgery along a trivial 2-knot. This lemma then follows from Lemma 6.

First, we slide C over C_K and call the new circle C', see Figure 11. The circle C' no longer links D_- or C_K, since they are 1-dimensional subsets in 4-dimensional space. Thus C' becomes a meridian circle of S, and its framing is changed from 0 to 1 with respect to a bounding 2-disk in B_5.

Next, we slide the annulus A over the 2-handle C', see Figure 10 again. Because of the non-triviality of the framing of C', A is twisted but it is absorbed in the 2-sphere K. Thus the resulting framed link is $(\{K, S\}, \{(C_K, 1), (C', 1)\})$.

Now, the pair consisting of the 2-sphere S and its meridian C' with 1-framing is split in a 4-ball from the other components of the framed link. Thus the pair contributes a connected summand of Gluck surgery along S, but this summand is trivial (S^4), because S is trivial and Gluck surgery along a trivial knot is trivial. Thus,

$$X(\mathcal{L}_U) = X(\{K, S\}, \{(C_K, 1), (C', 1)\}) = X(\{K\}, \{(C_K, 1)\}) = X(\mathcal{L}_{K,1}).$$

This proves the lemma. □

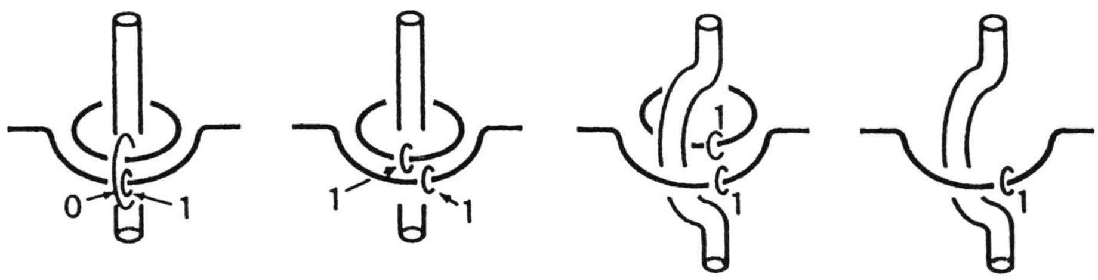

Fig. 11.

Acknowledgements

The first and third authors were partially supported by JSPS Research Fellowship for Young Scientists. They would like to express their sincere gratitude to the referee for helpful comments and suggestions.

Addendum. Here we list two additional facts on the relationship between ribbon moves and changing of bands (see Remark just after Theorem 2), which were verified in the process of revision.

Suppose that a 2-knot K in a 4-manifold X is obtained from K' by a ribbon move in a 4-ball U (see Section 4 for the definition and the notation A, D_- and S, see also Figure 9). The annulus A possibly connects two embedded 2-spheres with *non-zero* opposite self-intersection numbers. We take three points P_0, P_1 and Q on D_-, A and S, respectively, and two arcs c_0 connecting Q and P_0, and c_1 connecting Q and P_1 also in U. The 2-knot obtained from K and S by band sum along c_0 is K'. Regarding c_1, the following is proved (see, e.g. [7]):

The 2-knot obtained from K and S by band sum along c_1 is K itself.

Thus the original ribbon move in U can be achieved by changing of a band. Furthermore, the following is proved:

If the arc c_0 is homotopic to c_1 in the set of arcs that connects K and S in the ambient 4-manifold X, then the original ribbon move can be achieved by a sequence of ribbon moves whose corresponding annulus connects two embedded 2-spheres with trivial bundle.

References

[1] S. Akbulut, *On fake $S^3\tilde{\times}S^1\sharp S^2\times S^2$*, in "Combinatorial methods in topology and algebraic geometry", Contemporary Math. **44**, Amer. Math. Soc. (1985) 281–286.

[2] _________, *Constructing a fake 4-manifold by Gluck construction to a standard 4-manifold*, Topology **27** (1988)239–243.

[3] H. Gluck, *The embedding of two-spheres in the four-sphere*, Trans. Amer. Math. Soc. **104** (1962) 308–333.

[4] K. Habiro, T. Kanenobu and A. Shima, *Finite type invariants of ribbon 2-knots*, Univ. Tokyo Math. Sci. Preprint Ser. 98-22 (1998).

[5] A. Katanaga, O. Saeki, M. Teragaito and Y. Yamada, *Gluck surgery along a 2-sphere in a 4-manifold is realized by surgery along a projective plane*, to appear in Michigan Math. J.

[6] T. Kanenobu, *Cabling, twisting, and branching on higher dimensional knots*, in "Geometry and Algebra of 3:4 Dimensional Manifold", RIMS Kokyu-roku no.518, ed. F. Hosokawa (1984)1–9.

[7] T. Kanenobu and A. Shima, *Two filtrations of ribbon 2-knots*, preprint.

[8] A. Kawauchi, *A Survey of Knot Theory*, Birkhäuser Verlag, Basel-Boston-Berlin (1996) (originally published as "Knot Theory"(in Japanese), Springer Verlag Tokyo (1990)).

[9] Y. Marumoto, *Some higher dimensional knots* , Osaka J. Math., **24** no.4 (1987) 759–783.

[10] ___________, *Ribbon knots and Invariants of Theta-Curves*, J. Knot Theory and its Ram., **4** no.3 (1995) 481–491.

[11] P. Melvin, *Blowing up and down in 4-manifolds*, Ph. D. thesis, UC Berkeley (1977).

[12] E. Ogasa, *Ribbon-moves of 2-links*, Univ. Tokyo Math. Sci. Preprint Ser. 97-42 (1997).

[13] T. Yajima, *On simply knotted spheres in $\mathbf{R}^4$*, Osaka Math. J. **1** (1964)133–152.

[14] T. Yanagawa, *On ribbon 2-knots; the 3-manifold bounded by the 2-knot*, Osaka Math. J. **6** (1969) 447–464

The Planar Algebra of a bipartite graph.

Vaughan F.R. Jones

6th January 2000

Abstract

We review the definition of a general planar algebra $V = \cup V_k$. We show how to construct a general planar algebra from a bipartite graph by creating a specific model using statistical mechanical sums defined by labelled tangles. These planar algebras support a partition function for a closed tangle which is spherically invariant and defines a positive definite inner product on each V_k. We then describe how any planar algebra is naturally a cylic module in the sense of Connes and do some computations.

1 Introduction

The concept of planar algebra was introduced in [10] for many reasons, the most important of which was to help in the calculation of subfactors. As observed in [10], it is hardly surprising that such a natural concept arises in many other situations - notably in [1, 16, 14]. Our axioms for a planar algebra are in this respect rather special and could be criticised for being somewhat narrow because of the restrictions imposed by the shadings, but we would argue that our structure is to the more general ones as a group is to a semigroup. Indeed this could be made precise in the C^*-tensor category context, but the justification with the most content is that provided by the results of Popa. In a series of papers, [18, 19, 20] he came quite independently across a structure which he called a λ-lattice (shown in [10] to be equivalent to a planar algebra with positivity) from which he could construct a subfactor of a II_1 factor whose associated invariant is the λ-lattice one began with! The most recent paper in the series is a universal construction which should allow one to control the isomorphism classes of the II_1 factors in question. Popa's results should be viewed as a co-ordinatizaion theorem analogous to the fundamental theorem of projective geometry, in which the II_1 factors give the (non-commutative) co-ordinate ring associated to the combinatorico- geometrical structure defined by the planar algebra. In fact such a co-ordinatization theorem exists, and is much easier to prove, over an arbitrary field. It is the completeness properties associated with the II_1 factors that make Popa's results so compelling.

The planar algebra of a subfactor was approached in [10] in a somewhat abstract way as invariant vectors in the tensor powers of a bimodule. It would have also been possible to do it via an explicit model in a way similar in spirit to Ocneanu's paragroup approach. The first step in such a program would have been to construct a rather general kind of planar algebra using statistical mechanical sums on the principal graph of the subfactor. Then the actual planar algebra would be obtained as a planar subalgebra formed by certain "flat" elements. This approach was deliberately avoided as being long, clumsy and inelegant. However in more recent work the idea of constructing interesting planar algebras from the rather general ones based on graphs has turned out to be extremely useful and we find it necessary to give these general models. We will present such an application in a forthcoming paper. These general planar algebras may not turn out to be of much interest for their own sake as they can be formed from very general bipartite graphs. Their detailed construction does bring to light several interesting points however, such as the role of the choice of an eigenvector of the adjacency matrix of the graph in the statistical mechanical sums. In fact an arbitrary choice of vector(which we will call the "spin vector") would give a planar algebra but it would not in general have the property that a closed circle in a diagram contributes a simple multiplicative constant.

One feature of these general planar algebras is that there are no obstructions to obtaining graphs as "principal graphs" or at least connected components thereof. It is known ([8]) that graphs occuring as principal graphs of subfactors are quite rare so the role of the factor/connectedness condition becomes clear. This suggests the study of planar algebras intermediate between the rather simple kind constructed here and the restricted subfactor kind with connected principal graphs. We will present further results on this question, inculding an ABCDEFGHI classification for modulus less than 2, in a forthcoming paper.

We noticed some time ago that Connes' cyclic category appears in the annular or "affine" Temperley Lieb category. (Composition of morphisms in the Temperley Lieb category in general leads to closed circles, but they do not occur if one restricts to the annular tangles of Connes' category.) This means that any planar algebra is a cyclic module in the sense of Connes. In fact the natural adjoint map in the Temperley Lieb category defines a second copy of the cyclic category. Together with the first, they generate the Temperley Lieb category. If the planar algebra has the property that isolated circles in a tangle contribute a non-zero scalar multiplicative factor, the face maps of the second category provide homotopy contractions of the first so that the cyclic homology of such a planar algebra will always be that of the ground field. There are however many interesting planar algebras that do not have this property. We simply present these observations in this paper as we do not yet know how to use cyclic homology in planar algebras.

2 Definition of a general planar algebra.

A k−tangle for $k \geq 0$ is the unit disc D with $2k$ marked and numbered (clockwise) points on its boundary, containing a finite number of internal discs each with an even (possibly zero) number of marked and numbered points on their boundaries. All the marked points of all the discs are connected by smooth disjoint curves called the *strings* of the tangle. The strings lie between the internal discs and D and they meet all discs transversally. The strings must connect even-numbered boundary points to even-numbered ones and odd to odd. There may also be a finite number of closed strings (not connecting any discs) in the subset $\mathcal{I}$ which is the complement of the union of the internal discs in D. The connected components of $\mathcal{I}$ minus the strings are called the *regions* of the tangle and may be shaded black and white in a unique way with the convention that the region whose closure contains the interval on the boundary of D between the first and second marked points is shaded black. If necessary the strings of the tangle will be oriented so that black regions are on the right as one moves along strings. To indicate the first point on the boundary of a disc in a picture we will select the unshaded region immediately preceding the first point(in clockwise order) and place a * in that region near the relevant boundary component.

Tangles with no marked boundary points are called *closed* tangles and come in two kinds according to whether the region near the boundary is shaded or not.

Tangles are considered up to smooth isotopy of D. The isotopies may act non-trivially on the boundary of D. An example of a 4-tangle,with 7 internal discs is given in figure 1.1 below:

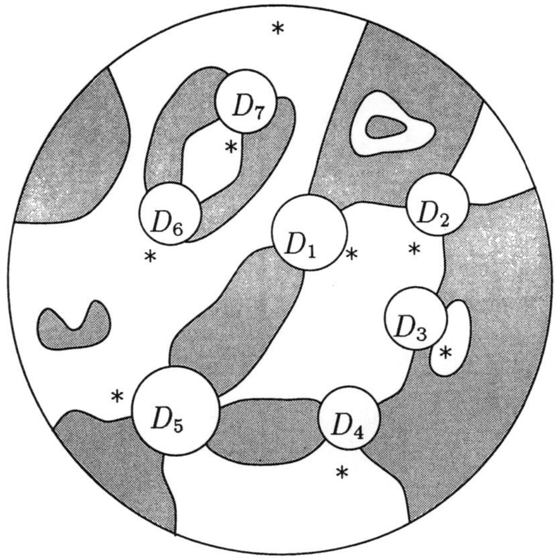

fig. 1.1

Tangles with the appropriate number of boundary points can be glued into the internal discs of another tangle making the set of all tangles into a coloured operad, the colour of an element being the number of boundary points (and the colour of the region near the boundary for 0-tangles). To perform the gluing operation, the tangle S to be glued to an internal disc(with the same colour as S) of another tangle T is first isotoped so that its boundary coincides with the boundary of the chosen internal disc $\mathcal{D}$, the marked points on each boundary disc being also made coincident by the isotpy. Some smoothing may need to be done near the marked points so that the strings of T and S meet smoothly. Finally the common boundary is removed. The result of the gluing is another tangle $T \circ_{\mathcal{D}} S$ with the same number of external marked points as S and having $n_T + n_S - 1$ internal discs, n_T and n_S being the numbers of internal discs of T and S respectively. It is clear that the isotopy class of $T \circ_{\mathcal{D}} S$ depends only on the isotopy classes of S and T and the choice of $\mathcal{D}$. An example of the gluing operation is depicted in figure 1.2:

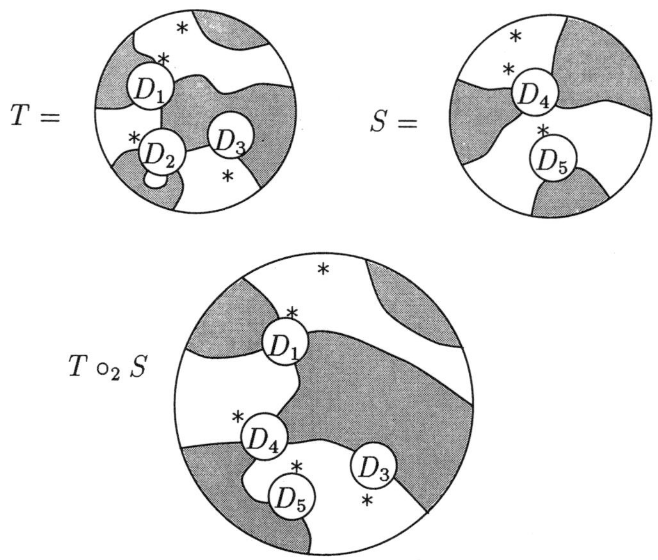

fig. 1.2

See [17] for the definition of an operad. Slight modifications need to be made to handle the colours.

Definition 2.1 *The planar operad $\mathcal{P}$ is the set of isotopy classes of planar tangles with colours and compositions defined above.*

One may construct another coloured operad with the same colours as $\mathcal{P}$ from vector spaces. Operad elements are then multilinear maps from "input" vector spaces, each vector space having a colour as in $\mathcal{P}$ to an "output" vector space. Composition is only permitted when the appropriate vector spaces have the same colour. One obtains another coloured operad $Vect$. See [17] for a preciese definition(without colours).

Definition 2.2 *A planar algebra is an operad homomorphism from $\mathcal{P}$ to $Vect$.*

What this means in more concrete terms is this: a planar algebra is a graded vector space V_k for $k > 0$ and two vector spaces V^+ and V^- (for $k = 0$) so that every element T of $\mathcal{P}$ determines a multilinear map from vector spaces, one for each internal disc of T, to the vector space of the boundary of T, vector spaces being required to have the same colour as the discs they are assigned to. Composition in $\mathcal{P}$ and $Vect$ correspond in the following sense (S (an r-tangle) and T as above): by singling out an internal k–disc $\mathcal{D}$, S determines a linear map from V_k to $Hom(W, V_r)$, W being the tensor product of the vector spaces corresponding to the internal discs of S other than $\mathcal{D}$. Composing this map with the multilinear map determined by T one obtains a multilinear map from the vector spaces of all the internal tangles of $T \circ_{\mathcal{D}} S$. This multilinear map must be the same as the one the planar algebra structure assigns to $T \circ_{\mathcal{D}} S$.

Here are three exercises to help understand this homomorphism property of a planar algebra.

(i) Show that V^+ and V^- are both commutative associative algebras with multiplication defined by the tangle of fig. 1.3 without any strings.

(ii) Show that each V_k becomes an associative algebra with multiplication being the bilinear map defined by the tangle below:

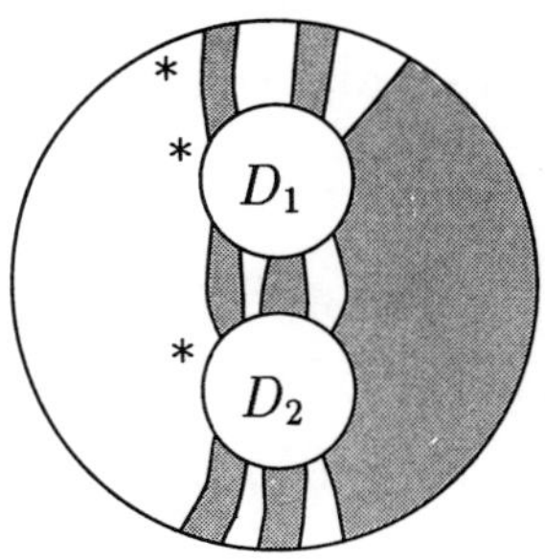

fig. 1.3 multiplication

(iii) Show that the V_k's for $k > 0$ and V^- for V_0 become an associative graded algebra over V_0 with multiplication being the bilinear map defined by the tangle below: (we will not show the shading any more-it is determined as soon as we know a region with a $*$, and given near the boundary if $k = 0$.)

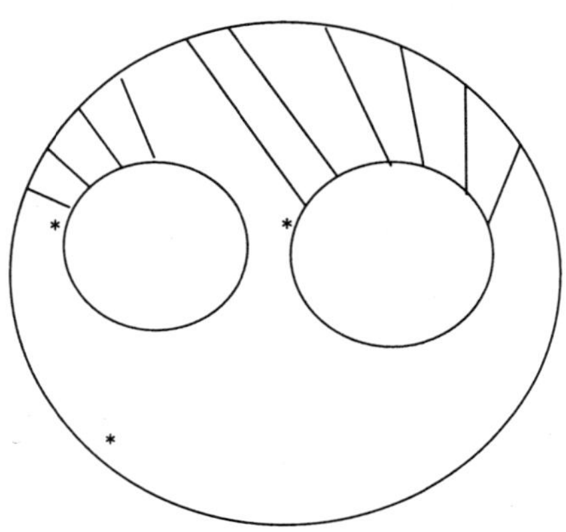

fig. 1.4 graded multiplication

Tangles without internal discs are required to give linear maps from the field to the output vector space. Thus the image of 1 under the 0-tangle with nothing inside is the identity for the algebras V_0. And in general the vector space spanned by k-tangles with no internal discs is a subalgebra of V_k.

The above definition of a planar algebra is quite general and one is interested in special cases. The following is rather commonly satisfied:

Definition 2.3 *The planar algebra will be said to have modulus δ if inserting a contractible circle inside a tangle causes its multilinear map to be multiplied by δ.*

There are maps from $\iota_k : V_k \to V_{k+1}$ defined by the "inclusion" tangle which consists of a single internal disc inside D with the first k boundary points of the internal disc connected to the first k boundary points of D, boundary points $k+1$ and $k+2$ of D connected by a string and boundary points $k+1$ to $2k$ of the internal disc connected to boundary points $k+3$ to $2k+2$ of D.

Proposition 2.4 *If V is a planar algebra with modulus $\delta \neq 0$, the ι_k are injective.*

Proof. Connecting the middle two boundary points and dividing by δ gives an inverse to ι_k. □

There may be examples of planar algebras where the ι_k are not injections but we have not looked in that direction. We will think of the V_k as being embedded one in the next via the ι_k.

Planar tangles possess the following involution: call the region preceeding the first boundary point of any disc the first region. Now reflect the tangle in a diameter passing through the first region on the boundary. Number all boundary points of all discs of the reflected tangle counting clockwise so that the image under the reflection of the first region becomes again the first region. The involution applied to the original tangle is the one obtained by this process.

If the field K possesses a conjugation and each V_k has a conjugate linear involution "$*$" we will say that V is a *planar *-algebra* if the involution on tangles and the involution on V commute in the obvious sense. If $K = \mathbb{R}$ or $\mathbb{C}$, V will be called a C^*-planar algebra if it is a planar *-algebra and each V_k becomes a C^*-algebra under its involution.

Planar algebras V and W are isomorphic if there are vector space isomorphisms $\theta_k : V_k \to W_k$ intertwining the actions of the planar operad. The isomorphisms are required to be *-isomorphisms in the planar *-algebra case.

There is a "duality" automorphism of the planar operad defined on a tangle by moving the first boundary point by one in a clockwise direction on every disc in the tangle and reversing the shading. Call this map $\Delta : \mathcal{P} \to \mathcal{P}$. It is clear that Δ preserves the composition of tangles.

Definition 2.5 *The dual $\hat{V}$ of the planar algebra V will be the planar algebra whose underlying vector spaces for $k > 0$ are those of V, $\hat{V}_0^{\pm} = V_0^{\mp}$, but for which the multilinear linear map corresponding to the tangle T is that of $\Delta(T)$.*

If V has modulus δ, so does $\hat{V}$. In general V is not isomorphic to $\hat{V}$. One may check for instance that the algebra structure defined on V_2 induced by multiplication in $\hat{V}_2$ is that defined by the tangle below:

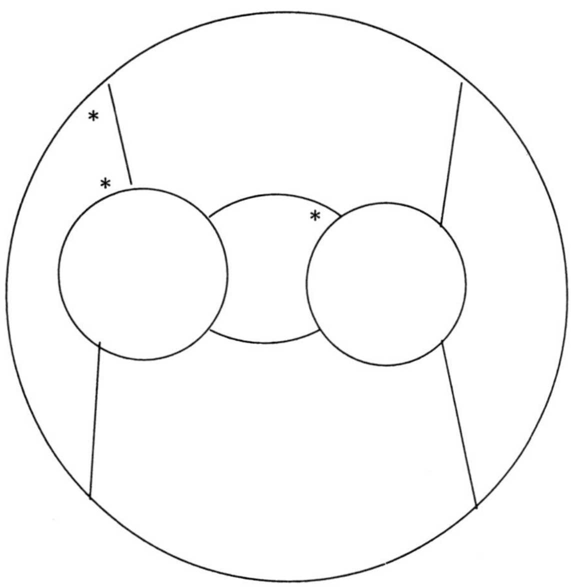

Comultiplication

We have somewhat abusively called this second multiplication "comultiplication" in [2]. For the planar algebra of a finite group as in [10], multiplication is that of the group algebra and comultiplication is that of functions on the group under pointwise multiplication.

On the other hand $\hat{\hat{V}}$ is isomorphic to V via the linear maps defined by the " rotation" tangles below:

The rotation tangle ρ.

3 The planar algebra of a bipartite graph.

Let Γ be a locally finite connected bipartite graph (possibly with multiple edges) with edge set $\mathcal{E}$, vertex set $\mathcal{U} = \mathcal{U}^+ \cup \mathcal{U}^-$, $\sharp(\mathcal{U}^+) = n_1$ and $\sharp(\mathcal{U}^-) = n_2$, $n_1 + n_2 = n = \sharp(\mathcal{U})$. No edges connect $\mathcal{U}_1$ to itself nor $\mathcal{U}_2$ to itself. The adjacency matrix of Γ is of the form $\begin{pmatrix} 0 & \Lambda \\ \Lambda^t & 0 \end{pmatrix}$ where there are Λ_{v^+,v^-} edges connecting v^+ to v^-.

The other piece of data we suppose given is a function $\mu : \mathcal{U} \to K$, $a \to \mu_a$ where K is the underlying field and μ_a is required to be different from zero for all a. In the cases of most interest so far (μ_a^2) has been an eigenvector for the adjacency matrix of Γ but that is only needed to guarantee that contractible circles inside pictures count as scalars.

The function μ will be called the *spin vector* and its value at a will be called the *spin* of a.

For each $k > 0$ let V_k be the vector space whose basis consists of loops of length $2k$ on Γ starting and ending at a point in $\mathcal{U}^+$. Such a loop will be represented by the pair (π, ϵ) of functions from $\{0, 1, 2, ..., 2k-1\}$ to $\Gamma \cup \mathcal{E}$ where the i-th. step in the loop goes from $\pi(i)$ to $\pi(i-1)$ along the edge $\epsilon(i)$ (the i's being counted modulo $2k$). Recording the vertices π of a path is redundant since the edges contain that information but in many examples of interest Γ will have no multiple edges in which case we would suppress the function ϵ. For $k = 0$ a loop of length 0 is just an element of $\mathcal{U}$, so we define ${V_0}^+$ (resp. ${V_0}^-$) to be the vector space with basis $\mathcal{U}^+$ (resp. $\mathcal{U}^-$).

We will make the V's into a general planar algebra in the sense of [10]. One may do this for discs and arbitrary smooth isotopies in the plane as in section 1 but consideration must be given to the angles at which the strings meet the boundaries of the discs. We prefer to use the equivalent picture where the discs are replaced by rectangles or "boxes" as in the second section of [10]. If we use the V's as a labelling set what we have to do is, given a k_0-tangle T in the sense of section 2 of [10], with an element of V_k assigned to each internal $k-$box of T, construct an "output" element $Z(T)$ in V_{k_0}, in such a way as to respect the compostion of tangles, be independent of isotopy and be multilinear in the "input" V elements. Many ingredients of the construction below were present in [13].

Definition of $Z(T)$. A *state* of the (unlabelled) tangle T will be a function σ: (regions of T)$\cup$ (strings of T) $\to \mathcal{U} \cup \mathcal{E}$ such that

(1) $\sigma(\{\text{shaded regions}\}) \subseteq \mathcal{U}^+$ and $\sigma(\{\text{unshaded regions}\}) \subseteq \mathcal{U}^-$.

(2) $\sigma(\{\text{strings}\}) \subseteq \mathcal{E}$.

(3) If the closures of the regions r_1 and r_2 both contain the string t then $\sigma(t)$ is an edge joining $\sigma(r_1)$ and $\sigma(r_2)$.

Now suppose T is labelled. Let $\{b\}$ be the set of internal boxes of T and let v_b be the vector in V_k be the vector assigned to the internal k-box b by the labelling.

To define the vector $Z(T)$ in V_{k_0} we must give the coefficient of a loop (π, ϵ) of length $2k_0$. Say that the state σ of T is *compatible* with (π, ϵ) in the following way. Consider the ith boundary segment of the external box of T (the one between the ith and (i+1)th boundary points according to the numbering convention of [10]). Then this segment is part of the boundary of a region r of T. The compatibility requirement of (π, ϵ) with σ is that $\pi(i) = \sigma(r)$, $\epsilon_i = \sigma(s)$, $\epsilon_{i+1} = \sigma(s')$ where s and s' are the strings of T meeting the boundary at the ith and $(i+1)$th points respectively. Compatibility of a state with a loop at an internal box of T is defined in the same way (with inside replaced by outside) so that every state assigns a loop (π_b, ϵ_b) to an internal box b of T as the only loop compatible with the state at b.

In the case $k = 0$ the whole boundary is a single segment and the loop is just a single vertex of $\mathcal{U}$. A state is compatible with the loop a on the boundary if it assigns a to the region near the boundary.

Now rotate the internal boxes of T so that they are all horizontal with the first boundary point at the top left. Isotope the strings if necessary so that any singularities of the y-coordinate function are isolated local maxima or minima.

We define the vector $Z(T)$ in V_{k_0} by its coefficient of the basis element (π, ϵ) which is:

$$(Z(T))_{(\pi,\epsilon)} = \sum_{\{\sigma \text{ compatible with}(\pi,\epsilon)\}} \prod_b (v_b)_{(\pi_b,\epsilon_b)} \prod_{\{\text{singularities } \alpha \text{ of y on strings}\}} \mu_\alpha$$

where (i) $(v)_{(\pi_b,\epsilon_b)}$ denotes the coefficient of the vector v in the basis $\{(\pi,\epsilon)\}$. (ii) The spin factor μ_α is the ratio $\frac{\mu_x}{\mu_y}$ where the state σ assigns x to the concave region near α and y to the region on the other side of the curve (regardless of the shading) as below:

This ends the definition of $Z(T)$. Note that the sum is finite even though Γ may be infinite because we only consider states compatible with the boundary state.

Note that the spin term

$$\prod_{\{\text{singularities } \alpha \text{ of y on strings}\}} \mu_\alpha$$

can be replaced by $\int \mu(\sigma)d(\theta)$ where σ denotes the locally constant function on the strings defined by σ and the spin vector whose value on a string segment is determined by the ratio of the spin values to the left and right of that string. This would allow one to define the "singularity" factor without considering singularities or arranging the tangle so that the boxes are all horizontal. See [12] for this in a simpler knot-theoretical context.

Theorem 3.1 *The above definition of $Z(T)$ for any planar tangle T makes the vector spaces of linear combinations of loops on Γ into a planar algebra with $dim(V_0^+) = n_1$ and $dim(V_0^-) = n_2$. This planar algebra will have modulus δ if (μ_a^2) is an eigenvector of the adjacency matrix with eigenvalue δ. It will be a planar *-algebra if K has a conjugation, the spin values are fixed by conjugation, and each V_k is equipped with the involution defined on the basis by reading loops backwards, and a C^*-planar algebra if, as well, $K = \mathbb{R}$ or $\mathbb{C}$.*

Proof. The first thing to show is that $Z(T)$ does not change if the labelled tangle is changed by isotopy of the disc. As in section 4 of [10] this follows from the fact that isotopy is generated by simple moves. Isotopies that do not change the maxima and minima are irrelevant. The first move is a cup-cap simplification which leaves each term in the sum invariant by the definition of the spin term. The second move required to generate isotopy is a 360° rotation of an internal box, since two ways of making the boxes horizontal with the first boundary point at the top left must differ by such rotations.

To see that such rotations do not affect Z, consider the spin term for a given state on a tangle that has a box surrounded by strings effecting a 360° rotation in the most obvious way with the maxima above the box and the minima below. Reading from the top down, cancellation occurs in the factors $\frac{\mu_x}{\mu_y}$ for all the maxima, and similarly for all the minima. So the overall spin term is exactly as if the rotation were not there.

Multilinearity is obvious, as is the compatibility of the operad gluing with $Z(T)$-just do the gluing operation with the inputs as horizontal boxes and the spin factors for the singularities of the y coordinate behave in the right way.

That contractible circles contribute a multiplicative factor of δ follows from the eigenvalue condition and the choice of the spin factors at local maxima and minima.

To see the *-algebra property, note that in the definition of the spin term, the thing that determines the numerator and denominator is the concavity of the region near the vertex. This is unaltered under reflection. The C^* property is clear since the * is just the transpose on a basis of matrix units. □

Definition 3.2 *With notation as in the previous theorem, we will call the planar algebra V (or $V(\Gamma)$ if necessary) the planar algebra of the bipartite graph Γ with respect to the spin vector μ.*

Notes. (i) The algebra structure on V_0^+ and V_0^- is such that the basis paths of length zero are idempotents, self adjoint in the *-case.

(ii) In general the loops with given first point and midway point form matrix units for a simple summand of V_k which is thus a multimatrix algebra with matrix summands indexed by pairs of vertices. Any pair of vertices will occur for appropriate k. The Bratteli diagram for the increasing sequence of multimatrix algebras V_k thus consists of a connected component for each vertex in $\mathcal{U}^+$ and each component is erected on a "principal graph" equal to Γ with starting vertex being the element of $\mathcal{U}^+$. See ([9, 6]) for a detailed treatment of such Bratteli diagrams. We give an example below of a graph Γ and the Bratteli diagram of its planar algebra:

The graph Γ

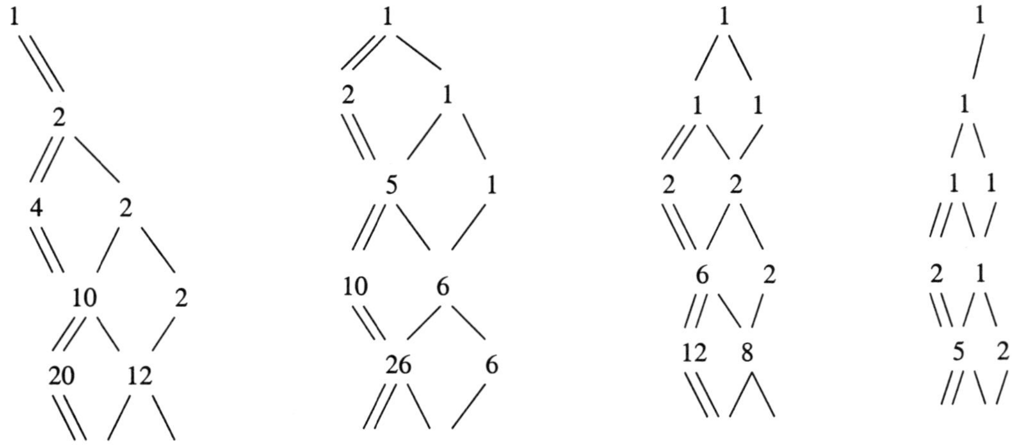

The Bratteli diagram of V(Γ)

(iii) The sum over all basis loops is the identity of V_k.

We shall now define linear functions on $V_0^{\pm}$. When applied to a labelled tangle with no boundary points, it will be called the *partition function* of the labelled tangle.

Definition 3.3 *The linear functions from* $Z : V_0^{\pm} \to K$ *are defined as the linear extensions of the function which takes the basis path a to* μ_a^4.

Thus for the partition function of a closed tangle one sums over states taking all possible values in all the regions, including the external one. There is an extra multiplicative spin factor of μ_a^4 for the external region.

Proposition 3.4 *The partition function of a closed labelled tangle* T *depends only on* T *up to isotopies of the 2-sphere.*

Proof. Spherical isotopy is generated by planar isotopy and isotopies that change a closed tangle by sending a string that meets the external region to a string that encloses the whole tangle(thus changing the shading of the external region). Invariance of the partition function under this move is easy to check. □

Up to this point the normalisation of the spin vector has been irrelevant. It is desirable that the partition function of an empty closed tangle be equal to 1. This suggests the following.

Definition 3.5 *We will say that the planar algebra of a graph is normalised if*

$$\sum_{a \in \mathcal{U}^+} \mu_a^4 = 1.$$

Note that this is the same as requiring $\sum_{a\in\mathcal{U}^-}\mu_a^4 = 1$.

Theorem 3.6 *Let K be $\mathbb{R}$ or $\mathbb{C}$ and let V be the normalised planar algebra of the finite graph Γ with respect to the appropriately normalised Perron Frobenius eigenvector of the adjacency matrix of Γ. Then $tr(x) = \delta^{-n+1}Z(\hat{x})$ defines a normalised trace on the union of the V's (with inclusion of V_k in V_{k+1} by adding a string to the right as usual-see [10]) where $\hat{x}$ is any 0-tangle obtained from x by connecting the first $k/2$ boundary points to the last $k/2$. The scalar product $< x, y >= tr(x^*y)$ is positive definite.*

Proof. Normalisation is a simple calculation which also shows that the definition of the trace is consistent with the inclusions. The property $tr(ab) = tr(ba)$ is a consequence of planar isotopy when all the strings added to x to get $\hat{x}$ go round x in the same direction, and spherical invariance reduces the general case to this one.

Positive definiteness follows from the fact that the loops, which form a basis of the V_k, are mutually orthogonal elements of positive length. In fact the square of the norm of a loop $(\pi, \epsilon) \in V_k$ is $\mu_{\pi(0)}^2\mu_{\pi(k)}^2$. □

Proposition 3.7 *The rotation tangle ρ is an isometry for the Hilbert space structure defined above by the trace and it acts on a basis path in V_k by:*

$$\rho(\pi, \epsilon) = \frac{\mu_{\pi(0)}\mu_{\pi(k)}}{\mu_{\pi(2k-2)}\mu_{\pi(k-2)}}(\alpha, \beta)$$

where $\alpha(i) = \pi(i-2)$ and $\beta(i) = \epsilon(i-2)$, with indices mod $2k$.

Proof. This is an exercise in using the definition of $Z(T)$. □

4 Examples

Some of the simplest examples were already present in [10].

Example 4.1 *Tensors.*

The tensor planar algebra of [10] is just the planar algebra of the bipartite graph with 2 vertices and n edges.

Example 4.2 *Spin models and discrete string theory.*

A so-called spin model of [10] is the planar algebra P^σ for the bipartite graph Γ with $\#(\mathcal{U}^+) = 1$ and $\#(\mathcal{U}^-) = n$. The somewhat mysterious normalisations of [10] are due to the spin vector which is the Perron Frobenius eigenvector of the adjacency matrix. This spin vector is particularly simple in this case which is why it was possible to complete the discussion of spin models in [10] without using the formalism developed here.

Observe that action of a permutation of the elements of $\mathcal{U}^-$ on loops preserves the planar algebra structure so that any group of such permutations defines a group of automorphisms of P^σ.

In [10] we considered the planar subalgebra $P^\mathcal{G}$ of P^σ generated by the single element of V_2 defined by the adjacency matrix of some arbitrary graph $\mathcal{G}$ with n vertices. The vertices of $\mathcal{G}$ form the set $\mathcal{U}^-$ in the spin model. The partition function of a labelled k- tangle T is then the number of graph homomorphisms from $col(T)$ to $\mathcal{G}$ where $col(T)$ is the planar graph obtained from T by taking as vertices the shaded regions of T and as edges the 2-boxes. The path (π, ϵ) is just a choice of vertices of $\mathcal{G}$ which specifies where these boundary vertices are sent by the graph homomorphism. As the tangle T becomes bigger we could imagine it filling up the inside of the disc so that we are exploring the graph $\mathcal{G}$ by counting larger and larger (singular) discs inside it. If the adjacency matrix were replaced by another symmetric matrix with the same pattern of zero entries we could think of its entries as interaction coefficients so we would have some kind of metric on the discs inside $\mathcal{G}$. Of course all this is the genus zero case. One might want to consider higher genus "surfaces" inside $\mathcal{G}$. For this one could try to extend the planar operad to allow systems of strings and internal discs on a surface of higher genus. It is not at all clear when this is possible. For vertex and spin models it should be straightforward enough though the spin term in the partition function will already cause a problem for spin models on higher genus surfaces.

It remains to be seen just how useful it is to explore a graph by counting homomorphisms of planar subgraphs into it. For a random graph one would expect that $P^\mathcal{G}$ is all of P^σ. But if $\mathcal{G}$ has automorphisms it is obvious that $P^\mathcal{G}$ will be contained in the planar subalgebra of P^σ given by fixed points for the action of the automorphism group of $\mathcal{G}$.

We record the following result giving a sufficient condition for any planar $*$−subalgebra of P^σ to be the fixed point algebra for the automorphism group coming from a group of permutations of the spins. We offer a proof that is curious in that the result is finite dimensional but we shall use type II_1 factors in our argument. Recall from [10] that the algebra V_k of P^σ is faithfully represented on the vector space $\otimes^{[\frac{k+1}{2}]}W$ where W is a vector space with basis equal to $\mathcal{U}^-$. We call an element of P^σ a *transposition* if it acts on

$\otimes^{[\frac{k+1}{2}]}W$ as a transposition between adjacent tensor product components.

Theorem 4.3 *A planar *-subalgebra $P = \{P_k\}$ of P^σ which contains a transposition is equal to the fixed points of P^σ under the action of some group of permutations of $\mathcal{U}^-$.*

Proof. The union of the increasing sequence of finite dimensional C^*−algebras P_k^σ admits a faithful trace. Complete the algebra using the GNS consruction ([21],page 41) to obtain the hyperfinite type II_1 factor R. The group S_n of all permutations of $\mathcal{U}^-$ acts on R so that every non-trivial permuation is an outer automorphism. It was shown in [11] that the fixed point algebra for the action of S_n is generated by the Temperley Lieb algebra and the "other" symmetric group-the representation of $S_{[\frac{k+1}{2}]}$ on $\otimes^{[\frac{k+1}{2}]}W$ which is obviously generated in P^σ by the transpostions. We see that the algebra R_0 generated in R by the P_k contains the fixed point algebra for the S_n action. Thus by the Galois theory for type II_1 factors R_0 is the fixed point algebra for some subgroup G of S_n.

The only thing left to show is that each individual P_k is the set of fixed points for the action of G. It is obvious that P_k is pointwise fixed. If x is in P_k^σ and is fixed by G and $\varepsilon > 0$ is given, there is a y in $P_{\hat{k}}$ for some sufficiently large $\hat{k}$ with $\|x-y\|_2 < \varepsilon$. But the conditional expectation $E_{P_k^\sigma}(y)$ is defined by a multiple of a tangle applied to y and so belongs to P_k. But x is in P_k^σ and so is fixed by the condtional expectation. We thus get an element of P_k within ε of x for every x, and P_k is finite dimensional. □

The above result can give an extremely rapid calculation of $P^{\mathcal{G}}$. In joint work with Curtin ([5]) we have shown that the planar algbebra thus associated to the Petersen graph is in fact the fixed point algebra for the automorphism group of that graph. It would be desirable to have an effective way of deciding if the first transposition is in $P^{\mathcal{G}}$.

Example 4.4 *Graphs with $\delta < 2$*

It is well known (see for instance [6]) that graphs the norm of whose adjacency matrix is < 2 are given by the Coxeter graphs A, D and E. These are the simplest examples where the formalism we have developed is necessary. The Perron Frobenius eigenvectors of the adjacency matrices can be found in [6]. It is of course possible to use an eigenvector other than the one of largest eigenvalue to obtain planar algebras with $\delta < 2$ from most graphs but the partition function will only give a positive definite form is one uses A, D or E.

The A, D, E graphs were used in the first construction of irreducible subfactors of non-integer index by finding what we would recognize as a connected planar algebra inside the general one defined above. In a future paper we will present a construction which gives all connected planar algebras with positivity and $\delta < 2$ in this way.

5 Towers of Algebras

An inclusion $A_0 \subseteq A_1$ of finite dimensional multimatrix algebras with the same identity defines a bipartite graph Γ with $\mathcal{U}^+$ and $\mathcal{U}^-$ being the sets of irreducible representations of A_0 and A_1 respectively. If $a \in \mathcal{U}^+$ and $b \in \mathcal{U}^-$ there are $n(a,b)$ edges between a and b where the restriction of b to A_0 contains a $n(a,b)$ times. We say the inclusion is *connected* if Γ is. Choose an eigenvector of the adjacency matrix of Γ with all entries non-zero and with non-zero eigenvalue (if there is one), and suppose the a entry of the eigenvector is a square, say ν_a^2. We can then define, as in [9] and [6], the Markov trace Tr of modulus δ^2 on A_1 by setting the trace of a minimal idempotent p_b in the matrix algebra direct summand of A_1 corresponding to $b \in \mathcal{U}^-$ to be ν_b^2, normalising the eigenvector so that

$$\sum_{b \in \mathcal{U}^-} dim(p_b A_1)\nu_b^2 = 1.$$

Performing the "basic construction" of [9] we obtain $A_2 =< A_1, e_1 >$, the algebra of linear endomorphisms of A_1 generated by left multiplication by A_1 and the "conditional expectation" $e_1 : A_1 \to A_0$ defined by $Tr(e_1(y)x) = Tr(yx)$ for $y \in A_1, x \in A_0$. The centre of A_2 is naturally identified with that of A_0 and the bipartite graph for $A_1 \subseteq A_2$ is the transpose of that for $A_0 \subseteq A_1$. If we use the same eigenvector of the adjacency matrix as for $A_0 \subseteq A_1$ to define a trace on A_2, the restriction of that trace to A_1 is Tr. Continuing in this way one gets a tower A_n for $n \geq 0$ with $A_{n+1} =< A_n, e_n >$ and a coherent trace Tr on $\cup_n A_n$ defined by $Tr(xe_n) = \delta^{-2}Tr(x)$ for $x \in A_n$.

On general principles we expect the centralisers $A_0' \cap A_k = B_k$ to form a planar algebra and this could no doubt be done by the method of section 4 of [10]. We give an alternative, explicit realisation of B_k as the planar algebra of Γ with spin vector equal to (ν_a). Before proceeding to construct the linear isomorphism between B and $V(\Gamma)$ we give two warnings.

(i) The centraliser inclusions $B_k \subseteq B_{k+1}$ are not connected - the planar algebra is not "connected" in the sense of [10].

(ii) The trace Tr on $V(\Gamma)$ does *not* correspond to the restriction of Tr to B.

The isomorphism will give an "intrinsic" meaning to the planar algebra of a bipartite graph though it is not quite canonical since there are many pairs $A_0 \subseteq A_1$ which realise a given Γ. To make the choice unique we shall now assume that $dim(pA_0) = 1$ for all minimal central idempotents p in A_0. This has the desirable consequence that the algebra A_1 itself can be realised as having a basis consisting of paths (π, ϵ) of length 2 on Γ starting at a point in $\mathcal{U}^+$. In this model the centre of A_1 can be identified with elements b of $\mathcal{U}^-$ thought of as sums of all loops from points in $\mathcal{U}^+$ to b and back.

So with the data Γ, ν form the planar algebra $V(\Gamma)$ and the pair $A_0 \subseteq A_1$ as above. Define the maps $\Theta : V_0^{\pm} \to A_1$ in the obvious way - a loop (π, ϵ) is a special kind of path so automatically defines an element of A_1.

In the following theorem E_i will denote the tangle having all strings vertically connecting the top to the bottom except two which connect the boundary points numbered i and $i+1$, $2k-i$ and $2k-i+1$ as below:

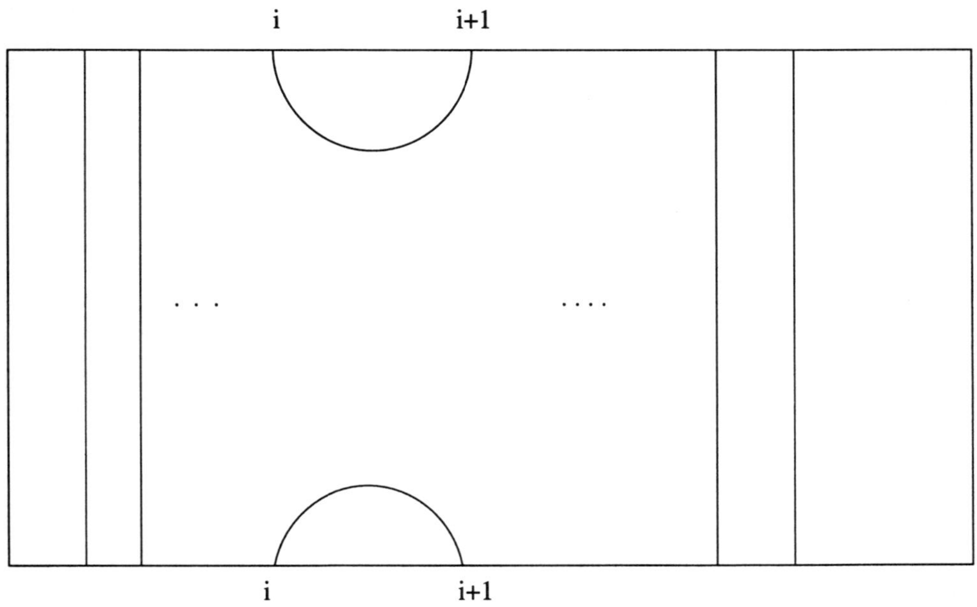

Theorem 5.1 *There is a linear extension of Θ to all of $V(\Gamma)$ such that:*
(i) Θ is an injective algebra homorphism.
(ii) $\Theta(E_i) = \delta e_i$.
(iii) $\Theta(V_k) = B_k$.

Proof. Paths of length $2k$ will be given by pairs (π, ϵ) of functions where $\pi : \{0, 1, ..., 2k\} \to \mathcal{U}$ and $\epsilon : \{0, 1, ..., 2k-1\} \to \mathcal{E}$ and the path goes from $\pi(i)$ to $\pi(i+1)$ along $\epsilon(i)$. The multiplication of paths is

$$(\pi_1, \epsilon_1)(\pi_2, \epsilon_2) = (\pi, \epsilon)$$

if $\epsilon_1(k+i) = \epsilon_2(k-i-1)$ for $0 \leq i < k$ in which case $\pi(i) = \pi_1(i)$ for $0 \leq i \leq k$ and $\epsilon(i) = \epsilon_1(i)$ for $0 \leq i < k$, $\pi(i) = \pi_2(i)$ $k \leq i \leq 2k$ and $\epsilon(i) = \epsilon_2(i)$ for $k \leq i < 2k$.

If the conditions are not satisfied, i.e. the "returning" path of (π_1, ϵ_1) is different from the "outgoing" path of (π_2, ϵ_2), the product is zero.

It follows from [6] that there is a unique algebra isomorphism Π between A_k and the algebra of paths of length $2k$ on Γ with initial vertex in $\mathcal{U}^+$ which is the extension of an identification of A_1 with linear combinations of paths of length two, and such that $\Pi(Z(E_i)) = \delta e_i$. Note that $Z(E_i)$ is a linear combination of loops which are also paths so $Z(E_i)$ can be viewed as an element of A_k. The formula at the top of page 86 of [6] is exactly what Z does to E_i if we choose ν as the square roots of the entries in the eigenvector of the adjacency matrix of Γ.

That multiplication of loops in V corresponds to multiplication of paths described above is just the definition of $Z(T)$ when T is the multiplication tangle in section 2 drawn with rectangular boxes rather than discs - all strings are vertical so there are no singularities and no spin term.

It is simple to check that a path is a loop if and only if the corresponding element of A_k is in the commutant B_k of A_0. Thus for Θ we may just take the restriction of Π to loops.$\Box$

Using Θ we can thus transport all the planar algebra structure of V, and the spherically invariant partition function, to B which becomes a planar algebra with $B_0^+ = A_0$ and $B_0^- = A_0' \cap A_1$. The normalised trace thus defined on B_k will be written tr and as we have pointed out, it is not the restriction of the Markov trace ot B_k. In particular if a is a minimal idempotent in A_0, identified with an element of $\mathcal{U}^+$, $tr(a) = \mu_a^4$ where μ is the multiple of ν with $\sum_{a \in \mathcal{U}^+} \mu_a^4 = 1$. But $Tr(a) = \nu_a^2$ where $\sum_{a \in \mathcal{U}^+} \nu_a^2 = 1$.

6 Connes' cyclic category

Connes' cyclic category is defined in ([3, 4, 15]) alternatively as (i) the category with objects C_i, for $i \geq 0$ and generated by morphisms $d_i : C_n \to C_{n-1}$, for $i = 0, 1, ..., n$, $s_i : C_n \to C_{n+1}$ for $i = 0, 1, ..., n$ and $t_i : C_n \to C_n$ subject to the relations

$$d_i d_j = d_{j-1} d_i \text{ for } i < j$$
$$s_i s_j = s_{j+1} s_i \text{ for } i \leq j$$
$$d_i s_j = \begin{cases} s_{j-1} d_i, & \text{for } i < j \\ id & \text{for } i = j, i = j+1 \\ s_j d_{i-1} & \text{for } i > j+1. \end{cases}$$

$$d_i t_n = t_{n-1} d_{i-1} \text{ for } 1 \le i \le n, \qquad d_0 t_n = d_n$$
$$s_i t_n = t_{n+1} s_{i-1} \text{ for } 1 \le i \le n, \qquad s_0 t_n = t_{n+1}^2 s_n$$
$$t_n^{n+1} = id.$$

(ii) the category whose objects are the sets of $ith.$ roots of unity and whose morphisms are homotopy classes of monotone degree one maps from the unit circle to itself sending roots of unity to roots of unity.

Fix an $n > 0$ and define elements d_i,s_i for $i = 0, 1, ..., n$ and t_n of the planar operad P as follows:

d_i is the tangle having one internal disc with $2n+2$ boundary points, and $2n$ boundary points on the outside disc. Internal boundary points numbered $2i+1$ and $2i+2$ are joined by a string. All internal boundary points are connected to external ones, with the first internal one connected to the first external one except when $i = 0$ when the first external point is connected to the third internal boundary point. These conditions uniquely determine d_i since the strings do not cross. .

For $0 \le i \le n$, s_i is the tangle having one internal disc with $2n+2$ internal boundary points, and $2n+4$ boundary points on the outside disc. External boundary points numbered $2i+2$ and $2i+3$ are connected and all other external ones are connected to internal ones with the first external point connected to the first internal point.

The tangle t_n is the clockwise rotation by two: there are $2n+2$ boundary points on both the internal and external discs, all strings connect inside points to outside ones and the first outside one is connected to the third inside one, as below:

Note that these tangles d_i,s_i and t_n have exactly one input and one output so their composition makes perfect sense provided the d's and s's have the right numbers of internal and external boundary points.

Theorem 6.1 *The operad element tangles defined above satisfy the relations of the generators of (the opposite of)Connes' cyclic category. The map they define from the cyclic category to annular tangles (with one input and one output) is injective.*

Proof. Verification of the relations is simply a matter of drawing pictures. Injectivity is more interesting. We prove it by providing a concrete realisation of the cyclic category which is arguably simpler than the one provided in [4]: for integers m and $n \ge 0$ let $\mathcal{C}_{m,n}$ be the set of all annular tangles (modulo a twist of 2π near the boundary) with $2m+2$ internal points and $2n+2$ external ones, with the following properties:

(i) All strings either connect an internal boundary point to an external one or a boundary point to one of it's neighbours.

(ii)If a string connects an internal boundary point to its neighbour the region between the string and the internal boundary is shaded, if it connects external boundary points the region between the string and the external boundary is unshaded.

A typical element in $\mathcal{C}_{2,3}$ is depicted below:

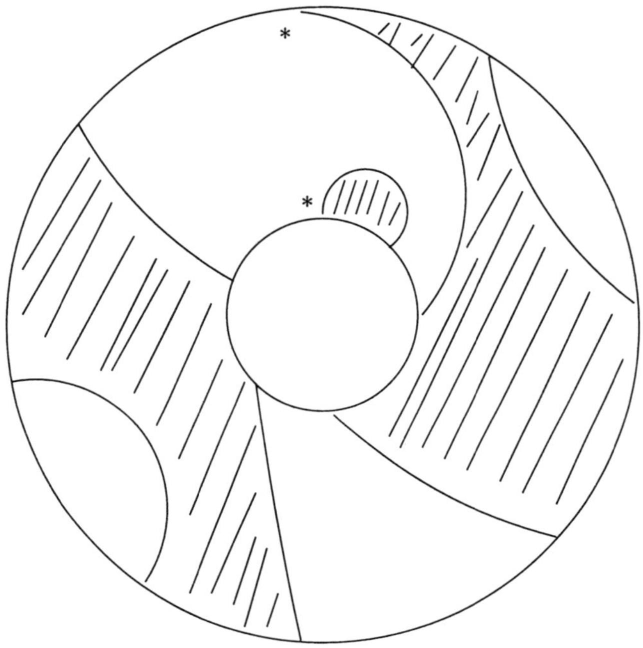

A morphism in the Cyclic Category

Construct the category with one object for each $n \geq 0$, the objects being the $2n+2$ boundary points up to isotopy and morphisms being the $\mathcal{C}_{m,n}$. One may check that composition of tangles makes this set into a small category (no closed loops are formed in composing such tangles) and that it is generated by the d_i,s_i and t_n . But to each element of this category there is a well defined homotopy class of degree one monotone maps from the circle to itself defined by isotoping the tangle so that the each root of unity is in exactly one unshaded region on the inside and outside. The degree one map can then be constructed as follows. First contract any regions enclosed by non through-strings to their segments on the internal or external boundary so that the shading near that boundary interval will reverse. Each unshaded interval on the outside boundary then contains exactly one root of unity and all the regions are topologically rectangles with internal and external boundary segments as opposite edges. These rectangles determine mappings from the inner circle to the outer one and after a litlle isotopy each internal root of unity may be sent onto the unique external root of unity in the same shaded region. Tangle composition obviously corresponds to composition of homotopy classes so we have a section of the map from the cyclic category to the $\mathcal{C}_{m,n}$'s which sends our d_i,s_i and t_n to the corresponding generators of the cyclic category. This is more than enough to show injectivity (and identify the image).□

Now define the tangles $\partial_i = \Delta(d_i)$, $\sigma_i = \Delta(s_i)$ where Δ is the duality map of the first secion. It follows from the previous result that ∂_i , σ_i and t_n $(= \Delta(t_n))$ define another copy of Connes' cyclic category in annular tangles. Note also that there is a natural involutory antiautomorphism of annular tangles (reflect with respect to a circle half way between the internal and external boundaries) which exchanges ∂ with s and d with σ.

Thus the vector spaces V_{n+1} become a cyclic module in two different ways. The two structures interact as follows.

Proposition 6.2 *If* V *has modulus* δ *then* $D_n\sigma_0 - \sigma_0 D_{n+1} = \delta id$ *where* $D_n = \sum_{i=0}^{n}(-1)^i d_i$ *is the Hochschild boundary*

Corollary 6.3 *If* V *has modulus* $\delta \neq 0$ *then it is acyclic so the cyclic homology is that of the ground field.*

The same result holds for any representations of the category of annular tangles as in [7].

Thus the cyclic homology is of no immediate interest in planar algebras coming from subfactors. On the other hand there are very interesting planar algebras of modulus 0. The first is the one coming from the skein theory of the Alexander polynomial. It can be obtained by specialising the planar algebra of example 2.5 in [10] to the values $x = \sqrt{t} - \frac{1}{\sqrt{t}}$. The planar algebra involved in a conjecture of Kashaev concerning asymptotics of the $sl(2)$ knot polynomials also has modulus 0. At this stage we have no results concerning the cyclic homology of these planar algebras. It is also true that the planar algebras of section 2 do not have any modulus at all if the spin vector is not an eigenvector for the adjacency matrix. Once again we have no results.

The appearance of cyclic homology in this context is not understood. One should note that, in the isomorphism of section 2 between the planar algebra of a finite graph and the centraliser tower of a finite dimensional multimatrix inculsion the cyclic module structure defined above corresponds to the usual one used in calculating the homology of the $A - A$ bimodule B.

Since the category or annular tangles is generated by the two copies of the cyclic category we have defined, and one of those categories is typically the adjoint of the other for some invariant inner product. It might be interesting to see if there are known natural examples of cyclic modules for which the cyclic category extends to an action of the whole annular structure and even to a planar algebra.

References

[1] J. Barrett and B. Westbury, *Spherical Categories*, hep-th preprint #9310164.

[2] D. Bisch and V.F.R. Jones,*Singly generated planar algebras of small dimension*, to appear, Duke Math. Journal

[3] A.Connes, *Noncommutative geometry*, Academic Press, 1994.

[4] ———, *Cohomologie cyclique et foncteurs* Ext^n, C.R.Acad.Sci.Paris **296**(1983),953–958.

[5] B. Curtin and V.F.R. Jones, *Planar algebras and graph automorphisms*, work in progress.

[6] F.M. Goodman, P. de la Harpe, and V.F.R. Jones, *Coxeter graphs and towers of algebras*, Springer-Verlag, 1989.

[7] J.J. Graham and G.I. Lehrer,*The representation theory of affine Temperley Lieb algebras*,L'Enseignement Mathématique **44**(1998),1–44.

[8] U. Haagerup *Principal graphs of subfactors in the index range* $4 < [M : N] < 3 + \sqrt{2}$ in: "Subfactors", World Scientific, Singapore-New Jersey-London-Hong Kong (1994) 1–39.

[9] V.F.R. Jones, *Index for subfactors*, Invent. Math **72** (1983), 1–25.

[10] ———, *Planar Algebras, I*, NZ Journal Math, to appear.

[11] ———, *The Potts model and the symmetric group*, in *Subfactors*,ed. Araki and Kawahigashi. (1994), 259–270

[12] ———, *On knot invariants related to some statistical mechanical models*, Pacific Journal of Mathematics **30**(1988), 311–334.

[13] V.F.R. Jones and V.S.Sunder,*Introduction to Subfactors*LMS lecture note series **234**(1997)

[14] G. Kuperberg *Spiders for rank 2 Lie algebras*, Commun. Math. Phys. **180**(1996),109–151.

[15] J-L. Loday, *Cyclic Homology*, Springer,Grundlehren,2nd.edition 1998.

[16] R. Longo and J. Roberts,*A theory of dimension*,Journal of K-theory **11** (1997), 103–159.

[17] J.P. May *Definitions: operads, algebras and modules*, Contemporary Mathematics **202**(1997),1–7.

[18] S.Popa, *Markov traces on universal Jones algebras and subfactors of finite index*, Invent. Math. **111** (1993), 375–405.

[19] ———,*An axiomatization of the lattice of higher relative commutants*, Invent. Math **120**(1995),427–445.

[20] ———.*Universal construction of subfactors*, preprint 1999.

[21] M. Takesaki, *Theory of Operator Algebras I*, Springer-Verlag (1979).

CERTAIN RACKS ASSOCIATED WITH THE BRAID GROUPS

SEIICHI KAMADA†
Department of Mathematics, Osaka City University
Osaka 558-8585, Japan

YUKIO MATSUMOTO
Graduate School of Mathematical Sciences, University of Tokyo
Tokyo 153-8914, Japan

ABSTRACT

A rack is a non-empty set with a binary operation such that right multiplication is an automorphism. New racks closely related to the braid groups and the mapping class groups will be introduced using "cords" on surfaces, and their structure will be investigated. Emphasis will be laid on the racks of cords on the plane and the sphere. In particular, their presentations and the center of the associated group of the rack of cords on the sphere will be given.

Keywords: Rack, braid group, mapping class group.

1. Introduction

A rack is a non-empty set with a binary operation such that right multiplication is an automorphism. In general, the binary operation is not associative and a rack is very different from a group. A typical example of a rack is the quandle of a knot due to D. Joyce [1], which classifies the knot up to mirror image. This is generalized to the fundamental racks of codimension-two links by R. Fenn and C. Rourke [2].

In this paper, we introduce certain racks closely related to the braid groups and the mapping class groups, and investigate their structure. They can be expected to play a fundamental role in the monodromy calculus appearing in 4-dimensional topology; in particular, in surface braids and in Lefschetz fibrations.

Let Σ be a connected oriented 2-manifold and let Q_{n+1} be a set of $n+1$ interior points of Σ: $Q_{n+1} = \{q_1, q_2, \ldots, q_{n+1}\}$, where $n+1 \geq 2$. A *cord* on (Σ, Q_{n+1})

†Present address: Department of Mathematics and Statistics, University of South Alabama, Mobile, AL 36688, USA.

is a simple curve (un-directed) on $\mathrm{int}(\Sigma)$ whose endpoints are distinct points of Q_{n+1}. We assume that the curve is disjoint from Q_{n+1} except at the endpoints (see Figure 1).

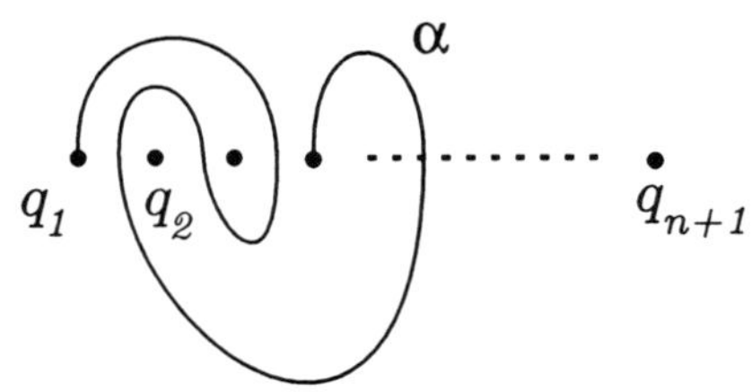

Figure 1

We denote by $M_{n+1}(\Sigma)$ the mapping class group of (Σ, Q_{n+1}), that is, the group of isotopy classes of orientation-preserving homeomorphisms of (Σ, Q_{n+1}) onto itself whose restrictions to the boundary $\partial\Sigma$ are the identity map.

For a cord α on (Σ, Q_{n+1}), we denote by $\tau[\alpha] \in M_{n+1}(\Sigma)$ the *disk twist* on α, which means the isotopy class of a homeomorphism of (Σ, Q_{n+1}) whose support is a disk neighbourhood of α in which it gives an anticlockwise "180°-twist" interchanging the endpoints of α.

Two cords α and α' are *isotopic* if they are ambient isotopic by an isotopy of Σ relative to Q_{n+1} and $\partial\Sigma$. Let $X_{n+1}(\Sigma)$ be the set of isotopy classes of cords on (Σ, Q_{n+1}). We will use the same symbol for a cord and its isotopy class. Notice that the element $\tau[\alpha] \in M_{n+1}(\Sigma)$ is well-defined for $\alpha \in X_{n+1}(\Sigma)$.

The notion of racks and similar notions have been studied by several authors from various points of view, see [1], [2], [3], [4], [5], [6]. We follow the terminology of Fenn and Rourke [2].

Definition 1.1 A non-empty set X with a binary operation $\wedge$ is a *rack* if the following two axioms are satisfied:

(i) For each $b \in X$, the mapping $X \to X$ defined by $a \mapsto a \wedge b$ is bijective.
(ii) $(a \wedge b) \wedge c = (a \wedge c) \wedge (b \wedge c)$ for all $a, b, c \in X$.

Let X be a rack, and let $F(X)$ denote the free group generated by all elements of X. Let N be the normal subgroup of $F(X)$ consisting of those elements $w \in F(X)$ such that $a \wedge w = a$ for all $a \in X$. Note that $a \wedge w$ makes sense for $a \in X$ and $w \in F(X)$, see [2]. The *operator group of* X is the quotient group $F(X)/N$, which is denoted by $Op(X)$. Let K be the normal subgroup of $F(X)$ generated by $(a \wedge b)b^{-1}a^{-1}b$ for all $a, b \in X$. The *associated group of* X is the quotient group $F(X)/K$, which is denoted by $As(X)$. The axioms of a rack imply that $K \subset N$, and we have a natural surjection $As(X) \to Op(X)$ (for details, see [2]).

Let $X_{n+1}(\Sigma)$ be the set of isotopy classes of cords on (Σ, Q_{n+1}) as defined above. We make the set $X_{n+1}(\Sigma)$ a rack by giving a binary operation $\wedge$ as follows:

$$\alpha \wedge \beta = (\alpha)\tau[\beta],$$

where we assume that $M_{n+1}(\Sigma)$ acts on $X_{n+1}(\Sigma)$ from the *right* and that for $\tau_1, \tau_2 \in M_{n+1}(\Sigma)$ the composition $\tau_1\tau_2$ means the mapping class which is obtained by applying τ_1 first and then τ_2. This operation $\wedge$ satisfies the two axioms of a rack. For any cord β, $\tau[\beta] \in M_{n+1}(\Sigma)$ gives a bijection from $X_{n+1}(\Sigma)$ to itself. So the first axiom is obvious. For the second, we notice that $\tau[(\beta)\tau[\gamma]] = \tau[\gamma]^{-1}\tau[\beta]\tau[\gamma]$ in $M_{n+1}(\Sigma)$ for any cords β, γ. Then we have $(\alpha \wedge \gamma) \wedge (\beta \wedge \gamma) = ((\alpha)\tau[\gamma])\tau[(\beta)\tau[\gamma]] = (\alpha)\tau[\beta]\tau[\gamma] = (\alpha \wedge \beta) \wedge \gamma$. We will investigate this rack $X_{n+1}(\Sigma)$.

2. Operator groups of $X_{n+1}(\mathbb{R}^2)$ and $X_{n+1}(S^2)$

Let Σ be $\mathbb{R}^2$ or S^2. Take a system of cords $\{\gamma_1, \dots, \gamma_n\}$ on (Σ, Q_{n+1}) such that the endpoints of γ_i are q_i and q_{i+1} and such that $\gamma_i \cap \gamma_j \subset Q_{n+1}$ for $i \neq j$. We will fix this system in what follows.

Lemma 2.1 *If $n+1 \geq 3$ and $\Sigma = \mathbb{R}^2$ or S^2, then the operator group $Op(X_{n+1}(\Sigma))$ is isomorphic to the mapping class group $M_{n+1}(\Sigma)$.*

Proof. Let $\tau : F(X_{n+1}(\Sigma)) \to M_{n+1}(\Sigma)$ be a homomorphism defined by

$$\alpha_1^{\epsilon_1} \dots \alpha_s^{\epsilon_s} \mapsto \tau[\alpha_1]^{\epsilon_1} \dots \tau[\alpha_s]^{\epsilon_s},$$

where $\epsilon_i \in \{\pm 1\}$ and $\alpha_i \in X_{n+1}(\Sigma)$. This is a surjective homomorphism because $\tau[\gamma_1], \dots, \tau[\gamma_n]$ generate the mapping class group $M_{n+1}(\Sigma)$. We will prove that $\ker \tau$ coincides with the normal subgroup N of the free group $F(X_{n+1}(\Sigma))$ defined in Section 1.

If $w = \alpha_1^{\epsilon_1} \dots \alpha_s^{\epsilon_s}$ is in $\ker \tau$, then $\alpha \wedge w = (\alpha)\tau[\alpha_1]^{\epsilon_1} \dots \tau[\alpha_s]^{\epsilon_s} = \alpha$ in $X_{n+1}(\Sigma)$ for any $\alpha \in X_{n+1}(\Sigma)$. So we have $\ker \tau \subset N$.

Conversely, if $w = \alpha_1^{\epsilon_1} \dots \alpha_s^{\epsilon_s}$ is in N, then for any $\alpha \in X_{n+1}(\Sigma)$, $\alpha = \alpha \wedge w = (\alpha)\tau[\alpha_1]^{\epsilon_1} \dots \tau[\alpha_s]^{\epsilon_s}$. In particular, $\tau[\alpha_1]^{\epsilon_1} \dots \tau[\alpha_s]^{\epsilon_s}$ preserves $\gamma_1, \dots, \gamma_n \in X_{n+1}(\Sigma)$. If $n+1 \geq 3$ and $\Sigma = \mathbb{R}^2$ or S^2, this means in turn that $\tau[\alpha_1]^{\epsilon_1} \dots \tau[\alpha_s]^{\epsilon_s}$ is isotopic to the identity through an isotopy of (Σ, Q_{n+1}). Thus, $w \in \ker \tau$. Hence $\ker \tau = N$, and we have $M_{n+1}(\Sigma) \cong F(X_{n+1}(\Sigma))/N = Op(X_{n+1}(\Sigma))$. □

3. Presentations of $X_{n+1}(\mathbb{R}^2)$ and $X_{n+1}(S^2)$

In this section we give presentations of the racks $X_{n+1}(\mathbb{R}^2)$ and $X_{n+1}(S^2)$.

Theorem 3.1 *The rack $X_{n+1}(\mathbb{R}^2)$ of cords on $(\mathbb{R}^2, Q_{n+1})$ has the following presentation:*

(i) *Generating set:*

$$\gamma_1, \gamma_2, \dots, \gamma_n$$

(ii) *Defining relations:*

$$\gamma_i \wedge \gamma_i = \gamma_i \qquad (i = 1, 2, \dots, n) \tag{1}$$

$$\gamma_i \wedge \gamma_j = \gamma_i \qquad (|i-j| > 1) \tag{2}$$

$$(\gamma_i \wedge \gamma_j) \wedge \gamma_i = \gamma_j \qquad (|i-j| = 1). \tag{3}$$

Theorem 3.2 *The rack $X_{n+1}(S^2)$ of cords on (S^2, Q_{n+1}) has the following presentation:*

(i) *Generating set:*

$$\gamma_1, \gamma_2, \ldots, \gamma_n$$

(ii) *Defining relations:*

$$\gamma_i \wedge \gamma_i = \gamma_i \qquad (i = 1, 2, \ldots, n) \tag{4}$$

$$\gamma_i \wedge \gamma_j = \gamma_i \qquad (|i-j| > 1) \tag{5}$$

$$(\gamma_i \wedge \gamma_j) \wedge \gamma_i = \gamma_j \qquad (|i-j| = 1) \tag{6}$$

$$\gamma_1 \wedge (\gamma_2 \ldots \gamma_n) = \gamma_n \wedge (\gamma_{n-1} \ldots \gamma_1). \tag{7}$$

Lemma 3.3 *Each $\alpha \in X_{n+1}(\mathbb{R}^2)$ is equal to an element of the form $\gamma_1 \wedge w$ where w is a word of $\gamma_1, \ldots, \gamma_n$.*

Proof. Since $\tau[\gamma_1], \ldots, \tau[\gamma_n]$ generate the mapping group $M_{n+1}(\mathbb{R}^2)$, we have $\alpha = (\gamma_1)\tau[\alpha_1]^{\epsilon_1} \ldots \tau[\alpha_s]^{\epsilon_s}$ for some $\alpha_1, \ldots, \alpha_s \in \{\gamma_1, \ldots, \gamma_n\}$ and $\epsilon_1, \ldots, \epsilon_s \in \{\pm 1\}$. Put $w = \alpha_1^{\epsilon_1} \ldots \alpha_s^{\epsilon_s}$. □

Corollary 3.4 *The rack $X_{n+1}(\mathbb{R}^2)$ is generated by $\gamma_1, \ldots, \gamma_n$.*

Corollary 3.5 *The group $As(X_{n+1}(\mathbb{R}^2))$ is generated by $\eta(\gamma_1), \ldots, \eta(\gamma_n)$, where $\eta : X_{n+1}(\mathbb{R}^2) \to As(X_{n+1}(\mathbb{R}^2))$ is the natural map.*

By Artin's theorem [7], [8], the $n+1$-string braid group B_{n+1} can be identified with the mapping class group $M_{n+1}(D^2)$ of (D^2, Q_{n+1}) with a sufficiently large disk D^2 which contains Q_{n+1}. For a cord α on $(\mathbb{R}^2, Q_{n+1})$, we define an $n+1$-string braid $\sigma[\alpha]$ to be the braid that is identified with the disk twist $\tau[\alpha]$ in $M_{n+1}(D^2)$ under the identification $B_{n+1} = M_{n+1}(D^2)$. The mapping $\sigma : X_{n+1}(\mathbb{R}^2) \to B_{n+1}$ satisfies $\sigma[\alpha \wedge \beta] = \sigma[(\alpha)\tau[\beta]] = (\sigma[\beta])^{-1}\sigma[\alpha]\sigma[\beta]$. Thus $\sigma : X_{n+1}(\mathbb{R}^2) \to (B_{n+1})_{\mathrm{conj}}$ is a rack homomorphism, where $(B_{n+1})_{\mathrm{conj}}$ is the conjugation rack of B_{n+1} (cf. [2]). Notice that $\sigma[\gamma_1], \ldots, \sigma[\gamma_n]$ are the standard generators, $\sigma_1, \ldots, \sigma_n$, of the braid group B_{n+1}. Artin proved that B_{n+1} has the following group presentation:

(i) Generating set:

$$\sigma_1, \ldots, \sigma_n$$

(ii) Defining relations:

$$\sigma_i\sigma_j = \sigma_j\sigma_i \qquad (|i-j| > 1)$$

$$\sigma_i\sigma_j\sigma_i = \sigma_j\sigma_i\sigma_j \qquad (|i-j| = 1).$$

By a *rack word of* $\gamma_1, \ldots, \gamma_n$, we mean a word of the form $c_0 \wedge (c_1^{\epsilon_1} \ldots c_m^{\epsilon_m})$, where c_i $(i = 0, \ldots, m)$ is an element from $\gamma_1, \ldots, \gamma_n$ and $\epsilon_i \in \{\pm 1\}$ $(i = 1, \ldots, m)$.

A word of $\sigma_1, \ldots, \sigma_n$ is called a *positive symmetric word* if it is

$$(s_1^{\epsilon_1} \ldots s_m^{\epsilon_m})^{-1} s_0 (s_1^{\epsilon_1} \ldots s_m^{\epsilon_m}),$$

where s_i $(i = 0, \dots, m)$ is an element from $\sigma_1, \dots, \sigma_n$ and $\epsilon_i \in \{\pm 1\}$ $(i = 1, \dots, m)$.

Rack words of $\gamma_1, \dots, \gamma_n$ are naturally in one-to-one correspondence to positive symmetric words of $\sigma_1, \dots, \sigma_n$ by

$$c_0 \wedge (c_1^{\epsilon_1} \dots c_m^{\epsilon_m}) \longleftrightarrow (s_1^{\epsilon_1} \dots s_m^{\epsilon_m})^{-1} s_0 (s_1^{\epsilon_1} \dots s_m^{\epsilon_m})$$

where $\sigma(c_i) = s_i$ for $i = 0, \dots, m$. Notice that

$$\sigma(c_0 \wedge (c_1^{\epsilon_1} \dots c_m^{\epsilon_m})) = (s_1^{\epsilon_1} \dots s_m^{\epsilon_m})^{-1} s_0 (s_1^{\epsilon_1} \dots s_m^{\epsilon_m}),$$

in the braid group B_{n+1}.

Proof of Theorem 3.1. By Lemma 3.3, $\gamma_1, \dots, \gamma_n$ generate $X_{n+1}(\mathbb{R}^2)$. It is easily seen that the relations (1)–(3) of the theorem hold. It remains for us to prove that if two rack words w and w' of $\gamma_1, \dots, \gamma_n$ represent the same element of $X_{n+1}(\mathbb{R}^2)$, then there exists a finite sequence of rack words of $\gamma_1, \dots, \gamma_n$ between w and w' such that each rack word is obtained from the previous one by a consequence of the relations.

Consider positive symmetric words corresponding to w and w'. Since $\sigma(w) = \sigma(w')$, the positive symmetric words represent the same element of B_{n+1}. By Theorem 1.4 of [9], there exists a finite sequence of positive symmetric words between the two positive symmetric words such that each word is obtained from the previous one by one of the following transformations (T0)–(T3):

(T0)

$$x^{-1} \sigma_k x \longleftrightarrow y^{-1} \sigma_k y$$

where x, y are words of $\sigma_1, \dots, \sigma_n$ which differ by the trivial relation $\sigma_i \sigma_i^{-1} = \sigma_i^{-1} \sigma_i = 1$, or one of the braid relations $\sigma_i \sigma_j = \sigma_j \sigma_i$ $(|i-j| > 1)$ and $\sigma_i \sigma_j \sigma_i = \sigma_j \sigma_i c_j$ $(|i-j| = 1)$,

(T1)

$$x^{-1} \sigma_i^{-\epsilon} \sigma_i \sigma_i^{\epsilon} x \longleftrightarrow x^{-1} \sigma_i x$$

where x is a word of $\sigma_1, \dots, \sigma_n$ and $\epsilon \in \{\pm 1\}$,

(T2)

$$x^{-1} \sigma_j^{-\epsilon} \sigma_i \sigma_j^{\epsilon} x \longleftrightarrow x^{-1} \sigma_i x$$

where x is a word of $\sigma_1, \dots, \sigma_n$, $|i-j| > 1$ and $\epsilon \in \{\pm 1\}$,

(T3)

$$x^{-1} \sigma_j^{-\epsilon} \sigma_i \sigma_j^{\epsilon} x \longleftrightarrow x^{-1} \sigma_i^{\epsilon} \sigma_j \sigma_i^{-\epsilon} x$$

where x is a word of $\sigma_1, \dots, \sigma_n$, $|i-j| = 1$ and $\epsilon \in \{\pm 1\}$.

Consider the sequence of rack words of $\gamma_1, \dots, \gamma_n$ corresponding to the above sequence of the positive symmetric words. This sequence is the desired one, because the transformation (T0) corresponds to the trivial relation of the free group $F(\gamma_1, \dots, \gamma_n)$ generated by $\gamma_1, \dots, \gamma_n$ or $\gamma_i \gamma_j = \gamma_j \gamma_i$ $(|i-j| > 1)$, $\gamma_i \gamma_j \gamma_i = \gamma_j \gamma_i \gamma_j$ $(|i-j| = 1)$ in the operator level. This is a consequence of the relations (1)–(3) of the theorem. The transformation (T1) corresponds to $\gamma_i \wedge \gamma_i^\epsilon = \gamma_i$, the transformation (T2) corresponds to $\gamma_i \wedge \gamma_j^\epsilon = \gamma_i$ for $|i-j|>1$, and the transformation (T3) corresponds to $\gamma_i \wedge \gamma_j^\epsilon = \gamma_j \wedge \gamma_i^{-\epsilon}$ for $|i-j|=1$. These are also consequences of the relations. □

Let α be a cord on $(\mathbb{R}^2, Q_{n+1})$. Consider a sufficiently large simple closed curve c in $\mathbb{R}^2$ surrounding Q_{n+1} and α. Let β be a simple curve in $\mathbb{R}^2 - Q_{n+1}$ whose endpoints are an interior point of α and a point of c such that β is disjoint from $\alpha \cup c$ except at the endpoints. By surgery along β and c, we have another cord α' on $(\mathbb{R}^2, Q_{n+1})$ from α. In Figure 2, we illustrate an example where $\alpha = \gamma_1$ and β is a trivial arc. We say that α' is obtained from α by a *flip* along β and c.

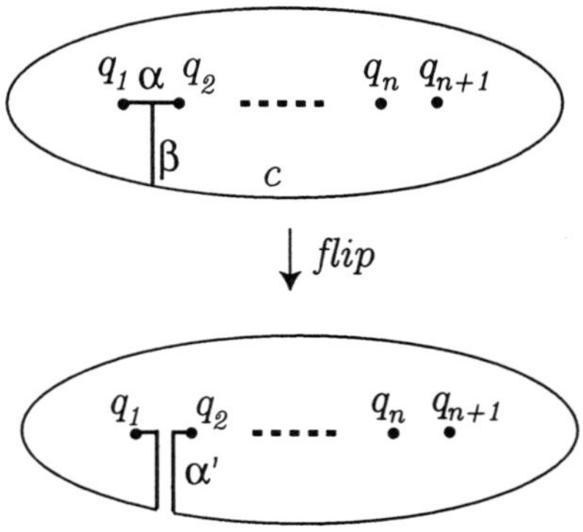

Figure 2

Let $\iota : X_{n+1}(\mathbb{R}^2) \to X_{n+1}(S^2)$ be the rack homomorphism induced by the inclusion map $(\mathbb{R}^2, Q_{n+1}) \to (S^2, Q_{n+1})$, where we assume that S^2 is the one-point compactification of $\mathbb{R}^2$. The element α' of $X_{n+1}(\mathbb{R}^2)$ obtained from α by a flip along β and c depends on β and c, however the image $\iota(\alpha')$ is the same as $\iota(\alpha)$.

We say that two rack words w and w' of $\gamma_1, \ldots, \gamma_n$ are *congruent modulo the relations* (4)–(7) if there exists a finite sequence of rack words of $\gamma_1, \ldots, \gamma_n$ between w and w' such that each rack word is obtained from the previous one by a consequence of the relations (4)–(7).

Proof of Theorem 3.2. We assume that S^2 is the one-point compactification of $\mathbb{R}^2$. Every cord on (S^2, Q_{n+1}) is isotopic to a cord on $(\mathbb{R}^2, Q_{n+1})$. Thus, by Lemma 3.3, $\gamma_1, \ldots, \gamma_n$ generate $X_{n+1}(S^2)$, where we use the same symbol γ_i for $\gamma_i \in X_{n+1}(\mathbb{R}^2)$ and $\iota(\gamma_i) \in X_{n+1}(S^2)$. It is easily checked that the relations (4)–(7) hold. It remains for us to prove that if two rack words w and w' of $\gamma_1, \ldots, \gamma_n$ represent the same element of $X_{n+1}(S^2)$, then they are congruent modulo the relations (4)–(7). Let α and α' be cords on $(\mathbb{R}^2, Q_{n+1})$ corresponding to the rack words w and w', respectively. Since $\iota(\alpha)$ and $\iota(\alpha')$ are the same element of $X_{n+1}(S^2)$, there exists a finite sequence of cords $\alpha_0, \alpha_1, \ldots, \alpha_s$ on $(\mathbb{R}^2, Q_{n+1})$ such that $\alpha = \alpha_0$, $\alpha' = \alpha_s$ and each α_i is obtained from α_{i-1} by a flip.

Let $w_0, w_1, \ldots, w_s$ be a sequence of rack words for $\alpha_0, \alpha_1, \ldots, \alpha_s$ with $w_0 = w$ and $w_s = w'$.

We consider a special case that α_{i-1} and α_i are the cords α and α' in Figure 2, respectively. Then α_{i-1} is represented by the word γ_1 and α_i is represented by the word $\gamma_n \wedge (\gamma_{n-1} \ldots \gamma_1 \gamma_n^{-1} \ldots \gamma_2^{-1})$. The word $\gamma_n \wedge (\gamma_{n-1} \ldots \gamma_1 \gamma_n^{-1} \ldots \gamma_2^{-1})$ is obtained from the word γ_1 by a consequence of the relation $\gamma_1 \wedge (\gamma_2 \ldots \gamma_n) = \gamma_n \wedge (\gamma_{n-1} \ldots \gamma_1)$. Since two rack words representing the same element of $X_{n+1}(\mathbb{R}^2, Q_{n+1})$ are congruent modulo the relations (4)–(6) by Theorem 3.1, we have that w_{i-1} and w_i are congruent modulo the relations (4)–(7).

Now we consider a general case where α_i is obtained from α_{i-1} by a flip along β and c. There exists an orientation-preserving homeomorphism h of $(\mathbb{R}^2, Q_{n+1})$ to itself such that $(\alpha_{i-1})h = \gamma_1$ and the images of β and c are as in Figure 2. Let v be an element of $F(X_{n+1}(\mathbb{R}^2))$ such that $\tau(v) = [h]$ where $\tau : F(X_{n+1}(\mathbb{R}^2)) \to Op(X_{n+1}(\mathbb{R}^2)) = M_{n+1}(\mathbb{R}^2)$ is the natural projection induced by the map $\tau : X_{n+1}(\mathbb{R}^2) \to M_{n+1}(\mathbb{R}^2)$. Then α_{i-1} is represented by the word $\gamma_1 \wedge (v^{-1})$ and α_i is represented by $(\gamma_n \wedge (\gamma_{n-1} \ldots \gamma_1 \gamma_n^{-1} \ldots \gamma_2^{-1})) \wedge (v^{-1})$. Since these words are congruent modulo the relation (4)–(7), we see that w_{i-1} and w_i are congruent modulo the relation (4)–(7). □

Corollary 3.6 *The associated group $As(X_{n+1}(\mathbb{R}^2))$ is isomorphic to the $n+1$-string braid group B_{n+1}.*

Proof. If a rack X has a rack presentation with generating set S and the defining relation set R, then the associated group $As(X)$ has a group presentation whose generating set is S and the defining relation set is R in which the rack operator $\wedge$ is interpreted as conjugation, cf. Lemma 4.3 of [2]. Thus, by Theorem 3.1, we have the result. □

Theorem 3.7 *The associated group $As(X_{n+1}(S^2))$ is isomorphic to the quotient group of B_{n+1} divided by the relations*

$$u\sigma_i = \sigma_i u \qquad (i = 1, 2, \ldots, n),$$

where $u = \sigma_1 \sigma_2 \ldots \sigma_{n-1} \sigma_n^2 \sigma_{n-1} \ldots \sigma_2 \sigma_1$.

Proof. By Theorem 3.2, we have a group presentation of the associated group $As(X_{n+1}(S^2))$ whose generating set is $\{\sigma_1, \ldots, \sigma_n\}$ and whose defining relation set is R:

$$\sigma_i \sigma_j = \sigma_j \sigma_i \qquad (|i-j| > 1) \tag{8}$$

$$\sigma_i \sigma_j \sigma_i = \sigma_j \sigma_i \sigma_j \qquad (|i-j| = 1) \tag{9}$$

$$\sigma_n^{-1} \ldots \sigma_2^{-1} \sigma_1 \sigma_2 \ldots \sigma_n = \sigma_1^{-1} \ldots \sigma_{n-1}^{-1} \sigma_n \sigma_{n-1} \ldots \sigma_1. \tag{10}$$

This is the quotient group of the braid group B_{n+1} divided by the relation (10).

In the braid group B_{n+1}, we have the relation

$$\sigma_1^{-1} \ldots \sigma_{n-1}^{-1} \sigma_n \sigma_{n-1} \ldots \sigma_1 = \sigma_n \ldots \sigma_2 \sigma_1 \sigma_2^{-1} \ldots \sigma_n^{-1}. \tag{11}$$

Using (10) and (11), we have

$$u\sigma_1 = \sigma_1 u, \tag{12}$$

where $u = \sigma_1\sigma_2\ldots\sigma_{n-1}\sigma_n^2\sigma_{n-1}\ldots\sigma_2\sigma_1$. Conversely the relations (11) and (12) imply the relation (10). Thus we may replace the relation (10) in the defining relation set R by (12). By considering geometric braids, we see easily that the element u commutes with $\sigma_2,\ldots,\sigma_n$ in the braid group B_{n+1}. Thus we may add the relations $u\sigma_i = \sigma_i u$ for $i = 2,\ldots,n$ to the set R. □

Remark. In [10] it is proved that the $n+1$-string braid group $B_{n+1}(S^2)$ of the 2-sphere S^2 is isomorphic to the quotient group of B_{n+1} divided by the relation

$$u = 1.$$

On the other hand in the associated group $As(X_{n+1}(S^2))$, the element u has infinite order, because by Theorem 3.7 the abelianization of $As(X_{n+1}(S^2))$ is an infinite cyclic group in which u is not the identity. Thus there exists a central extension

$$1 \to \langle u\rangle \to As(X_{n+1}(S^2)) \to B_{n+1}(S^2) \to 1.$$

4. Some general results

In this section, we consider a general connected oriented 2-manifold Σ, not necessarily $\mathbb{R}^2$ or S^2.

Lemma 4.1 *Let α and β be cords on (Σ, Q_{n+1}). If $\tau[\alpha]$ and $\tau[\beta]$ represent the same element of $M_{n+1}(\Sigma)$, then α and β are isotopic. (The map $\tau : X_{n+1}(\Sigma) \to M_{n+1}(\Sigma)$ is injective.)*

This seems to be well-known. Since the authors do not know a reference, we will give a proof.

Since $X_{n+1}(\Sigma) = X_{n+1}(\mathrm{int}(\Sigma))$ and there is a natural homomorphism $M_{n+1}(\Sigma) \to M_{n+1}(\mathrm{int}(\Sigma))$, we may assume $\partial\Sigma = \emptyset$ in what follows.

For a cord $\alpha \in X_{n+1}(\Sigma)$, we denote by $\tau^2[\alpha] \in M_{n+1}(\Sigma)$ a "360°-twist" along α. If $\tau[\alpha] = \tau[\beta]$ for $\alpha,\beta \in X_{n+1}(\Sigma)$, then $\tau^2[\alpha] = \tau^2[\beta]$.

Proof of Lemma 4.1. Let α, β be elements of $X_{n+1}(\Sigma)$ such that $\tau[\alpha] = \tau[\beta]$. The endpoints of α coincide with those of β. There exists a word w of $\gamma_1,\ldots,\gamma_n$ such that the endpoints of $\alpha \wedge w$ (and $\beta \wedge w$) are q_1 and q_2 of Q_{n+1} (Lemma 3.3). Since $\alpha = \beta$ if and only if $\alpha \wedge w = \beta \wedge w$ and since $\tau[\alpha] = \tau[\beta]$ if and only if $\tau[\alpha \wedge w] = \tau[\beta \wedge w]$, we may assume without loss of generality that the endpoints of α (and β) are q_1 and q_2.

Let $\alpha \in X_{n+1}(\Sigma)$ be a cord whose endpoints are q_1 and q_2. We call a disk $D(\alpha)$ in Σ a *tadpole with spine* α if $\alpha \subset D$, $\alpha \cap \partial D = \{q_1\}$ and $Q_{n+1} \cap D = \{q_1, q_2\}$ (see Figure 3). Let $\ell(\alpha)$ be the boundary simple closed curve $\partial D(\alpha)$ in the positive direction based at q_1.

Here we make a digression to the braid theory on Σ. Let Q be a finite set of points of Σ, and let $P(\Sigma, Q)$ denote the pure braid group on Σ with "base point" Q. Then we have the *fundamental exact sequence*, where $Q_{n+1} = \{q_1,\ldots,q_{n+1}\}$:

$$1 \to \pi_1(\Sigma - \{q_2,\ldots,q_{n+1}\}, q_1) \xrightarrow{j_*} P(\Sigma, Q_{n+1}) \xrightarrow{\pi_*} P(\Sigma, \{q_2,\ldots,q_{n+1}\}) \to 1.$$

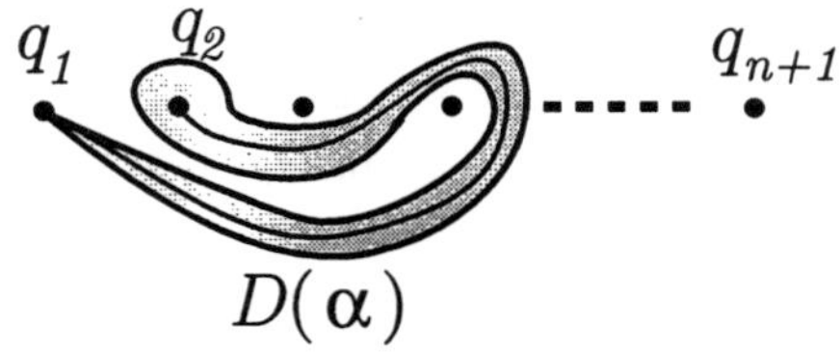

Figure 3

Here π_* is the homomorphism obtained by killing the first string (cf. [11]).

For $\Sigma \neq S^2$, the exactness of this sequence is proved in [8](Thm 1.4). For $\Sigma = S^2$ and $n+1 \geq 4$, the exactness is proved in [10]. Since $P(\Sigma, Q_{n+1}) \subset B_{n+1}(\Sigma)$, we have the following lemma.

Lemma 4.2 *The natural map*

$$j_* : \pi_1(\Sigma - \{q_2, \ldots, q_{n+1}\}, q_1) \to B_{n+1}(\Sigma)$$

is injective for $\Sigma \neq S^2$, or for $\Sigma = S^2, n+1 \geq 4$.

Given a braid $b \in B_{n+1}(\Sigma)$, there exists an isotopy $\{h_t\}_{0 \leq t \leq 1}$ of Σ onto itself such that $h_0 = \mathrm{id}$ and $\{h_t(Q_{n+1})\}_{0 \leq t \leq 1}$ represents the braid b. (See [7], [8].) A natural map $d_* : B_{n+1}(\Sigma) \to M_{n+1}(\Sigma)$ is defined as sending b to the mapping class $\in M_{n+1}(\Sigma)$ represented by the final stage h_1 of the isotopy $\{h_t\}_{0 \leq t \leq 1}$, cf. [8](p.153).

Lemma 4.3

$$\ker d_* \subset \mathrm{Center} B_{n+1}(\Sigma).$$

The proof is essentially the same as that of Lemma 4.2.1 in Birman's book [8].

Lemma 4.4 *If $\Sigma \neq \mathbb{R}^2, S^2, n+1 \geq 2$, or $\Sigma = \mathbb{R}^2, n+1 \geq 3$, or $\Sigma = S^2, n+1 \geq 4$, then the composition $d_* \circ j_* : \pi_1(\Sigma - \{q_2, \ldots, q_{n+1}\}, q_1) \to M_{n+1}(\Sigma)$ is injective.*

Proof. If $\Sigma \neq \mathbb{R}^2, S^2, n+1 \geq 2$, or $\Sigma = \mathbb{R}^2, n+1 \geq 3$, or $\Sigma = S^2, n+1 \geq 4$, the fundamental group $\pi_1(\Sigma - \{q_2, \ldots, q_{n+1}\}, q_1)$ is a free group of rank ≥ 2. In particular, it is centerless, and

$$j_*(\pi_1(\Sigma - \{q_2, \ldots, q_{n+1}\}, q_1)) \cap \mathrm{Center} B_{n+1}(\Sigma) = \{1\}. \tag{13}$$

By Lemma 4.2, the map $j_* : \pi_1(\Sigma - \{q_2, \ldots, q_{n+1}\}, q_1) \to B_{n+1}$ is injective, and by Lemma 4.3, $\ker d_* \subset \mathrm{Center} B_{n+1}(\Sigma)$. Hence the injectivity of $d_* \circ j_* : \pi_1(\Sigma - \{q_2, \ldots, q_{n+1}\}, q_1) \to M_{n+1}(\Sigma)$ follows from (13). □

Now we return to the proof of Lemma 4.1. Let $D(\alpha)$ be a tadpole with the spine α. Here α is a cord on (Σ, Q_{n+1}) whose endpoints are q_1 and q_2. It is not difficult to see that $d_* \circ j_* : \pi_1(\Sigma - \{q_2, \ldots, q_{n+1}\}, q_1) \to M_{n+1}(\Sigma)$ sends the boundary loop $\ell(\alpha)$ of $D(\alpha)$ to the 360°-twist $\tau^2[\alpha]$. Suppose we have two cords $\alpha, \beta \in X_{n+1}(\Sigma)$ whose endpoints are q_1 and q_2. If $\tau[\alpha] = \tau[\beta]$, then $\tau^2[\alpha] = \tau^2[\beta]$. Thus

$$d_* \circ j_*(\ell(\alpha)) = d_* \circ j_*(\ell(\beta)).$$

If $\Sigma \neq \mathbb{R}^2, S^2, n+1 \geq 2$, or $\Sigma = \mathbb{R}^2, n+1 \geq 3$, or $\Sigma = S^2, n+1 \geq 4$, then by Lemma 4.4, $d_* \circ j_*$ is injective, and we see that $\ell(\alpha)$ and $\ell(\beta)$ represent the same element of $\pi_1(\Sigma - \{q_2, \ldots, q_{n+1}\}, q_1)$. Then by Theorem 4.1 of [12], the closed curves $\ell(\alpha)$ and $\ell(\beta)$ are isotopic in $\Sigma - \{q_2, \ldots, q_{n+1}\}$ relative to $\{q_1\}$, and hence the tadpoles $D(\alpha)$ and $D(\beta)$ are isotopic in $\Sigma - \{q_3, \ldots, q_{n+1}\}$ relative to $\{q_1, q_2\}$. This implies that the cords α and β are isotopic in Σ relative to Q_{n+1}. Thus $\tau : X_{n+1}(\Sigma) \to M_{n+1}(\Sigma)$ is injective in this generic case.

If $\Sigma = \mathbb{R}^2$ or $S^2, n+1 = 2$, then $X_2(\Sigma)$ consists of a single element, and the injectivity of $\tau : X_2(\Sigma) \to M_2(\Sigma)$ is trivial.

Finally if $\Sigma = S^2, n+1 = 3$, then $X_3(S^2)$ is the dihedral rack R_3, [2](p.351), and $M_3(S^2)$ is the symmetric group on 3 letters, [11](p.288). The injectivity of $\tau : X_3(S^2) \to M_3(S^2)$ is proved easily in this special case. □

Corollary 4.5 *The natural map $\eta : X_{n+1}(\Sigma) \to As(X_{n+1}(\Sigma))$ is injective.*

Proof. The mapping $\tau : X_{n+1}(\Sigma) \to M_{n+1}(\Sigma)_{\text{conj}}$ is a rack homomorphism, where $M_{n+1}(\Sigma)_{\text{conj}}$ is the conjugation rack of $M_{n+1}(\Sigma)$ (cf. [2]). Thus by universality of the associated group [2], τ splits through η:

$$\tau : X_{n+1}(\Sigma) \xrightarrow{\eta} As(X_{n+1}(\Sigma)) \xrightarrow{\tau_\sharp} M_{n+1}(\Sigma).$$

The injectivity of η follows from the injectivity of τ. □

The following lemma is a general observation.

Lemma 4.6 *Let X be an arbitrary rack.*

(1) *If the natural map $\eta : X \to As(X)$ is injective, then we have*

$$Op(X) \cong As(X)/\text{Center}As(X).$$

(2) *If the natural map $\xi : X \to Op(X)$ is injective, then $Op(X)$ is centerless.*

Proof. Suppose that $\eta : X \to As(X)$ is injective. If a word $w \in F(X)$ represents an element $\eta(w)$ of $\text{Center}As(X)$, then for any $x \in X$,

$$\eta(x \wedge w) = \eta(w)^{-1}\eta(x)\eta(w) = \eta(x).$$

By the assumption, η is injective, so $x \wedge w = x$ for any $x \in X$. Thus the word w represents the identity of $Op(X)$. This proves that $\text{Center}As(X)$ is contained in $\ker(As(X) \to Op(X))$. The opposite inclusion $\ker(As(X) \to Op(X)) \subset \text{Center}As(X)$ is proved similarly but without assuming the injectivity of η. Thus

$$\ker(As(X) \to Op(X)) = \text{Center}As(X).$$

Since the natural map $As(X) \to Op(X)$ is surjective, we have the former assertion.

The proof of (2) is similar to the above argument. □

Theorem 4.7

$$Op(X_{n+1}(\Sigma)) \cong As(X_{n+1}(\Sigma))/\text{Center}As(X_{n+1}(\Sigma)).$$

Proof. By Corollary 4.5, the natural map $\eta : X_{n+1}(\Sigma) \to As(X_{n+1}(\Sigma))$ is injective. Then by Lemma 4.6(1), we have the result. □

5. The group $As(X_{n+1}(S^2))$

In this section we study the group $As(X_{n+1}(S^2))$ and give the structure of the center. Here we assume the presentation of $As(X_{n+1}(S^2))$ as in Theorem 3.7.

Theorem 5.1 *Suppose $n+1 \geq 3$. Center$As(X_{n+1}(S^2))$ is generated by $u = \sigma_1\sigma_2 \dots \sigma_{n-1}\sigma_n^2\sigma_{n-1} \dots \sigma_2\sigma_1$ and $v = (\sigma_1\sigma_2 \dots \sigma_n)^{n+1}$. Both the generators have infinite order, and satisfy a single relation $u^{n+1} = v^2$.*

Proof. By the remark below Theorem 3.7, the kernel of the natural surjection

$$As(X_{n+1}(S^2)) \to B_{n+1}(S^2) \tag{14}$$

is an infinite cyclic group generated by a central element

$$u = \sigma_1\sigma_2 \dots \sigma_{n-1}\sigma_n^2\sigma_{n-1} \dots \sigma_2\sigma_1.$$

Thus Center$As(X_{n+1}(S^2))$ certainly contains u.

Let v denote the word $(\sigma_1\sigma_2 \dots \sigma_n)^{n+1}$. Then under the assumption $n+1 \geq 3$, Artin [7] and Chow [13] show that v generates the center of the $n+1$-string braid group B_{n+1}, which is an infinite cyclic group. v belongs to Center$As(X_{n+1}(S^2))$, because $As(X_{n+1}(S^2))$ is a quotient group of B_{n+1} by Theorem 3.7. Thus we have seen that Center$As(X_{n+1}(S^2))$ contains u and v.

By Gillette and Van Buskirk [11](p.281), v represents in $B_{n+1}(S^2)$ an element of order 2, called the *Dirac braid*, and the kernel of the natural surjection

$$B_{n+1}(S^2) \to M_{n+1}(S^2) \tag{15}$$

is generated by the Dirac braid.

Note that, if $n+1 \geq 3$, $M_{n+1}(S^2)$ is isomorphic to $Op(X_{n+1}(S^2))$ by Lemma 2.1, and it is easy to see that the composite of the homomorphisms (14) and (15) is identified with the natural surjection

$$As(X_{n+1}(S^2)) \to Op(X_{n+1}(S^2)). \tag{16}$$

Theorem 4.7 identifies the kernel of (16) as the center of the associated group. Therefore, Center$As(X_{n+1}(S^2))$ is contained in the subgroup generated by u and v. Combining this with the opposite inclusion, we see that Center$As(X_{n+1}(S^2))$ coincides with the subgroup generated by u and v.

Next, we will study the relation between the two generators. In the braid group B_{n+1}, we have the following relation:

$$(\sigma_1\sigma_2 \dots \sigma_n)^{n+1} = (\sigma_n\sigma_{n-1} \dots \sigma_1)^{n+1}. \tag{17}$$

This relation would be most easily seen by drawing the diagrams of the braids. By the definition of u,

$$u(\sigma_n\sigma_{n-1}\dots\sigma_1)^{-1} = \sigma_1\sigma_2\dots\sigma_n.$$

Since u is a central element of $As(X_{n+1}(S^2))$,

$$u^{n+1}(\sigma_n\sigma_{n-1}\dots\sigma_1)^{-(n+1)} = (\sigma_1\sigma_2\dots\sigma_n)^{n+1}. \tag{18}$$

Combining (18) with (17) and the definition of v, we have the following relation in $As(X_{n+1}(S^2))$:

$$u^{n+1} = v^2. \tag{19}$$

Since u is of infinite order, v is also of infinite order in $As(X_{n+1}(S^2))$. If we move to $B_{n+1}(S^2)$ by imposing $u = 1$, the element v becomes the Dirac braid of order 2. Thus in $As(X_{n+1}(S^2))$, (19) is the only non-trivial relation between u and v. This completes the proof of Theorem 5.1. □

Corollary 5.2 *Suppose $n+1 \geq 3$. If $n+1$ is even,* Center$As(X_{n+1}(S^2)) \cong C_\infty \times C_2$. *If $n+1$ is odd,* Center$As(X_{n+1}(S^2)) \cong C_\infty$. *Here C_k denotes the cyclic group of order k.*

Proof. If $n+1$ is even, $u^{-\frac{n+1}{2}}v$ is an element of order 2. The factor C_∞ is generated by u, and the factor C_2 by $u^{-\frac{n+1}{2}}v$. If $n+1$ is odd, then $(u^{-\frac{n}{2}}v)^2 = u$ and $(u^{-\frac{n}{2}}v)^{n+1} = v$. Thus the infinite cyclic group generated by $u^{-\frac{n}{2}}v$ coincides with Center$As(X_{n+1}(S^2))$. □

We summarize the relationship of the group $As(X_{n+1}(S^2))$ to the mapping class groups and the braid groups.

(1) The braid group $B_{n+1}(S^2)$ is obtained from $As(X_{n+1}(S^2))$ by imposing $u = 1$. We already remarked this after Theorem 3.7.

(2) If $n+1 \geq 3$, we obtain the mapping class group $M_{n+1}(S^2)$ from $As(X_{n+1}(S^2))$ by imposing $u = v = 1$.

(3) If $n+1 \geq 3$, $M_{n+1}(\mathbb{R}^2)$ and $M_{n+1}(S^2)$ are centerless [11](p.288). We can give a rack theoretic proof of this fact using Lemmas 2.1, 4.1 and 4.6(2).

(4) If $n+1$ is even, we can take a double branched covering of S^2 with branch points Q_{n+1}. The covering space is an oriented closed surface Σ_g with genus $g = \frac{n-1}{2}$. Let $M_{\rm sym}(\Sigma_g)$ denote the "symmetric" subgroup of the mapping class group $M(\Sigma_g)$ consisting of those elements which commute with the covering transformation. Then for $n+1 \geq 6$, $M_{\rm sym}(\Sigma_g)$ is obtained from $As(X_{n+1}(S^2))$ by imposing $u^2 = 1, v = 1$. See [14](p.110).

6. Remarks

The racks $X_{n+1}(\mathbb{R}^2)$ and $X_{n+1}(S^2)$ introduced in this paper can be expected to play a fundamental role in the monodromy calculus appearing in 4-dimensional topology, for example, in surface braids [15] or in Lefschetz fibrations [16]. They are considered to provide a natural mathematical framework for the "quantum inverse" introduced in [17]. This will be fully discussed in a forthcomming paper [18].

After finishing the last draft of this paper, Luis Paris informed us of P. Dehornoy's paper [19], in which Dehornoy finds an interesting connection between free left distritutive systems (closely related to free racks) and braids which is different from ours. The authors thank L. Paris for this information and useful comments.

Acknowledgements

This research is supported by Grant-in-Aid for Scientific Research (No. 10440015), Ministry of Education, Science and Culture, Japan.

References

[1] D. Joyce, *A classifying invariant of knots; the knot quandle,* J. Pure and Applied Algebra **23** (1982) 37–65.
[2] R. Fenn and C. Rourke, *Racks and links in codimension two,* J. Knot Theory Ramifications **1** (1992) 343–406.
[3] M. Takasaki, *Abstraction of symmetric transformations (in Japanese),* Tohoku Math. J. **49** (1942/43), 145–207; Math. Rev. 9. p.8.
[4] S. V. Matveev, *Distributive groupoids in knot theory,* Math. USSR-Sbornik **47** (1984) 73-83.
[5] E. Brieskorn, *Automorphic sets and braids and singularities,* Contemp. Math. **78** (1988) 45–115.
[6] L. H. Kauffman, *Knots and Physics* (Series on Knots and Everything, 1, World Scientific Publ.,1991).
[7] E. Artin, *Theory of braids,* Ann. of Math. **48** (1947) 101–126.
[8] J. S. Birman, *Braids, links, and mapping class groups.* (Ann. Math. Studies 82, Princeton Univ. Press, Princeton, N. J., 1974).
[9] S. Kamada, *An observation of surface braids via chart description,* J. Knot Theory Ramifications **4** (1996) 517–529.
[10] E. Fadell and J. Van Buskirk, *The braid groups of E^2 and S^2,* Duke Math. J. **29** (1962) 243–258.
[11] R. Gillette and J. Van Buskirk, *The word probrem and consequences for the braid groups and mapping class groups of the 2-sphere,* Trans. Amer. Math. Soc. **131** (1968) 277–296.
[12] D. B. A. Epstein, *Curves on 2-manifolds and isotopies,* Acta Math. **115** (1966) 83–107.
[13] W.-L. Chow, *On the algebraic braid group,* Ann. of Math. **49** (1948) 654–658.
[14] J. S. Birman and H. M. Hilden, *On the mapping class groups of closed surfaces as covering spaces,* in: *Advances in the theory of Riemann surfaces* (Ann. Math. Studies 66, Princeton Univ. Press, Princeton, N. J., 1971) 81–115.
[15] S. Kamada, *On braid monodromies of non-simple braided surfaces,* Math. Proc. Camb. Phil. Soc. **120** (1996) 237–245.
[16] Y. Matsumoto, *Lefschetz fibrations of genus two—A topological approach—,* in: S. Kojima *et al*, eds., Proc. the 37-th Taniguchi Sympo. (World Scientific Publ., 1996) 123–148.
[17] Y. Matsumoto, *On quantum inverse in a free group,* in: *The fifth Korea-Japan School of Knots and Links* (Proc. Applied Math. Workshop, 8, KAIST, (Taejon, Korea 1997), 1998), 187–198.
[18] K. Habiro, S. Kamada, Y. Matsumoto and K. Yoshikawa, in preparation.
[19] P. Dehornoy, *Braid groups and left distributive operations,* Trans. Amer. Math. Soc. **345** (1994) 115–150.

AN EVALUATION OF THE COEFFICIENT POLYNOMIALS OF THE HOMFLY POLYNOMIAL OF A LINK

TAIZO KANENOBU
Department of Mathematics, Osaka City University
Sugimoto, Sumiyoshi-ku, Osaka 558-8585, Japan
kanenobu@sci.osaka-cu.ac.jp

ABSTRACT
We show that the values at $t=1$ of the first r coefficient polynomials of the HOMFLY polynomial of an r-component link L are determined by the linking numbers. This generalizes a formula of Hoste and one of Lickorish and Millett. The formula was proved before, using a different method, by J. Przytycki and related to homotopy skein modules [15].

Keywords: Link, HOMFLY polynomial, linking number, Vassiliev invariant.

1. Introduction

The *HOMFLY polynomial* $P(L;t,z) \in \mathbb{Z}[t^{\pm 1}, z^{\pm 1}]$ ([6, 16]) is an invariant of the isotopy type of an oriented link L, which is defined, as in [9], by the following formulas:

$$P(U;t,z) = 1; \tag{1.1}$$

$$t^{-1}P(L_+;t,z) - tP(L_-;t,z) = zP(L_0;t,z). \tag{1.2}$$

where U is the unknot and L_+, L_-, L_0 are three links that are identical except near one point where they are as in Fig. 1. We call (L_+, L_-, L_0) a *skein triple*. By [14, Proposition 22], the HOMFLY polynomial of an r-component link L is of the form

$$P(L;t,z) = \sum_{n\geq 0} P_{2n-r+1}(L;t) z^{2n-r+1}, \tag{1.3}$$

where each $P_{2n-r+1}(L;t) \in \mathbb{Z}[t^{\pm 1}]$ is called the *coefficient polynomial*; the powers of t which appear in it are all even or odd, depending on whether $2n-r+1$ is even or odd.

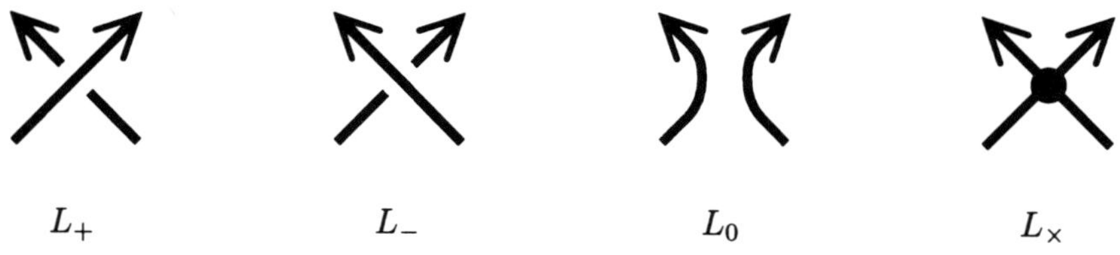

Fig. 1.

Let $L = K_1 \cup K_2 \cup \cdots \cup K_r$ be an oriented r-component link. We denote by λ_{ij} the linking number of K_i and K_j, which we may consider a map $L \mapsto \lambda_{ij}(L) \in \mathbb{Z}$. Let Λ^n be the set of the products of n linking numbers;

$$\Lambda^n = \left\{ \prod_{i<j} \lambda_{ij}^{e(i,j)} \;\middle|\; \sum_{i<j} e(i,j) = n \right\}.$$

For $\ell = \prod_{i<j} \lambda_{ij}^{e(i,j)} \in \Lambda^n$, we define a graph $\Psi(\ell)$ with r vertices $p_1, p_2, \ldots, p_r$ and n edges so that the number of the edges joining p_i to p_j is $e(i,j)$. Let T^n be the set of the products of n linking numbers whose associated graphs are unions of trees;

$$T^n = \{\, \ell \in \Lambda^n \mid \Psi(\ell) \text{ is a union of trees. } \}.$$

Main Theorem. *Suppose* $0 \le n \le r-1$. *Then*

$$\left[(t^{-1} - t)^{n-r+1} P_{2n-r+1}(L;t)\right]_{t=1} = \sum_{\ell \in T^n} \ell.$$

In particular, if $n = 0$, 1, 2, then

$$\left[(t^{-1} - t)^{1-r} P_{1-r}(L;t)\right]_{t=1} = 1; \tag{1.4}$$

$$\left[(t^{-1} - t)^{2-r} P_{3-r}(L;t)\right]_{t=1} = \sum_{i<j} \lambda_{ij}; \tag{1.5}$$

$$\left[(t^{-1} - t)^{3-r} P_{5-r}(L;t)\right]_{t=1} = \sum_{\substack{i<j,\ k<l \\ \text{either } i<k \text{ or } i=k,\ j<l}} \lambda_{ij}\lambda_{kl}, \tag{1.6}$$

where $r \ge 1$, $r \ge 2$ and $r \ge 3$, respectively. The equation (1.4) also follows from the explicit formula for $P_{1-r}(L;t)$ [14, Proposition 22], and (1.5) is [14, Proposition 24 (i)]. Also (1.5) and (1.6) follow from the explicit formulas for $P_{3-r}(L;t)$ and $P_{5-r}(L;t)$ given in [11]. Further, if $n = r-1$, then

$$a_{r-1}(L) = \sum_{\ell \in T^{r-1}} \ell, \tag{1.7}$$

where $a_{r-1}(L)$ is the coefficient of z^{r-1} of the Conway polynomial of L. This is given by Hoste [8, Theorem 2] (cf. [7]), and the new proof that uses a Vassiliev invariant is given in [12, Theorem 8.1]. Similarly, we prove Main Theorem using a Vassiliev invariant. The formula was proved before (but not yet published) by J. Przytycki [15] and A. Sikora [17, 18]. Przytycki generalized it, relating it to the homotopy skein module and the dichromatic polynomial.

2. Preliminaries

The *Conway polynomial* of L ([5]), $\nabla_L(z) \in \mathbb{Z}[z]$, is obtained from the HOMFLY polynomial by

$$\nabla_L(z) = P(L; 1, z), \tag{2.1}$$

which is of the form

$$\nabla_L(z) = \sum_{n \geq 0} a_{2n+r-1}(L) z^{2n+r-1}, \tag{2.2}$$

with $a_{2n+r-1}(L) = P_{2n+r-1}(L; 1) \in \mathbb{Z}$. In particular, $P_{2n-r+1}(L; 1) = 0$ for $0 \leq n \leq r-2$. Moreover, Kawauchi [13, Lemma 1.7] proves:

Lemma 1. *If* $0 \leq n \leq r-2$, *then* $P_{2n-r+1}(L;t)$ *is divisible by* $(t^{-1}-t)^{r-n-1}$.

We define the invariant of an r-component oriented link, b^r_n, $0 \leq n \leq r-1$, by

$$b^r_n(L) = \left[(t^{-1}-t)^{n+1-r} P_{2n+1-r}(L;t)\right]_{t=1} \in \mathbb{Z}. \tag{2.3}$$

In particular, $b^r_{r-1}(L) = a_{r-1}(L)$. Letting U^r be the trivial r-component link, we have

$$P_{2n+1-r}(U^r; t) = \begin{cases} (t^{-1}-t)^{r-1} & \text{if } n = 0; \\ 0 & \text{if } n \geq 1, \end{cases} \tag{2.4}$$

which implies

$$b^r_n(U^r) = \begin{cases} 1 & \text{if } n = 0; \\ 0 & \text{if } 1 \leq n \leq r-1. \end{cases} \tag{2.5}$$

An r-component *singular link* is the image of oriented ordered r circles under an immersion into S^3 whose only singularities are transverse double points. We assume that a double point on a singular link is a rigid (or flat) vertex. Let v be an isotopy invariant of an ordered r-component link, which takes values in the rational numbers. Then v can be uniquely extended to an r-component singular link invariant by the *Vassiliev skein relation*:

$$v(L_\times) = v(L_+) - v(L_-), \tag{2.6}$$

where $L_\times$ is a singular link with x a double point and L_+, L_- are ones obtained from $L_\times$ by replacing x by a positive crossing and a negative crossing, respectively; see Fig. 1. We call v a *Vassiliev* (*finite type*) *link invariant* if it satisfies the following axiom: there exists an integer n such that $v(L) = 0$ for any singular link L with more than n double points. The smallest such an integer n is the *order* of v.

The *n-configuration* which an r-component singular ordered oriented link with n double points respects is the n pairs of points on oriented r ordered circles. We use a *chord-diagram of order* n to represent it, that is, ordered oriented r circles with n chords joining the paired points. We shall not distinguish strictly a chord-diagram from a configuration. If v is a Vassiliev link invariant of order n, then v-value of a singular link with n double points depends only on its configuration. We may define a Vassiliev invariant for an unordered oriented link in the same way. Similarly we can define a Vassiliev invariant for an ordered oriented link as one for an unordered oriented link; cf. [19, p. 252]. The reader is referred to [1, 2, 3, 4, 10, 12, 19, 20, 21] for a more detailed exposition.

Lemma 2. *Suppose* $1 \leq n \leq r-1$. *Let* L^m *be a singular* r*-component link with* m *double points and* L_0^{m-1} *a singular link obtained from* L^m *by smoothing one of the double points of* L^m. *Then we have*

$$b_n^r(L^m) = \begin{cases} b_{n-1}^{r-1}(L_0^{m-1}) & \text{if } L_0^{m-1} \text{ has } r-1 \text{ components;} \\ 0 & \text{if } L_0^{m-1} \text{ has } r+1 \text{ components.} \end{cases}$$

Proof. Let $x_1, \ldots, x_{m-1}, x_m$ be the double points of L^m, and let L_0^{m-1} be obtained from L^m by smoothing x_m. Let $L_{x_1(\epsilon_1),\ldots,x_{m-1}(\epsilon_{m-1}),x(\epsilon_m)}$, $\epsilon_i = \pm$, be a non-singular link obtained from L^m by replacing each double point x_i by a positive crossing $x_i(+)$ or a negative crossing $x_i(-)$. Using the Vassiliev skein relation (2.6), we obtain:

$$\begin{aligned} b_n^r(L^m) &= \sum_{\epsilon_i=\pm}(-1)^{\mu(\epsilon)} b_n^r(L_{x_1(\epsilon_1),\ldots,x_{m-1}(\epsilon_{m-1}),x(\epsilon_m)}) \\ &= \sum_{\epsilon_i=\pm}(-1)^{\mu(\epsilon)} \left[(t^{-1}-t)^{n-r+1} P_{2n-r+1}(L_{x_1(\epsilon_1),\ldots,x_{m-1}(\epsilon_{m-1}),x_m(\epsilon_m)};t)\right]_{t=1} \\ &= \sum_{\epsilon_i=\pm}(-1)^{\mu(\epsilon')} \Big(\left[(t^{-1}-t)^{n-r+1} P_{2n-r+1}(L_{x_1(\epsilon_1),\ldots,x_{m-1}(\epsilon_{m-1}),x(+)};t)\right]_{t=1} \\ &\qquad - \left[(t^{-1}-t)^{n-r+1} P_{2n-r+1}(L_{x_1(\epsilon_1),\ldots,x_{m-1}(\epsilon_{m-1}),x(-)};t)\right]_{t=1}\Big), \end{aligned}$$

where $\mu(\epsilon)$ and $\mu(\epsilon')$ are the numbers of minus signs in $\epsilon_1, \ldots, \epsilon_{m-1}, \epsilon_m$ and $\epsilon_1, \ldots, \epsilon_{m-1}$, respectively; cf. [3, (2)]. From (1.2), this becomes

$$\begin{aligned} &\sum_{\epsilon_i=\pm}(-1)^{\mu(\epsilon')} \left[(t^{-1}-t)^{n-r+1} P_{2n-r}(L_{x_1(\epsilon_1),\ldots,x_{m-1}(\epsilon_{m-1}),x(0)};t)\right]_{t=1} \\ &= \left[(t^{-1}-t)^{n-r+1} P_{2n-r}(L_0^{m-1});t)\right]_{t=1}. \end{aligned}$$

If L_0^{m-1} has $r-1$ components, this equals to $b_{n-1}^{r-1}(L_0^{m-1})$, and if L_0^{m-1} has $r+1$ components, this equals to $\left[t^{-1}-t\right]_{t=1} \cdot b_n^{r+1}(L_0^{m-1}) = 0$. This completes the proof. □

Lemma 3. *The invariant* b_n^r, $0 \leq n \leq r-1$, *is a Vassiliev invariant for an unordered oriented link of order* $\leq n$.

Proof. From (1.4), $b_0^r(L) = 1$ for any r-component link L, and so the lemma is true for $n = 0$. It follows from Lemma 2 that the proof is completed by induction on n. □

Lemma 4. *Suppose $2 \le i+1 \le n \le r-1$. Let α^i be any i-configuration for oriented r circles. Then there exists a singular link M^i respecting α^i such that $b_n^r(M^i) = 0$.*

Proof. If the i-configuration α^i has k connected components, then it may be respected by a singular link M^i which is a split union of k singular links. Thus in particular, $k \ge \max\{r-i, 1\}$.

On the other hand, let L be an r-component oriented link which is a split union of k links; $L = L_1 \sqcup L_2 \sqcup \cdots \sqcup L_k$. Then

$$(t^{-1}-t)^{n-r+1}P(L;t,z) = (t^{-1}-t)^{n-r+k}P(L_1;t,z)P(L_2;t,z)\cdots P(L_k;t,z)z^{1-k}.$$

Since $n-r+k \ge 1$,

$$b_n^r(M^i) = \left[(t^{-1}-t)^{n-r+1}P_{2n-r+1}(M^i;t)\right]_{t=1} = 0,$$

which completes the proof. □

For an n-configuration α with r circles $S_1, S_2, \ldots, S_r$, we define a graph $\Phi(\alpha)$ with r vertices $p_1, p_2, \ldots, p_r$ and n edges by deforming the circle S_k into the vertex p_k and a chord joining S_i to S_j into an edge joining p_i to p_j.

The linking numbers λ_{ij} and their products are known to be Vassiliev invariants for an ordered oriented link; see [12, Proposition 3.1]. For $v = \prod_{i<j}\lambda_{ij}^{e(i,j)} \in \Lambda^n$, $\sum_{i<j} e(i,j) = n$, we define a graph $\Psi(v)$ with r vertices $p_1, p_2, \ldots, p_r$ and n edges so that the number of the edges joining p_i to p_j is $e(i,j)$. We shall use the following lemma [12, Proposition 4.2].

Lemma 5. *If $v = \prod_{i<j}\lambda_{ij}^{e(i,j)}$, then*

$$v(\alpha) = \begin{cases} \prod_{i<j}(e(i,j)!) & \text{if } \Psi(v) = \Phi(\alpha); \\ 0 & \text{if } \Psi(v) \ne \Phi(\alpha). \end{cases}$$

3. Proof of Main Theorem

Let $\{\,\alpha_s^i \mid s = 1, 2, \ldots, s_i\,\}$ be the set of all i-configurations for ordered r circles, and M_s^i be a singular link with i singular points which respects α_s^i. Let v be a Vassiliev invariant of order $n(\ge 1)$, for an r-component ordered oriented link. Then

$$v(L) = v(U^r) + \sum_{i=1}^{n-1}\sum_{s=1}^{s_i} m_s^i v(M_s^i) + \sum_{k=1}^{s_n} m_k^n v(\alpha_k^n), \tag{3.1}$$

where m_s^i are integers determined by the choice of a diagram for the link L and crossing changes to the trivial link U^r; cf. [3, Proposition 2]. By Lemma 4, we may suppose that $b_n^r(M_s^i) = 0$ for $1 \le i \le n-1$. Further using, (2.5), we obtain

$$b_n^r(L) = \sum_{k=1}^{s_n} m_k^n b_n^r(\alpha_k^n). \tag{3.2}$$

Using Lemma 2 and (2.5), we obtain

$$b_n^r(\alpha_k^n) = \begin{cases} 1 & \text{if } \Phi(\alpha_k^n) \text{ is a union of trees;} \\ 0 & \text{otherwise.} \end{cases} \tag{3.3}$$

For simplicity, we assume that the associated graphs $\Phi(\alpha_k^n)$, $k = 1, 2, \ldots, q$, are unions of trees and the rest, $\Phi(\alpha_k^n)$, $j = q+1, \ldots, s_n$, are not. Then (3.2) and (3.3) imply

$$b_n^r(L) = \sum_{k=1}^{q} m_k^n. \tag{3.4}$$

For each n-configuration α_k^n, $k = 1, 2, \ldots, q$, we define v_k^n to be the product of linking numbers satisfying $\Psi(v_k^n) = \Phi(\alpha_k^n)$. Then by Lemma 5,

$$v_j^n(\alpha_k^n) = \begin{cases} 1 & \text{if } j = k; \\ 0 & \text{if } j \neq k. \end{cases} \tag{3.5}$$

Also by [12, Lemma 4.3 (i)],

$$v_j^n(M_s^i) = 0 \tag{3.6}$$

for $1 \leq s \leq s_i$ and $1 \leq i \leq n-1$, and $v_j^n(U^r) = 0$. Therefore from (3.1), we have

$$v_j^n(L) = m_j^n. \tag{3.7}$$

Combining this with (3.4), we have

$$b_n^r(L) = \sum_{k=1}^{q} v_k^n(L) \tag{3.8}$$

for any r-component link L, thereby completing the proof. □

Acknowledgements

The author was partially supported by Grant-in-Aid for Scientific Research (No. 09640125), Ministry of Education, Science and Culture, Japan.

References

[1] D. Bar-Natan, *On the Vassiliev knot invariants*, Topology **34** (1995) 423–472.

[2] J. S. Birman, *New points of view in knot theory*, Bull. Amer. Math. Soc. **28** (1993) 253–287.

[3] J. S. Birman, *On the combinatorics of Vassiliev invariants*, in *Braid group, knot theory and statistical mechanics II* eds. M. L. Ge and C. N. Yang, Advanced Series in Mathematical Physics, World Scientific, Singapore-New Jersey-London-Hong Kong (1994) 1–19.

[4] J. S. Birman and X. S. Lin, *Knot polynomials and Vassiliev's invariants*, Inv. Math. **111** (1993) 225–270.

[5] J. H. Conway, *An enumeration of knots and links*, in *Computational Problems in Abstract Algebra* ed. J. Leech, Pergamon Press, New York (1969) 329–358.

[6] P. Freyd, D. Yetter, J. Hoste, W. B. R. Lickorish, K. Millett and A. Ocneanu, *A new polynomial invariant of knots and links*, Bull. Amer. Math. Soc. **12** (1985) 239–246.

[7] F. Hosokawa, *On ∇-polynomials of links* , Osaka. Math. J. **10** (1958) 273–282.

[8] J. Hoste, *The first coefficient of the Conway polynomial*, Proc. Amer. Math. Soc. **95** (1985) 299–302.

[9] V. F. R. Jones, *Hecke algebra representations of braid groups and link polynomials*, Ann. Math. **126** (1987) 335–388.

[10] T. Kanenobu and Y. Miyazawa, *HOMFLY polynomials as Vassiliev link invariants*, in *Knot Theory*, Banach Center Publications **42**, eds. V. F. R. Jones, J. Kania-Bartoszyńska, J. H. Przytycki, P. Traczyk and V. G. Turaev, Institute of Mathematics, Polish Academy of Science, Warsawa (1998) 165–185.

[11] T. Kanenobu and Y. Miyazawa, *The second and third terms of the HOMFLY polynomial of a link*, Math. Sem. Notes, Kobe Univ., to appear.

[12] T. Kanenobu, Y. Miyazawa and A. Tani, *Vassiliev link invariants of order three*, J. Knot Theory Ramifications **7** (1998) 433–462.

[13] A. Kawauchi, *On coefficient polynomials of the skein polynomial of an oriented link*, Kobe J. Math. **11** (1994) 49–68.

[14] W. B. R. Lickorish and K. C. Millett, *A polynomial invariant of oriented links*, Topology **26** (1987) 107–141.

[15] J. H. Przytycki, *Homotopy and q-homotopy skein modules of 3-manifolds: an example in Algebra Situs*, in *Proceedings of a Conference in Low-Dimensional Topology in Honor of Joan Birman's 70th Birthday*, Columbia University/Barnard College, New York, March 14–15, 1998, to appear.

[16] J. H. Przytycki and P. Traczyk, *Invariants of links of Conway type*, Kobe J. Math. **4** (1987) 115–139.

[17] A. S. Sikora, Master Degree Thesis, Warsaw University (1994).

[18] A. S. Sikora, *Note on the Homfly polynomial and linking numbers*, preprint.

[19] T. Stanford, *The functoriality of Vassiliev-type invariants of links, braids, and knotted graphs*, J. Knot Theory Ramifications **3** (1994) 247–262.

[20] T. Stanford, *Finite-type invariants of knots, links, and graphs*, Topology **35** (1996) 1027–1050.

[21] V. A. Vassiliev, *Cohomology of knot spaces*, in *Theory of Singularities and its Applications*, Advances in Soviet Mathematics **1**, ed. V. I. Arnold, Amer. Math. Soc., Providence, RI (1990) 23–69.

ESTIMATING THE SIZE OF SKEIN HOMOLOGIES

JOANNA KANIA-BARTOSZYŃSKA, JÓZEF H. PRZYTYCKI, AND ADAM S. SIKORA

ABSTRACT. We give a new method of distinguishing elements of skein homologies. Using this method we show that each skein homology module is at least as big as the Kauffman bracket skein module of a given 3-manifold.

1. INTRODUCTION

Skein modules are constructed by considering a free module spanned by ambient isotopy classes of links (possibly with some sort of decoration - like framing) in a 3-manifold, and then dividing it by a submodule generated by some skein relations [P-1]. More generally, using the same skein relations, one can define a chain complex whose 0-th homology is the original skein module, and whose higher degree homologies are some more complicated invariants of a given 3-manifold. This concept was introduced and described for the Kauffman bracket skein module in [BFK-2]. In this note we generalize some techniques developed in [BFK-2] to construct cycles and to identify non-boundary elements[1]. We use these techniques to show that each skein homology module of an arbitrary 3-manifold is at least as big as the Kauffman bracket skein module of the manifold.

For completeness, we recall main definitions from [BFK-2]. Given an oriented 3-manifold M, a framed link is an embedding of a disjoint union of annuli in M. Let $\mathcal{L}(M)$ denote the set of ambient isotopy classes of framed, unoriented links, including the empty link. A crossing ball for a framed link is an embedding of the pair (B^3, D^2) so that inside the crossing ball the link looks like  with D^2 lying in the page. A framed link with n ordered crossing balls is one where the balls are ordered and disjoint. We number the crossing balls from 1 to n, corresponding to their ordering. Two such objects are equivalent

The first author was partially supported by NSF-DMS-9626818, the second by NSF-DMS-98089555.

[1]This note originated from the discussion which took place during the first author's lecture on skein homologies at GWU, April 23, 1998. We would like to thank all the students who participated in this discussion.

if there is an ambient isotopy of the links that carries crossing balls to crossing balls in an order preserving fashion. The set of ambient isotopy classes excluding links with trivial components is denoted by $\mathcal{L}^n(M)$.

Let R be a ring with identity and A an invertible element of R. The i-th ball operator

$$\partial_i : R\mathcal{L}^n(M) \to R\mathcal{L}^{n-1}(M)$$

is defined locally at the i-th crossing ball by

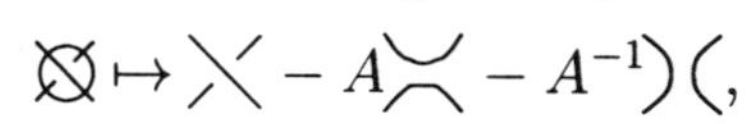

along with any necessary applications of $L \cup \bigcirc = -(A^2 + A^{-2})L$. The remaining crossing balls inherit the original ordering.

Define $d_0 : \mathcal{L}^0(M) \to R$ to be the zero map, and for $n \geq 1$ let

$$d_n : R\mathcal{L}^n(M) \to R\mathcal{L}^{n-1}(M)$$

be given by

$$d_n = \sum_{i=1}^{n} (-1)^i \partial_i.$$

It is easy to see that $(R\mathcal{L}^n(M), d_n)$ is a chain complex. Its homology is called the Kauffman bracket homology of a manifold M, and it is denoted by $K_*(M)$. The 0-th homology, $K_0(M)$, is the Kauffman bracket skein module of M. This module has been studied in several papers, e.g. [B], [BFK-1], [HP], [P-2], [PS].

2. The main result

In order to understand homologies one must be able to identify non-boundary cycles in the chain complex. We expand the methods introduced in [BFK-2]. In particular, we prove the following result.

Theorem 1. *Let M be an oriented 3-manifold. If $\frac{1}{2} \in R$ then for any $n = 0, 1, 2...$ there is an epimorphism of R-modules $\phi : K_n(M) \to K_0(M)$.*

Recently, there has been developed a connection between character varieties and the Kauffman bracket skein modules, [B, BFK-1, P-2, PS]. This connection implies an interesting corollary to the above theorem.

Corollary 2. *If one of the following conditions holds then $K_n(M)$ is infinitely generated for every n :*

- *$\frac{1}{2} \in R$ and $H_1(M, \mathbb{Z})$ is infinite;*

- *R is an algebraically closed field of characteristic* 0 *and the* SL_2*-character variety of* $\pi_1(M)$ *has at least one component of positive dimension.*

Proof. Each of these conditions guarantees that K_0 is infinitely generated. □

Proof of Theorem 1: Consider a map $\phi : \mathcal{L}^n(M) \to \mathcal{L}(M)$ defined by forgetting the crossing balls. This map induces a homomorphism of R-modules $\phi : R\mathcal{L}^n(M) \to K_0(M)$. Since each element of $im\ d_n$ is a linear combination of skein relations, the above homomorphism yields $\phi : R\mathcal{L}^n(M)/im(d_n) \to K_0(M)$. For brevity of notation, we denote the restriction of this homomorphism to homology by the same symbol, $\phi : K_n(M) \to K_0(M)$. We will prove that this map is an epimorphism. Let L be an arbitrary unoriented framed link in M without trivial components. We construct an element $\hat{L} \in \mathcal{L}^n(M)$ in the manner described in the proof of Proposition 2 [BFK-2]: Let $L_{\varepsilon_1,\ldots,\varepsilon_n} \in \mathcal{L}^n(M)$ be obtained from L by inserting n crossing balls in the pattern shown below.

$L_{\varepsilon_1,\ldots,\varepsilon_n} =$ 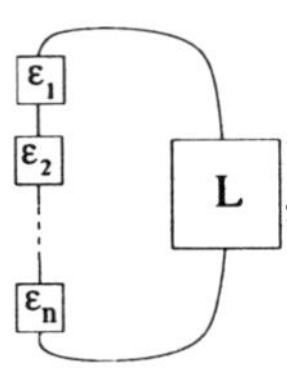 , $\varepsilon_1, \ldots, \varepsilon_n = \pm 1$,

where 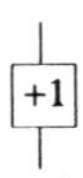 $= -A^3$ 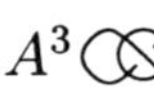, 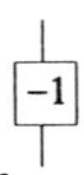 $=$ .

It does not matter which component of L has been selected for this operation. Let $\hat{L} = \sum_{\varepsilon_1,\ldots,\varepsilon_n=\pm 1} L_{\varepsilon_1,\ldots,\varepsilon_n}$. An easy argument (identical to that given in [BFK-2]) shows that $\partial_i \hat{L} = 0$, for $i = 1, 2, \ldots$ Therefore $d_n(\hat{L}) = 0$ and $\hat{L}$ is an n-cycle. Observe that $\phi(L_{\varepsilon_1,\ldots,\varepsilon_n}) = L$ for any choice of $\varepsilon_1, \ldots, \varepsilon_n = \pm 1$. Thus $\phi(\frac{1}{2^n}\hat{L}) = L \in K_0(M)$. This implies that ϕ is an epimorphism. □

There is yet another method of constructing n-cycles in $K_n(M)$. Observe that the symmetric group S_n acts on the space $\mathcal{L}^n(M)$ by permuting the crossing balls in each link. Let $L^{sym} = \Sigma_{\pi \in S_n} \pi L$ be the symmetrization of a link L, with n crossing balls. Similarly, we can define L^{cyc} to be $\Sigma_{\pi \in Z_n} \pi L$, where $Z_n \subset S_n$ is generated by the cycle $(1, 2, \ldots, n)$.

Proposition 3. (i) L^{sym} *is an* n*-cycle for* $n > 1$.
(ii) L^{cyc} *is an* n*-cycle for any even* n.
(iii) $\phi(L^{sym}) = n! L \in K_0(M)$

(iv) $\phi(L^{cyc}) = nL \in K_0(M)$. □

Example 4. *Let L be a positive trefoil with three balls, where the balls are placed around the three crossings in the alternating projection with the blackboard framing, and ordered in an arbitrary way. An easy calculation shows that L is a* 3*-cycle* [2]. *Since* $\phi(L) \neq 0$, *L represents a nontrivial element of* $K_3(S^3)$.

We have an exact sequence

$$0 \to Ker\ \phi \to K_n(S^3) \xrightarrow{\phi} K_0(S^3) \to 0.$$

Since $K_0(S^3) = R$ is a free R-module, the sequence splits and R is a direct summand of $K_n(S^3)$, $K_n(S^3) = R \oplus ker\ \phi$. The next example shows that $ker\ \phi$ is non-trivial for $n = 2$.

Example 5. *Let* $L =$ 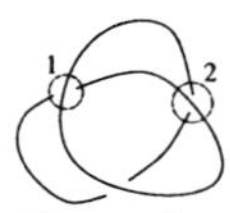 $+ A^3$ 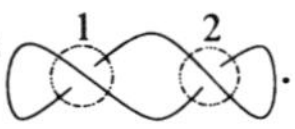.

An elementary calculation shows that L is a 2-cycle and that $\phi(L) = 0 \in K_0(M)$. However, L is a non-zero element of the 2-nd skein homology of S^3. This can be shown by considering a modification of ϕ defined as follows. Let $\psi : R\mathcal{L}^n(S^3) \to \mathcal{L}(S^3)$ be the map replacing each crossing ball with the opposite crossing (without the ball). Exactly as before, ψ induces a homomorphism from $K_n(S^3)$ to $K_0(S^3)$. Moreover, it is easy to see that $\psi(L) \neq 0$.

In this paper we have studied an ordered skein homology. Alternatively, one can consider an oriented skein homology, constructed out of links with crossing balls ordered up to an even permutation. We do not know if these homology theories coincide.

The methods of distinguishing nontrivial cycles presented in this paper can be sharpened by considering refinements of the Kauffman bracket skein module based on singular links.

References

[B] D. Bullock, *Rings of $SL_2(\mathbb{C})$-characters and the Kauffman bracket skein module*, Comment. Math. Helv. **72** (1997), 521–542.

[HP] J. Hoste, J. H. Przytycki, *A survey of skein modules of 3-manifolds*, in *Knots 90*, Proceedings of the International Conference on Knot Theory and Related Topics, Osaka (Japan), August 15-19, 1990, Editor A. Kawauchi, Walter de Gruyter 1992, 363–379.

[BFK-1] D. Bullock, C. Frohman, J. Kania-Bartoszyńska, *Understanding the Kauffman bracket skein module*, JKTR **8** (1999), 265–277.

[2] If 6 is not a zero divisor in R then this also follows from Proposition 3 applied to L, since $L^{sym} = 6L$.

[BFK-2] D. Bullock, C. Frohman, J. Kania-Bartoszyńska, *Skein Homology*, Canad. Math. Bull. **41** (1998), 140–144.

[P-1] J. H. Przytycki, *Skein modules of 3-manifolds*, Bull. Pol. Acad. Sci. **39** (1991), 91–100.

[P-2] J. H. Przytycki, *Algebraic topology based on knots: an introduction*, in: *Knots 96*, Proceedings of the Fifth International Research Institute of MSJ, edited by S. Suzuki, World Scientific Publishing Co., 1997, 279–297.

[PS] J. H. Przytycki, A. S. Sikora, *On skein algebras and $SL_2(\mathbb{C})$-character varieties*, Topology **39(1)** (2000), 115–148.

Department of Mathematics, Boise State University, Boise, ID 83725, USA

E-mail address: kania@math.idbsu.edu

Department of Mathematics, George Washington University, Washington, D. C. 20052 and University of Maryland, College Park, MD 20742

E-mail address: przytyck@gwu.edu

Department of Mathematics, University of Maryland, College Park, MD 20742

E-mail address: asikora@math.umd.edu

A Survey of Virtual Knot Theory

Louis H. Kauffman
Department of Mathematics, Statistics and Computer Science
University of Illinois at Chicago
851 South Morgan Street
Chicago, IL, 60607-7045

Abstract

This paper is a survey of the theory of virtual knots. It is a modified version of the lectures given by the author in the "Knots in Hellas" meeting in Delphi, Greece.

1 Introduction

This paper is a survey of virtual knot theory, a generalisation of classical knot theory [5], [6]. This paper gives the basic definitions, some fundamental properties,a collection of examples and discussion of the work that I and other people have so far contributed to this idea. This paper is an expanded version of the lectures that I gave at the "Knots in Hellas" conference in the summer of 1998. It is also an expanded version of the paper [6] by this author that introduces this theory of virtual knots.

Throughout this paper I shall refer to knots and links by the generic term "knot". In referring to a trivial fundamental group of a knot, I mean that the fundamental group is isomorphic to the integers.

The paper is organised as follows. Section 2 gives the definition of a virtual knot in terms of diagrams and moves on diagrams. Section 3 discusses both the motivation from knots in thickened surfaces and the abstract properties of Gauss codes. Section 3 proves basic results about virtual knots by using reconstruction properties of Gauss codes. In particular, we show how virtual knots can be identified as virtual by examining their codes. Section 4 discusses the fundamental group and the quandle extended for virtual knots. Examples are given of non-trivial virtual knots with trivial (isomorphic to the integers) fundamental group. An example shows that some virtual knots are distinguished from their mirror images by the fundamental group, a very non-classical effect. Section 5 shows how the bracket polynomial (hence the Jones polynomial) extends naturally to virtuals and gives examples of non-trivial virtual knots with trivial Jones polynomial. Examples of infinitely many distinct virtuals with the same fundamental group are verified by using the bracket polynomial. An example is given of a knotted virtual with trivial fundamental group and unit Jones polynomial. It is conjectured that this phenomenon cannot happen with virtuals whose shadow code is classical. In section 6 we show how to extend quantum link invariants and introduce the concept of virtual framing. This yields a virtually framed bracket polynomial distinct from the model in the previous section and to generalisation of this model to an invariant, $\overline{Z}(K)$, of virtual regular isotopy depending on infinitely many variables. Section 6 also introduces an example (See Figure 21) of a virtual knot that we are unable to prove to be non-trivial in the original sense of ignoring virtual framing. In fact the flat projection of this virtual is not yet proved to be non-trivial under the (unframed) flat moves defined in this section. This example, and its relatives, provides a challenge to the development of this theory. Section 7 discusses Vassiliev invariants, defines graphical finite type and proves that the weight systems are finite for the virtual Vassiliev invariants arising from the Jones polynomial. Section 8 discusses an application to Vassiliev invariants via semivirtual crossings and the Polyak algebra that was pioneered by Goussarov, Polyak and Viro in [19]. We show how the concept of semivirtual is motivated by the structure of the R-matrix in a quantum link invariant. Section 9 is a discussion of open problems.

Acknowledgment. It gives the author pleasure to thank the National Science Foundation for support of this research under NSF Grant DMS-9205277,the NSA for partial support under grant number MSPF-96G-179 and the Mittag-Leffler Institute for hospitality during the writing of this paper.

2 Defining Virtual Knots and Links

A classical knot [1] can be represented by a diagram. The diagram is a 4-regular plane graph with extra structure at its nodes. The extra structure is classically intended to indicate a way to embed a circle in three dimensional space. The shadow of a projection of this embedding is the given plane graph. Thus we are all familiar with the usual convention for illustrating a crossing by omitting a bit of arc at the node of the plane graph. The bit omitted is understood to pass underneath the uninterrupted arc. See Figure 1 .

From the point of view of a topologist, a knot diagram represents an "actual" knotted (possibly unknotted) loop embedded in three space. The crossing structure is an artifact of the projection to the plane.

I shall define a virtual knot (or link) diagram. The definition of a virtual diagram is just this: We allow a new sort of crossing, denoted as shown in Figure 1 as a 4-valent vertex with a small circle around it.

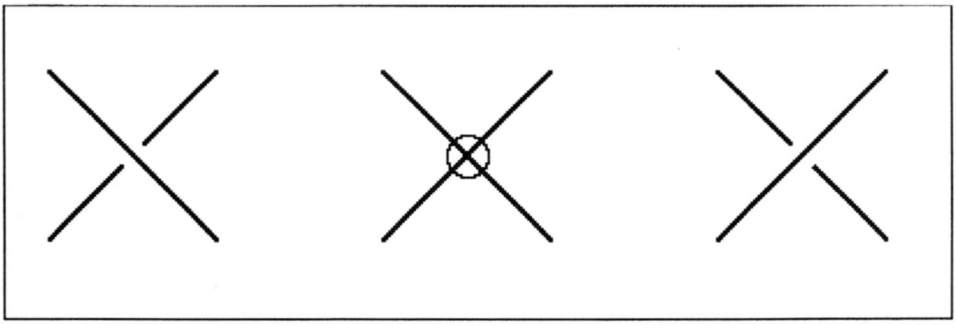

Figure 1 — Crossings and Virtual Crossings

This sort of crossing is called virtual. It comes in only one flavor. You cannot switch over and under in a virtual crossing. However the idea is not that a virtual crossing is just an ordinary graphical vertex. Rather, the idea is that the virtual crossing is not really there.

If I draw a non-planar graph in the plane it necessarily acquires virtual crossings. These crossings are not part of the structure of the graph itself.

They are artifacts of the drawing of the graph in the plane. The graph theorist often gets rid of a crossing in the plane by making it into a knot theorist's crossing, thereby indicating a particular embedding of the graph in three dimensional space. This is just what we do not do with our virtual knot crossings, for then they would be indistinct from classical crossings. The virtual crossings are not there. We shall make sense of that property by the following axioms generalising classical Reidemeister moves. See Figure 2.

The moves fall into three types: (A) Classical Reidemeister moves relating classical crossings; (B) Shadowed versions of Reidemeister moves relating only virtual crossings; (C) A triangle move that relates two virtual crossings and one classical crossing.

The last move (type C) is the embodiment of our principle that the virtual crossings are not really there. Suppose that an arc is free of classical crossings. Then that arc can be arbitrarily moved (holding its endpoints fixed) to any new location. The new location will reveal a new set of virtual crossings if the arc that is moved is placed transversally to the remaining part of the diagram. See Figure 3 for illustrations of this process and for an example of unknotting of a virtual diagram.

The theory of virtual knots is constructed on this combinatorial basis - in terms of the generalised Reidemeister moves. We will make invariants of virtual knots by finding functions well-defined on virtual diagrams that are unchanged under the application of the virtual moves. The remaining sections of this paper study many instances of such invariants.

3 Motivations

While it is clear that one can make a formal generalisation of knot theory in the manner so far described, it may not be yet clear why one should generalize in this particular way. This section explains two sources of motivation. The first is the study of knots in thickened surfaces of higher genus (classical knot theory is actually the theory of knots in a thickened two-sphere). The second is the extension of knot theory to the purely combinatorial domain of Gauss

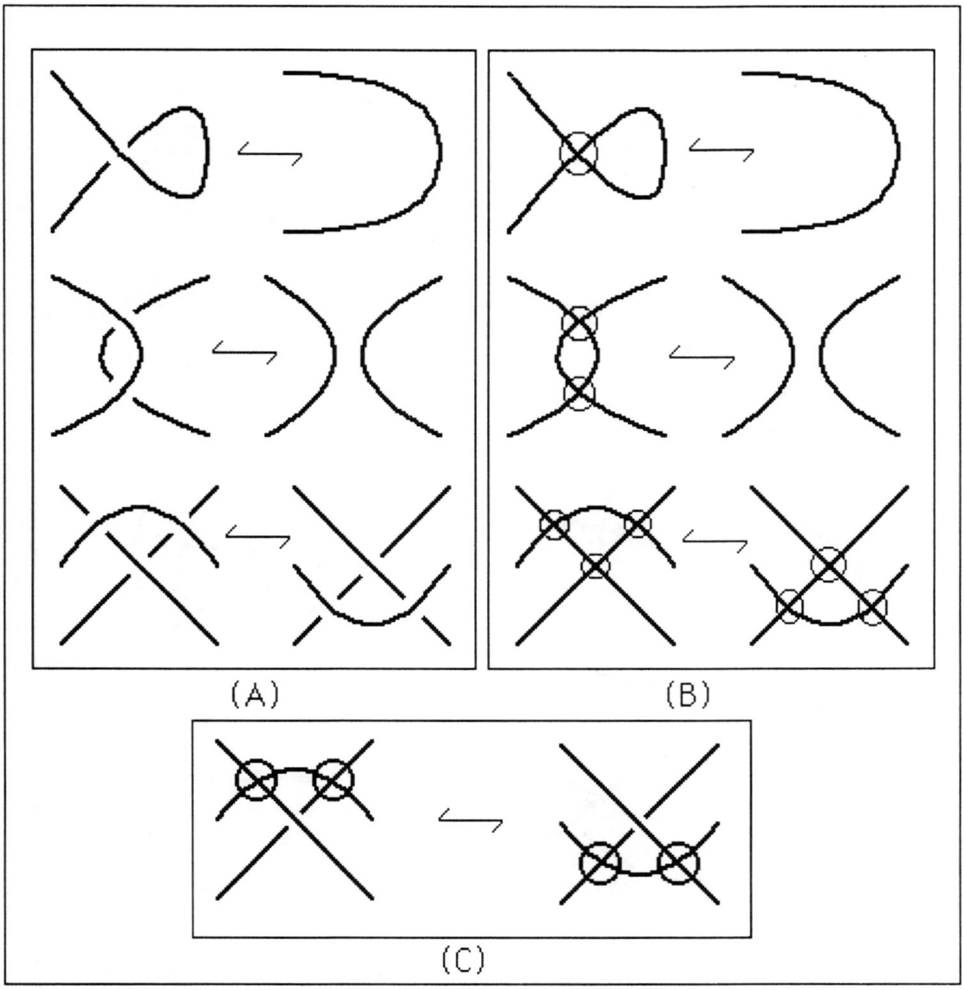

Figure 2 — Generalised Reidemeister Moves for Virtual Knot Theory

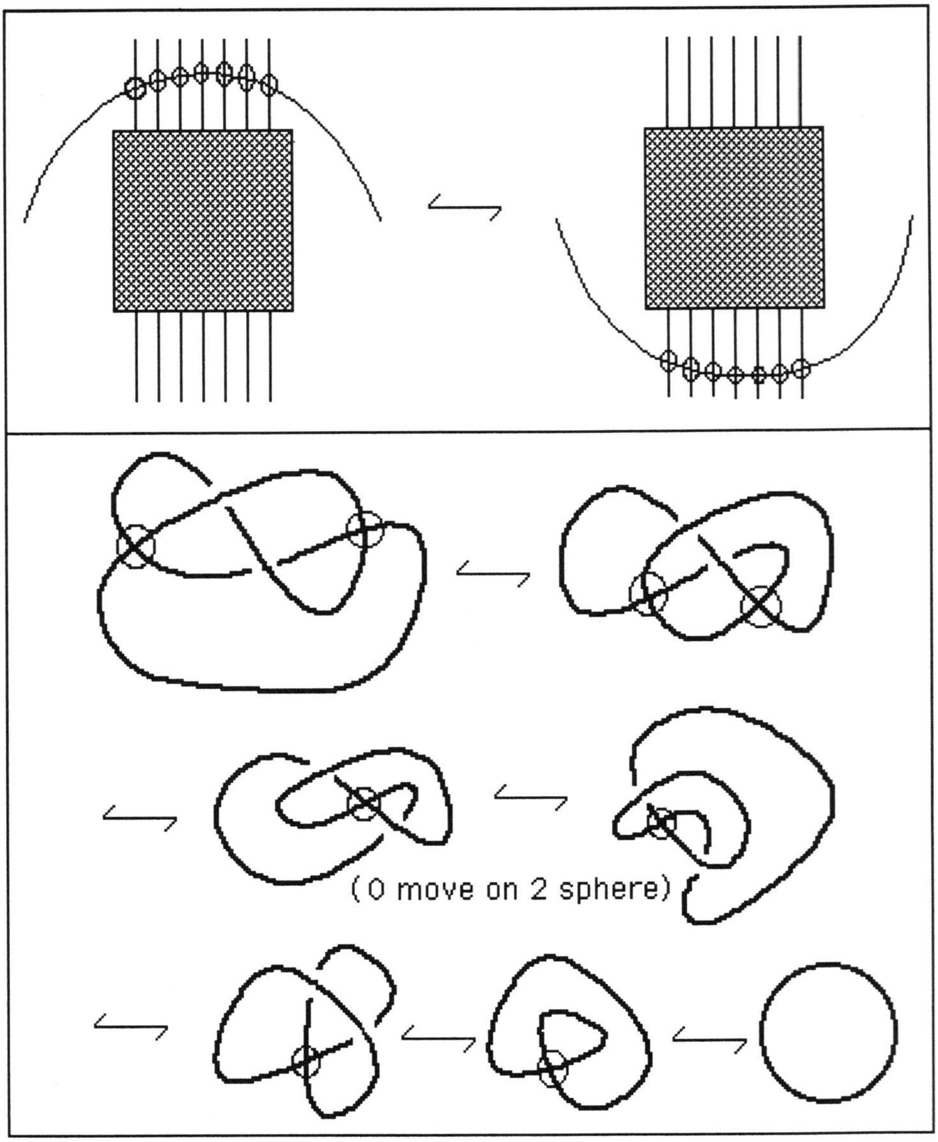

Figure 3 — Virtual Moves

codes and Gauss diagrams. It is in this second domain that the full force of the virtual theory comes into play.

3.1 Surfaces

Consider the two examples of virtual knots in Figure 4. We shall see later in this paper that these are both non-trivial knots in the virtual category. In Figure 4 we have also illustrated how these two diagrams can be drawn (as knot diagrams) on the surface of a torus. The virtual crossings are then seen as artifacts of the projection of the torus to the plane.

The knots drawn on the toral surface represent knots in the three manifold $T \times I$ where I is the unit interval and T is the torus. If S_g is a surface of genus g, then the knot theory in $S_g \times I$ is represented by diagrams drawn on S_g taken up to the usual Reidemeister moves transferred to diagrams on this surface.

As we shall see in the next section, abstract invariants of virtual knots can be interpreted as invariants for knots that are specifically embedded in $S_g \times I$ for some genus g. The virtual knot theory does not demand the use of a particular surface embedding, but it does apply to such embeddings. This constitutes one of the motivations.

3.2 Gauss Codes

A second motivation comes from the use of so-called *Gauss codes* to represent knots and links. The Gauss code is a sequence of labels for the crossings with each label repeated twice to indicate a walk along the diagram from a given starting point and returning to that point. In the case of multiple link components we mean a sequence labels, each repeated twice and intersticed by partition symbols "/" to indicate the component circuits for the code.

A *shadow* is the projection of a knot or link on the plane with transverse self-crossings and no information about whether the crossings are overcrossings or undercrossings. In other words, a shadow is a 4-regular plane graph. On such a graph we can count circuits that always cross (i.e., they never use

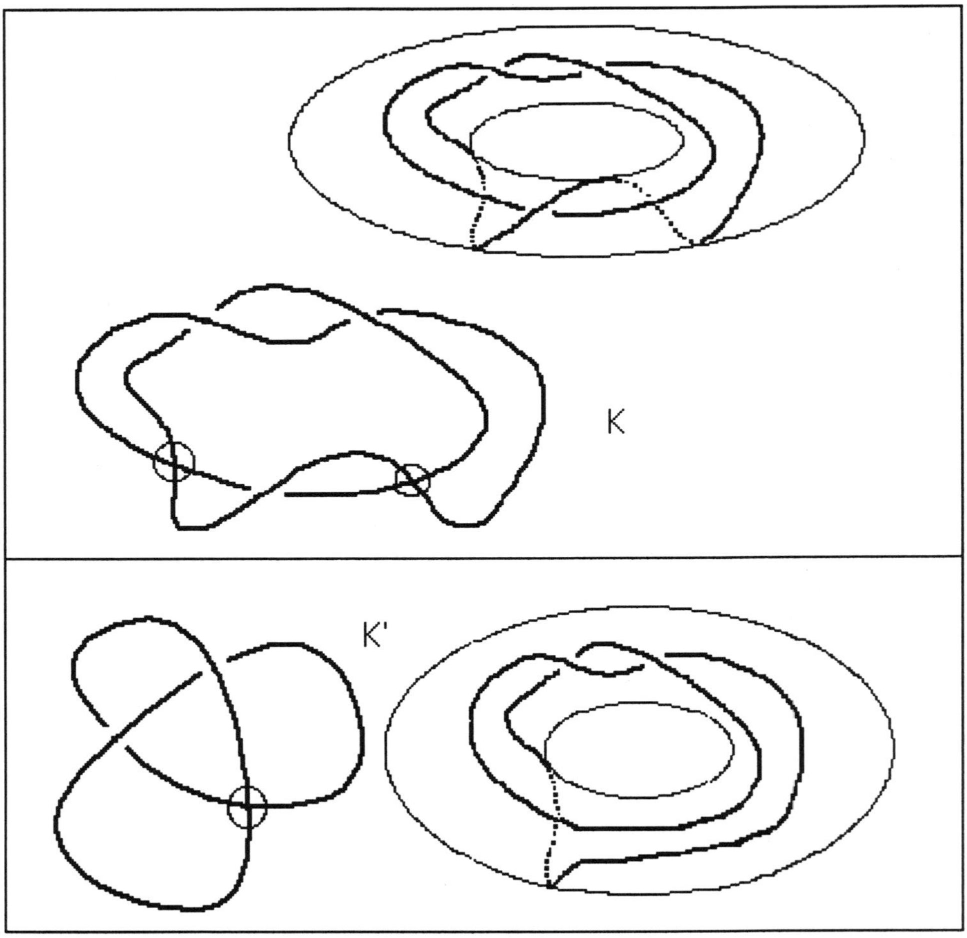

Figure 4 — Two Virtual Knots

two adjacent edges in succession at a given vertex) at each crossing that they touch. Such circuits will be called the *components* of the shadow since they correspond to the components of a link that projects to the shadow.

A single component shadow has a Gauss code that consists in a sequence of crossing labels, each repeated twice. Thus the trefoil shadow has code 123123. A multi-component shadow has as many sequences as there are components. For example 12/12 is the code for the Hopf link shadow.

Along with the labels for the crossings one can add the symbols O and U to indicate that the passage through the crossing was an overcrossing (O) or an undercrossing (U). Thus

$$123123$$

is a Gauss code for the shadow of a trefoil knot and

$$O1U2O3U1O2U3$$

is a Gauss code for the trefoil knot.

The Hopf link itself has the code $O1U2/U1O2$. See Figure 5.

Suppose that g is such a sequence of labels and that g is free of any partition labels. Every label in g appears twice. The first necessary criterion for the planarity of g is given by the following definition and Lemma.

Definition. A single component Gauss code g is said to be *evenly intersticed* if there is an *even* number of labels in between the two appearances of any label.

Lemma 1. If g is a single component planar Gauss code, then g is evenly intersticed.

Proof. This follows directly from the Jordan curve theorem in the plane.//

Example. The necessary condition for planarity in this Lemma is not sufficient. The code $g = 1234534125$ is evenly intersticed but not planar as is evident from Figure 5.

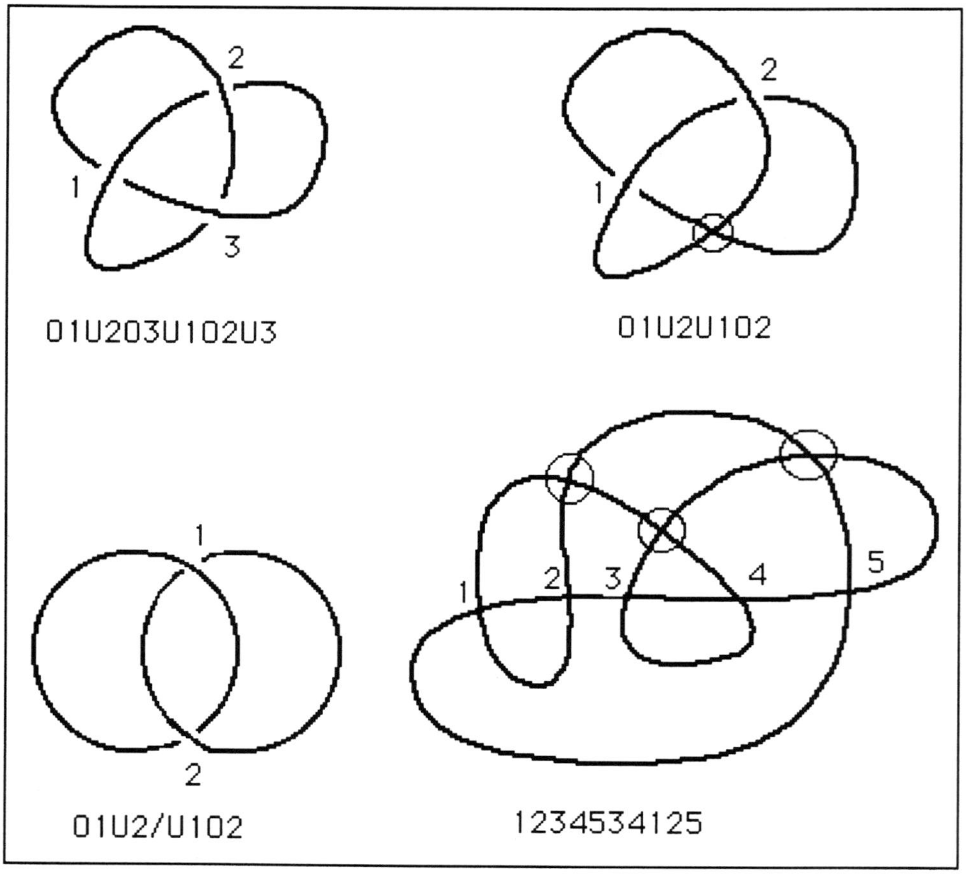

Figure 5 — Planar and Nonplanar Codes

Non-planar Gauss codes give rise to an infinite collection of virtual knots.

Local orientations at the crossings give rise to another phenomenon: virtual knots whose Gauss codes have planar realisations with different local orientations from their classical counterparts.

By orienting the knot, one can give orientation signs to each crossing relative to the starting point of the code—using the convention shown in Figure 6. This convention designates each oriented crossing with a *sign* of $+1$ or -1. We say that the crossing has positive sign if the overcrossing line can be turned through the smaller angle (of the two vertical angles at the crossing) to coincide with the direction of the undercrossing line. The signed code for the standard trefoil is

$$t = O1 + U2 + O3 + U1 + O2 + U3+,$$

while the signed code for a figure eight knot is

$$f = O1 + U2 + O3 - U4 - O2 + U1 + O4 - U3 - .$$

Here we have appended the signs to the corresponding labels in the code. Thus, crossing number 1 is positive in the figure eight knot, while crossing number 4 is negative. See Figure 6 for an illustration corresponding to these codes.

Now consider the effect of changing these signs. For example let

$$g = O1 + U2 + O3 - U1 + O2 + U3 - .$$

Then g is a signed Gauss code and as Figure 6 illustrates, the corresponding diagram is forced to have virtual crossings in order to accommodate the change in orientation. The codes t and g have the same underlying (unsigned) Gauss code $O1U2O3U1O2U3$, but g corresponds to a virtual knot while t represents the classical trefoil.

Definition of Virtuality. Carrying this approach further, we *define* a virtual knot as an equivalence class of oriented Gauss codes under abstractly defined Reidemeister moves for these codes—with no mention of

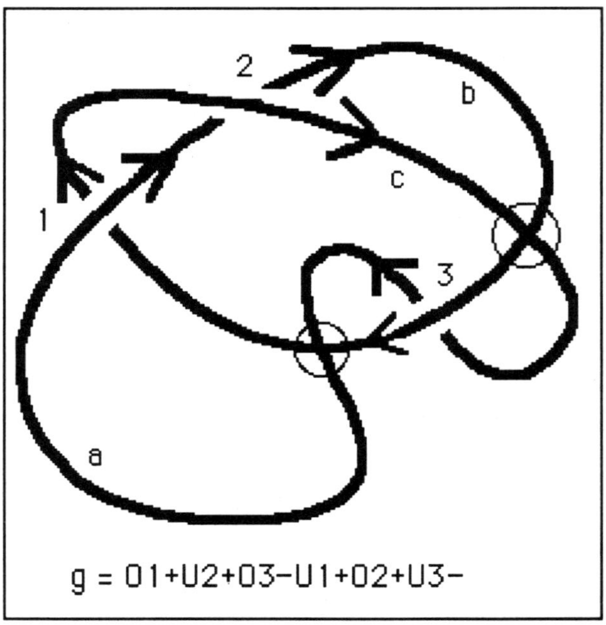

Figure 6 — Signed Gauss Codes

virtual crossings. (We omit here the enumeration of the abstract moves on the Gauss codes. They are easily obtained by translation from the diagrams of the oriented Reidemeister moves.) The virtual crossings become artifacts of a planar representation of the virtual knot. The move sets of type B and C for virtuals are diagrammatic rules that make sure that this representation of the oriented Gauss codes is faithful. Note, in particular, that the move of type C does not alter the Gauss code. With this point of view we see that the signed codes are knot theoretic analogues of the set of all graphs, and that the classical knot (diagrams) are the analogues of the planar graphs. This is the fundamental combinatorial motivation for our definitions of virtual knots and their equivalences.

Since it is useful to have a few more facts about the reconstruction of planar Gauss codes, we conclude this section with a quick review of that subject.

3.3 Gauss Codes and Reconstruction

In this section we recall an algorithm for reconstructing a planar diagram from its Gauss code. This algorithm also detects non-planar codes. We shall see that for a planar oriented Gauss code, the orientation signs in the code sequence are determined up to a small number of choices. Such sign sequences will be called *standard* (with the more technical definition to follow).

We shall prove the following Theorem.

Theorem 2. If K is a virtual knot whose underlying Gauss code is planar and whose sign sequence is standard, then K is equivalent to a classical knot.

The fundamental problem in Gauss codes is to give an algorithm for determining whether a given code can be realized by a planar shadow.

We will explain the detection and reconstruction algorithms for single component codes. The first necessary condition for planarity for a single component code is that it be evenly intersticed, as we have already remarked in Lemma 1.

If a code g is planar then a corresponding code for such an evenly paired

Jordan curve can be produced as follows: Let the labels in g be $1, 2, ..., n$. Starting with $i = 1$ *reverse the order of labels in between the two appearances of* i. Do this successively using $i = 1, 2, ..., n$. Let g^* be the resulting code.

In Figure 7 we see that the crossings of a planar shadow E can be smoothed to obtain a single Jordan curve in the plane. This Jordan curve can be seen as a circle with doubly repeated labels around its circumference so that some labels are paired by arcs inside the circle, and the remaining labels are paired by arcs outside the circle. The corresponding code is g^* as defined above. In this form of pairing *no two pairing arcs intersect one another.*

Remark. In the case of multiple component codes the algorithm for constructing g^* is modified as follows: Suppose that the two appearances of i occur in different components of the code, so that the code to be modified has the form

$$h = i\alpha/i\beta/R$$

where we have written the two components as adjacent code segments and started each with i (possible by rearrangement and cyclic permutation of the segments). Here R denotes the rest of the code sequences. Then replace h by

$$h' = i\alpha i\overline{\beta}/R$$

where $\overline{\beta}$ denotes the rewrite of β in reverse order. Note that the two components are amalgamated into one as a result of this process. Thus after applying this procedure successively to the labels in the code, we obtain a single code sequence g^* from a given multi-component code sequence g. For example, if $g = 1234/1536/2546$ then we get the following sequence of partial codes on the way to g^* :

$$g = 1234/1536/2546$$

$$g' = 12341635/2546 = 23416351/2546$$

$$g'' = 234163512645$$

$$g''' = 236143512645$$

$$g'''' = 236146215345$$

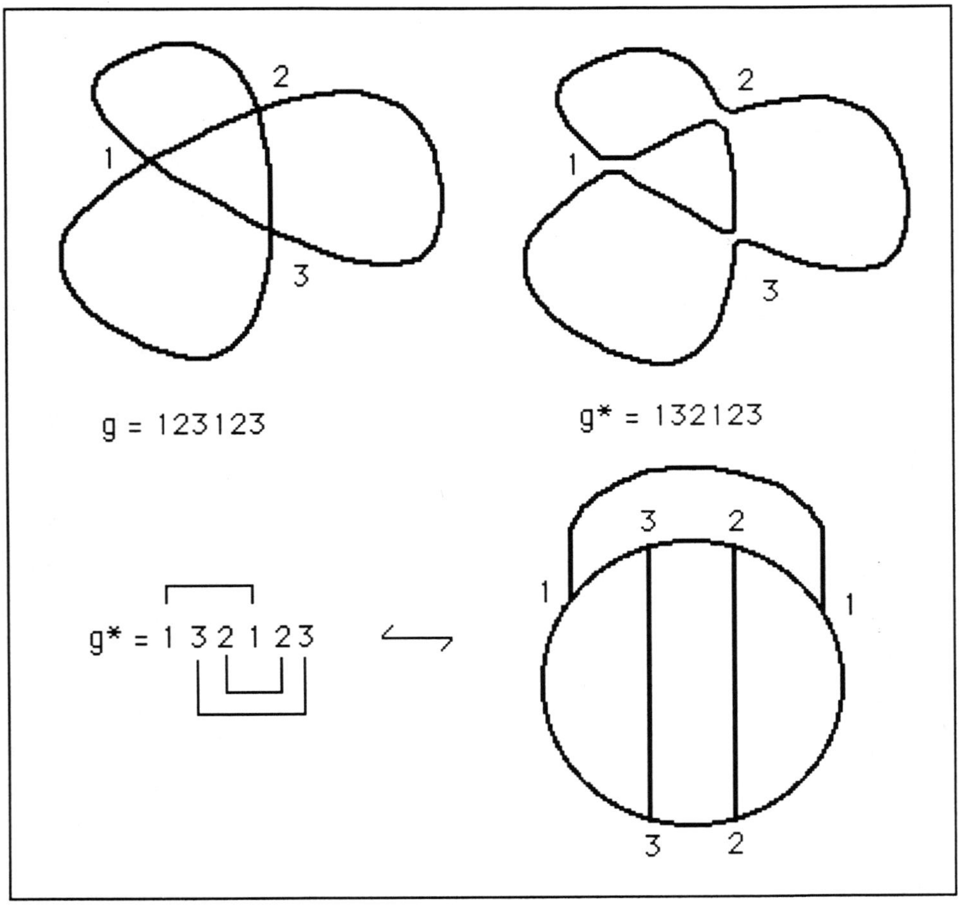

Figure 7 — If g is Planar, then g^* is Dually Paired.

$$g''''' = 236146215435$$

$$g^* = 236416215435$$

We leave it for the reader to check that g^* is dually paired.

We have the Lemma below.

Lemma 3. If g is a planar Gauss code, then g^* is dually paired.

Proof. The (easy) proof is omitted. See [9].

Figure 7 illustrates this situation and shows how the desired pairing can be written directly on the code g^* by pairing labels above and below the typographical line.

Lemma 4. If an evenly intersticed Gauss code g has g^* dually paired, then g is the Gauss code of a planar shadow.

Proof. Figure 8 shows how to reconstruct a shadow from any g satisfying the hypotheses of the Lemma.

These lemmas form the essentials of the reconstruction theory for planar Gauss codes.

Definition. A Gauss code g is said to be *prime* if it cannot be written as the juxtaposition of two Gauss codes on disjoint collections of labels. A non-prime code is said to be *composite.* For example, 123123 is prime but 121234543 is composite since it is the juxtaposition of 1212 and 34543.

In reconstructing a shadow from a Gauss code there is a choice of local orientation of the first crossing in the code. From then on the local orientations are determined by the reconstruction algorithm. See Figure 8 for an example of the procedure. Once we specify the local orientations in the code, the corresponding signs of the crossings are determined by whether there is an O or a U in the code. Thus up to these initial choices of orientation, the signs in a O/U code are all determined if the code is planar. It is this result that gives the proof of Theorem 2.

Proof of Theorem 2. Note that the reconstruction algorithm will give a planar embedding for this code with the same local orientations as those

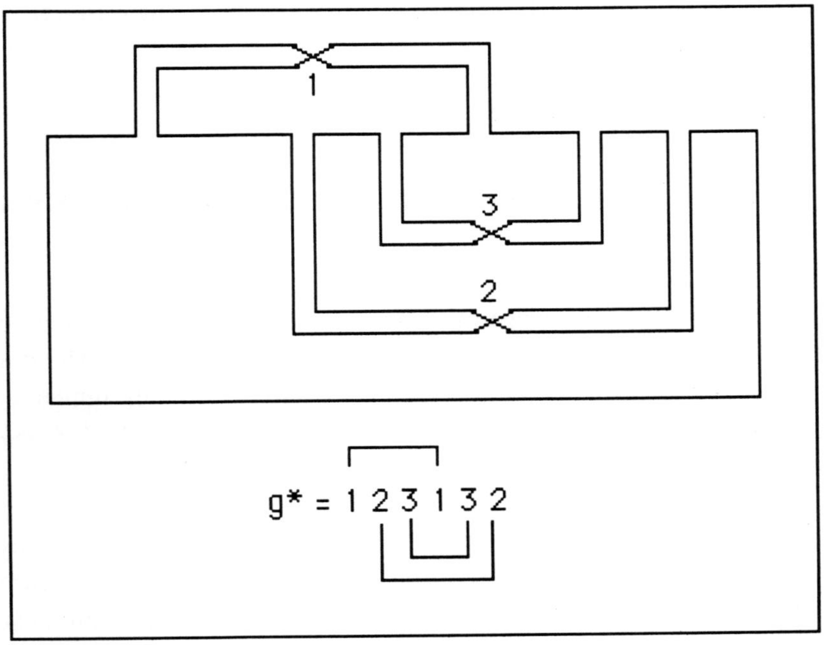

Figure 8 — Reconstruction from a Gauss Code

specified in the virtual diagram. In fact, we can assume that the planar positions of the crossings in the embedded diagram and the virtual diagram are identical (up to a global translation if comparison is desired). Now locate those arcs in the original diagram that involve virtual crossings and move them one-by-one into the positions indicated by the embedding. To accomplish this, start at the beginning of the code. Say the code reads $g = a_1 a_2 ... a_i a_{i+1} ... a_n$. In the virtual diagram there may be a series of virtual crossings between a_1 and a_2 but there will be no real crossings since the code is given by g. Therefore, the arc from a_1 to a_2 can be replaced (by virtual equivalence) to its position in the embedded diagram. Continue this process sequentially for $a_i a_{i+1}$ and the result is an equivalence through the virtual category of the original diagram with the embedded classical diagram. This completes the proof of Theorem 2. //

4 Fundamental Group, Crystals, Racks and Quandles

The fundamental group of the complement of a classical knot can be described by generators and relations, with one generator for each arc in the diagram and one relation for each crossing. The relation at a crossing depends upon the type of the crossing and is either of the form $c = b^{-1}ab$ or $c = bab^{-1}$ as shown in Figure 9.

We define the group $G(K)$ of an oriented virtual knot or link by this same scheme of generators and relations. An *arc* of a virtual diagram proceeds from one classical undercrossing to another (possibly the same) classical undercrossing. Thus no new generators or relations are added at a virtual crossing. It is easy to see that $G(K)$ is invariant under all the moves for virtuals and hence is an invariant of virtual knots.

There are virtual knots that are non-trivial but have trivial fundamental group. (We say that the fundamental group of a knot is trivial if it is isomorphic to the infinite cyclic group.) The virtual K' in Figure 4 is such an example.. We shall show that K' is a non-trivial virtual in the next section by using a generalisation of the bracket polynomial.

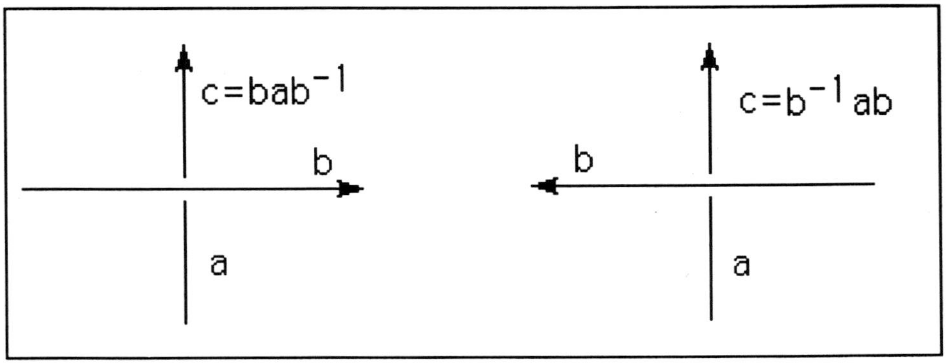

Figure 9 — Generators and Relations for the Fundamental Group

A generalization of the fundamental group called the quandle, rack or crystal (depending on notations and history) also assigns relations (in a different algebra) to each crossing. The quandle generalises to the virtual category. We first discuss the involutory quandle, $IQ(K)$, for a (virtual) knot or link K. The $IQ(K)$ does not depend upon the local orientations of the diagram and it assigns to each crossing the relation $c = a * b$ as in Figure 10.

The operation $a*b$ is a non-associative binary operation on the underlying set of the quandle, and it satisfies the following axioms:

1. $a * a = a$ for all a.
2. $(a * b) * b = a$ for all a and b.
3. $(a * b) * c = (a * c) * (b * c)$ for all a, b, c.

The algebra under these axioms with generators and relations as defined above is called the involutory quandle, $IQ(K)$. It is easy to see that the $IQ(K)$ is well-defined for K virtual.

An important special case of $IQ(K)$ is the operation $a * b = 2b - a$ where a

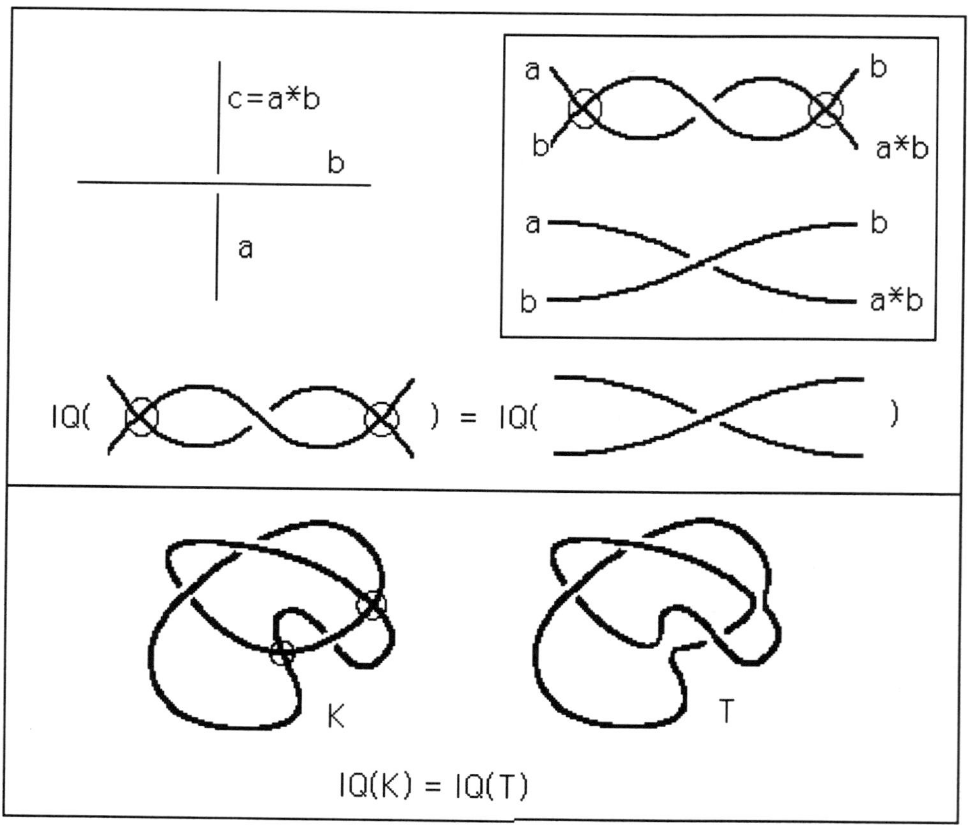

Figure 10 — The Involutory Quandle

and b are elements of a cyclic group Z/nZ for some modulus n. In the case of a classical knot K, there is a natural choice of modulus $D(K) = Det(M(K))$ where $M(K)$ is a minor of the matrix of relations associated with the set of equations $c = 2b - a$. This is called the determinant of the knot, in the classical case. More generally, one can define the determinant for any finite group presentation whose abelianization is the integers (as is the case with virtual knot groups). In our case one takes the greatest common divisor of the set of $(n-1) \times (n-1)$ minors of the $n \times n$ coloring relation matrix associated with the knot. In this way we can define the determinant $|D(K)|$ of any virtual knot. If K is virtual then $|D(K)|$ is an invariant of K. The virtual knot labelled K in Figure 4 has determinant equal to 3. The non-triviality of the determinant shows that this knot is knotted and in fact that it has non-trivial fundamental group. For more information on virtual knot groups see [15].

Another example of an involutory quandle is the operation $a * b = ba^{-1}b$. In classical knot theory this yields the fundamental group of the two-fold branched covering along the knot.

Here is a useful lemma about the IQ for virtuals.

Lemma 5.

$$IQ(K_{vxv}) = IQ(K_{\overline{x}})$$

where x denotes a crossing in the diagram K, vxv denotes that x is flanked by virtuals, and $K_{\overline{x}}$ denotes the diagram obtained by smoothing the flanking virtuals, and switching the intermediate crossing.

In other words the IQ for a classical crossing flanked by two virtual crossings is the same as the IQ of the diagram where the two virtual crossings are smoothed and the classical crossing is switched.

Proof. See Figure 10. //

Remark. In Figure 10 we illustrate that $IQ(K) = IQ(T)$ where K is the virtual knot also shown in Figure 6 and T is the trefoil knot.

Finally we discuss the full quandle of a knot and its generalization to virtuals. For this discussion the exponential notation of Fenn and Rourke [10] is convenient. Instead of $a * b$ we write a^b and assume that there is an operation of order two

$$a \longrightarrow \overline{a}$$

so that

$$\overline{\overline{a}} = a,$$

and for all a and b

$$\overline{a^b} = \overline{a}^b.$$

This operation is well-defined for all a in the underlying set Q of the quandle.

By definition

$$a^{bc} = (a^b)^c$$

for all a,b and c in Q.
The operation of exponentiation satisfies the axioms

1. $a^a = a$
2. $a^{b\overline{b}} = a$
3. $a^{(b^c)} = a^{\overline{c}bc}$

It follows that the set of the quandle acts on itself by automorphisms

$$x \longrightarrow x^a.$$

This group of automorphisms is a representation of the fundamental group of the knot. Note that if we define a^b by the formula

$$a^b = bab^{-1}$$

and

$$\overline{b} = b^{-1},$$

then we get the fundamental group itself as an example of a quandle. The *rack* [10] or *crystal* [11] is obtained by eliminating the first axiom. This makes the rack/crystal an invariant of framed knots and links. The three axioms correspond to invariance under the three Reidemeister moves.

If we now compare Lemma 5 with its possible counterpart for the full fundamental group or the quandle, we see that it no longer holds. Figure 11 shows the new relations in the quandle that are obtained after smoothing the two virtual crossings and switching the classical crossing. While the quandle of the simplified diagram is no longer isomorphic to the original quandle, the fact that we can articulate the change is often useful in computations.

Example. Consider the virtual knot K of Figure 6. We have seen that K has the same IQ as the trefoil knot. However, the quandle and fundamental group of K are distinct from those of the trefoil knot, and K is not equivalent to any classical knot. To see this consider the *Alexander quandle* [11] defined by the equations

$$a^b = ta + (1-t)b$$

and

$$a^{\bar{b}} = t^{-1}a + (1-t^{-1})b.$$

This quandle describes a module (the Alexander module) M over $Z[t,t^{-1}]$. In the case of the virtual knot K in Figure 6, we have the generating quandle relations $a^c = b$, $b^a = c$, $c^{\bar{b}} = a$. This results in the Alexander module relations $b = ta + (1-t)c$, $c = tb + (1-t)a$, $a = t^{-1}c + (1-t^{-1})b$. From this it is easy to calculate that the module $M(K) = \{0, m, 2m\}$ for a non-zero element m with $3m = 0$ and $tm = 2m$. Thus the Alexander module for K is cyclic of order three. Since no classical knot has a finite cyclic Alexander module, this proves that K is not isotopic through virtuals to a classical knot.

It should be remarked that the Alexander polynomial of a knot can be regarded as a generalization of the determinant of a knot. The colors are taken to be elements of the Alexander moddule (See [11] and [14]). Just as with the determinant of a knot, the Alexander polynomial is defined to be a greatest common divisor the the set of $(n-1)\times(n-1)$ minors of the $n \times n$ relation matrix for the Alexander module. Since, as we shall see below, there are actually *two* fundamental quandles associated with a virtual knot (The second,in

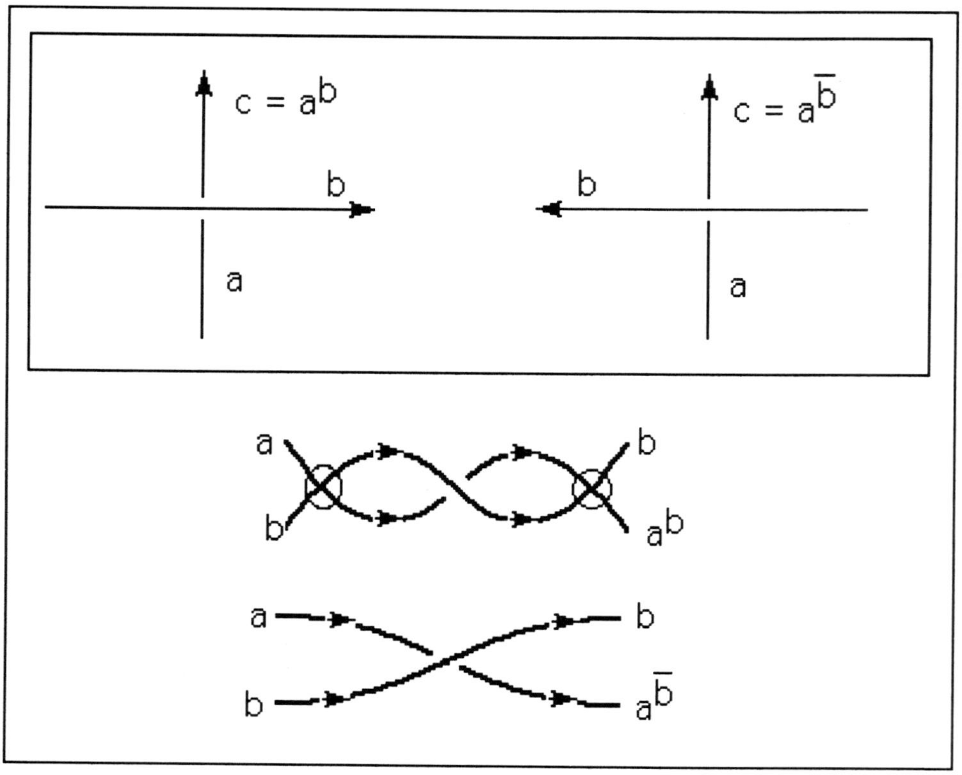

Figure 11 — Change of Relations for the Full Quandle

our language, is the quandle of the mirror image of the original knot.), there are two Alexander polynomials associated with a virtual. There are many questions related to the Alexander polynomials of virtual knots. We shall treat these matters elsewhere.

Finally, it should be remarked that the full quandle $Q(K)$ classifies a classical prime unoriented knot K up to mirror images. By keeping track of a *longitude* for the knot, one gets a complete classification. In the context of the quandle, the longitude can be described as the automorphism

$$\lambda : Q(K) \longrightarrow Q(K)$$

defined by the formula

$$\lambda(x) = x^{a_1^{\epsilon_1} a_2^{\epsilon_2} \ldots a_k^{\epsilon_k}}$$

where $\{a_1, a_2, \ldots, a_k\}$ is an ordered list of quandle generators encountered (as one crosses underneath) as overcrossing arcs as one takes a trip around the diagram. The ϵ denotes whether the generator is encountered with positive or negative orientation, and x^ϵ denotes x if $\epsilon = 1$ and $\overline{x}$ if $\epsilon = -1$. For a given diagram the longitude is well-defined up to cyclic re-ordering of this list of encounters. Exactly the same definition applies to virtual knots. It is no longer true that the quandle plus longitude classifies a virtual knot, as our examples of knotted virtuals with trivial fundamental group show.

On the other hand, we can use the quandle to prove the following result. This proof is due to Goussarov, Polyak and Viro [19].

Theorem 6. If K and K' are classical knot diagrams such that K and K' are equivalent under extended virtual Reidemeister moves, then K and K' are equivalent under classical Reidemeister moves.

Proof. Note that longitudes are preserved under virtual moves (adding virtual crossings to the diagram does not change the expression for a longitude). Thus an isomorphism from $Q(K)$ to $Q(K')$ induced by extended moves preserves longitudes. Since the isomorphism class of the quandle plus longitudes classifies classical knots, we conclude that K and K' are classically equivalent. This completes the proof. //

Remark. We would like to see a purely combinatorial proof of Theorem 6.

4.1 The GPV Example and a Generalisation.

We end this section with a variation of an example [16], [19] that shows that it is possible to have a virtual knot K with $Q(K)$ not isomorphic with $Q(K^*)$ where K^* is the mirror image of K. In other words, there are *two* quandles, or two fundamental groups associated with any given virtual knot!

This example is a slightly different take on an observation in [19]. Let K be a given (virtual) diagram, drawn in the plane. Pick the diagram up and turn it over (note that the crossings change diagrammatically, but correspond to the result of physically turning over the layout of criss-crossing strands with welds at the virtual crossings). Let $Flip(K)$ denote this overturned diagram. Define a new quandle $Q^*(K)$ by the formula $Q^*(K) = Q(Flip(K))$. Goussarov, Polyak and Viro take their "other" fundamental group to be the one defined by generators and relations obtained by "looking at the knot from the other side of the plane." At the quandle level this is the same as taking $Q^*(K)$. It is easy to see that $Q^*(K)$ *is isomorphic to* $Q(K^*)$. (Just note that if $c = a^{\bar{b}}$ then $\bar{c} = \overline{a^{\bar{b}}} = \bar{a}^{b}$. Use this to check that the two quandles are isomorphic through the mapping $a \longrightarrow \bar{a}$ taking one to the other.) Thus our version of this example is mathematically equivalent to the GPV version.

In Figure 12 the reader will find K and K^* with labelled arcs a,b,c,d. In K the quandle relations are $a = b^d$, $b = c^d$, $c = d^b$, $d = a^b$. The three-coloring of K in $Z/3Z$ with $a = 0$, $b = 2$, $c = 0$, $d = 1$ demonstrates that this quandle, and hence the fundamental group of K is non-trivial. On the other hand, K^* has quandle relations $a = a^a$, $c = b^c$, $d = c^c$, $a = d^a$, giving a trivial quandle. Thus K is distinguished from K^* by the quandle. This example also shows that K has non-trivial Alexander polynomial (using the fundamental group to define the Alexander polynomial - there is more than one Alexander polynomial for virtuals) but K^* has Alexander polynomial equal to 1.

We generalise this example by considering the 1–1 tangle W shown in Figure 13. Replacing a straight arc in a knot diagram by W does not affect the quandle, while replacing by its mirror image W^* changes the quandle relations in a generally non-trivial way (as in the GPV example). Insertion of

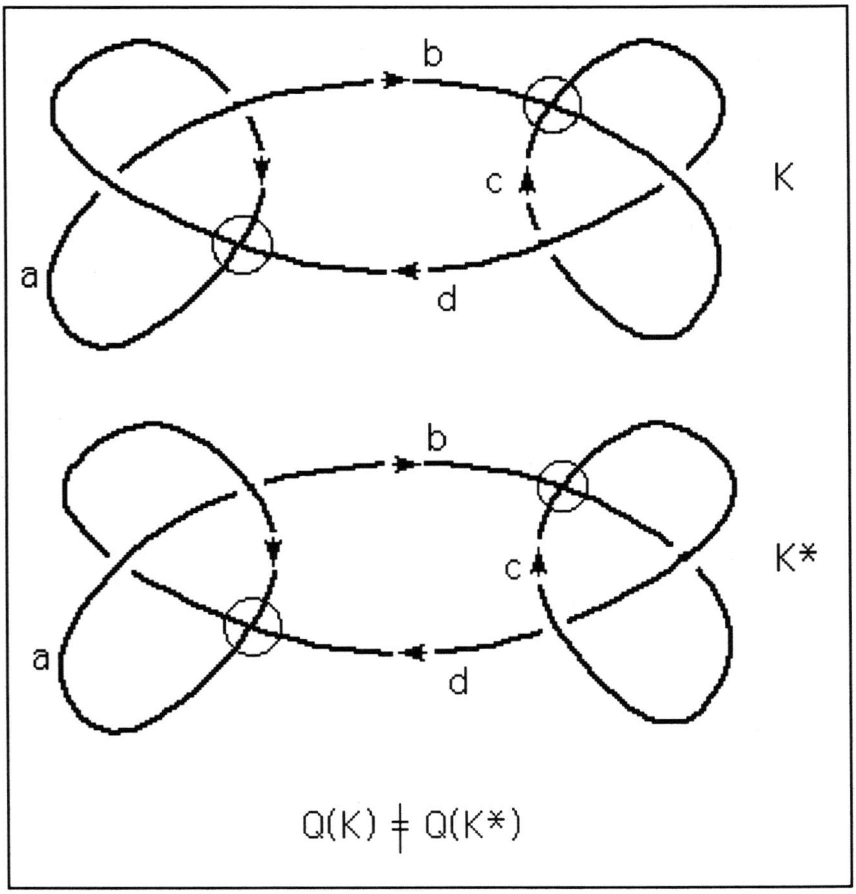

Figure 12 — $Q(K)$ is distinct from $Q(K^*)$.

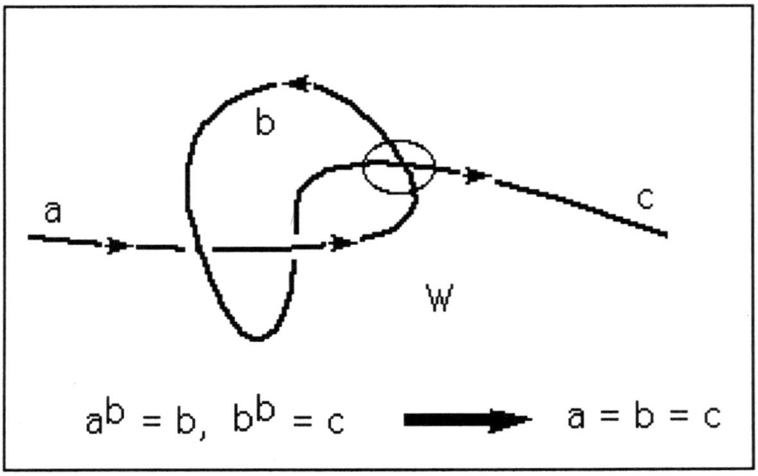

Figure 13 — The 1–1 Tangle W

W into knot diagrams produces infinitely many examples of pairs of virtual knots with the same quandle but different Jones polynomials. This last statement will be verified in the next section.

5 Bracket Polynomial and Jones Polynomial

The bracket polynomial [17] extends to virtual knots and links by relying on the usual formula for the state sum of the bracket, but allowing the closed loops in the state to have virtual intersections. Each loop is still valued at $d = -A^2 - A^{-2}$ and the expansion formula

$$< K >= A < K_a > +A^{-1} < K_b >$$

still holds where K_a and K_b denote the result of replacing a single crossing in K by smoothings of type a and type b as illustrated in Figure 14.

We must check that this version of the bracket polynomial is invariant under all but the first Reidemeister move (See the moves shown in Figure 2). Certainly the usual arguments apply to the moves of type (A). Moves of type

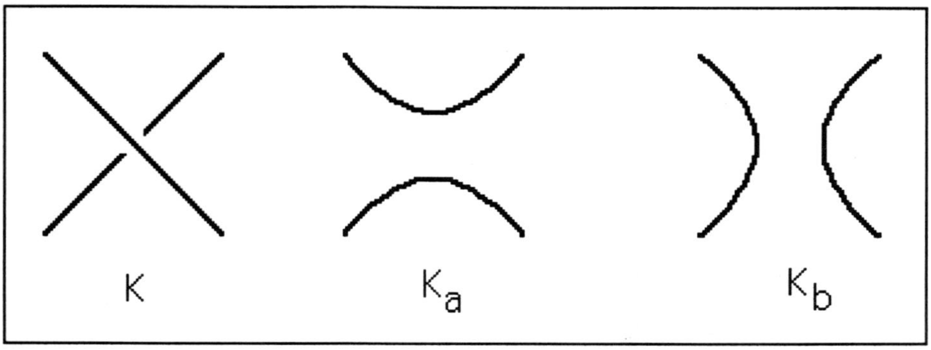

Figure 14 — Bracket Smoothings

(B) do not disturb the loop counts and so leave bracket invariant. Finally the move of type (C) receives the verification illustrated in Figure 15. This completes the proof of the invariance of the generalised bracket polynomial under move (C).

We define the writhe $w(K)$ for an oriented virtual to be the sum of the crossing signs—just as in the classical case.

The f-polynomial is defined by the formula

$$f_K(A) = (-A^3)^{-w(K)} < K > (A).$$

The Laurent polynomial, $f_K(A)$ is invariant under all the virtual moves including the classical move of type I.

Remark. It is worth noting that f_K can be given a state summation of its own. Here we modify the vertex weights of the bracket state sum to include a factor of $-A^{-3}$ for each crossing of positive sign, and a factor of A^{+3} for each factor of negative sign. It is then easy to see that

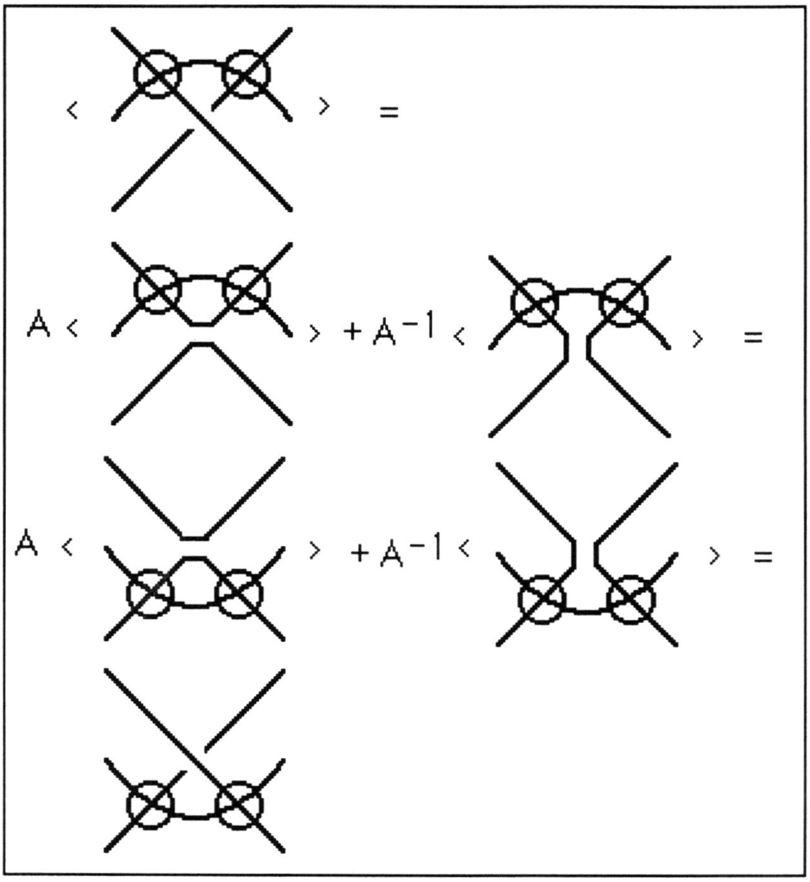

Figure 15 — Type (C) Invariance of the Bracket

$$f_{K_+} = -A^{-2} f_{K_0} - A^{-4} f_{K_\infty}$$
$$f_{K_-} = -A^{+2} f_{K_0} - A^{+4} f_{K_\infty}$$

where K_+ denotes K with a selected positive crossing, K_- denotes the result of switching only this crossing, K_0 denotes the result of making the oriented smoothing of this crossing, and K_∞ denotes the result of making an unoriented smoothing at this crossing. The states in this oriented state sum acquire sites with unoriented smoothings, but the procedure for evaluation is the same as before. For each state we take the product of the vertex weights multiplied by $d^{||S||-1}$ where $d = -A^2 - A^{-2}$ and $||S||$ denotes the number of loops in the state. Then f_K is the sum of these products, one for each state.

The following Lemma makes virtual calculations easier.

Lemma 7. $< K_{vxv} >=< K_x >$ where x denotes a crossing in the diagram K, vxv denotes that x is flanked by virtuals, and K_x denotes the diagram obtained by smoothing the flanking virtuals, and leaving the crossing the same.

Proof. The proof is shown in Figure 16.//

Note that this result has a different form from our corresponding Lemma about the involutory quandle $IQ(K)$. As a result we get an example of a virtual knot that is non-trivial (via the IQ) but has $f_K = 1$. *Hence we have a virtual knot K with Jones polynomial equal to 1.* The example is shown in Figure 17. Note that in Figure 10 we illustrated that this K has the same involutory quandle as the trefoil knot. We will see in Section 6 that K is not equivalent to a classical knot.

We now compute the bracket polynomial for our previous example with trivial fundamental group and we find that $< K' >= A^2 + 1 - A^{-4}$ and $f_{K'} = (-A^3)^{-2} < K' >= A^{-4} + A^{-6} - A^{-10}$. Thus K' has a non-trivial Jones polynomial. See Figure 18.

In Figure 18 we also indicate the result of placing the tangle W, discussed in Figure 13, into another knot or link. Since this is the same as taking a

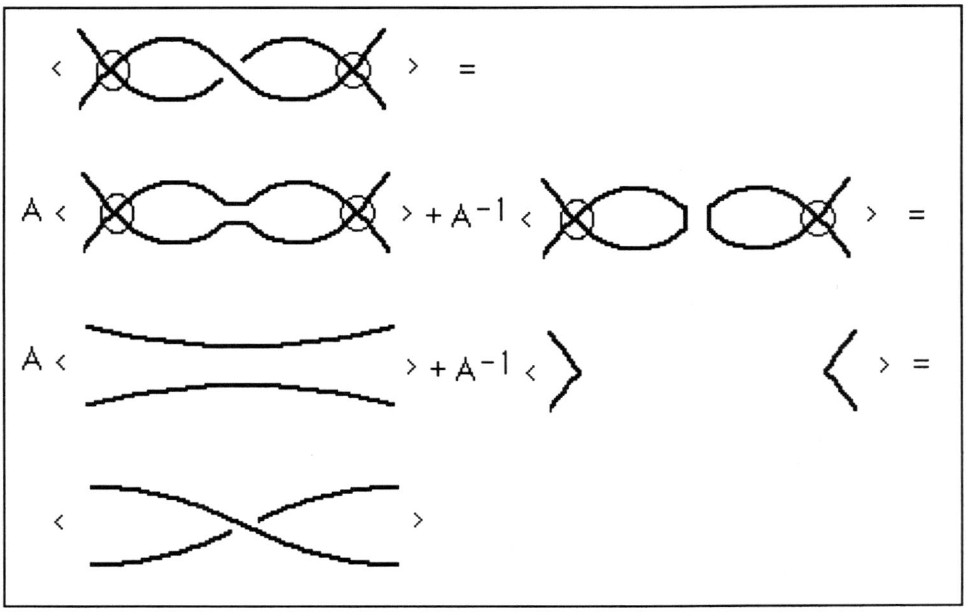

Figure 16 — Removal of Flanking Virtual Crossings

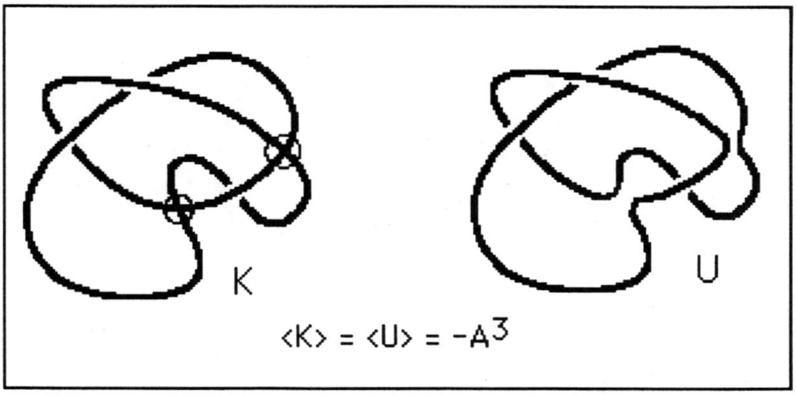

Figure 17 — A Knotted Virtual with Trivial Jones Polynomial

connected sum with K' it has the effect of multiplying the bracket polynomial by A^2+1-A^{-4}. Thus if L is any knot or link and $K'+L$ denotes the connected sum of K' along some component of L then $< K'+L >= (A^2+1-A^{-4}) < L >$ while $Q(K'+L) = Q(L)$ (as we verified in the last section). Thus for any knot L, successive connected sums with K' produces an infinite family of distinct virtual knots, all having the same quandle (hence same fundamental group).

Finally, we note that if the knot is given as embedded in $S_g \times I$ for a surface of genus g, and if its virtual knot diagram K is obtained by projecting the diagram on S_g into the plane, then $< K >$ computes the value of the extension of the bracket to the knots in $S_g \times I$ where all the loops have the same value $d = -A^2 - A^{-2}$. This is the first order bracket for link diagrams on a surface.

In Figure 4 we illustrated the non-trivial knot K with trivial Jones polynomial as embedded in $S_1 \times I$. This knot in $S_1 \times I$ is actually not trivial as can be seen from the higher Jones polynomials that discriminate loops in different isotopy classes on the surface.

In Figure 19 is another example of a virtual knot E and a corresponding

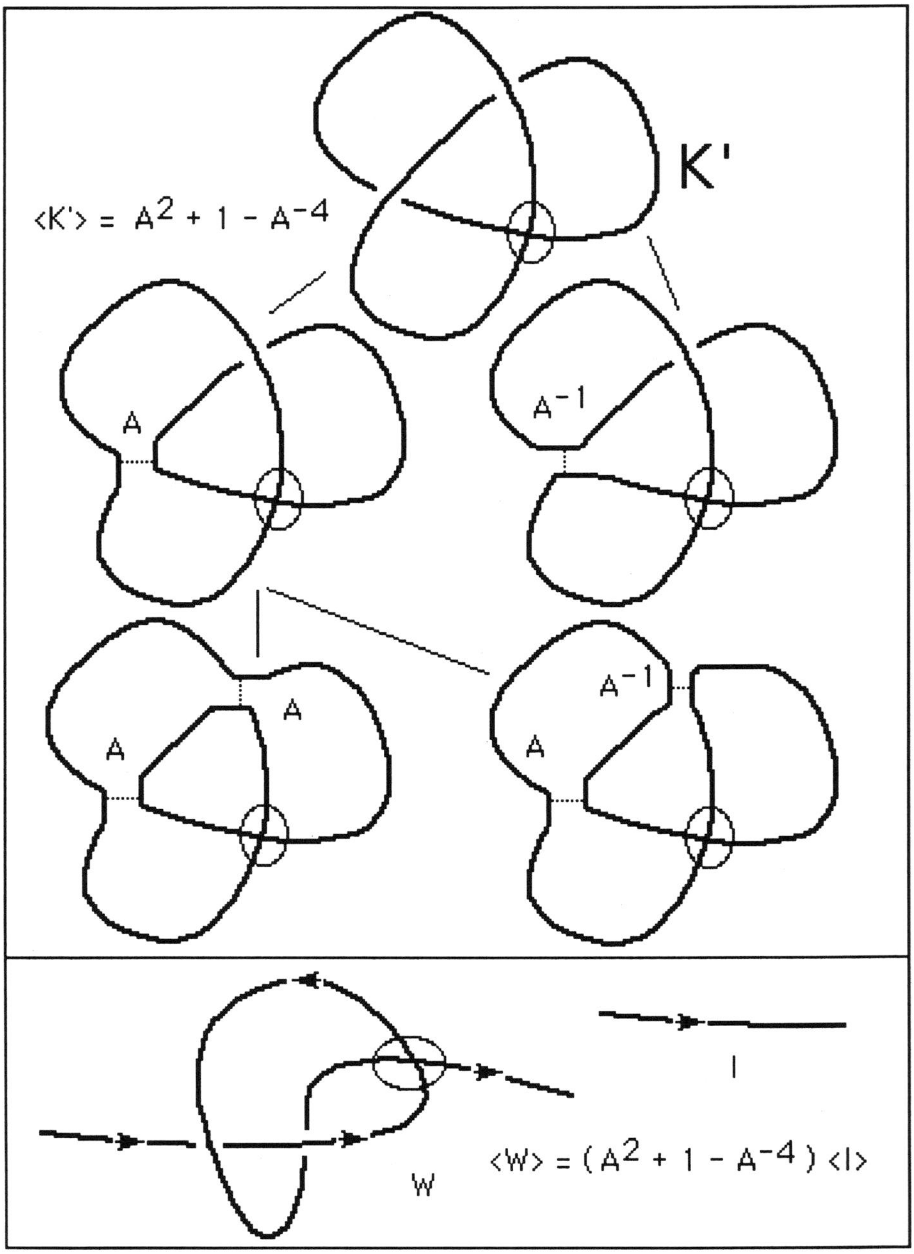

Figure 18 — Calculation of $< K' >$

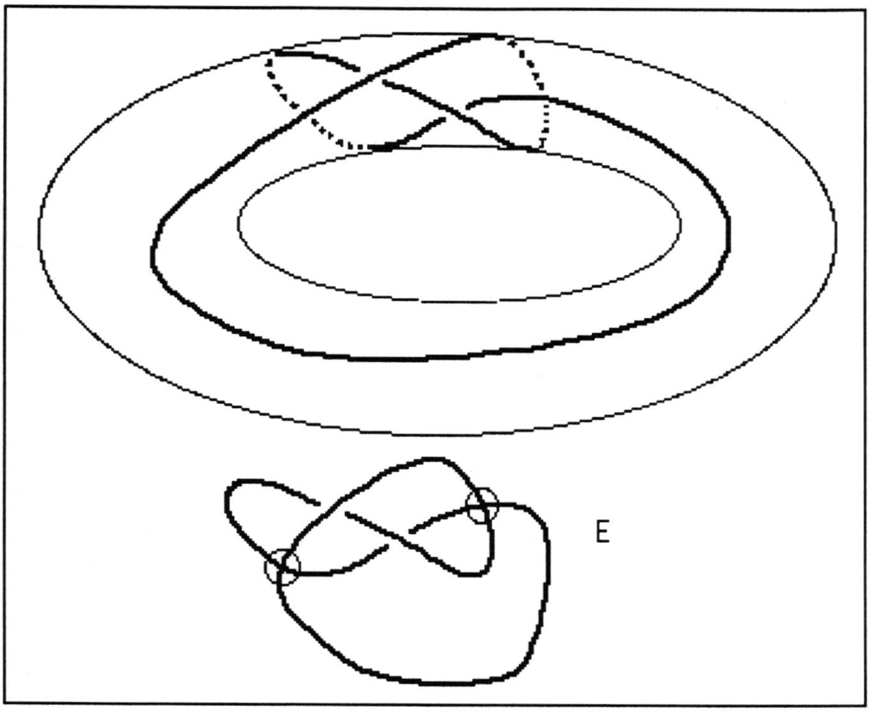

Figure 19 — A Knot in $S_1 \times I$ with Trivial Jones Polynomial

embedding in $S_1 \times I$. In this case, E is a trivial virtual knot (as is shown in Figure 3), but the embedding of E in $S_1 \times I$ is non-trivial (even though it has trivial fundamental group and trivial bracket polynomial). The non-triviality of this embedding is seen by simply observing that it carries a non-trivial first homology class in the thickened torus. In fact, if you expand the state sum for the bracket polynomial and keep track of the isotopy classes of the curves in the states, then the bracket calculation also shows this non-triviality by exhibiting as its value a single state with a non-contractible curve.

Virtual knot theory provides a convenient calculus for working with knots in $S_g \times I$. The virtuals carry many properties of knots in $S_g \times I$ that are independent of the choice of embedding and genus. This completes our quick survey of the properties of the bracket polynomial and Jones polynomial for virtual knots and links. Just as uncolorable graphs appear when one goes beyond the plane (for planar graph coloring problems) so knots of unit Jones polynomial appear as we leave the diagrammatic plane into the realm of the Gauss codes.

6 Unoriented Quantum Link Invariants

There are virtual link invariants corresponding to every quantum link invariant of classical links. However this must be said with a caveat: We do not assume invariance under the first classical Reidemeister move (hence these are invariants of regular isotopy) and we do not assume invariance under the flat version of the first Reidemeister move in the (B) list of virtual moves. Otherwise the usual tensor or state sum formulas for quantum link invariants extend to this generalized notion of regular isotopy invariants of virtual knots and links. In this section we illustrate this method by taking a different generalisation of the bracket that includes virtual framing. We apply this new invariant to distinguish a virtual knot that has Jones polynomial equal to one and a trivial fundamental group.

In order to carry out this program, we quickly recall how to construct quantum link invariants in the unoriented case. See [11] for more details. The link diagram is arranged with respect to a given "vertical" direction in the plane so that perpendicular lines to this direction intersect the diagram transversely or tangentially at maxima and minima.. In this way the diagram can be seen as constructed from a pattern of interconnected maxima, minima and crossings—as illustrated in Figure 20.

As illustrated in Figure 20, we associate symbols M_{ab} and M^{ab} to minima and maxima respectively, and symbols R^{ab}_{cd} and $\overline{R}^{ab}_{cd}$ to the two types of crossings. The indices on these symbols indicate how they are interconnected. Each maximum or minimum has two lines available for connection corresponding to the indices a and b. Each R , $\overline{R}$ has four lines available for connection. Thus the symbol sequence

$$T(K) = M_{ad}M_{bc}M^{ek}M^{lh}R^{ab}_{ef}R^{cd}_{gh}\overline{R}^{fg}_{kl}$$

represents the trefoil knot as shown in Figure 20. Since repeated indices show the places of connection, there is no necessary order for this sequence of symbols. I call $T(K)$ an *abstract tensor* expression for the trefoil knot K.

By taking matrices (with entries in a commutative ring) for the M's and the R's it is possible to re-interpret the abstract tensor expression as a summation of products of matrix entries over all possible choices of indices in the expression. Appropriate choices of matrices give rise to link invariants. If K is a knot or link and $T(K)$ its associated tensor expression, let $Z(K)$ denote the evaluation of the tensor expression that corresponds to the above choice of matrices. We will assume that the matrices have been chosen so that $Z(K)$ is an invariant of regular isotopy.

The generalisation of the quantum link invariant $Z(K)$ to virtual knots and links is quite straightforward. We simply ignore the virtual crossings in the diagram. Another way to put this is that we take each virtual crossing to be represented by crossed Kronecker deltas as in Figure 20. The virtual crossing is represented by the tensor

$$V^{ab}_{cd} = \delta^a_d \delta^b_c.$$

Here δ^a_b is the Kronecker delta. It is equal to 1 if $a = b$ and is equal to 0 otherwise. (Note that the Kronecker delta is well defined as an abstract tensor.)

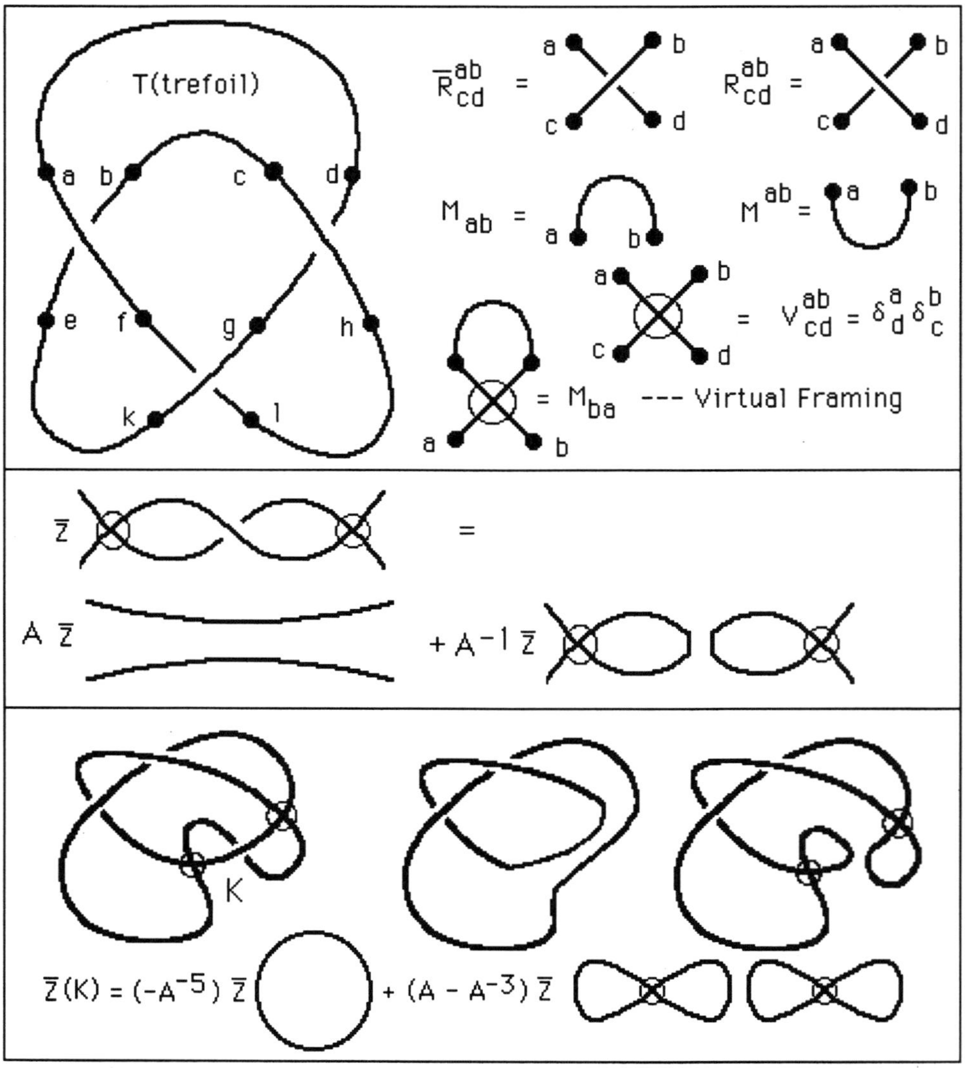

Figure 20 — Quantum Link Invariants

In extending $Z(K)$ to virtual knots and links by this method we cannot hope to obtain invariance under the type I virtual move. In fact, as Figure 20 shows, the presence of a virtual curl is indexed by the transpose M_{ba} of the tensor M_{ab}. Thus we define *virtual regular isotopy* to be invariance under all the extended Reidemeister moves for virtuals except type (A)I and (B)I. It is easy to see that $Z(K)$ extends in this way when $Z(K)$ is an invariant of regular isotopy for classical links.

In particular the bracket polynomial for classical knots is obtained by letting the indices run over the set $\{1,2\}$ with $M^{ab} = M_{ab}$ for all a and b and $M_{11} = M_{22} = 0$ while $M_{12} = iA$ and $M_{21} = -iA^{-1}$ where $i^2 = -1$. The R's are defined by the equations

$$R^{ab}_{cd} = AM^{ab}M_{cd} + A^{-1}\delta^a_c\delta^b_d$$

$$\overline{R}^{ab}_{cd} = A^{-1}M^{ab}M_{cd} + A\delta^a_c\delta^b_d$$

These equations for the R's are the algebraic translation of the smoothing identities for the bracket polynomial. Then we have

Theorem 8. With $Z(K)$ defined as above and K a classical knot or link, $Z(K) = d < K >$ where $d = -A^2 - A^{-2}$.

Proof. See [11].//

For this extension of $Z(K)$ to virtuals there is a state summation similar to that of the bracket polynomial. For this let C be a diagram in the plane that has only virtual crossings. View this diagram as an immersion of a circle in the plane. Let $rot(C)$ denote the absolute value of the Whitney degree of C as a immersion in the plane. (Since C is unoriented only the absolute value of the Whitney degree is well-defined.). The Whitney degree of an oriented plane immersion is the total algebraic number of 2π turns of the unit tangent vector to the curve as the curve is traversed once. Let $d(C)$ be defined by the equation

$$d(C) = (-1)^{rot(C)}(A^{2rot(C)} + A^{-2rot(C)}).$$

Let S be a state of a virtual diagram K obtained by smoothing each classical crossing in K. Let $C \in S$ mean that C is one of the curves in S. Let $< K|S >$

denote the usual product of vertex weights (A or A^{-1}) in the bracket state sum. Then

Proposition 9. The invariant of virtual regular isotopy $Z(K)$ is described by the following state summation.

$$Z(K) = \sum_{S} < K|S > \prod_{C \in S} d(C)$$

where the terms in this formula are as defined above. Note that $Z(K)$ reduces to $d < K >$ when K is a classical diagram.

Proof. The proof is a calculation based on the tensor model explained in this section. The details of this calculation are omitted.//

Remark. The state sum in Proposition 9 generalises to an invariant of virtual regular isotopy with an infinite number of polynomial variables, one for each regular homotopy class of unoriented curve immersed in the plane. To make this generalisation, let A_n for $n = 0, 1, 2, 3, ...$ denote a denumerable set of commuting independent variables. If C is an immersed curve in the plane, define $Var(C) = A_n$ where $n = rot(C)$, the absolute value of the Whitney degree of C. We take $A_1 = -A^2 - A^{-2}$ as before, but the other variables are independent of each other and of A.

Now define the generalisation of $Z(K)$, denoted $\overline{Z}(K)$, by the formula

$$\overline{Z}(K) = \sum_{S} < K|S > \prod_{C \in S} Var(C).$$

In this definition we have replaced the evaluation $d(C)$ by the corresponding variable $Var(C)$. In Figure 20 we illustrate the result of calculating $\overline{Z}(K)$ for a knot K with unit Jones polynomial. The result is

$$\overline{Z}(K) = (-A^{-5})A_1 + (A - A^{-3})A_0^2.$$

Since the coefficients of A_1 and A_0 are themselves invariants of virtual regular isotopy it follows, as we already knew, that K is a non-trivial virtual. This non-triviality is detected by our refinement $\overline{Z}(K)$ of the bracket polynomial. A similar phenomenon of refinement of invariants happens with other quantum link invariants. This will be the subject of a separate paper.

We end this section with an application of Proposition 9. Let D be the virtual knot diagram shown in Figure 21.

It is easy to see that D has f-polynomial equal to 1, hence Jones polynomial equal to 1. Use Lemma 7 to show that $< D >= -A^3$. D also has trivial fundamental group and quandle. Is D a non-trivial virtual knot? The answer is yes! It follows from the calculation of $Z(D)$. We omit the calculation, but give the result

$$Z(D) = A^7 - A^5 - 4A^3 + 2A + A^{-1} - A^{-3}.$$

It follows from this that D cannot be regularly isotopic to a standard virtual curl form. Hence D must be virtually knotted. This example shows the power of using the quantum invariants to study virtual knots.

There should be a direct way to see that D is knotted. Let E denote the *shadow* of the diagram D. That is, replace the classical crossings in D with flat crossings. Regard the flat crossings as *distinct* from virtual crossings, so that we get the rules for virtual isotopy of flat diagrams shown in Figure 22. By these rules a flat diagram corresponds to an oriented Gauss code without over or under crossing specifications. The virtual moves preserve the Gauss code just as before.

E is illustrated in Figure 21. Is E flat virtually equivalent to a circle with curls and virtual curls? I conjecture that the answer is no. Simpler examples of this sort of irreducibility are easy to produce. The diagram F in Figure 21 is irreducible because $\overline{Z}(F)(1) = -2A_0 + A_2$, as is easy to compute.

These examples lead us to the following definition. We call the *shadow code* of a diagram the underlying Gauss code of that diagram without any specifications of orientation or over/under crossing. We say that a virtual diagram is *almost classical* if its shadow code is planar.

Conjecture. There does not exist a non-trivial almost classical virtual knot with both trivial fundamental group and trivial Jones polynomial.

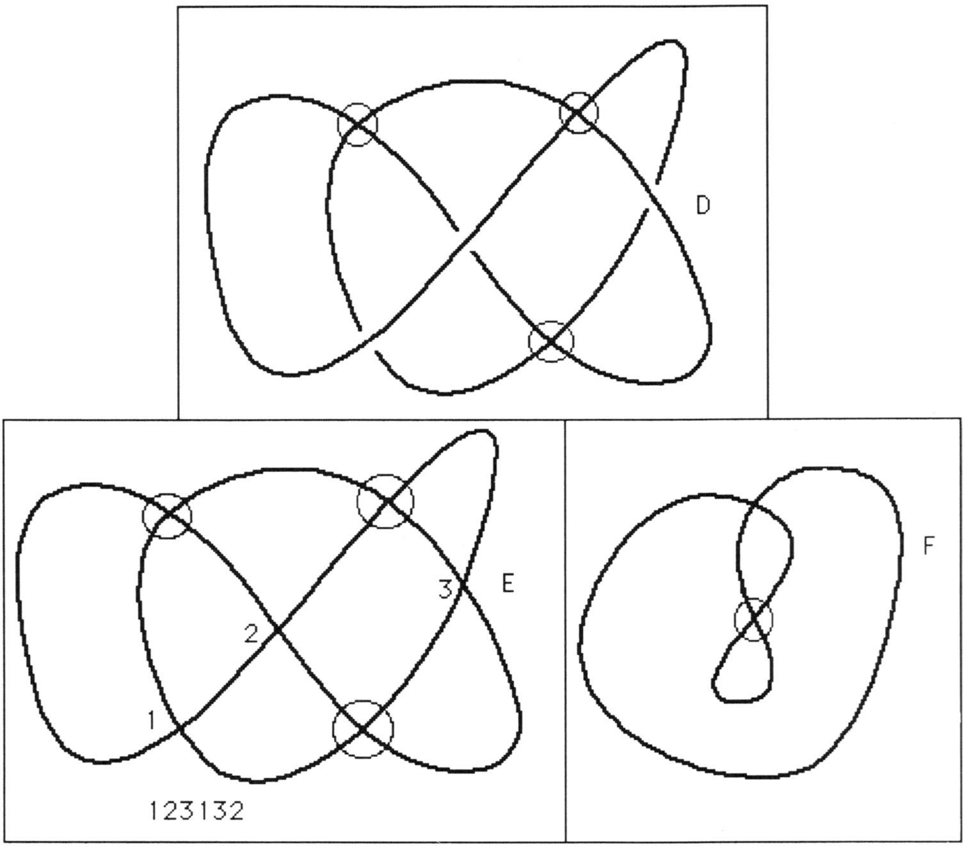

Figure 21 — D has Unit Jones Polynomial and Trivial Fundamental Group

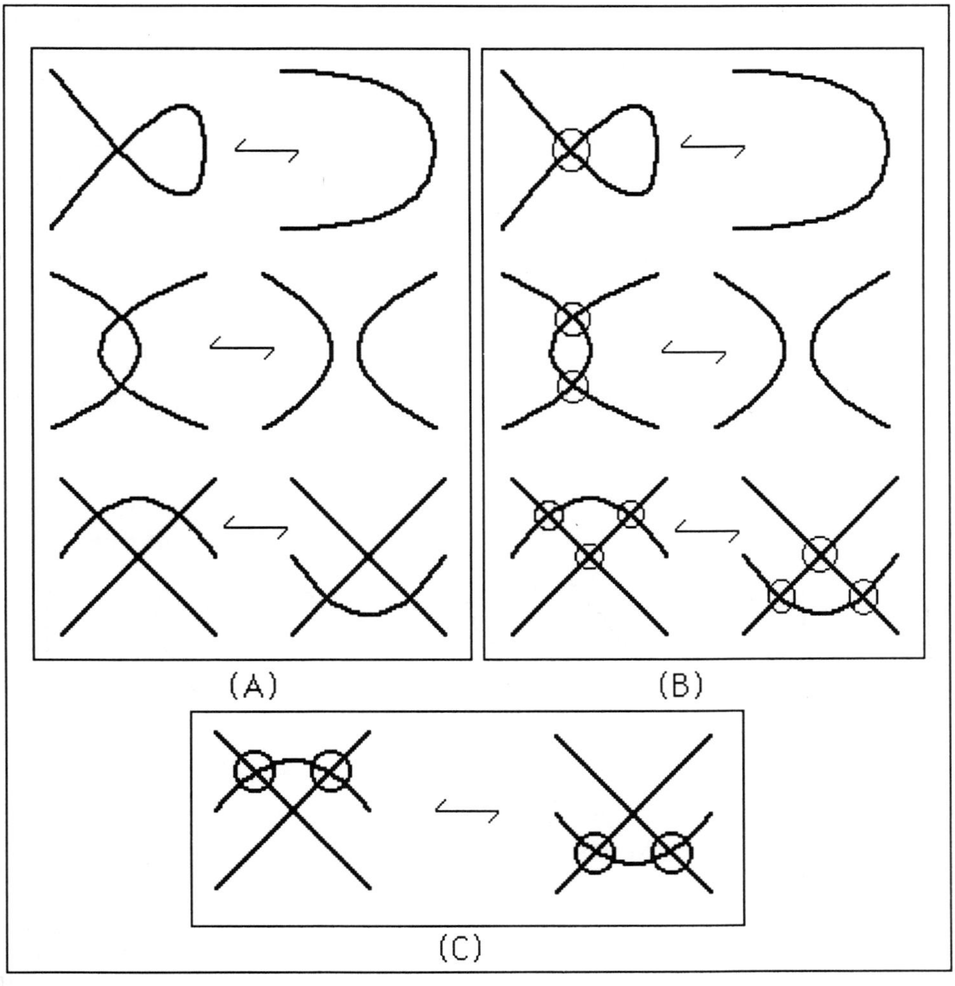

Figure 22 — Flat Virtual Moves

7 Virtual Vassiliev Invariants

We now study embeddings into R^3 (Euclidean three-space) of 4-valent graphs up to *rigid vertex isotopy*. In rigid vertex isotopy one can think of each graphical vertex as a rigid disk. The four graphical edges incident to this vertex are attached to the boundary of the disk at four specific points. In a rigid vertex isotopy the embedded edges of the graph can be isotoped freely, but the disks must move without deformation in the course of the isotopy. A consequence of this definition [18] is that the diagrammatic moves shown in Figure 23 capture rigid vertex isotopy just as the Reidemeister moves capture ambient isotopy. Figure 23 shows only the move types that are added to the usual list of Reidemeister moves. Two graphs with diagrammatic projections G_1 and G_2 are rigid vertex isotopic if and only if there is a series of moves of this type joining the two diagrams.

In Figure 23 there is also illustrated the one addition to virtual moves that is needed to complete the move set for rigid vertex isotopy of virtual knotted graphs. In this addition an arc with consecutive virtual crossings is moved to a new position across a rigid vertex. Here we must make a distinction between the graphical rigid vertices and the virtual vertices in the diagrams. Once this is done we directly extend discussions of invariants of rigid vertex graphs to invariants of virtual rigid vertex graphs. To see how this is done we will discuss invariants obtained by insertion into the vertices of a graph. We shall always mean the extension to virtual equivalence when we refer to ambient isotopy or to rigid vertex isotopy.

If one replaces each node of a (virtual) rigid vertex graph G with a tangle (possibly virtual) to form a virtual link K , then any rigid equivalence of G induces a corresponding equivalence of K. The consequence of this remark is that we can obtain invariants of rigid vertex graphs from any invariant of (virtual) knots and links by taking a systematic choice of tangle insertion. That is, if we have chosen tangle insertions $T_1, \ldots, T_n$ let $\{v_1, \ldots, v_m\}$ denote the set of vertices of G and let $a = (a_1, \ldots, a_m)$ with $1 \leq a_i \leq n$ denote a choice of tangle insertion for each vertex of G. Then let G^a denote the result of inserting tangle T_{a_i} at node i in G. Suppose that $R(K)$ is an ambient isotopy invariant of virtual knots and links K. Then define an extension of R to graphical imbeddings by the formula

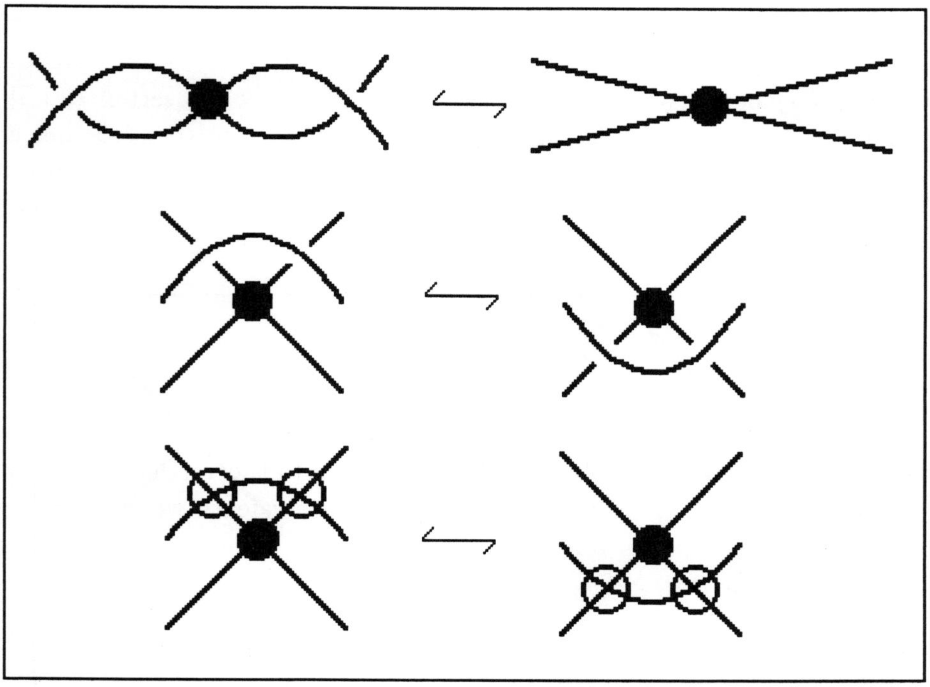

Figure 23 — Moves for Rigid Vertex Embeddings

$$R(G) = \sum_{a} x_{a_1} ... x_{a_m} R(G^a)$$

where $\{x_j | j = 1, ..., m\}$ is a new set of variables (or constants) independent of the variables already present in the invariant R. Our discussion shows that $R(G)$ is an invariant of virtual graph embeddings G.

While it is of interest to explore this larger class of induced invariants, we shall restrict ourselves to the generalisation of *Vassiliev invariants.* A Vassiliev invariant v is an invariant of rigid vertex (virtual) 4-valent graphs that satisfies

$$v(G|*) = v(G|+) - v(G|-)$$

where $(G|*)$ denotes an oriented graph G with a chosen vertex $*$. $(G|+)$ denotes the result of replacing the vertex $*$ with a positive crossing and $(G|-)$ is the result of replacing it with a negative crossing. See Figure 24. This is the traditional definition of a Vassiliev invariant and we adopt it verbatim for virtuals.

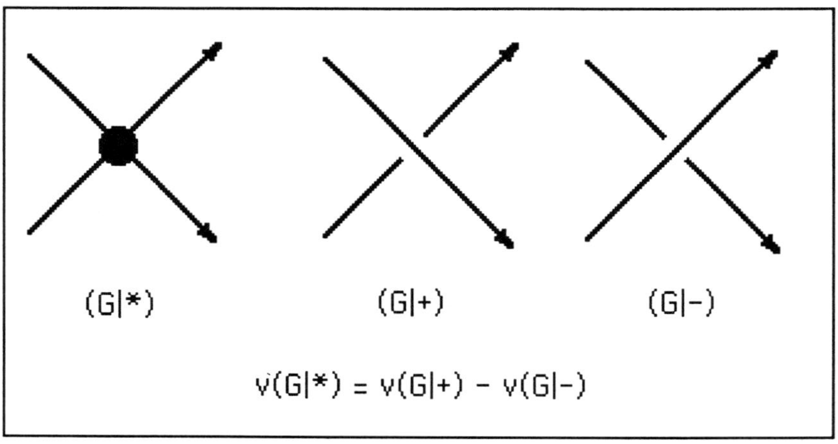

Figure 24 — Vassiliev Invariant Identity

Definition. Let $N(G)$ denote the number of vertices in the 4-valent graph

G. A Vassiliev invariant v is said to be of *graphical finite type* n if $v(G) = 0$ whenever $N(G) > n$. Note that this definition says nothing about the number of virtual crossings in the graph G.

Useful examples of virtual invariants of graphical finite type are obtained by taking the coefficients of x^m in

$$F_K(x) = f_K(e^x)$$

where $F_K(x)$ is extended to 4-valent graphs by the difference formula

$$F_K(x)(G|*) = F_K(x)(G|+) - F_K(x)(G|-).$$

The corresponding formula then holds for the coefficients of x^m in the power series expansion of $F_K(x)$.

Lemma 10. Let $F_K(x) = f_K(e^x)$ denote the power series resulting from substitution of e^x for the variable A in the Laurent polynomial $f_K(A)$ (defined in section 2). Write this power series in the form

$$F_K(x) = \sum_{m=0}^{\infty} v_m(K)x^m.$$

Then the numerical invariants $v_k(K)$ are of finite graphical type k.

Proof. Recall from section 5 that

$$f_{K_+} = -A^{-2}f_{K_0} - A^{-4}f_{K_\infty}$$
$$f_{K_-} = -A^{+2}f_{K_0} - A^{+4}f_{K_\infty}$$

It follows that

$$F_{K_*} = F_{K_+} - F_{K_-}$$

$$= f_{K_+}(e^x) - f_{K_-}(e^x)$$

is divisible by x. Thus if G has m nodes then F_G is divisible by x^m. This implies, for any G, that $v_k(G) = 0$ if $k < m = N(G)$. This is exactly the statement that v_k is of finite graphical type k.//

Proposition 11. Let G be a graph with n vertices so that $N(G) = n$, configured as a virtual diagram in the plane. Let $(G|+)$ denote the diagram G with a specific crossing of positive type and $(G|-)$ the diagram identical to G except that the crossing has been switched to one of negative type. Let $(G|*)$ denote the result of replacing this crossing by a graphical vertex. Let v be a Vassiliev invariant of type $n = N(G)$. Then $v(G|+) = v(G|-)$. Thus a Vassiliev invariant of type n is independent of the settings of the crossings (plus or minus) in a diagram for G.

Proof. $v(G|+) - v(G|-) = v(G|*)$ by the definition of a Vassiliev invariant. But $v(G|*) = 0$ since $(G|*)$ has $(n+1)$ vertices and v is of type n. This completes the proof.//

Corollary 12. If G and v are as in Proposition 11 , and G is a classical diagram (free of virtual crossings) then $v(G)$ does not depend upon the classical embedding of G in R^3 that is indicated by the diagram.

Proof. This follows directly from the switching independence shown in Proposition 11. //

For virtual Vassiliev invariants one should not expect the analog of this corollary to hold, but in fact it does hold for the virtual Vassiliev invariants induced from $f_K(A)$. That is, we shall show that the Vassiliev invariants $v_n(G)$ in the series $F_K(x)$ depend only on the chord diagram associated with G when G is a virtual diagram with n graphical nodes. This is the subject of the following subsection.

7.1 The Vassiliev Invariants Induced by the Jones-Polynomial

We shall use the Vassiliev invariants that arise from the bracket polynomial and the f-polynomial. This is equivalent to using the Vassiliev invariants that arise from the Jones polynomial. Let f(A) be any Laurent polynomial with coefficients as in the formula below

$$f(A) = c_1 A^{d_1} + c_2 A^{d_2} + ... + c_k A^{d_k},$$

where the degrees are integers arranged so that

$$d_1 \leq d_2 \leq ...d_k.$$

Then

$$f(e^x) = c_1 e^{xd_1} + c_2 e^{xd_2} + ... + c_k e^{xd_k}$$

$$= \sum_{n=0}^{\infty} (c_1 d_1^n + c_2 d_2^n + ... + c_k d_k^n) x^n / n!.$$

Thus if

$$F(x) = f(e^x) = \sum_{n=0}^{\infty} v_n x^n,$$

then

$$v_n = (c_1 d_1^n + c_2 d_2^n + ... + c_k d_k^n)/n!.$$

This gives a direct formula for the Vassiliev invariants v_n associated with f.

In particular, this gives us a direct method to read off the Vassiliev invariants associated with a given evaluation of the normalized bracket polynomial f_K. The invariant $v_n(G)$ is determined by the coefficients of $f_G(A)$ and the exponents of A in this Laurent polynomial.

Notational Discussion. Let $v_n(K)$ denote the n-th Vassiliev invariant induced from $f_K(A)$ as described in this section. Let G_*, G_+, G_- denote a triple of (virtual) graph diagrams that differ at the site of one rigid vertex (denoted $*$) by replacement by either a positive crossing (denoted $+$) or a negative crossing (denoted $-$). Let G_0 and G_∞ denote the oriented and unoriented smoothings of this crossing. Note that since we can speak of the evaluation of $f_{G_\infty}(A)$, it follows that v_n is defined for diagrams with non-oriented smoothings - one just evaluates the state sum in the usual way with single reverse-oriented loops taking the usual loop value of $-A^2 - A^{-2}$.

Theorem 13. With notation as above, the following recursion formula holds for the Vassiliev invariants $v_n(G)$.

$$v_n(G_*) = \sum_{k=0}^{n-1} c_{n,k}(v_k(G_0) + 2^{n-k} v_k(G_\infty))$$

where

$$(2^{n-k}(1+(-1)^{n-k+1})/(n-k)! = c_{n,k}.$$

The value of $v_0(K)$ on a virtual diagram without graphical nodes depends only on the number of components in the diagram, and is independent of the configuration of virtual crossings. Specifically,

$$v_0(K) = (-2)^{\mu(K)-1}$$

where $\mu(K)$ denotes the number of link components in K.

Corollary 14. The Vassiliev invariants $v_n(G)$ in the series $F_K(x)$ depend only on the chord diagram associated with G when G is a virtual diagram with n graphical nodes. Hence the weight systems for the invariants $v_n(G)$ do not depend upon virtual crossings.

Proof of Corollary. This follows directly from Theorem 13 since the recursion formula in that theorem computes $v_n(G)$ for a graph G with n nodes in terms of $v_0(K)$ for a collection of virtual knots $\{K\}$. Since v_0 is independent of virtual crossings, so is $v_n(G)$. This completes the proof.//

Proof of Theorem 13.

Recall from section 5 that

$$f_{G_+} = -A^{-2} f_{G_0} - A^{-4} f_{G_\infty}$$

$$f_{G_-} = -A^{+2} f_{G_0} - A^{+4} f_{G_\infty}$$

Hence

$$f_{G_*} = (A^2 - A^{-2}) f_{G_0} + (A^4 - A^{-4}) f_{G_\infty}.$$

Now suppose that

$$f_{G_0} = \sum_i a_i A^{n_i}$$

and that

$$f_{G_\infty} = \sum_i b_i A^{m_i}.$$

Then

$$f_{G_*} = \sum_i a_i A^{n_i+2} - a_i A^{n_i-2} + b_i A^{m_i+4} - b_i A^{m_i-4}.$$

Therefore

$$v_n(G_*) = (1/n!) \sum_i a_i((n_i+2)^n - (n_i-2)^n) + b_i((m_i+4)^n - (m_i-4)^n).$$

The first part of Theorem 13 follows from this formula by a direct application of the binomial theorem. For the second part, note that

$$v_0(K) = f_K(1) = (-1)^{w(K)} < K > (1).$$

For $A = 1$ it is easy to see that the only effect of the matrix model of section 6 on the bracket calculation is to multiply by $(-1)^{cv(K)}$ where $cv(K)$ is the number of virtual crossings in K. That is,

$$d < K > (1) = (-1)^{cv(K)} Z(K)(1)$$

where $Z(K)(1)$ is the matrix model of section 6 evaluated at $A = 1$.

In this matrix model there is no difference (at $A = 1$) between the crossings and the virtual crossings. They are both algebraically crossed Kronecker deltas. Consequently, $Z(K) = Z(K')$ where K' is the same diagram as K with all the virtual crossings replaced by (flat) classical crossings. Then it follows from the regular isotopy invariance of Z that $Z(K')(1) = (-2)^{\mu(K)}(-1)^{c(K')}$ where $c(K')$ is the total number of crossings in K'. Note that the value of $d = -2$ when $A = 1$. Hence

$$v_0(K) = (-1)^{w(K)}(-1)^{cv(K)}(-1)^{c(K')}(-2)^{\mu(K)-1}$$

Now we know that $w(K) + cv(K)$ is congruent modulo 2 to $c(K')$. Therefore

$$v_0(K) = (-2)^{\mu(K)-1}.$$

This completes the proof of the Theorem.//

Remark. An example for this Theorem is that v_2 gives the value -48 for both of the graphs shown in Figure 25. Each graph has two graphical nodes.

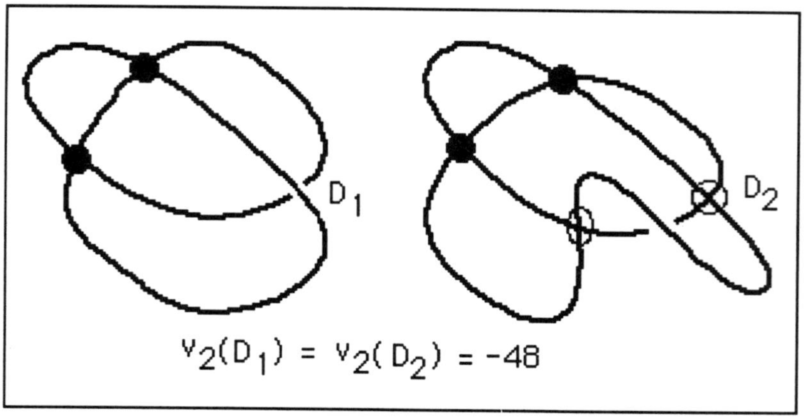

Figure 25 — v_2 dependence

One graph represents a virtual diagram inequivalent to any embedding of the other. The invariants $v_n(G)$ themselves depend on virtual crossings for the graphs with less than n nodes. In fact, in these intermediate ranges there is a dependency on infinitely many virtual diagrams so that these invariants are no longer as "finite" as the classical Vassiliev invariants. In [19] there is formulated a more restrictive notion of finite type virtual invariants. Our v_2 from the Jones polynomial is a first example of a finite graphical type invariant that is outside the scheme proposed by Goussarov, Polyak and Viro. More work needs to be done to have a complete theory of virtual Vassiliev invariants.

8 Quantum Link Invariants and the Polyak Algebra of Semi-Virtuals

Classical Vassiliev invariants can be used as building blocks for all the presently known quantum link invariants.

Suppose that the R-matrices in the model for a given quantum link invariant can be written as power series in h with matrix coefficients so that

$PR = I + r_+ h + O(h^2)$ and $PR^{-1} = I + r_- h + O(h^2)$ where P denotes the map of $V \otimes V$ that interchanges the tensor factors. Let $Z(K)$ denote the value of the (oriented) amplitude described by this choice of R (and the related augmentation structure corresponding to the cups and caps in the model). Then we can write

$$Z(K) = Z_0(K) + Z_1(K)h + Z_2(K)h^2 + \ldots$$

where each $Z_n(K)$ is an invariant of regular isotopy of the link K. Furthermore, we see easily that h *divides the series for* $Z(K_+) - Z(K_-)$. By the definition of the Vassiliev invariants this implies that h^k divides $Z(G)$ if G is a graph with k nodes. Therefore $Z_n(G)$ vanishes if n is less than the number of nodes of G. Therefore Z_n is a Vassiliev invariant of finite type n. Hence the quantum link invariant is built from an infinite sequence of interlocked Vassiliev invariants.

We can extend the oriented quantum invariant to an invariant of virtual knots in the same way that was explained in section 6 for unoriented invariants. The resulting invariants of virtual knots associate the permutation mapping P to each virtual crossing and they are invariants of virtual regular isotopy (with a virtual framing dependence that depends upon the choice of cup and cap mappings in the model). We shall treat virtual oriented quantum invariants in more detail elsewhere, but the point of the remark here is that all such invariants induce virtual Vasiliev invariants by the same argument that we have indicated above in the classical case. Thus there is a very rich vein of virtual Vassiliev invariants to be investigated.

Returning to classical invariants, recall that Kontsevich [12], [2] proved that a weight assignment for a Vassiliev top row that satisfies the 4-term relation actually extends to a Vassiliev invariant defined on all knots. His method is motivated by the perturbative expansion of the Witten integral and by Witten's interpretation of the integral in terms of conformal field theory.

The key to this approach is to see that the 4-term relations for the Vassiliev invariants are a kind of "infinitesimal braid relations". That is, one can re-write the 4-term relations in the form of tangle operators t^{ij} so that the commutator equation

$$[t^{12}, t^{13}] + [t^{13}, t^{23}] = 0$$

is an algebraic form of the 4-term relation. The 4-term relations translate exactly into these infinitesimal braid relations studied by Kohno [13].

The upshot of the Kontsevich work is a very specific integral formula for the Vassiliev invariants. See [2], [3] and [7] for the specifics. It is clear from the nature of the construction that the Kontsevich formula captures the various orders of perturbative terms in the Witten integral. However, it is not at all clear how to use this approach to extend to virtuals by interpreting the virtual crossing as a permutation operator. Such an interpretation goes outside the context of the integration procedure.

On the other hand, the infinitesimal braiding equation is also implicated in the formalism for the series expansion of a solution to the Yang Baxter equation. Hence this formalism fits into the context of the oriented quantum link invariants and their associated Vassiliev invariants. Lets return to the series form of the solution to the Yang-Baxter equation, as we have indicated it above.

$$PR = I + r_+ h + O(h^2)$$
$$PR^{-1} = I + r_- h + O(h^2).$$

Since $RR^{-1} = P$, it follows that $r_- = -r_+'$ where a' denotes the transpose of a. In the case that R can be augmented to produce a quantum link invariant, it follows that

$$t = r_+ - r_- = r + r'$$

(letting r denote r_+) satisfies the infinitesimal braiding relations

$$[t^{12}, t^{13}] + [t^{13}, t^{23}] = 0.$$

It is interesting to contemplate this fact, since r, being the coefficient of h in the series for R, necessarily satisfies the *classical Yang Baxter Equation*

$$[r^{13}, r^{23}] + [r^{12}, r^{23}] + [r^{12}, r^{13}] = 0.$$

(The classical Yang-Baxter equation for r is a direct consequence of the fact that R is a solution of the (quantum) Yang-Baxter equation.)

Thus via the quantum link invariants, we have provided a special condition (the assumption that r is the coefficient of h in a power series solution of the quantum Yang-baxter equation R, and that R can be augmented to produce a quantum link invariant) ensuring that a solution r of the classical Yang Baxter equation will produce a solution $t = r + r'$ of the infinitesimal braiding relation, whence a weight system for Vassiliev invariants. More work needs to be done to fully understand the relationship between solutions of the classical Yang-Baxter equation and the construction of Vassiliev invariants. We can now take a wider view and see how this context is related to a number of current developments about virtual invariants.

In [19] the authors consider *semivirtual crossings* and an algebra associated to the category of diagrams with semivirtual crossings. Semivirtual crossings are defined in the following way.

$$K_+ = K_0 + K_\oplus$$
$$K_- = K_0 + K_\ominus$$
$$K_+ - K_- = K_\odot = K_\oplus - K_\ominus$$

where $+$ denotes the presence of a positive crossing, $-$ denotes the presence of a negative crossing, $\oplus$ and $\ominus$ are designations for semivirtual crossings, and $\odot$ denotes a graphical node (Vassiliev vertex). Here the K denotes a knot or link diagram that is changed only at site of one crossing in each equation. In this interpretation, diagram sums occur in a free module with coefficients appropriate to the problem at hand. K can also denote the value of an invariant that is applied to a given diagram. It should be mentioned that all these matters can be expressed in terms of *Gauss diagrams* where one indicates an abstract circle for each knot component and uses arrows with signs to indicate the paired points of a classical crossing, dotted arrows with signs for semivirtual crossings and (just as with Gauss codes) the virtual crossings need not be indicated.

Reidemeister moves on diagrams with only classical crossings, then translate to *equations* in the category of diagrams with virtual and semivirtual crossings. In [19] the authors study Vassiliev invariants of *finite semivirtual degree n*. That is, after translation (by the above equations) into the semivirtual category, it is assumed that the invariant vanishes on all diagrams with more than n semivirtual crossings. Note that since

$$K_{\odot} = K_{\oplus} - K_{\ominus}$$

it follows that an invariant of semivirtual degree n is also of at least degree n in the standard Vassiliev sense of finite type. As a result, it is possible to use the theory of semivirtual finite degree invariants to study a significant subset of the the theory of Vassiliev invariants. It is conceivable that all finite type Vassiliev invariants of classical knots can be expressed as invariants of semivirtual finite degree. In the theory of semivirtual invariants the theory of virtual knots features in a necessary way through the expansion equations that translate classical knots to sums of semivirtual diagrams. It remains to be seen how this theory will contribute both to the theories of classical and virtual knots.

Finally, we point out how the expansion of an R-matrix in powers of h motivates the definition of semivirtual crossings. In the beginning of this section we wrote

$$PR = I + r_+ h + O(h^2)$$
$$PR^{-1} = I + r_- h + O(h^2).$$

Removing P and revising the notation so that R is replaced by R_+ and R^{-1} is replaced by R_-, we have

$$R_+ = R_0 + \rho_+ h + O(h^2)$$
$$R_- = R_0 + \rho_- h + O(h^2)$$

where $R_0 = P$ is the algebraic form of a virtual crossing. Thus

$$R_{\oplus} = \rho_+ h + O(h^2)$$
$$R_{\ominus} = \rho_- h + O(h^2)$$

so that

$$R_+ = R_0 + R_{\oplus}$$

$$R_{-} = R_0 + R_{\ominus}.$$

The semivirtual crossings here correspond to all of the R-matrix except for the permutation at the start. The algebraic equations for semivirtual crossings that derive from the Reidemeister moves are properties of these matrices $R_{\oplus}$ and $R_{\ominus}$. Thus the study of semivirtual invariants suggests that it will be of significance to study these truncated R- matrices in the context of quantum invariants and associated Vassiliev invariants. Clearly much more remains to be done in this field.

9 Discussion

This completes our introduction to virtual knot theory. There is much that begs for further investigation. We leave the following topics for sequels to this paper: the Alexander polynomial (there are a diversity of definitions that differ on virtual knots), virtual braids, virtual 3-manifolds, Vassiliev invariants induced from oriented quantum link invariants, more general structure of Vassiliev invariants.

Virtual braids is a subject very close to the "welded braids" of Fenn, Rimanyi and Rourke [20]. In fact, their welded braids are a quotient of the category of virtual braids that are defined through our approach. I am indebted to Tom Imbo for pointing out this connection. This topic will be the subject of a separate paper.

While we are on the subject of virtual braids it is worthwhile pointing out that the virtual braid group is an extension of the classical braid group by the symmetric group. If V_n denotes the n–strand virtual braid group, then V_n is generated by braid generators σ_1, ..., σ_{n-1} and virtual gnerators c_1,..., c_n where each virtual generator c_i has the form of the braid generator σ_i with the crossing replaced by a virtual crossing. Among themselves the braid generators satisfy the usual braiding relations. Among themselves, the virtual generators are a presentation for the symmetric group S_n. The relations that relate virtual generators and braiding geneerators are as follows:

$$\sigma_i^{\pm} c_{i+1} c_i = c_{i+1} c_i \sigma_{i+1}^{\pm},$$

$$c_i c_{i+1} \sigma_i^{\pm} = \sigma_{i+1}^{\pm} c_i c_{i+1},$$

$$c_i \sigma_{i+1}^{\pm} c_i = c_{i+1} \sigma_i^{\pm} c_{i+1}.$$

It is easy to see from this description of the virtual braid groups that all the braiding generators can be expressed in terms of the first braiding generator σ_1 (and its inverse) and the virtual generators. One can also see that Alexander's Theorem generalizes to virtuals: Every virtual knot is equivalent to a virtual braid. We shall discusss the possibility of a Markov Theorem for virtual braids elsewhere.

We also mention the following papers about virtual knots that have appeared (or sprouted) over the last year: [14], [15], [4], [8]. They include discussions of the fundamental groups of virtual knots, virtual tangles and relationships with surfaces and more elaborate Gauss codes. There is more to come!

This work began with an attempt to understand the Jones polynomial for classical knots by generalising that category. I hope that these considerations will lead to deeper insight into the Jones polynomial and its relationship with the fundamental group and quandle of a classical knot.

References

[1] C. W. Ashley, *The Ashley Book of Knots*, Doubleday, New York, London, Sydney, Auckland (1944).

[2] D. Bar-Natan, On the Vassiliev knot invariants, *Topology* **34** (1995), 423-472.

[3] Dror Bar-Natan, *Perturbative Aspects of the Chern-Simons Topological Quantum field Theory*, Ph. D. Thesis, Princeton University, June 1991.

[4] N. Kamada and S. Kamada, Abstract Link Diagrams and Virtual Knots, (to appear in *JKTR*)

[5] L. H. Kauffman, Talks at MSRI Meeting in January 1997, AMS Meeting at University of Maryland, College Park in March 1997, Isaac Newton Institute Lecture in November 1997, Knots in Hellas Meeting in

Delphi, Greece in July 1998, APCTP-NANKAI Symposium on Yang-Baxter Systems, Non-Linear Models and Applications at Seoul, Korea in October 1998.

[6] L. H. Kauffman, Virtual Knot Theory, *European J. Comb.* Vol. 20 (1999), 663-690.

[7] L. H. Kauffman, Witten's Integral and the Kontsevich Integrals, in *Particles, Fields, and Gravitation*, Proceedings of the Lodz, Poland (April 1998) Conference on Mathematical Physics edited by Jakub Rembliens-ki, AIP Conference Proceedings 453 (1998), pp. 368 -381.

[8] L. H. Kauffman and V. O. Manturov, Atoms and Virtual Knots, (in preparation)

[9] R. C. Read and P. Rosenstiehl, On the Gauss Crossing Problem, *Colloquia Mathematica societatis Janos Bolyai Vol. 18 - Combinatorics*, Keszthely, Hungary (1976), 843–876.

[10] R. Fenn and C. Rourke, Racks and Links in Codimension Two, *Journal of Knot Theory and its Ramifications* Vol. 1 (1992), 343–406.

[11] L. H. Kauffman, *Knots and Physics* World Scientific Pub. Co., Singapore (1991 and 1994).

[12] M. Kontsevich, Graphs, homotopical algebra and low dimensional topology, (preprint 1992).

[13] T. Kohno, Linear representations of braid groups and classical Yang-Baxter equations, *Contemporary Mathematics*, Vol. 78, Amer. Math. Soc., 1988, pp. 339-364.

[14] D. S. Silver and S. G. Williams, Virtual Tangles and a Theorem of Krebes, *Journal of Knot Theory and its Ramifications* Vol. 7, (1999), 941–945.

[15] D. S. Silver and S. G. Williams, Virtual Knot Groups, (these proceedings).

[16] O. Viro, Private Conversation (1998).

[17] L. H. Kauffman, State Models and the Jones Polynomial, *Topology*, Vol. 26 (1987), 395–407.

[18] L. H. Kauffman, Knots and Diagrams, *Lectures at Knots '96*, ed. S. Suzuki, World Sci. Pub. (1997), 123–194.

[19] M. Goussarov, M. Polyak and O. Viro, Finite type Invariants of Classical and Virtual Knots, (preprint October 1998 – math.GT/9810073).

[20] R. Fenn, R. Rimanyi and C. Rourke, The Braid Permutation Group, *Topology*, Vol. 36, No. 1 (1997), 123–135.

LOWER BOUNDS FOR THE UNKNOTTING NUMBERS OF THE KNOTS OBTAINED FROM CERTAIN LINKS

TOMOMI KAWAMURA
Graduate School of Mathematical Sciences, University of Tokyo
3-8-1 Komaba Meguro-ku, Tokyo 153-8914, Japan
email: kawamura@ms.u-tokyo.ac.jp

ABSTRACT

In this paper, by applying the results in [4], we shall compute lower bounds for the unknotting numbers of the knots obtained from certain links, for example, by band surgery.
Keywords: Knots, links, the unknotting number, band surgery.

1. Introduction and the statements of main results

A *link* is a closed oriented 1-manifold smoothly embedded in the 3-sphere S^3 and a *knot* is a link with one component. Let K be a knot. The *unknotting number* $u(K)$ of K is the minimal number of crossing changes needed to create the trivial knot. In this paper we shall estimate the unknotting numbers of the knots obtained from certain links.

Let $L = K_1^1 \cup \cdots \cup K_{m_1+n_1}^1 \cup K_1^2 \cup \cdots \cup K_{m_2+n_2}^2$ be an $((m_1+n_1, 0), (m_2+n_2, 0))$-cable on the Hopf link. We reverse orientations of components of L, $K_{m_i+j}^i$, $i = 1, 2$, $j = 1, \cdots, n_i$. We denote these new components by L_j^i, $i = 1, 2$, $j = 1, \cdots, n_i$. and this new link by $L_{(m_1,n_1),(m_2,n_2)}$.

Theorem 1.1 *Let K be an oriented knot in S^3 and $L_{(m_1,n_1),(m_2,n_2)}$ the above link in S^3, where $m_1 \geq n_1, m_2 \geq n_2$. If there is a compact connected oriented surface F in $S^3 \times [0,1]$ such that*

$$\partial F = F \cap (S^3 \times \{0,1\}) \tag{1.1}$$

$$F \cap (S^3 \times \{0\}) = L_{(m_1,n_1),(m_2,n_2)} \times \{0\} \tag{1.2}$$

$$F \cap (S^3 \times \{1\}) = K \times \{1\} \tag{1.3}$$

then

$$u(K) \geq (m_1 - n_1)(m_2 - n_2) - (m_1 + m_2) - g(F) + 1 \tag{1.4}$$

where $g(F)$ denotes the genus of F.

We shall give an example of the above surface F. We can obtain such surface from band surgery. Let L be an oriented link. Let $B_1, \cdots B_\nu$ be mutually disjoint oriented bands in S^3 such that $B_i \cap L = \partial B_i \cap L = \alpha_i \cup \alpha_i'$, where $\alpha_1, \alpha_1', \cdots, \alpha_\nu, \alpha_\nu'$ are disjoint connected arcs. The closure of $L \cup \partial B_1 \cup \cdots \cup \partial B_\nu - \alpha_1 \cup \alpha_1' \cup \cdots \cup \alpha_\nu \cup \alpha_\nu'$ is also the link. We will write it by L'. If L' has the orientation compatible with the orientation of $L - \bigcup_{i=1,\cdots,\nu} \alpha_i \cup \alpha_i'$ and $\bigcup_{i=1,\cdots,\nu}(\partial B_i - \alpha_i \cup \alpha_i')$, L' is called the link obtained from L by the *band surgery* along the bands $B_1, \cdots, B_\nu$.

Corollary 1.2 *If an oriented knot K in S^3 is obtained from $L_{(m_1,n_1),(m_2,n_2)}$ by the band surgery along the ν bands, then*

$$u(K) \geq (m_1 - n_1)(m_2 - n_2) - \frac{1}{2}(\nu + m_1 - n_1 + m_2 - n_2 - 1). \tag{1.5}$$

Prior to this paper, the author gave a proof of the following result in [4].

Theorem 1.3 *([4]) Let K be an oriented knot in S^3 and $L_{(m,0),(n,0)}$ the above link in S^3. If there is a compact connected oriented surface F in $S^3 \times [0,1]$ such that $\partial F = F \cap (S^3 \times \{0,1\})$, $F \cap (S^3 \times \{0\}) = L_{(m,0),(n,0)} \times \{0\}$, and $F \cap (S^3 \times \{1\}) = K \times \{1\}$, then*

$$u(K) \geq (m-1)(n-1) - g(\hat{F}) \tag{1.6}$$

where $g(\hat{F})$ denotes the genus of $\hat{F}$.

This theorem is the case $n_1 = n_2 = 0$ of Theorem 1.1, and we shall prove the case $n_1 \neq 0$ or $n_2 \neq 0$ by applying Theorem 1.3 in the next section. The proof of Theorem 1.3 is almost same as the argument in Auckly's lecture note [3]. This argument is based on the so-called generalized adjunction formula, which was proved using Seiberg-Witten theory by Kronheimer-Mrowka [7] and Morgan-Szabó-Taubes [10]. Auckly [3] commented that this technique gave an alternative proof of $u((p,q)\text{-torus knot}) = (p-1)(q-1)/2$, conjectured by Milnor [9], which was proved by Kronheimer-Mrowka [5][6]. We shall explain this alternative proof after proving the main results in this paper.

The author proved that the unknotting numbers of 10_{139} and 10_{152} are 4 applying Theorem 1.3 [4]. In fact, they are closed positive braids, so we can give an alternative proof of it by using the result of Rudolph which implies the slice-Bennequin inequality for braid [11], or the one of Menasco which gave an inequality with respect to the positive unknotting numbers [8]. Stoimenow also gave the unknotting numbers of 10_{139} and 10_{152} in [12], where he showed that the unknotting number of any braid positive knot is equal to its genus. Recently, A'Campo proved that the unknotting number of the fibered knot which corresponds to a generically immersed curve in the unit disk is equal to the number of the double points of this curve [1]. He also found that 10_{139} is obtained from such a curve with four double points [2] and that the fact $u(10_{139}) = 4$ is proved by his result.

A'Campo [1] and Rudolph [11] mainly observed links which appear as the intersections of embedded 3-spheres and algebraic curves in the 2-dimensional complex plane. Theorem 1.1 may possibly give sharper estimation for knots which do not appear in any algebraic curves.

2. The proof of main results

To prove Theorem 1.1 by applying Theorem 1.3, we show the following lemma.

Lemma 2.1 *Let K be an oriented knot in S^3 and $L_{(m_1,n_1),(m_2,n_2)}$ the link in S^3 which we defined in the previous section. We suppose $m_1 \geq n_1, m_2 \geq n_2$. We also consider the link $L_{(m_1-n_1,0),(m_2-n_2,0)}$ in S^3. If there is a compact connected oriented surface F in $S^3 \times [0,1]$ such that $\partial F = F \cap (S^3 \times \{0,1\})$, $F \cap (S^3 \times \{0\}) = L_{(m_1,n_1),(m_2,n_2)} \times \{0\}$, and $F \cap (S^3 \times \{1\}) = K \times \{1\}$, then there is a compact connected oriented surface F' in $S^3 \times [-1,1]$ such that*

$$\partial F' = F' \cap (S^3 \times \{-1,1\}) \tag{2.1}$$

$$F' \cap (S^3 \times \{-1\}) = L_{(m_1-n_1,0),(m_2-n_2,0)} \times \{-1\} \tag{2.2}$$

$$F' \cap (S^3 \times \{1\}) = K \times \{1\} \tag{2.3}$$

and

$$g(F') = g(F) + n_1 + n_2. \tag{2.4}$$

Proof. We can suppose that each $K^i_{m_i-n_i+j}$ lies in the tubular neighborhood of L^i_j. Let B^i_j be a trivial band connecting $K^i_{m_i-n_i+j}$ and L^i_j in the above tubular neighborhood. The link obtained from $L_{(m_1,n_1),(m_2,n_2)}$ by the band surgery along the bands B^i_j, $i = 1, 2$, $j = 1, \cdots n_i$ is the link $L_{(m_1-n_1,0),(m_2-n_2,0)} \cup \bigcup_{i=1,2,\ 1\leq j\leq n_i} O^i_j$ where each O^i_j is a split trivial component. Identifying $F' \cap (S^3 \times \{t\})$ with this band surgery in S^3 at time t ($t \in [-1,0]$) and $F' \cap (S^3 \times [0,1])$ with F, we get a proper surface $F' \subset S^3 \times [-1,1]$ with (2.1), (2.2), (2.3), and(2.4). Here we put

$$F' \cap (S^3 \times \{0\}) := L_{(m_1,n_1),(m_2,n_2)} \times \{0\} \tag{2.5}$$

$$F' \cap (S^3 \times \{-\tfrac{1}{4}\}) := (L_{(m_1,n_1),(m_2,n_2)} \cup \bigcup_{i,j} B^i_j) \times \{-\tfrac{1}{4}\} \tag{2.6}$$

$$F' \cap (S^3 \times \{-\tfrac{1}{2}\}) := (L_{(m_1-n_1,0),(m_2-n_2,0)} \cup \bigcup_{i,j} O^i_j) \times \{-\tfrac{1}{2}\} \tag{2.7}$$

$$F' \cap (S^3 \times \{-\tfrac{3}{4}\}) := (L_{(m_1-n_1,0),(m_2-n_2,0)} \cup \bigcup_{i,j} D^i_j) \times \{-\tfrac{3}{4}\} \tag{2.8}$$

$$F' \cap (S^3 \times \{-1\}) := L_{(m_1-n_1,0),(m_2-n_2,0)} \times \{-1\} \tag{2.9}$$

where D^i_j's are the disjoint disks with $\partial D^i_j = O^i_j$. Thus we obtain the desired surface F'. □

Proof of Theorem 1.1. The proof in the case $n_1 = n_2 = 0$ follows from Theorem 1.3. Hence in the case $n_1 \neq 0$ or $n_2 \neq 0$, we have

$$u(K) \geq (m_1 - n_1)(m_2 - n_2) - (m_1 - n_1 + m_2 - n_2) - g(F') + 1 \tag{2.10}$$

by means of Lemma 2.1 and Theorem 1.3. Then we obtain the inequality

$$u(K) \geq (m_1 - n_1)(m_2 - n_2) - (m_1 - n_1 + m_2 - n_2)$$

$$
\begin{aligned}
& -(g(F)+n_1+n_2)+1 \\
= \;& (m_1-n_1)(m_2-n_2)-(m_1+m_2)-g(F)+1. \qquad (2.11)
\end{aligned}
$$

Thus, we obtain the desired inequality (1.4). □

Proof of Corollary 1.2. Let $B_1, \cdots, B_\nu$ be the surgery bands. To apply Theorem 1.1, we construct a suitable surface F in $S^3 \times [0,1]$ similarly to the proof of Lemma 2.1. Identifying $F \cap (S^3 \times \{t\})$ with the band surgery in the time t ($t \in [0,1]$), we get a proper surface F with (1.1), (1.2), and (1.3). Here we put

$$
\begin{aligned}
F \cap (S^3 \times \{0\}) &:= L_{(m_1,n_1),(m_2,n_2)} \times \{0\} && (2.12) \\
F \cap (S^3 \times \{\tfrac{1}{2}\}) &:= (L_{(m_1,n_1),(m_2,n_2)} \cup \bigcup_{1 \le i \le \nu} B_i) \times \{\tfrac{1}{2}\} && (2.13) \\
F \cap (S^3 \times \{1\}) &:= K \times \{1\}. && (2.14)
\end{aligned}
$$

Then $g(F) = \frac{1}{2}(\nu - m_1 - n_1 - m_2 - n_2 + 1)$.

By Theorem 1.1, we have

$$
\begin{aligned}
u(K) \;\geq\; & (m_1-n_1)(m_2-n_2)-(m_1-n_1+m_2-n_2) \\
& -\{\tfrac{1}{2}(\nu - m_1 - n_1 - m_2 - n_2 + 1) + n_1 + n_2\} + 1 \\
= \;& (m_1-n_1)(m_2-n_2) - \frac{1}{2}(\nu + m_1 - n_1 + m_2 - n_2 - 1). \qquad (2.15)
\end{aligned}
$$

Thus, we obtain the desired inequality (1.5). □

Theorem 1.3 gives an alternative proof of $u((p,q)\text{-torus knot}) = (p-1)(q-1)/2$ as Auckly commented in his lecture note [3]. We show this equality by using Corollary 1.2.

Example. Let k, p, q be positive integers. We suppose that p and q are relatively prime and $q < kp$. A torus link of type $(p, 2kp)$ is obtained from a link $L_{(p,0),(kp,0)}$ by the band surgery along the kp bands as Yamamoto showed in [13]. A torus knot of type (p,q) is obtained from a torus link of type $(p, 2kp)$ by smoothing $(2kp-q)(p-1)$ crossings, that is the surgery along the $(2kp-q)(p-1)$ bands. Hence the (p,q)-torus knot is obtained from $L_{(p,0),(kp,0)}$ by the surgery along the $2kp^2 - kp - pq + q$ bands. Then by Corollary 1.2, we have

$$
\begin{aligned}
u((p,q)\text{-torus knot}) \;\geq\; & kp^2 - \frac{1}{2}(2kp^2 - kp - pq + q + p + kp - 1) \\
= \;& \frac{1}{2}(p-1)(q-1). \qquad (2.16)
\end{aligned}
$$

Then we have the equality $u((p,q)\text{-torus knot}) = (p-1)(q-1)/2$ since it is well known that the (p,q)-torus knot is unknotted by changing $(p-1)(q-1)/2$ crossings.

As the above example, Theorem 1.1 and Corollary 1.2 estimate the unknotting numbers at the same time for infinite sequences of the knots obtained from the links of type $L_{(m_1,n_1),(m_2,n_2)}$.

Acknowledgements

The author is partially supported by JSPS Research Fellowships for Young Scientists. She would like to thank Professor Toshitake Kohno for his encouragement. She also would like to thank Professor Norbert A'Campo and Doctor Alexander Stoimenow for their useful comments.

References

[1] N. A'Campo, *Generic immersions of curves, knots, monodromy and gordian number*, preprint.

[2] N. A'Campo, private communication.

[3] D. Auckly, *Surgery, knots, and the Seiberg-Witten equations*, Lectures for the 1995 TGRCIW, preprint.

[4] T. Kawamura, *The unknotting numbers of* 10_{139} *and* 10_{152} *are 4*, Osaka J. Math. **35** (1998), 539-546.

[5] P. Kronheimer and T. Mrowka, *Gauge theory for embedded surfaces, I*, Topology **32**(1993), 773-826.

[6] P. Kronheimer and T. Mrowka, *Gauge theory for embedded surfaces, II*, Topology **34**(1995), 37-97.

[7] P. Kronheimer and T. Mrowka, *The genus of embedded surfaces in the projective plane*, Math. Res. Lett. **1**(1994), 797-808.

[8] W. W. Menasco, *The Bennequin-Milnor unknotting conjectures*, C. R. Acad. Sci. Paris, **t.318**(1994), Série I, 831-836.

[9] J. Milnor, *Singular points of complex hypersurfaces*, Ann. of Math. Studies **61**, Princeton Univ. Press, (1968).

[10] J. Morgan, Z. Szabó and C. Taubes, *A product formula for the Seiberg-Witten invariants and the generalized Thom conjecture*, J. Differential Geom. **44**(1996), 706-788.

[11] L. Rudolph, *Quasipositivity as an obstruction to sliceness*, Bull. Amer. Math. Soc. **29**(1993), 51-59.

[12] A. Stoimenow, *Positive knots, closed braids and the Jones polynomial*, preprint, math.gt/9805078.

[13] M. Yamamoto, *Lower bounds for the unknotting numbers of certain torus knots*, Proc. Amer. Math. Soc. **86**(1982), 519-524.

TORSION LINKING FORMS ON SURFACE-KNOTS AND EXACT 4-MANIFOLDS

AKIO KAWAUCHI
Department of Mathematics, Osaka City University
Sumiyoshi-ku, Osaka 558-8585, Japan

ABSTRACT

We focus an interest on the torsion linking of a surface-knot. It is a knot invariant independent of the surface-knot group and its peripheral subgroup. It is identified with the torsion linking of any associated closed 4-manifold with infinite cyclic first homology. In the case of such a 4-manifold with an exact leaf, the linking of the leaf is identified with an orthogonal sum of it and a hyperbolic linking .

Keywords : torsion linking , surface-knot, exact 4-manifold, exact leaf

0. An aspect to surface-knot theory

A *surface-knot* is a closed connected surface F embedded in the 4-sphere S^4 by a smooth (or locally flat PL) embedding. A *surface-link* is a disjoint union of a finite number of surface-knots in S^4. In this paper, we mainly consider a surface-knot of an oriented surface F particularly with a positive genus. Then the genus of the surface F is called the *genus* of the surface-knot. A surface-knot of genus zero (i.e., an S^2-knot) is usually called a *2-knot*. Two surface-knots F_1 and F_2 are said to *belong to the same knot type* and denoted by $F_1 \cong F_2$, if there is a diffeomorphism (or equivalently, a PL auto-homeomorphism) f of S^4 with $f(F_1) = F_2$ such that both f and $f|_{F_1} : F_1 \to F_2$ are orientation-preserving. A surface-knot F is *trivial* or *TOP-trivial* respectively if it bounds a handlebody embedded in S^4 by a smooth (locally flat PL) embedding or a locally flat topological embedding. It is observed by F. Hosokawa and the author [6] that all the trivial surface-knots of the same genus belong to the same knot type. When we recognize two surface-knots of the same genus, it is a basic problem to determine whether or not they belong to the same knot type. In the case of a positive genus surface-knot F, the most powerful knot invariant known to this problem until now is the surface-knot group $\pi_1(S^4 - F) \cong \pi_1(E_F, b_0)$ and its peripheral subgroup, that is, the image of the natural homomorphism $\pi_1(\partial E_F, b_0) \to \pi_1(E_F, b_0)$ for the compact exterior E_F of F and a point $b_0 \in \partial E_F$ (cf. J. Boyle [1]). In this paper, as a knot type invariant which is independent of the surface-knot group and its peripheral subgroup, we shall emphasize the *torsion linking* ℓ_F of the surface-knot F, a Q/Z-valued *non-singular* symmetric bilinear form on a finite abelian group (called here a *linking*), which is derived from the infinite cyclic covering homology of E_F by applying the second duality in [**8**]. In [**9**] we used the relative torsion linking $\ell_{\emptyset,\partial E_F}$ of $(E_F;\emptyset,\partial E_F)$ which will be seen in this paper to be equivalent to the torsion linking ℓ_F of F to detect a relationship between the surface-knot group $\pi_1(S^4 - F)$ and the genus of F. In this paper, we shall first observe that *there exists a pair of genus one surface-knots whose surface-knot groups are isomorphic by a meridian-preserving and peripheral-subgroup-preserving isomorphism, but one of whose torsion linkings is the zero form and the other, a non-zero form* (see Example 1.7). To clarify a geometric meaning of ℓ_F, we consider a closed connected oriented 4-manifold M such that $H_1(M;Z) \cong Z$ which we call simply a Z^{H_1}*-manifold.* In this case, unless otherwise stated, we specify a generator of $H_1(M;Z)$ to give a unique cohomology class $\gamma \in H^1(M;Z)$ sending this generator to $1 \in Z$. This γ is needed to define the torsion linking ℓ_M of M. In §2, such a Z^{H_1}-manifold, denoted by M_F is constructed (but non-uniquely) from E_F and the 4-manifold $S^1 \times V_0 \# nS^2 \times S^2$ by pasting along their boundaries, where n is any non-negative integer and V_0 is a handlebody with $\partial V_0 \cong F$. For our purpose, we shall consider an *exact* Z^{H_1}-manifold, that is, a Z^{H_1}-manifold which admits an *exact* leaf, a special leaf satisfying a certain exactness condition (see §3). Then we show that *the linking ℓ_V (defined by Poincaré duality) of any exact leaf V of any exact Z^{H_1}-manifold M_F of F is isomorphic to the orthogonal sum of the torsion linking ℓ_F and a hyperbolic linking* (see Theorem 4.2) after observing that ℓ_F is isomorphic to the torsion linking of any Z^{H_1}-manifold M_F of F (see Lemma 2.2) and M_F is exact for a large n (see Theorem 3.5). This is a natural generalization of Farber's result on a Seifert hypersurface of the Farber-Levine linking of an S^2-knot (see M. Farber [**3**]). Further, using the t-isometry of ℓ_F, we can define a Z_4-valued invariant of ℓ_F which we call the *linking signature invariant* of the surface-knot F. This invariant will be discussed in [**13, 14**] as an analogy of the Arf invariant of a classical knot and in [**15**] from a viewpoint of a Rochlin type invariant of a surface-knot. A Z^{H_1}-manifold M is called a Z^{π_1}-*manifold* if $\pi_1(M) \cong Z$. A Z^{π_1}-manifold M is *split* or *TOP-split* respectively if it is diffeomorphic or homeomorphic to the connected sum $S^1 \times S^3 \# M_1$ with M_1 the simply connected 4-manifold obtained uniquely from M by a 2-handle surgery killing $\pi_1(M)$. The concept of an exact Z^{H_1}-manifold is motivated by a trial to correcting the author's error in [**10**] on the TOP-splitness for Z^{π_1}-manifolds (the error is in the statement "the lifting parts of E_s coincide"(p. 494, line 22) which is false when E_s is disconnected). Concerning this error, I. Hambleton-P. Teichner [**5**] gave an example of a non-TOP-split topological Z^{π_1}-manifold of Witt index zero and showed that every Z^{π_1}-manifold of Witt index ≥ 3 is necessarily TOP-split, where the *Witt index* of a Z^{H_1}-manifold M is meant by the integer $(\beta_2(M) - |\text{sign}(M)|)/2$. (Incidentally, it appears unknown whether there exists a non-TOP-split *smooth* Z^{π_1}-manifold.) In Corollary 3.4, we shall show a criterion of the TOP-splitness that *a Z^{π_1}-manifold is exact if and only if it is TOP-split*, meaning that the exactness on

Z^{H_1}-manifolds is a homological extension of the TOP-splitness on Z^{π_1}-manifolds. Some different criteria of the TOP-splitness for a Z^{π_1}-manifold are given by A. Cavicchioli-F. Hegenbarth [**2**] and M. Freedman-F. Quinn [**4**]. In J. Hillman-the authors' paper [**7**], it is shown that a surface-knot F of genus g is TOP-trivial if and only if the surface-knot group $\pi_1(S^4 - F)$ is isomorphic to Z, under the hypothesis that any zero signature spin Z^{π_1}-manifold M with $\beta_2(M) = 2g$ or $4g$ is TOP-split. Combining this result with I. Hambleton-P. Teichners' result above, we can conclude that *a surface-knot F of genus $g \geq 3$ is TOP-trivial if and only if* $\pi_1(S^4 - F) \cong Z$. For $g = 0$ the assertion is well-known as Freedman's theorem (cf. [**4**]), and for $g = 1, 2$ at present it remains open.

In §1, we define the torsion linking of a Z^{H_1}-manifold and then the torsion linking of a surface-knot. In §2, we explain how to construct a Z^{H_1}-manifold from a given surface-knot and show that the torsion linking ℓ_F of a surface-knot F is equal to the torsion linking of any Z^{H_1}-manifold M_F of F. In §3, we explain three basic results on exactness, namely an algebraic characterization of an exact leaf in Theorem 3.1, an algebraic characterization of an exact Z^{H_1}-manifold in Theorem 3.3, and the stable existence of an exact Z^{H_1}-manifold in Theorem 3.5. The proof of Corollary 3.4 is given here, but the proofs of Theorems 3.1, 3.3, and 3.5 are given in [**12**]. In §4, a relationship between the torsion linking of an exact Z^{H_1}-manifold M and the linking of any exact leaf V of M is established.

1. The torsion linking of a surface-knot

Let $p : \tilde{W} \to W$ be an infinite cyclic connected covering over a compact connected oriented 4-manifold W. Let $\gamma \in H^1(W; Z)$ be the associated (indivisible) cohomology class of the covering p. For a compact submanifold A of W, let $\tilde{A} = p^{-1}(A)$. Let $\Lambda = Z[t, t^{-1}]$. Then the homology $H_*(\tilde{W}, \tilde{A}; Z)$ is naturally regarded as a finitely generated Λ-module (see [**8**] for compact topological manifolds), where t denotes a generator of the infinite cyclic covering group of p specified by γ. For a Λ-module H, we denote the Λ-torsion part of H by TH and the Λ-torsion-free part H/TH by BH. Let $E^q H = Ext^q_\Lambda(H, \Lambda)$. A *leaf* V of γ is a bi-collared 3-submanifold $V \subset W$ such that $[V] = \gamma \cap [W] \in H_3(W, \partial W; Z)$. We can always take V connected (cf. [**11**]). A canonical lift of V to $\tilde{W}$ represents an element μ of $H_3(\tilde{W}, \partial\tilde{W}; Z)$ with $t\mu = \mu$. It turns out that μ is determined uniquely by the orientation of W and γ, which is called the *fundamental class* of the covering p (see [**8**,**11**]). We use the fact that every finitely generated Λ-module H admits a unique maximal finite Λ-submodule $DH (\cong E^2E^2H)$ (see [**8**, §3]). Let A and A' be compact 3-submanifolds of ∂W or $\emptyset$ such that $A' = \mathrm{cl}(\partial W - A)$. The Poincaré duality on $\tilde{W}$ over Λ and the universal coefficient exact sequence over Λ define a natural t-anti epimorphism

$$\theta_{A,A'} : DH_1(\tilde{W}, \tilde{A}; Z) \longrightarrow E^1BH_2(\tilde{W}, \tilde{A}'; Z)$$

which is an invariant of $(W; A, A'; \gamma)$ (see [8]). Let $DH_1(\tilde{W}, \tilde{A}; Z)^\theta$ be the kernel of $\theta_{A,A'}$. Similarly, let $DH_1(\tilde{W}, \tilde{A}'; Z)^\theta$ be the kernel of $\theta_{A',A}$ interchanging the roles of A and A'. Let $\tau H^*(\bullet; Z)$ be the image of the natural monomorphism

$$Ext^1_Z(H_{*-1}(\bullet; Z), Z) \longrightarrow H^*(\bullet; Z)$$

in the usual universal coefficient theorem. The following duality theorem between the finite Λ-modules $DH_1(\tilde{W}, \tilde{A}; Z)$ and $DH_1(\tilde{W}, \tilde{A}'; Z)$ is also given in [8]:

Lemma 1.1 (Duality Theorem). The cap product $\cap\mu : \tau H^2(\tilde{W}, \tilde{A}; Z) \to H_1(\tilde{W}, \tilde{A}'; Z)$ induces an isomorphism

$$\cap\mu : \mathrm{Hom}_Z(DH_1(\tilde{W}, \tilde{A}; Z)^\theta, Q/Z) \cong DH_1(\tilde{W}, \tilde{A}'; Z)^\theta.$$

By using Lemma 1.1, we define

$$\ell_{A,A'}(x, y) = ((\cap\mu)^{-1}(y))(x).$$

Since $\tau H^2(\bullet; Z)$ is the image of the Bockstein homomorphism $\delta_{Q/Z} : H^1(\bullet; Q/Z) \to H^2(\bullet; Z)$, we see from Lemma 1.1 that for any $x \in DH_1(\tilde{W}, \tilde{A}; Z)^\theta$ and $y \in DH_1(\tilde{W}, \tilde{A}'; Z)^\theta$, there are element $u_x \in H^1(\tilde{W}, \tilde{A}'; Q/Z)$ and $u_y \in H^1(\tilde{W}, \tilde{A}; Q/Z)$ such that $(\delta_{Q/Z} u_x) \cap \mu = x$ and $(\delta_{Q/Z} u_y) \cap \mu = y$. Then we have

$$\ell_{A,A'}(x, y) =< \delta_{Q/Z} u_x \cup u_y, \mu >=< u_x \cup \delta_{Q/Z} u_y, \mu >\in Q/Z.$$

Since $t\mu = \mu$, the following corollary is obtained:

Corollary 1.2. There is a t-isometric non-singular bilinear form

$$\ell_{A,A'} : DH_1(\tilde{W}, \tilde{A}; Z)^\theta \times DH_1(\tilde{W}, \tilde{A}'; Z)^\theta \longrightarrow Q/Z$$

which is an invariant of $(W; A, A'; \gamma)$.

Further, a symmetric property of cup product implies the identity

$$\ell_{A',A}(y, x) = \ell_{A,A'}(x, y).$$

In case $A = A' = \partial W = \emptyset$, the form $\ell_{\emptyset,\emptyset}$ is called the *torsion linking of* (W, γ) or simply *of* W if W is a Z^{H_1}-manifold. In case $\partial W \neq \emptyset$, this form $\ell_{A,A'}$ is called the *relative torsion linking of* $(W; A, A'; \gamma)$, or simply *of* $(W; A, A')$ if $H_1(W; Z) \cong Z$ and $H_1(W; Z)$ has a distinguished generator. A Λ-module H is said to be $(t-1)$-*divisible* if $t - 1$ is an automorphism of H. For an abelian group G, we denote the torsion part of G by tG and the torsion-free group $G/\mathrm{t}G$ by bG. It is well-known that every finitely generated $(t-1)$-divisible Λ-module H is a torsion Λ-module

$H = TH$ and the Z-torsion part $\mathrm{t}H$ is a finite Λ-module $\mathrm{t}H = DH$. This was first observed by M. A. Kervaire [**17**] (see also [**8**,§3]). For a surface-knot F in S^4, let N_F be a tubular neighborhood of F in S^4. *We choose a trivialization* $N_F = F \times D^2$ *of the trivial normal-disk bundle* N_F *over* F *so that every* 1*-cycle in* $F \times 1 (1 \in \partial D^2 = S^1)$ *is null-homologous in* E_F. Let $E_F = \mathrm{cl}(S^4 - N_F)$ be the *exterior* of the surface knot F in S^4. Then $H_1(E_F; Z) \cong Z$. The cohomology class $\gamma \in H^1(E_F; Z)$ needed for the infinite cyclic covering $\tilde{E}_F \to E_F$ is taken to send the meridian generator of $H_1(E_F; Z)$ to 1. By Wang exact sequence, the finitely generated Λ-module $H_1(\tilde{E}_F; Z)$ is $(t-1)$-divisible. The following lemma uses this fact:

Lemma 1.3. For a surface-knot F in S^4, the relative torsion linking $\ell_{\emptyset,\partial E_F}$ of $(E_F; \emptyset, \partial E_F)$ induces a t-isometric linking

$$\ell_F : \mathrm{t}H_1(\tilde{E}_F; Z)^\theta \times \mathrm{t}H_1(\tilde{E}_F; Z)^\theta \longrightarrow Q/Z.$$

We call this form ℓ_F the *torsion linking of* the *surface-knot* F.

Proof. By our choice of a trivialization $N_F = F \times D^2$, we have $\partial \tilde{E}_F = F \times R$ and $H_1(\partial \tilde{E}_F; Z) \cong [\Lambda/(t-1)]^{2g}$ where g denotes the genus of F. Since $H_1(\tilde{E}_F; Z)$ is $(t-1)$-divisible and $\partial \tilde{E}_F$ is connected, we obtain from the homology exact sequence of the pair $(\tilde{E}_F, \partial \tilde{E}_F)$ a natural Λ-isomorphism

$$\tilde{i}_* : H_1(\tilde{E}_F; Z) \cong H_1(\tilde{E}_F, \partial \tilde{E}_F; Z).$$

By the second duality of [**8**], $DH_2(\tilde{E}_F, \partial \tilde{E}_F; Z) = 0$. Then by the first duality of [**8**], $TH_2(\tilde{E}_F, \partial \tilde{E}_F; Z)$ is t-anti isomorphic to $E^1 H_1(\tilde{E}_F; Z)$ and hence is $(t-1)$-divisible. This implies that the homology exact sequence of the pair $(\tilde{E}_F, \partial \tilde{E}_F)$ induces a short exact sequence

$$0 \to BH_2(\tilde{E}_F; Z) \longrightarrow BH_2(\tilde{E}_F, \partial \tilde{E}_F; Z) \longrightarrow [\Lambda/(t-1)]^{2g} \to 0.$$

We note that $E^1 BH_2(\tilde{E}_F; Z)$ and $E^1 BH_2(\tilde{E}_F, \partial \tilde{E}_F; Z)$ are $(t-1)$-divisible, since $\theta_{\partial E_F, \emptyset}$ and $\theta_{\emptyset, \partial E_F}$ are t-anti epimorphisms. Thus, this short exact sequence induces the following commutative square with vertical natural isomorphisms:

$$\begin{array}{ccc} DH_1(\tilde{E}_F; Z) & \xrightarrow{\theta_{\emptyset,\partial E_F}} & E^1 BH_2(\tilde{E}_F, \partial \tilde{E}_F; Z) \\ \tilde{i}_* \cong \downarrow & & \cong \downarrow \\ DH_1(\tilde{E}_F, \partial \tilde{E}_F; Z) & \xrightarrow{\theta_{\partial E_F,\emptyset}} & E^1 BH_2(\tilde{E}_F; Z). \end{array}$$

By the definition of the relative torsion linking $\ell_{\emptyset,\partial E_F}$ and the naturality of cup product, we can define ℓ_F by the identity

$$\ell_F(x, y) = \ell_{\emptyset,\partial E_F}(x, \tilde{i}_*(y))$$

for $x, y \in \mathrm{t}H_1(\tilde{E}_F; Z)^\theta$. □

The following addition lemma follows from the definition and a property of cap product:

Lemma 1.4. Let F be the connected sum of surface-knots F_i $(i = 1, 2)$ in S^4. Then the torsion linking ℓ_F of F is Λ-isomorphic to the orthogonal sum of the torsion linkings ℓ_{F_i} of F_i $(i = 1, 2)$.

Proof. We can consider that $\tilde{E}_F$ is a union of $\tilde{E}_{F_1}$ and $\tilde{E}_{F_2}$ such that $\tilde{E}_{F_1} \cap \tilde{E}_{F_2} \cong D^2 \times R$ for a 2-disk D^2. By the excision isomorphism and the isomorphism $H_1(\tilde{E}_F; Z) \cong H_1(\tilde{E}_F, \partial\tilde{E}_F; Z)$ for every surface-knot F, we have a natural isomorphism

$$\begin{aligned} j_* : H_1(\tilde{E}_F, \partial\tilde{E}_F; Z) &\xrightarrow{\cong} H_1(\tilde{E}_F, \tilde{E}_{F_2} \cup \partial\tilde{E}_{F_1}; Z) \oplus H_1(\tilde{E}_F, \tilde{E}_{F_1} \cup \partial\tilde{E}_{F_2}; Z) \\ &\cong H_1(\tilde{E}_{F_1}, \partial\tilde{E}_{F_1}; Z) \oplus H_1(\tilde{E}_{F_2}, \partial\tilde{E}_{F_2}; Z). \end{aligned}$$

Also, the natural homomorphism

$$j_* : H_2(\tilde{E}_F, \partial\tilde{E}_F; Z) \longrightarrow H_2(\tilde{E}_{F_1}, \partial\tilde{E}_{F_1}; Z) \oplus H_2(\tilde{E}_{F_2}, \partial\tilde{E}_{F_2}; Z)$$

sends the fundamental class μ of the covering $\tilde{E}_F \to E_F$ to to the sum $\mu_1 + \mu_2$ for the fundamental class μ_i of the covering $\tilde{E}_{F_i} \to E_{F_i}$ $(i = 1, 2)$ because μ and μ_i are respectively represented by Seifert hypersurfaces V and V_i for F and F_i with some boundary-collars removed (see **[11]**) and we can take V as a disk-sum of V_1 and V_2. Further, using the natural isomorphism $i^* : \tau H^2(\tilde{E}_F; Z) \cong \tau H^2(\tilde{E}_{F_1}; Z) \oplus \tau H^2(\tilde{E}_{F_2}; Z)$, we obtain from the naturality of cap product the following commutative square:

$$\begin{array}{ccc} \tau H^2(\tilde{E}_F; Z) & \xrightarrow{\cap\mu} & H_1(\tilde{E}_F, \partial\tilde{E}_F; Z) \\ {\scriptstyle i^*}\downarrow\cong & & {\scriptstyle j_*}\downarrow\cong \\ \tau H^2(\tilde{E}_{F_1}; Z) \oplus \tau H^2(\tilde{E}_{F_2}; Z) & \xrightarrow{\cap\mu_1 + \cap\mu_2} & H_1(\tilde{E}_{F_1}, \partial\tilde{E}_{F_1}; Z) \oplus H_1(\tilde{E}_{F_2}, \partial\tilde{E}_{F_2}; Z). \end{array}$$

By Lemma 1.1 and the definition of the torsion linking of a surface-knot, we have the desired result. □

To compare with our torsion linking, we often consider a symmetric bilinear form

$$\ell_V : \mathrm{t}H_1(V; Z) \times \mathrm{t}H_1(V; Z) \to Q/Z$$

of a compact oriented 3-manifold V which is defined by Poincaré duality. This form ℓ_V is called the *linking of* V if it is non-singular. We note that if $\partial V = \emptyset$, then ℓ_V is a linking, but in general it may be a singular form which we do not call "linking". When $\partial V \neq \emptyset$, let B and B' be $\emptyset$ or compact subsurfaces of ∂V such that $B' = \mathrm{cl}(\partial V - B)$. Then by Poincaré duality we have a non-singular bilinear form

$$\ell_{B,B'} : \mathrm{t}H_1(V,B;Z) \times \mathrm{t}H_1(V,B';Z) \longrightarrow Q/Z,$$

which we call the *relative linking of* $(V;B,B')$. Then $\ell_{B,B'}(x,y) = \ell_{B',B}(y,x)$ for all $x \in \mathrm{t}H_1(V,B;Z)$ and $y \in \mathrm{t}H_1(V,B';Z)$. The following example gives the first non-trivial example on the torsion-linkings of surface-knots:

Example 1.5. Let K be a S^2-knot in S^4. Then $H_1(\tilde{E}_K;Z)$ is $(t-1)$-divisible and $H_2(\tilde{E}_K;Z)$ is Λ-torsion, i.e., $BH_2(\tilde{E}_K;Z) = 0$, by the Wang exact sequence. Hence the torsion linking ℓ_K of K is given as

$$\ell_K : \mathrm{t}H_1(\tilde{E}_K;Z) \times \mathrm{t}H_1(\tilde{E}_K;Z) \longrightarrow Q/Z.$$

This linking coincides with the Farber-Levine pairing (cf. M. Farber [**3**], J. Levine [**18**]). If F is the connected sum of K and a trivial surface-knot F^0 in S^4 of genus g, Lemma 1.4 says that $\ell_F : DH_1(E_F;Z)^\theta \times DH_1(E_F;Z)^\theta \to Q/Z$ is Λ isomorphic to ℓ_K. Thus, if $\mathrm{t}H_1(\tilde{E}_K;Z) \neq 0$, then we have lots of examples on the non-trivial torsion linkings of high genus surface-knots. Farber's result of [**3**] says that if V is a Seifert hypersurface for an S^2-knot K, then the linking ℓ_V : $\mathrm{t}H_1(V;Z) \times \mathrm{t}H_1(V;Z) \to Q/Z$ of V is isomorphic to the orthogonal sum of the linking ℓ_K and a hyperbolic linking, i.e., a linking represented by the block sum of the matrices

$$\begin{pmatrix} 0 & \frac{1}{n} \\ \frac{1}{n} & 0 \end{pmatrix}$$

with a finite number of integers $n \geq 2$. In Theorem 4.2, we shall give a certain generalization of this fact to a positive genus surface-knot. Here, we observe that a direct generalization of this fact to a positive genus surface-knot is impossible. For example, for any integer $n \geq 2$, we consider a closed oriented 3-manifold V' such that V' is smoothly embedded in S^4 and $H_1(V';Z) \cong Z_n \oplus Z_n$. [For example, embed the mapping cone $C(f_n)$ of a piecewise-linear map $f_n : S^1 \to S^1$ of degree n into S^4 and then take as V' the boundary of a regular neighborhood of the 2-dimensional polyhedron $C(f_n)$ in S^4.] Let T be a solid torus in V' whose core circle represents a generator of a direct summand Z_n in $H_1(V';X) \cong Z_n \oplus Z_n$. Let $V = \mathrm{cl}(V' - T)$. Then we have

$$\mathrm{t}H^2(V;Z) \cong \mathrm{t}H_1(V,\partial V;Z) \cong \mathrm{t}H_1(V',T;Z) \cong Z_n,$$

so that $H_1(V;Z) \cong Z \oplus Z_n$. This implies that a trivial surface F^0 of genus one admits a Seifert hypersurface V with $H_1(V;Z) \cong Z \oplus Z_n$. In case n is odd, we have an orthogonal splitting $Z_n \oplus Z_n$ of $H_1(V';Z)$ with respect to $\ell_{V'}$. Then for the 3-manifold V constructed above, the form $\ell_V : \mathrm{t}H_1(V;Z) \times \mathrm{t}H_1(V;Z) \to Q/Z$

is seen to be a linking. Since $\ell_{F^0} = 0$, this means that Farber's result does not directly generalize to a positive genus surface-knot. On the other hand, a weaker result for every surface-knot F is known by M. Sekine **[19]** (see Lemma 4.1 later for a Z^{H_1}-manifold version). Namely, the relative torsion-linking

$$\ell_{\emptyset,\partial E_F} : DH_1(\tilde{E}_F; Z)^\theta \times DH_1(\tilde{E}_F, \partial\tilde{E}_F; Z)^\theta \longrightarrow Q/Z$$

for every surface-knot F is isomorphic to the induced non-singular form of a restricted form of the relative linking $\ell : \mathrm{t}H_1(V; Z) \times \mathrm{t}H_1(V, \partial V; Z) \to Q/Z$ for every Seifert hypersurface V of F. Our generalization of Farber's result uses this fact. □

In the following example, we explain some properties of a ribbon surface-knot:

Example 1.6. A *ribbon surface-knot* F_R in S^4 is a surface-knot obtained from a trivial S^2-link $\cup_{i=1}^r S_i^2$ in S^4 by surgery along 1-handles with core arcs β_j $(j = 1, 2, \ldots s)$. Let x_i $(i = 1, 2, \ldots, r)$ be a basis of the free group $\pi_1(S^4 - (\cup_{i=1}^r S_i^2)\ b_0)$ such that x_i is represented by a path consisting of a meridian circle S_i^1 and an arc α_i from the base point b_0 to a point $p_i \in S_i^1$. If β_j joins a point $q_{k_j} \in S_{k_j}^2$ with a point $q_{h_j} \in S_{h_j}^2$, then we can construct a closed path β_j' with base point b_0 in $S^4 - (\cup_{i=1}^r S_i^2)$ from α_{k_j}, β_j, and α_{h_j} by slightly pushing out the end points q_{k_j} and q_{h_j} of β_j into $N_{k_j} - S_{k_j}^2$ and $N_{h_j} - S_{h_j}^2$, respectively, and then joining the point resulting from q_{k_j} with p_{k_j} by an arc α_{k_j}' in $N_{k_j} - S_{k_j}^2$ and the point resulting from q_{h_j} with p_{h_j} by an arc α_{h_j}' in $N_{h_j} - S_{h_j}^2$, where N_i $(i = 1, 2, \ldots, r)$ are mutually disjoint regular neighborhoods of S_i $(i = 1, 2, \ldots, r)$ in S^4. Let $w_j = [\beta_j'] \in \pi_1(S^4 - (\cup_{i=1}^r S_i^2), b_0)$. When the orientation of β_j' indicates the orientation of β_j with the initial point q_{k_j} and the end point q_{h_j} for each j, we obtain the following Wirtinger presentation of $\pi_1(S^4 - F_R, b_0)$ from the van Kampen theorem:

$$P = \langle x_1, x_2, \ldots, x_r | R_1, R_2, \ldots, R_s \rangle, \quad R_j = x_{h_j}^{-1} w_j^{-1} x_{k_j} w_j (j = 1, 2, \ldots, s).$$

We note that the group $G(P)$ of the presentation P does not depend on a choice of the arcs α_{h_j}' and α_{k_j}'. T. Yajima [21] first made this calculation and observed in principle that every Wirtinger presentation P with $H_1(G(P); Z) \cong Z$ is realizable by the surface-knot group of a ribbon surface-knot F_R in S^4 in this way. The ribbon surface-knot F_R has a Seifert hypersurface V_R diffeomorphic to the connected sum of a handlebody and some number of $S^1 \times S^2$ (see **[16]**). Since $\mathrm{t}H_1(V_R; Z) = 0$, the relative torsion linking $\ell_{\emptyset,\partial E_{F_R}} = 0$ by Sekine' result stated in Example 1.5 and hence the torsion linking

$$\ell_{F_R} : DH_1(\tilde{E}_{F_R}; Z)^\theta \times DH_1(\tilde{E}_{F_R}; Z)^\theta \longrightarrow Q/Z$$

is the zero form by Lemma 1.3. □

The following example shows that the torsion linking of a surface-knot is independent of the surface-knot group and its peripheral subgroup.

Example 1.7. Let K be the 2-twist spun (S^2-)knot of the 2-bridge knot of type (p, q) for any odd $p > 1$ which is a fibered S^2-knot with fiber a punctured manifold of the lens space $L(p, q)$ (see E. C. Zeeman [**22**]). This 2-sphere-knot group $\pi_1(S^4 - K)$ has the presentation $< x, a|a^p = 1, x^{-1}ax = a^{-1} >$ where x denotes a meridian element of K (see M. Teragaito [**20**, Lemma 1]). By taking $a = x^{-1}y$, this presentation changes into the following Wirtinger presentation:

$$< x, y|y = (x^{-1}y)^m x(x^{-1}y)^{-m}, y = ((x^{-1}y)^m x^{-1})^{-1} x((x^{-1}y)^m x^{-1}) >,$$

where x and y are meridian elements of K. By Example 1.6, there is a ribbon surface-knot F_R of genus one in S^4 with surface-knot group $\pi_1(S^4 - F_R)$ has the same presentation with x and y meridian elements. Let F be a surface-knot of genus one which is the connected sum of K and a trivial surface-knot F^0 of genus one. Then $\pi_1(S^4 - F)$ and $\pi_1(S^4 - F_R)$ are meridian-preservingly isomorphic. Since the commutator subgroups of these groups are cyclic groups of order p which are abelian, each peripheral subgroup is an infinite cyclic group with a meridian generator. Thus, *there is a meridian-preserving and peripheral-subgroup-preserving isomorphism* $\pi_1(S^4 - F) \cong \pi_1(S^4 - F_R)$. By Example 1.5, the torsion linking ℓ_F is a non-zero form since $DH_1(\tilde{E}_K; Z)^\theta = H_1(\tilde{E}_K; Z) \cong \Lambda/(p, t+1) \cong Z_p$, and by Example 1.6 $\ell_{F_R} = 0$, meaning that *the torsion linkling is independent of the surface-knot group and its peripheral subgroup.* In particular, *F_R and F belong to distinct unoriented knot types.* Another interesting property of these surface-knots F_R and F is that *F_R and F admit diffeomorphic Seifert hypersurfaces in S^4*. By Example 1.6, F_R has a Seifert hypersurface diffeomorphic to $V_d = (S^1 \times D^2)\#d(S^1 \times S^2)$ for some d, so that F_R has a Seifert hypersurface diffeomorphic to $V_d\#L(p, q)\#L(p, -q)$, for $L(p, q)\#L(p, -q)$ is smoothly embedded in S^4. Removing a Heegaard solid torus from the connected summand $L(p, q)$ of $L(p, q)\#L(p, -q)(\subset S^4)$, we see that the trivial surface-knot F^0 of genus one has a Seifert hypersurface diffeomorphic to $(S^1 \times D^2)\#L(p, -q)$, so that $F = K\#F^0$ has also a Seifert hypersurface diffeomorphic to $V_d\#L(p, q)\#L(p, -q)$ as desired. A higher genus example can be easily obtained by taking the connected sums of F_R and F with a trivial surface-knot of any non-zero genus, respectively. The knot types of F in the case when we fix p and vary q are discussed in [**13**]. □

2. Constructing a $\mathbf{Z}^{H_1}$-manifold from a surface-knot

For the exterior E_F of a surface-knot F in S^4 and a handlebody V_0 with $\partial V_0 = F$, let ϕ be an orientation-reversing diffeomorphism from $\partial V_0 \times S^1$ onto $\partial E_F = F \times S^1$ preserving the S^1-factors by counting our choice of a trivialization $N_F = S^1 \times D^2$. Since $H_1(E_F; Z) \cong Z$, we see from the Mayer-Vietoris sequence and the van Kampen theorem that the closed smooth 4-manifold $M_\phi = E_F \cup_\phi V_0 \times S^1$ is a Z^{H_1}-manifold and the inclusion $E_F \subset M_\phi$ induces an epimorphism $\pi_1(S^4 - F) = \pi_1(E_F) \to \pi_1(M_\phi)$. The following example shows that in general the group $\pi_1(M_\phi)$ and the topological type of M_ϕ depend on a choice of ϕ:

Example 2.1. Let k be a non-trivial knot in the 3-disk $D^3 \subset S^3$. Let $F = k \times S^1 \subset D^3 \times S^1 \subset \partial(D^3 \times D^2) = S^4$ be a surface-knot of genus 1. We take the simple loops $c = p \times S^1 (p \in k)$ and $c' = k \times 1$ on F. Let $V_0 = S^1 \times D^2$, $c_0 = 1 \times S^1$, and $c'_0 = S^1 \times 1$. Let ϕ and ϕ' be orientation-reversing diffeomorphisms $\partial V_0 \times S^1 \to \partial N_F = F \times S^1$ such that $\phi_1(c_0) = c$, $\phi_1(c'_0) = c'$, $\phi'_1(c_0) = c'$, and $\phi'_1(c'_0) = c$. Then by the van Kampen theorem the fundamental groups of the resulting Z^{H_1}-manifolds M_ϕ and $M_{\phi'}$ are given as follows:

$$\pi_1(M_\phi) \cong \pi_1(S^3 - k) \quad \textit{and} \quad \pi_1(M_{\phi'}) \cong \pi_1(\chi(k,0)),$$

where $\chi(k,0)$ denotes the 0-surgery manifold of S^3 along the knot k. Then we have

$$H_2(\pi_1(M_\phi); Z) \cong H_2(\pi_1(S^3 - k); Z) = 0.$$

On the other hand, since $\chi(k,0)$ is an irreducible manifold by the property R, we have

$$H_2(\pi_1(M_{\phi'}); Z) \cong H_2(\pi_1(\chi(k,0)); Z) \cong H_2(\chi(k,0); Z) \cong Z.$$

Thus, we see that $\pi_1(M_\phi)$ and $\pi_1(M_{\phi'})$ are not isomorphic. □

In spite of Example 2.1, we shall observe in the following lemma that the torsion linking of M_ϕ for every attachment $\phi : \partial V_0 \times S^1 \cong \partial N_F = F \times S^1$ preserving the S^1-factor is inherited from the torsion linking of the surface-knot F:

Lemma 2.2. For every orientation-reversing diffeomorphism

$$\phi : \partial V_0 \times S^1 \cong \partial E_F = F \times S^1$$

preserving the S^1-factor and every closed connected oriented 4-manifold W_0 with $H_1(W_0; Z) = 0$, the torsion linking of the Z^{H_1}-manifold $M_\phi \# W_0$ is Λ-isomorphic to ℓ_F.

Any Z^{H_1}-manifold $M_\phi \# W_0$ in Lemma 2.2 with $W_0 = \# n S^2 \times S^2$ for any $n \geq 0$ is simply called, in this paper, *a* Z^{H_1}*-manifold of the surface-knot* F and denoted by M_F.

Proof. Since $H_d(\tilde{M}_\phi, \tilde{E}_F; Z) = H_d((V_0, F) \times R; Z)$ is 0 for $d = 1$ and $[\Lambda/(t-1)]^g$ if $d = 2$, where g denotes the genus of F, and $H_1(\tilde{E}_F; Z)$ is $(t-1)$-divisible, we obtain from the homology exact sequence of the pair $(\tilde{M}_\phi, \tilde{E}_F)$ a natural isomorphism

$$H_1(\tilde{E}_F; Z) \cong H_1(\tilde{M}_\phi; Z).$$

By the homology exact sequence of the pair $(\tilde{M}_\phi, V_0 \times R)$ induces a natural short exact sequence

$$0 \to BH_2(\tilde{M}_\phi; Z) \longrightarrow BH_2(\tilde{M}_\phi, V_0 \times R; Z) = BH_2(\tilde{E}_F, \partial \tilde{E}_F; Z) \longrightarrow [\Lambda/(t-1)]^g \to 0.$$

Since $E^1BH_2(\tilde{E}_F, \partial\tilde{E}_F; Z)$ and $E^1BH_2(\tilde{M}_\phi; Z)$ are $(t-1)$-divisible by the t-anti epimorphisms $\theta_{\emptyset,\partial E_F}$ and $\theta_{\emptyset,\emptyset}$, we have the following commutative square with vertical natural isomorphisms :

$$\begin{array}{ccc} DH_1(\tilde{E}_F; Z) & \xrightarrow{\theta_{\emptyset,\partial E_F}} & E^1BH_2(\tilde{E}_F, \partial\tilde{E}_F; Z) \\ \cong\downarrow & & \cong\downarrow \\ DH_1(\tilde{M}_\phi; Z) & \xrightarrow{\theta_{\emptyset,\emptyset}} & E^1BH_2(\tilde{M}_\phi; Z), \end{array}$$

which implies by the definition of linking that the torsion linking of M_ϕ is Λ-isomorphic to ϕ_F. Writing $W = M_\phi \# W_0$, we also have the following commutative square with vertical natural isomorphisms :

$$\begin{array}{ccc} DH_1(\tilde{M}_\phi; Z) & \xrightarrow{\theta_{\emptyset,\emptyset}} & E^1BH_2(\tilde{M}_\phi; Z) \\ \cong\downarrow & & \cong\downarrow \\ DH_1(\tilde{W}) & \xrightarrow{\theta_{\emptyset,\emptyset}} & E^1BH_2(\tilde{W}; Z), \end{array}$$

which also implies that the torsion linking of W is Λ-isomorphic to the torsion linking of M_ϕ. □

We also note that there exists an orientation-reversing diffeomorphism ϕ such that the resulting Z^{H_1}-manifold M_ϕ is a spin 4-manifold (that is, the integral intersection form s of $H_2(M_\phi; Z)$ is an even form) (see [**7**]).

3. Three basic results on an exact $\mathbf{Z}^{H_1}$-manifold

For a leaf V of a Z^{H_1}-manifold M, we denote by M_V the 4-manifold obtained from M by splitting it along the leaf V. The boundary ∂M_V is the disjoint union $(-V^+) + V^-$ of the two copies $V^\pm$ of V under the natural identifications $i^\pm$: $V \cong V^\pm (\subset \partial M_V)$. For the infinite cyclic connected covering $p : \tilde{M} \to M$, we may consider that $p^{-1}(M_V, (-V^+) + V^-) = (\tilde{M}_V, (-\tilde{V}^+) + \tilde{V}^-)$ is the disjoint union of the covering translations $(t^i M_V, (-t^i V^+) + t^i \tilde{V}^-)$ $(i = 0, \pm 1, \pm 2, \ldots)$ of a pair $(M_V, (-V^+) + V^-)$ lifted to $\tilde{M}$, and $p^{-1}(V) = \tilde{V}$ is the disjoint union of $t^i V$ $(i = 0, \pm 1, \pm 2, \ldots)$ of V lifted to $\tilde{M}$. From a technical reason, we often regard M_V as $M_V = \mathrm{cl}(M - V \times [-1, 1])$ for a bi-collar neighborhood $V \times [-1, 1]$ of V in M. For a leaf V of a Z^{H_1}-manifold M, we call the image of the natural homomorphism

$$\tilde{i}_* : H_2(\tilde{V}; Z) \longrightarrow BH_2(\tilde{M}; Z)$$

the *leaf submodule* of $BH_2(\tilde{M}; Z)$ on $\tilde{V}$ and denote it by $X(\tilde{V})$. The Λ-intersection form $S : H_2(\tilde{M}; Z) \times H_2(\tilde{M}; Z) \to \Lambda$ defined by $S(x, y) = \sum_{i=-\infty}^{+\infty} s(t^i x, y) t^i$ (where s denotes the integral intersection form) induces a non-degenerate (but not necessarily non-singular) Λ-Hermitian form $BH_2(\tilde{M}; Z) \times BH_2(\tilde{M}; Z) \to \Lambda$ (cf. [**8**]), which we call also the Λ-intersection form and denote also by the same symbol S. Then we can state our first basic result:

Theorem 3.1. For a leaf V of a Z^{H_1}-manifold M, the following statements (1)-(3) are mutually equivalent:

(1) The natural semi-exact sequence

$$0 \to \mathrm{t}H_2(\tilde{M},\tilde{V};Z) \xrightarrow{\tilde{\partial}} \mathrm{t}H_1(\tilde{V};Z) \xrightarrow{\tilde{i}_*} \mathrm{t}H_1(\tilde{M};Z)$$

induced from the homology exact sequence of the pair $(\tilde{M},\tilde{V})$ is exact.

(2) The natural semi-exact sequence

$$0 \to \mathrm{t}H_2(\tilde{M},\tilde{M}_V;Z) \xrightarrow{\tilde{\partial}'} \mathrm{t}H_1(\tilde{M}_V;Z) \xrightarrow{\tilde{i}'_*} \mathrm{t}H_1(\tilde{M};Z)$$

induced from the homology exact sequence of the pair $(\tilde{M},\tilde{M}_V)$ is exact.

(3) There is a splitting $BH_2(\tilde{M};Z) \cong P \oplus X(\tilde{V}) \oplus Y$ such that $X(\tilde{V})$ is the leaf submodule on $\tilde{V}$ and the following conditions (3.1)-(3.3) are all satisfied:

(3.1) The direct summands P and $X(\tilde{V})\oplus Y$ are orthogonal, i.e., $P\perp(X(\tilde{V})\oplus Y)$ with respect to $S : BH_2(\tilde{M};Z) \times BH_2(\tilde{M};Z) \to \Lambda$.

(3.2) There is a Λ-basis for P which are represented by 2-cycles in M_V and on which the restriction $S|_P : P \times P \to \Lambda$ is represented by an integer matrix with determinant ± 1.

(3.3) The direct summand Y is Λ-free, and $X(\tilde{V})^{\perp} = X(\tilde{V})$, i.e., $X(\tilde{V})$ is self-orthogonal, with respect to the restriction $S|_{X(\tilde{V})\oplus Y} : X(\tilde{V}) \oplus Y \times X(\tilde{V}) \oplus Y \to \Lambda$.

In Theorem 3.1, we note that the leaf submodule $X(\tilde{V})$ need not be a free Λ-module although it is Λ-torsion-free.

Definition 3.2. A Z^{H_1}-manifold M is *exact* if there is an *exact* leaf V of M, that is, a leaf V of M satisfying one of the statements (1)-(3) of Theorem 3.1.

The following theorem shows that the exactness on a Z^{H_1}-manifold can be algebraically characterized:

Theorem 3.3. A Z^{H_1}-manifold M is exact if and only if there is a splitting $P \oplus X \oplus Y$ of $BH_2(\tilde{M};Z)$ such that

(1) $P\perp(X \oplus Y)$ with respect to $S : BH_2(\tilde{M};Z) \times BH_2(\tilde{M};Z) \to \Lambda$,

(2) there is a Λ-basis for P on which the restriction $S|_P : P \times P \to \Lambda$ is represented by an integer matrix with determinant ± 1, and

(3) the direct summand Y is a free Λ-module and $X^{\perp} = X$ with respect to the restriction $S|_{X\oplus Y} : X \oplus Y \times X \oplus Y \to \Lambda$.

In this case, we have a connected exact leaf V of M with $X(\tilde{V}) = X$ and a Λ-basis for P consisting of 2-cycles in $M_V \subset \tilde{M}$.

We note that Theorems 3.1 and 3.3 hold for topological Z^{H_1}-manifolds as well as the following corollary holds for topological Z^{π_1}-manifolds because the punctured manifold of every closed connected 4-manifold is smoothable (cf. [**4**]):

Corollary 3.4. A Z^{π_1}-manifold is exact if and only if it is TOP-split.

Proof. Let M be a Z^{π_1}-manifold. Then $H_2(\tilde{M};Z)$ is a free Λ-module by the first and second dualities of [**8**] (In fact, by the second duality $E^1BH_2(\tilde{M};Z) = E^2E^2H_2(\tilde{M};Z) = 0$ and by the first duality $E^1E^1H_2(\tilde{M};Z) = 0$, which imply that $H_2(\tilde{M};Z) = BH_2(\tilde{M};Z)$ is Λ-free). If M is TOP-split, then we can take $P = H_2(\tilde{M};Z)$ and $X = Y = 0$. Conversely, assume that M is exact. Then we have a splitting $H_2(\tilde{M};Z) = P\perp(X \oplus Y)$ as in Theorem 3.3. Using that $H_2(\tilde{M};Z)$ is Λ-free, we see that X is Λ-projective and hence Λ-free. Then we have a Λ-basis for $H_2(\tilde{M};Z)$ on which the Λ-intersection form S is represented by an integer matrix A. Let M_1 be the simply connected 4-manifold obtained from M by a surgery killing a generator of $\pi_1(M) \cong Z$. It is proved in [**10**] that M and $S^1 \times S^3 \# M_1$ are homology cobordant. This implies that M and $S^1 \times S^3 \# M_1$ have the same Kirby-Seibenmann invariant (for the Kirby-Seibenmann invariant is a cobordism invariant) and have the same Λ-intersection form represented by the matrix A. By M. Freedman-F. Quinn [**4**], there is a homeomorphism $M \to S^1 \times S^3 \# M_1$ and M is TOP-split. □

By Corollary 3.4, a non-TOP-split topological Z^{π_1}-manifold with Witt index zero given by I. Hambleton-P. Teichner [**5**] is not exact. The following stable existence theorem on exact Z^{H_1}-manifolds also holds for topological Z^{H_1}-manifolds:

Theorem 3.5. For every Z^{H_1}-manifold M, there is a non-negative integer n such that the connected sum $M \# n(S^2 \times S^2)$ is exact.

4. A relationship between the torsion linking of an exact $\mathbf{Z}^{H_1}$-manifold and the linking of an exact leaf

For a closed oriented 3-manifold V the linking

$$\ell_V : \mathrm{t}H_1(V;Z) \times \mathrm{t}H_1(V;Z) \longrightarrow Q/Z$$

of V is defined by the identity

$$\ell_V(x^V, y^V) = < \delta_{Q/Z} u_x^V \cup u_y^V, [V] > = < u_x^V \cup \delta_{Q/Z} u_y^V, [V] >$$

for $u_x^V, u_y^V \in H^1(V;Q/Z)$ with $(\delta_{Q/Z}u_x^V)\cap[V] = x^V$, $(\delta_{Q/Z}u_y^V)\cap[V] = y^V$, where $[V]$ denotes the fundamental class on V and $\delta_{Q/Z}$ denotes the Bockstein homomorphism $\delta_{Q/Z} : H^1(V;Q/Z) \to \tau H^2(V;Z)$. We first show the following lemma where (2) is given by M. Sekine **[19]**:

Lemma 4.1. For a leaf V of a Z^{H_1}-manifold M lifted to $\tilde{M}$, let $\mathrm{t}H_1(V;Z)^\theta$ be the subgroup of $\mathrm{t}H_1(V;Z)$ given by $(\cap[V])i^*(\tau H^2(\tilde{M};Z))$ in the following commutative diagram:

$$\begin{array}{ccc} \tau H^2(\tilde{M};Z) & \xrightarrow{\cap\mu} & \mathrm{t}H_1(\tilde{M};Z) \\ {\scriptstyle i^*}\downarrow & & \uparrow{\scriptstyle i_*} \\ \tau H^2(V;Z) & \xrightarrow[\cong]{\cap[V]} & \mathrm{t}H_1(V;Z). \end{array}$$

Then we have the following (1) and (2):

(1) $\mathrm{t}H_1(V;Z)^\theta \perp \mathrm{Ker}(i_*)$ with respect to ℓ_V.

(2) The map i_* induces an isomorphism from the induced linking of the restricted symmetric bilinear form

$$\ell_V|_{\mathrm{t}H_1(V;Z)^\theta} : \mathrm{t}H_1(V;Z)^\theta \times \mathrm{t}H_1(V;Z)^\theta \longrightarrow Q/Z$$

to the torsion linking

$$\ell : \mathrm{t}H_1(\tilde{M};Z)^\theta \times \mathrm{t}H_1(\tilde{M};Z)^\theta \longrightarrow Q/Z.$$

Proof. By Lemma 1.1 and the commutativity of the diagram above, we have $i_*(\mathrm{t}H_1(V;Z)^\theta) = \mathrm{t}H_1(\tilde{M};Z)^\theta$. For any $z^V \in \mathrm{Ker}(i_*)$ and any $x, y \in \mathrm{t}H_1(\tilde{M};Z)^\theta$ and any $x^V, y^V \in \mathrm{t}H_1(V;Z)^\theta$ with $i_*(x^V) = x, i_*(y^V) = y$, it suffices to show that $\ell_V(x^V, z^V) = 0$ and $\ell_V(x^V, y^V) = \ell(x,y)$. To see this, we find $u_z^V \in H^1(V;Q/Z)$ such that $(\delta_{Q/Z}u_z^V) \cap [V] = z^V$ and $u_x, u_y \in H^1(\tilde{M};Q/Z)$ such that $(\delta_{Q/Z}i^*u_x) \cap [V] = x^V$ and $(\delta_{Q/Z}i^*u_y) \cap [V] = y^V$ where i^* denotes the natural homomorphism $i^* : H^1(\tilde{M};Q/Z) \to H^1(V;Q/Z)$. Then a relation between cup and cap products implies:

$$\begin{aligned} \ell_V(x^V, z^V) &= < i^*u_x \cup \delta_{Q/Z}u_z^V, [V] > \\ &= < i^*u_x, z^V > \\ &= < u_x, i_*(z^V) > = 0. \end{aligned}$$

By the identities $(\delta_{Q/Z}u_x) \cap \mu = x$, $(\delta_{Q/Z}u_y) \cap \mu = y$, we have

$$\begin{aligned} \ell_V(x^V, y^V) &= < \delta_{Q/Z}i^*u_x \cup i^*u_y, [V] > \\ &= < \delta_{Q/Z}u_x \cup u_y, \mu > = \ell(x,y). \quad \square \end{aligned}$$

The following theorem is our main theorem:

Theorem 4.2. Let V be an exact leaf of an exact Z^{H_1}-manifold M. Then the linking

$$\ell_V : \mathrm{t}H_1(V;Z) \times \mathrm{t}H_1(V;Z) \longrightarrow Q/Z$$

of V is isomorphic to the orthogonal sum of the torsion linking

$$\ell_M : \mathrm{t}H_1(\tilde{M};Z)^{\theta} \times \mathrm{t}H_1(\tilde{M};Z)^{\theta} \longrightarrow Q/Z$$

and a hyperbolic linking.

Proof. We regard V as a canonical lift of V to $\tilde{M}$. First we show that the image of the natural homomorphism $i_* : \mathrm{t}H_1(V;Z) \to \mathrm{t}H_1(\tilde{M};Z)$ is equal to $\mathrm{t}H_1(\tilde{M};Z)^{\theta}$. For this proof, we use an argument of **[8]**. It suffices to show that $\mathrm{Im}(\tilde{i}_*) \subset \mathrm{Ker}(\theta_{\emptyset,\emptyset})$ for the natural homomorphism $\tilde{i}_* : \mathrm{t}H_1(\tilde{V};Z) \to \mathrm{t}H_1(\tilde{M};Z)$ and $\theta_{\emptyset,\emptyset} : \mathrm{t}H_1(\tilde{M};Z) \to E^1B_2H_2(\tilde{M};Z)$, because we have

$$\mathrm{Im}(i_*) \supset \tau H^2(\tilde{M};Z) \cap \mu = \mathrm{Ker}(\theta_{\emptyset,\emptyset})$$

by Lemma 1.1 and the commutative diagram in Lemma 4.1. We consider the following commutative diagram:

$$\begin{array}{ccccc}
\mathrm{t}H_1(\tilde{V} \times [-1,1];Z) & \xrightarrow{\cong} & TH^3_{\Lambda}(\tilde{M},\tilde{M}_V;Z) & \xrightarrow{\rho} & E^1H_2(\tilde{M},\tilde{M}_V;Z) \\
\downarrow & & \downarrow & & \downarrow{\scriptstyle\psi} \\
TH_1(\tilde{M};Z) & \xrightarrow{\cong} & TH^3_{\Lambda}(\tilde{M};Z) & \xrightarrow{\rho} & E^1H_2(\tilde{M};Z),
\end{array}$$

where $\cong$ denotes the Poincaré duality isomorphism over Λ and ρ is a natural Λ-epimorphism defined in **[8]** so that the restriction of the composite bottom map to $\mathrm{t}H_1(\tilde{M};Z)$ is equal to $\theta_{\emptyset,\emptyset}$ and the other maps are induced from inclusions. Since the image of the left vertical homomorphism is equal to $\mathrm{Im}(\tilde{i}_*)$, it suffices to show that $\psi = 0$. To see this, we consider the homology exact sequence of the pair $(\tilde{M},\tilde{M}_V)$:

$$H_2(\tilde{M};Z) \xrightarrow{\tilde{j}'_*} H_2(\tilde{M},\tilde{M}_V;Z) \xrightarrow{\tilde{\partial}'_*} H_1(\tilde{M}_V;Z) \xrightarrow{\tilde{i}'_*} H_1(\tilde{M};Z).$$

Since V is an exact leaf, we see from (2) of Theorem 2.1 that this exact sequence induces the following exact sequence:

$$0 \to \mathrm{Im}(\tilde{j}'_*) \xrightarrow{\subset} \mathrm{b}H_2(\tilde{M},\tilde{M}_V;Z) \xrightarrow{\tilde{\partial}'_*} \mathrm{b}H_1(\tilde{M}_V;Z) \xrightarrow{\tilde{i}'_*} \hat{H}_1(\tilde{M};Z) \to 0,$$

where $\hat{H}_1(\tilde{M};Z) = H_1(\tilde{M};Z)/\tilde{i}'_*(\mathrm{t}H_1(\tilde{M}_V;Z))$. Since $\mathrm{b}H_2(\tilde{M},\tilde{M}_V;Z) = \mathrm{b}H_2(V \times [-1,1], V \times \{-1,1\};Z) \otimes \Lambda$ and $\mathrm{b}H_1(\tilde{M}_V;Z) = \mathrm{b}H_1(M_V;Z) \otimes \Lambda$ are Λ-free and

$\hat{H}_1(\tilde{M};Z)$ has Λ-projective dimension ≤ 2, it follows that $\mathrm{Im}(\tilde{j}'_*)$ is Λ-projective, i.e., Λ-free. This implies that $\psi = 0$ as desired, for ψ is the composite

$$E^1H_2(\tilde{M},\tilde{M}_V;Z) \longrightarrow E^1\mathrm{Im}(\tilde{j}'_*) \longrightarrow E^1H_2(\tilde{M};Z).$$

Next, by Lemma 1.1 the map $\cap\mu : \tau H^2(\tilde{M};Z) \to \mathrm{t}H_1(\tilde{M};Z)$ induces an isomorphism $\cap\mu : \mathrm{Hom}_Z(\mathrm{t}H_1(\tilde{M};Z)^\theta, Q/Z) \cong \mathrm{t}H_1(\tilde{M};Z)^\theta$. Since $i_* : \mathrm{t}H(V;Z) \to \mathrm{t}H_1(\tilde{M};Z)^\theta$ is onto, the induced map

$$i_*^\# : \mathrm{Hom}_Z(\mathrm{t}H_1(\tilde{M};Z)^\theta, Q/Z) \longrightarrow \mathrm{Hom}_Z(\mathrm{t}H_1(V;Z), Q/Z)$$

is injective and we have the following commutative square:

$$\begin{array}{ccc} \mathrm{Hom}_Z(\mathrm{t}H_1(\tilde{M};Z)^\theta, Q/Z) & \xrightarrow[\cong]{\cap\mu} & \mathrm{t}H_1(\tilde{M};Z)^\theta \\ {\scriptstyle i_*^\#}\downarrow & & \uparrow{\scriptstyle i_*} \\ \mathrm{Hom}_Z(\mathrm{t}H(V;Z), Q/Z) & \xrightarrow[\cong]{\cap[V]} & \mathrm{t}H_1(V;Z). \end{array}$$

In this commutative square, let $\iota : \mathrm{t}H_1(\tilde{M};Z)^\theta \to \mathrm{t}H_1(V;Z)$ be the composite map $(\cap[V])(i_*^\#)(\cap\mu)^{-1}$, whose image is denoted by $\mathrm{t}H_1(V;Z)_*^\theta$. Then we have $i_*\iota = 1$, so that $\iota : \mathrm{t}H_1(\tilde{M};Z)^\theta \to \mathrm{t}H_1(V;Z)_*^\theta$ is an isomorphism, and $\mathrm{t}H_1(V;Z) = \mathrm{t}H_1(V;Z)_*^\theta \oplus \mathrm{Ker}(i_*)$. By Lemma 4.1, we have $\mathrm{t}H_1(V;Z)_*^\theta \perp \mathrm{Ker}(i_*)$ and ι induces an isomorphism from the torsion linking ℓ of M to the restriction $\ell_V|_{\mathrm{t}H_1(V;Z)_*^\theta}$. Let $\tilde{M}^+$ and $\tilde{M}^-$ be the connected components of the manifold obtained from $\tilde{M}$ by splitting along V so that $tV \subset \tilde{M}^+$ and $t^{-1}V \subset \tilde{M}^-$. Let $\tilde{V}^\pm = \tilde{V} \cap \tilde{M}^\pm$. Using the assumption that V is an exact leaf and (1) of Theorem 2.1, we can write $\mathrm{Ker}(i_*)$ as follows: $\mathrm{Ker}(i_*) = K^+ + K^-$ where K^+ (or K^-, respectively) denotes the subgroup generated by elements of $\mathrm{t}H_1(V;Z)$ represented by the boundaries of 2-cycles in $\mathrm{t}H_2(\tilde{M}^+,\tilde{V}^+;Z)$ (or $\mathrm{t}H_2(\tilde{M}^-,\tilde{V}^-;Z)$, respectively). For 1-cycles c and c' in K^+, let C and C' be 2-chains in $\mathrm{t}H_2(\tilde{M}^+,\tilde{V}^+;Z)$ with $\partial C = c$ and $\partial C' = c'$. Let $\mathrm{cl}_Q(C)$ and $\mathrm{cl}_Q(C')$ be rational 2-cycles in $\tilde{M}^+$ obtained from C and C' by adding rational 2-chains C_V^Q and $C_V'^Q$ in V bounded by $-c$ and $-c'$, respectively. Then we have the Q-intersection number $s(\mathrm{cl}_Q(C),\mathrm{cl}_Q(C')) = 0$ in $\tilde{M}^+$ because $\mathrm{cl}_Q(C)$ and $\mathrm{cl}_Q(C')$ are rationally homologous in $\tilde{M}^+$ to 2-cycles in $\tilde{V}^+$. Since $\ell_V(c,c') = -s(\mathrm{cl}_Q(C),\mathrm{cl}_Q(C'))$ (mod 1), this implies that $\ell_V(c,c') = 0 \in Q/Z$. Thus, $\ell_V(K^+,K^+) = 0$. Similarly, $\ell_V(K^-,K^-) = 0$. Using that $\ell_V|_{\mathrm{t}H_1(V;Z)_*^\theta}$ is non-singular, we see that $\ell_V|_{\mathrm{Ker}(i_*)}$ is non-singular and hence $K^+ \cap K^- = 0$. Let x_i $(i = 1,2,\ldots,r)$ be a generator system of a cyclic splitting of K^+, and $f_i \in \mathrm{Hom}_Z(K^+, Q/Z)$ $(i = 1,2,\ldots,r)$ the homomorphisms given by

$$f_i(x_j) = \frac{\delta_{ij}}{n_j},$$

where n_j denotes the group order of x_j. Since the linking $\ell|_{\mathrm{Ker}(i_*)}$ induces now an isomorphism $K^- \cong \mathrm{Hom}_Z(K^+, Q/Z)$, we can take a generator system y_i $(i = 1, 2, \ldots, r)$ of K^- corresponding to f_i $(i = 1, 2, \ldots, r)$ by this isomorphism. Then the linking $\ell|_{\mathrm{Ker}(i_*)}$ represents the block sum of hyperbolic linking matrices with respect to x_i, y_i $(i = 1, 2, \ldots, r)$ and thus it is a hyperbolic linking. □

The following corollary is direct from Theorem 4.2:

Corollary 4.3. Let V be an exact leaf of an exact Z^{H_1}-manifold M. If $\mathrm{t}H_1(V; Z)$ is a cyclic group or if $\ell_M \neq 0$ and $\mathrm{t}H_1(V; Z) \cong Z_p \oplus Z_p$ for a prime p, then the torsion linking ℓ_M of M is isomorphic to the linking ℓ_V of V.

The following example explains a relationship between Farber's result on Seifert hypersurfaces of S^2-knots in [**3**] and Theorem 4.2:

Example 4.4. A Z^{H_1}-manifold M_K of an S^2-knot K in S^4 is uniquely constructed from K and the addition number n of $S^2 \times S^2$ because every diffeomorphism of $S^1 \times S^2$ extends to a diffeomorphism of $S^1 \times D^3$. Let V be a closed oriented 3-manifold obtained from any Seifert hypersurface for K by filling up with a 3-ball. Since $X(\tilde{V}) = 0$ and $BH_2(\tilde{M}_K; Z)$ has a Λ-basis of $2n$ elements on which the Λ-intersection form S is represented by an integer matrix, we see from (3) of Theorem 3.1 that V is an exact leaf of M_K. Thus, the linking ℓ_V of V is isomorphic to the orthogonal sum of the torsion linking ℓ_{M_K} of M_K and a hyperbolic linking. Counting Lemma 2.2, we see that this result is nothing but Farber's result in [**3**]. □

The referee of this paper suggested to include the proof of the following lemma which is implicitly used in the preprint version:

Lemma 4.5. Let V be a compact connected oriented 3-manifold such that ∂V is non-empty and connected, and the form ℓ_V of V is a linking. Then there is an orientation-reversing diffeomorphism $\psi : \partial V \to \partial V_0$ for a handlebody V_0 such that the inclusion $i : V \subset V' = V \cup_\psi V_0$ induces an isomorphism $i_* : H_1(V; Z) \cong H_1(V'; Z)$, by which ℓ_V and $\ell_{V'}$ are isomorphic.

Proof. In the following homology exact sequence of the pair $(V, \partial V)$:

$$H_2(V, \partial V; Z) \xrightarrow{\partial} H_1(\partial V; Z) \xrightarrow{i_*} H_1(V; Z) \xrightarrow{j_*} H_1(V, \partial V; Z),$$

the non-singularity of ℓ_V means that j_* induces an isomorphism

$$\mathrm{t}H_1(V; Z) \cong \mathrm{t}H_1(V, \partial V; Z),$$

so that $\mathrm{Im}(i_*)$ is Z-free. Since, by Poincaré duality, $\mathrm{Im}(\partial)$ is a self-orthogonal complement with respect to the intersection form s on ∂V, we have a Z-basis

x_i, x_i' $(i = 1, 2, \ldots, g)$ for $H_1(\partial V; Z)$ such that $s(x_i, x_j') = \delta_{ij}$ and the sets x_i $(i = 1, 2, \ldots, g)$ and $i_*(x_i')$ $(i = 1, 2, \ldots, g)$ are bases of $\mathrm{Im}(\partial) = \mathrm{Ker}(i_*)$ and $\mathrm{Im}(i_*)$, respectively. Represent x_i and x_i' $(i = 1, 2, \ldots, g)$ by mutually disjoint embedded circles S_i^1 $(i = 1, 2, \ldots, g)$ and $(S^1)_i'$ $(i = 1, 2, \ldots, g)$ in ∂V, respectively, such that $S_i^1 \cap (S^1)_j'$ is $\emptyset$ for $i \neq j$ or one point for $i = j$. Then attach 2-handles h_i^2 $(i = 1, 2, \ldots, g)$ to V along mutually disjoint annuli $A_i \subset \partial V$ such that the core of A_i is S_i^1. Then $V \cup_{i=1}^g h_i^2$ is a compact connected oriented 3-manifold whose boundary is a 2-sphere. Filling up it with a 3-ball, we obtain a closed 3-manifold V'. We show that $V_0 = \mathrm{cl}(V' - V)$ is a desired handlebody. Since $H_1(V', V; Z) = \oplus_{i=1}^g H_1(h_i^2, A_i; Z) = 0$ and the boundary operator $\partial : H_2(V', V; Z) = \oplus_{i=1}^g H_2(h_i^2, A_i; Z) \to H_1(V; Z)$ is the zero map, the isomorphism $i_* : H_1(V; Z) \cong H_1(V'; Z)$ is obtained from the exact sequence of the pair (V', V), which also implies an isomorphism $\ell_V \cong \ell_{V'}$. □

The following lemma concerns an exact leaf of a Z^{H_1}-manifold (because every exact leaf satisfies the assumption of this lemma):

Lemma 4.6. For a leaf V of a Z^{H_1}-manifold M and the intersection form $s : H_2(M; Z) \times H_2(M; Z) \to Z$, assume that there is a direct summand $X(V) \oplus Y$ of $H_2(M; Z)$ such that $X(V) = \mathrm{Im}(H_2(V; Z) \to H_2(M; Z))$, the restriction $s|_{X(V) \oplus Y}$ is non-singular, and $X(V)^\perp = X(V)$ with respect to $s|_{X(V) \oplus Y}$. Then the boundary operator $\partial : H_2(M, V; Z) \to H_1(V; Z)$ induces an isomorphism

$$\mathrm{t}H_2(M, V; Z) \cong \mathrm{t}H_1(V; Z).$$

Proof. By the homology exact sequence of the pair (M, V), we have the following natural short exact sequence

$$0 \to H_2(M; Z)/X(V) \longrightarrow H_2(M, V; Z) \xrightarrow{\partial} H_1(V; Z) \to 0.$$

Since $\mathrm{t}(H_2(M; Z)/X(V)) = 0$, the map $\partial : \mathrm{t}H_2(M, V; Z) \to \mathrm{t}H_1(V; Z)$ is injective. Let c be a 1-cycle in $\mathrm{t}H_1(V; Z)$, which bounds a 2-chain C in M and a non-zero integer multiple mc of which bounds a 2-chain C_V in V. For every $x \in X(V)$ and the homology class $[mC - C_V] \in H_2(M; Z)$, we see from taking a collar of V in M that the intersection number $s([mC - C_V], x)$ is the m-multiple of an integer and is independent of a choice of m and C_V. Thus, the intersection number $s(C, x)$ is well-defined by the identity

$$s(C, x) = \frac{s([mC - C_V], x)}{m}.$$

By taking bases $x_i (i = 1, 2, \ldots, g)$ and $y_i = [C_i^y]$ $(i = 1, 2, \ldots, g)$ for $X(V)$ and Y, respectively, such that $s(x_i, y_j) = \delta_{ij}$, we construct a 2-chain

$$C' = C - \sum_{i=1}^{g} s(C, x_i) C_i^y$$

in M. Then $\partial C' = c$ and $s(C', x_i) = 0$ for all i. Thus, we have $[mC' - C_V] \in X(V)$, meaning that $[C'] \in \mathrm{t}H_2(M, V; Z)$ and hence $\partial : \mathrm{t}H_2(M, V; Z) \to \mathrm{t}H_1(V; Z)$ is surjective. □

The following example is an explanation on an exact leaf of a Z^{H_1}-manifold of a surface-knot, specially of a ribbon surface-knot:

Example 4.7. Let F be a surface-knot in S^4 of genus g which admits a Seifert hypersurface V such that the form ℓ_V of V is a linking. We choose a diffeomorphism $\psi : \partial V \to \partial V_0$ as in Lemma 4.5. By taking the diffeomorphism $\phi = \psi \times 1 : \partial E_F = F \times S^1 \to \partial V_0 \times S^1$, the Z^{H_1}-manifold M_F of F taking $n = 0$ (i.e., M_ϕ) is constructed. From a symplectic basis $x_i = [S_i^1]$, $x_i' = [(S^1)_i']$ $(i = 1, 2, \ldots, g)$ for $H_1(\partial V; Z)$ as in the proof of Lemma 4.5, a basis $\bar{x}_i$, $\bar{x}_i'$ $(i = 1, 2, \ldots, g)$ for $H_2(M_F; Z)$ is obtained as follows: Namely, we represent $\bar{x}_i$ by a 2-cycle in V' obtained from a 2-chain in V and a meridian disk of V_0 which are bounded by S_i^1, and $\bar{x}_i'$ by the 2-cycle $(S^1)_i' \times S^1 \subset F \times S^1 = \partial E_F$. We have $H_2(M_F; Z) \cong Z^{2g}$ (since $\chi(M_F) = 2g$) and $s(\bar{x}_i, \bar{x}_j) = s(\bar{x}_i', \bar{x}_j') = 0$ for all i, j and $s(\bar{x}_i, \bar{x}_j') = \delta_{ij}$. This implies that the homology classes $\bar{x}_i$, $\bar{x}_i'$ $(i = 1, 2, \ldots, g)$ form a basis for $H_2(M_F; Z)$ with respect to which s represents the block sum of g hyperbolic planes. In particular, M_F is a spin 4-manifold. By construction, the image $X(V') = \mathrm{Im}(H_2(V'; Z) \to H_2(M_F; Z))$ is a self-orthogonal complement with respect to s. By Lemma 4.6, we have an isomorphism

$$(4.7.1) \qquad \partial : \mathrm{t}H_2(M_F, V'; Z) \cong \mathrm{t}H_1(V'; Z).$$

By the excision Λ-isomorphisms

$$\mathrm{t}H_2(\tilde{M}_F, \tilde{V}'; Z) \cong \mathrm{t}H_2(M_F, V'; Z) \otimes_Z \Lambda \quad \text{and} \quad \mathrm{t}H_1(\tilde{V}'; Z) \cong \mathrm{t}H_1(V'; Z) \otimes_Z \Lambda,$$

we also obtain from (4.7.1) the following (4.7.2):

$$(4.7.2) \qquad \tilde{\partial} : \mathrm{t}H_2(\tilde{M}_F, \tilde{V}'; Z) \longrightarrow \mathrm{t}H_1(\tilde{V}'; Z) \quad \text{is injective.}$$

We note that (4.7.1) and (4.7.2) hold for V' in every Z^{H_1}-manifold M_F of F adding n copies of $S^2 \times S^2$. Although these properties are included in the exact sequence of (1) of Theorem 3.1, they do not mean in general that V' is an exact leaf of M_F. For example, consider a Seifert hypersurface V of a trivial surface-knot F^0 of genus one such that $\mathrm{t}H_1(V; Z) \cong Z_n$ for an odd $n > 1$ and the form ℓ_V of V is a linking (see Examples 1.5, 1.7). Then the torsion linking $\ell_{M_{F^0}}$ of M_{F^0} is the zero form by Lemma 2.2. If a leaf V' of M_{F^0} constructed as above from V is exact, then $\ell_{V'} (\cong \ell_V)$ is isomorphic to $\ell_{M_{F^0}} = 0$ by Corollary 4.3, which is a contradiction. Thus, V' cannot be any exact leaf of M_{F^0}, although it satisfies (4.7.1) and (4.7.2) as shown above. To obtain an example on an exact Z^{H_1}-manifold of a surface-knot, we consider any surface-knot F which bounds a Seifert hypersurface V with $\mathrm{t}H_1(V;Z) = 0$. For example, every ribbon surface-knot bounds such a Seifert hypersurface V (see Example 1.6). Then V' is an exact leaf of M_F by (1) of Theorem 3.1 and hence M_F is an exact Z^{H_1}-manifold, because $\mathrm{t}H_2(\tilde{M}_F, \tilde{V}'; Z) = \mathrm{t}H_1(\tilde{V}'; Z) = 0$ by (4.7.2). □

In the following example, we give a closed 4-manifold version of Example 1.7:

Example 4.8. Let F and F_R be the surface-knots of genus one in Example 1.7. We note that their surface-knot groups are meridian-preservingly isomorphic and their peripheral subgroups are infinite cyclic groups with meridian generators. Further, they have respectively Seifert hypersurfaces V and V_R diffeomorphic to $(S^1 \times D^2)\#d(S^1 \times S^2)\#L(p,q)\#L(p,-q)$. Then it is direct to see that the Z^{H_1}-manifolds M_F and M_{F_R} (for every $n \geq 0$) constructed in Example 4.7 by using Lemma 4.5 for V and V_R respectively have the properties that *M_F and M_{F_R} have isomorphic fundamental groups, have isomorphic hyperbolic intersection forms, and have respectively leaves V' and V'_R diffeomorphic to the same 3-manifold $\#(d+1)(S^1 \times S^2)\#L(p,q)\#L(p,-q)$.* On the other hand, since $\ell_F \neq 0$ and $\ell_{F_R} = 0$, we see from lemma 2.2 that *the torsion linking ℓ_{M_F} of M_F is a non-zero form whereas the torsion linking $\ell_{M_{F_R}}$ of M_{F_R} is the zero form.* As another interesting property, we observe that *the leaf V' of M_F is not exact whereas the leaf V'_R of M_{F_R} is exact.* In fact, if V' is exact, then $\ell_{V'}$ is isomorphic to the orthogonal sum of ℓ_{M_F} and a hyperbolic linking by Theorem 4.2, which is impossible since $\mathrm{t}H_1(\tilde{M}_F; Z)^\theta \cong H_1(\tilde{E}_F; Z) \cong Z_p$ and $\mathrm{t}H_1(V'; Z) \cong Z_p \oplus Z_p$. On the other hand, the connected summand V''_R of V'_R corresponding to $\#(d+1)(S^1 \times S^2)$ is an exact leaf of M_{F_R} by Example 4.7. Since a punctured manifold of the connected summand $L(p,q)\#L(p,-q)$ of V'_R is taken in a 4-ball in M_{F_R}, we have that $X(\tilde{V}''_R) = X(\tilde{V}'_R)$ in $BH_2(\tilde{M}_{F_R}; Z)$. By (3) of Theorem 2.1, V'_R is an exact leaf of M_{F_R}. For every surface-knot F, we denote by $-F$ the same surface-knot as F but with only the surface orientation reversed. If we consider the connected sums $F\#-F$ and $F_R\#-F_R$ instead of F and F_R, respectively, then *analogous Z^{H_1}-manifolds $M_{F\#-F}$ and $M_{F_R\#-F_R}$ for every $n \geq 0$ have isomorphic fundamental groups, have isomorphic hyperbolic intersection forms, and have the same 3-manifold $\#(2d+2)(S^1 \times S^2)\#L(p,q)\#L(p,-q)$ as exact leaves of both them, although $\ell_{M_{F\#-F}} \neq 0$ and $\ell_{M_{F_R\#-F_R}} = 0$.*

References

[1] J. Boyle, *Classifying 1-handles attached to knotted surfaces*, Trans. Amer. Math. Soc. **306** (1988), 475-487.

[2] A. Cavicchioli and F. Hegenbarth, *A note on four-manifolds with free fundamental groups*, J. Math. Sci. Univ. Tokyo **4** (1997), 435-451.

[3] M. Farber, *Duality in an infinite cyclic covering and even-dimensional knots*, Math. USSR-Izvestia **11** (1977), 749-781.

[4] M. H. Freedman and F. Quinn, *Topology of 4-Manifolds*, Princeton Univ. Press (1990).

[5] I. Hambleton and P. Teichner, *A non-extended Hermitian form over* $Z[Z]$, Manuscripta Math. (to appear).

[6] F. Hosokawa and A. Kawauchi, *Proposals for unknotted surfaces in four-spaces*, Osaka J. Math. **16** (1979), 233-248.

[7] J. A. Hillman and A. Kawauchi, *Unknotting orientable surfaces in the 4-sphere*, J. Knot Theory Ramifications **4** (1995), 213-224.

[8] A. Kawauchi, *Three dualities on the integral homology of infinite cyclic coverings of manifolds*, Osaka J. Math. **23** (1986), 633-651.

[9] A. Kawauchi, *The first Alexander modules of surfaces in 4-sphere*, In: Algebra and topology (Taejon,1990), Proc. KAIST Math. Workshop **5** (1990), 81-89.

[10] A. Kawauchi, *Splitting a 4-manifold with infinite cyclic fundamental group*, Osaka J. Math. **31** (1994), 489-495.

[11] A. Kawauchi, *On the fundamental class of an infinite cyclic covering*, Kobe J. Math. **15** (1998), 103-114.

[12] A. Kawauchi, *Algebraic characterization of an exact 4-manifold with infinite cyclic first homology*, Atti Sem. Mat. Fis. Univ. Modena (to appear).

[13] A. Kawauchi, *An intrinsic Arf invariant of a link and its surface-link analogue*, preprint.

[14] A. Kawauchi, *An intrinsic Arf invariant of a link and its surface-link analogue (the spin version)*, preprint.

[15] A. Kawauchi, *On some invariants of a zero-signature spin 4-manifold with non-trivial first cohomology*, preprint.

[16] A. Kawauchi, T. Shibuya, and S. Suzuki, *Descriptions on surfaces in four-space II. Singularities and cross-sectional links,*, Math. Sem. Notes, Kobe Univ. **11** (1983), 31-69.

[17] M. A. Kervaire, *Les noeuds de dimensions supérieures*, Bull. Soc. Math. France **93** (1965), 225-271.

[18] J. Levine, *Knot modules. I.*, Trans. Amer. Math. Soc. **229** (1977), 1-50.

[19] M. Sekine, *Kawauchi's second duality and knotted surfaces in 4-sphere*, Hiroshima J. Math. **19** (1989), 641-651.

[20] M. Teragaito, *Fibered 2-knots and lens spaces*, Osaka J. Math. **26** (1989), 57-63.

[21] T. Yajima, *On a characterization of knot groups of some spheres in* R^4, Osaka J. Math. **6** (1969), 435-446.

[22] E. C. Zeeman, *Twisting spun knots*, Trans. Amer. Math. Soc. **115** (1965), 471-495.

Extended Braids and Links

Roger Fenn
Ebru Keyman

July 29, 1999

Abstract

Our aim in this paper is to introduce and study the properties of different kinds of singular braids. Our starting point being the singular braids with double points as introduced in [3] and [5], we define a much bigger class of singular braids which we call geometric singular braids. Geometric singular braids form a monoid and we give an explicit presentation for this monoid. In this paper, we consider only singular braids where the tangents of the strings around a singular point generate a plane. In general this need not be the case. Finally, geometric singular links are defined and it is shown that there is a one-to-one correspondence between geometric singular braids and geometric singular links up to Markŏv equivalence.

1 Introduction

We define *extended braid diagrams* as planar graphs. We obtain groups or monoids out of them by defining equivalences. This is done in two ways. The first method is to compare their diagrams. We say that two extended braid diagrams are equivalent if they are connected to each other by a series of moves on the diagrams. The second way is to translate all this information to relations amongst the generators of a monoid or a group presentation. We then, define the *geometric braid monoid* where the equivalence of diagrams correspond to the isotopies of the ambient space with some restrictions around multiple points. It is not a group because singular points do not have inverses. After describing this monoid in terms of generators and relations we prove that it embeds in its group and obtain the *group of geometric braids.* Although the group of geometric braids admit some "*unnatural*" geometric moves corresponding to the cancellation of singular points, it is still possible to represent them by diagrams without loss of any information.

Next we concentrate on the relations between braids and links. We apply the classical theory to singular theory. Alexander and Markov theorems are well-known results in the theory of non-singular links. It is easy to associate a link to any braid and the former of these theorems says it is possible to associate a braid to every link. The latter says which braids give the same link. We define *geometric singular links* and adapt these theorems for geometric singular braids and geometric singular links. We prove the theorems with the methods which were used to prove the original theorems.

2 Extended Braids

An **extended braid diagram** on n strings is a finite planar graph Γ in $[0,1] \times \mathbf{R}$ consisting of a collection of n piecewise linear arcs monotonic in the x–component and joining a set of n verticies on $\{0\} \times \mathbf{R}$ to a set of n vertices on $\{1\} \times \mathbf{R}$. The vertices of Γ are:

- **Crossings:** This is when two arcs cross each other and one of the arcs is indicated by a broken line at the crossing as in figure 1. The arc indicated by a broken line is called the **under edge** and the other is called the **over edge**. There are two kinds of crossings, **positive** and **negative**.

Figure 1: Positive and negative crossings

- **Multiple points:** These are points of valency $2k$ for some $k = 1, 2, \ldots$. The k–tuple points have k **in-edges** on the left and k **out-edges** on the right. This means that the multiple point s is the final vertex of k edges and initial vertex of k edges. Label the in-edges with integers from 1 to k from bottom to top. That is to say, the lower most in-edge has label 1, the next 2 and so on. Associated with a k–tuple point s is a permutation π_s of the set $\{1, 2, \ldots, k\}$. This indicates that the continuation of the ith in-edge is the $\pi_s(i)$th out-edge counting from bottom to top. We label it by i. If $k = 2, 3$ we shall call them **double, triple** points respectively. The multiple points are also called **singular points** or **singularities**. It is convenient to consider the k–tuple point s with a small closed neighbourhood $N(s)$ around it such that $N(s)$ does not contain any crossings, any other singular points or any edges except the in and out-edges of s.

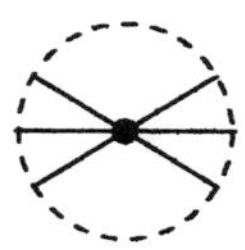

Figure 2: $N(s) \cap \Gamma$

- **Initial vertices:** These are the respective starting points of the n strings which are of valency 1 at $(0,1),(0,2),\ldots,(0,n)$.
- **Final vertices:** These are the respective final points of the n strings of valency 1 at $(1,1),(1,2),\ldots,(1,n)$.
- The vertex set contains $\partial N(s) \cap \Gamma$ for every multiple point s of Γ.
- All the other vertices are of valency 2.

A multiple point is determined by its valency and permutation. However when we introduce inverses, these will be distinct from the originals.

The edges of Γ are **arcs**, monotonic in the first coordinate x. An edge will be called **singular** if one of its end points is a singular point. Note that an edge can be singular at only one end since $\partial N(s) \cap \Gamma$ for every multiple point s.

We shall only be interested in diagrams up to an isotopy of $[0,1] \times \mathbf{R}$ fixing $\{0,1\} \times \mathbf{R}$ and preserving the monotonicity of the arcs. We will call such an isotopy a **geotopy**. During a geotopy the relative x values of two singular points or crossings on different strings may interchange.

The i-th string of Γ is a path which starts from $(0,i)$ and finishes at $(1,j)$. The diagram Γ defines a permutation $\pi = \pi_\Gamma$ in the natural way; $\pi(i) = j$ if the i string joins $(0,i)$ to $(1,j)$.

Example Figure 3 is a diagram of an extended braid on three strings where the permutation induced by the triple and the double point are (1 2 3) and (1 2) respectively and the permutation of the braid is (1 3 2).

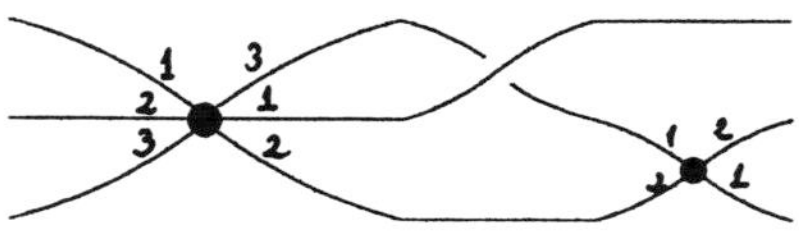

Figure 3: An extended braid diagram on 3 strings

There are many examples of extended braids in literature. Classical braids [2], [4] have only crossings. A positive braid on the other hand has only positive crossings.

A special type of extended braids, called *singular braids* [3], [5] and *welded braids* [9] are represented by diagrams which have crossings (both signs) and a double point with permutation (12). The double point has the same picture in both types of braids. However their movements through the same picture are different since singular braids and welded braids are defined with different equivalences. These will become clear in the next section.

Figure 4: A double point or a weld

Doodle braids, (twins) ([13]) and *permutation braids* have diagrams with only double points and no crossings.

Two extended braid diagrams Γ_1 and Γ_2 with the same number of strings can be composed to make a third diagram $\Gamma_1\Gamma_2$ by placing them side by side and shrinking the x coordinate by a half so that the result lies in $[0,1] \times \mathbb{R}$. This composition is clearly associative up to geotopy and has an identity which is the trivial diagram with no crossings or multiple points. Therefore braid diagrams on the same number of strings form a monoid up to geotopy. The generators of the monoid are the irreducible diagrams. We will adopt the general notation μ_{ik} for a k–tuple point $k \geq 2$. So μ_{ik} is an irreducible diagram consisting of $i-1$ horizontal strings as we travel upwards, a k–tuple point occupying the next k strings and finishing with $n-i-k+1$ horizontal strings where n is the total number of strings. The notation μ_{ik} does not specify the permutation of the singular point. If the permutation π of the multiple point is of relevance we write $\mu_{ik\pi}$.

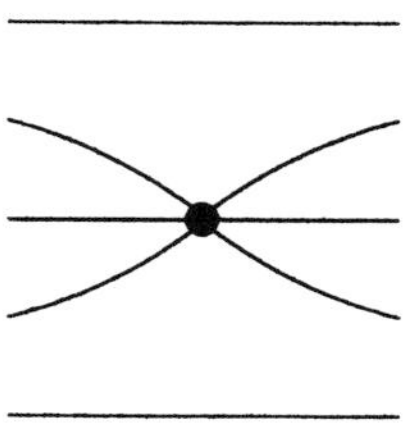

Figure 5: μ_{23} on a 5 string diagram.

An extended braid diagram is said to be in **general position** if all the crossings and multiple points have different x values. Any diagram can be put in general position by a geotopy. Moreover we can assume that the diagram is the product of irreducible diagrams each containing exactly one multiple point or one crossing point.

We reserve the symbols σ_i, (σ_i^{-1}) for a positive (negative) crossing between i and $i+1$ strings and τ_i for a double point with permutation (12) again on i and $i+1$ strings (although μ_{i2} includes τ_i).

We can also think of extended braid diagrams in 3-space. We can lift an extended braid diagram in such a way that the lift of a singular point s is planar within a small neighbourhood $N(s)$ of s and the interior of $N(s)$ is kept fixed. We will also require that the initial and final vertices are kept fixed and the diagram does not have any self intersections apart from the singularities. The strings at a positive or a negative crossing are also disjoint and the string represented by a broken arc will be lifted to an arc that passes under the other string at the crossing. We lift the diagram such that it is still strictly monotonic in the x-direction so there are no arcs parallel to the z-axis. The permutation induced by the singular points makes sense by the way the lift is defined. Conversely given any such diagram in general position it has a unique projection on to the plane. Throughout we will use the two equivalent notions interchangeably whichever is convenient at the time.

2.1 Equivalences of extended braid diagrams

We shall define the equivalence of diagrams as a sequence of allowable moves considered below. We will for convenience always allow geotopy. As for the other moves they may or may not be allowed. For example an arc is allowed to move over a weld but not under it, which is allowed for the double points of Baez and Birman. The equivalence classes of diagrams on n strings are called **extended**

braids on n strings.

The generators correspond to irreducible diagrams of crossings and multiple points and the relations correspond to certain allowable moves on the diagram.

Interchanges: Two generators may change their respective x values as a consequence of geotopy. In terms of generators this corresponds to

$$\begin{aligned} \sigma_i^\varepsilon\sigma_j^\delta &= \sigma_j^\delta\sigma_i^\varepsilon && \text{if} \quad |\mathrm{i}-\mathrm{j}|>1;\ \varepsilon,\delta=\pm 1 \\ \mu_{ik}\sigma_j^\varepsilon &= \sigma_j^\varepsilon\mu_{ik} && \text{if} \quad i+k<j+1 \quad \text{or} \quad j+1<i \\ \mu_{ik}\mu_{jl} &= \mu_{jl}\mu_{ik} && \text{if} \quad i+k<j+1 \quad \text{or} \quad j+l<i+1 \end{aligned} \tag{1}$$

where $\sigma_j^\varepsilon, \mu_{ik}$ represent the irreducible diagrams.

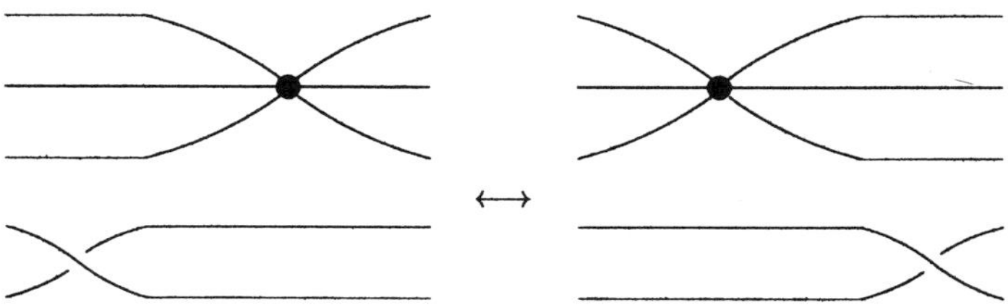

Figure 6: Commuting generators

Arc moves: Let D be a disk in $(0,1)\times\mathbf{R}$ such that its boundary is the union of two arcs γ_1, γ_2 which are monotonic in x and such that γ_1 is part of a string of a diagram Γ. Suppose that γ_1 does not run through any singular point with order greater than or equal to 3, and its end points are neither crossings nor multiple points. Suppose further that the interior of γ_2 "*meets*" Γ only in crossings or double points in a finite number of places. Then an arc move of γ_1 over D replaces γ_1 by γ_2 and the crossings and double points of γ_1 are replaced by new crossings and double points corresponding to the points where γ_2 meets the arcs of Γ.

There are some special kinds of arc moves. If γ_1 is an over arc and is replaced by γ_2 which is also an over arc then this is called an **over arc move**. When we say γ_1 is an over arc, then γ_1 is not necessarily over at *one* crossing but it is the over edge at *all* the crossings that it goes through. Similarly we can define an **under arc move**.

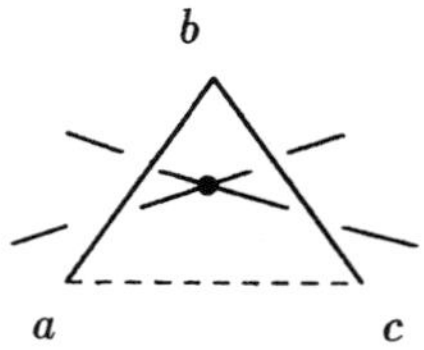

Figure 7: An over arc move

Twists: This corresponds physically to twisting a k–tuple point through 180 degrees about a horizontal axis parallel to the x-axis. The corresponding relation is of the form

$$\Delta_{ik}\mu_{ik\pi} = \mu_{ik\pi'}\Delta_{ik} \tag{2}$$

where $\Delta_{ik} = \sigma^{\varepsilon}_{i+k-2}(\sigma^{\varepsilon}_{i+k-3}\sigma^{\varepsilon}_{i+k-2})\dots(\sigma^{\varepsilon}_{i}\sigma^{\varepsilon}_{i+1}\dots\sigma^{\varepsilon}_{i+k-2})$ and

$$\pi' = \begin{pmatrix} 1 & \dots & k-1 & k \\ k+1-a_k & \dots & k+1-a_2 & k+1-a_1 \end{pmatrix} \text{ if } \pi = \begin{pmatrix} 1 & 2 & \dots & k \\ a_1 & a_2 & \dots & a_k \end{pmatrix}.$$

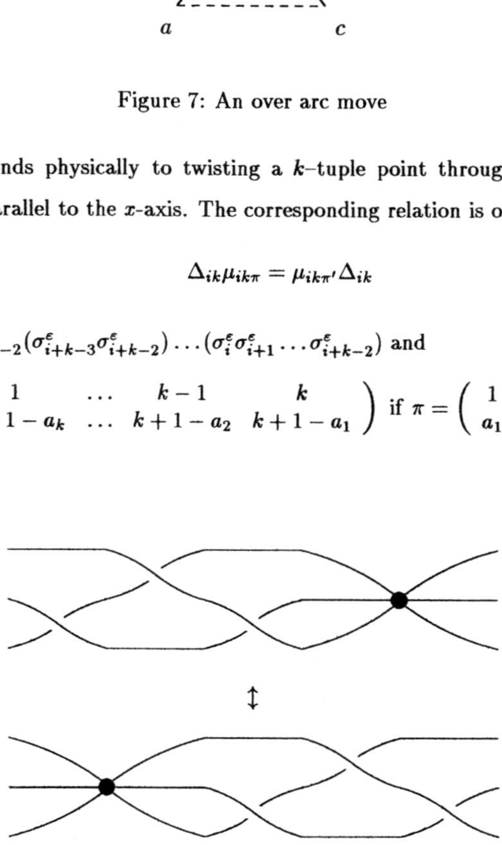

Figure 8: A twist around a triple point

Notice that interchanges preserve both the order and the permutation of the multiple point whereas a twist changes the permutation induced by the multiple point we are twisting but the order of this point remains unchanged.

Births and deaths: Let μ be a k–tuple point. An **inverse** to μ (if it exists) will also be a k–tuple point with inverse permutation which we will denote by $\bar{\mu}$. So $\pi_{\bar{\mu}} = \pi_{\mu}^{-1}$. A birth of μ and $\bar{\mu}$ is the replacement of k horizontal strings by a side by side appearance of μ and $\bar{\mu}$ in either order. A death is the reverse process.

We will find it convenient to represent μ by a heavy blob and $\bar{\mu}$ by a hollow blob.

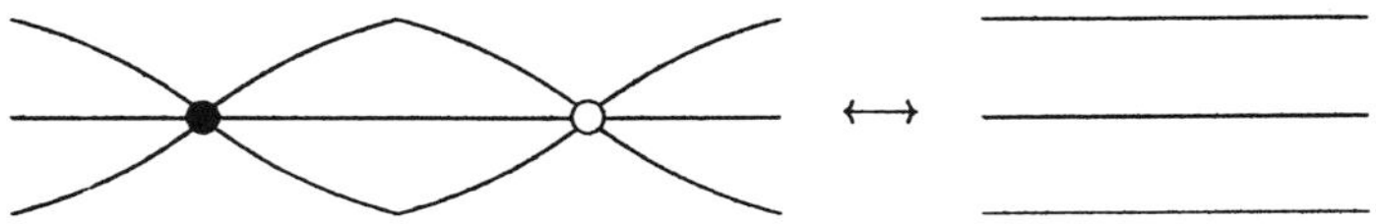

Figure 9: Birth and death of a triple point

2.2 Geometric braids

In this section we will construct a specific kind of extended singular braid monoid, which we will call the monoid of *geometric braids*. We will also show that this monoid embeds in a group.

Consider braid diagrams on n strings where all the singular crossings are represented by heavy blobs. We say that two diagrams are geometrically equivalent if one can be obtained from the other by a finite sequence of the moves:

1. Interchanges.
2. Twists.
3. Over and under arc moves.
4. Births and deaths if inverses exist.

Consider the monoid with generators $\{\sigma_j^\varepsilon, \mu_{ik} \mid i,j = 1,\ldots,n-1,\ 2 \le k \le n+1-i,\ \varepsilon = \pm 1\}$. Interchanges and twists correspond to relations 1 and 2 respectively and the arc moves correspond to a finite application of the relations:

$$
\begin{aligned}
\sigma_i\sigma_i^{-1} &= \sigma_i^{-1}\sigma_i = 1 \\
\sigma_i\sigma_{i\pm 1}\sigma_i &= \sigma_{i\pm 1}\sigma_i\sigma_{i\pm 1} \\
\mu_{ik}\sigma_{i+k-1}\sigma_{i+k-2}\cdots\sigma_i &= \sigma_{i+k-1}\sigma_{i+k-2}\cdots\sigma_i\mu_{i+1,k} \\
\mu_{ik}\sigma_{i-1}\cdots\sigma_{i+k-2} &= \sigma_{i-1}\cdots\sigma_{i+k-2}\mu_{i-1,k}
\end{aligned}
\tag{3}
$$

Suppose that Γ_1 and Γ_2 are geometrically equivalent diagrams. Then the lift of the deformation from one to the other in the 3-space corresponds to an ambient isotopy of $\mathbf{R}^3$ mapping Γ_1 to Γ_2

leaving the neighbourhood of the multiple points planar and fixing the initial and final vertices. Conversely if there is an ambient isotopy of $\mathbf{R}^3$ mapping Γ_1 to Γ_2 by keeping the end points fixed and the neighbourhoods of the singularities planar then, as diagrams on the plane, Γ_2 can be obtained from Γ_1 by finite application of the allowable moves. Notice that if we drop the condition on $N(s)$ not all the istopies of $\mathbf{R}^3$ can be represented as a sequence of the lift of allowable moves on the diagram since we exclude some isotopies around singular points of order ≥ 3, see [1]. The extended braid monoid defined by the above relations is called the **geometric braid monoid**. The relations are motivated by the relations of singular braids SB_n as defined by Baez and Birman which have only double points as multiple points and in each case the permutation is (12). We will drop all (or some) adjectives in front of braid when there is no ambiguity. Denote the monoid of geometric braids by MB_n, then SB_n is a submonoid of MB_n. It has been shown in [8] that SB_n embeds in a group by showing that in can be embedded in a monoid which satisfies the diamond property. In [12], a monoid presentation which can be embedded is given. We show that the standard presentation of MB_n satisfies the conditions of the main theorem of [12]. First we repeat the main theorem of this paper.

Theorem 2.1 *Consider a monoid $\mathcal{M}$ with the following presentation:*
Generators $\mathbf{X} = \mathbf{A} \cup \mathbf{B}$ where $\mathbf{A} = \{a_1, \ldots, a_n\}$, $\mathbf{B} = \{b_1, \ldots, b_m\}$.
Relations $\mathbf{R} = \mathbf{R}_1 \cup \mathbf{R}_2 \cup \mathbf{R}_3 \cup \mathbf{R}_4$ where
$\mathbf{R}_1$ consists of relations of the form $u = v,\ u, v \in F^+(\mathbf{A})$.
$\mathbf{R}_2 = \{a_i u_i = u_i a_i = 1 \mid \text{for some } u_i \in F^+(\mathbf{A}), \text{for all } i = 1, \ldots, n\}$.
$\mathbf{R}_3$ consists of relations of the form $u b_j = b_k u$ for some $j, k = 1, \ldots, m$ and $u \in F^+(\mathbf{A})$.
$\mathbf{R}_4$ consists of relations of the form $b_j b_k = b_k b_j$ for some $j, k = 1, \ldots, m$.
Then $\mathcal{M}$ embeds in a group G.

Proof We refer to [12] for the proof.

Theorem 2.2 *The monoid of geometric braids MB_n embeds in its group GB_n.*

Proof Let MB_n be given by the standard presentation where the generators are $\{\sigma_i, \bar{\sigma}_i, \mu_{jk} \mid i, j = 1, \ldots, n-1\}$ and the relations are equations 1–3. This presentation of MB_n satisfies the conditions of theorem 2.1. With the notation of that theorem let the generators be $\mathbf{A} = \{\sigma_i, \bar{\sigma}_i\}$, $\mathbf{B} = \{\mu_{jk}\}$.

Then:

The first relation of 1, 2 and second of 3 belong to $\mathbf{R}_1$.

$\mathbf{R}_2$ is satisfied by the first relation of 3.

The second of 1, third and fourth of 3 belong to $\mathbf{R}_3$.

And finally the last of 1 belong to $\mathbf{R}_4$. □

Let GB_n be the groupification of MB_n. Geometrically we will represent an element of GB_n by a braid diagram where μ_{ik}^{-1} corresponds to a hollow blob $\bar{\mu}_{ik}$. Then the allowable moves will be the same as MB_n with the addition of births and deaths which correspond to $\mu_{ik}\bar{\mu}_{ik} = \bar{\mu}_{ik}\mu_{ik} = 1$ in terms of generators and relations.

We can now consider elements of GB_n as geometric braids in which double points and anti-double points can appear. In view of the above this does not lead to any loss of information.

We can make the following geometric observations. Let us assume that we have the birth of a pair $s\bar{s}$ of order k. Let γ_j be the arc which starts from the jth vertex on the right side of $\partial N(s)$ and ends on a vertex on the left side of $\partial N(\bar{s})$, for $j = 1, \ldots, k$. Now if we fix the end points of $\gamma_1, \ldots, \gamma_k$ and consider the diagram just consisting of these arcs, then this diagram is geometrically equivalent to the identity braid on k strings. Let us call the set $\{\gamma_1, \ldots, \gamma_k\}$ the **joining strings** of s and $\bar{s}$. Moreover if we join the initial (terminal) points of γ_j and γ_{j+1} with an arc that lies on $\partial N(s)$ ($\partial N(\bar{s})$), which does not go through any other vertex on $\partial N(s)$ ($\partial N(\bar{s})$) then we obtain a simple closed curve C_j which bounds a disc D_j which can be chosen to be disjoint from the braid apart from its boundary in $\mathbf{R}^3$. Moreover in every equivalent diagram we can find such discs. A k-tuple pair ss' is called a **cancelling pair** if

- if one of them is represented by a black blob and the other by a hollow blob,
- they have a set of joining strings,
- the joining strings bound discs D_j, $j = 1, \ldots, k-1$ which are disjoint from the braid apart from their boundaries,
- $\partial D_j = C_j$ where C_j is as described in birth and does not have any singular point lying on it,
- $\pi_{s'} = \pi_s^{-1}$ or $\pi_{s'} = \begin{pmatrix} 1 & \ldots & k-1 & k \\ k+1-a_k & \ldots & k+1-a_2 & k+1-a_1 \end{pmatrix}$ where

$$\pi_s^{-1} = \begin{pmatrix} 1 & 2 & \dots & k \\ a_1 & a_2 & \dots & a_k \end{pmatrix} \text{ and}$$

- if the first case holds then the set of joining strings is geometrically equivalent to $\Delta^{2r} \in B_k$ for some $r \in \mathbb{Z}$,
- if the second case holds, the set of joining strings is geometrically equivalent to $\Delta^{2r+1} \in B_k$ for some $r \in \mathbb{Z}$.

We will use this definition of *cancelling* to prove Markov's theorem which is the main theorem in the next section.

2.3 Geometric Singular Links

A **singular link diagram** Λ in $\mathbb{R}^2$ is a finite collection of piecewise linear closed curves. The intersection points are either **singular points** or **crossings** –a crossing can occur only if *two* arcs are meeting. At a crossing one of the arcs is indicated by a broken line which is called the **under** arc. The other is the **over** arc. A singular point is indicated by a heavy blob or a hollow blob. Two singular link diagrams are equivalent if they are connected by a finite sequence of moves on the diagrams. An equivalence class of diagrams is called a **singular link**. Different allowable moves give different classes of singular links. In this paper, we will mainly concentrate on *geometric singular links.*

A **geometric singular link diagram** Λ satisfies the following:

1. Λ is piecewise linear.
2. Λ is oriented.
3. Λ has a vertex set and an edge set. The vertex set satisfies some properties which will be listed separately. An edge $|ab|$ is a polygonal arc in Λ which joins a to b where a, b belong to the vertex set $\mathcal{V}$. The initial and terminal vertices of $|ab|$ are a and b respectively and it respects the orientation of the link. A **path** is a collection of edges, $e_1 \dots e_l$ where the initial vertex of e_{i+1} is the terminal vertex of e_i for $i = 1, \dots, l-1$. If $e_i = |a_i a_{i+1}|$ then we denote the path $e_1 \dots e_l$ by $[a_1 a_{l+1}]$.

4. A **singular edge** is an edge which has a singular point as its initial or final vertex. The singular edge whose final (initial) vertex is a singular point is called an **in-edge (out-edge)**. For any singular point s the in-edges (and hence the out-edges) are adjacent.

5. We consider the diagrams up to geotopy. That is an isotopy on $\mathbf{R}^2$. Here we do not require the geotopy to preserve the monotonicity of arcs as we did in braid diagrams.

The origin O will be our reference point. All the link diagrams will be assumed to avoid the origin. We can assume this without loss of generality since if Λ passes through O, we can shift it slightly by a geotopy so that it does not pass through the origin. Suppose that for an edge $|ab|$ directed from a to b which does not contain any singular points, travelling from $|Oa|$ to $|Ob|$ through a continuous family of paths $|Op|$, $p \in |ab|$ is entirely in counterclockwise or clockwise direction with respect to the origin then $|ab|$ is called **positive** or **negative** respectively. If there is a point $p \in |ab|$ such that $|ap|$ is positive (negative) and $|pb|$ is negative (positive) then p is called a **turning point.** The vertex set $\mathcal{V}$ of Λ is required to satisfy the following:

- It contains all the crossing points and singular points of the diagram.
- Between every over and under crossing point there is a vertex $p \in \mathcal{V}$.
- All the turning points are included in $\mathcal{V}$.
- We consider all the singular points s with a small neighbourhood $N(s)$ around them. For all multiple points s, $\partial N(s) \cap L \in \mathcal{V}$. This indicates that a singular edge can not be both in and out.

By geotopy if necessary we will choose $N(s)$ sufficiently small, so that all the singular edges in $N(s)$ are of the same sign. Call s **positive (negative)** if the singular edges in $N(s)$ are positive (negative). In $N(s)$ the edges are labelled so that we know which out-edge follows which in-edge. Start from an out-edge and travel in the clockwise direction around s. The first in-edge encountered is labelled 1, the next 2 and so on. If π is the permutation that s induces then the continuation of the ith in-edge (which is labelled by i) is the $\pi(i)$th out-edge. We count the out-edges starting from the one nearest to the first in-edge and travelling counterclockwise. We label the $\pi(i)$th out-edge by i.

Definition Two geometric singular link diagrams are equivalent if they are connected to each other by the three Reidemeister moves **R1**, **R2**, **R3**, **R3'** (as shown in figure 10), twist, birth and death where a death can take place only among cancelling pairs and a birth takes place such that if s, s' are the two new multiple points then the in-edges of s are adjacent and so are the in-edges of s'. The last two moves will be explained in greater detail in the next section.

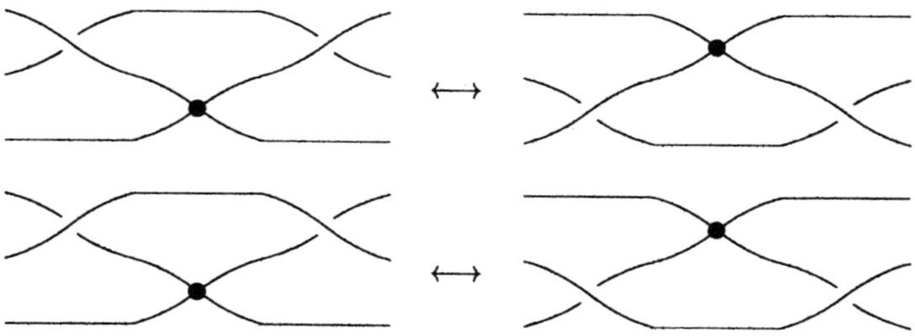

Figure 10: An **R3'** move

A **closed braid** is a link diagram all of whose edges and singular points are positive. We can obtain a geometric singular link by joining the end points of a singular braid $\beta \in GB_n$. Like geometric braids, we can also talk about geometric links in the 3-space as a lift of their diagrams leaving the neighbourhoods of the multiple points planar. Moreover we may always assume that in the lift there are no edges parallel to a chosen axis of rotation through the reference point O. Let us fix the axis of rotation to be the z-axis. We may further assume that no edge cuts the z-axis. Conversely, given any diagram in the 3-space where the edges around singular points are planar around a small neighbourhood, with a slight isotopy if necessary it will satisfy these conditions. Hence we can talk about the projection on the xy -plane and the lift of a diagram interchangeably without losing any information.

Theorem 2.3 *Every geometric singular link L can be obtained by closing some braid $\beta \in GB_n$.*

Proof Let Λ be the projection of L on the plane. To prove theorem 2.3, we will get a closed braid Λ' from Λ via allowable moves and cut it along an axis to get a braid diagram. We may assume without loss of generality that Λ lies in the lower half plane $H^- = \{(x,y) \in \mathbf{R}^2 \mid y < 0\}$. Let s be a singular point of Λ and $N(s)$ be its closed neighbourhood. The arcs in $N(s)$ may be positive or not. By applying a geotopy leaving $\Lambda - N(s)$ fixed, rotate $N(s)$ so that all edges are positive in a closed neighbourhood $N'(s) \subset N(s)$. This operation is called **correcting** a neighbourhood of a singularity. Apply this to all the singular points in Λ. After this correction we may still assume that Λ lies in H^-. This is the first step to get a closed braid from a link. The second step is to replace all the non-singular negative edges by positive edges via arc moves.

Let $|ac|$ be a non-singular negative edge. Take a point $b \in H^+$ which is not on the diagram such that $|ab|$, $|bc|$ are positive. Observe that the point b is a new vertex. Replace $|ac|$ by $|ab|$, $|bc|$ such that $|ab|$, $|bc|$ have the same *type* as $|ac|$ (see figure 7). By saying the same type we mean that if $|ac|$ is an over (under) arc then so are $|ab|$ and $|bc|$. Note that $|ac|$ is either over or under but it cannot be singular since all the singularities have been corrected and hence every singular edge lies in a corrected neighbourhood. It is possible that $|ac|$ does not contain any crossing points. In this case we are free to choose the types of $|ab|$, $|bc|$ but they should be the same type with each other. Let $< a, b, c >$ denote the convex hull of $\{a, b, c\}$ then in the 3-space we can find a disc D bounded by $< a, b, c >$ such that $D^o \cap L = \emptyset$ where D^o denotes the interior of D. This is possible because all three edges are of the same type. We visualise this arc move as the edge $|ac|$ deforming into two edges over the disc D in $\mathbb{R}^3$. In this way, discard all the negative edges. After all the negative edges are replaced by positive edges we have a closed braid Λ' equivalent to Λ. Let $P = \{(0, y) \in \mathbb{R}^2 \mid y > 0\}$. By applying geotopy if necessary we can assume that P does not intersect any crossing or singular point of Λ'. Now we can cut Λ' along P to get a diagram of a braid β such that $\hat{\beta} \cong L$. □

Birman proved the Alexander theorem for singular links which contain only double points as singular points and no cancellation is allowed. In other words, every singular link is equivalent to a closed singular braid $\beta \in SB_n$ in [5]. The above proof is the extension of her proof to our case. Notice that this is not the most efficient way of finding the braid of a given link. In the classical case Vogel [16] describes a method to find the braid with the lowest index and his method can be adapted to the singular case. However this is not necessary for our purposes.

Example of Alexander's theorem:

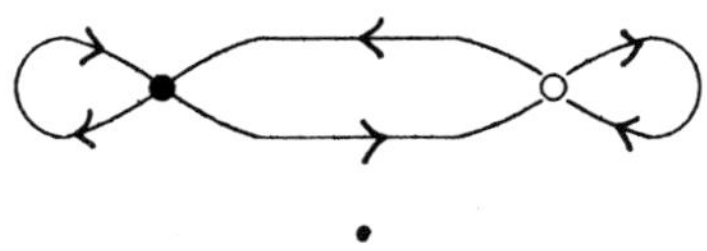

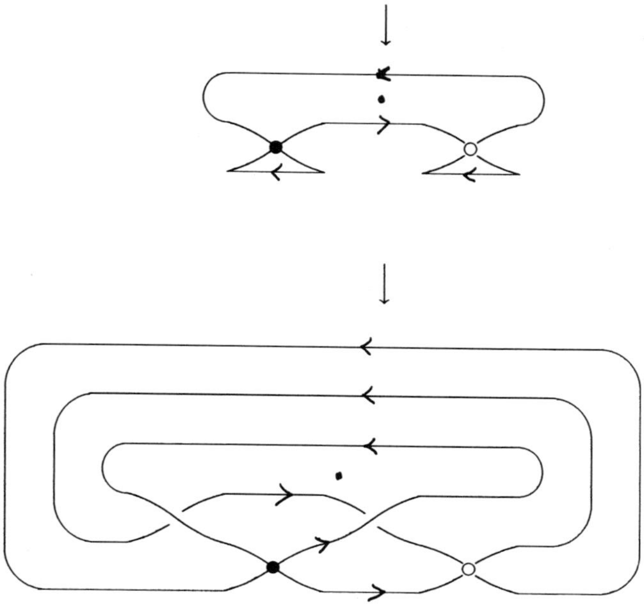

So 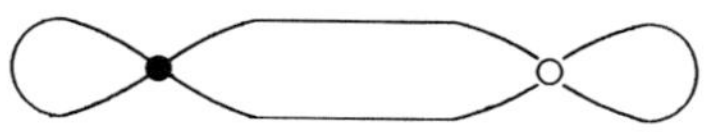 $\cong \widehat{\sigma_2\tau_1\sigma_2^{-1}\tau_1^{-1}}$

Remember that the orientation on our links is such that the in-edges of multiple points are adjacent. If this were not the case, then the above theorem would be false.

Example

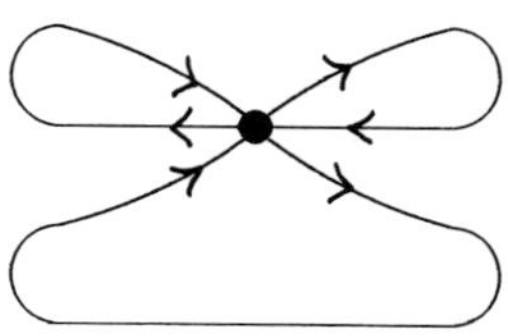

Figure 11: A singular knot which is not the closure of a braid

2.4 Markov's Theorem

In this section we will prove Markov's theorem for geometric singular closed braids. In [10], Gemein proves Markov's theorem for links obtained by taking the closure of elements of SB_n, by extending Birman's proof in [4] for the non-singular case. We will also follow Birman's proof. Another possibility might be to generalise the proof of Lambropoulou and Rourke in [14]. Assume that everything is piecewise linear. Then two non-singular link diagrams are called combinatorially equivalent if one can be obtained from the other by a sequence of the following two moves or their inverses.

1. Subdivision.

2. An over or an under arc move.

Theorem 2.4 *Two link diagrams are combinatorially equivalent if and only if they are equivalent links.*

This is a well-known result and the proof may be found in various books, for example [6]. Let us now consider oriented geometric singular links. This result can be extended to the singular case by adding two more moves as below. However by a subdivision we can only add vertices which are not crossings or singular points and by an inverse to a subdivision we can remove vertices which are not crossings or singular points. The extra moves are:

3. Twist around a singular point.

4. Insertion or deletion of a cancelling pair of multiple points.

Theorem 2.5 *If two geometric singular links are isotopic to each other by an isotopy keeping a neighbourhood of every singular point fixed then they are equivalent links.*

Proof By an isotopy a k–tuple point can travel along k disjoint untangled arcs, be turned, twisted and rotated. Fix the starting point of these arcs at the beginning of the journey. Now we have a family of arcs satisfying the properties of a set of joining strings of a cancelling pair. Hence to achieve the travel of the singular point s, we can insert a pair $\bar{s}s'$ where $\bar{s}$ is near s and s' is placed at the end of the family of arcs. Then cancel $s\bar{s}$. □

Birman in [4] showed that given two equivalent classical oriented closed braids Λ_1 and Λ_2, one can be obtained from the other by a finite sequence of **B1** and **B2** moves and their inverses as shown in figure 12.

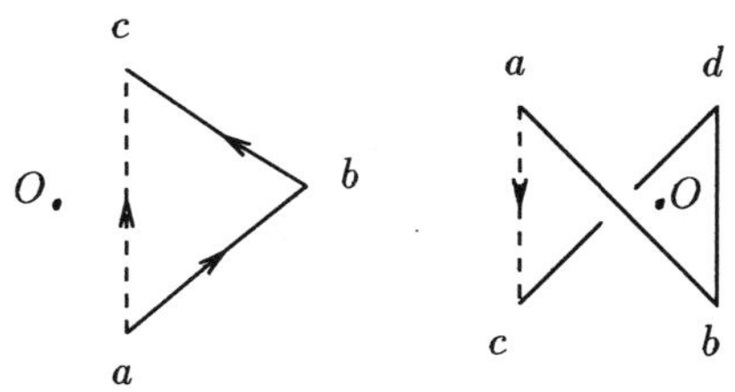

Figure 12: a: **B1** move   b: **B2** move

B1 is a single arc move where the arc deleted and the two arcs that replace it are all positive. **B2** is a collection of some arc moves. The result is a replacement of a positive arc by three positive arcs. The significance of **B1** and **B2** moves is that when they are applied to a closed braid the resulting link is also a closed braid.

We will generalise her proof of Markov's theorem to prove a general Markov theorem. Given a link diagram Λ, let the **height** $h(\Lambda)$ be the number of its non-singular negative edges plus the number of its negative singular points counted by multiplicity. By this we mean that a negative singular point of order k contributes k to the height. So if Λ is a closed braid then $h(\Lambda) = 0$. We will show that if Λ_1 and Λ_2 are combinatorially equivalent closed braids we can find a sequence of moves from Λ_1 to Λ_2 where after each move the new diagram is also a closed braid diagram.

B1 and **B2** are the two non-singular moves. Since we have singular links we have to consider type 3 and 4 moves as described in the beginning of the section. Now let us define these moves combinatorially to suit our purposes.

3. Twist around a singular point. We will allow a twist only when all the edges involved are of the same sign as the singular point. We can ensure this by applying a sequence of arc moves to the non-singular edges involved so this does not lead to any loss of information. In the combinatorial set up, it looks as follows. We can imagine all the non-singular edges that are involved to lie in a rectangle, which we will call the the rectangle of the twist. On one side of this rectangle is the multiple point of order k. Now also suppose that on the other side of the rectangle are k free edges - non-singular edges with no crossings - emanating from the end points of the edges within the rectangle. This can be guaranteed by a subdivision. By choosing the rectangle small enough we may assume these free edges have the same sign as the multiple point. To apply the twist remove the multiple point and replace it by the free edges as follows: Remove the free edges and replace them by a singular point s' of order k where the permutation $\pi_{s'}$ is

$$\begin{pmatrix} 1 & \dots & k-1 & k \\ k+1-a_k & \dots & k+1-a_2 & k+1-a_1 \end{pmatrix} \text{ if } \pi_s = \begin{pmatrix} 1 & 2 & \dots & k \\ a_1 & a_2 & \dots & a_k \end{pmatrix}.$$

Remove the singualr point s together with the edges attached to it and replace it by k free edges. A twist does not change the height. It will be denoted by W. We will sometimes say a positive twist or a negative twist depending on the sign of the multiple point.

4. A pair ss' can be deleted only if it is a cancelling pair (as defined in section 2.2). Suppose that ss' is a cancelling pair of order k. Then to delete the pair, remove both singularities together with the singular edges attached to them. In other words remove the interiors of $N(s)$ and $N(s')$. Now put in edges e_i, $i = 1, \dots, k$ inside $\partial N(s)$ such that e_i joins the initial vertex of ith in-edge to the terminal vertex of the ith out-edge (this is *not* the edge labelled by i) which were deleted. Similarly insert edges f_i, $i = 1, \dots, k$ inside $\partial N(s')$. Now the the signs of e_i are the same as the sign of s because if a_i, a_i' are the respective initial and final vertices of e_i then the angle from $|Oa_i|$ to $|Oa_i'|$ is the sum of the angles from $|Oa_i|$ to $|Os|$ and from $|Os|$ to $|Oa_i'|$. Since a multiple point of order k is replaced by k edges of the same sign with the multiple point, deletion is height preserving. Insertion is the reverse of cancellation. We denote insertion by I or $I(k_{ss'})$ when we want to be more precise.

We will call any arc move, **B2**, $I(k)$, W and their inverses **elementary** moves and will denote the set of them by $\mathcal{E}$. In [4], Birman proved the two following lemmas.

Lemma 2.6 *Let Λ be a link diagram with at least one non-singular negative edge and let $\Lambda' = A(\Lambda)$ where A is a height preserving arc move or a* **B2** *move. Then there is a chain of diagrams $\Lambda \to \Lambda_1 \to \dots \to \Lambda_s \to \Lambda'$ such that $s \geq 1$ and $h(\Lambda_i) < h(\Lambda)$ for every $i = 1, \dots, s$ and every transformation is a single arc move or a* **B2** *move (or their inverses).*

Lemma 2.7 *Let $\Lambda, \Lambda', \Lambda''$ be link diagrams with $h(\Lambda), h(\Lambda'') < h(\Lambda')$ and for some non-singular elementary transformations A, B $\Lambda \xrightarrow{A} \Lambda' \xrightarrow{B} \Lambda''$. Then there is another chain $\Lambda \to \Lambda_1 \to \ldots \to \Lambda_s \to \Lambda''$ such that $h(\Lambda_i) < h(\Lambda')$ for every $i = 1, \ldots, s$ and every transformation is a single arc move or a* **B2** *move (or their inverses).*

Lemmas 2.6 and 2.7 are the core of the proof of Markov's theorem both in singular and non-singular cases. Although the link diagrams Birman considers do not have singular points, the proofs are not affected by the existence of multiple points since the transformtions in the lemmas are non-singular.

Theorem 2.8 *Let Λ, Λ' be two combinatorially equivalent closed braids. Then there is a chain of link diagrams from one to the other each obtained from the previous by an elementary move and such that none of the links in the sequence contains a negative singular point.*

Proof Let $\Lambda \to \Lambda_1 \to \ldots \to \Lambda_l \to \Lambda'$ be any chain of elementary transformations. If none of the links contain a negative singular point then this is the required sequence. If not say Λ_i is the first diagram that a negative singular point s appears. It can only appear by the insertion $\Lambda_{i-1} \xrightarrow{I} \Lambda_i$ and we know that it has to be deleted at some point since $h(\Lambda') = 0$. Before the insertion, replace all the negative edges (that s will be inserted on) of Λ_{i-1} by positive edges via arc moves. Applying I now will result as the insertion of a positive singular point. There are two problems that need to be dealt with. The first is that if there were a twist that involved s, then it would be negative. To overcome this we apply a sequence of arc moves to make all the edges that were involved in the twist positive, apply the twist, -now a positive one- and apply the inverse sequence of the arc moves. We may need a geotopy to make the rectangle of the new twist disjoint from the singular point. The second problem is that we had a set of negative edges in Λ_{i-1} which became singular in Λ_i. Now we have two sets of positive edges and only one set of them becomes singular. The remaining set of positive edges stays attached to $N(s)$ unless there is a twist that involves s. Whenever we apply a twist to s we carry the extra set of positive edges by a trivial twist so that they stay attached to $N(s)$. And after we delete s, we again have two sets of positive edges. Now apply a sequence of inverse arc moves to get back to the negative set of edges. □

Theorem 2.9 *Let Λ, Λ' be two combinatorially equivalent closed braids. Then one can be obtained* *from the other by a sequence of* **B1**, **B2**, *W, I moves and their inverses where the insertions do not create any negative singular points.*

Proof If Λ, Λ' are combinatorially equivalent then there is a sequence of links from one to another, obtained by elementary moves. By theorem 2.8 we can replace this sequence by $\Lambda \to \Lambda_1 \to \Lambda_2 \to \ldots \to \Lambda_l \to \Lambda'$ where no Λ_i contains a negative singular point. The rest follows easily from lemmas 2.6 and 2.7 and can also be found in [4]. □

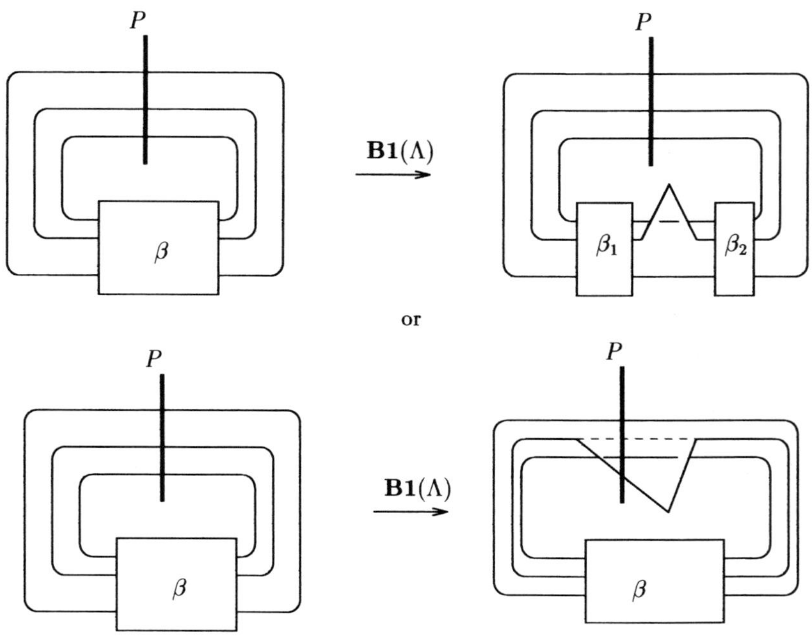

Figure 13: The effect of **B1** on a closed braid Λ

Lemma 2.10 *Let Λ be a closed braid in general position and let β be the braid obtained by cutting Λ open along P. Then β' which is the braid obtained by cutting* $\mathbf{B1}(\Lambda)$ *open along P is a conjugate of β in GB_n.*

Proof Let Λ be a closed braid. Suppose we have $\mathbf{B1}(\Lambda)$ replacing an edge $|ac|$ by two edges $|ab|, |bc|$. If $|ac| \cap P = \emptyset$ then $|ab|, |bc|$ also do not intersect P. Because if they do we must get a negative edge.

So Λ and $\mathbf{B1}(\Lambda)$ come from the same braid. If $|ac| \cap P \neq \emptyset$ then at least one of $|ab| \cap P, |bc| \cap P$ is non-empty. Hence if Λ comes from $\beta \in GB_n$ then $\mathbf{B1}(\Lambda)$ comes from $\alpha^{-1}\beta\alpha \in GB_n$ where $\alpha = \sigma_j^{\varepsilon_j}\sigma_{j+1}^{\varepsilon_{j+1}}\ldots\sigma_k^{\varepsilon_k}$ or $\alpha = \sigma_j^{\varepsilon_j}\sigma_{j-1}^{\varepsilon_{j-1}}\ldots\sigma_i^{\varepsilon_i}$ for some $i \leq j \leq k$ and $\varepsilon = \pm 1$. See figure 13. $\square$

Lemma 2.11 *If* **B2** *is applied to a closed braid* Λ *in general position and if cutting* Λ *open along* P *gives a braid* $\beta \in GB_n$ *then cutting* $\mathbf{B2}(\Lambda) = \Lambda'$ *open along* P *will give a braid* $\beta' \in GB_{n+1}$ *which is related to* β *by a sequence of* **M1** *and* **M2** *moves where* **M1** *is conjugation by the elements of* GB_n.

Proof Suppose that Λ is the diagram of $\hat{\beta}$, $\Lambda' = \mathbf{B2}(\Lambda)$ with $|ac|$ the edge to be replaced and Λ' comes from the braid β'. If $|ac| \cap P = \emptyset$ and if $\beta = \beta_1\beta_2$ with $\beta_1, \beta_2 \in GB_n$, then up to a conjugate $\beta'\alpha_1\beta_1\alpha_2\sigma_n^{\pm 1}\alpha_2^{-1}\beta_2\alpha_1^{-1}$ where $\alpha_1 \in GB_{n+1}$ and $\hat{\beta}' = \mathbf{B2}(\Lambda)$, see figure 14. So we have the following sequence, $\beta' \overset{\mathbf{M1}}{\to} \beta_1\alpha_2\sigma_n^{\pm 1}\alpha_2^{-1}\beta_2 \overset{\mathbf{M1}}{\to} \alpha_2^{-1}\beta_2\beta_1\alpha_2\sigma_n^{\pm 1} \overset{\mathbf{M2}}{\to} \alpha_2^{-1}\beta_2\beta_1\alpha_2 \overset{\mathbf{M1}}{\to} \beta_2\beta_1 \overset{\mathbf{M1}}{\to} \beta$. If $|ac| \cap P \neq \emptyset$ then up to a conjugate $\beta' = \alpha\beta\alpha^{-1}\sigma_n^{\pm 1} \overset{\mathbf{M2}}{\to} \alpha\beta\alpha^{-1} \overset{\mathbf{M1}}{\to} \alpha^{-1}\alpha\beta = \beta$. $\square$

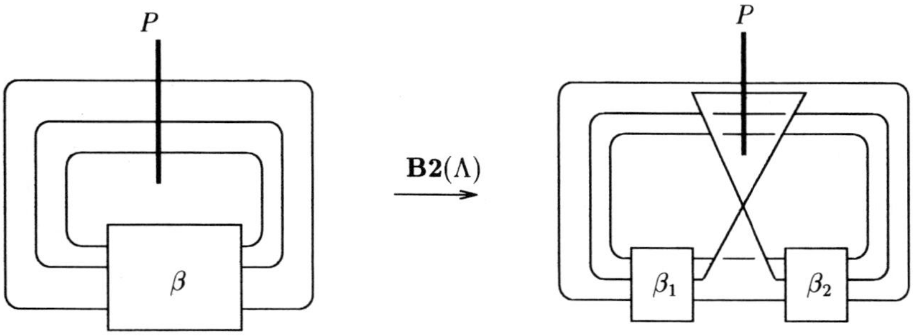

Figure 14: The effect of **B2** on a closed braid Λ

Lemma 2.12 *If a singular link diagram* Λ' *is obtained from* Λ *by a twist or insertion or deletion of singular points then they arise from conjugate braids in* GB_n. $\square$

Thus we can conclude:

Theorem 2.13 *Two braids* $\beta_1 \in GB_n$, $\beta_2 \in GB_m$ *define equivalent links if and only if they are related by a sequence of the following moves:*

M1 $\beta_2 = \alpha^{-1}\beta_1\alpha$ *for some* $\alpha \in GB_n$ *and* $n = m$.

M2 $\beta_2 = \beta_1\sigma_n^{\pm 1}$ *and* $m = n + 1$.

References

[1] D. Armand-Ugon, R. Gambini and P. Mora, *Intersecting braids and intersecting knot theory.* J. of Knot Theory and Its Ramifications **4** No:1 (1995), 1–12.

[2] E. Artin, *Theory of Braids.* Ann. of Math. (2) **48** (1947), 101–128.

[3] J. Baez, *Link invariants and perturbation theory.* Lett. Math. Phys. **2** (1992), 43–51.

[4] J. Birman, *Braids, links and mapping class groups.* Ann. Math. Studies **82**, Princeton University Press, Princeton, (1974).

[5] J. Birman, *New points of view in knot theory.* Bull. AMS **28** No:2 (1993), 253–287.

[6] G. Burde, H. Zieschang, *Knots.* De Gruyter Studies in Mathematics **5**, New York, (1985).

[7] R. Fenn, *PicTEX macros for drawing braids and links.* Preprint.

[8] R. Fenn, E. Keyman, C. Rourke *Singular braids embed in a group.* J. of Knot T. and Ram. **7** (1998), 881–892.

[9] R. Fenn, R. Rimányi, C. Rourke, *The braid-permutation group.* Topology **36** (1997), 123–135.

[10] B. Gemein, *Singular braids and Markov's theorem* J. of Knot T. and Ram. **4** (1997), 441–454.

[11] E. Keyman, *Singular braids and links.* D. Phil. Thesis. University of Sussex, (1997).

[12] E. Keyman, *A new criterion for embedding monoids.* Preprint.

[13] M. Khovanov, *Doodle groups.* Preprint.

[14] S.Lambropoulou, C.Rourke, *On Markov's theorem.* University of Warwick, preprint series (1992).

[15] P. Traczyk, *A new proof of Markov's braid theorem.* Banach Center Publications **42** (1998), 409–419.

[16] P.Vogel, *Representation of links by braids - a new algorithm.* Com. Math. Helvetici **65** No: 1 (1990), 104–113.

Roger Fenn
SMS,
University of Sussex,
Falmer, Brighton, BN1 9QH, UK
rogerf@sussex.ac.uk

Ebru Keyman
Osmangazi Üniversitesi
Matematik Blm.
26480 Eskişehir TURKEY
ekeyman@ogu.edu.tr

On Spaces of connected Graphs I

Properties of Ladders

Jan A. Kneissler

Abstract

We examine spaces of connected tri-/univalent graphs subject to local relations which are motivated by the theory of Vassiliev invariants. It is shown that the behaviour of ladder-like subgraphs is strongly related to the parity of the number of rungs: there are similar relations for ladders of even and odd lengths, respectively. Moreover, we prove that - under certain conditions - an even number of rungs may be transferred from one ladder to another.

1 Introduction

This note is the first in a series of three papers and its main goal is to provide relations that will be employed by the following ones. Nevertheless, we have the feeling that the results are interesting enough to stand alone. The objects we deal with are combinatorial multigraphs in which all vertices have valency 1 or 3, equipped with some additional data: at each trivalent vertex a cyclic ordering of the three incoming edges is specified and every univalent vertex carries a colour. In pictures we assume a counter-clockwise ordering at every trivalent vertex and indicate the colour of univalent vertices by integers. The graphs have to be connected and must contain at least one trivalent vertex, but we do not insist that there are univalent vertices. For simplicity, such graphs will be called *diagrams* from now on.

The motivation is given by the combinatorial access to Vassiliev invariants. We briefly mention the most relevant facts about this comprehensive family of link invariants. Vassiliev invariants with values in a ring R form a filtered algebra. A deep theorem of Kontsevich (see [4] and [1]) states that for $\mathbf{Q} \subset R$ the filtration quotients are isomorphic to the graded dual of the Hopf-algebra of chord diagrams. By the structure theorem of Hopf-algebras it suffices to examine the primitive elements.

The algebra of chord diagrams is rationally isomorphic ([1]) to an algebra of tri-/univalent graphs (sometimes called Chinese characters), defined by relations named (AS) and (IHX) (see Definition 2.1). The coproduct is given by distributing the components into two groups, so the primitives are spanned by connected graphs. The colours of the univalent vertices represent the link-components and half the total number of vertices corresponds to the degree of Vassiliev invariants. The number of univalent vertices allows a second grading that corresponds to the eigenspace-decomposition of primitive Vassiliev invariants with respect to the cabling operation.

The most important open question in Vassiliev theory is whether all invariants of finite degree taken together form a complete invariant of knots and links. For knots there is a weaker (but equally essential) question:

Question 1.1 Are there any (rational valued) Vassiliev invariants that are able to detect non-invertibility of knots?

There is a weak hope that this question might be settled in the combinatorial setting, where it translates into the question whether all diagrams with an odd number of univalent vertices vanish.

A progress in the combinatorics of connected graphs ([6]) was initiated by the observation that replacing a trivalent vertex by the graph that is shown in relation (x) in the next section is a well-defined operation (i.e. independent of the choice of a trivalent vertex and the orientation). This operation called x_n is an element of an algebra called Λ that acts on spaces of connected graphs. One purpose of our investigations (and in fact the initial one) was to obtain a family of relations in Λ. The result is presented in [2].

Our philosophy is to declare diagrams that are coming from lower degrees by the action of Λ as uninteresting, which explains why we factor by the relation (x) in $\hat{B}^u$ (see Definition 2.1). This can be justified in two ways: First, note that Question 1.1 is equivalent to asking whether $\hat{B}^u = 0$ for all odd u. Second, any good upper bound for the dimensions of $\hat{B}^u$ will lead to good upper bounds for $\dim B^u$ as we demonstrate in [3].

2 Results

First let us introduce notations for certain subgraphs of diagrams. An edge that connects a trivalent and a univalent vertex is called *leg*. For $n \geq 2$, a subgraph consisting of $3n+2$ edges of the following type

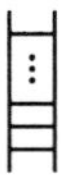

is called a *n-ladder*. The two uppermost and the two lowermost edges are called *ends* of the ladder. They may be connected to univalent or trivalent vertices in the rest of the diagram. The $3n-2$ other edges are called *interior* of the ladder. A ladder is said to be *odd* or *even* according to the parity of n. A 2-ladder is also called *square*. Finally, for a partition $u = (u_1, \ldots, u_n)$, $u_1 \leq u_2 \leq \cdots u_n$, we say a diagram is u-coloured if its univalent vertices carry the colours 1 to n and there are exactly u_i univalent vertices of colour i for $1 \leq i \leq n$.

Definition 2.1 For any partition u let

$$B(u) \;:=\; \mathbf{Q}\langle\ u\text{-coloured diagrams}\ \rangle \;/\; (\mathrm{AS}), (\mathrm{IHX}) \quad \text{and} \quad \hat{B}(u) \;:=\; B(u) \;/\; (\mathrm{x}),$$

where (AS), (IHX) and (x) are the following local relations:

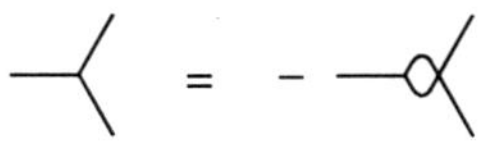

(AS)

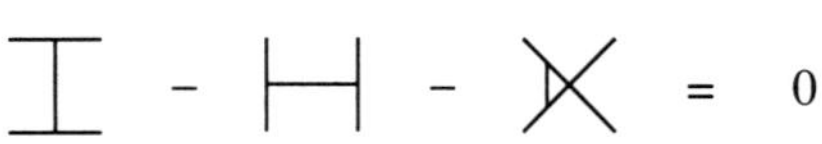

(IHX)

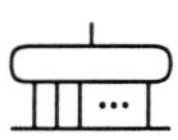 $= 0$ (x)

The extremal partitions $u_1 = \cdots = u_n = 1$ and $n = 1$ are most important, so they get their own names: $F(n) := B\big((1, \cdots, 1)\big)$, $\hat{F}(n) := \hat{B}\big((1, \cdots, 1)\big)$ and $B^u := B\big((u)\big)$, $\hat{B}^u := \hat{B}\big((u)\big)$.

Remark 2.2 The careful reader will notice that in our calculations we have to divide by 2 and 3 only, so all statements remain valid if one works with $\mathbf{Z}[\frac{1}{6}]$-modules instead of $\mathbf{Q}$-vectorspaces.

Let us mention an important consequence of the relations (AS), (IHX) and (x):

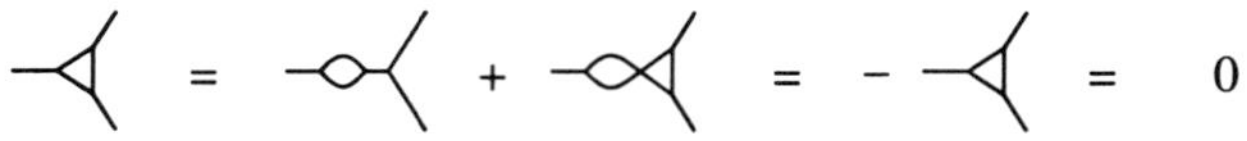

(t)

By (IHX) and (t) the ends of a square may be permuted:

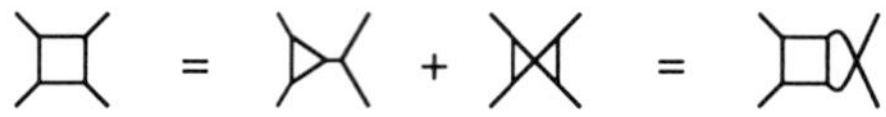

This obviously implies that neighbouring ends of any n-ladder can be interchanged. But it is a little surprise that for all even ladders any permutation of the ends yields the same element, i.e. for $(2m)$-ladders ($m \geq 1$) we have the following relation called LS ("ladder symmetry"):

(LS)

Here and later on, when making statement about n-ladders, we draw the corresponding picture for a generic but small value of n. This should not lead to confusion.
For any $m \geq 1$, we have the following relation of the IHX-type, involving odd ladders of length $2m+1$:

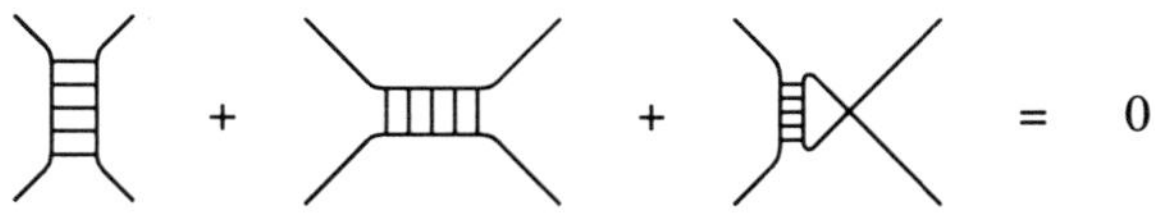

(LIHX)

We present a further relation named LI, in which an edge has been glued to non-neighbouring ends of an odd ladder:

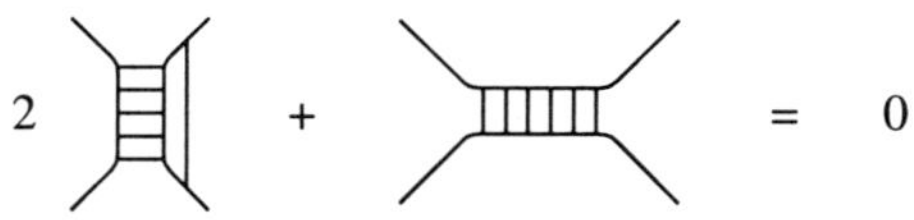

(LI)

Finally, we have the following relation LL that replaces two parallel ladders of lengths $2n+1$ and $2m+1$ by a single $(2n+2m+2)$-ladder:

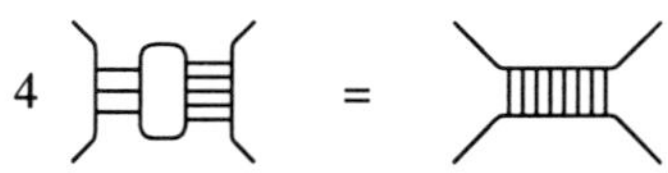

(LL)

Theorem 1 *The relations (LS), (LIHX), (LI) and (LL) are valid in $\hat{B}(u)$ for any partition u.*

Definition 2.3 Suppose we have a diagram D containing two ladders L_1, L_2 which have disjoint interiors. If we remove the interiors of L_1 and L_2 from D we get a – possibly disconnected – graph D'. Let $D'_1, \ldots D'_k$ denote the components of D' that contain at least one end of L_1 *and* at least one end of L_2. If $k = 1$ and D'_1 is a tree and D'_1 contains *exactly* one end of L_1 and *exactly* one end of L_2, then we say that L_1 and L_2 are *weakly connected*. Otherwise we call L_1 and L_2 *strongly connected*. If the intersection of the interiors of L_1 and L_2 is not empty (which implies that L_1 and L_2 are sub-ladders of a single longer ladder) then L_1 and L_2 are also called strongly connected.

Example: In the following diagrams the 3-ladders and 4-ladders are strongly connected to the square but weakly connected to each other (so being strongly connected is not a transitive relation):

For the rest of this section let us assume that D is a u-coloured diagram with two specified 3-ladders L_1 and L_2. For $a, b \geq 2$ let $D_{a,b}$ denote the diagram that is obtained by replacing L_1 and L_2 by two ladders (in the same orientation) of length a and b, respectively.

Theorem 2 (Square-Tunnelling relation) *If L_1 and L_2 are strongly connected in D then $D_{2,4} = D_{4,2}$ in $\hat{B}(u)$.*

If either L_1 or L_2 is subladder of a longer ladder, then L_1 or L_2 are automatically strongly connected, which allows to make a nice statement:

Corollary 2.4 *For any $a \geq 2$, $b \geq 4$ with $a+b \geq 7$ we have $D_{a,b} = D_{a+2,b-2}$ in $\hat{B}(u)$.*

One might ask whether the strong connectivity condition is essential in Theorem 2, in other words:

Question 2.5 Are there any u-coloured diagrams D with *weakly* connected ladders such that $D_{2,4} \neq D_{4,2}$?

At least for the case length$(u) = 1$, we may give the answer:

Theorem 3 (Square-Tunnelling relation in $\hat{B}^u$) *Let D be a diagram of B^u (i.e. all univalent vertices carry the same colour), then $D_{2,4} = D_{4,2}$.*

Compared to the other relations, which are local, the square-tunnelling relation has a completely different character: it relates ladders that might be located arbitrarily far apart in a diagram. We already mentioned in the introduction a situation where a global structure (the action of Λ on $B(u)$) emerges from local relations ((IHX) and (AS)). There, it is easy to understand how the subgraphs x_n (shown in relation (x)) move around, because they can go from one vertex to a neighbouring one. But here, we are not able to trace the way of the square from the ladder in which it disappears to the other ladder. It is just like the quantum-mechanical effect where an electron tunnels through a classically impenetrable potential barrier: one can calculate that it is able to go from one place to the other, but we cannot tell how it actually *does* it.

3 The local ladder relations

Let us start with two well-known and frequently-used implications of (IHX), (AS) and (t). We assume that we are given a diagram D of $\hat{F}(n)$ together with an arbitrary grouping of the legs of D into two classes called "entries" and "exits". Say there are p entries and q exits (thus $p+q=n$). For $1 \leq i \leq p$ $(1 \leq j \leq q)$ let D_i (D^j) denote the elements of $\hat{F}(n+1)$ one obtains when the $(n+1)$-th leg is glued to the i-th entry (the j-th exit) of D. Furthermore for $1 \leq i < j \leq p$ $(1 \leq k < l \leq q)$ let D_{ij} (D^{kl}) denote the elements of $\hat{F}(n)$ having an additional edge between the i-th and j-th entry (the k-th and l-th exit) of D. D_i, D^j, D_{ij}, D^{kl} look typically like this:

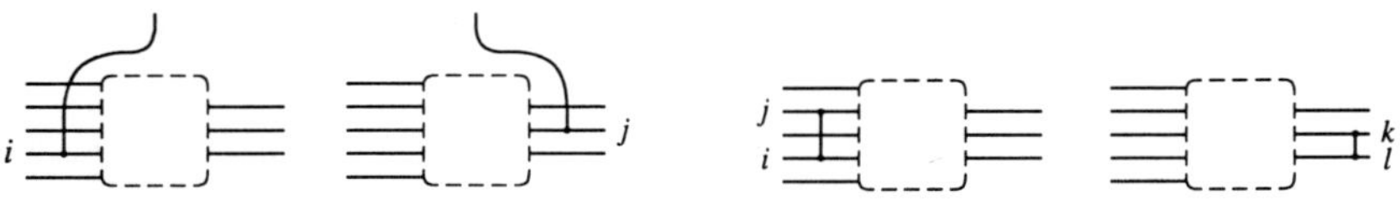

Lemma 3.1 *For any $D \in \hat{F}(n)$ with p entries and q exits, we have*

a) $\sum_{i=1}^{p} D_i = \sum_{j=1}^{q} D^j$

b) $\sum_{1\leq i<j\leq p} D_{ij} = \sum_{1\leq k<l\leq q} D^{kl}.$

Proof We cut the box into little slices, such that each slice is of one of the following types:

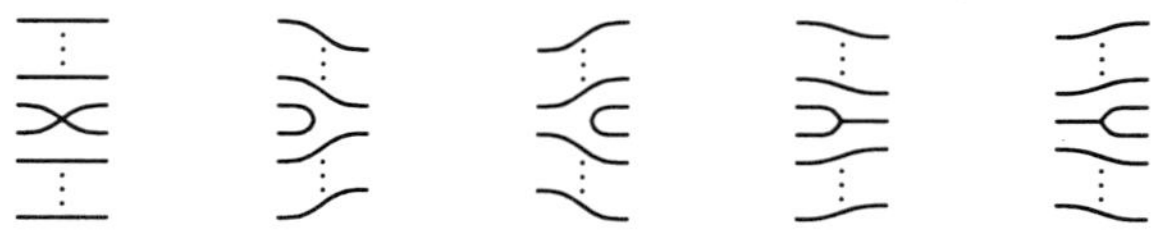

It is easy to verify that a) and b) are valid for slices of these types. □

Let us introduce a notation for certain elements of $\hat{F}(4)$. For a word w in the letters c_1, c_2, c_3, let $\langle w\rangle$ be a diagram of $\hat{F}(4)$ that is constructed as follows: Take three strings, put edges between the strings according to w, and then glue string 1 and string 2 together on the right side. c_1, c_2 and c_3 correspond to 1-2, 2-3 and 1-3 edges, respectively. For example $\langle c_2c_3c_1^2c_3\rangle$, $\langle c_2^4\rangle$, $\langle c_3^4\rangle$ are the following diagrams:

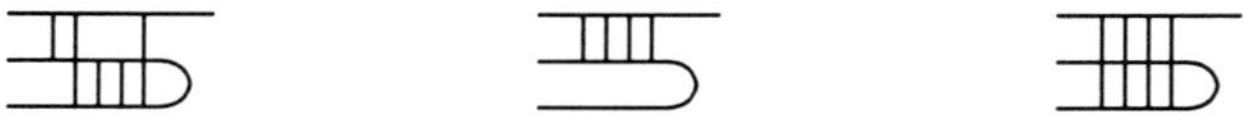

Proposition 3.2 *For $m\geq 1$ and any word u in the letters c_1, c_2, c_3, we have* $\langle uc_2^{2m}\rangle = \langle uc_3^{2m}\rangle$.

Proof The statement is implicitly hidden in section 5 of [6], but to the readers convenience we will give a simple inductive proof of the statement. We know already that it is true for $m=1$. We have the following equalities for arbitrary words u, v:

$$\langle uc_1v\rangle + \langle uc_2v\rangle + \langle uc_3v\rangle = 0 \tag{i}$$

$$\langle uc_1c_2^n\rangle = \langle uc_1c_3^n\rangle \tag{ii}$$

$$\langle uc_2c_1c_2^n\rangle = \langle uc_2c_3c_2^n\rangle \quad \text{for } n\geq 2 \tag{iii}$$

$$\langle uc_3c_1c_3^n\rangle = \langle uc_3c_2c_3^n\rangle \quad \text{for } n\geq 2 \tag{iv}$$

(i) is due to Lemma 3.1 b). (ii) and (iii) are implications of (IHX) and (x):

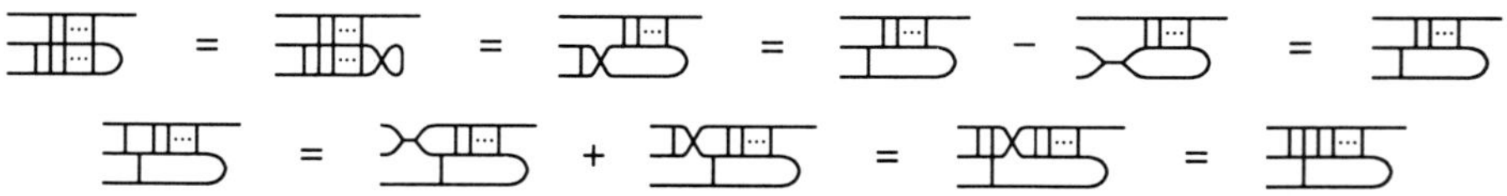

(iv) is shown similarly to (iii), interchanging string 1 with string 2. (i), (iii) and (iv) imply

$$2\langle uc_2c_3c_2^n\rangle = -\langle uc_2^{n+2}\rangle \quad\text{and}\quad 2\langle uc_3c_2c_3^n\rangle = -\langle uc_3^{n+2}\rangle \quad \text{for } n\geq 2. \tag{v}$$

Assuming $m\geq 2$ and $\langle wc_2^{2m-2}\rangle = \langle wc_3^{2m-2}\rangle$ by induction hypothesis, we obtain

$$\begin{aligned}\langle uc_2^{2m}\rangle - \langle uc_3^{2m}\rangle &\overset{\text{(i)}}{=} -\langle uc_3c_2^{2m-1}\rangle - \langle uc_1c_2^{2m-1}\rangle + \langle uc_2c_3^{2m-1}\rangle + \langle uc_1c_3^{2m-1}\rangle \\ &\overset{\text{(ii)}}{=} -\langle uc_3c_2^{2m-1}\rangle + \langle uc_2c_3^{2m-1}\rangle \\ &\overset{\text{I.H.}}{=} -\langle uc_3c_2c_3^{2m-2}\rangle + \langle uc_2c_3c_2^{2m-2}\rangle \overset{\text{(v)}}{=} \frac{1}{2}\langle uc_3^{2m}\rangle - \frac{1}{2}\langle uc_2^{2m}\rangle.\end{aligned}$$

□

Proof of Theorem 1 The first equality of the (LS)-relation is true by Proposition 3.2. To prove the second equality, we simply swap the upper ends of the ladder and use the first equality of (LS):

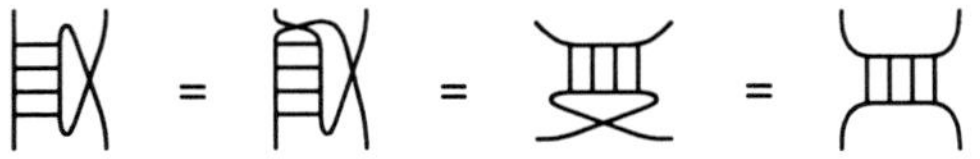

If we take three ends of a $(2m)$-ladder as entries and the forth as exit, apply Lemma 3.1 b) and (LS), we get the (LIHX)-relation:

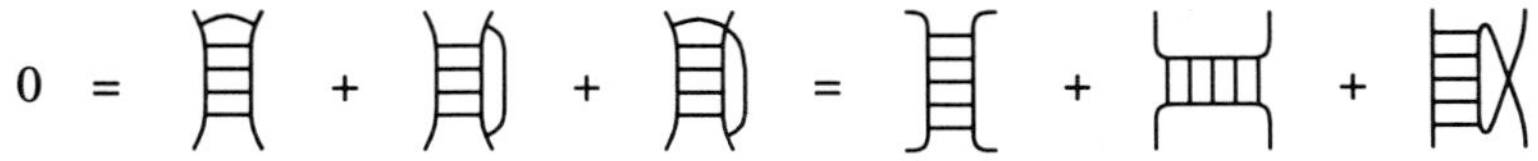

Relation (LI) is obtained when we append an edge between the ends on the right side of (LIHX) and use (IHX) and (x):

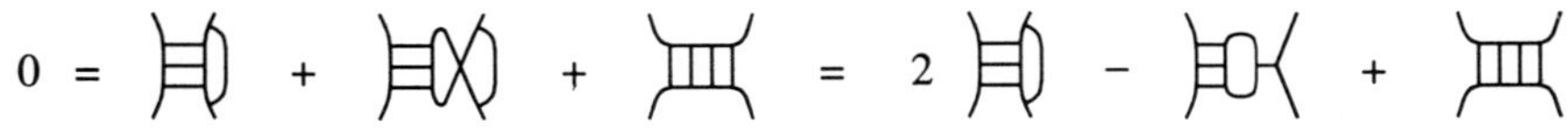

(LL) is the hardest one; we apply (IHX) and use Lemma 3.1 a):

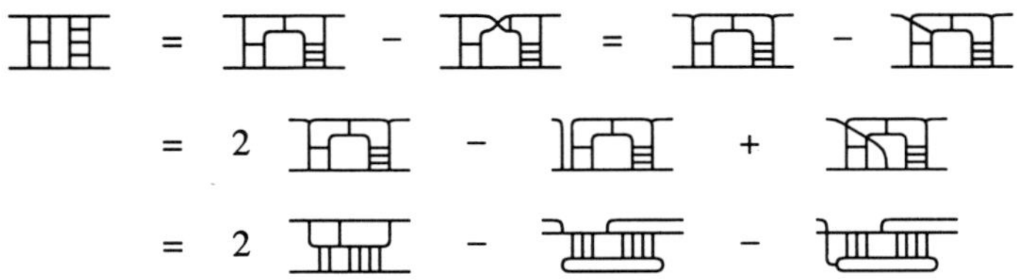

To each of the three resulting diagrams we apply the following identity

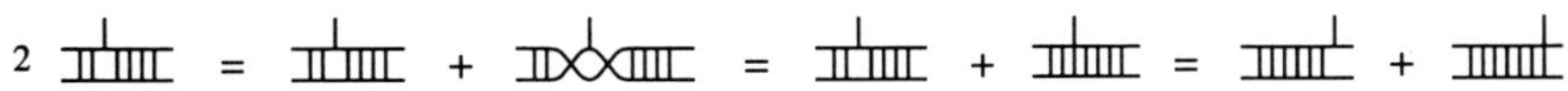

to obtain by (t), (IHX), (x), (LS) and (LI) the promised result:

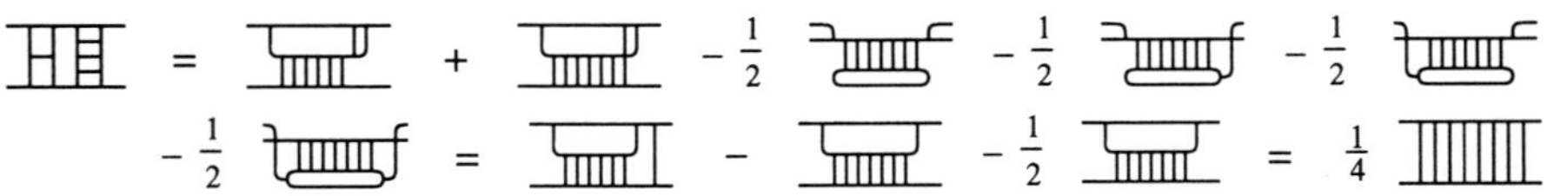

□

4 Relations in $\hat{F}(6)$

The key to Theorem 1 is a certain equation in $\hat{F}(6)$. Written in the notation we will introduce in this section, it appears misleadingly simple:

$$[y]_1^3 = 0.$$

Nevertheless, it requires a lot of calculation. We have spent a considerable amount of time trying to take the computation into a bearable form. We hope – for the readers sake – that this has not been a vain endeavour.

4.1 The algebra Ξ

We do not like to draw pictures all the time, so we introduce an algebra that enables us to write down sufficiently many elements of $\hat{F}(6)$ in terms of a few number of symbols.

Let Ξ be the space of tri-/univalent graphs with exactly six numbered univalent vertices quotiented by the (AS), (IHX) and (x). The only difference between Ξ and $\hat{F}(6)$ is that in Ξ we do not require the graphs to be connected. Also we do not insist that there actually *are* any trivalent vertices.

All graphs in an (AS)- or (IHX)-relation have same number of components. This implies that $\Xi = \hat{F}(6) \oplus \Delta$ where Δ is the subspace of Ξ that is spanned by disconnected graphs. So we can project any relation that we find in Ξ to $\hat{F}(6)$.

Let us introduce a multiplication in Ξ: For any two graphs D_1, D_2 let $D_1 D_2$ denote the graph that is obtained when the univalent vertices 6, 5, 4 of D_1 are glued to univalent vertices 1, 2, 3 of D_2, respectively. This extends to a well defined map $\Xi \otimes \Xi \to \Xi$, so we may view Ξ as algebra. It is a graded algebra, but we make no explicit use of this. In pictures we always arrange the ends always like this:

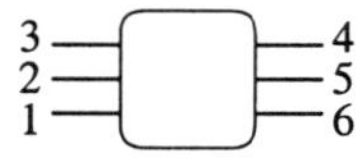

Let c_1, c_2, c_3, s_1, s_2, d_1, d_2, y denote the elements of Ξ that are represented by the following graphs:

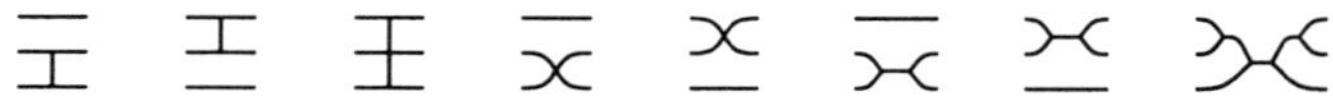

Furthermore let $l_1, l_2, l_3, l_4, \ldots$ be the family of elements of Ξ that is represented by the following graphs (l_n contains a n-ladder for $n \geq 2$ and $l_1 = c_1 c_2$):

Let $z := c_1 + c_2 + c_3$, then by Lemma 3.1 a), z is central in Ξ. We will continually substitute c_3 by $z - c_1 - c_2$ and collect all z's at the beginnings of words.

Proposition 4.1 *The following relations are fulfilled in Ξ:*

$$\begin{aligned}
s_1^2 &= s_2^2 = (s_1 s_2)^3 = 1 \\
d_i c_i &= c_i d_i = 0 \quad \textit{for } i = 1,2 \\
s_i c_i &= c_i s_i \quad \textit{for } i = 1,2 \\
s_i c_j &= c_k s_i \quad \textit{for } \{i,\, j,\, k\} = \{1,\, 2,\, 3\} \textit{ and } i \neq 3 && (1)\\
d_i &= c_i (1 - s_i) \quad \textit{for } i = 1,2 && (2)\\
c_1 d_2 c_1 &= c_2 d_1 c_2 && (3)\\
d_2 c_1^{2n+1} d_2 &= d_2 c_1 c_2^{2n-1} c_1 d_2 = z d_2 c_1^{2n} d_2 = -\frac{1}{2} z^{2n+2} d_2 \quad \textit{for } n \geq 1 && (4)\\
l_n (1 + s_1) &= d_1 c_2 c_1^{n-1} + c_3 c_1^n \quad \textit{for } n \geq 2 && (5)\\
s_2 l_n &= -l_n + c_2^{n-1} c_1 d_2 + c_2^n c_3 \quad \textit{for } n \geq 2 && (6)\\
c_1 y c_1 &= s_1 c_1 y c_1 = z c_1 d_2 c_1 - z c_1 s_1 s_2 d_1 c_2 && (7)\\
z^2 d_2 c_1 &= z^2 s_1 s_2 d_1 c_2 && (8)
\end{aligned}$$

Proof The first six equalities are obvious: $d_i c_i$ and $c_i d_i$ contain triangles, $s_i c_i = c_i s_i$ is a double application of (AS) and $d_i = c_i - s_i c_i$ is just (IHX). $c_1 d_2 c_1$ and $c_2 d_1 c_2$ are identical graphs:

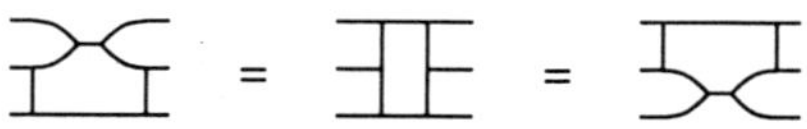

We will use in the sequel that due to Lemma 3.1 b), $z^k d_2$ is represented by the following graph:

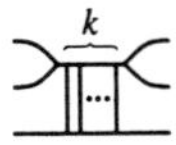

Then all equalities in (4) are simple consequences of (LI) and (LS):

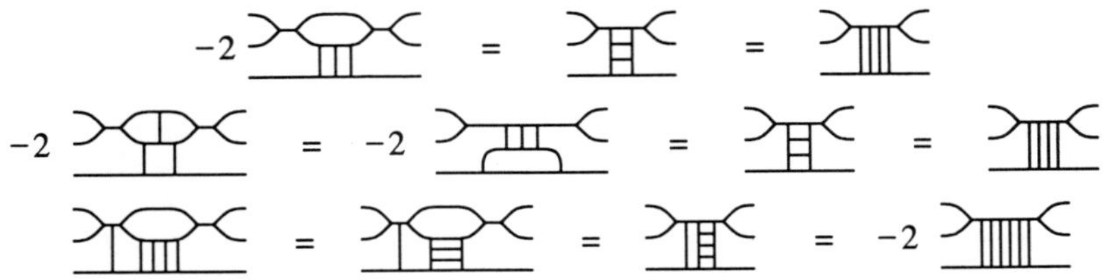

To show (5) and (6), we have to use Lemma 3.1 a) (for (6) simply rotate the picture by π):

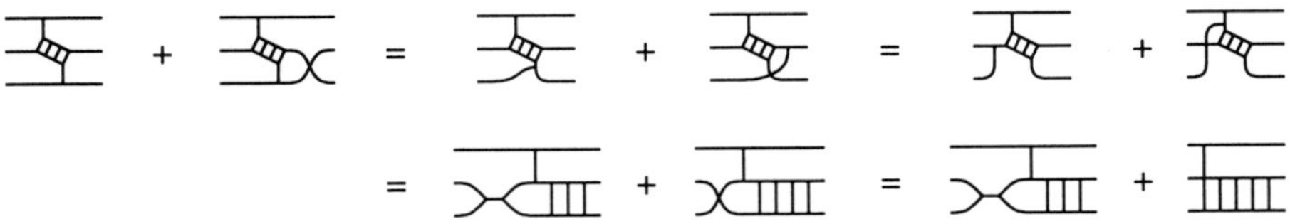

Finally we use (t) and (IHX) to show (7) and (8):

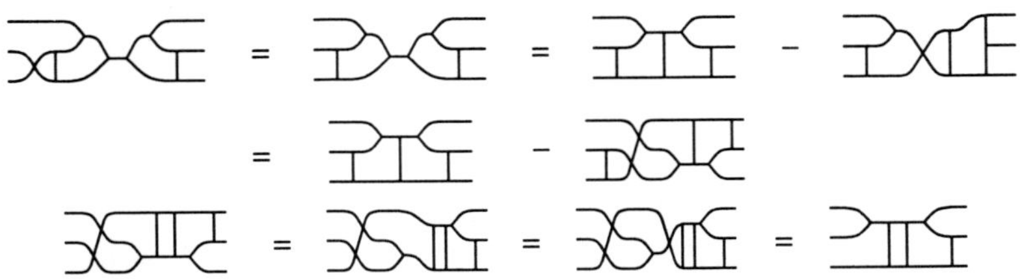

□

Let us elaborate some consequences of the relations, which we will use later. (Note that the application of a previous equation is indicated by stacking its number atop the equality sign but we do not mention the use of equations without number.)

$$
\begin{aligned}
s_i d_i &= d_i s_i \overset{(2)}{=} s_i(1-s_i)c_i = (s_i-1)c_i \overset{(2)}{=} -d_i && (9)\\
d_i^2 &\overset{(2)}{=} d_i c_i - d_i s_i c_i = 2c_i d_i = 0 && (10)\\
d_2 c_1 d_2 &\overset{(9)}{=} \frac{1}{2} d_2 c_1 d_2 + \frac{1}{2} d_2 s_2 c_1 s_2 d_2 \overset{(1)}{=} \frac{1}{2} d_2(c_1+c_3)d_2 = \frac{1}{2} d_2(z-c_2)d_2 \overset{(10)}{=} 0 && (11)\\
d_2 d_1 c_2 &\overset{(2)}{=} (1-s_2)c_2 d_1 c_2 \overset{(3)}{=} (1-s_2)c_1 d_2 c_1 \overset{(1,9)}{=} c_1 d_2 c_1 + c_3 d_2 c_1 = z d_2 c_1 && (12)\\
d_2 d_1 d_2 &\overset{(2)}{=} d_2 d_1 c_2(1-s_2) \overset{(12)}{=} z d_2 c_1 (1-s_2) \overset{(1,9)}{=} z d_2(c_1+c_3) = z^2 d_2 && (13)
\end{aligned}
$$

$$
\begin{aligned}
c_2^3 + 2c_2c_1c_2 - zc_2^2 &= c_2(2c_1+c_2-z)c_2 = c_2c_1c_2 - c_2c_3c_2 \overset{(1)}{=} c_2c_1c_2 - c_2s_2c_1s_2c_2\\
&\overset{(2)}{=} c_2c_1c_2 - (c_2-d_2)c_1(c_2-d_2) \overset{(11)}{=} d_2c_1c_2 + c_2c_1d_2 && (14)
\end{aligned}
$$

$$
\begin{aligned}
z^2 d_2(z-c_1) &= z^2 d_2 c_3 \overset{(13)}{=} d_2 d_1 d_2 c_3 \overset{(2,11)}{=} -d_2 s_1 c_1 d_2 c_3 = -d_2 s_1 (z-c_2-c_3) d_2 c_3\\
&\overset{(1)}{=} -z d_2 s_1 d_2 c_3 + d_2 c_2 s_1 d_2 c_3 \overset{(9)}{=} z d_2 s_2 s_1 d_2 c_3 \overset{(2,1)}{=} z d_2 d_1 s_2 s_1 c_3\\
&\overset{(1)}{=} z d_2 d_1 s_2 c_2 s_1 \overset{(2)}{=} z d_2 d_1 (c_2-d_2) s_1 \overset{(12,13)}{=} z^2 d_2 c_1 s_1 - z^3 d_2 s_1 && (15)
\end{aligned}
$$

$$\begin{aligned}(c_2-d_2)c_1^2d_2 &\overset{(2)}{=} c_2s_2c_1^2d_2 \overset{(1,9)}{=} -c_2c_3^2d_2 = -c_2(z-c_1-c_2)^2d_2\\ &= 2zc_2c_1d_2-c_2^2c_1d_2-c_2c_1^2d_2\end{aligned} \tag{16}$$

4.2 The Subspace $[\Xi_0]$

The symmetric group S_6 operates on Ξ by permutation of the univalent vertices. This allows us to regard Ξ as a $\mathbf{Z}[S_6]$-module. To prevent confusion, we use a dot to indicate this operation (so $\sigma \in \mathbf{Z}[S_6]$ applied to $\xi \in \Xi$ will be written as $\sigma \cdot \xi$). Note that in general $\sigma \cdot \xi_1\xi_2 \neq (\sigma \cdot \xi_1)\xi_2$.

The elementary transpositions in S_6 will be named $\tau_i := (i \;\; i{+}1)$. Obviously, $\tau_1 \cdot \xi = s_1\xi$, $\tau_2 \cdot \xi = s_2\xi$, $\tau_4 \cdot \xi = \xi s_2$ and $\tau_5 \cdot \xi = \xi s_1$. Furthermore let $\mu := (1\;\;6)(2\;\;5)(3\;\;4)$, then μ operates by mirroring along the y-axis (there is always an even number of trivalent vertices, so (AS) causes no change of sign). For abbreviation we introduce the symbol $[*]_*^*$:

$$[\xi]_i^j \;:=\; (1-\mu)(1+\tau_1)(1+\tau_3)(1+\tau_5)\cdot(c_1^i\xi c_1^j) \quad \text{for any } \xi\in\Xi \text{ and } i,j\geq 0.$$

The maps $\xi \to [\xi]_i^j$ are well-defined $\mathbf{Q}$-endomorphisms of Ξ. Obviously $[c_1\xi]_i^j = [\xi]_{i+1}^j$ and $[\xi c_1]_i^j = [\xi]_i^{j+1}$. A useful feature of this notation is that whenever $\mu \cdot w = w$ then $[z^k w]_i^i = 0$. For instance, this is the case if w is a palindrome in the letters s_*, c_*, d_*. Other pleasing properties of $[*]_*^*$ are given in the following proposition.

Proposition 4.2 *For any $\xi \in \Xi$, $i,j \geq 0$ we have*

$$[\xi]_i^j \;=\; -[\mu\cdot\xi]_j^i \;=\; [s_1\xi]_i^j \;=\; [\tau_3\cdot\xi]_i^j \;=\; [\xi s_1]_i^j \tag{17}$$

$$c_1[\xi]_i^j c_1 \;=\; [\xi]_{i+1}^{j+1} \tag{18}$$

$$c_1[\xi]_i^j + [\xi]_i^j c_1 \;=\; [\xi]_{i+1}^j + [\xi]_i^{j+1}. \tag{19}$$

Moreover, if $\mu\cdot\xi = \xi$ and $[\xi]_0^1 = 0$, then $[\xi]_i^j = 0$ for all i, $j \geq 0$.

Proof The last three factors of $\pi := (1-\mu)(1+\tau_1)(1+\tau_3)(1+\tau_5)$ commute with each other and their product commutes with $(1-\mu)$. Since $\tau_i^2 = \mu^2 = 1$, we have $\pi = -\pi\mu = \pi\tau_1 = \pi\tau_3 = \pi\tau_5$, which implies (17). Equality (18) becomes clear if one realises that for $i \in \{1, 3, 5\}$

$$\begin{aligned} c_1(\tau_i\cdot\xi) &= \tau_i\cdot c_1\xi, & (\tau_i\cdot\xi)c_1 &= \tau_i\cdot\xi c_1\\ c_1(\mu\cdot\xi) &= \mu\cdot\xi c_1, & (\mu\cdot\xi)c_1 &= \mu\cdot c_1\xi. \end{aligned}$$

Let $x := (1+\tau_1)(1+\tau_3)(1+\tau_5)\cdot c_1^i\xi c_1^j$, then $[\xi]_i^j = x - \mu\cdot x$ and

$$\begin{aligned} c_1[\xi]_i^j + [\xi]_i^j c_1 &= c_1x - c_1(\mu\cdot x) + xc_1 - (\mu\cdot x)c_1 \;=\; c_1x - \mu\cdot xc_1 + xc_1 - \mu\cdot c_1x\\ &= (1-\mu)\cdot c_1x + (1-\mu)\cdot xc_1 \;=\; [\xi]_{i+1}^j + [\xi]_i^{j+1}. \end{aligned}$$

Finally, the following identity shows by induction that $[\xi]_0^0 = [\xi]_0^1 = 0$ implies $[\xi]_0^j = 0$ for $j \geq 2$:

$$[\xi]_0^j \;=\; [\xi]_0^j + [\xi]_1^{j-1} - [\xi]_1^{j-1} \overset{(19,18)}{=} c_1[\xi]_0^{j-1} + [\xi]_0^{j-1}c_1 - c_1[\xi]_0^{j-2}c_1$$

By (18) we obtain $[\xi]_i^j = c_1^i[\xi]_0^{j-i}c_1^i = 0$ for $i \leq j$. If $i > j$ we have $[\xi]_i^j \overset{(17)}{=} -[\mu\cdot\xi]_j^i = [\xi]_j^i = 0$.
□

Definition 4.3 Let Ξ_0 denote the subalgebra of Ξ that is generated by $\{c_1, c_2, c_3, d_1, d_2\}$. Let $[\Xi_0]$ denote the subspace of Ξ that is spanned by $\{\, [\xi]_i^j \mid \xi \in \Xi_0;\; i,j \geq 0 \,\}$.

Remark 4.4 Equations (9) and (17) imply that $[z^k w]_i^j = 0$ if w begins or ends with d_1. (3), (12), (13) and $d_1c_1 = c_1d_1 = d_1^2 = 0$ allow to eliminate all d_1-s in the middle of words. Thus $[\Xi_0]$ is spanned by elements of the form $[z^k\, w]_i^j$ where w is a word in the letters c_1, c_2, d_2.

Let us do some auxiliary calculations in $[\Xi_0]$ based on Proposition 4.2.

$$[z^k c_2]_i^j \overset{(17)}{=} \frac{1}{2}[z^k c_2]_i^j + \frac{1}{2}[z^k s_1 c_2 s_1]_i^j \overset{(1)}{=} \frac{1}{2}[z^k(c_2+c_3)]_i^j = \frac{1}{2}[z^k(z-c_1)]_i^j = 0 \tag{20}$$

$z^2 d_2 c_1(1+s_1) \overset{(15)}{=} z^3 d_2(1+s_1)$ implies $[z^k d_2]_0^1 = [z^{k+1} d_2]_0^0 = 0$, so by Proposition 4.2

$$[z^k d_2]_i^j = 0 \quad \text{for } k \geq 2. \tag{21}$$

We can generalise this a little more:

$$[z^k d_2 w]_i^j = 0 \quad \text{if } k \geq 2 \text{ and } w \text{ is a word in the letters } c_1 \text{ and } c_2. \tag{22}$$

This is shown by induction on the length of w. If w is empty, then the statement is (21), otherwise if w begins with c_2 then $d_2 w = 0$, so let us assume $w = c_1 u$.

$$2[z^k d_2 c_1 u]_i^j \overset{(2)}{=} [z^k d_2 c_1(1+s_1)u]_i^j + [z^k d_2 d_1 u]_i^j \overset{(15)}{=} [z^{k+1} d_2 u]_i^j + [z^{k+1} d_2 s_1 u]_i^j + [z^k d_2 d_1 u]_i^j$$

By (1) and (17), the second term is equal to $[z^{k+1} d_2 u']_i^j$ where $u' = s_1 u s_1$ is obtained by replacing in u every c_2 by $z - c_1 - c_2$. If u is empty or begins with c_1 then $[z^k d_2 d_1 u]_i^j = 0$, otherwise $u = c_2 v$ and $[z^k d_2 d_1 u]_i^j \overset{(12)}{=} [z^{k+1} d_2 c_1 v]_i^j$. So the induction hypothesis applies to all three terms on the right side of the equation above and we are done.

We are interested in $[\Xi_0]$ because it contains the element $[y]_1^3$ we want to calculate.

$$\begin{aligned}
[y]_{i+1}^{j+1} - [zd_2]_{i+1}^{j+1} &= [c_1 y c_1 - z c_1 d_2 c_1]_i^j \overset{(7)}{=} -[z c_1 s_1 s_2 d_1 c_2]_i^j \overset{(1)}{=} -[z s_1 s_2 c_3 d_1 c_2]_i^j \\
&\overset{(17)}{=} -[z^2 s_1 s_2 d_1 c_2]_i^j + [z s_2 c_2 d_1 c_2]_i^j \overset{(8,2)}{=} -[z^2 d_2 c_1]_i^j + [z(c_2 - d_2) d_1 c_2]_i^j \\
&\overset{(3,12)}{=} -[z^2 d_2]_i^{j+1} + [z c_1 d_2 c_1]_i^j - [z^2 d_2 c_1]_i^j \overset{(21)}{=} [zd_2]_{i+1}^{j+1}.
\end{aligned}$$

This allows us to express $[y]_i^j$ as element of $[\Xi_0]$:

$$[y]_i^j = 2[zd_2]_i^j \text{ for } i, j \geq 1 \tag{23}$$

Next, we show that for $i,\ j,\ k \geq 0$ and arbitrary $\xi \in \Xi$, we have

$$2[\xi\, c_1^2 c_2]_i^j = [z\xi]_i^{j+2} - [\xi]_i^{j+3} \tag{24}$$
$$2[\xi\, d_2 c_1 c_2]_i^j = 2[z\xi\, d_2]_i^{j+1} - [\xi\, d_2]_i^{j+2} \tag{25}$$
$$2[\xi\, c_2 c_1 c_2]_i^j = [z\xi\, c_2]_i^{j+1} - [\xi\, c_2]_i^{j+2} + [\xi\, c_1 d_2]_i^{j+1}. \tag{26}$$

Equations (24)-(26) are shown simultaneously:

$$[u(c_1 - d_1)c_2]_i^j \overset{(2)}{=} [u c_1 s_1 c_2]_i^j \overset{(1,17)}{=} [u c_1 c_3]_i^j = [u c_1(z - c_1 - c_2)]_i^j = [zu]_i^{j+1} - [u]_i^{j+2} - [u c_1 c_2]_i^j$$

This implies $2[u\, c_1 c_2]_i^j = [zu]_i^{j+1} - [u]_i^{j+2} + [u\, d_1 c_2]_i^j$. If we set $u = \xi c_1$, $u = \xi d_2$, $u = \xi c_2$ and apply $c_1 d_1 = 0$, $d_2 d_1 c_2 \overset{(12)}{=} z d_2 c_1$, $c_2 d_1 c_2 \overset{(3)}{=} c_1 d_2 c_1$ we obtain (24), (25), (26), respectively.

Remark 4.5 Using the relations we have found so far, one can show that $[\Xi_0]$ is spanned by the elements of the following forms:

- $[z^k c_2^2]_i^{i+1}$ with $k \geq 1,\ i \geq 0$,
- $[z^k d_2]_i^j$ with $k \in \{0, 1\}$ and $0 \leq i < j$,
- $[d_2 c_1^{2n} d_2]_i^j$ with $0 \leq i < j$ and $n \geq 1$.

We only give a few hints how this can be shown. Let $\sigma_n := \frac{1}{2}(1+s_1)c_1^n$, then $\sigma_n\sigma_m = \sigma_{n+m}$ and $c_1\sigma_n = \sigma_n c_1 = \sigma_{n+1}$. By (2) and Remark 4.4, $[\Xi_0]$ then is spanned by elements of the form $[z^k\sigma_0 w_1 \sigma_{n_1} w_2 \cdots \sigma_{n_{l-1}} w_l \sigma_0]_p^q$ with $n_i \geq 1$ and $w_i = d_2$ or $w_i = c_2^j$; this will be called *normal form* from now on. The w_i are called *segments* and we define the simplicity of an element in normal form by: k + number of d_2-segments. For a linear combination x of words of the same simplicity we say $x \sim 0$ if any element $[z^k uxw]_p^q$ in normal form can be expressed by a sum of elements with simpler normal forms (i.e. elements that contain more z-s or more d_2-s). Of course, by $x \sim y$ we mean $x - y \sim 0$. It is not hard to show that

$$\begin{aligned}
\sigma_n c_2 \sigma_m &\sim -\frac{1}{2}\sigma_{n+m+1} && \text{(a)}\\
\sigma_n c_2^{i+3}\sigma_m &\overset{(14)}{\sim} -2\sigma_n c_2^{i+1}\sigma_1 c_2 \sigma_m \overset{(a)}{\sim} \sigma_n c_2^{i+1}\sigma_{m+2} && \text{(b)}\\
\sigma_n c_2^2 \sigma_{m+1} + \sigma_{n+1} c_2^2 \sigma_m &\sim \frac{5}{4}\sigma_{n+m+3} && \text{(c)}\\
\sigma_n c_2^2 \sigma_1 c_2^2 \sigma_m &\sim -\frac{1}{2}\sigma_n c_2^5 \sigma_m && \text{(d)}\\
\sigma_n c_2^2 \sigma_1 d_2 \sigma_m &\overset{(16)}{\sim} -2\sigma_n c_2 \sigma_2 d_2 \sigma_m \overset{(a)}{\sim} \sigma_{n+3} d_2 \sigma_m. && \text{(e)}
\end{aligned}$$

Using (a) and (b), we may eliminate all c_2^j-segments with $j \neq 2$. Due to (c), we can reduce the σ_n-s between two c_2^2-segments or between a c_2^2-segment and a d_2-segment to become σ_1 and then apply (d) or (e). Like this we are left only with two types of normal forms: the ones that contain only d_2-segments and those which consist of a single c_2^2-segment. In the first case we may eliminate $d_2\sigma_{2n-1}d_2$ and $zd_2\sigma_{2n}d_2$ by (4) and (11). If a normal form contains $d_2\sigma_{2n}d_2\sigma_{2m}d_2$ then after rotating the even ladders with (LS), we may apply the relation (LL) to transform it into $\frac{1}{4}z^{n+m+2}d_2$. In the second case we may use (c) to put the c_2^2 in the middle to get $[z^k c_2^2]_i^i = 0$ or $[z^k c_2^2]_i^{i+1}$. If $k = 0$ we may apply equation (36). This shows that any element of $[\Xi_0]$ is a linear combination of elements of the types in Remark 4.5.

Remark 4.6 Using the relations that will be found in the next section, the statement of Remark 4.5 still can be improved a little bit: $[\Xi_0]$ is spanned by $[z^k c_2^2]_{i-1}^i$, $[d_2]_0^i$, $[zd_2]_0^i$, $[zd_2]_1^i$, $[zd_2]_i^{i+1}$ with $i, k \geq 1$.

4.3 Additional Relations in $[\Xi_0]$

Let us consider a family $(H_n)_{n\geq 1}$ of elements of Ξ, where H_n is the sum of the four possible ways of attaching the lower ends of a n-ladder to the first or second strand of 1, and the upper ends to the third strand. For example, H_3 looks like this:

Let $R_n := [H_n]_0^1$. Applying Lemma 3.1 a) twice, we see that $H_n c_1 = c_1 H_n$. Obviously $\mu \cdot H_n = H_n$, so $(1-\mu)\cdot H_n c_1 = H_n c_1 - c_1 H_n = 0$, thus

$$R_n = 0 \tag{27}$$

We now present another way to write R_n as element of $[\Xi_0]$ in order to obtain new relations. Let $H_{n,i}$ denote the i-th term of H_n in the order of the picture above, then $H_{n,2} = s_1 H_{n,1} s_1$ and $H_{n,4} = s_1 H_{n,3} s_1$. Thus by (17) we have

$$R_n = 2[H_{n,1}]_0^1 + 2[H_{n,3}]_0^1.$$

$H_{n,3}$ contains a $(n+2)$-ladder, so by (LS) and (LIHX)

$$H_{n,3} = \left\{ \begin{array}{ll} c_2^{n+2} & \text{if } n \text{ is even} \\ -c_2^{n+2} - \tau_3 \cdot c_2^{n+2} & \text{if } n \text{ is odd} \end{array} \right\}$$

and by (17) we get

$$2[H_{n,3}]_0^1 = \left(3\cdot(-1)^n - 1\right)[c_2^{n+2}]_0^1.$$

Lemma 3.1 a) implies

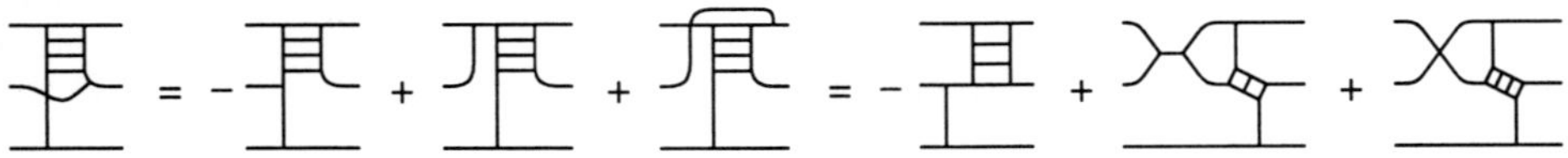

$$H_{n,1} = -c_1 H_{n-1,3} + d_2\, l_n + s_2\, l_{n+1} \quad \text{for } n \geq 1. \tag{28}$$

Now $[c_1 H_{n-1,3}]_0^1 = [H_{n-1,3}]_1^1 = 0$ because of $\mu \cdot H_{n-1,3} = H_{n-1,3}$. The second and third term of $2[H_{n,1}]_0^1$ are according to (28)

$$\begin{aligned}
2[d_2 l_n]_0^1 &\overset{(17)}{=} [d_2 l_n(1+s_1)]_0^1 \overset{(5)}{=} [d_2 d_1 c_2 c_1^{n-1} + d_2 c_3 c_1^n]_0^1 \overset{(12)}{=} [z d_2 c_1^n + d_2(z - c_1 - c_2)c_1^n]_0^1 \\
&= 2[z d_2 c_1^n]_0^1 - [d_2 c_1^{n+1}]_0^1 = 2[z d_2]_0^{n+1} - [d_2]_0^{n+2} \\
2[s_2 l_{n+1}]_0^1 &\overset{(6)}{=} 2[-l_{n+1} + c_2^n c_1 d_2 + c_2^{n+1} c_3]_0^1 = 2[c_2^n c_1 d_2]_0^1 + 2[z c_2^{n+1}]_0^1 - 2[c_2^{n+1}]_0^2 - 2[c_2^{n+2}]_0^1.
\end{aligned}$$

The first equation is valid only for $n \geq 2$; for $n = 1$ we use $l_1 = c_1 c_2$. In the last equation we used $[l_n]_i^j = 0$, which is a consequence of $\tau_3 \cdot l_n = -l_n$ (relation (AS)). Summarising the calculations of this subsection so far, we may state that

$$\begin{aligned}
R_1 &= 2[d_2 c_1 c_2]_0^1 + 2[c_2 c_1 d_2]_0^1 + 2[z c_2^2]_0^1 - 2[c_2^2]_0^2 - 6[c_2^3]_0^1 \quad \text{and} && (29)\\
R_n &= 2[z d_2]_0^{n+1} - [d_2]_0^{n+2} + 2[c_2^n c_1 d_2]_0^1 + 2[z c_2^{n+1}]_0^1 - 2[c_2^{n+1}]_0^2 + 3\left((-1)^n - 1\right)[c_2^{n+2}]_0^1. && (30)
\end{aligned}$$

with $n \geq 2$ are trivial elements of $[\Xi_0]$. In the spirit of Remark 4.5 one is able to present R_n as linear combination of simple elements. We need this for $n \leq 3$ which requires a lengthy computation that is done in the appendix. The results are

$$\begin{aligned}
R_1 &= 3[d_2]_1^2 - 6[z c_2^2]_0^1 && (31)\\
R_2 &= 2[d_2 c_1^2 d_2]_0^1 - 3[z d_2]_0^3 && (32)\\
R_3 &= 4[d_2 c_1^2 d_2]_0^2 - 6[z d_2]_0^4. && (33)
\end{aligned}$$

To simplify the calculation of R_2 and R_3, the following formula has been used

$$[d_2]_i^j = 2[z c_2^2]_{i-1}^{j-1} \quad \text{for } i, j \geq 1. \tag{34}$$

Because of $[c_1 d_2 c_1 - 2 z c_2^2]_0^1 \overset{(31)}{=} \frac{1}{3} R_1 \overset{(27)}{=} 0$ Proposition 4.2 implies (34).
Now we finally may combine (32), (33) and (23) to obtain the desired result:

$$\begin{aligned}
0 \overset{(27)}{=} &\frac{1}{3} R_3 - \frac{2}{3}(c_1 R_2 + R_2 c_1) \overset{(19)}{=} \frac{4}{3}[d_2 c_1^2 d_2]_0^2 - 2[z d_2]_0^4 - \frac{4}{3}[d_2 c_1^2 d_2]_1^1 + 2[z d_2]_1^3 \\
&- \frac{4}{3}[d_2 c_1^2 d_2]_0^2 + 2[z d_2]_0^4 = 2[z d_2]_1^3 \overset{(23)}{=} [y]_1^3
\end{aligned} \tag{35}$$

4.4 Calculations for parts 2 and 3

Now, having the relation $[y]_1^3 = 0$ in hand, we may stop messing around in Ξ and proceed with the proof of the square-tunnelling relation. But now that we have gone through so much trouble, we first take profit out of our current knowledge to provide some more equations that are urgently needed in [2] and [3].
We start with the observation that $\tau_3 \cdot c_2 s_2 c_1 c_2 = c_2 s_2 c_2 c_1 - c_2 s_2 c_1 c_2$:

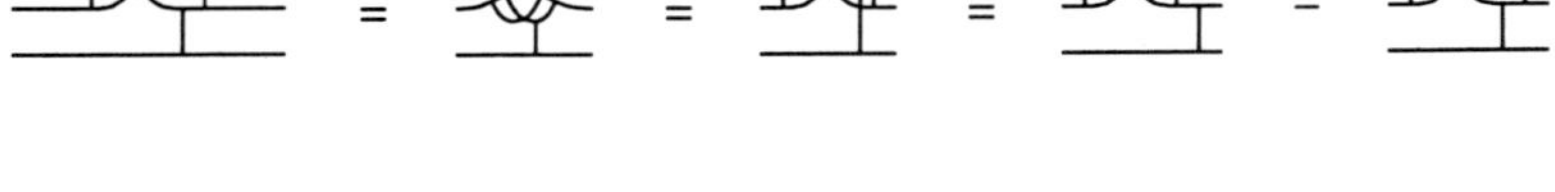

We obtain for any $i, j \geq 0$

$$\begin{aligned} [c_2^2]_i^{j+1} &= [c_2 s_2 c_2 c_1]_i^j = [(1+\tau_3)\cdot c_2 s_2 c_1 c_2]_i^j \overset{(17,2)}{=} 2[(c_2-d_2)c_1c_2]_i^j \overset{(26,25)}{=} [zc_2]_i^{j+1} - [c_2]_i^{j+2} \\ &\quad + [c_1 d_2]_i^{j+1} - 2[zd_2]_i^{j+1} + [d_2]_i^{j+2} \overset{(20)}{=} [d_2]_{i+1}^{j+1} + [d_2]_i^{j+2} - 2[zd_2]_i^{j+1} \end{aligned} \tag{36}$$

We use this for $i = 1$ and $j = 1$ to get an expression for $c_1^2 c_2^2 c_1^2$:

$$\begin{aligned} 0 &= c_1\big([c_2^2]_1^2 - [d_2]_2^2 - [d_2]_1^3 + 2[zd_2]_1^2\big) \overset{(23)}{=} c_1\big([c_2^2]_1^2 - [d_2]_1^3 + [y]_1^2\big) \\ &= (1+\tau_1)(1+\tau_3)(1+\tau_5)\cdot\big(c_1^2c_2^2c_1^2 - c_1^3c_2^2c_1 - c_1^2d_2c_1^3 + c_1^4d_2c_1 + c_1^2yc_1^2 - c_1^3yc_1\big) \end{aligned}$$

Using $\tau_1 \cdot c_1^2\xi = c_1^2\xi$ and $\tau_3 \cdot c_1^2c_2^2c_1^2 = \tau_5 \cdot c_1^2c_2^2c_1^2 = c_1^2c_2^2c_1^2$, we obtain

$$c_1^2c_2^2c_1^2 = \frac{1}{4}(1+\tau_3)(1+\tau_5)\cdot\big(c_1^3yc_1 - c_1^4d_2c_1 + c_1^3c_2^2c_1 - c_1^2yc_1^2 + c_1^2d_2c_1^3\big). \tag{37}$$

Another relation that will be needed in [3] is the following: $(n \geq 1)$

$$\begin{aligned} d_1 H_{n,1} &\overset{(28)}{=} -d_1c_1H_{n-1,3} + d_1d_2l_n + d_1s_2l_{n+1} \\ &\overset{(6,1)}{=} d_1d_2l_n - d_1l_{n+1} + d_1c_2^n c_1 d_2 + d_1 c_2^{n+1} c_1 s_2 \end{aligned} \tag{38}$$

Equations (18) and (35) imply $[y]_n^{n+2} = 0$ in $\hat{F}(6)$. In [2] we need a similar relation in $F(6)$. This means that the whole calculation should be repeated in $F(6)$, i.e. using the following relation instead of (x):

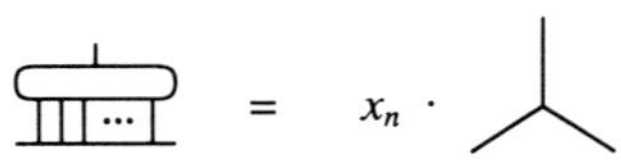

Interpreting $[y]_i^j$ as element of $F(6)$ we would then obtain a relation of the following form (note that only x_n-s with $n \leq 4$ can occur and that $x_1 = 2t$, $x_2 = t^2$, $x_4 = \frac{4}{3}t\,x_3 - \frac{1}{3}t^4$, see [6]):

$$[y]_n^{n+2} = \sum_i \lambda_i\,\xi_i \quad \text{with } \lambda_i \in \mathbf{Q}[t, x_3] - \mathbf{Q} \text{ and } \xi_i \in F(6)$$

For $n = 2$ we may even take the relation into a form in which all ξ_i are of the form $[y]_*^*$ (which is not possible for $n = 1$):

$$\begin{aligned} 3[y]_2^4 &= 9t\,[y]_2^3 + 3t\,[y]_1^4 - 9t^2\,[y]_1^3 + 6t^2\,[y]_0^4 + (4t^3 + 2x_3)\,[y]_1^2 \\ &\quad - 18t^3\,[y]_0^3 + 4t(4t^3 - x_3)\,[y]_0^2 + 8t^2(x_3 - t^3)\,[y]_0^1 \end{aligned} \tag{39}$$

4.5 The Square-Tunnelling Relation

First, we use $\tau_1 \cdot c_1yc_1^3 = s_1c_1yc_1^3 \overset{(7)}{=} c_1yc_1^3$ and $\tau_5 \cdot c_1yc_1^3 = c_1yc_1^3s_1 \overset{(2)}{=} c_1yc_1^3$ to reformulate equation (35):

$$(1+\tau_3)\cdot(c_1yc_1^3 - c_1^3yc_1) = \frac{1}{4}(1-\mu)(1+\tau_1)(1+\tau_3)(1+\tau_5)\cdot c_1yc_1^3 = \frac{1}{4}[y]_1^3 \overset{(35)}{=} 0 \tag{40}$$

We turn our attention towards $\hat{F}(n+4)$. For $n \geq 0$ let $D_{n,1}$, $D'_{n,1}$, $D_{n,2}$ and $D'_{n,2}$ denote the elements of $\hat{F}(n+4)$ that are represented by:

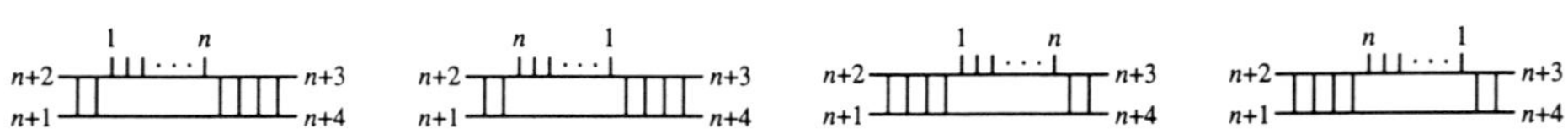

Furthermore let $D_n := D_{n,1} - D_{n,2}$ and $D'_n := D'_{n,1} - D'_{n,2}$.

Lemma 4.7 $D_n = 0$ *for all* $n \geq 0$.

Proof The assertion is true for $n = 0$, since $D_{0,1} = D_{0,2}$. We proceed by induction and assume that $n \geq 1$ and $D_i = 0$ for $i < n$. We glue a line with n legs with labels from 1 to n between the ends 3 and 4 of $c_1yc_1^3$ and rename the ends $1, 2, 4, 6$ to $n+1, n+2, n+3, n+4$; the result is $D_{n,1}$:

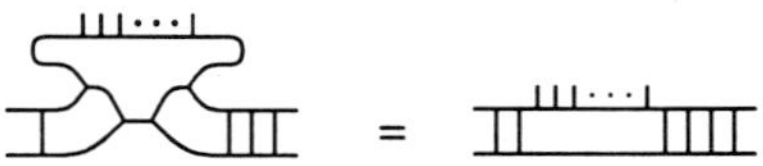

If we do the same with $\tau_3 \cdot c_1yc_1^3$ and apply (AS) at the n legs, we get $(-1)^n D'_{n,1}$.
A similar statement can be done for $c_1yc_1^3$, $\tau_3 \cdot c_1yc_1^3$ and $D_{n,2}$, $(-1)^n D'_{n,2}$, respectively, thus equation (40) implies:

$$D_n + (-1)^n D'_n \;=\; D_{n,1} - D_{n,2} + (-1)^n D'_{n,1} - (-1)^n D'_{n,2} \;=\; 0. \tag{41}$$

By permuting neighbouring ends of both ladders, we get the following representation of $D'_{n,1}$:

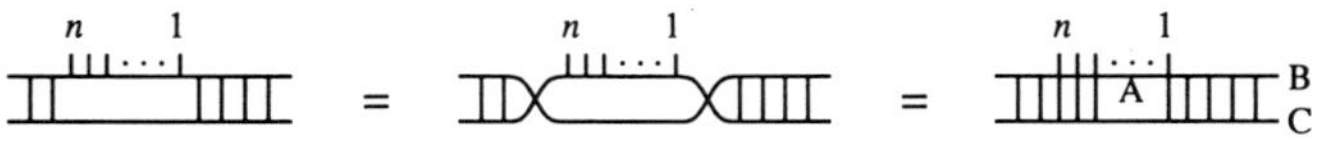

By Lemma 3.1 a) we can push all n legs off the lower strand, one after another. Let us push them all to the right, then we get 3^n terms which have legs at the edges indicated by the letters A, B, C in the picture above. The coefficient of each term is $(-1)^{\text{number of legs in position A}}$.

We do the same trick for $D'_{n,2}$ and get a similar sum with the 2-ladder and the 4-ladder exchanged. If there are legs in position B or C, then there must be less than n legs between the two ladders. By induction assumption all these terms in $D'_{n,1}$ are equal to those in $D'_{n,2}$. So in $D'_n = D'_{n,1} - D'_{n,2}$ only the two terms having all n legs in position A remain. The process reverses the order of the legs, so the remaining two diagrams are $D_{n,1}$ and $D_{n,2}$:

$$D'_n \;=\; (-1)^n D_{n,1} - (-1)^n D_{n,2} \;=\; (-1)^n D_n \tag{42}$$

The equations (41) and (42) imply the assertion of the lemma. □

Proof of Theorem 2 We consider a graph containing two ladders with disjoint interiors. Let us remove the interiors of these ladders and call the components of the remaining graph that contain ends of both ladders *joining*. A sequence of different consecutive edges from one ladder to the other is called *joining path*. Remember that the condition of Theorem 2 is satisfied iff

- either there are two or more joining components, or
- there is a single joining component that contains two ends of one of the ladders or
- there is a single joining component that contains a circle.

The last case splits up into two subcases: either there are at least two different paths joining the ladders or there is a unique joining path. So to prove Theorem 2 we have to show that in the following four situations we may exchange the square and the 4-ladder:

a) 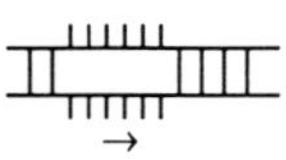

b) 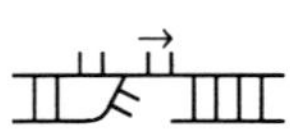

c) 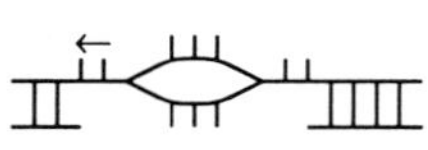

d) 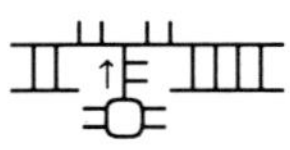

In situation a), we choose a joining path p and push all edges arriving at p by Lemma 3.1 a) through one of the ladders, as indicated by the little arrow. We can do the same for the graph in which both ladders have been exchanged. Both times we get a linear combination of diagrams in which p has become a single edge. The only difference is that in each term the ladders have been exchanged. By Lemma 4.7 the two expressions are identical.

In situation b) we have two joining paths p_1 and p_2 that meet each other and then coincide. We first push away all edges arriving at the common part of p_1 and p_2, to get diagrams in which p_1 and p_2 have only an edge e in common (e is the end of one of the ladders L). By Lemma 4.7 a) we may express each of these diagram by the three graphs we obtain if we cut off p_2 from e and glue it to one of the three other ends of L. So we may express any graph in situation b) by a sum of graphs of situation a).

Exactly the same trick allows to reduce case c) to b). Finally in situation d), we have a joining path p_1, a circle l and a path p_2 connecting l and p_1. We assume that p_2 is a single edge (otherwise proceed as above to clean p_2 as indicated by the arrow) and apply (IHX) at p_2. The result is a difference of two graphs of type c). This completes the proof of Theorem 2. □

Proof of Theorem 3 The univalent vertices in the following pictures are marked by bullets. If a graph in B^u has weakly connected ladders of lengths 2 and 4, it typically looks like this (the omitted parts of the graph are supposed to be located inside the boxes):

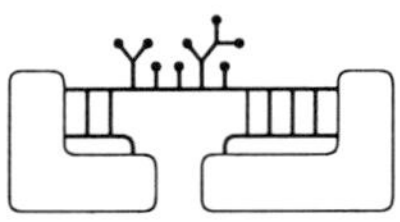

Since all univalent vertices carry the same colour, the (AS)-relation implies:

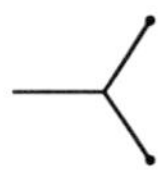 $= \quad 0$ $(\star)$

Lemma 3.1 a) yields

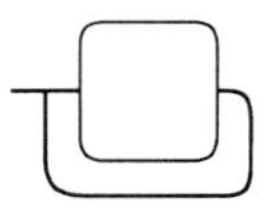 $=$ 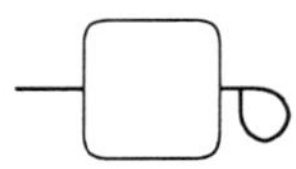 $= \quad 0$

if the box contains no univalent vertices. So we assume that there is at least one univalent vertex inside each box and consider diagrams that look like this:

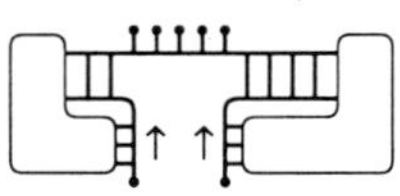

As before, we push all disturbing edges one by one through the ladders to obtain graphs that are trivial because of $(\star)$, graphs that are of type b) or d) (in the proof of Theorem 2) and graphs that look like this:

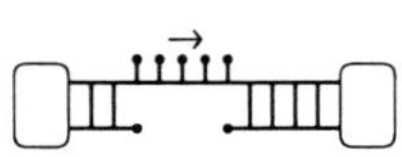

In the latter case, we push away all legs between the ladders. Because of $(\star)$ we only have to consider the terms in which all these legs end up in the two boxes. Thus to prove the theorem, it remains to show that we may exchange ladders in graphs of this type:

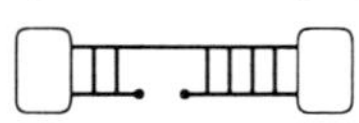

This is done by gluing univalent vertices of the same colour to the ends 3 and 4 in equation (40) so that ξ and $\tau_3 \cdot \xi$ become the same element.

So in the case of unicoloured univalent vertices we may drop the somehow artificial condition of strong connectivity and state that the square-tunnelling relation holds in all connected graphs. □

References

[1] D. Bar-Natan, *On the Vassiliev knot invariants,* Topology **34** (1995), 423-472.

[2] J. A. Kneissler, *On Spaces of Connected Graphs II: Relations in* Λ, this volume.

[3] J. A. Kneissler, *On Spaces of Connected Graphs III: The ladder filtration*, this volume.

[4] M. Kontsevich, *Vassiliev's knot invariants,* Adv. in Sov. Math., **16(2)** (1993), 137-150.

[5] V. A. Vassiliev, *Cohomology of knot spaces,* Theory of Singularities and its Applications (ed. V. I. Arnold), Advances in Soviet Math., **1** (1990) 23-69.

[6] Pierre Vogel, *Algebraic structures on modules of diagrams,* Université Paris VII preprint, July 1995.

Mathematisches Institut
Beringstrasse 1
D-53115 Bonn
e-mail:jk@math.uni-bonn.de
http://math.uni-bonn.de/people/jk

Appendix

Here we give a detailed calculation that allows to express the elements R_1, R_2 and R_3 (given by (29) and (30) in section 4.3) as linear combination of basic elements (according to Remark 4.5). In every step, we use one of the following equations ($i, j \geq 0$ and u, v arbitrary words):

$$
\begin{aligned}
[u\, c_1^2 c_2]_i^j &= \tfrac{1}{2}[zu]_i^{j+2} - \tfrac{1}{2}[u]_i^{j+3} && \text{(A)}\\
[u\, d_2 c_1 c_2]_i^j &= [zu\, d_2]_i^{j+1} - \tfrac{1}{2}[u\, d_2]_i^{j+2} && \text{(B)}\\
[u\, c_2 c_1 c_2]_i^j &= \tfrac{1}{2}[zu\, c_2]_i^{j+1} - \tfrac{1}{2}[u\, c_2]_i^{j+2} + \tfrac{1}{2}[u\, c_1 d_2]_i^{j+1} && \text{(C)}\\
[u\, c_2^3\, v]_i^j &= [zu\, c_2^2\, v]_i^j - 2[u\, c_2 c_1 c_2\, v]_i^j + [u\, d_2 c_1 c_2\, v]_i^j + [u\, c_2 c_1 d_2\, v]_i^j && \text{(D)}\\
[u\, c_2^2 c_1 d_2\, v]_i^j &= 2[zu\, c_2 c_1 d_2\, v]_i^j - 2[u\, c_2 c_1^2 d_2\, v]_i^j + [u\, d_2 c_1^2 d_2\, v]_i^j && \text{(E)}\\
[u\, c_2]_i^{j+2} &= [zu\, c_2]_i^{j+1} - 2[u\, c_2 c_1 c_2]_i^j + [u\, c_1 d_2]_i^{j+1} && \text{(C}'\text{)}\\
[u\, c_2 c_1 c_2\, v]_i^j &= \tfrac{1}{2}[zu\, c_2^2\, v]_i^j - \tfrac{1}{2}[u\, c_2^3\, v]_i^j + \tfrac{1}{2}[u\, d_2 c_1 c_2\, v]_i^j + \tfrac{1}{2}[u\, c_2 c_1 d_2\, v]_i^j && \text{(D}'\text{)}\\
[z\, c_2^2]_i^j &= \tfrac{1}{2}[d_2]_{i+1}^{j+1} && \text{(F)}
\end{aligned}
$$

The equations (A)-(F) are (24), (25), (26), (14), (16), (34); (C′) and (D′) are equivalent to (C) and (D). Note that equation (F) is not used in the calculation of R_1, because the value of R_1 is an essential ingredient in the proof of (34).

Elements of the form $[z^k\, w]_i^j$ (where w is a word in the letters c_1, c_2, d_2) are trivial in each of the following cases: (see (11), (20), (22), (4))

a) w contains $d_2 c_2$ or $c_2 d_2$ or $d_2 c_1 d_2$ as subword,

b) $w = c_2$,

c) $k \geq 2$ and $w = d_2$ or $w = d_2 c_1 c_2^n$,

d) $w = d_2 c_1 c_2 c_1 d_2$,

e) $k \geq 1$ and $w = d_2 c_1^2 d_2$,

f) w is a palindrome and $i = j$.

The basic elements are named $e_1, \ldots, e_k$. In the calculations each term t of the left side of equations (A) to (F) has one of the following three properties:

- t or $\mu \cdot t$ is one of the basic elements, or
- t or $\mu \cdot t$ belongs to one of the classes a) to f) and is henceforth 0, or
- t or $\mu \cdot t$ has been calculated in a previous equation.

We indicate the second case by writing 0_x (x one of the letters a to f), and the third case by the equation number in brackets. If $\mu \cdot t$ (the reversed word) instead of t satisfies the condition, the sign of the coefficient of t has to be changed according to (17):

$$[w]_i^j \;=\; -\mu \cdot [w]_i^j \;=\; -[\mu \cdot w]_j^i.$$

In each step the result is a linear combination of basic elements which is written as k-dimensional row-vector.

Calculation of R_1

Basic elements: $e_1 = [z\,c_2^2]_0^1$, $e_2 = [d_2]_0^3$, $e_3 = [d_2]_1^2$, $e_4 = [z\,d_2]_0^2$.

$$
\begin{array}{rcllr}
[d_2c_1^2c_2]_0^0 & \overset{(A)}{=} & \frac{1}{2}[z\,d_2]_0^2 - \frac{1}{2}[d_2]_0^3 & & \\
& = & \frac{1}{2}e_4 - \frac{1}{2}e_2 & = \quad (0, -\frac{1}{2}, 0, \frac{1}{2}) & (1) \\
[z\,d_2c_1c_2]_0^0 & \overset{(B)}{=} & [z^2\,d_2]_0^1 - \frac{1}{2}[z\,d_2]_0^2 & & \\
& = & 0_c - \frac{1}{2}e_4 & = \quad (0, 0, 0, -\frac{1}{2}) & (2) \\
[c_2^2c_1d_2]_0^0 & \overset{(E)}{=} & 2[z\,c_2c_1d_2]_0^0 - 2[c_2c_1^2d_2]_0^0 + [d_2c_1^2d_2]_0^0 & & \\
& = & -2(2) + 2(1) + 0_f & = \quad (0, -1, 0, 2) & (3) \\
[c_2c_1c_2]_0^1 & \overset{(C)}{=} & \frac{1}{2}[z\,c_2]_0^2 - \frac{1}{2}[c_2]_0^3 + \frac{1}{2}[d_2]_1^2 & & \\
& = & 0_b + 0_b + \frac{1}{2}e_3 & = \quad (0, 0, \frac{1}{2}, 0) & (4) \\
[c_2^2c_1c_2]_0^0 & \overset{(D')}{=} & \frac{1}{2}[z\,c_2^3]_0^0 - \frac{1}{2}[c_2^4]_0^0 + \frac{1}{2}[c_2d_2c_1c_2]_0^0 + \frac{1}{2}[c_2^2c_1d_2]_0^0 & & \\
& = & 0_f + 0_f + 0_a + \frac{1}{2}(3) & = \quad (0, -\frac{1}{2}, 0, 1) & (5) \\
[d_2c_1c_2]_0^1 & \overset{(B)}{=} & [z\,d_2]_0^2 - \frac{1}{2}[d_2]_0^3 & & \\
& = & e_4 - \frac{1}{2}e_2 & = \quad (0, -\frac{1}{2}, 0, 1) & (6) \\
[d_2c_1c_2]_1^0 & \overset{(B)}{=} & [z\,d_2]_1^1 - \frac{1}{2}[d_2]_1^2 & & \\
& = & 0_f - \frac{1}{2}e_3 & = \quad (0, 0, -\frac{1}{2}, 0) & (7) \\
[c_2^2]_0^2 & \overset{(C')}{=} & [z\,c_2^2]_0^1 - 2[c_2^2c_1c_2]_0^0 + [c_2c_1d_2]_0^1 & & \\
& = & e_1 - 2(5) - (7) & = \quad (1, 1, \frac{1}{2}, -2) & (8) \\
[c_2^3]_0^1 & \overset{(D)}{=} & [z\,c_2^2]_0^1 - 2[c_2c_1c_2]_0^1 + [d_2c_1c_2]_0^1 + [c_2c_1d_2]_0^1 & & \\
& = & e_1 - 2(4) + (6) - (7) & = \quad (1, -\frac{1}{2}, -\frac{1}{2}, 1) & (9)
\end{array}
$$

$$
\begin{array}{rcl}
R_1 & = & 2[d_2c_1c_2]_0^1 + 2[c_2c_1d_2]_0^1 + 2[z\,c_2^2]_0^1 - 2[c_2^2]_0^2 - 6[c_2^3]_0^1 \\
& = & 2(6) - 2(7) + 2e_1 - 2(8) - 6(9) \; = \; (-6, 0, 3, 0)
\end{array}
$$

Calculation of R_2

We use equation (F) at one place (line (9)). The basic elements are:

$$e_1 = [z^2\, c_2^2]_0^1,\quad e_2 = [d_2]_0^4,\quad e_3 = [z\, d_2]_0^3,\quad e_4 = [z\, d_2]_1^2,\quad e_5 = [d_2 c_1^2 d_2]_0^1.$$

$$\begin{aligned}
[z\, d_2 c_1^2 c_2]_0^0 &\stackrel{(A)}{=} \tfrac{1}{2}[z^2\, d_2]_0^2 - \tfrac{1}{2}[z\, d_2]_0^3 \\
&= 0_c - \tfrac{1}{2} e_3 &&= (0,\ 0,\ \text{-}\tfrac{1}{2},\ 0,\ 0) && (1) \\
[z\, c_2^2 c_1 d_2]_0^0 &\stackrel{(E)}{=} 2[z^2\, c_2 c_1 d_2]_0^0 - 2[z\, c_2 c_1^2 d_2]_0^0 + [z\, d_2 c_1^2 d_2]_0^0 \\
&= 0_c + 2(1) + 0_e &&= (0,\ 0,\ \text{-}1,\ 0,\ 0) && (2) \\
[z\, c_2^2 c_1 c_2]_0^0 &\stackrel{(D')}{=} \tfrac{1}{2}[z^2\, c_2^3]_0^0 - \tfrac{1}{2}[z\, c_2^4]_0^0 + \tfrac{1}{2}[z\, c_2 d_2 c_1 c_2]_0^0 + \tfrac{1}{2}[z\, c_2^2 c_1 d_2]_0^0 \\
&= 0_f + 0_f + 0_a + \tfrac{1}{2}(2) &&= (0,\ 0,\ \text{-}\tfrac{1}{2},\ 0,\ 0) && (3) \\
[d_2 c_1 c_2]_2^0 &\stackrel{(B)}{=} [z\, d_2]_2^1 - \tfrac{1}{2}[d_2]_2^2 \\
&= -e_4 + 0_f &&= (0,\ 0,\ 0,\ \text{-}1,\ 0) && (4) \\
[d_2 c_1 c_2]_0^2 &\stackrel{(B)}{=} [z\, d_2]_0^3 - \tfrac{1}{2}[d_2]_0^4 \\
&= e_3 - \tfrac{1}{2} e_2 &&= (0,\ \text{-}\tfrac{1}{2},\ 1,\ 0,\ 0) && (5) \\
[z\, d_2 c_1 c_2]_0^1 &\stackrel{(B)}{=} [z^2\, d_2]_0^2 - \tfrac{1}{2}[z\, d_2]_0^3 \\
&= 0_c - \tfrac{1}{2} e_3 &&= (0,\ 0,\ \text{-}\tfrac{1}{2},\ 0,\ 0) && (6) \\
[z\, c_2 c_1 c_2]_0^1 &\stackrel{(C)}{=} \tfrac{1}{2}[z^2\, c_2]_0^2 - \tfrac{1}{2}[z\, c_2]_0^3 + \tfrac{1}{2}[z\, d_2]_1^2 \\
&= 0_b + 0_b + \tfrac{1}{2} e_4 &&= (0,\ 0,\ 0,\ \tfrac{1}{2},\ 0) && (7) \\
[z\, d_2 c_1 c_2]_1^0 &\stackrel{(B)}{=} [z^2\, d_2]_1^1 - \tfrac{1}{2}[z\, d_2]_1^2 \\
&= 0_c - \tfrac{1}{2} e_4 &&= (0,\ 0,\ 0,\ \text{-}\tfrac{1}{2},\ 0) && (8) \\
\tfrac{1}{2}[d_2]_1^3 \stackrel{(F)}{=} [z\, c_2^2]_0^2 &\stackrel{(C')}{=} [z^2\, c_2^2]_0^1 - 2[z\, c_2^2 c_1 c_2]_0^0 + [z\, c_2 c_1 d_2]_0^1 \\
&= e_1 - 2(3) - (8) &&= (1,\ 0,\ 1,\ \tfrac{1}{2},\ 0) && (9) \\
[c_2 c_1 c_2]_0^2 &\stackrel{(C)}{=} \tfrac{1}{2}[z\, c_2]_0^3 - \tfrac{1}{2}[c_2]_0^4 + \tfrac{1}{2}[d_2]_1^3 \\
&= 0_b + 0_b + (9) &&= (1,\ 0,\ 1,\ \tfrac{1}{2},\ 0) && (10) \\
[d_2 c_1^2 c_2]_1^0 &\stackrel{(A)}{=} \tfrac{1}{2}[z\, d_2]_1^2 - \tfrac{1}{2}[d_2]_1^3 \\
&= \tfrac{1}{2} e_4 - (9) &&= (\text{-}1,\ 0,\ \text{-}1,\ 0,\ 0) && (11) \\
[c_2^2 c_1 d_2]_0^1 &\stackrel{(E)}{=} 2[z\, c_2 c_1 d_2]_0^1 - 2[c_2 c_1^2 d_2]_0^1 + [d_2 c_1^2 d_2]_0^1 \\
&= -2(8) + 2(11) + e_5 &&= (\text{-}2,\ 0,\ \text{-}2,\ 1,\ 1) && (12) \\
[z\, c_2^3]_0^1 &\stackrel{(D)}{=} [z^2\, c_2^2]_0^1 - 2[z\, c_2 c_1 c_2]_0^1 + [z\, d_2 c_1 c_2]_0^1 + [z\, c_2 c_1 d_2]_0^1 \\
&= e_1 - 2(7) + (6) - (8) &&= (1,\ 0,\ \text{-}\tfrac{1}{2},\ \text{-}\tfrac{1}{2},\ 0) && (13) \\
[c_2^3]_0^2 &\stackrel{(D)}{=} [z\, c_2^2]_0^2 - 2[c_2 c_1 c_2]_0^2 + [d_2 c_1 c_2]_0^2 + [c_2 c_1 d_2]_0^2 \\
&= (9) - 2(10) + (5) - (4) &&= (\text{-}1,\ \text{-}\tfrac{1}{2},\ 0,\ \tfrac{1}{2},\ 0) && (14)
\end{aligned}$$

$$\begin{aligned}
R_2 &= 2[z\, d_2]_0^3 - [d_2]_0^4 + 2[c_2^2 c_1 d_2]_0^1 + 2[z\, c_2^3]_0^1 - 2[c_2^3]_0^2 \\
&= 2e_3 - e_2 + 2(12) + 2(13) - 2(14) \;=\; (0,\ 0,\ \text{-}3,\ 0,\ 2)
\end{aligned}$$

Calculation of R_3

Relation (F) will be used in line (24); furthermore it allows to use the basic element e_1 in two different forms:

$$e_1 = [z\,c_2^2]_1^2 = \frac{1}{2}[d_2]_2^3, \quad e_2 = [z^3\,c_2^2]_0^1, \quad e_3 = [d_2]_0^5,$$

$$e_4 = [z\,d_2]_0^4, \qquad e_5 = [z\,d_2]_1^3, \qquad e_6 = [d_2c_1^2d_2]_0^2.$$

$$\begin{aligned}
[d_2c_1c_2]_0^3 &\overset{(B)}{=} [z\,d_2]_0^4 - \tfrac{1}{2}[d_2]_0^5 \\
&= e_4 - \tfrac{1}{2}e_3 \quad = \quad (0,\,0,\,\text{-}\tfrac{1}{2},\,1,\,0,\,0) && (1)\\
[d_2c_1c_2]_2^1 &\overset{(B)}{=} [z\,d_2]_2^2 - \tfrac{1}{2}[d_2]_2^3 \\
&= 0_{\mathrm{f}} - e_1 \quad = \quad (\text{-}1,\,0,\,0,\,0,\,0,\,0) && (2)\\
[c_2c_1c_2]_1^2 &\overset{(C)}{=} \tfrac{1}{2}[z\,c_2]_1^3 - \tfrac{1}{2}[c_2]_1^4 + \tfrac{1}{2}[d_2]_2^3 \\
&= 0_{\mathrm{b}} + 0_{\mathrm{b}} + e_1 \quad = \quad (1,\,0,\,0,\,0,\,0,\,0) && (3)\\
[z^2\,c_2^2c_1c_2]_0^0 &\overset{(D')}{=} \tfrac{1}{2}[z^3\,c_2^3]_0^0 - \tfrac{1}{2}[z^2\,c_2^4]_0^0 + \tfrac{1}{2}[z^2\,c_2d_2c_1c_2]_0^0 + \tfrac{1}{2}[z^2\,c_2^2c_1d_2]_0^0 \\
&= 0_{\mathrm{f}} + 0_{\mathrm{f}} + 0_{\mathrm{a}} + 0_{\mathrm{c}} \quad = \quad (0,\,0,\,0,\,0,\,0,\,0) && (4)\\
[z\,d_2c_1^2c_2]_1^0 &\overset{(A)}{=} \tfrac{1}{2}[z^2\,d_2]_1^2 - \tfrac{1}{2}[z\,d_2]_1^3 \\
&= 0_{\mathrm{c}} - \tfrac{1}{2}e_5 \quad = \quad (0,\,0,\,0,\,0,\,\text{-}\tfrac{1}{2},\,0) && (5)\\
[z^2\,c_2c_1c_2]_0^1 &\overset{(C)}{=} \tfrac{1}{2}[z^3\,c_2]_0^2 - \tfrac{1}{2}[z^2\,c_2]_0^3 + \tfrac{1}{2}[z^2\,d_2]_1^2 \\
&= 0_{\mathrm{b}} + 0_{\mathrm{b}} + 0_{\mathrm{c}} \quad = \quad (0,\,0,\,0,\,0,\,0,\,0) && (6)\\
[z\,d_2c_1c_2]_1^1 &\overset{(B)}{=} [z^2\,d_2]_1^2 - \tfrac{1}{2}[z\,d_2]_1^3 \\
&= 0_{\mathrm{c}} - \tfrac{1}{2}e_5 \quad = \quad (0,\,0,\,0,\,0,\,\text{-}\tfrac{1}{2},\,0) && (7)\\
[z\,d_2c_1^2c_2]_0^1 &\overset{(A)}{=} \tfrac{1}{2}[z^2\,d_2]_0^3 - \tfrac{1}{2}[z\,d_2]_0^4 \\
&= 0_{\mathrm{c}} - \tfrac{1}{2}e_4 \quad = \quad (0,\,0,\,0,\,\text{-}\tfrac{1}{2},\,0,\,0) && (8)\\
[z\,d_2c_1c_2]_2^0 &\overset{(B)}{=} [z^2\,d_2]_2^1 - \tfrac{1}{2}[z\,d_2]_2^2 \\
&= 0_{\mathrm{c}} + 0_{\mathrm{f}} \quad = \quad (0,\,0,\,0,\,0,\,0,\,0) && (9)\\
[z\,c_2c_1c_2]_0^2 &\overset{(C)}{=} \tfrac{1}{2}[z^2\,c_2]_0^3 - \tfrac{1}{2}[z\,c_2]_0^4 + \tfrac{1}{2}[z\,d_2]_1^3 \\
&= 0_{\mathrm{b}} + 0_{\mathrm{b}} + \tfrac{1}{2}e_5 \quad = \quad (0,\,0,\,0,\,0,\,\tfrac{1}{2},\,0) && (10)\\
[z^2\,c_2^2]_0^2 &\overset{(C')}{=} [z^3\,c_2^2]_0^1 - 2[z^2\,c_2^2c_1c_2]_0^0 + [z^2\,c_2c_1d_2]_0^1 \\
&= e_2 - 2(4) + 0_{\mathrm{c}} \quad = \quad (0,\,1,\,0,\,0,\,0,\,0) && (11)\\
[d_2c_1^2c_2]_0^2 &\overset{(A)}{=} \tfrac{1}{2}[z\,d_2]_0^4 - \tfrac{1}{2}[d_2]_0^5 \\
&= \tfrac{1}{2}e_4 - \tfrac{1}{2}e_3 \quad = \quad (0,\,0,\,\text{-}\tfrac{1}{2},\,\tfrac{1}{2},\,0,\,0) && (12)
\end{aligned}$$

$$
\begin{aligned}
[z\,d_2c_1c_2]_0^2 &\overset{(\mathrm{B})}{=} [z^2\,d_2]_0^3 - \tfrac{1}{2}[z\,d_2]_0^4 \\
&= 0_{\mathrm{c}} - \tfrac{1}{2}e_4 &&= (0,\ 0,\ 0,\ \text{-}\tfrac{1}{2},\ 0,\ 0) && (13)\\
[d_2c_1c_2c_1c_2]_0^1 &\overset{(\mathrm{C})}{=} \tfrac{1}{2}[z\,d_2c_1c_2]_0^2 - \tfrac{1}{2}[d_2c_1c_2]_0^3 + \tfrac{1}{2}[d_2c_1^2d_2]_0^2 \\
&= \tfrac{1}{2}(13) - \tfrac{1}{2}(1) + \tfrac{1}{2}e_6 &&= (0,\ 0,\ \tfrac{1}{4},\ \text{-}\tfrac{3}{4},\ 0,\ \tfrac{1}{2}) && (14)\\
[c_2^2c_1d_2]_2^0 &\overset{(\mathrm{E})}{=} 2[z\,c_2c_1d_2]_2^0 - 2[c_2c_1^2d_2]_2^0 + [d_2c_1^2d_2]_2^0 \\
&= -2(13) + 2(12) - e_6 &&= (0,\ 0,\ \text{-}1,\ 2,\ 0,\ \text{-}1) && (15)\\
[z\,c_2^3]_0^2 &\overset{(\mathrm{D})}{=} [z^2\,c_2^2]_0^2 - 2[z\,c_2c_1c_2]_0^2 + [z\,d_2c_1c_2]_0^2 + [z\,c_2c_1d_2]_0^2 \\
&= (11) - 2(10) + (13) - (9) &&= (0,\ 1,\ 0,\ \text{-}\tfrac{1}{2},\ \text{-}1,\ 0) && (16)\\
[z\,c_2^2c_1d_2]_1^0 &\overset{(\mathrm{E})}{=} 2[z^2\,c_2c_1d_2]_1^0 - 2[z\,c_2c_1^2d_2]_1^0 + [z\,d_2c_1^2d_2]_1^0 \\
&= 0_{\mathrm{c}} + 2(8) + 0_{\mathrm{e}} &&= (0,\ 0,\ 0,\ \text{-}1,\ 0,\ 0) && (17)\\
[d_2c_1c_2^3]_0^1 &\overset{(\mathrm{D})}{=} [z\,d_2c_1c_2^2]_0^1 - 2[d_2c_1c_2c_1c_2]_0^1 + [d_2c_1d_2c_1c_2]_0^1 + [d_2c_1c_2c_1d_2]_0^1 \\
&= -(17) - 2(14) + 0_{\mathrm{a}} + 0_{\mathrm{d}} &&= (0,\ 0,\ \text{-}\tfrac{1}{2},\ \tfrac{5}{2},\ 0,\ \text{-}1) && (18)\\
[z\,c_2^2c_1c_2]_1^0 &\overset{(\mathrm{C})}{=} \tfrac{1}{2}[z^2\,c_2^2]_1^1 - \tfrac{1}{2}[z\,c_2^2]_1^2 + \tfrac{1}{2}[z\,c_2c_1d_2]_1^1 \\
&= 0_{\mathrm{f}} - \tfrac{1}{2}e_1 - \tfrac{1}{2}(7) &&= (\text{-}\tfrac{1}{2},\ 0,\ 0,\ 0,\ \tfrac{1}{4},\ 0) && (19)\\
[z^2\,c_2^3]_0^1 &\overset{(\mathrm{D})}{=} [z^3\,c_2^2]_0^1 - 2[z^2\,c_2c_1c_2]_0^1 + [z^2\,d_2c_1c_2]_0^1 + [z^2\,c_2c_1d_2]_0^1 \\
&= e_2 - 2(6) + 0_{\mathrm{c}} + 0_{\mathrm{c}} &&= (0,\ 1,\ 0,\ 0,\ 0,\ 0) && (20)\\
[z\,c_2^2c_1d_2]_0^1 &\overset{(\mathrm{E})}{=} 2[z^2\,c_2c_1d_2]_0^1 - 2[z\,c_2c_1^2d_2]_0^1 + [z\,d_2c_1^2d_2]_0^1 \\
&= 0_{\mathrm{c}} + 2(5) + 0_{\mathrm{e}} &&= (0,\ 0,\ 0,\ 0,\ \text{-}1,\ 0) && (21)\\
[z\,c_2^4]_0^1 &\overset{(\mathrm{D})}{=} [z^2\,c_2^3]_0^1 - 2[z\,c_2c_1c_2^2]_0^1 + [z\,d_2c_1c_2^2]_0^1 + [z\,c_2c_1d_2c_2]_0^1 \\
&= (20) + 2(19) - (17) + 0_{\mathrm{a}} &&= (\text{-}1,\ 1,\ 0,\ 1,\ \tfrac{1}{2},\ 0) && (22)\\
[z\,c_2^2c_1c_2]_0^1 &\overset{(\mathrm{D}')}{=} \tfrac{1}{2}[z^2\,c_2^3]_0^1 - \tfrac{1}{2}[z\,c_2^4]_0^1 + \tfrac{1}{2}[z\,c_2d_2c_1c_2]_0^1 + \tfrac{1}{2}[z\,c_2^2c_1d_2]_0^1 \\
&= \tfrac{1}{2}(20) - \tfrac{1}{2}(22) + 0_{\mathrm{a}} + \tfrac{1}{2}(21) &&= (\tfrac{1}{2},\ 0,\ 0,\ \text{-}\tfrac{1}{2},\ \text{-}\tfrac{3}{4},\ 0) && (23)\\
\tfrac{1}{2}[d_2]_1^4 &\overset{(\mathrm{F})}{=} [z\,c_2^2]_0^3 \overset{(\mathrm{C}')}{=} [z^2\,c_2^2]_0^2 - 2[z\,c_2^2c_1c_2]_0^1 + [z\,c_2c_1d_2]_0^2 \\
&= (11) - 2(23) - (9) &&= (\text{-}1,\ 1,\ 0,\ 1,\ \tfrac{3}{2},\ 0) && (24)\\
[d_2c_1^2c_2]_1^1 &\overset{(\mathrm{A})}{=} \tfrac{1}{2}[z\,d_2]_1^3 - \tfrac{1}{2}[d_2]_1^4 \\
&= \tfrac{1}{2}e_5 - (24) &&= (1,\ \text{-}1,\ 0,\ \text{-}1,\ \text{-}1,\ 0) && (25)\\
[d_2c_1c_2]_1^2 &\overset{(\mathrm{B})}{=} [z\,d_2]_1^3 - \tfrac{1}{2}[d_2]_1^4 \\
&= e_5 - (24) &&= (1,\ \text{-}1,\ 0,\ \text{-}1,\ \text{-}\tfrac{1}{2},\ 0) && (26)\\
[c_2^2c_1d_2]_1^1 &\overset{(\mathrm{E})}{=} 2[z\,c_2c_1d_2]_1^1 - 2[c_2c_1^2d_2]_1^1 + [d_2c_1^2d_2]_1^1 \\
&= -2(7) + 2(25) + 0_{\mathrm{f}} &&= (2,\ \text{-}2,\ 0,\ \text{-}2,\ \text{-}1,\ 0) && (27)
\end{aligned}
$$

$$[c_2^3]_1^2 \stackrel{(D)}{=} [z\,c_2^2]_1^2 - 2[c_2c_1c_2]_1^2 + [d_2c_1c_2]_1^2 + [c_2c_1d_2]_1^2$$
$$= e_1 - 2(3) + (26) - (2) \quad = \quad (1, \text{-}1, 0, \text{-}1, \text{-}\tfrac{1}{2}, 0) \qquad (28)$$

$$[c_2^3c_1c_2]_1^0 \stackrel{(C)}{=} \tfrac{1}{2}[z\,c_2^3]_1^1 - \tfrac{1}{2}[c_2^3]_1^2 + \tfrac{1}{2}[c_2^2c_1d_2]_1^1$$
$$= 0_{\mathrm{f}} - \tfrac{1}{2}(28) + \tfrac{1}{2}(27) \quad = \quad (\tfrac{1}{2}, \text{-}\tfrac{1}{2}, 0, \text{-}\tfrac{1}{2}, \text{-}\tfrac{1}{4}, 0) \qquad (29)$$

$$[c_2^2c_1c_2]_2^0 \stackrel{(C)}{=} \tfrac{1}{2}[z\,c_2^2]_2^1 - \tfrac{1}{2}[c_2^2]_2^2 + \tfrac{1}{2}[c_2c_1d_2]_2^1$$
$$= -\tfrac{1}{2}e_1 + 0_{\mathrm{f}} - \tfrac{1}{2}(26) \quad = \quad (\text{-}1, \tfrac{1}{2}, 0, \tfrac{1}{2}, \tfrac{1}{4}, 0) \qquad (30)$$

$$[d_2c_1c_2c_1c_2]_1^0 \stackrel{(C)}{=} \tfrac{1}{2}[z\,d_2c_1c_2]_1^1 - \tfrac{1}{2}[d_2c_1c_2]_1^2 + \tfrac{1}{2}[d_2c_1^2d_2]_1^1$$
$$= \tfrac{1}{2}(7) - \tfrac{1}{2}(26) + 0_{\mathrm{f}} \quad = \quad (\text{-}\tfrac{1}{2}, \tfrac{1}{2}, 0, \tfrac{1}{2}, 0, 0) \qquad (31)$$

$$[c_2^3c_1d_2]_0^1 \stackrel{(D)}{=} [z\,c_2^2c_1d_2]_0^1 - 2[c_2c_1c_2c_1d_2]_0^1 + [d_2c_1c_2c_1d_2]_0^1 + [c_2c_1d_2c_1d_2]_0^1$$
$$= (21) + 2(31) + 0_{\mathrm{d}} + 0_{\mathrm{a}} \quad = \quad (\text{-}1, 1, 0, 1, \text{-}1, 0) \qquad (32)$$

$$[c_2^4]_0^2 \stackrel{(D)}{=} [z\,c_2^3]_0^2 - 2[c_2c_1c_2^2]_0^2 + [d_2c_1c_2^2]_0^2 + [c_2c_1d_2c_2]_0^2$$
$$= (16) + 2(30) - (15) + 0_{\mathrm{a}} \quad = \quad (\text{-}2, 2, 1, \text{-}\tfrac{3}{2}, \text{-}\tfrac{1}{2}, 1) \qquad (33)$$

$$[c_2^5]_0^1 \stackrel{(D)}{=} [z\,c_2^4]_0^1 - 2[c_2c_1c_2^3]_0^1 + [d_2c_1c_2^3]_0^1 + [c_2c_1d_2c_2^2]_0^1$$
$$= (22) + 2(29) + (18) + 0_{\mathrm{a}} \quad = \quad (0, 0, \text{-}\tfrac{1}{2}, \tfrac{5}{2}, 0, \text{-}1) \qquad (34)$$

$$R_3 = 2[z\,d_2]_0^4 - [d_2]_0^5 + 2[c_2^3c_1d_2]_0^1 + 2[z\,c_2^4]_0^1 - 2[c_2^4]_0^2 - 6[c_2^5]_0^1$$
$$= 2e_4 - e_3 + 2(32) + 2(22) - 2(33) - 6(34) \quad = \quad (0, 0, 0, \text{-}6, 0, 4)$$

Braid structures in knot complements, handlebodies and 3–manifolds

Sofia Lambropoulou
Mathematisches Institut, Göttingen Universität

Abstract

We consider braids on $m+n$ strands, such that the first m strands are trivially fixed. We denote the set of all such braids by $B_{m,n}$. Via concatenation $B_{m,n}$ acquires a group structure. The objective of this paper is to find a presentation for $B_{m,n}$ using the structure of its corresponding pure braid subgroup, $P_{m,n}$, and the fact that it is a subgroup of the classical Artin group B_{m+n}. Then we give an irredundant presentation for $B_{m,n}$. The paper concludes by showing that these braid groups or appropriate cosets of them are related to knots in handlebodies, in knot complements and in c.c.o. 3–manifolds.

1 Introductory notions and motivations

Definition 1 *The set of all elements of the classical Artin group B_{m+n} for which, if we remove the last n strands we are left with the identity braid on m strands, shall be denoted by $B_{m,n}$ (see figure 1(a) below for an example in $B_{3,3}$). The elements of $B_{m,n}$ are special cases of 'mixed braids' (cf. section 6).*

Figure 1:

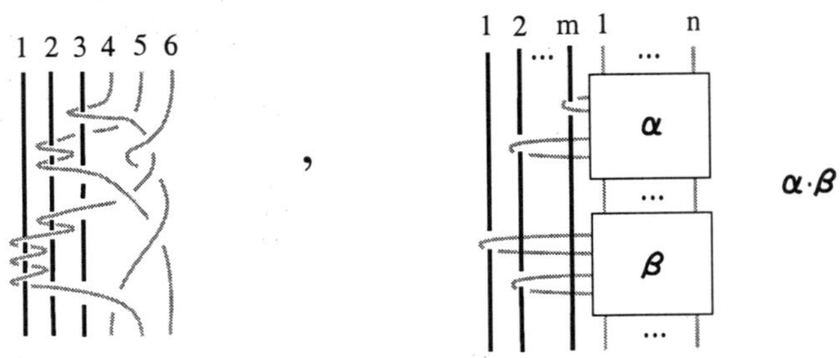

(a) (b)

Concatenation is a closed operation in $B_{m,n}$: the product $\alpha \cdot \beta$ of two elements $\alpha, \beta \in B_{m,n}$ is also an element of $B_{m,n}$ (see figure 1(b)). Thus $B_{m,n} \leq B_{m+n}$. Our purpose is to obtain a simple presentation for $B_{m,n}$.

The motivation for studying these braids comes from studying oriented knots and links in knot complements, in c.c.o. 3-manifolds and in handlebodies, since these spaces may be represented by a fixed braid or a fixed integer-framed braid in S^3. Then knots and links in these spaces may be represented by elements of the above braid groups $B_{m,n}$ or of appropriate cosets of these groups. More precisely, if M denotes the complement of the m-unlink or a connected sum of m lens spaces of type $L(p,1)$ or a handlebody of genus m, then knots and links in these spaces may be represented precisely by the mixed braids in $B_{m,n}$, for $n \in \mathbb{N}$. In the case $m = 1$, $B_{1,n}$ is the Artin group of type $\mathcal{B}$ (cf. [4], [5], [6]). If M is generic, concatenation is no more a closed operation of mixed braids, but as we show in section 6, knots and links in M may be represented by mixed braids in $B_{m,n}$, for $n \in \mathbb{N}$, followed by a fixed part associated to M, i.e. by elements of a coset of $B_{m,n}$.

We recall now some facts about braids and pure braids. For more details and a complete study of the classical theory of braids the reader is referred to [1]. The pure braid group, P_n, corresponding to the classical Artin group on n strands, B_n, consists of all elements in B_n that induce the identity permutation in S_n, thus $P_n \lhd B_n$ and P_n is generated by the elements

$$\begin{aligned} a_{ij} &= {\sigma_i}^{-1}{\sigma_{i+1}}^{-1}\ldots{\sigma_{j-2}}^{-1}{\sigma_{j-1}}^{2}\sigma_{j-2}\ldots\sigma_{i+1}\sigma_i \\ &= \sigma_{j-1}\sigma_{j-2}\ldots\sigma_{i+1}{\sigma_i}^{2}{\sigma_{i+1}}^{-1}\ldots{\sigma_{j-2}}^{-1}{\sigma_{j-1}}^{-1}, \quad 1 \le i < j \le n. \end{aligned}$$

The generators a_{ij} may be pictured geometrically as an elementary loop between the ith and jth strand (cf. figure 2).

The most important property of pure braids is that they have a canonical form, the so-called *'Artin's canonical form'*, which says that every element, A, of P_n may be written uniquely in the form:

$$A = U_1 U_2 \cdots U_{n-1}$$

where each U_i is a uniquely determined product of powers of the a_{ij} using only those with $i < j$. Geometrically, this means that any pure braid can be 'combed' i.e. can be written canonically as the pure braiding of the first strand with the rest, then keep the first strand fixed and uncrossed and have the pure braiding of the second strand and so on (cf. figure 3).

The main idea for finding a presentation for P_n is the following: The combing of a strand may be regarded as a loop in the complement space of the other strands, and as such is an element of a free group since the fundamental group of a punctured disc is free. Thus,

$$P_n = F_{n-1} \rtimes \cdots \rtimes F_2 \rtimes F_1 = F_{n-1} \rtimes P_{n-1},$$

where each F_i is a free group on the generators $a_{1,i+1}, \ldots, a_{i,i+1}$ (the elementary loops between the $(i+1)$st strand and all its previous ones), and where the action is induced by conjugation. It turns out that P_n has $\frac{n(n-1)}{2}$ generators and $\frac{1 \cdot 2^2 + 2 \cdot 3^2 + \ldots + (n-2)(n-1)^2}{2}$ relations, the following:

$$a_{ij}{}^{-1} a_{rs} a_{ij} = \begin{cases} a_{rs} & \text{if } i<j<r<s \quad \text{or } r<i<j<s, \\ a_{is}a_{js}a_{is}{}^{-1} & \text{if } i<j<s, \\ a_{is}a_{js}a_{is}a_{js}{}^{-1}a_{is}{}^{-1} & \text{if } i<j<s, \\ a_{is}a_{js}a_{is}{}^{-1}a_{js}{}^{-1}a_{rs}a_{js}a_{is}a_{js}{}^{-1}a_{is}{}^{-1} & \text{if } i<r<j<s. \end{cases}$$

Based on these ideas, we introduce in section 2 the pure braid group $P_{m,n}$ and in section 3 we find a presentation for it. Then in section 4 we put together a presentation for $B_{m,n}$, which we simplify in section 5. In section 5 we also give a Dynkin-diagram related to $B_{m,n}$. Finally, in section 6 we explain that elements of $B_{m,n}$ represent oriented knots and links in certain spaces and that appropriate cosets of $B_{m,n}$ represent knots and links in the generic cases of knot complements and c.c.o. 3-manifolds.

The results here have been preliminary studied by the author in [4] and have been presented in various mathematical meetings since 1995. A. Sossinsky, independently, motivated by the same topological considerations, studies these groups in [8] and he conjectures the irredundant presentation for $B_{m,n}$. Moreover, V. Vershinin in [9] studies the groups $B_{m,n}$ in connection to handlebodies of genus g, taking a configuration-spaces approach. Back in 1993 Alastair Leeves had found a presentation for $B_{m,n}$, which was presented in [4], but a proof was never published. The author is thankfull to A. Leeves for inspiring discussions at the time. Also, her grateful thanks are due to Bernard Leclerc for his careful reading through this work and his very valuable comments.

This is the first paper in a sequel of three. The next one gives expressions for algebraic equivalence of braids reflecting knot isotopy in arbitrary knot complements and c.c.o. 3-manifolds. The case of handlebodies is joint work with Reinhard Häring-Oldenburg.

2 The pure braid group $P_{m,n}$

Definition 2 *The corresponding pure braid group $P_{m,n}$ of $B_{m,n}$ is defined as $P_{m,n} = B_{m,n} \cap P_{m+n}$, i.e. $P_{m,n} \leq P_{m+n}$ and it does not contain pure braiding among the first m strands.*

By its definition, $P_{m,n}$ is generated by the pure braid generators a_{ij} for $i \in \{1, \ldots, m+n-1\}$ and $j \in \{m+1, \ldots, m+n\}$ of P_{m+n} (see figure 2). Then, $B_{m,n}$

is clearly generated by the elementary mixed braids (drawn below) a_{ij} for $i \in \{1,\ldots,m+n-1\}$ and $j \in \{m+1,\ldots,m+n\}$ together with $\sigma_{m+1},\ldots,\sigma_{m+n-1}$, the elementary crossings among the last n strands. Note that the inverses of the a_{ij}'s and σ_k's are represented by the same geometric pictures, but with the opposite crossings.

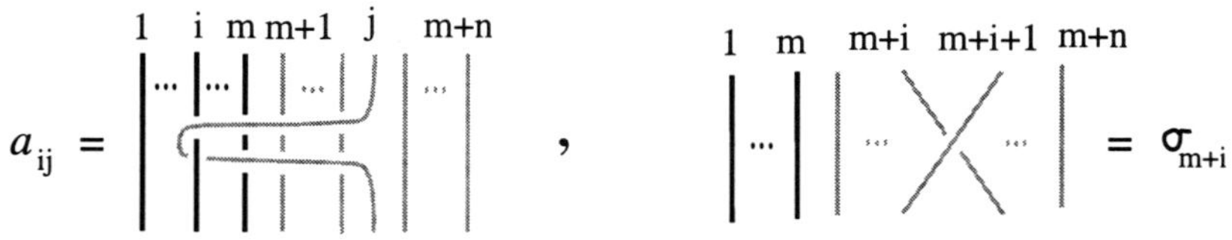

Figure 2

Also, by definition we have an exact sequence

$$1 \longrightarrow P_{m,n} \longrightarrow B_{m,n} \longrightarrow S_n \longrightarrow 1.$$

In particular, $P_{m,n} \lhd B_{m,n}$. More precisely, we have the following relations:

$$\sigma_k{}^{-1} a_{ij}{}^{\pm} \sigma_k = \begin{cases} a_{ij}{}^{\pm} & \text{if } k \le i-2 \text{ or } i+1 \le k \le j-2 \text{ or } k \ge j+1, \\ a_{i-1,j}{}^{\pm} & \text{if } k = i-1, \\ a_{ij} a_{i+1,j}{}^{\pm} a_{ij}{}^{-1} & \text{if } k = i, \\ a_{i,j-1}{}^{\pm} & \text{if } k = j-1, \\ a_{ij} a_{i,j+1}{}^{\pm} a_{ij}{}^{-1} & \text{if } k = j. \end{cases} \qquad \square$$

We shall call these *mixed relations* and we shall denote them by M_1, M_2, M_3, M_4 and M_5 in the order they are written.

Thus $B_{m,n}$ is a group extension of $P_{m,n}$ by S_n. This will yield a presentation for $B_{m,n}$, conditionally to knowing a presentation for $P_{m,n}$.

3 A presentation for $P_{m,n}$

Theorem 1 *The pure braid group $P_{m,n}$ is generated by the elements a_{ij} for $i \in \{1,\ldots,m+n-1\}$, $j \in \{m+1,\ldots,m+n\}$ and $i < j$, which are subject to the relations:*

$$\begin{array}{llll} (P_1) & a_{ij}{}^{-1} a_{rs} a_{ij} = a_{rs} & \textit{if } i<j<r<s & \textit{or } r<i<j<s, \\ (P_2) & a_{ij}{}^{-1} a_{js} a_{ij} = a_{is} a_{js} a_{is}{}^{-1} & & \textit{if } i<j<s, \\ (P_3) & a_{ij}{}^{-1} a_{is} a_{ij} = a_{is} a_{js} a_{is} a_{js}{}^{-1} a_{is}{}^{-1} & & \textit{if } i<j<s, \\ (P_4) & a_{ij}{}^{-1} a_{rs} a_{ij} = a_{is} a_{js} a_{is}{}^{-1} a_{js}{}^{-1} a_{rs} a_{js} a_{is} a_{js}{}^{-1} a_{is}{}^{-1} & & \textit{if } i<r<j<s. \end{array}$$

Relations P_1, P_2, P_3 and P_4 shall be called *pure braid relations.* Note that P_1 and P_4 involve the strands i, r, j, s, whilst P_2 and P_3 involve the strands i, j, s.

Proof By its definition and by the fact that the a_{ij}'s (for all indices) generate P_{m+n} follows that the above set of elements is indeed a set of generators for $P_{m,n}$. Since $P_{m,n} \leq P_{m+n}$ we can apply on its elements Artin's combing. As for P_n, the combing of a strand can be regarded as a loop in the complement space of the strands with smaller index (including the m fixed ones) and as such it is an element of a free group. Therefore we have:

$P_{m,1} = F_m = \langle a_{1,m+1}, a_{2,m+1}, \dots, a_{m,m+1} \rangle$, the free group on m generators.

Further is: $P_{m,2} = F_{m+1} \rtimes P_{m,1} = \langle a_{1,m+2}, \dots, a_{m,m+2}, a_{m+1,m+2} \rangle \rtimes P_{m,1}$, i.e. F_{m+1} is the free group on $m+1$ generators, and $P_{m,1}$ acts on F_{m+1} by conjugation, via the relations of the pure braid group P_{m+1} for appropriate indices. We proceed inductively to obtain:

$$\begin{aligned} P_{m,n} &= F_{m+n-1} \rtimes \cdots \rtimes F_{m+1} \rtimes F_m = F_{m+n-1} \rtimes P_{m,n-1} \\ &= \langle a_{1,m+n}, \dots, a_{m,m+n}, \dots, a_{m+n-1,m+n} \rangle \rtimes P_{m,n-1}, \end{aligned}$$

where F_{m+n-1} is the free group on $m+n-1$ generators, and where $P_{m,n-1}$ acts on F_{m+n-1} by conjugation, via the relations of the pure braid group P_{m+n} for appropriate indices, i.e. via the relations P_1, P_2, P_3 and P_4. □

Some remarks are now due.

Remark 1 The groups $P_{m,n}$ and P_{m+n} have seemingly the same presentation. For $m \neq 1$ is, though, $P_{m,n} \neq P_{m+n}$. The difference lies in the restriction of the indices of the generators. In fact, $P_{m,n}$ has $\frac{n(n+2m-1)}{2}$ generators, which is the number of generators of P_{m+n} less the number of generators of P_m. Moreover, $P_{m,n}$ has $\frac{(m-1)\cdot m^2+m\cdot(m+1)^2+\cdots+(m+n-2)(m+n-1)^2}{2} - \frac{(m-1)\cdot m\cdot n\cdot(n+2m-1)}{4}$ relations, which is the number of relations of P_{m+n} less the number of relations of P_m. In the case $m = 1$ holds $P_{1,n} = P_{1+n}$, which follows immediately from the definition of $P_{m,n}$ or can be observed from its presentation for $m = 1$.

Remark 2 In [4] there is a discussion about the groups $B_{m,n}$ and a different line of proof is given for finding a presentation. There, by $P_{m,n}$ we denoted some smaller pure braid subgroups, for which it is rather complicated to find a presentation. But the case $m = 1$ was extensively treated, also in the sequel papers [5, 6]. In all these previous results $P_{1,n}$ denoted the free group $F_n = \langle a_{12}, a_{13}, \dots, a_{1,n+1} \rangle$ and *not* the corresponding pure braid group of $B_{1,n}$. That's why we had then $B_{1,n} = F_n \rtimes B_n$. We hope that the readers familiar with those results will not be in confusion.

4 A presentation for $B_{m,n}$

In section 2 we showed that $1 \longrightarrow P_{m,n} \longrightarrow B_{m,n} \longrightarrow S_n \longrightarrow 1$ and in section 3 we found a presentation for $P_{m,n}$. Recall that S_n has the presentation:

$$\langle s_1, ..., s_{n-1} \mid s_i s_j = s_j s_i \ \text{ for } \ |i-j| > 1, \ \ s_i s_{i+1} s_i = s_{i+1} s_i s_{i+1}, \ \ {s_i}^2 = 1 \rangle.$$

We are now ready to put together a presentation for $B_{m,n}$. Namely, we can apply a result from the theory of group presentations (see [3], p.139), that gives a presentation for a group extension of two groups with known presentations. Indeed, the following is then a presentation for $B_{m,n}$.

$$\left\langle \begin{array}{l|l} a_{1,m+1}, \ldots, a_{1,m+n}, \ldots, a_{m,m+1}, \ldots, a_{m,m+n}, & P_1, P_2, P_3, P_4, \\ a_{m+1,m+2}, \ldots, a_{m+1,m+n}, \ldots, a_{m+n-1,m+n}, & M_1, M_2, M_3, M_4, M_5, \\ \sigma_{m+1}, \sigma_{m+2}, \ldots, \sigma_{m+n-1} & \Sigma_1, \Sigma_2, \Sigma_3. \end{array} \right\rangle,$$

where the relations Σ_1, Σ_2 and Σ_3 are satisfied by the $\sigma_{m+1}, \sigma_{m+2}, \ldots, \sigma_{m+n-1}$ and they are the following:

$$\begin{array}{lrcll} (\Sigma_1) & \sigma_i \sigma_j & = & \sigma_j \sigma_i & \text{if} \quad |i-j| > 1, \\ (\Sigma_2) & \sigma_i \sigma_{i+1} \sigma_i & = & \sigma_{i+1} \sigma_i \sigma_{i+1} & \text{if} \quad m+1 \leq i \leq m+n-2, \\ (\Sigma_3) & {\sigma_i}^2 & = & a_{i,i+1} & \text{if} \quad m+1 \leq i \leq m+n-2. \end{array}$$

Σ_1 and Σ_2 are the *'braid relations'*.

Notice now that relations Σ_3 for $i \in \{m+1, \ldots, m+n-1\}$ and $j \in \{m+2, \ldots, m+n\}$ do not involve any mixed braiding and so they may be taken as defining relations, namely:

$$\begin{array}{lcl} {a_{m+1,m+2}}^{\pm} & := & {\sigma_{m+1}}^{\pm 2}, \\ {a_{m+1,m+3}}^{\pm} & := & \sigma_{m+2} {\sigma_{m+1}}^{\pm 2} {\sigma_{m+2}}^{-1}, \ldots \\ {a_{ij}}^{\pm} & := & \sigma_{j-1} \ldots \sigma_{i+1} {\sigma_i}^{\pm 2} {\sigma_{i+1}}^{-1} \ldots {\sigma_{j-1}}^{-1}, \\ & \vdots & \\ {a_{m+n-1,m+n}}^{\pm} & := & {\sigma_{m+n-1}}^{\pm 2}. \end{array}$$

Therefore, we want to omit eventually these a_{ij}'s from the list of generators of $B_{m,n}$ and subsequently to eliminate or simplify all relations involving these elements, applying Tietze transformations. Indeed, we examine one by one the relations and we have:

$\boldsymbol{P_1}$: ${a_{ij}}^{-1} a_{rs} a_{ij} = a_{rs}$ for the case $r < i < j < s$. If all $r, i, j, s \in \{m+1, \ldots, m+n\}$, the relations follow from Σ_1 and Σ_2 and so we only keep the ones where $r \in \{1, \ldots, m\}$ and $i \in \{m+1, \ldots, m+n-1\}$ or $r, i \in \{1, \ldots, m\}$.

P_2 : $a_{ij}^{-1}a_{js}a_{ij} = a_{is}a_{js}a_{is}^{-1}$ for $i < j < s$. Since $j, s \in \{m+1, \ldots, m+n\}$ the only case to be kept is when $i \in \{1, \ldots, m\}$, as the relations for $i \in \{m+1, \ldots, m+n-1\}$ follow from the braid relations.

P_3 : $a_{ij}^{-1}a_{is}a_{ij} = a_{is}a_{js}a_{is}a_{js}^{-1}a_{is}^{-1}$ for $i < j < s$. As in the previous case, the only relations that do not follow from Σ_1 and Σ_2 are the ones where $i \in \{1, \ldots, m\}$.

P_4 : $a_{ij}^{-1}a_{rs}a_{ij} = a_{is}a_{js}a_{is}^{-1}a_{js}^{-1}a_{rs}a_{js}a_{is}a_{js}^{-1}a_{is}^{-1}$ for $i < r < j < s$. Since $j, s \in \{m+1, \ldots, m+n\}$ the only relations to be kept are those where either $i \in \{1, \ldots, m\}$ and $r \in \{m+1, \ldots, m+n-1\}$ or $i, r \in \{1, \ldots, m\}$.

M_1 : $\sigma_k^{-1}a_{ij}^{\pm}\sigma_k = a_{ij}^{\pm}$ for $k \leq i-2$ or $i+1 \leq k \leq j-2$ or $k \geq j+1$. Here also we have $j \in \{m+1, \ldots, m+n\}$ and $k \in \{m+1, \ldots, m+n-1\}$. Now, if $i \in \{m+1, \ldots, m+n-1\}$, all these relations follow from Σ_1 and Σ_2, whilst for $i \in \{1, \ldots, m\}$ it only makes sense to consider $k \leq j-2$ or $k \geq j+1$.

M_2 : $\sigma_{i-1}^{-1}a_{ij}^{\pm}\sigma_{i-1} = a_{i-1,j}^{\pm}$. Since $i-1 \in \{m+1, \ldots, m+n-1\}$ it must be $i > m+1$ and so all these relations follow from Σ_1 and Σ_2.

M_3 : $\sigma_i^{-1}a_{ij}^{\pm}\sigma_i = a_{ij}a_{i+1,j}^{\pm}a_{ij}^{-1}$. This is analogous to the above case, since $i \geq m+1$.

M_4 : $\sigma_{j-1}^{-1}a_{ij}^{\pm}\sigma_{j-1} = a_{i,j-1}^{\pm}$. Here also the only cases that do not follow from the braid relations are the ones with $i \in \{1, \ldots, m\}$.

M_5 : $\sigma_j^{-1}a_{ij}^{\pm}\sigma_j = a_{ij}a_{i,j+1}^{\pm}a_{ij}^{-1}$. As above, the only relations surviving are the ones where $i \in \{1, \ldots, m\}$.

Remark 3 The remaining relations of P_1 for $i \in \{m+1, \ldots, m+n-1\}$ follow from the simpler relations: $a_{ij}\sigma_k = \sigma_k a_{ij}$ for $k \leq j-2$ or $k \geq j+1$, which coincides with the remaining of M_1 above.

To summarize, we showed that the following is a presentation for $B_{m,n}$.

$$B_{m,n} = \left\langle \begin{array}{l} a_{1,m+1}, \ldots, a_{1,m+n}, \ldots, a_{m,m+1}, \ldots, a_{m,m+n}, \\ a_{m+1,m+2}, \ldots, a_{m+1,m+n}, \ldots, a_{m+n-1,m+n}, \\ \sigma_{m+1}, \sigma_{m+2}, \ldots, \sigma_{m+n-1} \end{array} \middle| \begin{array}{l} P_1', P_2', P_3', P_4', \\ M_1', M_2', M_3', \\ \Sigma_1, \Sigma_2 \end{array} \right\rangle,$$

where we have:

$$
\begin{array}{llll}
(P_1') & a_{ij}a_{rs} & = & a_{rs}a_{ij} \quad \text{for } r < i < j < s,\ 1 \le i, r \le m, \\
(P_2') & a_{ij}{}^{-1}a_{js}a_{ij} & = & a_{is}a_{js}a_{is}{}^{-1} \quad \text{for } i < j < s,\ 1 \le i \le m, \\
(P_3') & a_{ij}{}^{-1}a_{is}a_{ij} & = & a_{is}a_{js}a_{is}a_{js}{}^{-1}a_{is}{}^{-1} \quad \text{for } i < j < s,\ 1 \le i \le m, \\
(P_4') & a_{ij}{}^{-1}a_{rs}a_{ij} & = & a_{is}a_{js}a_{is}{}^{-1}a_{js}{}^{-1}a_{rs}a_{js}a_{is}a_{js}{}^{-1}a_{is}{}^{-1} \quad \text{for } i < r < j < s, \\
 & & & 1 \le i \le m,\ 1 \le r \le m+n-1, \\
(M_1') & \sigma_k{}^{-1}a_{ij}{}^{\pm}\sigma_k & = & a_{ij}{}^{\pm} \quad \text{for } k \le j-2 \text{ or } k \ge j+1 \text{ and } 1 \le i \le m, \\
(M_2') & a_{ij}{}^{\pm} & = & \sigma_{j-1}a_{i,j-1}{}^{\pm}\sigma_{j-1}{}^{-1} \quad \text{for } 1 \le i \le m, \\
(M_3') & \sigma_j{}^{-1}a_{ij}{}^{\pm}\sigma_j & = & a_{ij}a_{i,j+1}{}^{\pm}a_{ij}{}^{-1} \quad \text{for } 1 \le i \le m.
\end{array}
$$

Having now done the first, 'obvious' clearing in the original presentation of $B_{m,n}$, we observe that many of the above relations are redundant or they simplify further, and that we may omit the a_{ij}'s with $i \ge m+1$.

Theorem 2 *The following is a presentation for $B_{m,n}$:*

$$
B_{m,n} = \left\langle \begin{array}{l} a_{1,m+1}, \ldots, a_{1,m+n}, \cdots, \\ a_{m,m+1}, \ldots, a_{m,m+n}, \\ \sigma_{m+1}, \sigma_{m+2}, \ldots, \sigma_{m+n-1} \end{array} \;\middle|\; \begin{array}{l} \Sigma_1, \Sigma_2, (1), (2), (3), (4), \\ \textit{for all appropriate indices} \end{array} \right\rangle,
$$

where we have:

$$
\begin{array}{llll}
(1) & \sigma_k{}^{-1}a_{ij}{}^{\pm}\sigma_k & = & a_{ij}{}^{\pm} \quad \textit{for } k \le j-2 \textit{ or } k \ge j+1, \\
(2) & a_{ij}{}^{\pm} & = & \sigma_{j-1}a_{i,j-1}{}^{\pm}\sigma_{j-1}{}^{-1}, \\
(3) & \sigma_j{}^{-1}a_{ij}{}^{\pm}\sigma_j & = & a_{ij}a_{i,j+1}{}^{\pm}a_{ij}{}^{-1} \\
(4) & a_{ij}{}^{\pm}a_{r,j+1}{}^{\pm} & = & a_{r,j+1}{}^{\pm}a_{ij}{}^{\pm} \quad \textit{for } r < i.
\end{array}
$$

Proof Relations (1), (2) and (3) are precisely M_1', M_2' and M_3', whilst relations (4) are a special case of relations P_1'. So we have to show that P_2', P_3', P_4' and P_5' as well as the rest cases of P_1' follow from $\Sigma_1, \Sigma_2, (1), (2), (3)$ and (4). Before continuing we note that, using (2), relation (3) is equivalent to

$$\sigma_j a_{ij} \sigma_j a_{ij}{}^{\pm} = a_{ij}{}^{\pm}\sigma_j a_{ij}\sigma_j$$

and relation (4) is equivalent to

$$a_{ij}{}^{\pm}(\sigma_j a_{rj}{}^{\pm}\sigma_j{}^{-1}) = (\sigma_j a_{rj}{}^{\pm}\sigma_j{}^{-1})a_{ij}{}^{\pm}.$$

We shall also use these forms in the proof. We proceed now case by case. The underlining indicates the expressions involved in each step of the proof.

$$
\begin{aligned}
a_{ij}\underline{a_{rs}} &\overset{M_2'}{=} \underline{a_{ij}\sigma_{s-1}\cdots\sigma_{j+1}a_{r,j+1}}\sigma_{j+1}^{-1}\cdots\sigma_{s-1}^{-1}\\
&\overset{M_1'}{=} \sigma_{s-1}\cdots\sigma_{j+1}\underline{a_{ij}(\sigma_j a_{rj}\sigma_j^{-1})}\sigma_{j+1}^{-1}\cdots\sigma_{s-1}^{-1}\\
&\overset{(4)}{=} \sigma_{s-1}\cdots\sigma_{j+1}(\sigma_j a_{rj}\sigma_j^{-1})\underline{a_{ij}\sigma_{j+1}^{-1}\cdots\sigma_{s-1}^{-1}}\\
&\overset{M_1'}{=} \underline{\sigma_{s-1}\cdots\sigma_j a_{rj}\sigma_j^{-1}\cdots\sigma_{s-1}^{-1}}a_{ij}\\
&\overset{M_2'}{=} a_{rs}a_{ij}.
\end{aligned}
$$

$$
\begin{aligned}
a_{ij}^{-1}\;a_{js}\;a_{ij} &\overset{M_2'}{=} a_{ij}^{-1}(\underline{\sigma_{s-1}\cdots\sigma_{j+1}\sigma_j^{2}\sigma_{j+1}^{-1}\cdots\sigma_{s-1}^{-1}})a_{ij}\\
&\overset{M_1'}{=} (\sigma_{s-1}\cdots\sigma_{j+1})\underline{a_{ij}^{-1}\sigma_j^{2}}a_{ij}(\sigma_{j+1}^{-1}\cdots\sigma_{s-1}^{-1})\\
&\overset{M_3'}{=} (\sigma_{s-1}\cdots\sigma_{j+1})\sigma_j a_{ij}a_{i,j+1}^{-1}\underline{a_{ij}^{-1}\sigma_j}a_{ij}(\sigma_{j+1}^{-1}\cdots\sigma_{s-1}^{-1})\\
&\overset{M_3'}{=} (\sigma_{s-1}\cdots\sigma_{j+1})\sigma_j a_{ij}\underline{a_{i,j+1}^{-1}}\sigma_j a_{ij}a_{i,j+1}^{-1}a_{ij}^{-1}a_{ij}(\sigma_{j+1}^{-1}\cdots\sigma_{s-1}^{-1})\\
&\overset{M_2'}{=} (\sigma_{s-1}\cdots\sigma_{j+1})\sigma_j a_{ij}\sigma_j^{2}a_{ij}^{-1}\sigma_j^{-1}(\sigma_{j+1}^{-1}\cdots\sigma_{s-1}^{-1})\\
&= \underline{(\sigma_{s-1}\cdots\sigma_j)a_{ij}(\sigma_j^{-1}\cdots\sigma_{s-1}^{-1}}\;\underline{\sigma_{s-1}\cdots\sigma_j)\sigma_j^{2}(\sigma_j^{-1}\cdots\sigma_{s-1}^{-1}}\\
&\quad\times \underline{\sigma_{s-1}\cdots\sigma_j)a_{ij}^{-1}(\sigma_j^{-1}\cdots\sigma_{s-1}^{-1})}\\
&\overset{M_2'}{=} a_{is}a_{js}a_{is}^{-1}.
\end{aligned}
$$

For P_3' we have:

$$
\begin{aligned}
a_{is}\;\underline{a_{js}}\;a_{is}\;\underline{a_{js}^{-1}}\;a_{is}^{-1} &\overset{M_2'}{=} \underline{a_{is}}(\sigma_{s-1}\cdots\sigma_{j+1}\sigma_j^{2}\sigma_{j+1}^{-1}\cdots\sigma_{s-1}^{-1})\underline{a_{is}}\\
&\quad\times(\sigma_{s-1}\cdots\sigma_{j+1}\sigma_j^{-2}\sigma_{j+1}^{-1}\cdots\sigma_{s-1}^{-1})\underline{a_{is}^{-1}}\\
&\overset{M_2'}{=} (\sigma_{s-1}\cdots\sigma_j a_{ij}\sigma_j^{-1}\cdots\sigma_{s-1}^{-1})(\sigma_{s-1}\cdots\sigma_{j+1}\sigma_j^{2}\sigma_{j+1}^{-1}\cdots\sigma_{s-1}^{-1})\\
&\quad\times(\sigma_{s-1}\cdots\sigma_j a_{ij}\sigma_j^{-1}\cdots\sigma_{s-1}^{-1})(\sigma_{s-1}\cdots\sigma_{j+1}\sigma_j^{-2}\sigma_{j+1}^{-1}\cdots\sigma_{s-1}^{-1})\\
&\quad\times(\sigma_{s-1}\cdots\sigma_j a_{ij}^{-1}\sigma_j^{-1}\cdots\sigma_{s-1}^{-1})\\
&= (\sigma_{s-1}\cdots\sigma_j)a_{ij}\sigma_j^{2}\underline{a_{ij}\sigma_j^{-2}a_{ij}^{-1}}(\sigma_j^{-1}\cdots\sigma_{s-1}^{-1}).
\end{aligned}
$$

On the other hand:

$$
\begin{aligned}
a_{ij}^{-1}\;\underline{a_{is}}\;a_{ij} &\overset{M_2'}{=} \underline{a_{ij}^{-1}\sigma_{s-1}\cdots\sigma_{j+1}}\sigma_j a_{ij}\sigma_j^{-1}\underline{\sigma_{j+1}^{-1}\cdots\sigma_{s-1}^{-1}a_{ij}}\\
&\overset{M_1'}{=} \sigma_{s-1}\cdots\sigma_{j+1}\underline{a_{ij}^{-1}\sigma_j a_{ij}\sigma_j^{-1}}a_{ij}\sigma_{j+1}^{-1}\cdots\sigma_{s-1}^{-1}\\
&\overset{M_3'}{=} \sigma_{s-1}\cdots\sigma_{j+1}\sigma_j a_{ij}\sigma_j(\sigma_j\sigma_j^{-1})a_{ij}^{-1}\sigma_j^{-2}a_{ij}(\sigma_j\sigma_j^{-1})\sigma_{j+1}^{-1}\cdots\sigma_{s-1}^{-1}\\
&= (\sigma_{s-1}\cdots\sigma_j)a_{ij}\sigma_j^{2}\underline{\sigma_j^{-1}a_{ij}^{-1}\sigma_j^{-2}a_{ij}\sigma_j}(\sigma_j^{-1}\cdots\sigma_{s-1}^{-1}).
\end{aligned}
$$

Therefore, it suffices to show that the underlined expressions in the last equation of either side are equal. Indeed we have:

$$
\begin{aligned}
a_{ij}\ \ \sigma_j^{-1}\ \ & \underline{\sigma_j{}^{-1}a_{ij}{}^{-1}(\sigma_j{}^{-1}a_{ij}{}^{-1}}\sigma_j{}^2a_{ij}\sigma_j) \\
\overset{M_3'}{=}\ & a_{ij}\underline{\sigma_j{}^{-1}a_{ij}{}^{-1}\sigma_j{}^{-1}a_{ij}{}^{-1}}\sigma_j{}^{-1}\sigma_j{}^2a_{ij}\sigma_j \\
\overset{M_3'}{=}\ & a_{ij}a_{ij}{}^{-1}\sigma_j{}^{-1}a_{ij}{}^{-1}\sigma_j{}^{-1}\sigma_ja_{ij}\sigma_j \\
=\ & 1.
\end{aligned}
$$

Finally, for P_4' and $r \in \{1, \ldots, m\}$ we have:

$$
\begin{aligned}
a_{is}\ \ a_{js}\ \ & a_{is}{}^{-1}\ a_{js}{}^{-1}\ a_{rs}\ a_{js}\ a_{is}\ a_{js}{}^{-1}\ a_{is}{}^{-1} \\
\overset{M_2'}{=}\ & (\sigma_{s-1}\ldots\sigma_ja_{ij}\sigma_j{}^{-1}\ldots\sigma_{s-1}{}^{-1})\cdot(\sigma_{s-1}\ldots\sigma_{j+1}\sigma_j{}^2\sigma_{j+1}{}^{-1}\ldots\sigma_{s-1}{}^{-1}) \\
& \times(\sigma_{s-1}\ldots\sigma_ja_{ij}{}^{-1}\sigma_j{}^{-1}\ldots\sigma_{s-1}{}^{-1})\cdot(\sigma_{s-1}\ldots\sigma_{j+1}\sigma_j{}^{-2}\sigma_{j+1}{}^{-1}\ldots\sigma_{s-1}{}^{-1}) \\
& \times(\sigma_{s-1}\ldots\sigma_ja_{rj}\sigma_j{}^{-1}\ldots\sigma_{s-1}{}^{-1})\cdot(\sigma_{s-1}\ldots\sigma_{j+1}\sigma_j{}^2\sigma_{j+1}{}^{-1}\ldots\sigma_{s-1}{}^{-1}) \\
& \times(\sigma_{s-1}\ldots\sigma_ja_{ij}\sigma_j{}^{-1}\ldots\sigma_{s-1}{}^{-1})\cdot(\sigma_{s-1}\ldots\sigma_{j+1}\sigma_j{}^{-2}\sigma_{j+1}{}^{-1}\ldots\sigma_{s-1}{}^{-1}) \\
& \times(\sigma_{s-1}\ldots\sigma_ja_{ij}{}^{-1}\sigma_j{}^{-1}\ldots\sigma_{s-1}{}^{-1}) \\
=\ & \sigma_{s-1}\ldots\sigma_{j+1}\underline{\sigma_ja_{ij}\sigma_j{}^2a_{ij}{}^{-1}\sigma_j{}^{-2}a_{rj}\sigma_j{}^2a_{ij}\sigma_j{}^{-2}a_{ij}{}^{-1}\sigma_j{}^{-1}}\sigma_{j+1}{}^{-1}\ldots\sigma_{s-1}{}^{-1}.
\end{aligned}
$$

On the other hand:

$$
\begin{aligned}
a_{ij}{}^{-1}\ \ \underline{a_{rs}}\ \ a_{ij}\ & \overset{M_2'}{=}\ \underline{a_{ij}{}^{-1}\sigma_{s-1}\ldots\sigma_{j+1}\sigma_ja_{rj}\sigma_j{}^{-1}\sigma_{j+1}{}^{-1}\ldots\sigma_{s-1}{}^{-1}a_{ij}} \\
\overset{M_1'}{=}\ & \sigma_{s-1}\ldots\sigma_{j+1}\underline{a_{ij}{}^{-1}\sigma_ja_{rj}\sigma_j{}^{-1}a_{ij}}\sigma_{j+1}{}^{-1}\ldots\sigma_{s-1}{}^{-1}.
\end{aligned}
$$

Again, it suffices to show that the underlined expressions in the last equation of either side are equal. Indeed we have:

$$
\begin{aligned}
\sigma_j\ \ a_{ij}\ \ & \sigma_j{}^2\,a_{ij}{}^{-1}\,\sigma_j{}^{-2}\,a_{rj}\,\sigma_j{}^2\,a_{ij}\,\underline{\sigma_j{}^{-2}\,a_{ij}{}^{-1}\,\sigma_j{}^{-1}\,(a_{ij}{}^{-1}}\,\sigma_j\,a_{rj}{}^{-1}\,\sigma_j{}^{-1}\,a_{ij}) \\
\overset{M_3'}{=}\ & \sigma_ja_{ij}\sigma_j{}^2a_{ij}{}^{-1}\sigma_j{}^{-2}a_{rj}\sigma_j{}^2a_{ij}a_{ij}{}^{-1}\sigma_j{}^{-1}a_{ij}{}^{-1}\sigma_j{}^{-2}\sigma_ja_{rj}{}^{-1}\sigma_j{}^{-1}a_{ij} \\
=\ & \sigma_ja_{ij}\sigma_j{}^2a_{ij}{}^{-1}\sigma_j{}^{-2}\underline{a_{rj}\sigma_ja_{ij}{}^{-1}\sigma_j{}^{-1}a_{rj}{}^{-1}}\sigma_j{}^{-1}a_{ij} \\
\overset{(4)}{=}\ & \sigma_ja_{ij}\sigma_j{}^2\underline{a_{ij}{}^{-1}\sigma_j{}^{-1}a_{ij}{}^{-1}\sigma_j{}^{-2}}a_{ij} \\
\overset{M_3'}{=}\ & \sigma_ja_{ij}\sigma_j{}^2\sigma_j{}^{-2}a_{ij}{}^{-1}\sigma_j{}^{-1}a_{ij}{}^{-1}a_{ij} \\
=\ & 1.
\end{aligned}
$$

The case where $r \in \{m+1, \ldots, m+n-1\}$ is completely analogous. □

5 Irredundant presentation for $B_{m,n}$

Looking now at the last presentation for $B_{m,n}$ we observe that relations (2) may be seen as defining relations for $1 \le i \le m$ and $j \ge m+2$, namely:

$$\begin{aligned} a^{\pm}_{i,m+2} &:= \sigma_{m+1} a^{\pm}_{i,m+1} \sigma^{-1}_{m+1}, \\ a^{\pm}_{i,m+3} &:= \sigma_{m+2}\sigma_{m+1} a^{\pm}_{i,m+1} \sigma^{-1}_{m+1}\sigma^{-1}_{m+2}, \\ &\vdots \\ a^{\pm}_{i,m+n} &:= \sigma_{m+n-1} \cdots \sigma_{m+1} a^{\pm}_{i,m+1} \sigma^{-1}_{m+1} \cdots \sigma^{-1}_{m+n-1}. \end{aligned}$$

Therefore, we want to omit further these a_{ij}'s from the list of generators, and subsequently to eliminate or simplify all relations involving these elements. Indeed, we have:

Theorem 3 *The following is a presentation for $B_{m,n}$:*

$$B_{m,n} = \left\langle \begin{array}{l} a_{1,m+1}, a_{2,m+1}, \ldots, a_{m,m+1}, \\ \sigma_{m+1}, \sigma_{m+2}, \ldots, \sigma_{m+n-1} \end{array} \middle| \begin{array}{l} \Sigma_1, \Sigma_2, (1'), (2'), (3'), \\ \textit{for all appropriate indices} \end{array} \right\rangle,$$

where:

$$\begin{array}{lrcl} (1') & {\sigma_k}^{-1} {a_{i,m+1}}^{\pm} \sigma_k &=& {a_{i,m+1}}^{\pm}, \quad k \ge m+2, \\ (2') & {a_{i,m+1}}^{\pm} \sigma_{m+1} a_{i,m+1} \sigma_{m+1} &=& \sigma_{m+1} a_{i,m+1} \sigma_{m+1} {a_{i,m+1}}^{\pm}, \\ (3') & {a_{i,m+1}}^{\pm} (\sigma_{m+1} {a_{r,m+1}}^{\pm} {\sigma_{m+1}}^{-1}) &=& (\sigma_{m+1} {a_{r,m+1}}^{\pm} {\sigma_{m+1}}^{-1}) {a_{i,m+1}}^{\pm}, \quad r < i. \end{array}$$

Proof Relations (1′) are a special case of relations (1), relations (2′) are a special case of relations (3) and relations (3′) are a special case of relations (4). Note that relations (3′) are equivalent to

$${a_{i,m+1}}^{\pm} {a_{r,m+2}}^{\pm} = {a_{r,m+2}}^{\pm} {a_{i,m+1}}^{\pm}, \quad r < i.$$

We show the sufficiency of the new relations by examining each case.

For relations (1) and for $m+1 \le k \le j-2$ we have:

$$\begin{aligned} {\sigma_k}^{-1} \underline{{a_{ij}}^{\pm}} \sigma_k &\overset{(2)}{=} \underline{{\sigma_k}^{-1} (\sigma_{j-1} \cdots \sigma_{k+2}} \sigma_{k+1} \cdots \sigma_{m+1} {a_{i,m+1}}^{\pm} \\ &\quad \times {\sigma_{m+1}}^{-1} \cdots {\sigma_{k+1}}^{-1} \underline{{\sigma_{k+2}}^{-1} \cdots {\sigma_{j-1}}^{-1}) \sigma_k} \\ &\overset{\Sigma_1}{=} (\sigma_{j-1} \cdots \sigma_{k+2}) \underline{{\sigma_k}^{-1} \sigma_{k+1} \sigma_k} \cdots \sigma_{m+1} {a_{i,m+1}}^{\pm} \\ &\quad \times {\sigma_{m+1}}^{-1} \cdots \underline{{\sigma_k}^{-1} {\sigma_{k+1}}^{-1} \sigma_k} ({\sigma_{k+2}}^{-1} \cdots {\sigma_{j-1}}^{-1}) \\ &\overset{\Sigma_2}{=} (\sigma_{j-1} \cdots \sigma_{k+2}) \sigma_{k+1} \sigma_k \underline{{\sigma_{k+1}}^{-1} \sigma_{k-1} \cdots \sigma_{m+1}} {a_{i,m+1}}^{\pm} \\ &\quad \times \underline{{\sigma_{m+1}}^{-1} \cdots {\sigma_{k-1}}^{-1} \sigma_{k+1}} {\sigma_k}^{-1} {\sigma_{k+1}}^{-1} ({\sigma_{k+2}}^{-1} \cdots {\sigma_{j-1}}^{-1}) \\ &\overset{\Sigma_1}{=} (\sigma_{j-1} \cdots \sigma_{m+1}) \underline{{\sigma_{k+1}}^{-1} {a_{i,m+1}}^{\pm} \sigma_{k+1}} ({\sigma_{m+1}}^{-1} \cdots {\sigma_{j-1}}^{-1}) \\ &\overset{(1')}{=} \sigma_{j-1} \cdots \sigma_{m+1} {a_{i,m+1}}^{\pm} {\sigma_{m+1}}^{-1} \cdots {\sigma_{j-1}}^{-1}, \ (k+1 \ge m+2) \\ &\overset{(2)}{=} {a_{ij}}^{\pm}. \end{aligned}$$

Further, for relations (1) and for $k \geq j+1$ we have:

$$\begin{aligned}
\sigma_k{}^{-1}\underline{a_{ij}}{}^{\pm}\sigma_k &\overset{(2)}{=} \underline{\sigma_k{}^{-1}(\sigma_{j-1}\ldots\sigma_{m+1}a_{i,m+1}{}^{\pm}\sigma_{m+1}{}^{-1}\ldots\sigma_{j-1}{}^{-1})\sigma_k} \\
&\overset{\Sigma_1}{=} (\sigma_{j-1}\ldots\sigma_{m+1})\underline{\sigma_k{}^{-1}a_{i,m+1}{}^{\pm}\sigma_k}(\sigma_{m+1}{}^{-1}\ldots\sigma_{j-1}{}^{-1}) \\
&\overset{(1')}{=} \sigma_{j-1}\ldots\sigma_{m+1}a_{i,m+1}{}^{\pm}\sigma_{m+1}{}^{-1}\ldots\sigma_{j-1}{}^{-1}) \\
&\overset{(2)}{=} a_{ij}{}^{\pm}.
\end{aligned}$$

In the last two equations we used that $k \geq m+2$. For relations (3) we have:

$$\begin{aligned}
\sigma_j\underline{a_{ij}}\sigma_j\underline{a_{ij}{}^{\pm}} &\overset{(2)}{=} \sigma_j(\sigma_{j-1}\ldots\sigma_{m+1}a_{i,m+1}\underline{\sigma_{m+1}{}^{-1}\ldots\sigma_{j-1}{}^{-1})\sigma_j} \\
&\quad\times \underline{(\sigma_{j-1}\ldots\sigma_{m+1}}a_{i,m+1}{}^{\pm}\sigma_{m+1}{}^{-1}\ldots\sigma_{j-1}{}^{-1}) \\
&\overset{\Sigma_2}{=} \sigma_j\ldots\sigma_{m+1}a_{i,m+1}\underline{\sigma_j}\ldots\underline{\sigma_{m+2}}\sigma_{m+1}\underline{\sigma_{m+2}{}^{-1}}\ldots\underline{\sigma_j{}^{-1}} \\
&\quad\times a_{i,m+1}{}^{\pm}\sigma_{m+1}{}^{-1}\ldots\sigma_{j-1}{}^{-1} \\
&\overset{\Sigma_1,(1')}{=} \underline{\sigma_j(\sigma_{j-1}\sigma_j)\ldots(\sigma_{m+1}\sigma_{m+2})a_{i,m+1}\sigma_{m+1}a_{i,m+1}{}^{\pm}} \\
&\quad\times (\sigma_{m+2}{}^{-1}\sigma_{m+1}{}^{-1})\ldots(\sigma_j{}^{-1}\sigma_{j-1}{}^{-1}) \\
&\overset{\Sigma_2}{=} (\sigma_{j-1}\sigma_j)\ldots(\sigma_{m+1}\sigma_{m+2})\underline{\sigma_{m+1}a_{i,m+1}\sigma_{m+1}a_{i,m+1}{}^{\pm}} \\
&\quad\times (\sigma_{m+2}{}^{-1}\sigma_{m+1}{}^{-1})\ldots(\sigma_j{}^{-1}\sigma_{j-1}{}^{-1}) \\
&\overset{(2')}{=} (\sigma_{j-1}\sigma_j)\ldots(\sigma_{m+1}\sigma_{m+2})a_{i,m+1}{}^{\pm}\sigma_{m+1}a_{i,m+1} \\
&\quad\times \underline{\sigma_{m+1}(\sigma_{m+2}{}^{-1}\sigma_{m+1}{}^{-1})\ldots(\sigma_j{}^{-1}\sigma_{j-1}{}^{-1})} \\
&\overset{\Sigma_2}{=} (\sigma_{j-1}\underline{\sigma_j})\ldots(\sigma_{m+1}\underline{\sigma_{m+2}})a_{i,m+1}{}^{\pm}\sigma_{m+1}a_{i,m+1} \\
&\quad\times (\underline{\sigma_{m+2}{}^{-1}}\sigma_{m+1}{}^{-1})\ldots(\underline{\sigma_j{}^{-1}}\sigma_{j-1}{}^{-1})\sigma_j
\end{aligned}$$

$$\begin{aligned}
&\overset{\Sigma_1,(1')}{=} (\sigma_{j-1}\ldots\sigma_{m+1})a_{i,m+1}{}^{\pm}\underline{\sigma_j\ldots\sigma_{m+2}\sigma_{m+1}\sigma_{m+2}{}^{-1}\ldots\sigma_j{}^{-1}} \\
&\quad\times a_{i,m+1}(\sigma_{m+1}{}^{-1}\ldots\sigma_{j-1}{}^{-1})\sigma_j \\
&\overset{\Sigma_2}{=} \underline{(\sigma_{j-1}\ldots\sigma_{m+1})a_{i,m+1}{}^{\pm}\sigma_{m+1}{}^{-1}\ldots\sigma_{j-1}{}^{-1}}\sigma_j \\
&\quad\times \underline{\sigma_{j-1}\ldots\sigma_{m+1}a_{i,m+1}(\sigma_{m+1}{}^{-1}\ldots\sigma_{j-1}{}^{-1})}\sigma_j \\
&\overset{(2)}{=} a_{ij}{}^{\pm}\sigma_j a_{ij}\sigma_j.
\end{aligned}$$

Finally, for relations (4) we have:

$$\begin{aligned}
a_{ij}^{\pm}\ (\sigma_j\, a_{rj}^{\pm}\, \sigma_j^{-1}) &\overset{(2)}{=} (\sigma_{j-1}\ldots\sigma_{m+1}a_{i,m+1}^{\pm}\sigma_{m+1}^{-1}\ldots\sigma_{j-1}^{-1})\sigma_j \\
&\quad \times (\sigma_{j-1}\ldots\sigma_{m+1}a_{r,m+1}^{\pm}\sigma_{m+1}^{-1}\ldots\sigma_{j-1}^{-1})\sigma_j^{-1} \\
&\overset{\Sigma_2}{=} \sigma_{j-1}\ldots\sigma_{m+1}a_{i,m+1}^{\pm}\sigma_j\ldots\sigma_{m+2}\sigma_{m+1}\sigma_{m+2}^{-1}\ldots\sigma_j^{-1}a_{r,m+1}^{\pm} \\
&\quad \times \sigma_{m+1}^{-1}\ldots\sigma_j^{-1} \\
&\overset{\Sigma_1,(1')}{=} (\sigma_{j-1}\sigma_j)\ldots(\sigma_{m+1}\sigma_{m+2})a_{i,m+1}^{\pm}\sigma_{m+1}a_{r,m+1}^{\pm} \\
&\quad \times (\sigma_{m+2}^{-1}\sigma_{m+1}^{-1})\ldots(\sigma_j^{-1}\sigma_{j-1}^{-1})\sigma_j^{-1} \\
&\overset{\Sigma_2}{=} (\sigma_{j-1}\sigma_j)\ldots(\sigma_{m+1}\sigma_{m+2})a_{i,m+1}^{\pm}(\sigma_{m+1}a_{r,m+1}^{\pm}\sigma_{m+1}^{-1}) \\
&\quad \times (\sigma_{m+2}^{-1}\sigma_{m+1}^{-1})\ldots(\sigma_j^{-1}\sigma_{j-1}^{-1}) \\
&\overset{(3')}{=} (\sigma_{j-1}\sigma_j)\ldots(\sigma_{m+1}\sigma_{m+2})\sigma_{m+1}a_{r,m+1}^{\pm}\sigma_{m+1}^{-1}a_{i,m+1}^{\pm} \\
&\quad \times (\sigma_{m+2}^{-1}\sigma_{m+1}^{-1})\ldots(\sigma_j^{-1}\sigma_{j-1}^{-1}) \\
&\overset{\Sigma_2}{=} \sigma_j(\sigma_{j-1}\sigma_j)\ldots(\sigma_{m+1}\sigma_{m+2})a_{r,m+1}^{\pm}\sigma_{m+1}^{-1}a_{i,m+1}^{\pm} \\
&\quad \times (\sigma_{m+2}^{-1}\sigma_{m+1}^{-1})\ldots(\sigma_j^{-1}\sigma_{j-1}^{-1}) \\
&\overset{\Sigma_1,(1')}{=} \sigma_j\ldots\sigma_{m+1}a_{r,m+1}^{\pm}\sigma_j\ldots\sigma_{m+2}\sigma_{m+1}^{-1}\sigma_{m+2}^{-1}\ldots\sigma_j^{-1} \\
&\quad \times a_{i,m+1}^{\pm}\sigma_{m+1}^{-1}\ldots\sigma_{j-1}^{-1} \\
&\overset{\Sigma_2}{=} \sigma_j\sigma_{j-1}\ldots\sigma_{m+1}a_{r,m+1}^{\pm}\sigma_{m+1}^{-1}\ldots\sigma_{j-1}^{-1}\sigma_j^{-1} \\
&\quad \times \sigma_{j-1}\ldots\sigma_{m+1}a_{i,m+1}^{\pm}\sigma_{m+1}^{-1}\ldots\sigma_{j-1}^{-1} \\
&\overset{(2)}{=} \sigma_j(a_{rj}^{\pm}\sigma_j^{-1}a_{ij}^{\pm}). \qquad \square
\end{aligned}$$

The system of generators in the last presentation of $B_{m,n}$ is irredundant, in the sense that no proper subset of it can generate $B_{m,n}$. In order now to simplify the notation we will relabel the generators $a_{1,m+1}, \ldots, a_{m,m+1}, \sigma_{m+1}, \ldots, \sigma_{m+n-1}$ to $a_1, \ldots, a_m, \sigma_1, \ldots, \sigma_{n-1}$ accordingly, to obtain the following, final presentation for $B_{m,n}$:

$$B_{m,n} = \left\langle \begin{array}{c} a_1, \ldots, a_m, \\ \sigma_1, \ldots, \sigma_{n-1} \end{array} \left| \begin{array}{l} \sigma_k\sigma_j = \sigma_j\sigma_k, \quad |k-j| > 1 \\ \sigma_k\sigma_{k+1}\sigma_k = \sigma_{k+1}\sigma_k\sigma_{k+1}, \quad 1 \le k \le n-1 \\ a_i\sigma_k = \sigma_k a_i, \quad k \ge 2,\ 1 \le i \le m, \\ a_i\sigma_1 a_i\sigma_1 = \sigma_1 a_i\sigma_1 a_i, \quad 1 \le i \le m \\ a_i(\sigma_1 a_r\sigma_1^{-1}) = (\sigma_1 a_r\sigma_1^{-1})a_i, \quad r < i. \end{array} \right. \right\rangle.$$

Remark 4 It is worth mentioning that the above presentation of $B_{m,n}$ is very similar to that of the Artin braid group associated to the Dynkin diagram below.

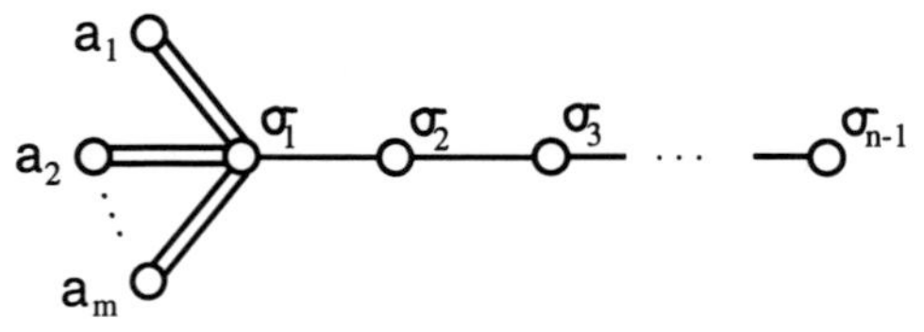

In the diagram the single bonds mean relations of degree 3, the double bonds relations of degree 4, and if two generators are not connected by a bond they commute. The two presentations differ only in the last relation, which in the case of the Artin group (cf. [2]) is a mere commutation relation between a_i and a_r. Nevertheless, in the case $m = 1$, $B_{1,n}$ is the Artin group of type $\mathcal{B}$.

6 The cosets $C_{m,n}$

In this section the word *knot* will be used to mean *knots and links.*

Let now $S^3 \backslash K$ be the complement of the oriented knot K in S^3. Obviously, $S^3 \backslash K$ can be represented in S^3 by the knot K. By classical results of Lickorish and Wallace a closed, connected, orientable 3-manifold can be obtained (not uniquely) by surgery along an integer-framed knot in S^3, so it can be represented in S^3 by this knot. We shall denote by M either a knot complement or a c.c.o. 3-manifold. Then, by fixing M we may also fix a knot in S^3 representing M, and this knot may be assumed to be a closed braid, say $\widehat{B}$. It is shown in [7] that knots in these spaces can be represented by 'mixed braids', which contain the braid B as a fixed subbraid. More precisely, we have the following:

Definition 3 *A mixed braid is a special element of B_{m+n} consisting of two disjoint orbits of strands, such that the subbraid forming the one orbit consists of the first m strands and it is a fixed element of B_m.*

If now the manifold M is a handlebody of genus m, knots in M can clearly be represented by elements of $B_{m,n}$. Further, if M is the complement of the m-unlink or a connected sum of m lens spaces of type $L(p,1)$, then knots in M are represented by elements of $B_{m,n}$, for $n \in \mathbb{N}$. In the special case where M is the complement of the trivial knot or $L(p,1)$, knots in M are represented by elements of $B_{1,n}$, the Artin group of type $\mathcal{B}$ (cf. [4], [5], [6]). These are rather special cases of knot complements or c.c.o. 3-manifolds.

In the generic case the subbraid representing M will not be the identity braid (for an example see figure 3(a)). In the generic case the multiplication of two mixed braids in B_{m+n} related to M is not a closed operation, since concatenation does not preserve the structure of the manifold.

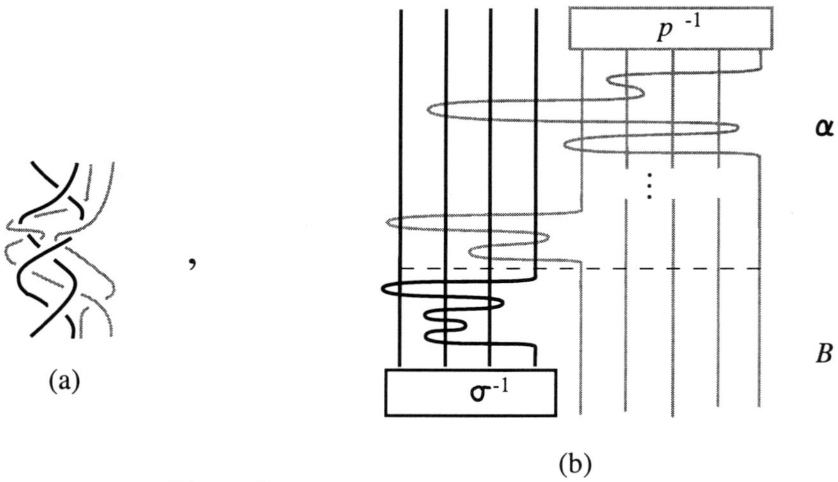

Figure 3

The following proposition shows that, nevertheless, we still have braid structures in M. Indeed, let $\bigcup_{n=1}^{\infty} C_{m,n}$ denote the disjoint union of the cosets of all mixed braids associated to a generic M. Then we have the following:

Proposition 1 *For a fixed M, $C_{m,n}$ is a coset of $B_{m,n}$ in B_{m+n}.*

Proof Let $A \in C_{m,n}$. We shall show that A can be written as a product $\alpha \cdot B$, where $\alpha \in B_{m,n}$ and B is the fixed braid representing M in S^3. Indeed, we notice first that by symmetry, Artin's combing for pure braids can be also applied starting from the last strand of a pure braid. So, we multiply A from the top with a braid p on the last n strands and with a braid σ on the first m strands, such that $pA\sigma$ is a pure braid in B_{m+n}. Then we apply to it Artin's canonical form from the end. This will separate A into two parts, one being an element of $B_{m,n}$ and the other being the fixed braid B embedded in B_{m+n} (see figure 3(b)). □

A final comment is now due.

Remark 5 The groups $B_{m,n}$ and their cosets $C_{m,n}$ will be used for yielding an algebraic version of Markov's theorem for isotopy of knots in knot complements and 3-manifolds. For the purpose of constructing knot invariants following the line of Jones-Ocneanu one can use the irredundant presentation of $B_{m,n}$ for considering appropriate quotient algebras that satisfy a quadratic skein relation for the σ_i's.

References

[1] J.S. Birman, "Braids, links and mapping class groups", Ann. of Math. Stud. **82**, Princeton University Press, Princeton, 1974.

[2] E. Brieskorn, K. Saito, Artin-Gruppen und Coxeter-Gruppen, *Inventiones Math.* **17**, 245–271 (1972).

[3] D.L. Johnson, "Presentations of Groups", LMS Student Texts **15**, 1990.

[4] S. Lambropoulou, "A study of braids in 3–manifolds", Ph.D. thesis, Warwick, 1993.

[5] S. Lambropoulou, Solid torus links and Hecke algebras of B-type, Proceedings of the Conference on Quantum Topology, D.N. Yetter ed., *World Scientific Press*, 1994.

[6] S. Lambropoulou, Knot theory related to generalized and cyclotomic Hecke algebras of type B, *J. Knot Theory and its Ramifications* **Vol. 8, No. 5**, 621–658 (1999).

[7] S. Lambropoulou, C.P. Rourke, Markov's theorem in 3-manifolds, *Topology and its Applications* **78**, 95–122 (1997).

[8] A.B. Sossinsky, Preparation theorems for isotopy invariants of links in 3-manifolds, Quantum Groups, Proceedings, *Lecture Notes in Math.* **1510**, Springer-Verlag Berlin a.o. 354–362 (1992).

[9] V.V. Vershinin, Homology of braid groups in handlebodies, Preprint No 96/06-2, Université de Nantes (1996).

S.L.: BUNSENSTRASSE 3-5, MATHEMATISCHES INSTITUT, GÖTTINGEN UNIVERSITÄT, 37073 GÖTTINGEN, GERMANY. E-MAIL: SOFIA@CFGAUSS.UNI-MATH.GWDG.DE

On the remarkable properties of the hyperbolic Whitehead link cone-manifold *

ALEXANDER D. MEDNYKH

Sobolev Institute of Mathematics, Novosibirsk, 630090, Russia

e-mail : mednykh@math.nsc.ru

Abstract

Denote by $W(m,n)$ the hyperbolic cone-manifold whose underlying space is the 3-sphere and singular geodesics are formed by two components of the Whitehead link with cone angles $2\pi/m$ and $2\pi/n$. The aim of the paper is to establish the Tangent and Sine Rules relating the complex lengthes of the singular geodesics and the cone angles of $W(m,n)$. An explicit upper bound for the real length of the singular geodesic is also given.

Keywords: hyperbolic orbifold, hyperbolic cone-manifold, complex length, the Tangent Rule, the Sine Rule

1. Introduction

A 3–dimensional *hyperbolic cone–manifold* is a 3–dimensional riemannian manifold of constant negative sectional curvature with cone-type singularity along simple closed geodesics. To each component of singular set we associate a real number $n \geq 1$ such that the cone-angle around the component is $2\pi/n$. The concept of the hyperbolic cone-manifold generalizes the hyperbolic manifold which appears in the partial case when all cone-angles are 2π. The hyperbolic cone-manifold is also a generalization of a hyperbolic 3–orbifold which arises when all associated numbers n are integers.

Denote by $W(m,n)$ the cone-manifold whose underlying space is the 3-sphere and whose singular set consists of two components of the Whitehead link with cone angles $2\pi/m$ and $2\pi/n$ (see Fig.1).

We remark that $W(n,m) = W(m,n)$ by definition and the properties of the Whitehead link.

In the partial case when $m,n \in \{3,4,\ldots\}$ the properties of $W(m,n)$ were investigated by many authors [1], [2], [3], [4]. In this case the orbifold $W(m,n)$ can be represented in the form $W(m,n) = \mathbb{H}^3/\Gamma$, where Γ is a discrete cocompact group of the isometries of the 3 –dimensional hyperbolic space $\mathbb{H}^3$ with the following presentation

$$\Gamma = \langle S,T : S^m = T^n = 1,\ SL = LS,\ L = STS^{-1}T^{-1}ST^{-1}S^{-1}T\rangle. \qquad (1.1)$$

*Supported by the RFFI (grant 99-01-00630)

H.Helling, A.Kim, J.Mennicke [1] had constructed a fundamental set for the group Γ in $\mathbb{H}^3$ in the case $m = n$ and had investigated the geometrical and arithmetical properties of 3 – dimensional hyperbolic manifolds arising as cyclic n–fold coverings of $W(n,n)$. H.M.Hilden, M.T.Lozano and J.M.Montesinos [2], [4] have investigated the hyperbolicity and arithmeticity of the orbifold $W(m,n)$ for all $m, n \in \mathbb{N}$. In the paper [5] a simple construction is given for the fundamental set of the cone-manifold $W(m,n)$; that is bounded by two pairs of curvilinear faces mutually identified by S and T.

In this paper we shall prove the following theorems.

Theorem A (The Tangent Rule). *Let $\gamma_m = \gamma_m(n)$ be a complex length of the singular geodesic of the hyperbolic cone-manifold $W(m,n)$ with cone angle $2\pi/m$. Then*

$$\frac{\tanh\frac{\gamma_m}{4}}{\tanh\frac{\gamma_n}{4}} = \frac{\tan\frac{\pi}{m}}{\tan\frac{\pi}{n}}.$$

Theorem B (The Sine Rule). *Let $\gamma_m = l_m + i\,\varphi_m$ be a complex length of the singular geodesic of the hyperbolic cone-manifold $W(m,n)$ with cone angle $2\pi/m$. Then*

$$\frac{\sin\frac{\varphi_m}{2}}{\sinh\frac{l_m}{2}} = \frac{\sin\frac{\varphi_n}{2}}{\sinh\frac{l_n}{2}}.$$

Theorem C. *Let $l_m = l_m(n)$ be a real length of the singular geodesic of the hyperbolic cone-manifold $W(m,n)$ with cone angle $2\pi/m$. Then*

$$\cosh l_m(n) < 17.$$

The upper bound is sharp and $\cosh l_m(n) \to 17$ *as* $m \to 2$ *and* $n \to \infty$.

2. The complex distances and lengthes

Let $\mathbb{H}^3 = \{(z,\xi) \in \mathbb{C} \times \mathbb{R} : \xi > 0\}$ be the upper half model of the 3 -dimensional hyperbolic space endowed by the Riemannian metric $ds^2 = \dfrac{dzd\overline{z} + d\xi^2}{\xi^2}$. We indentify the group of orientation preserving isometries of $\mathbb{H}^3$ with the group $PSL(2,\mathbb{C})$ consisting of linear fractional transformations

$$A : z \in \mathbb{C} \to \frac{az+b}{cz+d}.$$

By the canonical procedure the linear transformation A can be uniquely extended to the isometry of $\mathbb{H}^3$. We prefer to deal with the matrix $\widetilde{A} = \begin{pmatrix} a & b \\ c & d \end{pmatrix} \in SL(2,\mathbb{C})$ rather than the element $A \in PSL(2,\mathbb{C})$. The matrix $\widetilde{A}$ is uniquely up to sign determined by the element A. If there will be no confusions we shall use the same letter A for both A and $\widetilde{A}$.

Recall some definitions and results from the book [6]. A matrix $l \in SL(2,\mathbb{C})$ satisfying $tr(l) = 0$ will be called a (normalized) *line matrix*. We have from definition $l^2 = -I$, where I is the identity matrix. Hence any line matrix determines a half-turn about a line in $\mathbb{H}^3$, and this line determines matrix up to sign. According to [6, p.63] there exists a natural one-to-one correspondence between line matrices and oriented lines in $\mathbb{H}^3$. Hereby, if a line matrix l deternines an oriented line $[e, e']$ with and points e and e', then the line matrix $(-l)$ determines the line $[e', e]$. Moreover, if matrix $f \in SL(2,\mathbb{C})$ is considered as a motion of $\mathbb{H}^3$, then the matrix lfl^{-1} determines the line $[f(e), f(e')]$.

Let A and B be oriented lines determined by line matrices a and b. A complex number μ is called a *complex distance* from A to B if it real part $\Re\,\mu$ is the distance from A to B, and it imaginary part $\Im\,\mu$ is the angle from A to B in the obvious sense. We get [6, p.68]

$$\cosh \mu = -\frac{1}{2}\, tr(ab). \tag{2.1}$$

From now on, all lines in this paper are supposed to be oriented.

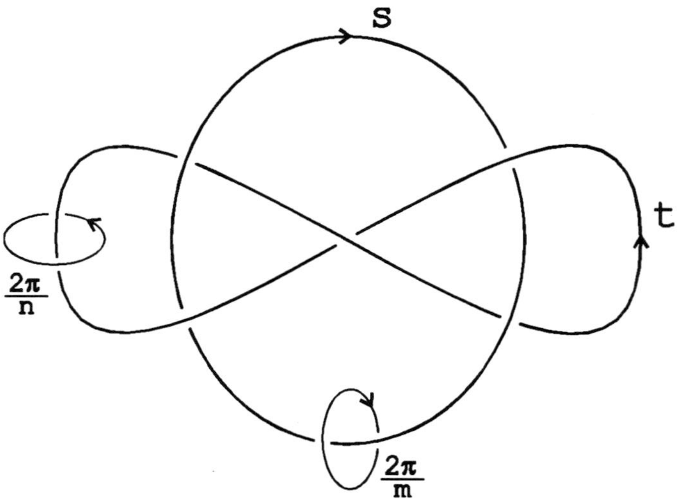

Fig.1. The Whitehead link cone-manifold $W(n, m)$

Let us investigate the hyperbolic structure of the cone-manifold $W = W(m, n)$. The singular set $\Sigma = \Sigma^1 \cup \Sigma^2$ of W consists of two components $\Sigma^1 = \Sigma^1(m)$ and $\Sigma^2 = \Sigma^2(n)$ with cone-angles $2\pi/m$ and $2\pi/n$ respectively. W defines a nonsingular but incomplete hyperbolic manifold $N = W - \Sigma$. The fundamental group of the manifold N has the following presentation

$$\Phi = \langle s, t : \; sl = ls, \; l = sts^{-1}t^{-1}st^{-1}s^{-1}t \rangle,$$

where s and t are meridians of the components Σ^1 and Σ^2 respectively, and l is a longitude of Σ^1.

The hyperbolic structure of N defines, up to congugation in $PSL(2,\mathbb{C})$, a holonomy homomorphism

$$\hat{h} = \hat{h}_{m,n} : \Phi \to PSL(2,\mathbb{C}).$$

The images $\hat{h}(s)$ and $\hat{h}(t)$ of meridians s and t are rotations in $\mathbb{H}^3$ on angles $2\pi/m$ and $2\pi/n$ respectively. It is shown in [7], see also [3] and [9], that the holonomy homomorphism of a compact orientable cone-manifold can be lifted to $SL(2,\mathbb{C})$. Denote by $h = h_{m,n} : \Phi \to SL(2,\mathbb{C})$ this lifting homomorphism and set $\Gamma = \Gamma_{m,n} = h_{m,n}(\Phi)$. The group $\Gamma = \langle S, T\rangle$ is generated by two matricies S and T with the following properties:

$$tr(S) = 2\cos\frac{\pi}{m},\ tr(T) = 2\cos\frac{\pi}{n},$$

$$SL = LS, L = STS^{-1}T^{-1}ST^{-1}S^{-1}T. \tag{2.2}$$

Recall that subgroup G of $SL(2,\mathbb{C})$ is called *elementary* if it has a finite orbit in $\mathbb{H}^3 \cup \overline{\mathbb{C}}$. We remark the group (2.2) is non-elementary and not necassary discrete. It will be shown below (see Remark 2.6) that there is a one-to-one correspondence between hyperbolic structures on the cone-manifold $W(m,n)$ and conjugacy classes of the presentations $h_{m,n} : \Phi \to SL(2,\mathbb{C})$, where the group $\Gamma = \Gamma_{m,n}$ is non-elementary and is not conjugated to a subgroup of $SL(2,\mathbb{R})$.

Introduce the following definitions:

Definition 2.1. Let A be an isometry of $\mathbb{H}^3$ different from identity. It has two fixed points u and v in $\overline{\mathbb{C}}$ which may coincide. Consider the oriented axis $[u,v]$. A complex number $\delta(A)$ is call the *displacement* of A if its real part is the distance which the axis is translated and its imaginary part is the angle through which the half–planes bounded by the axis are rotated.

The isometry A without an orientation of its axis determines $\delta(A)$ only up to sign. By [6, p.46] for the isometry given by matrix $A \in SL(2,\mathbb{C})$ we have

$$2\cosh\delta(A) = \mathrm{tr}(\mathrm{A}^2) = \mathrm{tr}^2(\mathrm{A}) - 2. \tag{2.3}$$

Definition 2.2. A *complex length* $\gamma_m = \gamma_m(n)$ of the singular component $\Sigma^1 = \Sigma^1(m)$ of the cone-manifold $W(m,n)$ is defined as displacement of the isometry L of $\mathbb{H}^3$, where $L = STS^{-1}T^{-1}ST^{-1}S^{-1}T$ is represented by the longitude of Σ^1.

Similary, a *complex lenght* $\gamma_n = \gamma_n(m)$ of the singular component $\Sigma^2 = \Sigma^2(n)$ of the cone-manifold $W(m,n)$ is defined as displacement of the isometry V of $\mathbb{H}^3$, where $V = TST^{-1}S^{-1}TS^{-1}T^{-1}S$ is represented by the longitude of Σ^2,

By [6, p.46] we have

$$\cosh\gamma_m = \frac{1}{2}\,\mathrm{tr}\,(L^2) \tag{2.4}$$

and

$$\cosh\gamma_n = \frac{1}{2}\,\mathrm{tr}\,(V^2). \tag{2.5}$$

We note [11, p.38] that the meridian-longitude pair {s, l} is uniquely determined up to a common conjugating element of the group Φ. Hence, the complex length

$\gamma_m = l_m + i\, \varphi_m$ is uniquely determined up to sign and (mod $2\pi i$) by the above definition.

We need two conventions to choose correctly real and imaginary parts of γ_m. The first convention is the following. Since $\Sigma^1(m)$ does not shrink to the point, $l_m \neq 0$. Hence, we choose γ_m in such a way that $l_m = \Re \gamma_m > 0$. The second convention is concerned with imaginary part $\varphi_m = \Im \gamma_m$. We want to choose φ_m such that the following identity holds

$$\cosh \frac{\gamma_m}{2} = -\frac{1}{2}\operatorname{tr}(L) \tag{2.6}$$

By virtue of identity $\operatorname{tr}(L)^2 - 2 = \operatorname{tr}(L^2)$ equality (2.4) is a consiquence of (2.6). The converse, in general, is true only up to sign. Under the second convention (2.4) and (2.6) are equivalent.

Remark 2.3. There is a natural geometrical reason for choosing minus in (2.6). Indeed, it will be shown in Lemma 2.7 that $L = -D^{-2}$, where D is a matrix in $SL(2, \mathbb{C})$ such that $DSD^{-1} = S$ and $DTD^{-1} = T^{-1}L$. Geometrically speaking, this means that D defines an order two isometry of the cone-manifold $W(m,n)$ corresponding to rotation along the component $\Sigma^1(m)$. Moreover, the axis of D coincides with common axis of S and L. Consider an orientable normal to axis D determined by line matrix a . The motion of $\mathbb{H}^3$ defined by matrix D moves it to the oriented line determined by DaD^{-1}. The displacement of this motion is $-\gamma_m/2$, where γ_m is displacement of the motion defined by matrix L. Hence, $-\gamma_m/2$ is the complex distance from oriented line determined by line matrix a to oriented line determined by line matrix DaD^{-1}. By formula (2.1) we get

$$\cosh(-\frac{\gamma_m}{2}) = -\frac{1}{2}\operatorname{tr}(a \cdot DaD^{-1}). \tag{2.7}$$

Since $a \cdot DaD^{-1} = aaD^{-1}D^{-1}$, $a^2 = -I$, and $(D^{-1})^2 = -L$ the equality (2.7) gives (2.6).

The two above conventions lead to convinient analytic formulas for calculation of γ_m and l_m. More precisely, there are simlple relations between these numbers and eigenvalues of matrix L. Recall that $\det L = 1$ and axis of L coincides with axis of S. Hence matrix L is not parabolic and has two eigenvalues f and $1/f$. We choose f so that $|f| > 1$. The case $|f| = 1$ is impossible because in this case the matrix L is elliptic and $l_m = 0$. We have from (2.6)

$$\frac{e^{\frac{\gamma_m}{2}} + e^{-\frac{\gamma_m}{2}}}{2} = \frac{-f - f^{-1}}{2},$$

where $|e^{\frac{\gamma_m}{2}}| = e^{\frac{l_m}{2}} > 1$ and $|f| > 1$. Hence

$$f = -e^{\frac{\gamma_m}{2}} \tag{2.8}$$

and

$$|f| = e^{\frac{l_m}{2}}. \tag{2.9}$$

Respectively, if g be an eigenvalue of matrix V such that $|g| > 1$, then

$$g = -e^{\frac{\gamma_n}{2}} \tag{2.10}$$

and

$$|g| = e^{\frac{l_n}{2}}. \tag{2.11}$$

Now we want to find a canonical representation for holonomy $h_{m,n}$ in each class of conjugacy.

Proposition 2.4. *Let $\Gamma = \langle S, T \rangle$ be a group whose generators S and T satisfy to conditions (2.2). Suppose that the group Γ is non-elementary and is not conjugated to a subgroup of $SL(2, \mathbb{R})$. Then, up to cojugation Γ in the group $SL(2, \mathbb{C})$, its generators S and T can be chosen thus that*

$$S = \begin{pmatrix} \cos\mu & i\, e^{\frac{\rho}{2}} \sin\mu \\ i\, e^{-\frac{\rho}{2}} \sin\mu & \cos\mu \end{pmatrix}, \quad T = \begin{pmatrix} \cos\nu & i\, e^{-\frac{\rho}{2}} \sin\nu \\ i\, e^{\frac{\rho}{2}} \sin\nu & \cos\nu \end{pmatrix}, \tag{2.12}$$

where $\mu = \pi/m$, $\nu = \pi/n$, ρ is a complex distance between axis of S and *T, and $u = \cosh\rho$ is a non-real root of the complex distance equation*

$$u^3 - MNu^2 + \frac{M^2N^2 + M^2 + N^2 - 1}{2} u + MN = 0, \tag{2.13}$$

$M = \cot\mu$, and $N = \cot\nu$.

Proof. After a suitable conjugation in the group $SL(2, \mathbb{C})$, one can assume that the axes of elliptic elements S and T are lines with endpoints $\{\pm e^{\frac{\rho}{2}}\}$ and $\{\pm e^{-\frac{\rho}{2}}\}$ respectively. Since $\mathrm{tr}(S) = 2\cos\mu$ and $\mathrm{tr}(T) = 2\cos\nu$, the matrices S and T are given by the formulas (2.5) . Check that ρ coincides with the complex distance $\rho(S, T)$ between oriented axes $[-e^{\frac{\rho}{2}}, e^{\frac{\rho}{2}}]$ and $[-e^{-\frac{\rho}{2}}, e^{-\frac{\rho}{2}}]$ of S and T. Indeed, the line matrices p, q, corresponding to these lines can be obtained from the matricies S,T by taking $\mu = \nu = \frac{\pi}{2}$.

We have $p = \begin{pmatrix} 0 & ie^{\frac{\rho}{2}} \\ ie^{-\frac{\rho}{2}} & 0 \end{pmatrix}$, $q = \begin{pmatrix} 0 & ie^{-\frac{\rho}{2}} \\ ie^{\frac{\rho}{2}} & 0 \end{pmatrix}$. By [6, p.68] we get $\cosh\rho(S, T) = -\frac{1}{2}\, tr(pq) = \cosh\rho$.

Now we show that $u = \cosh\rho$ is a non-real root of equation (2.13). For this purpose we have to use the condition $SL = LS$ in (2.2).

Let a be a line matrix corresponding to a common normal to axes of S and T. If matrices S and T are represented in the form (2.12) then one can take $a = \begin{pmatrix} i & 0 \\ 0 & -i \end{pmatrix}$. It is not difficult to verify that $aSa^{-1} = S^{-1}$ and $aTa^{-1} = T^{-1}$.

The following lemma gives simple criteria for matrices S and L to commute.

Lemma 2.5. *The following conditions are equivalent:* (i) $SL = LS$; (ii) $aLa^{-1} = L^{-1}$; (iii) $\mathrm{tr}(aL) = 0$.

Proof. At first we show that (i) and (ii) are equivalent. Indeed, since $L = STS^{-1}T^{-1}ST^{-1}S^{-1}T$ we get $aLa^{-1} = S^{-1}T^{-1}STS^{-1}TST^{-1} = S^{-1}L^{-1}S$. Hence

(ii) holds if and only if S and L^{-1} commute. The last property is equivalent to (i). Because of $a^2 = -I$ the condition (ii) can be rewritten in the form $aLaL = -I$; that is equivalent to (iii).

By lemma 2.5 and direct calculations we have

$$\mathrm{tr}(aL) = \frac{16 \sinh \rho}{(1+M^2)^2(1+N^2)^2} \tag{2.14}$$

$$\times(\cosh^3 \rho - MN\cosh^2 \rho + \frac{M^2N^2 + M^2 + N^2 - 1}{2}\cosh \rho + MN) = 0.$$

If $\sinh \rho = 0$ then the axes S and T coincide, and the group Γ is elementary. This contradicts to our assumption. Consiquently, $u = \cosh \rho$ is a root of equation (2.13).

Suppose that $u = \cosh \rho$ is a real root. Let

$$R(a,b,c,d) = \frac{(c-a)(d-b)}{(c-b)(d-a)}$$

be the cross ratio of four points. Then $R(-e^{\frac{\rho}{2}}, e^{\frac{\rho}{2}}, -e^{-\frac{\rho}{2}}, e^{-\frac{\rho}{2}}) = \dfrac{\cosh \rho - 1}{\cosh \rho + 1} \in \mathbb{R} \cup \{\infty\}$. We have that the axes $[-e^{\frac{\rho}{2}}, e^{\frac{\rho}{2}}]$ and $[-e^{-\frac{\rho}{2}}, e^{-\frac{\rho}{2}}]$ of S and T lie in a common plane. If the axes intersect then the group $\Gamma = \langle S, T\rangle$ has a fixed point and is elementary. If they do not intersect, Γ is conjugated to a subgroup of $SL(2,\mathbb{R})$. By assumption both cases are impossible.

We have shown that u is a non-real root of equation (2.13).

Remark 2.6. Set $z = \mathrm{tr}(S^{-1}T)$. ¿From the presentation (2.12) in Proposition 2.4 we have

$$z = 2\,(\cos \mu \cos \nu - \sin \mu \sin \nu \cosh \rho). \tag{2.15}$$

In terms of z the complex distance equation (2.13) is equivalent to the following trace equation

$$st - (s^2 + t^2 - 2)z + stz^2 - z^3 = 0, \tag{2.16}$$

where $s = \mathrm{tr}(S) = 2\cos \mu$ and $t = \mathrm{tr}(T) = 2\cos \nu$, which was found earlier in [2].

Now we want to show that the Whitehead link cone-manifold is hyperbolic if and only if the trace equation (2.16) and, hence, the complex distance equation (2.13), have a non-real root. This takes place in the region V outside of discriminant curve of the equation (2.16) which is shown in [2, Fig.2]. In particular, $W(m,n)$ is hyperbolic if $n > 2.507$ and $m > 2.507$.

Indeed, in [5] the fundamental set for cone-manifold $W(m,n)$ is constructed consisting of 16 tetrahedra having one common edge. Each of these tetrahedra does not collapse to a point because its edges can be expressed in terms of the real part $r > 0$ of the complex distance $\rho = r + i\phi$ between axis S and T and appear to be positive. Some additional arguments are needed to show that the fundamental set contains no flat or negatively oriented tetrahedra. Consider the case of equal parameters $m = n$. Based on the symmetries of the Whitehead link it was shown in [10] by elementary arguments that dihedral angles of eight tetrahedra at the common edge are equal to $\pi/2 - \theta$ while the other eight have the angles equal to $\theta - \pi/4$. As a result the sum of the angles equals

$$8(\pi/2 - \phi) + 8(\phi - \pi/4) = 2\pi.$$

We need only to show that $\pi/4 < \phi < \pi/2$. It easily follows from the fact that $t = \cos 2\phi$ is the real (and hence negative) root of the complex distance equation (2.13). In the case of m not equal to n the similar argument can be applied. The details can be found in the forthcoming paper [8].

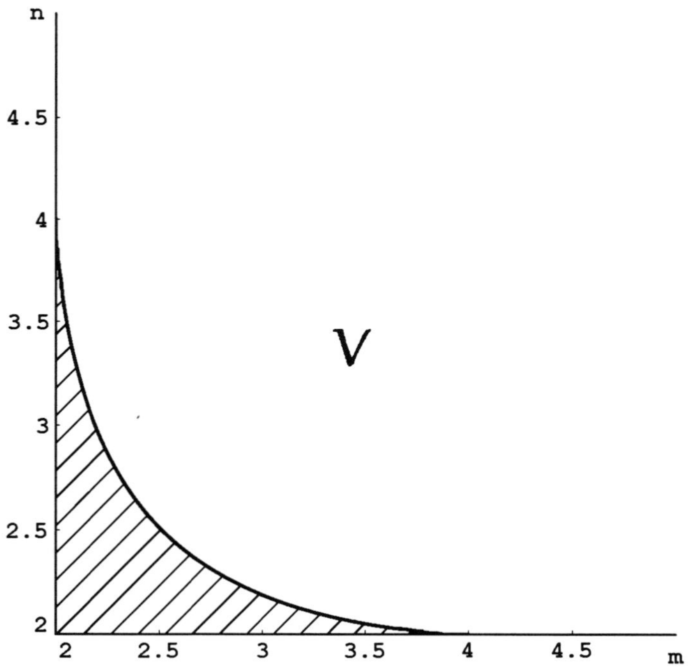

Fig.2. The region of hyperbolicity of the cone-manifold $W(m,n)$.

The next step is to find relations between values u, f and γ_m. They are given by the following two assertions.

Lemma 2.7 *Let $\Gamma = \langle S, T\rangle$ be the same as in Proposition 2.4. Suppose that $u = \cosh\rho$, where ρ is a complex distance between axes S and T chosen thus that $\Im\, u > 0$, f is an eigenvalue of the matrix L such that $|f| > 1$, and $N = \cot\frac{\pi}{n}$. Then u and f are related by the following formulas*

$$u = iN\frac{f-1}{f+1} \tag{2.17}$$

and

$$f = \frac{Ni+u}{Ni-u}. \tag{2.18}$$

Proof. After a suitable conjugation by isometry of $\mathbb{H}^3$ one can assume that the fixed points of S are 0 and ∞. In this case the generators S and T of the group Γ are given by the following matricies in $SL(2,\mathbb{C})$.

$$S = \begin{pmatrix} e^{i\frac{\pi}{m}} & 0 \\ 0 & e^{-i\frac{\pi}{m}} \end{pmatrix}, T = \begin{pmatrix} a & b \\ c & d \end{pmatrix}, \tag{2.19}$$

where $tr(T) = a + d = 2\cos\frac{\pi}{n}$.

We note [1] that the group Γ has the following outer automorphism of order 2

$$\theta : S \to S, \quad T \to T^{-1}L, \quad L \to L. \tag{2.20}$$

As all cone-angles corresponding to the region of hyperbolicity V are less than $\pi/2$ by the Rigidity theorem for cone-manifolds proved by S. Kojima [9] there exists an element $D \in SL(2,\mathbb{C})$ such that

$$DSD^{-1} = S, \quad DTD^{-1} = T^{-1}L, \quad DLD^{-1} = L. \tag{2.21}$$

Our next goal is to find matrix D explicitly. We get from (2.2), (2.20), and (2.21)

$$\begin{aligned} D^2\ S\ D^{-2} &= L^{-1}S\ L, \\ D^2\ T\ D^{-2} &= L^{-1}T\ L, \\ D^2\ L\ D^{-2} &= L^{-1}L\ L. \end{aligned} \tag{2.22}$$

We recall [12, p.90] that centralizer of non-elementary subgroup of $SL(2,\mathbb{C})$ consists of $\{\pm I\}$. Hence by virtue of (2.22) we have

$$D^2 = \pm\ L^{-1}. \tag{2.23}$$

Since $DSD^{-1} = S$ the matrix D has the form $\begin{pmatrix} k & 0 \\ 0 & k' \end{pmatrix}$ or the form $\begin{pmatrix} 0 & ki \\ k'i & 0 \end{pmatrix}$, where $kk' = 1$. It follows from (2.23) that $D^2 \neq \pm I$, hence

$$D = \begin{pmatrix} k & 0 \\ 0 & k' \end{pmatrix} \tag{2.24}$$

for some $k \in \mathbb{C}$ and $kk' = 1$.

We note that

$$DTD = \begin{pmatrix} k^2a & b \\ c & k'^2d \end{pmatrix}$$

and

$$-T^{-1} = \begin{pmatrix} -d & b \\ c & -a \end{pmatrix}. \tag{2.25}$$

From $DTD^{-1} = T^{-1}SL$ and (2.23) we have

$$DTD = \pm T^{-1}. \tag{2.26}$$

It is easy to check from (2.23) and (2.25) that the case $DTD = T^{-1}$ is impossible. Indeed, in this case $b = c = 0$ and the axes of S and T coinside. Hence

$$DTD = -T^{-1}. \tag{2.26}$$

Substituting (2.19) and (2.24) in (2.26) we get $k^2 a = -d$. The case $a = d = 0$ is excluded since $a + d = 2\cos\dfrac{\pi}{n} > 0$. So $k^2 = -d/a$ and we have from (2.23) and (2.24)

$$L^{-1} = -D^2 = \begin{pmatrix} d/a & 0 \\ 0 & a/d \end{pmatrix}.$$

Hence the eigenvalues of the matrix L are d/a and a/d.

Let ρ be a complex distance between axes of S and T. For given non-singular matrix A we set $norm(A) = \frac{1}{\sqrt{det(A)}} A$. Consider the normalized matricies $p = norm(S - S^{-1})$ and $q = norm(T - T^{-1})$ corresponding to half–turns about axes of S and T respectively. We have from (2.19) that $det(S - S^{-1}) = 4\sin^2\frac{\pi}{m}$ and

$$det(T - T^{-1}) = -(a+d)^2 + 4(ad - bc) = 4\sin^2\frac{\pi}{n}.$$

Hence $p = \begin{pmatrix} i & 0 \\ 0 & -i \end{pmatrix}$ and $q = \dfrac{1}{2\sin\frac{\pi}{n}} \begin{pmatrix} a-d & 2b \\ 2c & d-a \end{pmatrix}$. From (2.1) we obtain

$$\cosh\rho = -\frac{1}{2}\,tr(pq) = -\frac{i}{2\sin\frac{\pi}{n}}(a-d) = iN\frac{d-a}{2\cos\frac{\pi}{n}}, \tag{2.27}$$

where $N = \cot\frac{\pi}{n}$. Since $a + d = 2\cos\frac{\pi}{n}$ we get from (2.27)

$$\cosh\rho = iN\,\frac{d/a - 1}{d/a + 1}. \tag{2.28}$$

We remark that the linear fractional transformation $u = iN\dfrac{f-1}{f+1}$ maps the region $|f| > 1$ on the extended complex plane onto the upper half–plane $\Im\, u > 0$. Taking $u = \cosh\rho$ and $f = d/a$ we deduce from the formula (2.28) the assertion of the lemma.

As an immediate consequence of this lemma and formula (2.8) we have the following result.

Corollary 2.8 *Suppose that $u = \cosh\rho$, where ρ is a complex distance between axes S and T chosen thus that $\Im\, u > 0$, and γ_m is a complex length of the singular component of cone-manifold $W(m,n)$ with cone-angle $2\pi/m$. Then*

$$u = i\cot\frac{\pi}{n}\coth\frac{\gamma_m}{4}. \tag{2.29}$$

3. The Tangent Rule

In this section we shall prove the following

Theorem A (The Tangent Rule). *Let $\gamma_m = \gamma_m(n)$ be a complex length of the singular geodesic of the hyperbolic cone-manifold $W(m,n)$ with cone angle $2\pi/m$. Then*

$$\frac{\tanh \frac{\gamma_m}{4}}{\tanh \frac{\gamma_n}{4}} = \frac{\tan \frac{\pi}{m}}{\tan \frac{\pi}{n}}.$$

Proof. By Corollary 2.8 we get

$$u = i \cot \frac{\pi}{n} \coth \frac{\gamma_m}{4}.$$

Similarly, taking into account that $u = u(m,n) = u(n,m)$, we obtain

$$u = i \cot \frac{\pi}{m} \coth \frac{\gamma_n}{4}.$$

Comparing the last two formulas we have the assertion of the theorem.

4. The Sine Rule

The real and imaginary parts of singular lengthes of the cone-manifold $W(m,n)$ are related to each others by the following theorem.

Theorem B (The Sine Rule). *Let $\gamma_m = l_m + i\,\varphi_m$ be a complex length of the singular geodesic of the hyperbolic cone-manifold $W(m,n)$ with cone angle $2\pi/m$. Then*

$$\frac{\sin \frac{\varphi_m}{2}}{\sinh \frac{l_m}{2}} = \frac{\sin \frac{\varphi_n}{2}}{\sinh \frac{l_n}{2}}.$$

Proof. For given m, n we put for simplicity $l = l_m$, $\varphi = \varphi_m$, and $\gamma = \gamma_m = l + i\,\varphi$. By Corollary 2.8 we get

$$-i\frac{u}{N} = \coth \frac{\gamma}{4}, \qquad i\frac{\bar{u}}{N} = \coth \frac{\overline{\gamma}}{4}. \tag{4.1}$$

Since $l > 0$ we have $\cosh \frac{l}{2} > 1$ and $\cosh \frac{l}{2} - \cos \frac{\varphi}{2} > 0$. We set $z = \frac{\gamma}{4}$ and remark that

$$\coth \bar{z} - \coth z = \frac{\sinh(z - \bar{z})}{\sinh z \sinh \bar{z}} = \frac{2i \sin \frac{\varphi}{2}}{\cosh \frac{l}{2} - \cos \frac{\varphi}{2}} \tag{4.2}$$

and

$$\coth \bar{z} + \coth z = \frac{\sinh(\bar{z} + z)}{\sinh z \sinh \bar{z}} = \frac{2 \sinh \frac{l}{2}}{\cosh \frac{l}{2} - \cos \frac{\varphi}{2}} \tag{4.3}$$

We have also

$$i\frac{\bar{u}}{N} + i\frac{u}{N} = \frac{2i\Re\, u}{N} \tag{4.4}$$

and

$$i\frac{\bar{u}}{N} - i\frac{u}{N} = \frac{2\Im\, u}{N}. \tag{4.5}$$

Taking into account (4.1) we get from (4.2)-(4.5)

$$\frac{\sin\frac{\varphi}{2}}{\cosh\frac{l}{2}-\cos\frac{\varphi}{2}}=\frac{\Re\, u}{N} \tag{4.6}$$

$$\frac{\sinh\frac{l}{2}}{\cosh\frac{l}{2}-\cos\frac{\varphi}{2}}=\frac{\Im\, u}{N} \tag{4.7}$$

Dividing (4.6) on (4.7) we obtain

$$\frac{\sin\frac{\varphi}{2}}{\sinh\frac{l}{2}}=\frac{\sin\frac{\varphi_m}{2}}{\sinh\frac{l_m}{2}}=\frac{\Re\, u}{\Im\, u}. \tag{4.8}$$

By similar arguments observing $u=u(m,n)=u(n,m)$ we have

$$\frac{\sin\frac{\varphi_n}{2}}{\sinh\frac{l_n}{2}}=\frac{\Re\, u}{\Im\, u}. \tag{4.9}$$

Comparing the formulas (4.8) and (4.9) finish the proof of the theorem.

5. An example and further development

We illustrate the Theorems A and B by the following example.

Example 5.1. The orbifold $\mathbf{W}(3,4)$. In this case $m=3$ and $n=4$. The root $u, \Im u>0$ of the complex distance equation

$$u^3-\frac{1}{\sqrt{3}}\,u^2+\frac{1}{3}\,u+\frac{1}{\sqrt{3}}=0$$

is given by $u=\frac{1}{\sqrt{3}}+i\sqrt{\frac{2}{3}}$. By formulas (2.9) and (2.18) we have

$$e^{\frac{\gamma_3}{2}}=-f=\frac{u+Ni}{u-Ni}=i(\sqrt{3}+\sqrt{2}),$$

$$e^{\frac{\gamma_4}{2}}=-g=\frac{u+Mi}{u-Mi}=\frac{1+i}{\sqrt{2}}(\sqrt{2}+1).$$

Hence, taking into account that $\gamma=l+i\varphi$ we obtain

$$\frac{\tanh\frac{\gamma_3}{4}}{\tanh\frac{\gamma_4}{4}}=\frac{\tan\frac{\pi}{3}}{\tan\frac{\pi}{4}}=\sqrt{3},$$

$$\frac{\sin\frac{\varphi_3}{2}}{\sinh\frac{l_3}{2}}=\frac{\sin\frac{\varphi_4}{2}}{\sinh\frac{l_4}{2}}=\frac{1}{\sqrt{2}}.$$

Remark 5.2. The key lemma for proving of Theorems A and B is Lemma 2.7. For the proof of this lemma we have used only two facts— the existence of automorphism $\theta: S\to S,\, T\to T^{-1}L,\, L\to L$, induced by order two rotation of the singular link along the longitude L and Kojima's Rigidity theorem for cone-manifolds. Both these results hold for infinite family of the two bridge link cone-manifolds $\frac{p}{q}(m,n)$ with the slope

$$\frac{p}{q} = 2 + \frac{1}{r+1/2} = \frac{4r+1}{2r+1},$$

where r is a positive integer. Here, the underlying space of cone-manifold is the three-sphere, and the singular set consists of two components of $\frac{p}{q}$–link with cone-angles $2\pi/m$ and $2\pi/n$. In particular, for $r = 1$ we get the Whitehead link and for $r = 2,3,4,5$ the links 6^2_3 ,$7^2_3, 8^2_6$, and 9^2_{10} in notation of [13] respectively. Hence, the theorems similar to Theorems A and B can be obtaned also for cone-manifolds $\frac{p}{q}(m,n)$.

Remark 5.3. The Theorem A fails for the link 6^2_2 with the slope $\frac{10}{3}$. Indeed, denote by σ the singular geodesic of the orbifold $\frac{10}{3}(2,n)$ labeled by n. Then $\frac{10}{3}(2,n)$ is hyperbolic for $n \geq 4$ because its two–fold covering branched over σ is the Figure eight orbifold $\frac{5}{2}(n)$. An immediate consequence of Theorem A is that the complex length γ_n of σ is pure imaginary. This is a contradiction since σ is not a point.

6. The upper bound for the real length of singular geodesics

Denote by $W(m,\infty)$ the hyperbolic cone-manifold with a cone-angle $2\pi/m$ on the one component of the Whitehead link and with the complete hyperbolic cusp on the another.

Lemma 6.1. *For any $m > 2$ the cone-manifold $W(m,\infty)$ is hyperbolic and*

$$\cosh l_m(\infty) = \frac{M^4 + 10M^2 + 17}{(M^2+1)^2},$$

where $M = \cot\frac{\pi}{m}$.

Proof. Consider $W(m,\infty)$ as a geometric limit of cone-manifolds $W(m,n)$, where $n \to \infty$. We set $U = u/N$, where u and N are the same as in Proposition 2.4, By Lemma 2.7 the eigenvalue f of the matrix L is given by $f = \dfrac{i+U}{i-U}$

To find $l(m,\infty)$ we remark that the number $U = u/N$, $\Im U > 0$ is a complex root the equation

$$U^3 - MU^2 + (\frac{M^2+1}{2} + \frac{M^2-1}{2}N^2)\,U + \frac{M}{N^2} = 0, \tag{6.1}$$

$M = \cot\frac{\pi}{m}$ and $N = \cot\frac{\pi}{n}$. As $N \to \infty$ the last equation reduces to

$$U^3 - MU^2 + \frac{M^2+1}{2}U = 0, \tag{6.2}$$

Hence $U = \frac{1}{2}(M + i\sqrt{M^2+2})$ and

$$|f| = |\frac{i+U}{i-U}| = \frac{\sqrt{M^2+2}+1}{\sqrt{M^2+2}-1}. \tag{6.3}$$

Finaly, taking into account formula (2.4) we get from (6.3)

$$\cosh l_m(\infty) = \frac{|f|^2+|f|^{-2}}{2} = \frac{M^4+10M^2+17}{(M^2+1)^2}.$$

Lemma 6.2. *Given $m > 2$ the function $n \to \cosh l_m(n)$ is strictly increasing and*

$$\cosh l_m(n) < \cosh l_m(\infty) = \frac{M^4+10M^2+17}{(M^2+1)^2},$$

where $M = \cot\frac{\pi}{m}$.

Proof. By formulas (2.9) and (2.18) we have

$$l_m(n) = 2\,\log|f| = 2\,\log|\frac{Ni+u}{Ni-u}|.$$

The first assertion of the lemma is a routine consequence of this formula and analytic properties of linear fractional function $u \to \dfrac{Ni+u}{Ni-u}$. The contour plot of the function $\cosh l_m(n)$ is shown on the Fig.3.

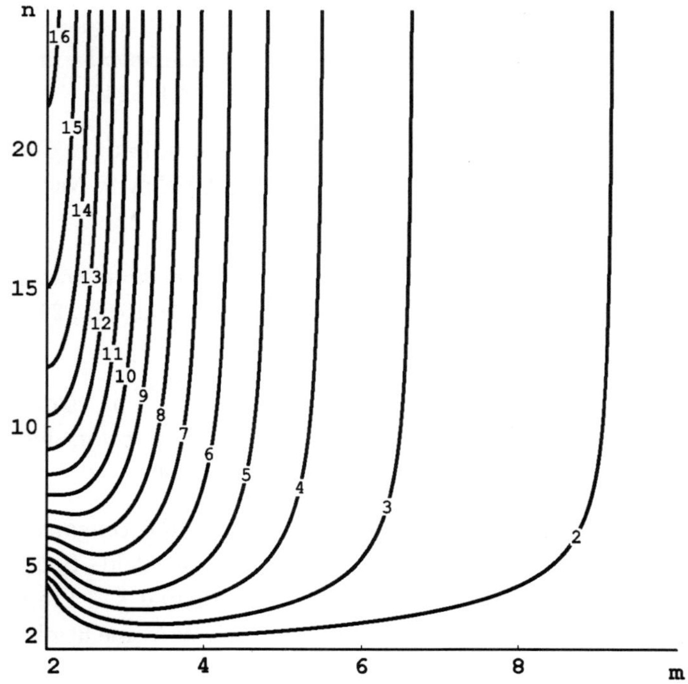

Fig.3. Contour plot of the function $\cosh l_m(n)$

The second assertion of the lemma immediately follows from the first one and Lemma 6.1.

The upper bound for the real length of singular geodesics of the cone-manifold $W(m,n)$ is given by the following theorem.

Theorem C. *Let $l_m = l_m(n)$ be a real length of the singular geodesic of the hyperbolic cone-manifold $W(m,n)$ with cone angle $2\pi/m$. Then*

$$\cosh l_m(n) < 17.$$

The upper bound is sharp and $\cosh l_m(n) \to 17$ *as $m \to 2$ and $n \to \infty$.*

Proof. By Corollary 2.8 we have

$$\cosh l_m(n) < \cosh l_m(\infty) = \frac{M^4 + 10M^2 + 17}{(M^2+1)^2} < 17,$$

Since $M = \cot\frac{\pi}{m}$, $\cosh l_m(\infty) \to 17$ as $m \to \infty$. Hence, by continuity of function $l = l_m(n)$, we obtain

$$\cosh l_m(n) \to 17$$

as $m \to 2$ and $n \to \infty$.

The author is grateful to referee for helpful remarks and suggestions.

References

[1] H. Helling, A. C. Kim, J. L. Mennicke, *Some Honey-combs in the hyperbolic 3-space,* Communications in Algebra **23** (1995), 5221–5242.

[2] H. M. Hilden, M. T. Lozano, J. M. Montesinos-Amilibia, *A Characterization of Arithmetic Subgroups of $SL(2,\mathbb{R})$ and $SL(2,\mathbb{C})$,* Math. Nach., **159** (1992), 245–270.

[3] F. González-Acuña, J. M. Montesinos-Amilibia, *On the character varieties of group representations in $SL(2,\mathbb{C})$ and $PSL(2,\mathbb{C})$,* Math. Z., **214** (1993), 627–652.

[4] H. M. Hilden, M. T. Lozano, J. M. Montesinos-Amilibia, *On the arithmetic 2-bridge knots and link orbifolds and a new knot invariant,* Journal of Knot Theory and Its Ramifications, **4** (1995), no.1, 81–114.

[5] A. Mednykh, A. Rasskazov, *On the structure of the canonical fundamental set for the 2-bridge link orbifolds,* Universität Bielefeld, Sonderforschungsbereich 343, "Discrete Structuren in der Mathematik", Preprint, 98–062, http://www.mathematik.uni-bielefeld.de/sfb343/Welcome.html.

[6] W. Fenchel, *Elementary Geometry in Hyperbolic Space,* Walter de Gruyter, Berlin–New York 1989.

[7] Qing Zhou, *The Moduli Space of Hyperbolic Cone Structures,* Preprint, 1998, math.GT/9805129, http://xxx.lanl.gov.

[8] Ann Chi Kim, A. Mednykh, *Hyperbolic structures on the Whitehead link cone-manifold,* (in preparation).

[9] S. Kojima, *Deformations of hyperbolic 3-cone-manifolds,* J. Differential Geometry, **49** (1998), no.3, 469–516.

[10] R. Shmatkov, *On a Cone-Manifold with the Euclidean Structure on the Whitehead Link,* Universität Bielefeld, Sonderforschungsbereich 343, "Discrete Structuren in der Mathematik", Preprint, 98–061, http://www.mathematik.uni-bielefeld.de/sfb343/Welcome.html.

[11] G. Burde, H. Zieschang, *Knots,* De Gruyter Studies in Mathematics, Berlin - New-York, 1985.

[12] A. F. Beardon, *The geometry of discrete groups,* Graduate Texts in Math 91, Springer-Verlag, 1983.

[13] D. Rolfsen, *Knots and links,* Publish of Perish Inc., Berkeley Ca., 1976.

MONTE CARLO EXPLORATIONS OF POLYGONAL KNOT SPACES

KENNETH C. MILLETT

Department of Mathematics
University of California, Santa Barbara
Santa Barbara, CA 93106, USA
Email: millett@math.ucsb.edu

ABSTRACT

Polygonal knots are embeddings of polygons in three space. For each n, the collection of embedded n-gons determines a subset of Euclidean space whose structure is the subject of this paper. Which knots can be constructed with a specified number of edges? What is the likelihood that a randomly chosen polygon of n-edges will be a knot of a specific topological type? At what point is a given topological type most likely as a function of the number of edges? Are the various orderings of knot types by means of "physical properties" comparable? These and related questions are discussed and supporting evidence, in many cases derived from Monte Carlo explorations, is provided.

Keywords: Monte Carlo, polygonal knots, energy, thickness, HOMFLY, geometric knots, physical knot theory

1. INTRODUCTION

The topological and geometric knotting of circles occurs in many contexts in the natural sciences: in the biology of DNA, in the chemistry of relatively small molecules, in the physics of macromolecules, in the spatial structure of trajectories of dynamical systems and line field conformations, and in complex movements of robots. Whereas topological knots have been studied from the time of Gauss, the study of geometric knots is a relatively recent event stimulated by these applications to the natural sciences. Geometric knotting, in contrast to topological knotting, concerns the imposition of geometric constraints on spatial polygons. Thus, constraints of length and of angle on allowed configurations of spatial polygons and their transformations define new notions of knotting and equivalence of knots. These constraints represent an effort to capture the local "stiffness" of molecular structures, as found in some mathematical models of DNA and other macromolecules, or other limitations on the physical system under consideration.

To systematize the study of polygonal knots it is convenient to consider functions taking the standard polygon of n sides into three-dimensional Euclidean space such that the edges go to line segments connecting the images of the corresponding vertices. Such maps are determined by a list of n distinct points, each corresponding to the image of the respective vertex of the polygon. The maps need not necessarily respect the lengths of edges nor the angles between successive edges. Each successive pair of points is therefore connected by a straight line segment, with the first and last points being connected to close the loop. For polygonal knots, care is necessary to insure that edges do not intersect except as necessary to connect adjacent edges. When a first and a second vertex are specified (giving an initial vertex and orientation of the polygon), the collection of all such maps corresponds to the product of n three-dimensional Euclidean or, equivalently, a 3n-dimensional Euclidean space. The entire collection such knots determines an open subset of this space, Geo(n). Requiring that each of the segments has length one determines a subspace of Geo(n), another knot space denoted by Equ(n), whose dimension is twice the number of vertices. This formulation of polygonal knot space and its topology is studied in the foundational papers of Dick Randell and, later, in those of Jorge Alberto Calvo.

Global properties of these knot spaces and the specific instances of knots within them provide information employed in their application to questions arising in the natural sciences mentioned earlier. The structure of these polygonal knot spaces is, however, extremely complex and resistant to direct study. In the case of lattice knot models, one method of gaining information and making quantitative estimations of underlying structure is the statistical sampling of the population of the space. The asymptotic occurrence of knotting has been a long time objective of research and is the subject of papers of Diao, Sumners, Pippenger, Whittington, Van Rensburg, and others. Prior to and concurrent with these theoretical advances, numerical studies have been undertaken.

Among the more successful methods used to investigate the asymptotic properties of knot spaces is the Metropolis Monte Carlo sampling of the space. The objective of this paper is the application of this and other similarly random methods in a new investigation of the nature of polygonal knotting. The estimations of knot types are accomplished by use of the HOMFLY polynomial. While this knot invariant does not, in general, provide a faithful representation of knot types, it is a much finer filter than those more commonly used, such as the Alexander polynomial or certain Vassiliev invariants arising as evaluations of knot polynomials. The second section of the paper reviews some of the fundamental definitions, concepts and results, and identifies some of the more central questions. It also describes some of the key relevant results of current research. The third section contains a short discussion of the key elements of the computer exploration of knot spaces. The fourth section presents the results of the Metropolis Monte Carlo studies. The fifth, and final section, focuses on conclusions derived from this study and discusses suggestions for directions of further research.

In addition to my own work some of the results reported here are due to Jorge Alberto Calvo, while others are the result of joint work with him. Still others reflect results of research underway in collaboration with Eric J. Rawdon.

2. Definitions, Fundamental Concepts, Questions and Results

The general context for studying spaces of polygonal knots is described in Randell [1, 2]. In this paper, I will employ the specific description used in Calvo [3] and in Calvo and Millett [4]. Let P_n be the standard regular polygon with n unit length edges in , equipped with a distinguished first vertex, called the base point, at $(r_n, 0)$, where $r_n = 1/\sin(\pi/n)$. Ordering the vertices in the counter clockwise direction places the mth vertex at $r_n e^{2i\pi(m-1)/n} = (r_n \cos(2\pi(m-1)/n), r_n \sin(2\pi(m-1)/n))$, thereby determining an orientation on P_n. Let Map(n) denote the space of maps from P_n into $\mathbf{R}^3$ that are linear on the edges of P_n. Such maps uniquely correspond to points in $\mathbf{R}^{3n}$ by associating to each map the list of the images of the successive vertices, beginning at the base point and proceeding in the direction of the orientation. This association provides an identification of Map(n) with the structure of a topological space, $\mathbf{R}^{3n}$, that we will use in this study. Polygonal knots are the images of those maps that are embeddings, i.e. homeomorphisms of the regular n-gon to its image. The requirement that it be an embedding adds constraints giving rise to a 3n-dimensional dense open subspace of $\mathbf{R}^{3n}$, denoted by Geo(n). The constraints determine the discriminant, the set of maps which are not embeddings, as the closure of a finite set of semi-algebraic varieties. This set separates Map(n) into a finite number of connected components. Each component defines a distinct geometric knot type. Imposing the extra condition that these embeddings be length-preserving, we obtain a 2n dimensional submanifold Equ(n) of Geo(n), whose components define equilateral knot types.

Knot spaces are typically non-compact manifolds. For example, the space of based oriented equilateral triangles is homeomorphic to $\mathbf{R}^3$x SO(3); here, the first factor corresponds to the position of the first vertex in space. In order to simplify the analysis we consider the subspaces in which the first vertex goes to the origin. These are denoted by $Map_0(n)$, $Geo_0(n)$, and $Equ_0(n)$ respectively. In this paper we will also study the subset of Geo(n) whose vertices lie within the cube $[0,1]^3$. It is denoted by Cube(n).

The path components of Geo(n), Cube(n), and Equ(n) define potentially distinct theories of geometric knotting as well as models of interest in different natural contexts. For each n, we know that there are only finitely many geometric knot types [1, 5]. Note also that Cube(n) and Equ_0 (n) have compact closures in Map(n) and $Map_0(n)$, respectively. In order to provide a compact model in which to study the structure of Geo(n) we observe that Geo(n) has a radial structure that respects geometric knot types. Any expanding or contracting homothety preserves the geometric knot type and is isotopic to the identity. Thus, any representative of a geometric knot type is equivalent to one whose vertices determine points lying on the unit sphere in $\mathbf{R}^{3n}$. Furthermore, the proportion of geometric knots in Geo(n), the proportion in a ball in Geo(n) centered at the origin, or the proportion lying on a sphere in Geo(n) centered at the origin, are equal.

The function which assigns to each point in Map(n) the sum of the lengths of the edges of the image polygon defines a family of "level hypersurfaces" consisting of polygons of constant total length. For any non-zero total length, the corresponding hypersurface meets each ray from the origin exactly once thereby defining a diffeomorphism of the level surface to the standard sphere. As a consequence, each geometric knot is equivalent to one lying on the smooth sphere of images of total edge length equal to one. With an appropriate normalization of edge lengths, Equ(n) can be realized as a 2n-dimensional subset of this (3n - 1)-dimensional smooth sphere of geometric knots.

An important consequence of the discussion in the previous two paragraphs is that, in Metropolis Monte Carlo explorations of Geo(n), one may require that knots correspond to a collection of vertices lying within a specified ball. The proportion of their occurrence is equally an estimation of the proportion of the ball or in its boundary sphere consisting of knots of this type.

As soon as one begins to look closely at geometric and equilateral knots, three initial questions are posed.

QUESTION 1. For each topological knot type, K, what is the smallest n for which there is a realization with n edges?

This has been variously called the “stick number,” “edge number,” “broken line number,” or “minimal polygon index” of the topological knot type. In this paper we will employ the term “polygon number” for this concept.

QUESTION 2. Is the polygon number the same for geometric knots and equilateral knots?

QUESTION 3. How many distinct geometric or equilateral knot types are there as a function of n, the number of edges?

Cube(n), Geo(n) and, Equ(n) are connected for $n \leq 5$. This is a "folk" result. One nice proof is attributed to Nicholas Kuiper [1, 2]. Stimulated by discussions with biologists, chemists, and, most directly, by participation in a 1991 NATO conference on “Topological Fluid Dynamics” organized by Moffat, Zaslavsky, Comte, and Tabor [6], I first constructed equilateral models of the hexagonal trefoil knot, the heptagonal figure eight knot, and several eight crossing knots as well as a variety of more complex knots. I also began to study “random knotting” with the assistance of undergraduate summer interns. One of these was Jorge Alberto Calvo whose doctoral research results are discussed later in this paper. Colin Adams has also directed summer undergraduate research projects which have provided important contributions to the determination of the polygon number [7, 8, 9], especially for composite knots. The polygon number has also been explored by Jin in connection with Kuiper’s super bridge number [10, 11].

In his 1998 UCSB Ph.D. dissertation research [3,12,13], Calvo showed that, for n = 6, each of these spaces have exactly five components. That the set of topological unknots in Equ(6) and Geo(6) is connected is an earlier result of unpublished joint work with Rosa Orellana during summer 1993. This was discussed in [14]. Calvo has provided a new proof of this basic result in his thesis as part of his complete analysis of these spaces. He shows that there is one component of topologically trivial knots and two each of right handed and left handed trefoils. He also shows that Geo(7) has exactly five components. There is one component each of trivial knots and right and left hand trefoils and two components of figure eight knots. For n = 8, by direct construction there are known to be at least twenty components since only the unknot, the figure eight, 6_3, and the sum of the left and right trefoil knots (the square knot) are achiral among the twelve constructed topological knot types. In his thesis, Calvo showed that these, and possibly, the knot 8_{18} are the only topological knot types that can be realized in Geo(8). Subsequently, he proved that 8_{18} cannot occur. Monte Carlo searches have provided realizations of all these knot types in Equ(8) with the exceptions of 3.1#3.1 and 8_{19}.

In her 1997 Ph.D. dissertation, Monica Meissen, [15, 16], constructed nine-gon realizations of all seven crossing prime knots as well as 8_{21}. In addition, nine-gon realizations have been achieved for 8_{16}, 8_{17}, 8_{18}, 8_{21}, 9_{40}, 9_{41}, 9_{42}, and 9_{46}. As a consequence of Calvo's results, all of these knots must have geometric polygon number equal to nine.

I have employed several random knot generation and sampling strategies in order to provide some data relevant to these questions and to estimations of the complexity of knot spaces for larger numbers of vertices, including estimates of asymptotic properties. These data also give one rough sketch of the gross structure of these spaces for smaller numbers of vertices [14]. In the case of Cube(n), randomly generated coordinates of the vertices were selected with respect to the uniform distribution on the cube. In the case of Geo(n), n points in the unit three ball are selected with respect to the uniform distribution and collectively rescaled to define a point on the unit sphere in R^{3n}. In the case of Equ(n), the pivot transformation with randomly selected pairs of vertices and randomly selected angles was used. The pivot transforms the map by fixing one of the pieces determined by the pair of vertices and rotating the image of the other piece about the axis determined by the images of the designated vertices by the given angle. Up to Euclidean rotation, about the same axis, the map is equivalent to one given by reversing the roles of the two pieces. If the map is an embedding and the angle is sufficiently small, the result of the pivot is an equivalent embedding. In order to insure that one could sample the entire equilateral knot space, a couple of theorems are required.

THEOREM 1. [14] For any two maps in Map(n) there is a finite sequence of translations, rotations, and pivots taking one map to the other.

The proof of this result can be modified to prove the following theorem.

THEOREM 2. Any path connecting two maps in Map(n) can be approximated as closely as desired by a sequence of pivots.

The first theorem insures that any equilateral knot can be connected to the standard regular polygon by a sequence of these operations. The second is used to seek optimal representations of a geometric knot type by random perturbations of an instance of the knot. It implies that any knot equivalent representative can be reached by some sequence of pivots.

QUESTION 4. What proportion of knot space consists of knots of a given topological or geometric type?

In effect, we are asking, "If a polygonal knot is chosen at random, what is the likelihood that a knot of this type will be selected?" In the context of ring polymers, the asymptotic behavior of the probability of knotting has had a long history. Frisch and Wasserman [17] and Delbruck [18] conjectured that the asymptotic probability of knotting is one. This conjecture has been the subject of both numerical and well as theoretical study in a variety of contexts. In 1988 Sumners and Whittington [19] and, independently, Pippenger [20], established the conjecture in the context of the simple cubic lattice. Diao, Pippenger, and Sumners established the conjecture in the case of Gaussian random polygons [21]. Diao later studied the structure of equilateral spatial knots as well as minimal knots in the cubic lattice [22, 23]. Data from the current study concerning the proportion of distinct HOMFLY [24, 25, 26] polynomials, corresponding to specific knot types or small families of knot types, is reported later in the paper. Beyond the cases studied by Calvo, the question of geometric knot types is currently intractable due to the absence of effective invariants distinguishing geometric knot types. In Calvo's study of hexagonal knots and heptagonal knots, volume preserving involutions of the knot space demonstrate that all components of the a given topological knot type occupy the same proportion of the knot space.

QUESTION 5. How does the likelihood of a knot type depend on the number of edges? For example, how do the probabilities of the unknot, of prime and of composite knots depend on the number of edges?

In the cubic lattice, the likelihood of having a trefoil summand, or perhaps any other prime knot, goes to one as the number of edges goes to infinity. Van Rensburg and Whittington [27, 28] have estimated the probability, P(n), that an n-gon is knotted is given by $P(n) = 1 - \exp(-\alpha n + o(n))$ where $\alpha = (7.6 \pm 0.9)\text{x}10^{-6}$. Another approach is to consider "Gaussian" random polygons in which the edge lengths satisfy a Gaussian distribution. Diao, Pippenger and Sumners [21] have shown that the probability of knotting tends to one exponentially with n. Specifically, they prove "There exists a constant $\varepsilon > 0$ such that P(n) is at least $1 - \exp(-n^{\varepsilon})$ provided that n is large enough."

In this study I have used the HOMFLY polynomial as a filter to attempt to identify (families of) topological knot types. In estimations of knot types I will not identify chiral presentations, as least to the extent that they are distinguished by their HOMFLY polynomials. Data will be presented concerning likelihood of a knot type for knots through the eight crossing knots as well as a selection of composite knots. The number of edges for which a given knot type occurs with its maximal relative probability as well as the numerical likelihood of the knot type occurring as a 16 edge knot will be also reported. An interesting analysis occurs when regarding these as providing parameters by which knot types can be given a linear order much in the same way that knots have been ordered by their minimal number of crossing representations. The optimal relative occurrence provides a measure suggesting "ideal" polygonal representations analogous to minimal crossing number or minimal polygon number representations. This line of study is the setting of the next question.

QUESTION 6. For each fixed n, what are their "ideal" configurations associated to each geometric knot type?

The question of what one means by an "ideal" configuration is one of great complexity and importance [29]. It is, in fact, the subject of an entire book [4]. One reason for the importance of this question is the possibility that the properties of ideal configurations correspond to physically observed behaviors of macromolecules under a wide range of conditions. An example of this is the electrophoretic mobility of DNA knots. Another is the desire to exhibit some intrinsic symmetries that have been observed physically. Thus, a central question is to provide an appropriate definition of the "ideal configuration".

One approach has been to define a function of the configuration such as an energy, a thickness, or another property of the specific spatial conformation to be optimized through deformations that respect the relevant structure determining the knot space. A key property of an energy function, for example, is being a positive function on the space of knots that it tends to infinity as singular maps are approached. Minimal energy configurations are discrete and finite in number. These can, in theory, be determined and compared in an effort to select one that is "ideal." An interesting historic objective was the creation of an energy such that a topological unknot would "flow" to a standard configuration by moving in a reverse gradient direction toward a minimal energy conformation, the standard unknot. While this approach has yet to bear definitive theoretical fruit, there have been many proposals for "energies" for smooth and polygonal knots. A particularly interesting family of functions have been proposed by O'Hara [30, 31, 32, 33] and Freedman, He, and Wang [34]. Kusner and Sullivan have proposed a mobius energy [35]. Another has been proposed for polygonal knots by Jonathon Simon, [36, 37]: the sum, over non-adjacent edges of the polygon, of the product of the lengths of the edges divided by the square of the minimum distance in three-space between the edges. By subtracting an intrinsic term from the Simon's energy, for example given by summing the product of the lengths of the edges divided by the square of the minimal arc distance along the portion of the knot connecting the two non-adjacent edges. This defines a discrete version of the Freedman, He, and Wang energy. The resulting energy function, denoted by $\varepsilon(K)$, has the attractive property that it is finite for polygonal knots.

However, it is not a "good" polygonal knot energy in the sense proposed by Diao, Ernst, and van Rensburg [38, 99] as it fails to be "asymptotically finite." This means that sequence of values $\varepsilon(K)$ fails to converge under subdivision of the polygonal knot K.

Another example of an "ideal property" is the thickness of the polygonal knot. The "thickness" of a smooth knot is the radius of the thickest tube, or tubular neighborhood consisting of normal non intersecting discs, whose axial curve is a representative of the knot type of unit length [40, 41, 42]. In this paper we will use Eric Rawdon's [43, 44, 45] definition of thickness for polygonal knots. It has the property that the thickness of inscribed polygonal approximations of a smooth curve approach the thickness of the curve. For a polygon of unit length, Rawdon's thickness, $\tau(K)$, is the minimum of a polygonal version of the doubly-critical self-distance function (distances between two points on the curve) and a minimal radial distance arising from adjacent edges due to curvature. For non unit length polygons, the value is divided by the length of the polygon. For many purposes the "length" of a knot, defined to be the reciprocal of the thickness, appears to possess a more natural structure and is preferred.

Our investigations of these quantities make an allowed random perturbation of the configuration, calculate the quantity to be optimized, compare it to the current value, and keep the new value and configuration if it represents an improvement over the currently held value. In order to improve the effectiveness of this simple method, we keep track of the nature of the perturbation and increase or decrease its magnitude and apply it again if an improvement has occurred. One might think of this as a choice of step size along a randomly selected path in knot space. We have not used some of the more complex methods associated with annealing approaches to optimization as this approach has proved to be adequately effective. Rather than seeking "ideal" knot conformations,
I search for "optimal" knot conformations with respect to a specific criterion. This search for optimal knots was be taken over conformations within a specific equilateral knot type or over families of related knot types, i.e. all representing a given topological knot type for varying numbers of edges. The question of optimization across all equilateral knot types representing the same topological knot type is being explored in ongoing research.

Thus, as the number of edges vary, another notion of optimization arises by seeking the maximum of the relative probability, for each fixed n, of the topological knot type. For a given optimal energy or thickness knot configuration we will also look at such physical knot theoretical properties as the average crossing number and the average writhe. These are averages over all projections of the given configuration of the number of crossings and, respectively, the signed sum of the crossing types. A crossing is given the value of 1 or -1 as shown in the Figure 1. Note that, for a knot, this value does not depend on the orientation of the knot.

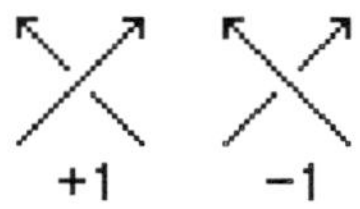

FIGURE 1. Crossing Convention

3. FOUNDATIONAL ASPECTS OF COMPUTER EXPLORATIONS OF KNOT SPACE

One important tool in the study of knot spaces has been the calculation of knot invariants. Historically, the Alexander polynomial has been the principal invariant used. It continues to be the most popular invariant, due to the relative ease of its calculation. Since 1984, however, other alternatives have become possible with the creation of the Jones polynomial and its successors, the HOMFLY polynomial, the Kauffman polynomial and, more recently, the "quantum" and Vassiliev finite type invariants [46, 47]. While, in theory, these are impractical due to the complexity of their computation, they have proven remarkably effective in our studies. The program used in this project was developed in collaboration with Bruce Ewing [48, 49]. It is able to calculate the HOMFLY polynomial for knots with up to 120 crossings. The HOMFLY polynomial is a finite Laurent polynomial in two variables, l and m, with integer coefficients, which is associated to each topological knot type. For the 2977 knot represented with fewer than 13 crossings there are only 76 cases that have the same first term as the trivial knot. By considering the entire invariant, these are easily eliminated. There are small families of knots having the same invariant and most, but not all, chiral knots are distinguished by the HOMFLY polynomial.

Many of the quantitative questions such as the probability of knotting or polygon number for a topological knot type are extremely difficult if not intractable through traditional geometric or algebraic topological methods. The Monte Carlo method [50], therefore, remains the most accessible. There are several obstacles to its application, however. First is the question of taking a uniformly distributed random sample of the space of knots.

In the case of Geo(n) or Cube(n), the Monte Carlo approach is quite straight forward. With respect to the uniform distribution on [0,1], one selects a list of 3n numbers to represent the coordinates of the n vertices of the polygon in Cube(n). For Geo(n), one randomly selects points n points on the unit sphere with respect to the uniform distribution and randomly selects n-1 radii randomly with respect to the uniform distribution on [0,1]. By allowing the first vertex to have magnitude one and the remaining vertices to lie within the unit ball with radii determined above, one has identified the vertex of a polygon in Geo(n) of largest magnitude, applied a homothety to place the polygon within the unit ball in 3-space, and used this largest vertex as the starting vertex. Direct calculation of the desired quantity, for example the curvature or the HOMFLY polynomial, provides the necessary data.

In the case of Equ(n) a Monte Carlo strategy is also possible. Using the pivot transformation with randomly selected axis vertices and rotation angle as described above, one can act upon the polygonal maps preserving the length and the linearity of the edges. Theorem 1 states that every polygonal knot can be reached (the transformation is irreducible) and every walk has the same probability of occurring (the transformation is reversible). Unfortunately, in order to insure that sampling by means of a pivot transformation is uniform with respect to the intrinsic measure, one needs to know that the pivot transformation is measure preserving. This, however, has not been demonstrated. So, although, I will later present results gained from the pivot sampling they will be representative of a limit measure that may differ from some more natural or intrinsic measure on the knot space.

QUESTION 7. What is the limit measure associated to the pivot transformation on Equ(n)?

Another context in which one can employ the pivot transformation is the study of ideal configurations. In this setting one uses a variation of the pivot transformation in which one dynamically selects, as a function of the current polygon, the range of the random rotation so as to insure that the geometric knot type is preserved throughout the rotation. In effect, one selects a short path determining a motion of the knot that does not change its knot type. Such transformations can be used to minimize energy, thickness, etc. This research is the subject of a study undertaken in collaboration with Eric Rawdon, [51]. Some preliminary elements of this research will be reported in the next section.

In all of the computer simulations and Metropolis Monte Carlo explorations extreme care must be exercised in the interpretation of the results. One example is the calculation of the HOMFLY polynomial in which integer algebra must be employed. A more delicate dimension is the determination of the topological knot type via the calculation of the HOMFLY polynomial or in the modification of an equilateral knot via the pivot transformation. All of these have implicit approximations as part of the method. Thus one must verify that the program is sufficiently accurate for the purpose of the study. A concrete example is provided by the creation of equilateral knots or their optimization with respect to some criterion. The computer approximation must be sufficiently accurate so as to insure that, although the precise data may not represent an equilateral knot, there is a nearby geometrically equivalent knot which is equilateral. Furthermore, transformations of "nearly equilateral" geometric knots imply that there exists a nearby transformation of the associated nearby equilateral knots. All this is necessary, of course, in order to be able to demonstrate that computer created nearly equilateral knots imply the existence of equivalent equilateral knots. These results will be found in the report the collaboration with Rawdon, [51].

4. The Results of the Metropolis Monte Carlo Studies

In this section, I will describe results of several recent simulations and numerical calculations as they concern the questions discussed above.

Question 1: What is the smallest number of edges required for an equilateral or geometric knot to realize a given topological knot type? A recent report is provided in Calvo and Millett [4]. Since then further progress has occurred, both Monte Carlo and theoretical. The consequences of this work is reflected in Tables 1 and 2.

Question 2: Is the polygon number the same for geometric and equilateral knots?
Question 3: How many distinct geometric or equilateral knot types are there as a function of n, the number of edges? The data represented in Figures 2 and 3 provide evidence in support of the conjecture that there are differences between the polygon numbers for geometric and equilateral knots. These data are also consistent with the exponential growth in distinct knots expected for the two knot theories. With respect to question 1, however, there is a remote possibility that the differences in numbers of knot types observed is an artifact of very small relative probabilities for equilateral representatives of the topological knot type rather than the non existence of representatives. An explicit example of this difference is required in order to settle this question.

These data are derived from a specific sample and are not a prediction of the actual number of distinct HOMFLY polynomials occurring for equilateral or geometric knots with a specific number of edges. The larger number of distinct polynomials observed for 24-gons is the result of a significantly larger sample size. The prediction of the total number of distinct polynomials is a more delicate question. An example of this later question is the effort to predict the number of distinct HOMFLY polynomials that arise for equilateral 50 edge knots. The growth in the number of distinct types can analyzed as a function on the number of observations and this sequence of data used to estimate the total number of types. The graph of the data is shown in Figure 4 from Calvo and Millett [4].

In that paper, we used this data to give an estimate of about 3,472 distinct knot polynomials occurring for knots in Equ(50). Of this, 2935 gave been observed. This is significantly smaller than the more than 124,359 HOMFLY polynomials of geometric knots that have been observed in Cube(32) but is comparable to the number observed for geometric knots with 18 to 19 edges. In order to give a credible prediction for either the equilateral or geometric knots a similar analysis would have to be undertaken for each of the sample points.

Question 4: What proportion of the knot space consists of knots of each type?" The estimation of an answer to this question, for any given knot type and knot space, requires a substantial collection of Monte Carlo studies and an analysis of the data for each knot type or HOMFLY polynomial, as is the case here. For Cube(n), the current status of calculation is shown in Figures 5, 6, 7, and 8 for selected knot types.

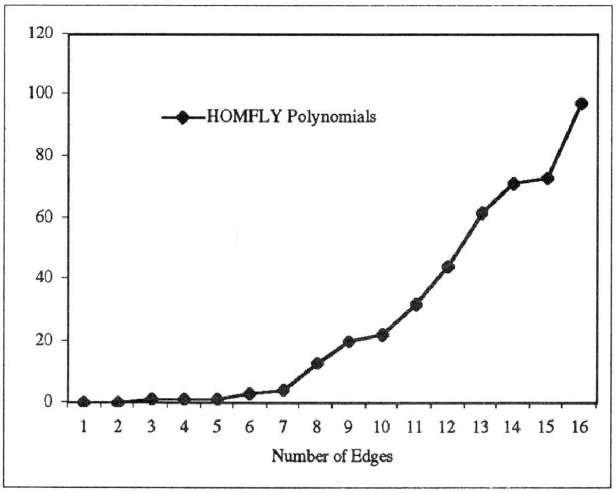

FIGURE 2. Knot Growth in Equ(n)

TABLE 1: Observed geometric polygon numbers p(.) for knots with nine or fewer crossings. Stars (*) indicate cases for which the minimal polygon number is shown.

K	p(K)	K	p(K)	K	p(K)
0	3*	8_{13}	10	9_{22}	12
		8_{14}	10	9_{23}	13
3_1	6*	8_{15}	12	9_{24}	12
		8_{16}	9*	9_{25}	13
4_1	7*	8_{17}	9*	9_{26}	12
		8_{18}	9*	9_{27}	12
5_1	8*	8_{19}	8*	9_{28}	12
5_2	8*	8_{20}	8*	9_{29}	12
		8_{21}	9*	9_{30}	13
6_1	8*	$3_1 + 5_1$	11	9_{31}	13
6_2	8*	$3_1 - 5_1$	11	9_{32}	12
6_3	8*	$3_1 + 5_2$	12	9_{33}	12
$3_1 + 3_1$	8*	$3_1 - 5_2$	11	9_{34}	12
$3_1 - 3_1$	8*	$4_1 + 4_1$	11	9_{35}	13
				9_{36}	14
7_1	9*	9_1	13	9_{37}	14
7_2	9*	9_2	14	9_{38}	14
7_3	9*	9_3	12	9_{39}	13
7_4	9*	9_4	14	9_{40}	9*
7_5	9*	9_5	13	9_{41}	9*
7_6	9*	9_6	12	9_{42}	9*
7_7	9*	9_7	12	9_{43}	10
$3_1 + 4_1$	9*	9_8	13	9_{44}	10
		9_9	13	9_{45}	10
8_1	11	9_{10}	13	9_{46}	9*
8_2	10	9_{11}	12	9_{47}	12
8_3	11	9_{12}	12	9_{48}	10
8_4	11	9_{13}	13	9_{49}	10
8_5	12	9_{14}	10	$3_1 + 6_1$	12
8_6	11	9_{15}	14	$3_1 - 6_1$	11
8_7	11	9_{16}	14	$3_1 + 6_2$	13
8_8	10	9_{17}	13	$3_1 - 6_2$	12
8_9	10	9_{18}	13	$3_1 + 6_3$	13
8_{10}	10	9_{19}	13	$4_1 + 5_1$	14
8_{11}	10	9_{20}	13	$4_1 + 5_2$	14
8_{12}	10	9_{21}	13	$3_1 + 3_1 \pm 3_1$	10*

TABLE 2: Observed equilateral polygon numbers ep(.) for knots with nine or fewer crossings. Stars (*) indicate cases for which the minimal polygon number is shown.

K	ep(K)	K	ep(K)	K	ep(K)
0	3*	8_{13}	12	9_{22}	??
		8_{14}	12	9_{23}	??
3_1	6*	8_{15}	16	9_{24}	??
		8_{16}	13	9_{25}	??
4_1	7*	8_{17}	13	9_{26}	19
		8_{18}	??	9_{27}	18
5_1	8*	8_{19}	9	9_{28}	??
5_2	8*	8_{20}	9	9_{29}	??
		8_{21}	10	9_{30}	15
6_1	8*	$3_1 + 5_1$	14	9_{31}	??
6_2	8*	$3_1 - 5_1$	15	9_{32}	16
6_3	8*	$3_1 + 5_2$	14	9_{33}	??
$3_1 + 3_1$	9	$3_1 - 5_2$	16	9_{34}	??
$3_1 - 3_1$	9	$4_1 + 4_1$	15	9_{35}	??
				9_{36}	18
7_1	9*	9_1	??	9_{37}	??
7_2	11	9_2	??	9_{38}	??
7_3	11	9_3	??	9_{39}	16
7_4	12	9_4	??	9_{40}	??
7_5	11	9_5	??	9_{41}	??
7_6	11	9_6	??	9_{42}	9*
7_7	11	9_7	??	9_{43}	13
$3_1 + 4_1$	11	9_8	??	9_{44}	13
		9_9	??	9_{45}	11
8_1	14	9_{10}	??	9_{46}	12
8_2	12	9_{11}	??	9_{47}	14
8_3	14	9_{12}	15	9_{48}	14
8_4	13	9_{13}	16	9_{49}	16
8_5	13	9_{14}	??	$3_1 + 6_1$	15
8_6	13	9_{15}	??	$3_1 - 6_1$	??
8_7	13	9_{16}	??	$3_1 + 6_2$	??
8_8	14	9_{17}	16	$3_1 - 6_2$	??
8_9	16	9_{18}	??	$3_1 + 6_3$	??
8_{10}	13	9_{19}	18	$4_1 + 5_1$	??
8_{11}	12	9_{20}	18	$4_1 + 5_2$	??
8_{12}	12	9_{21}	??	$3_1 + 3_1 \pm 3_1$	??

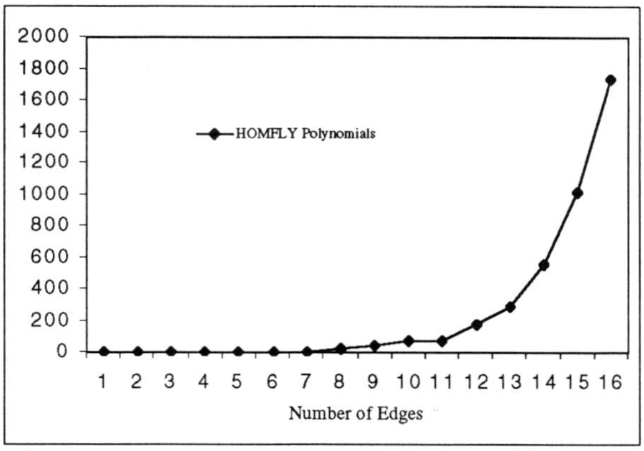

FIGURE 3. Knot Growth in Cube(n)

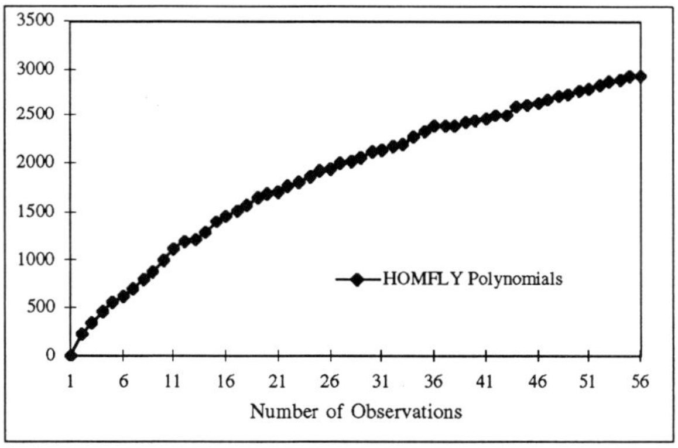

FIGURE 4. Distinct HOMFLY Polynomials of 50 Edge Knots

Question 5: How does the likelihood of a knot type depend on the number of edges? For example, how do the probabilities of randomly choosing an unknot, a prime or a composite knots occurring depend on the number of edges?

For Cube(n), the data shown in the previous figures are qualitatively consistent with the behavior exhibited in lattices and the case of Gaussian random polygons explored by Deguchi and Tsurusaki [52, 53]. Taking the perspective of scaling in statistical mechanics and considering the context of critical phenomena of second order phase transitions, they propose asymptotic formulae for random knotting, for a knot type K, of the form

$$P_K[n - 4] = C_K(n - 4)^{\upsilon K}\mathrm{Exp}[-\kappa_K(n - 4)].$$

An analysis of the data collected in this study shows that is model is not sufficient to faithfully represent the data. Instead, it appears that a quadratic term in the exponential is required. At this time, the reason for this term is unexplained. Thus, the model proposed in this paper is

$$P_K[n - 4] = C_K(n - 4)^{\upsilon_K}\mathrm{Exp}[\kappa_K(n - 4) + \delta_K(n - 4)^2].$$

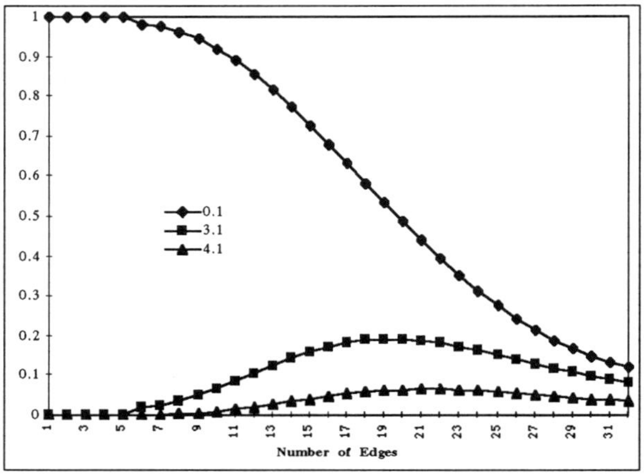

FIGURE 5. Proportion of Knot Types in Cube(n)

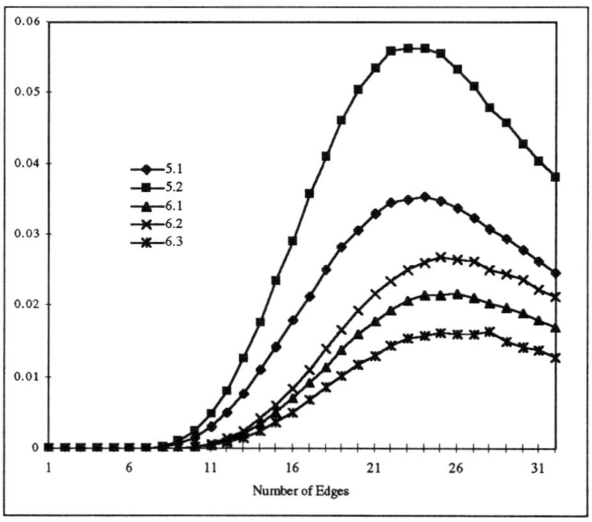

FIGURE 6. Proportion of Knot Types in Cube(n)

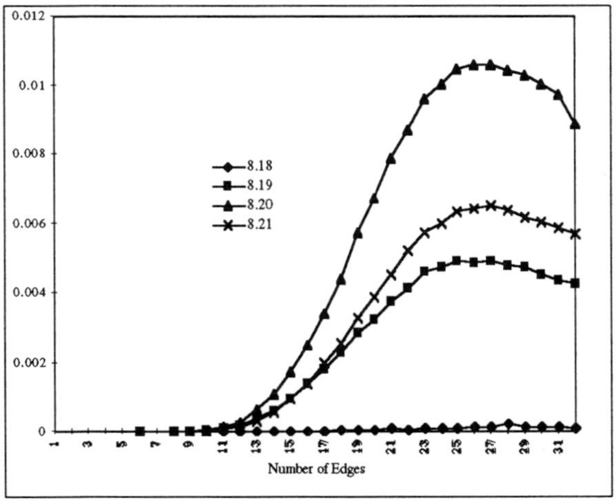

FIGURE 7. Proportion of Knot Types in Cube(n)

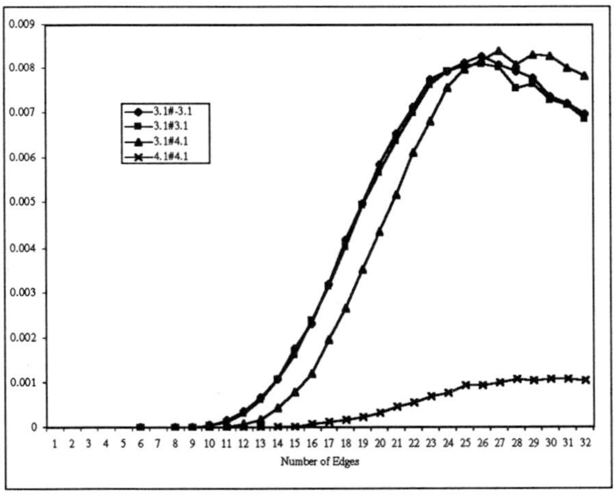

FIGURE 8. Proportion of Knot Types in Cube(n)

For the unknot, the predicted lattice dependence is of the form

$$P_0[n - 4] = \mathrm{Exp}[\, -\kappa(n - 4) + o(n)]$$

but the data for the case of geometric unknots, see Figure 9, gives the form

$$P_0[n - 4] = 0.988966(n - 4)^{-0.0270147}\mathrm{Exp}[0.0154788(n - 4) - 0.0034218\,(n - 4)^2].$$

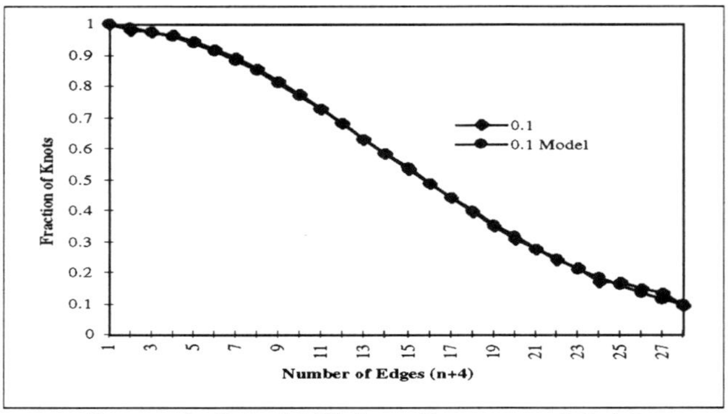

FIGURE 9. Cube(n) Unknot Occurrence Model

For the trefoil knot the data gives, see Figure 10,

$$P_{3.1}[n - 4] = 0.0135348(n - 4)^{0.523199}\text{Exp}[0.192969\,(n - 4) - 0.0034218\,(n - 4)^2].$$

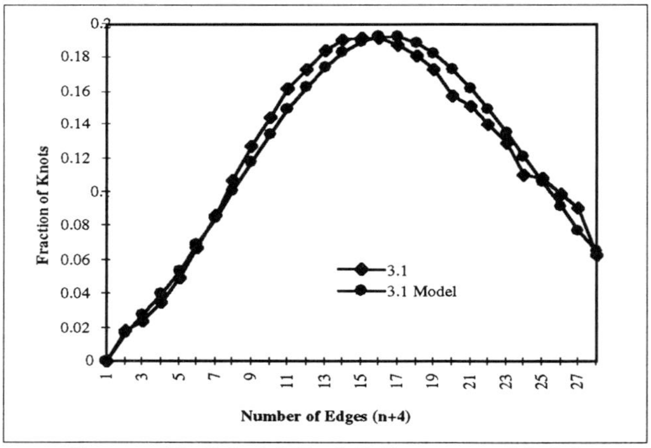

FIGURE 10. Cube(n) Trefoil Occurrence Model

For the 4.1, the figure 8 knot, see Figure 11, one has

$$P_{4.1}[n - 5] = 0.000712362(n - 5)^{1.69835}\text{Exp}[0.0828738(n - 5) - 0.005909\,(n - 5)^2].$$

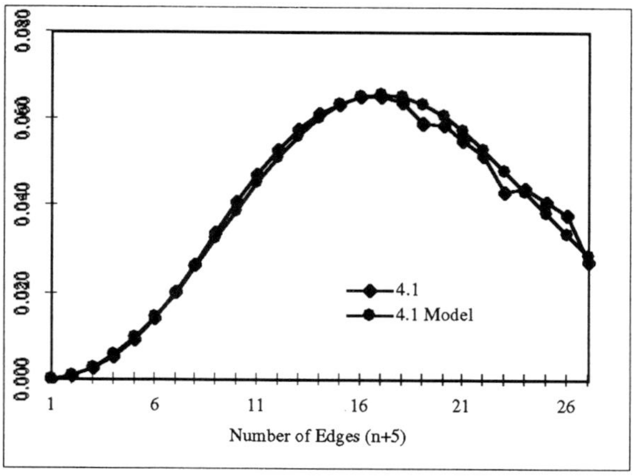

FIGURE 11. Cube(n) Figure Eight Occurrence Model

TABLE 3. Knotting as a function of the number of edges:

$$P_K[n] = C_K(n - n_0)^{\upsilon K}\text{Exp}[\upsilon_K(n - n_0) + \delta_K(n - n_0)^2]$$

where $n_0 + 1$ is the number of edges required to construct the geometric knot.

Knot	$\ln(C_K)$	κ_K	υ_K	δ_K
0.1	-0.01110	-0.027015	0.015479	-0.0034218
3.1	-4.30249	0.523199	0.192969	-0.0073583
4.1	-7.24692	1.69835	0.082874	-0.0059091
5.1	-8.85832	1.93224	0.121593	-0.00690878
5.2	-8.43491	1.9929	0.115512	-0.00698158
6.1	-12.502	3.62345	-0.00817	0.00500914
6.2	-11.705	2.94066	0.102771	-0.00680935
6.3	-12.2566	2.95392	0.105574	-0.00691127
8.19	-14.1467	3.63823	-0.016296	-0.00424356
8.20	-15.0832	4.96809	-0.191126	-0.00124664
8.21	-13.317	3.19543	0.0541537	-0.00582986
3.1#-3.1	-14.6166	4.64039	-0.163117	-0.00197747
3.1#3.1	-12.1329	2.96944	0.0226628	-0.00498713
3.1#4.1	-12.7173	3.10306	0.0379747	-0.00516541

Table 3 contains the various knot parameters that have been calculated for the cube knot data. They represent one manner in which the various knot types can be ordered in terms of their statistical occurrence in this knot space. In the exploration of the physical properties of the various knot types arising in the natural sciences, another important parameter appears to be the number of edges at which a knot type attains its maximal relative probability of occurrence. This has been the subject of applications in molecular biology, for example in the papers of Stasiak and collaborators [54, 55, 56, 57, 58, 59]. Another example of this sort is probability of occurrence for a specified number of edges. The number of edges giving the maximal likelihood of a topological knot type, together with its probability of occurrence in Cube(16) are reported in Table 4.

TABLE 4. Maximal Geometric Knotting Probabilities

Knot Type	maximal likelihood	16 edge probability
3.1	20	0.173646
4.1	21	0.0472002
5.1	26	0.0179719
5.2	24	0.0289013
6.1	26	0.0069933
6.2	28	0.00832214
6.3	28	0.00505106
8.19	30	0.00134359
8.20	27	0.00243887
8.21	27	0.00137309
3.1#-3.1	23	0.00226923
3.1#3.1	26	0.00238724

Question 6: For each fixed n, how do the various "ideal" configurations associated to each geometric knot type order the knots types? How do these ideal configuration orders vary with n?

Tables 5 and 6 report some initial results of joint work with Eric Rawdon on energy, rope length, and the average crossing number of the energy minimized configuration of geometric knots. Note that these tables report similar but distinct orderings of the various topological knot types depending on the number of edges and the characteristic.

5. CONCLUSIONS AND AREAS FOR FURTHER RESEARCH

The determination of the polygon number for a given topological knot type is a very challenging problem for any specific example. As demonstrated by the results of Calvo, even for small numbers of edges, the systematic determination of all knots types occurring for a specific number of edges is even harder. Calvo has determined the geometric octagonal topological knot types and all have been observed in the Metropolis Monte Carlo sampling of Geo(8). In the study of a much larger number of samples taken

TABLE 5: 16 Edge Equilateral Knot Orders

Position	Probability	Energy	Rope Length	Average Crossing Number
1	0.1	0.1	0.1	0.1
2	3.1	3.1	3.1	3.1
3	4.1	4.1	4.1	4.1
4	5.2	5.1	5.1	5.1
5	5.1	5.2	5.2	5.2
6	6.2	3.1#3.1	8.19	3.1#3.1
7	6.1	6.1	3.1#3.1	6.1
8	6.3	8.19	6.1	6.2
9	3.1#-3.1	6.2	6.3	6.3
10	3.1#3.1	6.3	3.1#-3.1	8.19
11	8.2	8.20	6.2	7.1
12	8.19	3.1#-3.1	7.1	8.20
13	8.21	7.1	9.46	3.1#-3.1
14	7.3	7.2	8.20	3.1#4.1
15	7.2	3.1#4.1	7.2	7.2
16	7.5	7.3	7.4	7.3
17	7.7	9.46	3.1#4.1	9.46
18	7.6	7.4	7.5	7.6
19	7.1	7.5	7.3	8.21
20	3.1#4.1	8.21	7.6	7.5
21	7.4	9.42	10.144	7.4
22	9.42	10.144	8.21	3.1#5.1
23	8.2	7.6	3.1#5.1	9.42
24	9.46	9.43	9.42	3.1#5.2
25	9.44	3.1#5.1	3.1#5.2	8.1
26	8.11	8.1	9.45	10.144
27	8.6	3.1#5.2	9.49	9.43
28	8.1	8.7	8.7	8.7
29	8.14	9.45	8.15	4.1#4.1
30	8.9	10.151	10.145	8.9
31	8.10	9.49	8.9	7.7

TABLE 6: 32 Edge Equilateral Knot Orders

Position	Probability	Energy	Rope Length	Average Crossing Number
1	0.1	0.1	0.1	0.1
2	3.1	3.1	3.1	3.1
3	4.1	4.1	4.1	4.1
4	5.2	5.1	5.1	5.1
5	5.1	5.2	5.2	5.2
6	6.2	3.1#-3.1	3.1#-3.1	3.1#-3.1
7	3.1#3.1	3.1#3.1	6.1	3.1#3.1
8	6.1	6.1	3.1#3.1	6.1
9	3.1#-3.1	6.2	6.3	6.2
10	6.3	6.3	6.2	6.3
11	3.1#4.1	7.1	8.19	7.1
12	7.5	8.19	8.2	3.1#4.1
13	3.1#-5.2	3.1#4.1	7.1	8.19
14	8.2	7.2	7.2	7.2
15	7.6	7.3	7.3	8.2
16	8.19	8.2	3.1#4.1	7.7
17	7.7	7.4	8.21	7.3
18	8.21	7.5	7.6	7.4
19	7.4	7.7	9.46	7.5
20	7.3	8.21	7.7	7.7
21	7.2	7.6	9.42	8.21
22	9.44	3.1#5.1	7.4	7.6
23	8.11	8.1	9.43	3.1#5.1
24	9.42	3.1#5.2	7.5	3.1#5.2
25	7.1	9.42	10.144	4.1#4.1
26	8.15	9.46	3.1#5.1	8.1
27	8.9	8.2	8.1	8.2
28	8.4	4.1#4.1	3.1#5.2	9.42
29	9.46	9.43	9.49	9.46
30	8.6	8.6	8.1	9.43
31	8.7	8.9	4.1#4.1	8.8

from Equ(8), the knots 8.19, 8.20 and 3.1 ± 3.1 have not yet been observed. Despite having constructed equilateral models for these knots and being quite convinced that equilateral instances of them exist, this can not be confirmed by the current Monte Carlo data. Stronger evidence suggesting that geometric and equilateral polygon numbers must differ for most knot types is provided by comparison of the substantial differences in the growths of observed HOMFLY polynomials shown in Figures 2 and 3. The determination of a specific case in which the equilateral and geometric polygon numbers differ, however, remains unachieved. As a consequence, the determination of growth rate for equilateral and geometric knots is also still open. The data suggest that, asymptotically, the number of distinct HOMFLY polynomials of geometric knots with n edges is about ten times the number of distinct HOMFLY polynomials of equilateral knots with n^2 edges.

The proportion of the space of geometric knots appears to be given by a function of the form

$$P_K[n - 4] = C_K(n - 4)^{\upsilon K} \mathrm{Exp}[\kappa_K(n - 4) + \delta_K(n - 4)^2]$$

where the specific coefficients vary with the associated knot type. This model differs from those proposed for lattice knots or appearing in Deguchi and Tsurusaki by the presence of a quadratic term. The presence of this term does not yet have a theoretical explanation.

The various quantities such as the relative probability of occurrence, the number of edges at which this is probability is maximal, any of the knot energies, or the thickness of the polygonal knot can be used to impose an order on the topological knot types. The data has shown that these orders, while similar in many respects, are not entirely equivalent. Of particular interest, in terms of the present research, is the relationship between the order arising from relative probability and that arising from the energy or thickness. Unfortunately, the data in this study suggest that the relative probability order may change with increasing numbers of edges. Further research is necessary to confirm this assertion is necessary. A deeper investigation into the properties of pairs or families of knots that exhibit this behavior could also be important. Another fundamental question to be explored would be whether the probability of the unknot is ever smaller than that of any other knot, e.g. the trefoil knot.

In this paper we have considered Monte Carlo data generated for Equ(n), Geo(n), and Cube(n) and have noted the occurrence of quantitative differences in the data despite qualitative similarities. A. Stasiak has suggested that an important factor in accounting for the quantitative differences may be the influence of spatial constraints on the distributions. Effectively, in the case of the cube knots, the observed distribution may be the superposition of two or more distributions reflecting distinctive contributions arising from spatial constraints or barriers as well as one reflecting the intrinsic knotting. Further work, whose consequences will be reported in a sequel to this paper, is underway to explore these possibilities.

A somewhat different perspective on these questions arises by exploring the geometric structure of the various polygonal knot spaces and the components that determine the associated polygonal knot types. While the Monte Carlo simulations provide average data on the extent of these components, their precise geometric nature is a mystery. Are these components "uniformly long, skinny, thread-like" entities or do they have "fat" or "thick" regions that account from the preponderance of the relative probability? This question and the relationship of the answers to the relative probability and to a wide range of physical knot invariants provides the stimulus for a new collection of questions. These are the subject of a research collaboration with Eric Rawdon and the subject of another paper.

Acknowledgments

During Fall 1998, I was on sabbatical leave at the Centre de Mathematiques et d'Informatique at the Universite de Provence, Marseille, France. I wish to thank the CMI for support of this research and my colleagues there for their hospitality. Computational work was undertaken at the CMI and at the University of California, Santa Barbara. I also thank Stefan Boeriu and Joel Rising, at UCSB, for their extraordinary assistance in seeking new and creative ways in which to make this work possible despite the severely limited computer resources available for this work. Thanks are also due to Jon Simon, Greg Buck, and Andrzej Stasiak for their interest in this work, their suggestions for improvement of various elements and, most especially, their excellent questions. Most especially, I would like to thank Bruce Ewing for his invaluable contributions to the HOMFLY knot polynomial program and Jorge Alberto Calvo for his enduring interest and contribution to this project.

References

[1] R. Randell, *A Molecular Conformation Space*, MATH/CHEM/COMP 1987, Studies in Physical and Theoretical Chemistry, 54, 1988, p 125-140

[2] R. Randell, *Conformation Spaces of Molecular Rings*, MATH/CHEM/COMP 1987, Studies in Physical and Theoretical Chemistry, 54, 1988, p 141-156.

[3] J. A. Calvo, *Geometric knot spaces and polygonal isotopy,* Knots 98 Proceedings, 1999.

[4] J. A. Calvo and K. C. Millett, *Minimal Edge Piecewise Linear Knots*, Ideal Knots (A. Stasiak, V. Katritch, and L. H. Kauffman, eds.), Series on Knots and Everything, vol. 19, World Scientific, Singapore, 1999.

[5] H. Whitney, *Elementary Structure of Real Algebraic Varieties*, Annals of Math 66 (1957), 545- 556.

[6] H.K. Moffat, G.M. Zaslavsky, P. Comte, M. Tabor, editors, *Topological Aspects of the Dynamics of Fluids and Plasmas*, Kluwer, Dordrecht,1992.

[7] C.C. Adams, *The Knot Book: An Elementary Introduction to the Mathematical of Knots*, W. H. Freeman and Co., New York, (1994).

[8] C.C. Adams, B. M. Brennan, D. L. Greilsheimer, and A. K. Woo, *Stick numbers and composition of knots and links*, Journal of Knot Theory and its Ramifications 6 no. 2 (1997) 149- 161.

[9] E. Furstenberg, J. Lie, & J. Schneider, *Stick knots*, preprint (1997).

[10] G.T. Jin and H. S. Kim, *Polygonal Knots*, Journal of the Korean Mathematical Society 30 no. 2 (1993), 371-383.

[11] G.T. Jin, *Polygon Indices and the superbridge indices of torus knots and links*, Journal of Knot Theory and its Ramifications 6 no. 2 (1997), 281-289.

[12] J. A. Calvo, *The Embedding Space of Hexagonal Knots*, to Journal of Knot Theory and its Ramifications.

[13] J. A. Calvo, *Geometric Knot Theory*, Ph.D. Thesis, University of California (1998).

[14] K. C. Millett, *Knotting of Regular Polygons in 3-Space*, Journal of Knot Theory and Its Ramifications 3 no. 3 (1994), 263 - 278.

[15] M. Meissen, *Edge number results for piecewise linear knots*, Knot Theory, Banach Center Publications Vol. 42, Polish Academy of Sciences, Institute of Mathematics, Warsaw (1998).

[16] M. Meissen, *Lower and upper bounds on edge numbers and crossing numbers of knots*, Ph.D. Thesis, University of Iowa (1997).

[17] H. L. Frisch and E. Wasserman, *Chemical topology*, J. Ame. Chem. Soc. 83 (1961), 3789-3795.

[18] M. Delbruck, *Knotting problems in biology*, Math. Problems in the Biological Sciences, Proc. Symp. Appl. Math. 14, Providence, RI: American Mathematical Society (1962), 55-63.

[19] D.W. Sumners and S. G. Whittington, *Knots in self-avoiding walks*, J. Phys. A: Math. Gen. 21 (1988), 1689-1694.

[20] N. Pippenger, *Knots in random walks*, Discrete Appl. Math. 25 (1989), 273 - 278.

[21] Y. Diao, N. Pippenger and D. W. Sumners, *On Random Knots*, Journal of Knot Theory and Its Ramifications, 3 no. 3 (1994) 419-429.

[22] Y. Diao, *The knotting of equilateral polygons in R^3*, Journal of Knot Theory and Its Ramifications 4 no. 2 (1995) 189-196.

[23] Y. Diao, *Minimal knotted polygons on the cubic lattice*, Journal of Knot Theory and its Ramifications 2 no. 4 (1993) 413-425.

[24] P. Freyd, D. Yetter, J. Hoste, W. B. R. Lickorish, K. C. Millett and A. Ocneanu, *A new polynomial invariant of knots and links*, Bulletin of the American Mathematical Society (New Series) 12 no. 2 (1985) 239 - 289.

[25] W.B.R. Lickorish and K. C. Millett, *A new polynomial invariant of oriented links*, Topology 26 no. 1 (1987), 107-141.

[26] W.B.R. Lickorish and K. C. Millett, *The new polynomial invariants fo knots and links*, Mathematics Magazine, 61 no. 1 (1988) 3-23.

[27] E. J. van Rensburg and S. G. Whittington, *The BFACF algorithm and knotted polygons, J. Phys. A 24 (1991), no. 23, 5553-5567.*

[28] E. J. van Rensburg and S. G. Whittington, *The knot probability in lattice polygons, J. Phys. A 23 (1990), no. 15, 3573-3590.*

[29] G. Buck and J. Orloff, *Computing canonical conformations for knots*, Topology Appl. 51 (1993), 246-253.

[30] J. O'Hara, *Energy of a knot*, Topology, 30 (1991), 241-247.

[31] J. O'Hara, *Energy functionals of knots*, Proccedings of the 1991 Topology Conference-Hawaii, K. H. Doverman (ed), 1992, 201-214.

[32] J. O'Hara, *Family of energy functionals of knots*, Topology and its Applications, 48 (1992) 147-161.

[33] J. O'Hara, *Energy Functionals of knots*, Topology and its Applications 56 (1994) 45-61.

[34] M.H. Freedman, Z-X He and Z. Wang, *Mobius energy of knots and unknots*, Annals of Mathematics, 139 (1994), 1-50.

[35] R. Kusner and J. Sullivan, *Mobius energies for knots and links, surfaces and manifolds*, Geometry Center Research Report GCG64, 1993, rev. 1994 to apppear in Proc. of 1993 Georgia Top. Conf.

[36] J. Simon, *Energy Functions for Polygonal Knots*, Journal of Knot Theory and its Ramifications, 3 no. 3 (1944), 299-320.

[37] J. Simon, *Energy Functions for Knots: Beginning to Predict Physical Behaviour*, IMA lecture and paper, 1994

[38] Y. Diao, C. Ernst and E. J. van Rensburg, *In search of a good polygonal knot energy*, Journal of Knot Theory and Its Ramifications vol 6 no 5 (1997) 633-657.

[39] Y. Diao, C. Ernst and E. J. van Rensburg, *Properties of knot energies*, to appear in Topology and Geometry in Polymer Science, (eds. S. G. Whittington, D. W. Sumners, & T. Lodge), IMA Volumes in Mathematics and its Applications.

[40] R. Litherland, J. Simon, O. Durmeric and E. Rawdon, *Thickness of knots*, Topology and its Applications, 91 (3), 1999, 233-244.

[41] D. Durumeric, R. Litherland, E. Rawdon, J. Simon, *Thickness of knots*, II, preprint.

[42] G. Buck, J. Simon and R. Scharein, *Physical knots: radiating tubes, rope-length, and crossing number*, preprint June 1997.

[43] E.J. Rawdon, *Thickness of polygonal knots*, Ph.D Thesis, University of Iowa, 1997.

[44] E.J. Rawdon, Approximating Smooth Thickness, to appear in the Journal of Knot Theory and its Ramifications.

[45] E.J. Rawdon, Approximating the Thickness of a Knot, , Ideal Knots (A. Stasiak, V. Katritch, and L. H. Kauffman, eds.), Series on Knots and Everything, vol. 19, World Scientific, Singapore, 1999.

[46] T.M. Przytycka and J. H. Przytycki, *Subexponentially computable truncations of Jones-type polynomials*, Graph Structure Theory, Contemporary Mathematics Vol. 147, American Mathematical Society, Providence, Rhode Island (1993) 63-108.

[47] F. Jaeger, D. L. Vertigan and D. J. A. Welsh, *On the computational complexity of the Jones and Tutte polynomials*, Math. Proc. Camb. Phi. Soc. 108 (1990) 35-53.

[48] B. Ewing and K. C. Millett, *A load balanced algorithm for the calculation of the polynomial knot and link invariants*, "The Mathematical Heritage of C. F. Gauss", World Scientific Publishing Co., Singapore, 1991.

[49] B. Ewing and K. C. Millett, *Computational algorithms and the complexity of link polynomials*, Progress in Knot Theory and related Topics, Travaux en Cours 56, Hermann, Paris, 1996, pp. 51 - 68.

[50] N. Metropolis, A. W. Rosenbluth, M. N. Rosenbluth, A. H. Teller and E. Teller, *Equation of state calculations by fast computing machines*, J. Chem. Phys. 21 (1953) 1087-1092.

[51] K.C. Millett and E.J. Rawdon, Energy, Thickness and Other Physical Aspects of Knots, in preparation 1999

[52] T. Deguchi and K. Tsurusaki, *Topology of closed random polygons*, Journal of the Physical Society of Japan, 62 no. 5 (1993) 1411-1414.

[53] T. Deguchi and K. Tsurusaki, *A statistical study of random knotting using Vassiliev Invariants*, Journal of Knot Theory and Its Ramifications 3 no. 3 (1994), 321-354.

[54] V. Katritch, J. Bednar, D. Michoud, R. G. Scharein, J. Dubochet and A. Stasiak, *Geometry and physics of knots*, Nature, 384 (1996) 142-145.

[55] V. Katritch, W. Olson, P. Pieranski, J. Dubochet and A. Stasiak, *Properties of ideal composite knots*, Nature 388 (1997) 148-151.

[56] B. Laurie, V. Katritch, J. Sogo, T. Holler, J. Dubochet and A. Stasiak, *Geometry and physics of catenanes applied to the study of DNA replication*, Biophysical Journal 74 (1998) 2815-2822.

[57] A. Stasiak, V. Katritch, J. Bednar, D. Michoud and J. Dubochet, *Electrophoretic mobility of DNA knots*, Nature 384 (1996) 122.

[58] A.V. Vologodksii, N. Crisona, B. Laurie, P. Pieranski, V. Katritch, J. Dubochet and A. Stasiak, *Sedimentation and electrophoretic migration of DNA knots and catenanes*, J. Mol. Biol. 278 (1998) 1 - 3.

[59] A. Dobay, P.E. Sottas, P. Pieranski, V. Katritch, J. Dubochet, A. Stasiak, *Predicting Optimal Lengths of Random Knots,* preprint 1999

MUTUAL BRAIDING AND THE BAND PRESENTATION OF BRAID GROUPS

HUGH R. MORTON
Department of Mathematical Sciences,
University of Liverpool,
Liverpool, L69 3BX

MARTA RAMPICHINI
Dipartimento di Matematica
Università degli Studi
via Saldini 50
20133 Milano, Italy.

ABSTRACT

This work is concerned with detecting when a closed braid and its axis are 'mutually braided' in the sense of Rudolph [7]. It deals with closed braids which are fibred links, the simplest case being closed braids which present the unknot. The geometric condition for mutual braiding refers to the existence of a close control on the way in which the whole family of fibre surfaces meet the family of discs spanning the braid axis. We show how such a braid can be presented naturally as a word in the 'band generators' of the braid group discussed by Birman, Ko and Lee [1] in their recent account of the band presentation of the braid groups. In this context we are able to convert the conditions for mutual braiding into the existence of a suitable sequence of band relations and other moves on the braid word, and thus derive a combinatorial method for deciding whether a braid is mutually braided.

Keywords: Fibred knot; open book; braid group; exchangeable braid; mutual braiding; band presentation.

1 Introduction

This is an account of part of the second author's doctoral dissertation [4]. It is an extension of the work of the first author on exchangeable braids dating from about 1981, reported at the Sussex Low-dimensional Topology meeting of 1982, [3]. Among its antecedents are a problem of Stallings on representing the unknot as a closed braid, and some constructions of fibred knots by Goldsmith using closed braids. Subsequent work has been done by Rudolph [7], and by John Salkeld [8] who showed that a braid β is exchangeable if and only if β^2 closes to a fibred link. A more extended account has been written, giving further details of the techniques and their applications to exchangeable braids [5].

2 Exchangeable braids

The original idea arose from the principle that many features of a braid β are best seen by looking at its closure $\hat{\beta}$ *along with its axis* A as a link $A \cup B$ with a distinguished unknotted component A, as in figure 1.

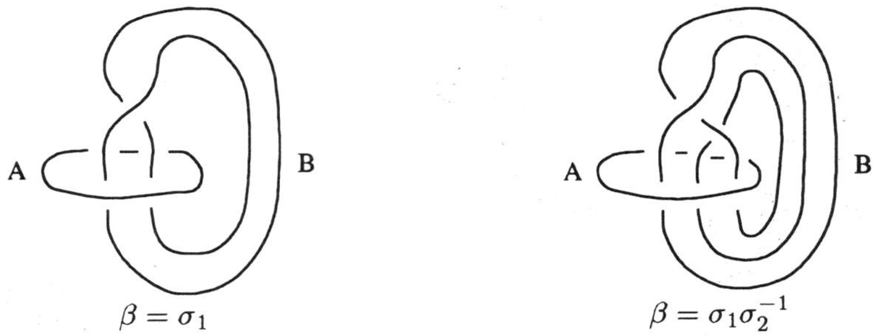

Figure 1.

Suppose conversely that we are given a link $A \cup B$ with A unknotted. Then there is a fibre projection $p_A : S^3 - A \to S^1$ whose fibres $A_\theta = p_A^{-1}(e^{i\theta})$ are discs.

Definition. We say that B is *braided rel* A if it can be isotoped, avoiding A, to meet all fibres transversely.

Then $A \cup B$ can be viewed as a closed braid $B = \hat{\beta}$ with axis A; the braid β is determined up to conjugacy in B_n where $n = |\mathrm{lk}(A, B)|$. The link $A \cup B$ is thus a good way to capture the conjugacy class of a braid β.

Exchangeable braids arise naturally when we look at such a link and ask whether it is also possible for B to be the axis and A the closed braid, in other words whether we can have B unknotted and $A = \hat{\alpha}$ braided rel B.

Definition. We say that $A \cup B$ is *exchangeably braided* if A and B are unknotted, B is braided rel A and A is braided rel B.

The examples of figure 1 can be redrawn to see that both are exchangeably braided, as in figure 2. In each of these cases there is even an isotopy of the 2-component link which interchanges A and B, so that the braids α and β are the same up to conjugacy.

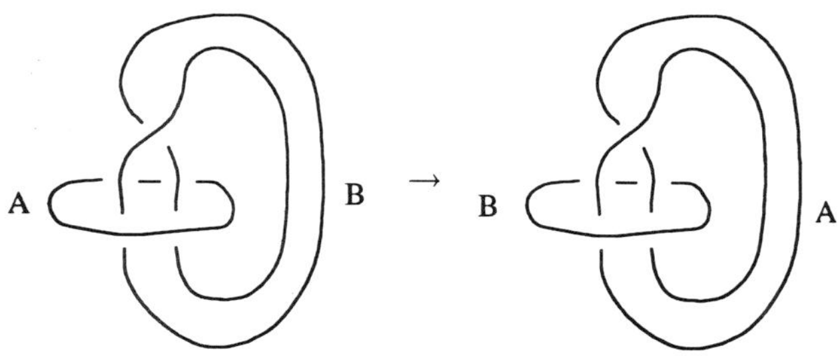

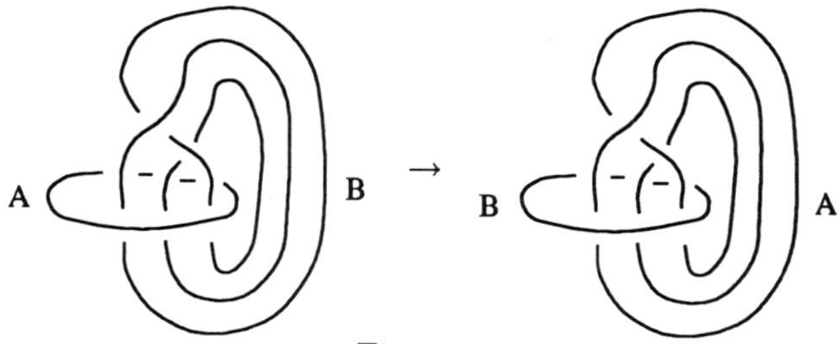

Figure 2.

This is not true in general. For example the braid β shown in figure 3, which is built as a satellite of $\sigma_1\sigma_2^{-1}$ with pattern σ_1, is exchangeable and the exchanged braid α, shown also in figure 3, is built as a satellite of σ_1 with pattern $\sigma_1\sigma_2^{-1}$. However there can not be an isotopy of the link $A \cup B$ interchanging the components, since its 2-variable Alexander polynomial $\Delta_{A\cup B}(a,b) = \sum c_{rs}a^r b^s$, defined up to a power of a and b, has coefficient matrix

$$C = \begin{pmatrix} 0 & 0 & 1 & 0 & 0 & 0 \\ 1 & 0 & 1 & 0 & 1 & 0 \\ 0 & 0 & 1 & 0 & 0 & 0 \\ 0 & 0 & 0 & 1 & 0 & 0 \\ 0 & 1 & 0 & 1 & 0 & 1 \\ 0 & 0 & 0 & 1 & 0 & 0 \end{pmatrix}.$$

Now C is not symmetric, and consequently α is not conjugate to β.

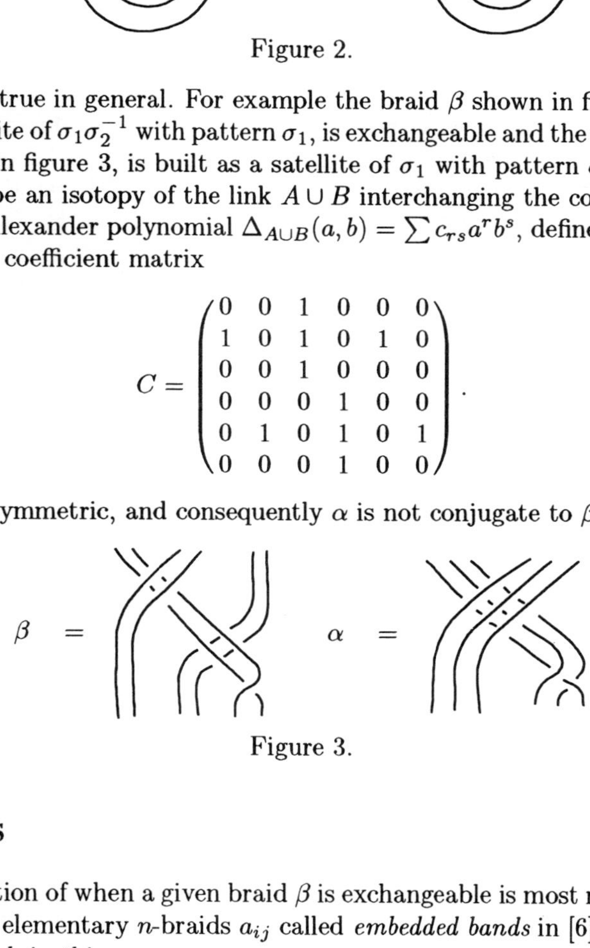

Figure 3.

3 Bands

The basic question of when a given braid β is exchangeable is most readily answered in terms of the elementary n-braids a_{ij} called *embedded bands* in [6], which we shall simply call *bands* in this paper.

Definition. The *band* $a_{ij} = a_{ji}$ is the braid illustrated in figure 4, in which the strings i and j interchange in the positive sense in front of the other strings.

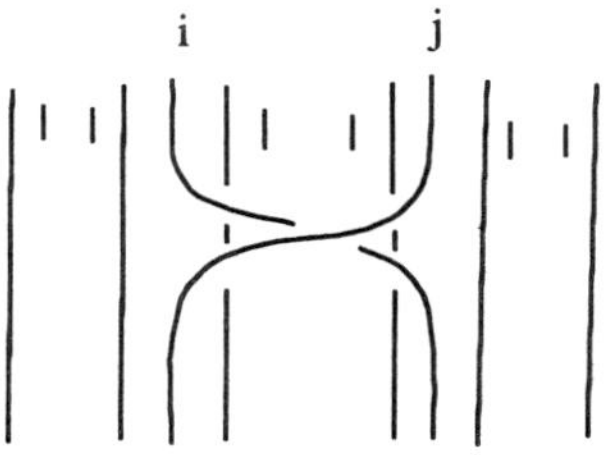

Figure 4.

Definition. A *Stallings* n-braid is a product of $n-1$ bands or their inverses whose closure is connected.

The closure of such a braid is readily seen to be spanned by a disc meeting the axis in n points made up of one disc for each string, joined by $n-1$ half-twisted bands. It is clearly necessary for an exchangeable n-braid β to have unknotted closure $B = \hat{\beta}$, and for B to be spanned by a disc which meets the axis A in n points. This property was shown in [3] to characterise Stallings braids up to conjugacy. It follows that any exchangeable braid is conjugate to a Stallings braid.

This condition alone is not enough, however; there has to be not just one spanning disc, but a whole family $\{B_\varphi\}, \varphi \in [0, 2\pi]$ which all meet A in n points. Examples of non-exchangeable Stallings braids are readily available for $n \geq 4$. While the Stallings 4-braids $a_{13}(a_{23})^{-1}a_{24}$, illustrated in figure 5, and $a_{13}a_{23}a_{24}$ are both exchangeable, the Stallings braid $\beta = a_{24}a_{23}a_{13}$ is *not*. This can be detected from the nature of the Alexander polynomial $\Delta_{A\cup\hat{\beta}}(a,b) = \sum d_{rs}a^r b^s$ of any closed braid and axis. Such a polynomial always has the form $\det(aI - B(b))$ where $B(b)$ is the $(n-1)\times(n-1)$ reduced Burau matrix of β, [3]. The extreme coefficients of powers of a are then monomials in b with coefficient ± 1. A similar constraint must apply to the extreme coefficients of powers of b when the braid is exchangeable. In the case of β above, the matrix D of coefficients of the Alexander polynomial is

$$D = \begin{pmatrix} 1 & 0 & 0 & 0 \\ 1 & -1 & 2 & -1 \\ -1 & 2 & -1 & 1 \\ 0 & 0 & 0 & 1 \end{pmatrix};$$

although the first and last rows have each a single entry ± 1, this does not hold for the first and last columns, so in this case A can not be braided rel $\hat{\beta}$.

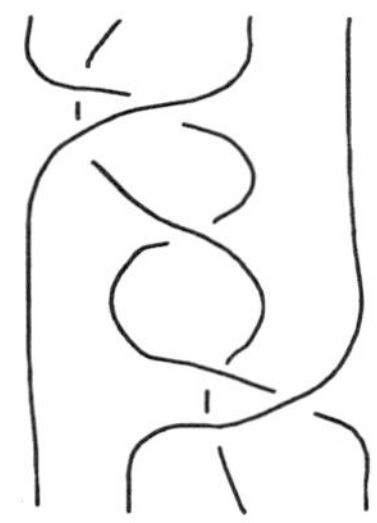

Figure 5.

Birman, Ko and Lee [1] have recently looked at the presentation of the braid groups using bands as generators. The relations in this presentation turn out to give a very good combinatorial method to help in determining when a braid is exchangeable.

The relations themselves can be stated very simply as follows.

- For $i > j > k$ we have $a_{ij}a_{jk} = a_{jk}a_{ki} = a_{ki}a_{ij}$.

  The common product of these pairs of bands can be visualised as a positive twist on the strings i, j and k through one-third of a full turn.

- The bands a_{pq} and a_{rs} commute if p, q, r and s are all different, and the pairs (p, q) and (r, s) do not interlace.

  Thus $a_{14}a_{23} = a_{23}a_{14}$ but $a_{13}a_{24} \neq a_{24}a_{13}$.

4 Generalised exchangeable braids

In trying to determine which Stallings braids are exchangeable we follow Goldsmith [2] in using the term *generalised axis* for an oriented fibred link A, not necessarily the unknot, and extending the definition of a closed braid to include a curve B which meets all the fibres A_θ of the fibration $p_A : S^3 - A \to S^1$ transversely. In this case we also extend our previous terminology to say that B is braided rel A when A is fibred and the fibration can be chosen so that B meets all the fibres transversely, and thus meets each fibre in $n = |\mathrm{lk}(A, B)|$ points.

Definition. Call $A \cup B$ a *(generalised) exchangeable link* if A and B are both fibred, B is braided rel A *and* A is braided rel B.

In the most general case, when both A and B are knotted, our analysis is still rather sketchy, even in the case of braid index 1.

In what follows we shall restrict attention to the case in which one component, A say, is unknotted, when B will determine a classical braid β up to conjugacy. We call β a *generalised exchangeable braid* if the axis A is braided rel $B = \hat{\beta}$ as above.

Theorem 1 *Every generalised exchangeable braid β is conjugate to a product of k bands, such that the resulting banded surface made up from one disc for each braid string joined by k half-twisted bands forms a fibre surface for $B = \hat{\beta}$.*

Proof : We may suppose that β is an n-braid such that $B = \hat{\beta}$ is fibred and whose axis A is braided rel B. Then A meets each fibre B_φ in n points, with all intersections in the same sense. Concentrate on just one of these fibre surfaces B_0 and its intersection with all the discs A_θ which span the axis A, or equivalently look at the function $p_A|B_0 - A \to S^1$. We can assume that the intersections are transverse except for finitely many values of θ where there is either a saddle or a centre. Now B_0 has minimal genus among spanning surfaces of B and all its intersections with A are in the same sense. Hence, following standard arguments as in [6], we can isotop it to eliminate centres. There are then $n - \chi(B_0)$ saddles, each of which lies on a component of $A_\theta \cap B_0$ as in figure 6.

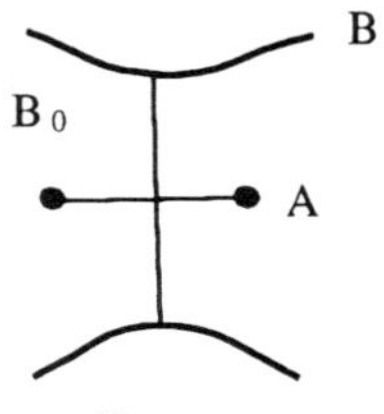

Figure 6.

The remainder of B_0 is foliated by the arcs of $A_\theta \cap B_0$ in a way which is determined up to isotopy by the position of the saddles. Each saddle determines two crossing arcs in B_0, one which joins two points of the boundary B and one which connects two of the n points of A lying in B_0. Within B_0 take a neighbourhood N of the points of A and the cross arcs of the $k = n - \chi(B_0)$ saddles which join them, whose boundary meets the foliation transversely. The resulting subsurface differs from B_0 only by a collar of the boundary B, so that B can be isotoped through B_0, avoiding A, to form the boundary of N. We may choose a sufficiently close neighbourhood N as a surface in S^3 made up of n disc neighbourhoods of the intersections with the axis together with neighbourhoods of the k cross-arcs each at a different level of θ. The close neighbourhood of each such arc within N lies as a half-twisted band when viewed with A as axis, while its boundary still lies as a closed braid. As a result, B_0 is isotopic to a banded surface N whose boundary can replace the original curve B without loss. When the points of intersection of B_0 with the axis are labelled $1, \ldots, n$ in order around A then the neighbourhood of a saddle arc joining points i and j becomes a band $(a_{ij})^{\pm 1}$, whose sign depends on the direction of tangency of the surfaces B_0 and A_θ at that saddle. The boundary can be presented as the closure of a braid consisting of the product of these $k = n - \chi(B_0)$ bands in the order of the value of θ at each saddle. □

We may thus assume that any exchangeable braid is, up to conjugacy, a product of bands and that a fibre surface for its closure is the resulting banded surface. In the case when B is unknotted this is the result from [3] about Stallings braids.

Corollary 1 *There are only a finite number of possible exchangeable links $A \cup B$ with unknotted A, and a given genus for B and linking number n.*

Proof : The component B can be written as the closure of an n-braid β with axis A. By the theorem we may assume that β is the product of k bands, where k is determined by the genus of B and the linking number n. There are only finitely many products of k bands. □

To help decide exactly which products of bands *are* exchangeable, we must look at the singular foliations induced on *every one* of the family of fibre surfaces B_φ. At the same time we will see singular foliations of each disc A_θ by the curves of $A_\theta \cap B_\varphi$.

We can best keep track of the foliations by plotting the values of $(\theta, \varphi) \in S^1 \times S^1$ for which A_θ and B_φ do not meet transversely. After a small isotopy we may assume that the singularities of any intersections are generic, so that the graphic of singular values may look something like figure 7.

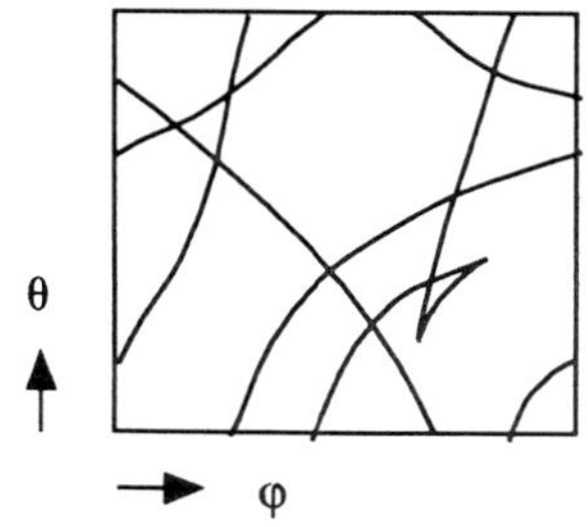

Figure 7.

At a general singular point (θ, φ) there is either a single saddle or centre tangency between A_θ and B_φ. For any φ we can calculate the Euler characteristic of the n-punctured surface $B_\varphi - A$ as the number of centres minus the number of saddles in the foliation of B_φ, which we can read from the graphic on the line of the chosen value of φ. Since this number must be constant ($= \chi(B_0) - n$) as φ varies we see that this sum remains fixed as the vertical line sweeps through the graphic. Similarly each horizontal line meets the graphic so as to contribute $\chi(A_\theta) - n = 1 - n$ to the Euler characteristic for the n-punctured disc $A_\theta - B$.

Thus in the generic case no smooth local maxima or minima can occur either vertically or horizontally on the lines of the graphic, which must consist of a number of monotone increasing or decreasing lines with some simple crossings and cusps. Each monotone line represents either a persistent saddle or centre, by constancy of the Euler characteristic. At a crossing, two tangencies generically occur at distinct places and the saddles or centres continue through the crossings, while at each cusp a line of centres and a line of saddles meet. The slope of a line is positive when the surfaces A_θ and B_φ have the same orientation at the point of tangency, and negative when they have reverse orientations.

5 Mutual braiding and labelled graphics

To simplify further analysis we now impose Rudolph's condition that the two fibrations form *mutually braided open books*, [7]. Equivalently we require all the local tangencies between fibres of the two families to be saddles, so that the graphic of singular values has no centres, and hence no cusps. Rampichini proves in [4, 5] that, for an exchangeable link with one unknotted component, this condition can be assumed without loss of generality.

Definition. A link $A \cup B$ with unknotted A which satisfies Rudolph's condition is called *mutually braided.*

The graphic of a mutually braided link then consists only of a number of monotone lines, with simple crossings. We can add some simple combinatorial information to the graphic of a mutually braided link to describe the positions of the saddles in each disc A_θ. The boundary of each disc A_θ is the fixed curve A which meets every fibre B_φ in n points. The n points of A where $\varphi = 0$ as $1, \ldots, n$ dissect A into n half-open segments. Label these $1, \ldots, n$ in order around A.

Where $A_\theta \cap B_\varphi$ contains a single saddle tangency the cross-arc of the saddle joins two points of A lying in different segments i and j. Label the corresponding point (θ, φ) of the graphic of singular values with the pair ij. Along any line of the graphic the label remains constant until we reach a point where two lines of saddles cross, or we get to the transition where φ increases through the level $\varphi = 2\pi$ to return to $\varphi = 0$. At this transition the corresponding arcs of the saddles will move from segments i and j to segments $i + 1$ and $j + 1$.

Before describing the labelling of the graphic at a crossing it is worth visualising the singular foliation of the disc A_θ for fixed θ. A typical configuration with $n = 4$ is shown in figure 8.

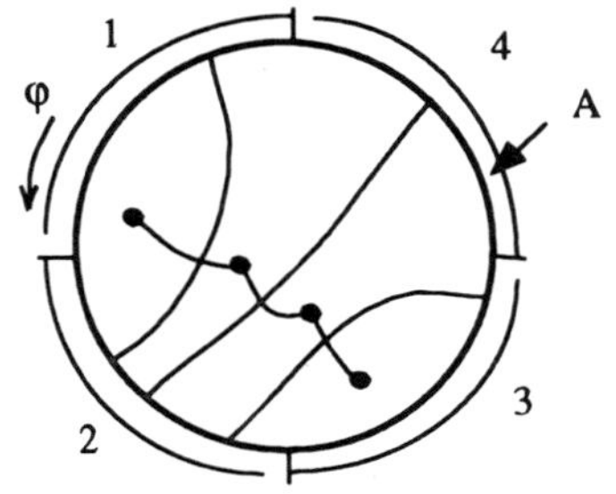

Figure 8.

The general line θ = constant meets the graphic in $n-1$ points, each labelled by some pair ij. These correspond to the cross arms of $n-1$ saddles, and the labelling tells us in which segments of the circle A their ends lie. They divide the disc A_θ into n regions, each of which contains one point of B; a choice of this point and the rest of the foliation is determined up to isotopy by the saddle arms. As θ increases the position of the saddles, and the foliation changes, initially by an isotopy in which the end points of the positive saddle arms rotate in one direction while those of the negative arms rotate in the other.

This evolution is viewed by Rampichini as a sort of cinematic film, in which the screen is a disc with fixed boundary on which is projected the foliation of A_θ with the parameter θ in the role of a time variable. The main part of the action is the movement of the saddle arms. This is punctuated by isolated instants where two lines of saddles cross in the graphic. In the film at each such critical time θ there are two saddles having the same value of φ and thus two singular points in the set $A_\theta \cap B_\varphi$. These singular points either belong to two disjoint components of $A_\theta \cap B_\varphi$, each consisting of simple saddle arms connecting non-interlocking pairs of segments p, q and r, s say, or the two saddles belong to the same component of $A_\theta \cap B_\varphi$ which lies in A_θ as shown in figure 9.

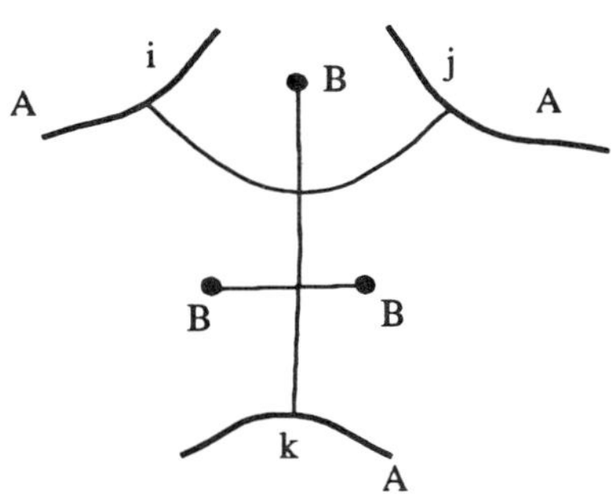

Figure 9.

Nothing very dramatic happens in the film when disjoint saddle arms are involved; they simply interchange levels of the value φ, while in the graphic the labels pq and rs continue through the crossing. Where the two saddles belong to the same component, however, we see the saddle arms in the film coming together at one end and then reforming with one of the two arms remaining essentially the same and the other changing radically to emerge from the other end of the unchanged saddle arm. The whole process is determined by the 'Christmas tree' configuration of the intermediate position as in figure 9. The essential information is the labelling of the three segments i, j and k of the end points, and the knowledge of which segment lies at the base of the tree. The labelling of the saddles on the graphic before and after the crossing depends on this combinatorial information. As a summary of the film itself it is useful to think of simply drawing the sequence of shots at the critical levels, with each Christmas tree truncated to form a 'T' and the signs of each saddle indicated so that the direction of its evolution in the film is known. Rampichini presents such data, with some extension, in what she terms a 'film script', with these shots to dictate the changes of scene.

The labelling on the graphic is then quite constrained in its nature at the crossing points. It is easier to describe the exact transitions of labelling at the crossings when we change our viewpoint and consider what happens as we change φ.

For a fixed φ the graphic provides a complete list of all the saddles in B_φ which arise in the foliation by the intersections with the level discs A_θ, and the labelling shows which of the segments of A are connected by their cross-arms. The labelling on the graphic, read in order of increasing θ, determines a word, w_φ, as a product of bands, using a_{ij} or its inverse according to the slope of the line on the graphic with label ij.

Theorem 2 *The band word w_φ determined by a labelled graphic arising from a mutually braided link changes only*

- *when two lines of the graphic cross, or*
- *when a line on the graphic passes between $\theta = 0$ and $\theta = 2\pi$.*

In the first case w_φ changes by the application of a band relation on passing the crossing. In the second case, which occurs exactly $n-1$ times, either a final band a_{ij} is transferred to the beginning of the band word w_φ, or an initial band a_{ij}^{-1} to the end, as φ increases.

Proof : The first case is shown by examining the possible generic behaviour close to the configuration in figure 9. Full details are given in [4, 5]. The second case is immediate from the monotonicity of the lines and the fact that there are $n-1$ saddles at $\theta = 0$. □

As a result, the labelled graphic arising from a mutally braided closed braid B determines a sequence of band words w_φ, each representing B and a spanning surface B_φ. The sequence starts from the band word w_0, which can be regarded as a given band word for the proposed exchangeable braid B along with its spanning surface, and finishes with the word $w_{2\pi}$, which is the original word w_0 with all its indices i, j reduced by 1 mod n. They are connected by a sequence of moves consisting of an unspecified number of applications of the band relations along with exactly $n-1$ cyclings as specified in the theorem.

By way of example, a labelled graphic for the exchangeable 4-braid with band word $a_{34}(a_{12})^{-1}a_{23}$ is shown in figure 10.

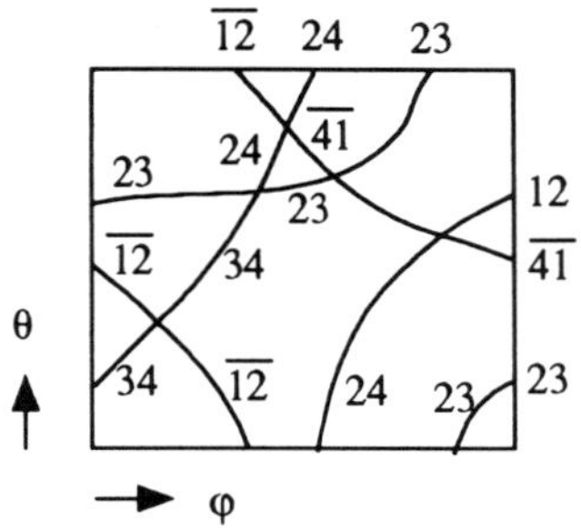

Figure 10.

The resulting sequence of band words, given simply by their indices, is listed below.

$$\begin{aligned}&(34)\overline{(12)}(23) \to \overline{(12)}(34)(23) \to (34)(23)\overline{(12)}\\ &\to (23)(24)\overline{(12)} \to (23)\overline{(14)}(24) \to (24)(23)\overline{(14)}\\ &\to (24)\overline{(14)}(23) \to (23)(24)\overline{(14)} \to (23)\overline{(41)}(12).\end{aligned}$$

6 Band algorithms for mutual braiding

The previous section shows that if a braid determines a mutually braided open book then there must be a labelled graphic giving rise to a suitably restricted sequence of band words. It can be shown, with a little more argument, that there are only finitely many possible sequences which need to be considered when asking if a given band word determines a mutually braided link, and if none of them fulfil all the necessary conditions for arising from a labelled graphic then the link is definitely not mutually braided.

The key to the converse of this result, which also covers Rampichini's extension to exchangeable braids, lies in the reconstruction of a family of spanning surfaces from an 'admissibly' labelled graphic, [4, 5].

Definition. A graphic of monotone increasing and decreasing lines in $S^1 \times S^1$, labelled by pairs of numbers i, j between 1 and n, is *admissibly labelled* if

- the labelled lines cross one basic circle $\theta = 0$ in $n - 1$ points,
- the band words w_φ read from the graphic following the basic circles $\varphi =$ constant of the other family satisfy the band relations at every crossing,
- the indices in the bands all change by 1 at the reference circle $\varphi = 0$,
- the $n - 1$ labels on the circle $\theta = 0$ contain no interlocking pairs.

Theorem 3 *Every admissibly labelled graphic arises from some exchangeable braid.*

Proof : Use the last condition in the definition to draw a set of saddles in the disc A_0 which correspond with the labelling when $\theta = 0$. The evolution of the graphic as θ increases then provides a film script, which is converted into a film by filling in suitably the non-singular parts of the foliation at each level. Use of standard models around the singular levels where there are crossings in the graphic, coupled with the monotonicity of the movement of the saddle arms with increasing θ, ensures that the family of surfaces in the solid torus generated in this way are non-singular everywhere, [4, 5]. □

Consequently an admissibly labelled graphic with a given band word at $\varphi = 0$ will give rise to a mutually braided fibration for the corresponding closed braid.

Rampichini also shows how the graphic for a general exchangeable braid, which includes cusps and lines of centres, can be used to construct an admissibly labelled graphic, and hence deduce that any exchangeable braid is mutually braided [4, 5].

We thus have an algorithm to decide whether a band word w_0 determines a mutually braided (and indeed an exchangeable) braid.

Algorithm. Look for a chain of moves to convert the word w_0 to the same word but with all indices reduced by 1 mod n. The moves must consist of a number of band relations and exactly $n-1$ cyclings in which a positive band is moved from back to front or a negative band from front to back of the word. It can be shown that no repetitions of words between cyclings need occur in an admissibly labelled graphic, so there are only a finite number of possible chains to consider [4, 5].

If no chains satisfy these conditions then the braid is *not* mutually braided.

Given a chain satisfying these conditions try to realise it by an admissibly labelled graphic. It may not be possible to do this, bearing in mind that the lines in the graphic, which are determined combinatorially by the chain as φ increases, are required to move strictly monotonically in θ, and must end at the same level that they begin. If any chain can be realised then the braid *is* mutually braided, otherwise again it is not.

As an example we apply this method to the Stallings braid $w = a_{24}a_{23}a_{13}$ mentioned earlier to show that its closure is not exchangeable.

We must consider possible chains of moves which lead to the word $w' = a_{13}a_{12}a_{24}$, and include exactly 3 cyclings. Now no band relations are possible in w so the first move in any chain must be a cycling, to get $a_{13}a_{24}a_{23}$. Again no relations are possible, so we must cycle to $a_{23}a_{13}a_{24}$. Once more there are no relations and we must cycle, returning to w. Since we have used all 3 cyclings there is nothing more we can do; as we have not reached w' we conclude that the braid is not exchangeable.

In contrast we *can* find a chain and an admissibly labelled graphic leading from the word $v = a_{13}a_{23}a_{24}$ to $v' = a_{24}a_{12}a_{13}$, by making use of band relations such as $a_{23}a_{24} = a_{24}a_{34}$.

Acknowledgments

The second author was partially supported by a British Council Fellowship Award during the academic year 1996-97.

References

[1] J.S.Birman, K.H.Ko and S.J.Lee, A new approach to the word and conjugacy problems in the braid groups. Adv. Math. 139 (1998), 322-353

[2] D.L.Goldsmith, Symmetric fibered links. In 'Knots, groups and 3-manifolds', ed. L.P.Neuwirth. Ann. Math. Studies 84, Princeton University Press 1975, 3-23.

[3] H.R.Morton, Exchangeable braids. In 'Low Dimensional Topology', London Mathematical Society Lecture Notes 95, ed. R. A. Fenn. Cambridge University Press (1985), 86-105.

[4] M.Rampichini, Exchangeable fibred links. Doctoral thesis, University of Milan, 1998.

[5] M.Rampichini, Mutually braided links, films and exchangeability. Quaderno 10/1999, Dipartimento di Matematica 'F. Enriques', via Saldini 50, 20133, Milano, Italy. Preprint 1999.

[6] L.Rudolph, Braided surfaces and Seifert ribbons for closed braids. Comm. Math. Helv. 58 (1983), 1-37.

[7] L.Rudolph, Mutually braided open books and new invariants of fibered links. Contemp. Math. 78 (1988), 657-673.

[8] J.O.Salkeld, Generalised exchangeable braids. PhD thesis, University of Liverpool, 1997.

QUANTUM $SU(2)$-INVARIANTS FOR THREE-MANIFOLDS ASSOCIATED WITH NON-TRIVIAL COHOMOLOGY CLASSES MODULO TWO

HITOSHI MURAKAMI

ABSTRACT. We show an integrality of the quantum $SU(2)$-invariant associated with a non-trivial first cohomology class modulo two.

For a given Lie group G and an integer (level) k, E. Witten gave an 'invariant' of three-manifolds by using the path integral [33]. Mathematically rigorous proof of its existence was given by several people. See for example [26, 7, 4, 32, 29]. There are also some refinements for these invariants corresponding to various structures of the three-manifolds, such as cohomology classes and spin structures [30, 7, 9, 18, 3, 2].

Some topological properties of the invariants were given especially for $G = SU(2)$. The first striking result was found by R. Kirby and P. Melvin stating that the quantum $SU(2)$-invariant with level two splits into the sum of the invariants associated with spin structures and each summand can be described in terms of the μ-invariant. They and X. Zhang obtained a topological interpretation for the quantum $SU(2)$-invariant with level four [8]. In this case the invariant splits into the sum of the invariants associated with the first cohomology classes modulo two and each summand can be described in terms of the Witt invariant and the cohomologies of both the three-manifold and its double cover defined by the cohomology class. Similarly the author gave a topological interpretation for the quantum $SU(3)$-invariant with level three [18]. It also decomposes into the sum of the invariants associated with the first cohomology classes modulo three and each summand can be described in terms of the cohomologies of the manifold and its triple cover.

If $G = SU(n)$ and n and the level k are coprime, then the invariant splits into the product of the $U(1)$-invariant [21, 5] and $PSU(n)$-invariant [7, 10]. If $n + k$ is odd prime and the three-manifold is a rational homology sphere, the invariant is a cyclotomic integer [17, 19, 15, 16, 28]. Similar integrality holds for a three-manifold with positive first Betti number. Moreover one can define the perturbative invariant from the $PSU(n)$-invariant for rational homology three-spheres [22, 24, 12]. Recently it is proved by T.T.Q. Le that every coefficient of the perturbative invariant for integral homology three-spheres is of finite type in T. Ohtsuki's sense [23]. (For $n = 2$ case, it is proved in by A. Kricker and B. Spence [11]. In particular its first non-trivial coefficient coincides with (a multiple of) the Casson invariant [1]. (For rational homology three-spheres, Le showed in [12] that this coincides with a

---

Date: August 3, 1999.

Partially supported by Waseda University Grant for Special Research Projects (No. 98A-623) and Grant-in-Aid for Scientific Research (C) (No. 09640135), the Ministry of Education, Science, Sports and Culture.

multiple of the Casson-Walker invariant [31].) See also [25] for recent developments of these invariants; especially their relation with the LMO invariant [13].

In this paper we consider the quantum $SU(2)$-invariant with level divisible by four. By V. Turaev, and Kirby and Melvin [7, 30] it decomposes into the sum of the invariants associated with the cohomology classes modulo two. We will consider the case where $(n+2)/2$ is an odd prime.

Let p be an odd prime and M a $\mathbf{Z}/p\mathbf{Z}$-homology three-sphere. For a cohomology class $\theta \in H^1(M;\mathbf{Z}/2\mathbf{Z})$, let $\tau_{2p}^{SU(2)}(M,\theta)$ be the quantum $SU(2)$-invariant with level $2p-2$ associated with the $2p$th root of unity $q=\exp(\pi\sqrt{-1}/p)$. In the previous paper [20] the author proved

Theorem 1. *For an odd prime p and a $\mathbf{Z}/p\mathbf{Z}$-homology three-sphere M, we have $\tau_{2p}^{SU(2)}(M,0) \in \mathbf{Z}[1/2,\xi]$ and the 'coefficient' of $(\xi-1)$ is congruent to the Casson-Walker invariant modulo p, where $\xi=\exp(2\pi\sqrt{-1}/p)$ and p is an odd prime.*

In this paper we will show a weaker result for non-trivial cohomology classes. We will show

Theorem 2. *Let p be an odd prime and M a $\mathbf{Z}/p\mathbf{Z}$-homology three-sphere. Then we have*

$$\tau_{2p}^{SU(2)}(M,\theta) \in \begin{cases} (\xi-1)\mathbf{Z}[1/2,\xi] & \text{if } \theta\cup\theta\cup\theta=0, \\ \sqrt{-1}(\xi-1)\mathbf{Z}[1/2,\xi] & \text{if } \theta\cup\theta\cup\theta\neq 0. \end{cases}$$

Acknowledgment. The author thanks the organizers, especially Sofia Lambropoulou, of the conference 'Knots in Hellas' for their hospitality.

1. PROOF

We put $p=4u+\varepsilon$ with $\varepsilon=\pm1$. Put $\xi=\exp(2\pi\sqrt{-1}/p)$, $q=\exp(\pi\sqrt{-1}/p)$ and $s=\exp(\pi\sqrt{-1}/2p)$. Therefore $\xi=q^2=s^4$. For a non-negative integer k, we denote by $[k]$ the quantum integer $(s^k-s^{-k})/(s-s^{-1})$.

Suppose that a closed three-manifold M is given by an algebraically split, framed link $\mathcal{L}$ with m components. We also assume that the framing of $\mathcal{L}$ is $(f_1,f_2,\ldots,f_m)$ with $f_i\not\equiv 0 \mod p$. Moreover we assume that $f_i\equiv 0 \mod 2$ for $i\leq b$ and $f_i\equiv 1 \mod 2$ for $b<i\leq m$. Let $\theta\in H^1(M;\mathbf{Z}/2\mathbf{Z})$ be presented by the sublink $\mathcal{L}_\theta=L_1\cup L_2\cup\cdots\cup L_c$ $(c\leq b)$. We put $f_i=2g_i$ for $1\leq i\leq b$.

By [7, (8.32) Theorem], $\tau_{2p}^{SU(2)}(M,\theta)=\alpha(\mathcal{L})\Sigma(\mathcal{L})$ where $\alpha(\mathcal{L})$ and $\Sigma(\mathcal{L})$ is given as follows.

$$\begin{aligned} \alpha(\mathcal{L}) &= \left(\sqrt{\frac{1}{p}}\sin\left(\frac{\pi}{2p}\right)\right)^m \exp\left(-\frac{3(p-1)}{8p}\times 2\pi\sqrt{-1}\right)^{\sigma(\mathcal{L})} \\ \Sigma(\mathcal{L}) &= \sum_{\substack{0<k_i\leq p \\ \mathcal{L}_{\mathbf{k}}=\mathcal{L}_\theta}} 2^{|T_{\mathbf{k}}|}[\mathbf{k}]J(\mathcal{L},\mathbf{k}), \end{aligned}$$

where $J(\mathcal{L},\mathbf{k})$ is the colored Jones polynomial of $\mathcal{L}$ with color $\mathbf{k}=(k_1,k_2,\ldots,k_m)$ $(0<k_i\leq p)$, $\mathcal{L}_{\mathbf{k}}$ is the sublink of $\mathcal{L}$ consisting of components with the color k_i even, $|T_{\mathbf{k}}|$ is the number of k_i which are less than p, $\sigma(\mathcal{L})$ is the number of positives minus that of negatives in $\{f_1,f_2,\ldots f_m\}$, and $[\mathbf{k}]=[k_1][k_2]\cdots[k_m]$.

After some calculations we can simplify $\alpha(\mathcal{L})$ and $\Sigma(\mathcal{L})$ as follows.

Lemma 1.1 (Lemma 2.2 in [20]).

$$\alpha(\mathcal{L}) = \frac{(\xi^u + \xi^{-u})^m}{(2G(\xi))^m}(-1)^{u\sigma(\mathcal{L})+\sigma_-(\mathcal{L})(\varepsilon-1)/2}\xi^{3\times\bar{8}\varepsilon\sigma(\mathcal{L})},$$

where $G(\xi)$ *is a Gaussian sum* $\sum_{k=0}^{p-1}\xi^{k^2}$, $\bar{8}$ *is the inverse of* 8 *in* $\mathbf{Z}/p\mathbf{Z}$ *and* $\sigma_-(\mathcal{L})$ *is the number of negatives in* $\{f_1, f_2, \dots f_m\}$. *Note that* $G(\xi)$ *is of the form* $\gamma(\xi - 1)^{(p-1)/2}$ *with* γ *invertible in* $\mathbf{Z}[\xi]$.

Lemma 1.2 (see Lemma 2.3 in [20]).

$$\Sigma(\mathcal{L}) = [2]2^c \sum_{n_1,n_2,\dots,n_m=1}^{(p-1)/2} (-1)^{\sum_i n_i} V(L^{(2n_1-1,2n_2-1,\dots,2n_c-1,2n_{c+1},2n_{c+2},\dots,2n_m)})$$

$$\times \prod_{l=1}^{c} C_{n_l}(g_l) \prod_{l=c+1}^{m} S_{n_l}(f_l)$$

with

$$S_n(f) = \sum_{k=n}^{p-n-1} (-1)^k[2k+1]\xi^{(k^2+k)f/2}\binom{k+n}{k-n}$$

and

$$C_n(g) = \sum_{k=n}^{(p-1)/2} (-1)^k[2k]s^{(4k^2-1)g}\binom{k+n-1}{k-n},$$

where $L^{(d_1,d_2,\dots,d_m)}$ *is the (unframed) link obtained from* $\mathcal{L}$ *by replacing ith component with* d_i *parallels respecting the zero-framing. For a link* L, $V(L)$ *is the Jones polynomial defined by the skein relation* $qV(L_+) - q^{-1}V(L_-) = (s - s^{-1})V(L_0)$ *for the usual skein triple* (L_+, L_-, L_0) *and by* $V(\textit{trivial knot}) = 1$.

Proof. The proof is similar to [20, Lemma 2.3] and so details are omitted. □

To prove Theorem 2, we prepare three lemmas.

Lemma 1.3 (Lemma 2.4 in [20]). $S_n(f) \in \mathbf{Z}[\xi]$ *is divisible by* $(\xi - 1)^{(p-1)/2-n}$ *in* $\mathbf{Z}[\xi]$.

Lemma 1.4. $\sqrt{-1}^{g+1}C_n(g) \in \mathbf{Z}[\xi]$ *is divisible by* $(\xi - 1)^{(p-1)/2-n+1}$ *in* $\mathbf{Z}[\xi]$.

Proof. First note that $s = \sqrt{-1}^{\varepsilon}\xi^{-\varepsilon u}$. Therefore an easily calculation shows that

$$C_n(g) = \frac{\sqrt{-1}^{-\varepsilon(g+1)}\xi^{\varepsilon ug}}{\xi^u + \xi^{-u}} \sum_{k=n}^{(p-1)/2} (\xi^{-2\varepsilon uk} - \xi^{2\varepsilon uk})\xi^{gk^2}\binom{k+n-1}{k-n}.$$

Now the proof follows from [14, Lemma 6.2] (see also [22]). □

The following lemma is a generalization of a result of M. Sakuma [27, Theorems 1 and 2], who proves the case where $c = 0$.

Lemma 1.5. *For an algebraically split link* L, *we have*

$$\sqrt{-1}^{c+1}V(L^{(2n_1-1,2n_2-1,\dots,2n_c-1,2n_{c+1},2n_{c+2},\dots,2n_m)}) \in \mathbf{Z}[\xi]$$

is divisible by $(\xi - 1)^{\sum_{l=1}^m n_l - c}$.

Proof. We will show more generally that for a link $L = L_1 \cup L_2$, $\sqrt{-1}^{\sharp(L_1)+1} V(L_1 \cup L_2^2) \in \mathbf{Z}[\xi]$ is divisible by $(\xi - 1)^{\sharp(L_2)}$, where L_2^2 is the two-parallel of L_2 with respect to the zero-framing and $\sharp(L_i)$ is the number of components in L_i $(i = 1, 2)$.

First note that for any link L with $\sharp(L)$ components, $s^{\sharp(L)+1} V(L) \in \mathbf{Z}[q] = \mathbf{Z}[\xi]$ from the skein relation. Since $s = \sqrt{-1}^{\varepsilon} \xi^{-4\varepsilon u}$, we see that $\sqrt{-1}^{\sharp(L_1)+1} V(L_1 \cup L_2^2) \in \mathbf{Z}[\xi]$.

We will use L.H. Kauffman's bracket polynomial [6] $\langle \mathcal{L} \rangle$ for a framed link $\mathcal{L}$ defined by

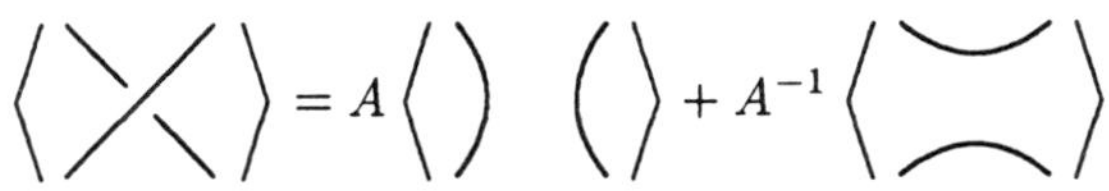

and

$$\langle \mathcal{O} \rangle = 1,$$

where $\mathcal{O}$ is the trivial knot with zero-framing. Since $V(L) = (-1)^{\sharp(L)+1} \langle \mathcal{L} \rangle \Big|_{A=s^{-1/2}}$, we only have to prove that $\langle \mathcal{L}_1 \cup \mathcal{L}_2^2 \rangle$ is divisible by $(A^2 + A^{-2})^{\sharp(\mathcal{L}_2)}$ (note that $s + s^{-1} = \sqrt{-1}^{\varepsilon}(\xi^{-4\varepsilon u} - \xi^{4\varepsilon u})$). Here $\mathcal{L} = \mathcal{L}_1 \cup \mathcal{L}_2^2$ is the zero-framed link obtained from $L = L_1 \cup L_2^2$.

By simple calculations we have

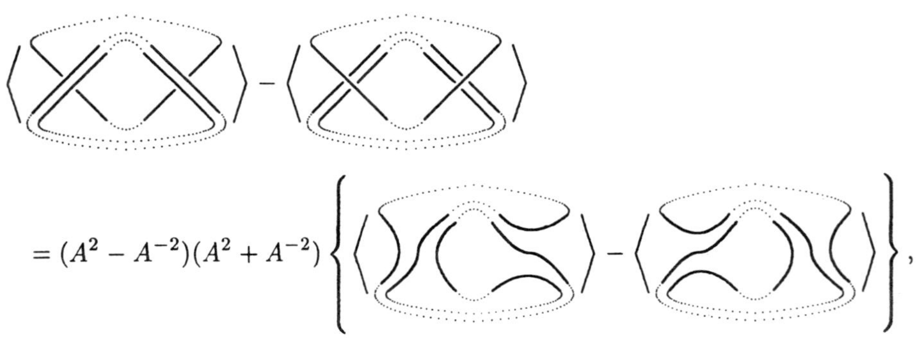

where dotted lines indicate the connectivity. Note that the term in the curly bracket is in $\sqrt{-1}^{\sharp(\mathcal{L}_1)+1} \mathbf{Z}[\xi]$ from the connectivity. By induction on $\sharp(\mathcal{L}_2)$, we see that the divisibility of $\langle \mathcal{L}_1 \cup \mathcal{L}_2^2 \rangle$ does not change by double crossing change between a single strand and a double strand. This shows that the divisibility of $\langle \mathcal{L}_1 \cup \mathcal{L}_2^2 \rangle$ is greater than or equal to that of $\langle \mathcal{L}_1 \rangle \langle \mathcal{L}_2^2 \rangle$ which is greater than or equal to that of $\langle \mathcal{L}_2^2 \rangle$, which is greater than or equal to $\sharp(\mathcal{L}_2)$ by [20, Lemma 4.1]. This completes the proof. □

Proof of Theorem 2. We only prove the case where M is obtained from an algebraically split, framed link since the general case follows from T. Ohtsuki's diagonalizing lemma [24, Lemma 2.4].

Now we assume that $\theta \in H^1(M; \mathbf{Z}/2\mathbf{Z})$ is given as described before. From the lemmas above, we easily see that $\sqrt{-1}^{\sum_{i=1}^{c} g_i} \alpha(\mathcal{L}) \Sigma(\mathcal{L})$ belongs to $\mathbf{Z}[1/2, \xi]$ and divisible by $(\xi - 1)$ (Note that $\sqrt{-1}[2] \in \mathbf{Z}[\xi]$ and divisible by $(\xi - 1)$ and that we need $1/2$ from $\alpha(\mathcal{L})$).

Now from [8], $\theta \cup \theta \cup \theta \equiv \sum_{l=1}^{c} g_l \mod 2$. Therefore $\tau_{2p}^{SU(2)}(M,\theta) \in \mathbf{Z}[\xi]$ if $\theta \cup \theta \cup \theta \equiv 0 \mod 2$ and $\sqrt{-1}\mathbf{Z}[\xi]$ otherwise, completing the proof. □

REFERENCES

[1] S. Akbulut and J.D. McCarthy, *Casson's Invariant for Oriented Homology 3-spheres - An Exposition*, Mathematical Notes, vol. 36, Princeton University Press, Princeton, 1990.

[2] C. Blanchet, *Hecke algebras, modular categories and 3-manifolds quantum invariants*, math.GT/9803114.

[3] ———, *Refined quantum invariants for three-manifolds with structure*, Knot theory (Warsaw, 1995) (Warsaw), Polish Acad. Sci., Warsaw, 1998, pp. 11–22.

[4] C. Blanchet, N. Habegger, G. Masbaum, and P. Vogel, *Three-manifold invariants derived from the Kauffman bracket*, Topology **31** (1992), 685–699.

[5] T. Gocho, *The topological invariant of three-manifolds based on the* u(1) *gauge theory*, Proc. Japan Acad. Ser. A Math. Sci. **66** (1990), no. 8, 237–239.

[6] L.H. Kauffman, *State models and the Jones polynomial*, Topology **26** (1987), 395–407.

[7] R. Kirby and P. Melvin, *The 3-manifold invariants of Witten and Reshetikhin–Turaev for* $sl(2,\mathbb{C})$, Invent. Math. **105** (1991), 473–545.

[8] R. Kirby, P. Melvin, and X. Zhang, *Quantum invariants at the sixth root of unity*, Comm. Math. Phys. **151** (1993), 607–617.

[9] T. Kohno and T. Takata, *Symmetry of Witten's 3-manifold invariants for* $sl(n,\mathbb{C})$, J. Knot Theory Ramifications **2** (1993), 149–169.

[10] ———, *Level-rank duality of Witten's 3-manifold invariants*, Progress in algebraic combinatorics (Fukuoka, 1993) (Tokyo), Math. Soc. Japan, Tokyo, 1996, pp. 243–264.

[11] A. Kricker and B. Spence, *Ohtsuki's invariants are of finite type*, J. Knot Theory Ramifications **6** (1997), no. 4, 583–597.

[12] T. T. Q. Le, *On perturbative PSU(n) invariants of rational homology 3-spheres*, math.GT/9802032.

[13] T. T. Q. Le, J. Murakami, and T. Ohtsuki, *On a universal perturbative invariant of 3-manifolds*, Topology **37** (1998), no. 3, 539–574.

[14] X.-S. Lin and Z. Wang, *On Ohtsuki's invariants of integral homology 3-spheres*, Acta Math. Sinica (Series B), to appear (math.GT/9509149).

[15] G. Masbaum and J.D. Roberts, *A simple proof of integrality of quantum invariants at prime roots of unity*, Math. Proc. Cambridge Philos. Soc. **121** (1997), 443–454.

[16] G. Masbaum and H. Wenzl, *Integral modular categories and integrality of quantum invariants at roots of unity of prime order*, J. Reine Angew. Math. **505** (1998), 209–235.

[17] H. Murakami, *Quantum* $SU(2)$*-invariants dominate Casson's* $SU(2)$*-invariant*, Math. Proc. Cambridge Philos. Soc. **115** (1993), 253–281.

[18] ———, *Quantum invariants for 3-manifolds*, The 3rd Korea-Japan School of Knots and Links (K.H. Ko and G.T. Jin, eds.), Proc. Appl. Math. Workshops, vol. 4, 1994, pp. 129–143.

[19] ———, *Quantum* $SO(3)$*-invariants dominate the* $SU(2)$*-invariant of Casson and Walker*, Math. Proc. Cambridge Philos. Soc. **117** (1995), 237–249.

[20] ———, *Quantum* $SU(2)$*-invariants of three-manifolds associated with the trivial first cohomology class modulo two*, Low Dimensional Topology (H. Nencka, ed.), Contemporary Mathematics, vol. 233, 1999.

[21] H. Murakami, T. Ohtsuki, and M. Okada, *Invariants of three-manifolds derived from linking matrices of framed links*, Osaka J. Math. **29** (1992), 545–572.

[22] T. Ohtsuki, *A polynomial invariant of integral homology 3-spheres*, Math. Proc. Cambridge Philos. Soc. **117** (1995), no. 1, 83–112.

[23] ———, *Finite type invariants of integral homology 3-spheres*, J. Knot Theory Ramifications **5** (1996), no. 1, 101–115.

[24] ———, *A polynomial invariant of rational homology 3-spheres*, Invent. Math. **123** (1996), no. 2, 241–257.

[25] ———, *A filtration of the set of integral homology 3-spheres*, Proceedings of the International Congress of Mathematicians, Vol. II (Berlin, 1998), 1998, pp. 473–482.

[26] N.Yu. Reshetikhin and V.G. Turaev, *Invariants of 3-manifolds via link polynomials and quantum groups*, Invent. Math. **103** (1991), 547–597.

[27] M. Sakuma, *An evaluation of the Jones polynomial of a parallel link*, Math. Proc. Cambridge Philos. Soc. **104** (1988), 105–113.
[28] T. Takata and Y. Yokota, *The $PSU(N)$ invariants of 3-manifolds are algebraic integers*, J. Knot Theory Ramifications **8** (1999), no. 4, 521–532.
[29] V. Turaev and H. Wenzl, *Quantum invariants of 3-manifolds associated with classical simple Lie algebras*, Internat. J. Math. **4** (1993), 323–358.
[30] V. G. Turaev, *State sum models in low-dimensional topology*, Proceedings of the International Congress of Mathematicians, (Kyoto, 1990), Vol. I, Math. Soc. Japan, Tokyo, 1991, pp. 689–698.
[31] K. Walker, *An Extension of Casson's Invariant*, Annals of Mathematics Studies, vol. 126, Princeton University Press, Princeton, 1992.
[32] H. Wenzl, *Braids and invariants of 3-manifolds*, Invent. Math. **114** (1993), 235–275.
[33] E. Witten, *Quantum field theory and the Jones polynomial*, Comm. Math. Phys. **121** (1989), 351–399.

Department of Mathematics, School of Science and Engineering, Waseda University, Ohkubo, Shinjuku, TOKYO 169-8555, JAPAN, and Mittag-Leffler Institute, Auravägen 17, S-182 62, Djursholm, SWEDEN
E-mail address: hitoshi@uguisu.co.jp

RELATIONS AMONG SELF DELTA-EQUIVALENCE AND SELF SHARP-EQUIVALENCES FOR LINKS

YASUTAKA NAKANISHI
Department of Mathematics, Kobe University, Nada-ku, Kobe 657-8501, Japan
E-mail: naknaisi@math.kobe-u.ac.jp

TETSUO SHIBUYA
Department of Mathematics, Osaka Institute of Technology, Asahi-ku, Osaka 535-8585, Japan
E-mail: shibuya@ge.oit.ac.jp

ABSTRACT

In this note, we will study on generalized unknotting operations for links, especially with the condition to restrict their deformations for the same component, and we will show their differences.

Keywords: self Δ-unknotting operation, self #-unknotting operation, Alexander polynomial

1. Introduction

Throughout this note, a link means a union of mutually disjoint tame and oriented circles in the oriented 3-dimensional Euclidean space $\mathbf{R}^3$.

For two link diagrams K and L which differ only in one place as in Fig. 1 (a) (or (b), (c)), a local move between K and L is called a #(I)-move (or a #(II)-move, a Δ-move, respectively). (A #(I)-move is originally called a #-move in [8]. A #(II)-move is originally called a pass-move in [5].) Furthermore, for two links κ and λ represented by K and L, κ and λ are said to be transformed into each other by a #(I)-move (or a #(II)-move, a Δ-move, respectively). It is known that any two knots can be transformed into each other by a finite sequence of #(I)-moves. Two knots can be transformed into each other by a finite sequence of #(II)-moves if and only if the Arf invariants of the two knots coincide. Any two knots can be transformed into each other by a finite sequence of Δ-moves. There are also corresponding results for links, for example, in [5], [6], [8], [9], [15], [16], [18].

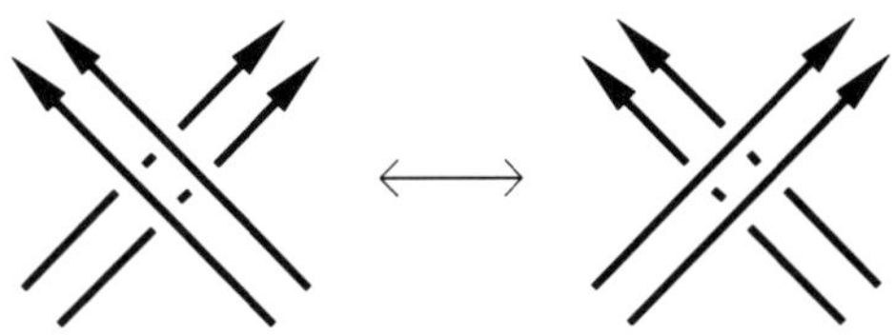

(a)

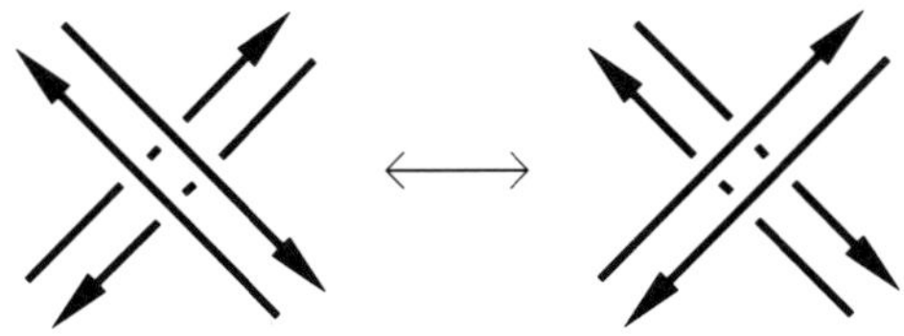

(b)

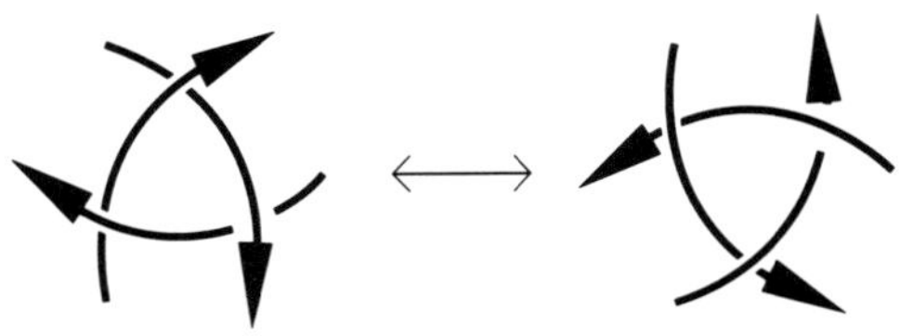

(c)

Figure 1

In this note, we shall study the special types of the above moves as follows. If all arcs illustrated in the Figure 1 are contained in the same component, the above move is called a self #(I)-move (or a self #(II)-move, a self Δ-move, respectively). Furthermore, for two links κ and λ represented by K and L, κ and λ are said to be transformed into each other by a self #(I)-move (or a self #(II)-move, a self Δ-move, respectively). For two links κ and λ, if κ can be transformed into λ by a finite sequence of self #(I)-moves (or self #(II)-moves, self Δ-moves, respectively), κ and λ are said to be *self #-equivalent (I)* (or *self #-equivalent (II)*, *self Δ-equivalent*, respectively). We shall consider the relations among those three concepts of equivalence relations for links.

The following is known in [9], [15].

Proposition. *For two links κ and λ, if κ and λ are self #-equivalent (II), then they are self #-equivalent (I).*

But the converse does not hold. That is because a self #(I)-move changes the Arf invariant, denoted by $\varphi(\lambda)$, of knots (see [13]). On the other hand, a self #(II)-move does not change the Arf invariant of knots (see [5], [6]). However, we shall show that, in the case that the Arf invariants of each corresponding components coincide, the converse holds as in Theorem 1. The proof will be given in the section 2.

Theorem 1. *For two n-component links $\kappa = \kappa_1 \cup \cdots \cup \kappa_n$ and $\lambda = \lambda_1 \cup \cdots \cup \lambda_n$, if κ and λ are self #-equivalent (I) and the Arf invariants of each corresponding components coincide; $\varphi(\kappa_i) = \varphi(\lambda_i)$ for each $i = 1, \ldots, n$, then κ and λ are self #-equivalent (II).*

Next we consider the relation between self #-equivalence (I) and self Δ-equivalence for links.

Theorem 2. *For two links κ and λ, if κ and λ are self Δ-equivalent, then they are self #-equivalent (I).*

The above Theorem 2 is already proved by the second author in [16]. Here we shall give an alternate proof by the results in [1] and [11] in the section 2.

We remark that the converse does not hold as in Theorem 3. The proof will be given in the sections 3 and 4.

Theorem 3. *For any integer n greater than 1, there are infinitely many n-component links which are self #-equivalent (I) but not self Δ-equivalent.*

2. Proofs of Theorems 1 and 2

In order to prove Theorem 1, we prepare the following Lemma 1.1.

The concepts of #-cobordisms (I) and (II) for n-component links are introduced by the second author in [15] and [17]. We consider two links κ in $\mathbf{R}^3 \times \{1\}$ and λ in $\mathbf{R}^3 \times \{0\}$, and a set of proper annuli $\mathcal{A}$ in $\mathbf{R}^3 \times [0,1]$ with $\mathcal{A} \cap \mathbf{R}^3 \times \{1\} = \kappa$ and $\mathcal{A} \cap \mathbf{R}^3 \times \{0\} = -r\lambda$, where $-r\lambda$ means the reflective inverse of λ. For each (annulus) component A of $\mathcal{A}$, if A is locally flat except a finite number of singular points in the interior *int* $\mathcal{A}$ of $\mathcal{A}$ such that for each singular point P the pair of boundaries of regular neighbourhoods $(\partial N(P : \mathbf{R}^3 \times [0,1]), \partial N(P : \mathcal{A}))$ is the link as illustrated in Figure 2 (a) (or (b)), $\mathcal{A}$ is called #(I)-annuli (or #(II)-annuli). Furthermore, if $\mathcal{A}$ is a set of mutually disjoint annuli, the two links κ and λ are said to be #-cobordant (I) (or #-cobordant (II), respectively).

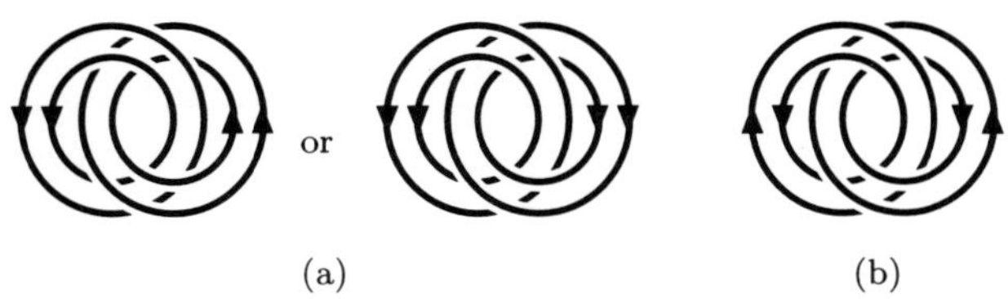

Figure 2

Lemma 1.1. *Two links κ and λ in $\mathbf{R}^3$ are self #-equivalent (I) (or self #-equivalent (II)) if and only if they are #-cobordant (I) (or #-cobordant (II), respectively).*

The proof of Lemma 1.1 can be obtained by a similar argument as in that of Lemma 1.17 in [14].

Proof of Theorem 1. Suppose that two n-component links $\kappa = \kappa_1 \cup \cdots \cup \kappa_n$ in $\mathbf{R}^3 \times \{1\}$ and $\lambda = \lambda_1 \cup \cdots \cup \lambda_n$ in $\mathbf{R}^3 \times \{0\}$ are self #-equivalent (I) with $\varphi(\kappa_i) = \varphi(\lambda_i)$ for each $i = 1, \ldots, n$. By Lemma 1.1, there is a union of mutually disjoint #(I)-annuli $\mathcal{A} = A_1 \cup \cdots \cup A_n$ in $\mathbf{R}^3 \times [0,1]$ between κ and λ with $\partial A_i = \kappa_i \cup \lambda_i$. Here the set $\mathcal{S}(A_i)$ of singular points of A_i consists of finite isolated points in *Int* A_i and, for each singular point P in $\mathcal{S}(A_i)$, $(\partial N(P : \mathbf{R}^3 \times [0,1]), \partial N(P : \mathcal{A}))$ is the link as illustrated in Figure 2 (a), whose Arf invariant is equal to 1. As $\varphi(\kappa_i) = \varphi(\lambda_i)$, $\mathcal{S}(A_i)$ consists of even number of points, say $P_{i1}, \ldots, P_{i2m_i}$, each of which is contained in *int* A_i. Let $\alpha_{i1}, \ldots, \alpha_{im_i}$ be mutually disjoint simple arcs on *int* A_i with $\partial\alpha_{ij} = P_{ij} \cup P_{im_i+j}$ for $j = 1, \ldots, m_i$. Then $\partial N(\alpha_{ij} : A_{ij})$ $(= L_{ij})$ in $\partial N(\alpha_{ij} : \mathbf{R}^3 \times [0,1])$ is obtained by a 1-fusion, [7], of two links as illustrated in Fig. 2 (a). Since the Arf invariant of each link in Fig. 2 (a) is 1, that of L_{ij} is 0. Therefore L_{ij} can be deformed into a trivial link $\mathcal{O}$ by a finite sequence of #(II)-moves (not self #(II)-moves) by Theorem A.2 in [9]. Hence there are proper disks $\mathcal{D}_{ij}$ (which are not mutually disjoint) in $N(\alpha_{ij} : \mathbf{R}^3 \times [0,1])$ with $\partial\mathcal{D}_{ij} = L_{ij}$ such that $\mathcal{S}(\mathcal{D}_{ij})$ consists of isolated points in $int\mathcal{D}_{ij}$ and, for a point Q of $\mathcal{S}(\mathcal{D}_{ij})$, $\partial N(Q : \mathcal{D}_{ij})$ is the link in Fig. 2 (b). Let $A_{i0} = (A_i - \bigcup_{j=1}^{m_i} N(\alpha_{ij} : A_i)) \cup (\bigcup_{j=1}^{m_i} \mathcal{D}_{ij})$ and $\mathcal{A}_0 = A_{10} \cup \cdots \cup A_{n0}$. Then $\mathcal{A}_0$ is a union of mutually disjoint #(II)-annuli between κ and λ in $\mathbf{R}^3 \times [0,1]$ by the construction of each A_{i0}. Therefore κ and λ are self #-equivalent (II) by Lemma 1.1. The proof is complete.

Proof of Theorem 2. A (self) Δ-move can be realized by two (self) local moves illustrated in Fig. 3 (a), (b) and each of which can be realized by a finite sequence of (self) #(I)- and (self) #(II)-moves, [1], [11]. Hence we obtain Theorem 2 by Proposition.

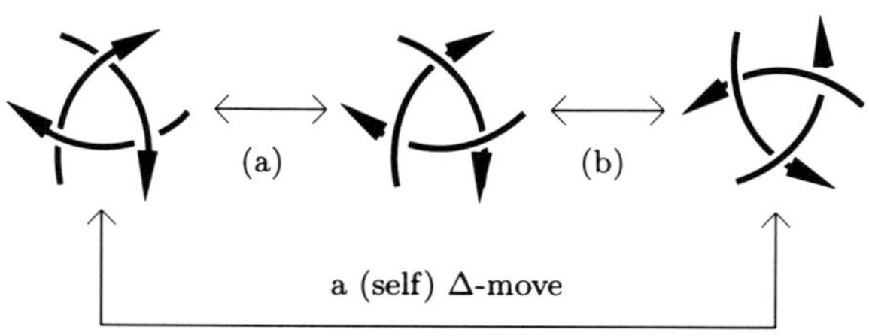

Figure 3

3. Proof of Theorem 3

In order to prove Theorem 3, we prepare the following Lemmas 3.1, 3.2, and 3.3.

Lemma 3.1. *For two links κ and λ, if κ and λ are self Δ-equivalent, then, for properly chosen Alexander matrices $A_\kappa(t_1, \ldots, t_n)$ and $A_\lambda(t_1, \ldots, t_n)$, we have*

$$A_\kappa(t_1, \ldots, t_n) \equiv A_\lambda(t_1, \ldots, t_n) \ mod\ \{(1-t_1)^2, \ldots, (1-t_n)^2\},$$

and hence, for the Alexander polynomials $\Delta_\kappa(t_1, \dots, t_n)$ *and* $\Delta_\lambda(t_1, \dots, t_n)$,

$$\Delta_\kappa(t_1, \dots, t_n) \equiv \pm t_1^{r_1} \cdots t_n^{r_n} \Delta_\lambda(t_1, \dots, t_n) \bmod \{(1-t_1)^2, \dots, (1-t_n)^2\}.$$

Further, we have similar statements for the elementary ideals of deficiency greater than 1.

In the above, $f(t_1, \dots, t_n) \equiv g(t_1, \dots, t_n) \bmod \{h_1(t_1, \dots, t_n), \dots, h_j(t_1, \dots, t_n)\}$ means that $f(t_1, \dots, t_n)$ and $g(t_1, \dots, t_n)$ are in the same class of the quotient $\mathbf{Z}\langle t_1, \dots, t_n\rangle/(h_1(t_1, \dots, t_n), \dots, h_j(t_1, \dots, t_n))$ of the n variable Laurent polynomial ring by the ideal generated by $h_1(t_1, \dots, t_n), \dots, h_j(t_1, \dots, t_n)$. The proof of Lemma 3.1 will be given in the section 4.

For a 2-component link $\lambda = \lambda_1 \cup \lambda_2$ with the linking number $lk(\lambda_1, \lambda_2) = 0$, the following fact is known (see [2], [10]).

Lemma 3.2. *The Alexander polynomial of a 2-component link* $\lambda = \lambda_1 \cup \lambda_2$ *with the linking number* $lk(\lambda_1, \lambda_2) = 0$ *is characterized by the following condition:*

$$\Delta_\lambda(t_1, t_2) = (1-t_1)(1-t_2)m(t_1, t_2) \text{ where } m(t_1^{-1}, t_2^{-1}) = m(t_1, t_2).$$

From Lemmas 3.1 and 3.2, we have the following.

Lemma 3.3. *For two 2-component links* $\kappa = \kappa_1 \cup \kappa_2$ *and* $\lambda = \lambda_1 \cup \lambda_2$ *with linking number* 0, *if* κ *and* λ *are self* Δ*-equivalent, then we have*

$$\Delta_\kappa(t_1, t_2) \equiv \pm t_1^{r_1} t_2^{r_2} \Delta_\lambda(t_1, t_2) \bmod \{(1-t_1)^2, (1-t_2)^2\}.$$

Further, for $m(t_1, t_2)$ *in the sense of Lemma 3.2, we have*

$$m_\kappa(1,1) = \pm m_\lambda(1,1).$$

For the link $\kappa(2,i)$ illustrated in Fig. 4, we have $m_{\kappa(2,i)}(1,1) = i$. Therefore, for two integers i and j with $|i| \neq |j|$, the links $\kappa(2,i)$ and $\kappa(2,j)$ are not self Δ-equivalent by Lemma 3.3. For any integer n greater than 2, we consider the n-component link $\kappa(n,i)$ which is the split union of $\kappa(2,i)$ and the $(n-2)$-component trivial link. For two integers i and j with $|i| \neq |j|$, the links $\kappa(n,i)$ and $\kappa(n,j)$ are not self Δ-equivalent. On the other hand, if $i \equiv j \bmod 2$, $\kappa(n,i)$ and $\kappa(n,j)$ are self #-equivalent (II) and so self #-equivalent (I) by Proposition. The proof of Theorem 3 is complete.

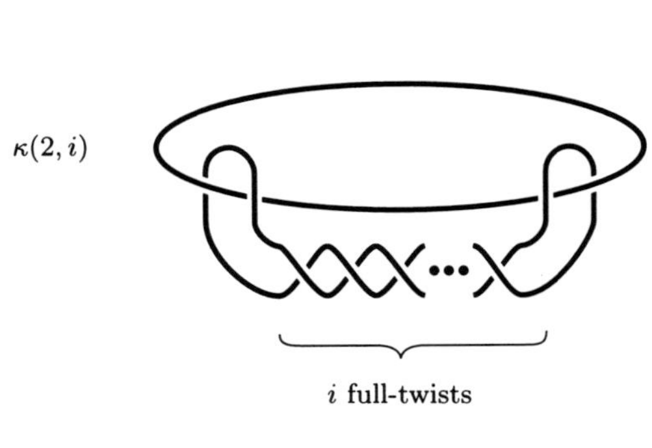

Figure 4

Note. We can see the four 2-component links $0_1^2, 5_1^2, 7_3^2, 9_{10}^2$ with the linking number 0, in the table of Rolfsen [12]. Here, $\kappa(0) = 0_1^2, \kappa(1) = 5_1^2, \kappa(2) = 7_3^2, \kappa(3) = 9_{10}^2$.

4. Proof of Lemma 3.1

To prove Lemma 3.1, it is sufficient to show the following Lemma 3.1a.

Lemma 3.1a. *Let $\kappa = \kappa_1 \cup \cdots \cup \kappa_n$ and $\lambda = \lambda_1 \cup \cdots \cup \lambda_n$ be two n-component links, whose diagrams differ only one place as in Fig. 1 (c), where all arcs belong to the i-th components κ_i and λ_i. Then, for properly chosen Alexander matrices $A_\kappa(t_1, \ldots, t_n)$ and $A_\lambda(t_1, \ldots, t_n)$, we have*

$$A_\kappa(t_1, \ldots, t_n) \equiv A_\lambda(t_1, \ldots, t_n) \bmod \{(1-t_i)^2\},$$

and hence, for the Alexander polynomials $\Delta_\kappa(t_1, \ldots, t_n)$ and $\Delta_\lambda(t_1, \ldots, t_n)$,

$$\Delta_\kappa(t_1, \ldots, t_n) \equiv \pm t_1^{r_1} \cdots t_n^{r_n} \Delta_\lambda(t_1, \ldots, t_n) \bmod \{(1-t_i)^2\}.$$

Further, we have similar statements for the elementary ideals of deficiency greater than 1.

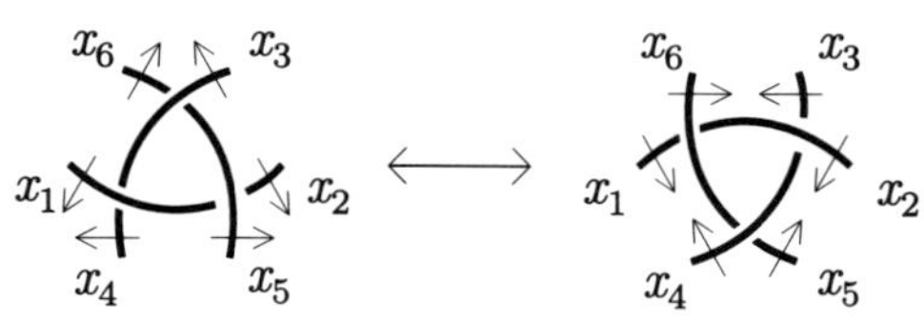

Figure 5

Proof of Lemma 3.1a. We can choose generators $x_1, \ldots, x_6$ of the fundamental group $\pi_1(S^3 - \kappa)$ as shown in the left figure of Fig. 5, and further we choose other generators $x_7, \ldots$ as usual. Similarly, we choose generators $x_1, \ldots, x_6$ of the fundamental group $\pi_1(S^3 - \lambda)$ as shown in the right figure of Fig. 5, and further we choose other generators $x_7, \ldots$ as usual. Then, we have group presentations of $\pi_1(S^3 - \kappa)$ and $\pi_1(S^3 - \lambda)$:

$$|x_1, x_2, x_3, x_4, x_5, x_6, x_7, \ldots \quad | \quad x_1x_5x_2^{-1}x_5^{-1}, x_3x_1x_4^{-1}x_1^{-1}, x_5x_3x_6^{-1}x_3^{-1}, r_j|,$$

$$|x_1, x_2, x_3, x_4, x_5, x_6, x_7, \ldots \quad | \quad x_1x_6x_2^{-1}x_6^{-1}, x_3x_2x_4^{-1}x_2^{-1}, x_5x_4x_6^{-1}x_4^{-1}, r_j|.$$

Here, r_j are relations corresponding to other crossings. We use Fox's free differential calculus ([3, 4]). We have the Alexander matrices $A_\kappa(t_1, \ldots, t_n)$ and $A_\lambda(t_1, \ldots, t_n)$ of κ and λ as follows:

$$\left|\begin{array}{ccccccc} 1 & -t_i & 0 & 0 & t_i - 1 & 0 & \mathbf{0} \\ t_i - 1 & 0 & 1 & -t_i & 0 & 0 & \mathbf{0} \\ 0 & 0 & t_i - 1 & 0 & 1 & -t_i & \mathbf{0} \\ \mathbf{a} & \mathbf{b} & \mathbf{c} & \mathbf{d} & \mathbf{e} & \mathbf{f} & \mathbf{M} \end{array}\right|,$$

$$\left|\begin{array}{ccccccc} 1 & -t_i & 0 & 0 & 0 & t_i - 1 & \mathbf{0} \\ 0 & t_i - 1 & 1 & -t_i & 0 & 0 & \mathbf{0} \\ 0 & 0 & 0 & t_i - 1 & 1 & -t_i & \mathbf{0} \\ \mathbf{a} & \mathbf{b} & \mathbf{c} & \mathbf{d} & \mathbf{e} & \mathbf{f} & \mathbf{M} \end{array}\right|.$$

Here, **a**, **b**, **c**, **d**, **e**, **f** are column-vectors and **M** is a matrix, and the submatrix $|\mathbf{a} \;\; \mathbf{b} \;\; \mathbf{c} \;\; \mathbf{d} \;\; \mathbf{e} \;\; \mathbf{f} \;\; \mathbf{M}|$ is corresponding to the result of calculation $\mathfrak{a}\phi(\partial r_j / \partial x_i)$. Each $\mathbf{0}$ is a row-vector with all entries 0.

By fundamental deformations of presentation matrices, the above two matrices become to the following forms:

$$\begin{vmatrix} 1 & -t_i & 0 & 0 & t_i-1 & 0 & \mathbf{0} \\ t_i-1 & 0 & 1 & -t_i & 0 & 0 & \mathbf{0} \\ 0 & 0 & t_i-1 & 0 & 1 & -t_i & \mathbf{0} \\ \mathbf{a} & \mathbf{b} & \mathbf{c} & \mathbf{d} & \mathbf{e} & \mathbf{f} & \mathbf{M} \end{vmatrix}$$

$$\sim \begin{vmatrix} t_i-1 & 1 & -t_i & 0 & 0 & \mathbf{0} \\ 0 & t_i-1 & 0 & 1 & -t_i & \mathbf{0} \\ \mathbf{a}+t_i^{-1}\mathbf{b} & \mathbf{c} & \mathbf{d} & \mathbf{e}+t_i^{-1}(t_i-1)\mathbf{b} & \mathbf{f} & \mathbf{M} \end{vmatrix}$$

$$\sim \begin{vmatrix} 0 & t_i-1 & 1 & -t_i & \mathbf{0} \\ \mathbf{a}+t_i^{-1}\mathbf{b}+t_i^{-1}(t_i-1)\mathbf{d} & \mathbf{c}+t_i^{-1}\mathbf{d} & \mathbf{e}+t_i^{-1}(t_i-1)\mathbf{b} & \mathbf{f} & \mathbf{M} \end{vmatrix}$$

$$\sim |\,\mathbf{a}+t_i^{-1}\mathbf{b}+t_i^{-1}(t_i-1)\mathbf{d} \quad \mathbf{c}+t_i^{-1}\mathbf{d}+t_i^{-1}(t_i-1)\mathbf{f} \quad \mathbf{e}+t_i^{-1}(t_i-1)\mathbf{b}+t_i^{-1}\mathbf{f} \quad \mathbf{M}\,|$$

$$\sim |\,t_i\mathbf{a}+\mathbf{b}+(t_i-1)\mathbf{d} \quad t_i\mathbf{c}+\mathbf{d}+(t_i-1)\mathbf{f} \quad t_i\mathbf{e}+(t_i-1)\mathbf{b}+\mathbf{f} \quad \mathbf{M}\,|,$$

$$\begin{vmatrix} 1 & -t_i & 0 & 0 & 0 & t_i-1 & \mathbf{0} \\ 0 & t_i-1 & 1 & -t_i & 0 & 0 & \mathbf{0} \\ 0 & 0 & 0 & t_i-1 & 1 & -t_i & \mathbf{0} \\ \mathbf{a} & \mathbf{b} & \mathbf{c} & \mathbf{d} & \mathbf{e} & \mathbf{f} & \mathbf{M} \end{vmatrix}$$

$$\sim \begin{vmatrix} t_i-1 & 1 & -t_i & 0 & 0 & \mathbf{0} \\ 0 & 0 & t_i-1 & 1 & -t_i & \mathbf{0} \\ \mathbf{b}+t_i\mathbf{a} & \mathbf{c} & \mathbf{d} & \mathbf{e} & \mathbf{f}-(t_i-1)\mathbf{a} & \mathbf{M} \end{vmatrix}$$

$$\sim \begin{vmatrix} 0 & t_i-1 & 1 & -t_i & \mathbf{0} \\ \mathbf{b}+t_i\mathbf{a}-(t_i-1)\mathbf{c} & \mathbf{d}+t_i\mathbf{c} & \mathbf{e} & \mathbf{f}-(t_i-1)\mathbf{a} & \mathbf{M} \end{vmatrix}$$

$$\sim |\,\mathbf{b}+t_i\mathbf{a}-(t_i-1)\mathbf{c} \quad \mathbf{d}+t_i\mathbf{c}-(t_i-1)\mathbf{e} \quad \mathbf{f}-(t_i-1)\mathbf{a}+t_i\mathbf{e} \quad \mathbf{M}\,|.$$

Here, we remark that

$$\mathbf{f}-(t_i-1)\mathbf{a}+t_i\mathbf{e}+(t_i-1)(\mathbf{b}+t_i\mathbf{a}-(t_i-1)\mathbf{c}) \equiv \mathbf{f}+t_i\mathbf{e}+(t_i-1)\mathbf{b} \bmod (t_i-1)^2,$$

$$\mathbf{d}+t_i\mathbf{c}-(t_i-1)\mathbf{e}+(t_i-1)(\mathbf{f}+t_i\mathbf{e}+(t_i-1)\mathbf{b}) \equiv \mathbf{d}+t_i\mathbf{c}+(t_i-1)\mathbf{f} \bmod (t_i-1)^2,$$

$$\mathbf{b}+t_i\mathbf{a}-(t_i-1)\mathbf{c}+(t_i-1)(\mathbf{d}+t_i\mathbf{c}+(t_i-1)\mathbf{f}) \equiv \mathbf{b}+t_i\mathbf{a}+(t_i-1)\mathbf{d} \bmod (t_i-1)^2.$$

So, the above matrix is congruent mod $(t_i-1)^2$ to the previous one. Hence we complete the proof.

For the consideration on orders of components, we have the proof of Lemma 3.1 from the above Lemma 3.1a.

References

[1] H. Aida, *The oriented Δ_{ij}-moves on links*, Kobe J. Math. **9** (1992), 163–170.

[2] J.L. Bailey, *Alexander invariants of links*, Ph.D. Thesis (1977), Univ. of British Columbia.

[3] R.H. Fox, *Free differential calculus, I*, Ann. of Math. **57** (1953), 547–560.

[4] R.H. Fox, *Free differential calculus, II*, Ann. of Math. **59** (1954), 196–210.

[5] L.H. Kauffman, *Formal knot theory*, Math. Notes, vol. 30, Princeton Univ. Press, Princeton, New Jersey, 1983.

[6] L.H. Kauffman and T.F. Banchoff, *Immersion and mod-2 quadratic forms*, Amer. Math. Monthly **84** (1977), 168–185.
[7] A. Kawauchi, T. Shibuya and S. Suzuki, *Descriptions on surfaces in four space, I*, Math. Sem. Notes Kobe Univ. **10** (1982), 75–125.
[8] H. Murakami, *Some metrics on classical knots*, Math. Ann. **270** (1985), 35–45.
[9] H. Murakami and Y. Nakanishi, *On a certain move generating link-homology*, Math. Ann. **284** (1989), 75–89.
[10] Y. Nakanishi, *A surgical view of Alexander invariants of links*, Math. Seminar Notes Kobe Univ. **8** (1980), 199–218.
[11] Y. Nakanishi, *Replacements in the Conway third identity*, Tokyo J. Math. **14** (1991), 197–203.
[12] D. Rolfsen, *Knots and links*, Publish or Perish Inc., Berkeley, 1976.
[13] R.A. Robertello, *An invariant of knot cobordism*, Comm. Pure Appl. Math. **18** (1965), 543–555.
[14] T. Shibuya, *On the cobordisms of links in 3-space*, Kobe J. Math. **1** (1984), 119 –131.
[15] T. Shibuya, *Self #-unknotting operation of links*, Mem. Osaka Insti. Tech. **34** (1989), 9 –17.
[16] T. Shibuya, *Two self #-equivalences of links in solid tori*, Mem. Osaka Insti. Tech. **35** (1990), 13 –24.
[17] T. Shibuya, *Self Δ-equivalence of ribbon links*, Osaka J. Math. **33** (1996), 751 –760.
[18] M. Yamasaki, *On a surface in S^3 (in Japanese)*, Sûrikaisekikenkyûsho Kôkyûroku **297** (1977), 92–99.

Towards a complexity measure theory for vortex tangles

Renzo L. Ricca

Department of Mathematics, University College London
Gower Street, London WC1E 6BT, UK
E-mail: ricca@math.ucl.ac.uk

Abstract

In this paper we address the problem of measuring structural complexity of generic tangles of vortex lines in a fluid domain, by using a combination of geometric and topological techniques. To this end new concepts based on the idea of structural 'tropicity' are introduced to determine 'tubeness', 'sheetness' and 'bulkiness' of a vortex tangle and to evaluate the degree of topological entanglement. A number of cases are considered: from highly organised, coherent vortex regions, given by the embedding of vortex coils, knots and links on nested tori, to less organised vortical flows, such as tangles of chaotic vortex lines. Various measures of linking (and helicity) are presented as well as estimates of writhing and crossing numbers based on geometric and topological information. Moreover, by using the concept of signature preserving flow we extend the definition of classical stability to include wilder vortex dynamics that during evolution preserve structural complexity. The tools and the new concepts presented in this paper are useful for the classification and study of general flow fields and can be employed to develop computational techniques for measuring structural complexity.

1 Vortex structures as complex tangles of vortex lines

Modelling and control of turbulent flows still defy mathematics for difficulty and complexity. Spontaneous formation and interaction of self-organized, coherent vortex regions are generic features of turbulent flows (Vincent & Meneguzzi, 1991; Jiménez *et al.*, 1993). Coherence (for a proper definition see §6 below) is essentially due to the strong correlation of dynamical properties (such as fluid pressure, velocity, vorticity, strain, etc.) confined to localized regions in the fluid domain $\mathcal{D}$. These regions are occupied by extended, filamentary 'structures' in the shape of vortex tubes and sheets (see, for example, Schwarz, 1988; She *et al.*, 1990; figure 1). These structures are present where vorticity $\boldsymbol{\omega}(\mathbf{X}, t) = \nabla \times \mathbf{u}(\mathbf{X}, t)$ (where $\mathbf{u}(\mathbf{X}, t)$ denotes

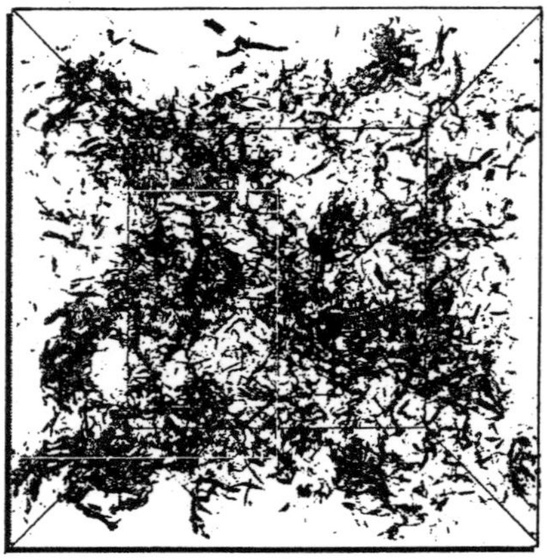

Figure 1: Localization of vortex structures in a visualization of data from 512^3-element turbulence simulation in a numerical box (from Fernandez *et al.*, 1996).

fluid velocity and everything is smooth function of the position vector $\mathbf{X}$ in $\mathbb{R}^3$ and time t) is highly localized in a compact domain $\Omega \subset \mathbb{R}^3$, of volume $V(\Omega)$ small compared to the total volume $V(\mathcal{D})$. We assume that the fluid in $\mathcal{D}$ is homogeneous and incompressible, i.e. $\nabla \cdot \mathbf{u} = 0$, and that there are no sources or sinks of vorticity, that is $\nabla \cdot \boldsymbol{\omega} = 0$.

Numerical solutions of the real fluid equations (the Navier-Stokes equations) and real-time simulations of complex flow patterns (see, for example, the papers edited by Jiménez, 1991; Barenghi *et al.*, 1998) show that vortex lines are fundamental constituents of complex vortical flows. Recent progress in computational fluid dynamics, visualization techniques (Zabusky *et al.*, 1993; Fernandez *et al.*, 1996), and fast data processing make it possible accurate identification and detailed description of vortical structures. Formation, interaction and re-structuring of complex tangles of vortex lines can now be traced with unprecedented resolution (figure 2). Evolution and interaction of large-scale vortices (produced naturally or artificially), as well as detailed mechanisms of braiding, reconnection and linking, are all fundamental mechanisms of vortex dynamics. Geometric and topological measures provide very important information. Recent applications of geometric and topological concepts to fluid mechanics in general, and vortex dynamics in particular, have led to the development of new tools in the mathematical study of fluid flows (see the papers edited by Moffatt & Tsinober, 1990 and Moffatt *et al.*, 1992; see also Ricca & Berger, 1996). A combination of differential geometric techniques and knot theoretical tools has proved to be particularly useful (Ricca, 1998a).

In the context of the Euler equations (inviscid limit of the Navier-Stokes equations), where there is no dissipation, vortex lines are material lines frozen in the

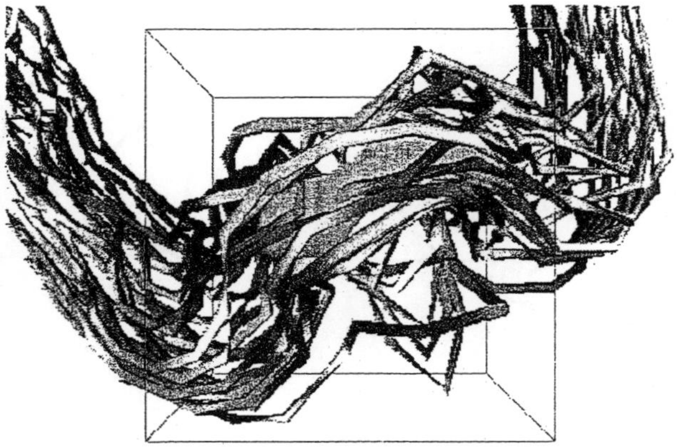

Figure 2: Close-up view of interaction and reconnection of vortex lines visualized by advanced visiometric techniques of computational fluid dynamics (from Fernandez *et al.*, 1996).

fluid (consequence of Hemholtz's theorem). Moreover vortex strength (intensity) and all topological properties are conserved (consequence of the Cauchy solutions to the Euler equations, see Saffman, 1992). Reconnections and consequential changes in vortex topology, determined by the action of dissipative effects at a microscopic physical scale, cannot take place in ideal fluids. Despite these limitations, study of ideal evolution allows a detailed understanding of the role played by geometric and topological quantities and provide us with important tools for measuring structural complexity of real vortex tangles. It is in this context that we want to present and discuss the concepts below.

2 Tropicity and linking of particle trajectories, streamlines and vortex lines

Fluid motion is induced by a smooth velocity field $\mathbf{u} = \mathbf{u}(\mathbf{X}, t)$ in $\mathcal{D}$, which satisfies the solenoidal condition $\nabla \cdot \mathbf{u} = 0$ in $\mathcal{D}$ and the condition to be at rest ($\mathbf{u} = 0$) at infinity (or on $\partial\mathcal{D}$). As time passes, fluid particles move from one position to another: if $\mathbf{a} = \mathbf{X}(\mathbf{a}, 0)$ denotes the initial position of a fluid particle at time $t = 0$, then we can define a flow map φ associated with $\mathbf{u}$, so that at each instant t a particle at $\mathbf{a}$ is sent to the final position $\mathbf{X}(\mathbf{a}, t)$ by the mapping

$$\varphi : \mathbf{a} \to \mathbf{X}, \qquad \forall t \in I \, , \tag{1}$$

where I denotes some finite time interval. A *particle trajectory* $\chi_{\mathbf{a}}$ is the collection of all particle positions from $\mathbf{a}$ to $\mathbf{X}$, and is labeled by the initial position from

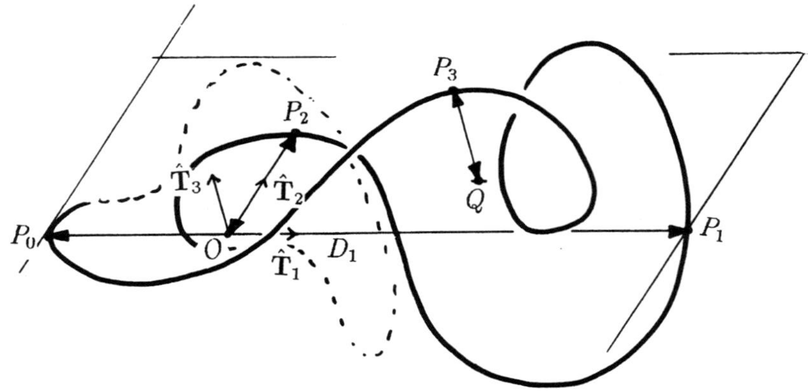

Figure 3: Directional tropic vectors of a curve in $\mathbb{R}^3$.

which it originates. Particle trajectories are the result of a time integration of positions, whereas material lines (such as streamlines or vortex lines) are instantaneous snapshots of vector field lines. In particular, at time t a *streamline* is a space curve $\Upsilon_t(s) = (x(s), y(s), z(s))$, whose directional tangent coincides with the velocity field $\mathbf{u} = (u_x, u_y, u_z)$ at each point of $\Upsilon_t(s)$, i.e.

$$\frac{dx/ds}{u_x} = \frac{dy/ds}{u_y} = \frac{dz/ds}{u_z} . \tag{2}$$

Similarly, a *vortex line* $\Gamma_t(s) = (x(s), y(s), z(s))$ is the embedding of the vorticity $\boldsymbol{\omega} = (\omega_x, \omega_y, \omega_z)$ onto the tangent field of $\Gamma_t(s)$, i.e.

$$\frac{dx/ds}{\omega_x} = \frac{dy/ds}{\omega_y} = \frac{dz/ds}{\omega_z} . \tag{3}$$

For steady flows, whose evolution is governed by autonomous velocity fields, particle trajectories and streamlines coincide.

Structural complexity of particle trajectories and material lines can be characterized according to their spatial distribution and degree of linking. The following concepts are of general applicability, regardless if referred to particle trajectories or material lines. To fix ideas let us consider a generic tangle[1] $\mathcal{T}$ of n vortex lines Γ_n, i.e. $\mathcal{T} = \bigcup_n \Gamma_n$. Let us introduce the concept of 'directional tropic vector' and 'tropic measure' to characterize the degree of tubeness, sheetness and bulkiness of $\mathcal{T}$. Take a vortex line Γ and two points P_i and P_j on Γ. The maximal extension of Γ in $\mathcal{D}$ (see figure 3) is given by

$$D_1 = \max \mathrm{d}(P_i, P_j) \equiv \overline{P_0P_1} = \sqrt{X^2 + Y^2 + Z^2} , \tag{4}$$

[1]In this paper we do not use the word 'tangle' in a strict topological sense.

where $P_0 = (x_0, y_0, z_0)$, $P_1 = (x_1, y_1, z_1)$ and

$$X = x_1 - x_0 \ , \quad Y = y_1 - y_0 \ , \quad Z = z_1 - z_0 \ . \tag{5}$$

The unit vector $\hat{\mathbf{T}}_1 = (P_1 - P_0)/D_1$ given by the director cosines X/D_1, Y/D_1, Z/D_1 provides the principal directional axis of Γ in $\mathcal{D}$, and we call it *first directional tropic vector*. Let us consider the width of Γ in $\mathcal{D}$. Take the maximal distance of another point P_i on Γ (not aligned with P_0 and P_1) from the line $l(P_0, P_1)$: the width of Γ is given by

$$D_2 = \max \mathrm{d}(P_i, l(P_0, P_1)) \equiv \overline{P_2 O} \ , \tag{6}$$

where $O \in l(P_0, P_1)$ (O being not necessarily a point on Γ) is the orthogonal projection of P_2 on $\overline{P_0 P_1}$. As above, we can define the *second directional tropic vector* as the unit vector $\hat{\mathbf{T}}_2 = (P_2 - O)/D_2$, which gives direction and orientation of the second principal directional axis of Γ in $\mathcal{D}$. Finally, take the plane $\pi = \pi(P_0, P_1, P_2)$ and consider the maximal distance from π of a fourth point P_i on Γ, but off the plane π. A measure of the three-dimensionality of the spatial distribution of Γ in $\mathcal{D}$ is given by

$$D_3 = \max \mathrm{d}(P_i, \pi(P_0, P_1, P_2)) \equiv \overline{P_3 Q} \ , \tag{7}$$

along the direction of $\hat{\mathbf{T}}_3 = \hat{\mathbf{T}}_1 \times \hat{\mathbf{T}}_2$ (*third directional tropic vector*).

Now let us consider the tangle $\mathcal{T} = \bigcup_n \Gamma_n$. The above concepts of tropicity are extended to $\mathcal{T}$, where $P_i \in \mathcal{T}$ and D_i ($i = 1, 2, 3$) denotes the maximal size of $\mathcal{T}$ along the i-th principal axis. Tropic measures are used to identify a particular shape of the flow structure. The tangle configuration can be characterized by one of the following tropic quantities:

- if $D_3 = O(D_2)$ and $D_2 \ll D_1$, then we have *tubeness* $\stackrel{\text{def}}{=} \Lambda_1 \equiv \dfrac{D_1}{D_2}$;
- if $D_3 \ll D_2$ and $D_2 = O(D_1)$, then we have *sheetness* $\stackrel{\text{def}}{=} \Lambda_2 \equiv \dfrac{D_2^2}{D_3^2}$;
- if $D_3 = O(D_2)$ and $D_2 = O(D_1)$, then we have *bulkiness* $\stackrel{\text{def}}{=} \Lambda_3 \equiv \dfrac{D_1 D_2 D_3}{D_3^3}$.

Other geometric measures can be employed to quantify alignment, rotation and stretching of particle trajectories and material lines, and for a discussion of these quantities we refer to the paper of Tabor (1992).

Directional tropic axes find an interesting application in the characterization of the topological entanglement of $\mathcal{T}$. A fundamental measure is given by the asymptotic linking number. The concept of linking number plays an important role in fluid dynamics (see the review article of Arnold & Khesin, 1998). For a tangle of particle trajectories (and material lines) we can define the asymptotic linking number as follows. Consider two particle trajectories $\chi_{\mathbf{a}}$ and $\chi_{\mathbf{b}}$, with $\mathbf{a} \neq \mathbf{b}$. After some long time T these trajectories are artificially closed on themselves by 'shortest paths',

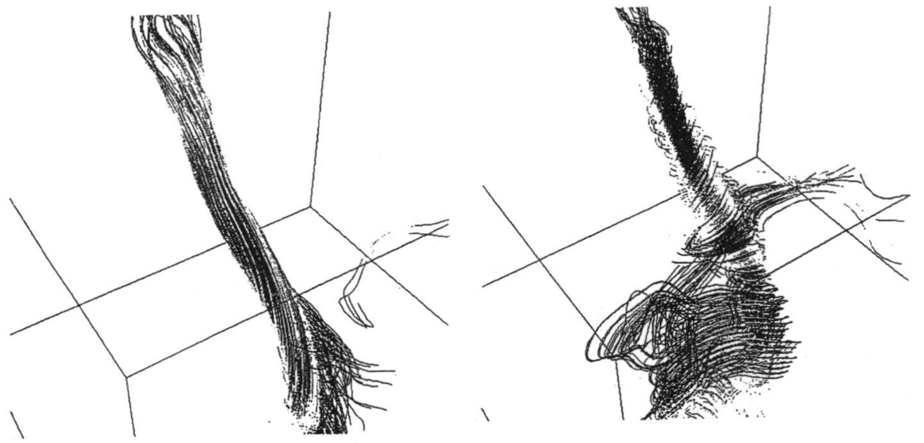

Figure 4: Close-up view of streamlines (on the left) and vortex lines (on the right) of a vortex tube in a numerical simulation of homogeneous, isotropic turbulent flow (from She *et al.*, 1990).

thereby defining two closed paths, say $\bar{\chi}_{\mathbf{a}}$ and $\bar{\chi}_{\mathbf{b}}$. The linking number $Lk(\bar{\chi}_{\mathbf{a}}, \bar{\chi}_{\mathbf{b}})$ is defined as the number of signed intersection points of $\bar{\chi}_{\mathbf{a}}$ with the surface bounded by $\bar{\chi}_{\mathbf{b}}$. The *asymptotic linking number* $\ell k(\mathbf{a}, \mathbf{b})$, associated with $\chi_{\mathbf{a}}$ and $\chi_{\mathbf{b}}$ is then defined as

$$\ell k(\mathbf{a}, \mathbf{b}) = \lim_{T \to \infty} \frac{Lk_T(\bar{\chi}_{\mathbf{a}}, \bar{\chi}_{\mathbf{b}})}{T^2} , \tag{8}$$

where $Lk_T(\bar{\chi}_{\mathbf{a}}, \bar{\chi}_{\mathbf{b}})$ is the linking number computed for orbits integrated for time T. The limit exists for almost every pair $\mathbf{a}$, $\mathbf{b}$ and $\ell k(\mathbf{a}, \mathbf{b})$ is of L^1 class. It is useful to define also the *average linking number* $\widetilde{\ell k} = <\ell k(\mathbf{a}, \mathbf{b})>$, where $< \cdot >$ denotes average over the entire fluid domain.

A fundamental connection between topology and fluid mechanics is provided by the relationship between linking number and helicity. The average linking of any two orbits of a vector field coincides with the Hopf invariant and is indeed a measure of the helicity $\mathcal{H}$ of that field. For the vorticity field $\boldsymbol{\omega}$ *kinetic helicity* is defined by

$$\mathcal{H} = \int_{\mathcal{D}} \mathbf{u} \cdot \boldsymbol{\omega} \, dV , \tag{9}$$

and is a measure of the average linking number of the tangle of vortex lines in $\mathcal{D}$.

A measure of relative linking of material lines about principal directional axes of a given flow pattern provides a useful estimate of the amount of winding localized in the flow. This measure is given by the *directional linking number* $\widetilde{\ell k}(\hat{\mathbf{T}}_i)$ defined by the degree of linking of material lines with a principal tropic axis $\vec{\lambda}_i$ (along $\hat{\mathbf{T}}_i$), i.e.

$$\widetilde{\ell k}(\hat{\mathbf{T}}_i) = <\ell k(\mathbf{a}, \hat{\mathbf{T}}_i)> , \tag{10}$$

where

$$\ell k(\mathbf{a}, \hat{\mathbf{T}}_i) = \lim_{T\to\infty} \frac{Lk_T(\bar{\chi}_{\mathbf{a}}, \vec{\lambda}_i)}{T^2} , \tag{11}$$

and where $\vec{\lambda}_i$ is kept fixed in $\mathcal{D}$. Figure 4 shows an example of localization, where there is a high degree of tubeness and winding of streamlines and vortex lines, measured by Λ_1 and $\widetilde{\ell k}(\hat{\mathbf{T}}_1)$.

3 Measures of organised structural complexity: vortex flow in a toroidal domain

Let us first recall a fundamental result of topological fluid mechanics (see the review paper by Ricca, 1998a). For a vortex tangle $\mathcal{T}$ total helicity $\mathcal{H}$ is given by

$$H(\mathcal{T}) = \sum_i Lk_i\, \Phi_i^2 + 2\sum_{i\neq j} Lk_{ij}\, \Phi_i\Phi_j , \tag{12}$$

where Lk_i is the (self-) linking number of the tube axis of the i-th vortex with respect to the framing induced by the embedding of $\boldsymbol{\omega}$ in $\mathcal{D}$, and Lk_{ij} denotes the linking number of the i-th vortex with the j-th vortex. All quantities in (12) are conserved during evolution. For each vortex component we have $Lk = Wr + Tw$, and so we need a way to estimate Wr and Tw. The writhe can be measured by the sum of the signed crossings of the diagram of the vortex axis onto some projection plane, averaged over all projections, that is

$$Wr = < n_+(\boldsymbol{\nu}) - n_-(\boldsymbol{\nu}) > , \tag{13}$$

where $<\cdot>$ denotes averaging over all directions $\boldsymbol{\nu}$ of projection, and $n_\pm$ denotes the number of apparent $\pm$ crossings, from the direction of projection $\boldsymbol{\nu}$. For a nearly plane curve (except small indentations to allow crossings) the writhe can be directly estimated by the sum of the signed crossings. The total twist Tw is given by the sum of the normalized total torsion

$$\frac{1}{2\pi}\int_\Gamma \tau(s)\, ds , \tag{14}$$

(τ torsion of the vortex axis Γ, s arc-length) and the normalized intrinsic twist of the $\boldsymbol{\omega}$-lines about Γ.

For organised vortex flows in a toroidal domain complexity can be evaluated by direct measures of topological entanglement. Let us identify Ω with a toroidal domain Π, with irrotational fluid in $\mathcal{D} - \Pi$. A vortex ring in a fluid at rest provides a good example. Let us assume that vorticity may be decomposed into a toroidal and poloidal component (along the longitudinal and meridian direction of Π, respectively), i.e. $\boldsymbol{\omega} = \boldsymbol{\omega}_t + \boldsymbol{\omega}_p$, and that $|\boldsymbol{\omega}_t| \gg |\boldsymbol{\omega}_p|$. The particular form

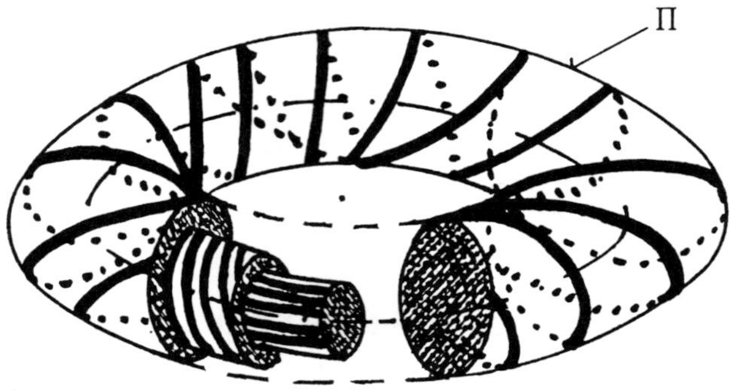

Figure 5: A collection of vortex lines uniformly coiled, knotted and linked on nested tori may provide an example of vortex ring modelling.

of the distribution, given by functional dependence of ω in Π, must satisfy a Poisson equation for vorticity, with appropriate boundary conditions on $\partial\Pi$ (Saffman, 1992). Hence, admissible distributions of vorticity must be solutions to the Poisson equation. Here we want to concentrate the attention on basic aspects of modelling, by simply assuming that the functional distribution ω (expression of the structural complexity of the flow field) is a solution of the Poisson equation. This assumption and the discussion that follows are consistent with the global theory of steady vortex rings omeomorphic to a torus (see Fraenkel & Berger, 1974). Vortex ring distributions of complex geometry and topology of the constituent vortex lines are indeed admissible as possible solutions. These include vortex lines uniformly coiled on nested tori, or in the shape of torus knots, or Hopf links on Π (figure 5).

We consider here three examples of vortex ring distributions.

Vortex coils

Take a uniform distribution of unknotted vortex coils, multi-covering a family of nested tori in $\mathcal{D}$. Complexity is given by the distribution of multi-coverings of toroidal and poloidal vortex coils, multiple foldings of the standard circle $\mathcal{U}_\circ$. From a geometric viewpoint we classify these coils into two categories: toroidal coils $\mathcal{U}_{m,1}$ and poloidal coils $\mathcal{U}_{1,m}$, where the first index denotes the number of times the coil wraps Π in the longitudinal direction (along the torus grand circle), and the second index denotes the number of times the coil wraps Π in the meridian direction (along the torus small circle; see figure 6).

All unknots are isotopic to one another, hence $\mathcal{U}_{m,1} \sim \mathcal{U}_{1,m} \sim \mathcal{U}_\circ$, and each coil contributes to the complexity of the system through individual contribution to the degree of linking Lk of the vortex ring. Here Lk denotes the limiting form of the Gauss linking number (self-linking) and is a measure of the linking of each coil with the torus axis. Since Lk is the sum of writhe Wr and total twist Tw, both decidedly

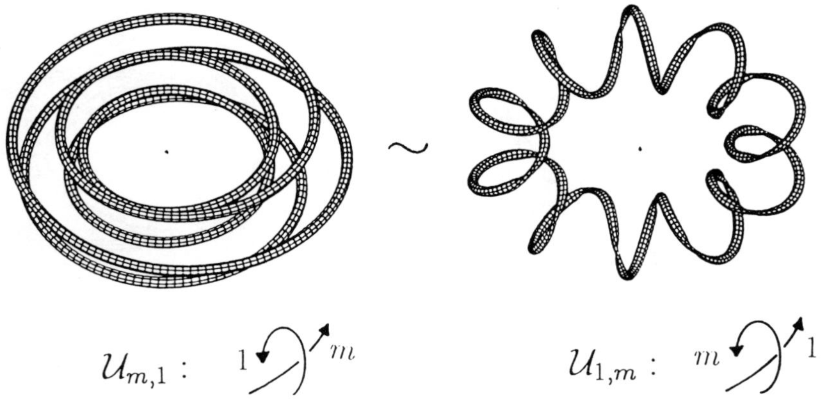

Figure 6: An example of toroidal coil $\mathcal{U}_{m,1}$ (on the left) and poloidal coil $\mathcal{U}_{1,m}$ (on the right). These coils wrap a mathematical torus Π (not visible in the figure) m times in the longitudinal (or meridian) direction and only once in the meridian (or longitudinal) direction.

non-zero, we have non-zero linking in the physical system. Hence, $Lk = m$ and the total helicity $\mathcal{H}$ (Moffatt & Ricca, 1992) amounts to

$$\mathcal{H} = \int_{\mathcal{D}} \mathbf{u} \cdot \boldsymbol{\omega} \, dV = (Wr + Tw)\, \Phi^2 = Lk\, \Phi^2 = m\, \Phi^2 \ , \tag{15}$$

where Φ is the total strength of the vortex ring. In this case m is an obvious measure of structural complexity.

Vortex knots

Vortex lines in the shape of torus knots can be taken as another example of elementary constituents of organised structural complexity. Torus knots $\mathcal{K}_{p,q}$ are closed curves wrapped around Π $p > 1$ times in the longitudinal direction, and $q > 1$ times in the meridian direction (p, q relatively prime integers; figure 7). Note that the two knots obtained by exchanging p and q (for given p and q) are topologically equivalent, i.e. $\mathcal{T}_{p,q} \sim \mathcal{T}_{q,p}$. When $p = 1$ (or $q = 1$) the curve is *not* knotted and reduces to the coil $\mathcal{U}_{1,m}$ (or $\mathcal{U}_{m,1}$). Winding number $w = q/p$ and self-linking number $Lk = pq$ provide a measure of topological complexity of the system.

Studies on vortex knots received new impetus recently, with the work of Kida (1981), Keener (1990) and Ricca (1993), who showed that vortex torus knots $\mathcal{T}_{p,q}$ that move under the so-called Localized Induction Approximation equation (LIA for short) are steady solutions of ideal fluid dynamics. LIA is however a cut-off of the Biot-Savart equation that governs the motion under the Euler equations. Numerical investigations based on the linearized equations have shown (Ricca *et al.*, 1999) that the motion consists of a simultaneous rotation and translation of the vortex filament about the symmetry axis. Aspects of stability profoundly influence the dynamics.

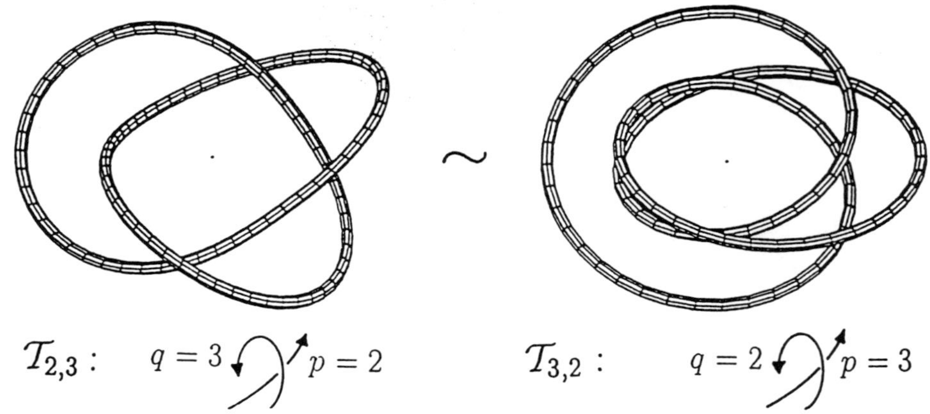

Figure 7: The standard trefoil knot $\mathcal{T}_{2,3}$ (on the left) and the isotope form $\mathcal{T}_{3,2}$ (on the right). The two knots have different geometry but are topologically equivalent.

The motion of the vortex system is studied when small-amplitude perturbations are superimposed on a steady state. If the amplitude of the perturbation decays with time, then the system is said to be stable, if it grows, then we have instability. Torus knots with winding number $w > 1$ are found to be LIA-stable, whereas the vortex knots are unstable if $w < 1$, in agreement with the stability analysis of the LIA theory (Ricca, 1993). The stable knot types tested can travel a considerable distance (compared to the maximal vortex size) in the fluid domain, without any appreciable change in shape and regardless of their complexity. Unstable knots unfold very rapidly and break down through reconnections. It is interesting to note also that poloidal and toroidal coils of the type discussed earlier are found to follow the same stability criterium: poloidal coils $\mathcal{U}_{1,m}$ are found to be LIA-stable, whereas toroidal coils $\mathcal{U}_{m,1}$ are LIA-unstable. The most interesting results, however, come from the investigation of the vortex evolution governed by the full Biot-Savart equation (no approximations involved). Torus coil and knot types are found to evolve stably, without visible change of shape, regardless of the value of their winding number, $w < 1$ or $w > 1$. This behaviour appears to be generic and independent of the complexity of the coil or knot type tested. The stabilizing action of distant parts of the vortex system, absent in the local analysis of LIA, indicates that the effects due to distant vortex elements considerably influence the long-term behaviour of the whole system.

Vortex links

Vortex lines could be linked on nested tori by Hopf links $\mathcal{L}_n^k$, where n denotes the number of components and k the degree of linking of each link type. For link systems with linking number zero (such as Borromean rings) k may be then a generalized, high-order linking number (Berger, 1990). In this case a vortex ring system is obtained by embedding n-component vortex links onto nested tori (see figure 8). In

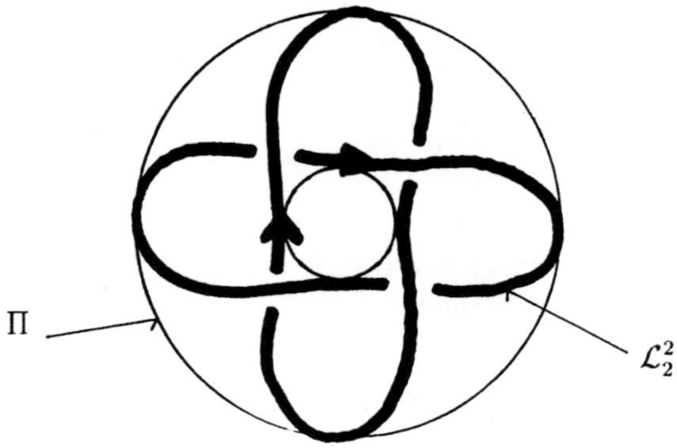

Figure 8: Example of a two-component vortex link on a mathematical torus Π: here $n = 2$ and $k = 2$.

this case a measure of complexity is given by nk.

Although studies of ideal vortex links date back to J.J. Thomson (1883; see also Ricca 1998b), no further progress has been done since. The classical results of J.J. Thomson regard motion and stability of vortex links of the type $\mathcal{L}_n^k$. These vortex link systems were found to be steady configurations, moving in the fluid without change of shape as rigid bodies. Their motion is the sum of rotation around the central axis of the torus and translation along this axis. Hence, the resulting dynamics is given by a screw motion of the whole system in $\mathcal{D}$ (as in the case of the torus knots above). Another interesting result regards stability. For n vortices $\mathcal{L}_n^k$ Thomson finds that the system is stable if $n \leq 6$, determining the period of vibration as a function of k, a result that provides an early and remarkable example of relationship between vortex dynamics and topology.

4 Geometric and topological measures for thin cored vortex filaments

Since thin cored vortex filaments are highly localized objects, their visualization and characterization is comparatively much easier than that of complex vortical flows. By identifying thin cored vortex filaments, of small circular cross section of radius σ, with their centreline Γ we can find estimates of the writhing number. An estimate based on the LIA theory (Ricca, 1995) gives

$$Wr(\Gamma) \sim \int_\Gamma (L^2c^2 - 1)\tau \, ds \; , \tag{16}$$

where L is the total length of the vortex axis Γ, of curvature c and torsion τ. From the previous equation and the definition of self-linking number $Lk(\Gamma)$ of Γ, we have

$$Lk(\Gamma) \sim L^2 \int_\Gamma c^2 \tau \, ds \,. \tag{17}$$

Useful relations between global geometric measures and topological information are given by Milnor. A way to estimate whether Γ is knotted or not is provided by the following inequality for the total curvature of Γ (Milnor, 1950): if

$$\int_\Gamma c \, ds \geq 2\pi\beta(\Gamma) \,, \tag{18}$$

where $\beta(\Gamma)$ is the bridge index of Γ (topological invariant), then Γ is a knot. For the unknot $\mathcal{U}_\circ$ we have $\beta(\mathcal{U}_\circ) = 1$. Clearly, if

$$\int c(s) \, ds < 4\pi\beta(\Gamma) \,, \tag{19}$$

then Γ is unknotted. Moreover, the following inequalities hold true (Milnor, 1953):

$$2\pi \leq \int_\Gamma c \, ds \leq \int_\Gamma (c^2 + \tau^2)^{1/2} \, ds \leq \int_\Gamma c \, ds + \int_\Gamma |\tau| \, ds \,, \tag{20}$$

and by applying a Milnor result (1953) to our definitions of tropicity, we have

$$\int_\Gamma c \, ds + \int_\Gamma |\tau| \, ds \geq 2\pi Lk(\Gamma, \vec{\lambda}_i) \,, \tag{21}$$

where $Lk(\Gamma, \vec{\lambda}_i)$ is the linking number of Γ with a directional tropic axis.

By considering the directional projection diagram of Γ, we have various estimates of the writhing number. The following results hold true:

1. (Cantarella, De Turck & Gluck, 1998):

$$|Wr(\Gamma)| \leq \frac{1}{4} \left(\frac{L}{\sigma} \right)^{4/3} \,. \tag{22}$$

   Remark: if $R = 1/c$ is the local radius of curvature of Γ, then $\dfrac{L}{\sigma} > \dfrac{R}{\sigma} \gg 1$ and bounds on the writhe have no practical interest.

2. (Fuller, 1972): Let Γ_0 be a reference curve and $\mathbf{t}_0(\xi)$ and $\mathbf{t}(\xi)$ $(a \leq \xi \leq b)$ be the tangents to Γ_0 and Γ, respectively. Then

$$|\sin \pi \left(Wr(\Gamma_0) - Wr(\Gamma) \right)| \leq 2 \int_a^b \left(\frac{3}{4(b-a)} + \left| \frac{d\mathbf{t}_0}{d\xi} \right| + \left| \frac{d\mathbf{t}}{d\xi} \right| \right) |\mathbf{t}_0 - \mathbf{t}| \, d\xi \,. \tag{23}$$

3. (Aldinger, Klapper & Tabor, 1998): Let π be an oriented plane of unit normal $\boldsymbol{\nu}$ such that $\boldsymbol{\nu}$ is never parallel to the tangent $\mathbf{t}$ to Γ, and Γ_0 be the $\boldsymbol{\nu}$-projection of Γ onto π (hence Γ_0 is a planar curve on π) of arc-length s_0, curvature c_0 and total length L_0. Then

$$|Wr(\Gamma) - Wr(\Gamma_0)| \leq \frac{|c_0|}{2\pi} L_0 \left(1 - \frac{1}{2}\left(\frac{L_0}{L}\right)^3\right) , \qquad (24)$$

and

$$|Wr(\Gamma) - Wr(\Gamma_0)| \leq \frac{1}{2\pi}\left(1 - \left(\frac{L_0}{L}\right)^2\right)^{1/2} \left(L_0 \int_{\Gamma_0} c_0^2 \, ds_0\right)^{1/2} . \qquad (25)$$

Remark: with the exception of L, obtained by direct estimates, the other quantities refer to the geometry of the planar curve Γ_0 (these quantities can be easily obtained as a by-product of the numerical code).

Another measure of topological complexity is given by the *asymptotic crossing number* $ac(\Gamma)$ introduced by Freedman & He (1991). By considering the projected diagram of Γ onto the plane of normal $\boldsymbol{\nu}$, we have

$$ac(\Gamma) = \frac{1}{4\pi}\int_\Gamma\int_\Gamma \frac{|(\mathbf{X}(s) - \mathbf{X}(s^*)) \cdot d\mathbf{X}(s) \wedge d\mathbf{X}(s^*)|}{|\mathbf{X}(s) - \mathbf{X}(s^*)|^3} = \langle n_+(\boldsymbol{\nu}) + n_-(\boldsymbol{\nu})\rangle , \qquad (26)$$

where s and s^* label two different points on Γ. As for the estimate of the writhe (see previous section) $n_\pm$ denotes the number of apparent $\pm$ crossings, from the direction of projection $\boldsymbol{\nu}$, and $\langle\cdot\rangle$ denotes average over all directions $\boldsymbol{\nu}$ of projection. We emphasize that whereas the writhe gives a purely geometric information of the complexity of the vortex pattern, Freedman & He claim that the asymptotic crossing number provides a topological measure of the entanglement of Γ. Through application of the asymptotic crossing number, they found lower bounds for minimum energy levels of magnetic knots. Similar applications in vortex dynamics might find interesting results for energy estimates associated with changes in vortex topology due to reconnection.

5 Measures of topological entanglement of open vortex structures based on reference fields

Measures of topological entanglement of open vortex structures (figure 9) are particularly useful for numerical diagnostics and can be based on reference test fields. Relative measures of topological complexity do not provide invariant quantities, but give indications of local entanglement of the flow field. To illustrate these concepts, we follow Polifke and Levich (1991). Let us decompose the fluid domain D into two simply connected regions $D = D^\circ \bigcup D^\sharp$. If D is unbounded, we assume that all

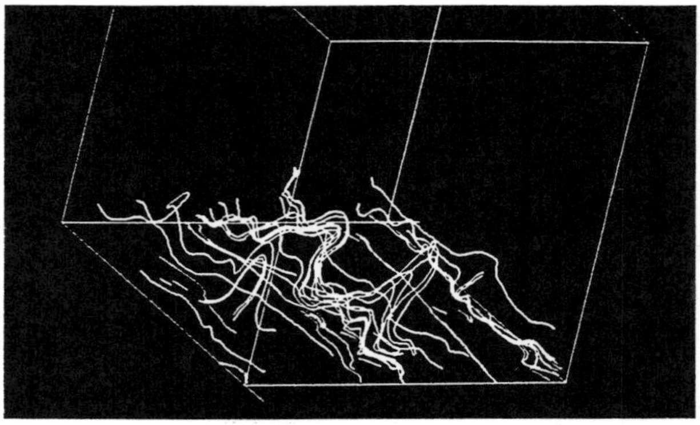

Figure 9: Hairpin vortex lines visualized in a numerical simulation of a boundary layer of a fluid flowing against a wall (from Lesieur (Ed.), 1991).

surface integrals vanish at infinity. Consider two fields $\boldsymbol{\omega}^1$ and $\boldsymbol{\omega}^2$, where $\boldsymbol{\omega}^1 \equiv \boldsymbol{\omega}$ is the given vorticity field and $\boldsymbol{\omega}^2$ is a reference vorticity field such that $\boldsymbol{\omega}^2 \equiv \boldsymbol{\omega}^1$ in D°. The corresponding velocity fields $\mathbf{u}^i$ are defined by $\nabla \times \mathbf{u}^i = \boldsymbol{\omega}^i$ $(i = 1, 2)$. All fields are assumed to be solenoidal and smooth in D, with their normal components continuous across the boundary $\partial D^\sharp$. Berger and Field (1984) have shown that the helicity difference $\Delta\mathcal{H} = \mathcal{H}^1 - \mathcal{H}^2$ of total helicities $\mathcal{H}^i$, given by

$$\Delta\mathcal{H} = \int_D \mathbf{u}^1 \cdot \boldsymbol{\omega}^1 \, dV - \int_D \mathbf{u}^2 \cdot \boldsymbol{\omega}^2 \, dV \ , \tag{27}$$

is gauge invariant and depends only on the vorticity fields inside $D^\sharp$. It can be shown that

$$\Delta\mathcal{H} = \int_{D^\sharp} \nabla\xi \cdot (\boldsymbol{\omega}^1 + \boldsymbol{\omega}^2) \, dV - \int_{\partial D^\sharp} \xi(\boldsymbol{\omega}^1 + \boldsymbol{\omega}^2) \cdot \boldsymbol{\nu} \, dA \ , \tag{28}$$

where $\nabla\xi = \mathbf{u}^1 - \mathbf{u}^2$, and $\boldsymbol{\nu}$ is the unit normal to the surface area A of $\partial D^\sharp$. Polifke and Levich suggest that the most suitable reference field should be a potential field given by $\boldsymbol{\omega}^2 = \nabla\phi$, where ϕ is a harmonic function ($\nabla^2\phi = 0$) determined uniquely by the Neumann boundary conditions

$$\nabla\phi \cdot \boldsymbol{\nu}|_{\partial D^\sharp} = \boldsymbol{\omega} \cdot \boldsymbol{\nu}|_{\partial D^\sharp} \ . \tag{29}$$

An alternative measure is the relative helicity $\mathcal{H}_R$ given by

$$\mathcal{H}_R \equiv \int_{D^\sharp} \mathbf{u}^1 \cdot (\boldsymbol{\omega}^1 - \boldsymbol{\omega}^2) \, dV \ , \tag{30}$$

which clearly requires less information than that needed for (28). Admittedly, the physical and topological interpretation of $\mathcal{H}_R$ is not so clear. Its relation with $\Delta\mathcal{H}$ is given by

$$\Delta\mathcal{H} = \mathcal{H}_R + \int_{D^\sharp} \mathbf{u}^2 \cdot (\boldsymbol{\omega}^1 - \boldsymbol{\omega}^2) \, dV \ , \tag{31}$$

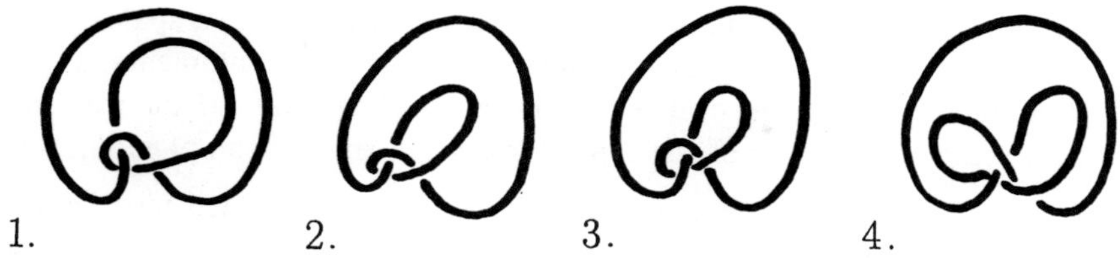

Figure 10: A fluid structure $\mathcal{T}$ evolves under a flow map φ from the initial configuration 1 ($\mathcal{T}_1 = \varphi_1(\mathcal{T})$) to the final configuration 4 ($\mathcal{T}_4 = \varphi_4(\mathcal{T})$). Only the evolution between 2 and 3 is governed by a signature-preserving flow φ_s; hence $\mathcal{T}_2 \approx \mathcal{T}_3$, but $\mathcal{T}_1 \not\approx \mathcal{T}_4$ (i.e. $\mathcal{T}_1$ and $\mathcal{T}_4$ have different geometric signature).

where the integral term in the r.h.s of (31) is generally non-zero. However, Polifke & Levich argue that since $\nabla^2 \mathbf{u}^2 = 0$ in $D^\sharp$, the integral term above should be negligible for most field configurations, implying that $\mathcal{H}_R$ would actually provide reliable relative measures of entanglement. A number of test cases were studied by these authors to support their case.

6 Dynamical measures of structural complexity

Vortex structures may evolve by preserving geometric and topological aspects of their shape, while allowing large-amplitude deformations. Even in the presence of dissipation, we have cases where geometric features are preserved for a relatively long time compared to the typical viscous dissipation time. It is therefore necessary to extend classical concepts of dynamical stability to describe a much richer family of dynamics. Another important issue in the study of complex flow systems is 'robustness', that may characterize certain fluid evolutions, where fluid properties (such as helicity) tend to resist from decaying under dissipation. To make progress in this direction we need to introduce new concepts based on dynamical measures of structural complexity. These concepts are also useful to develop a measure theory that is able to classify structured flow patterns dynamically. Let us introduce the following:

Definition: φ_s is a *(geometric) signature-preserving flow* if it admits finite, large-amplitude deformations (diffeomorphisms) that make the final configuration recognizable in terms of the initial pattern.

Let us denote by $\mathcal{T}_t = \varphi_t(\mathcal{T})$ the instantaneous mapping of $\mathcal{T}$ by φ. If a structure $\mathcal{T}_t$ evolves under φ_s for some time $t \in I$, then we write $\mathcal{T}_{t_1} \approx \mathcal{T}_{t_2}$ (where t_1 and t_2 take values in I) to denote conservation of geometric signature (figure 10). Another important issue is vortex coherency, that can be defined as follows:

Definition: A vortex structure is said to be *coherent* if there is a spatial/temporal statistical correlation over the mean values (or higher-order moments) of the dynamical characteristics taken over a length-scale of the order of the fluid domain and a time-scale of the order of the typical evolution time of the phenomenon.

Dynamics that preserve geometric features (such as tropicity, mean curvature, writhing number, twist, etc.) and topology (knot and link type, linking number, etc.), while allowing large-amplitude deformations, have not been studied so far and there is therefore a lack of rigorous definition and analysis. Upper bounds on tropic measures, alignment and Lyapunov exponents, global and local curvatures, etc. must be used to specify the diffeomorphisms that are signature preserving from the general flow maps. Here we propose a new definition of dynamical stability in relation to these specific evolutions. By taking into account large-amplitude perturbations we can extend the usual concept of Lyapunov stability to include the dynamics governed by geometric signature-preserving flows. The corresponding motion is characterized by finite, large-amplitude, (Lyapunov) stable evolutions that conserve geometric coherency and signature for some (finite) time of physical interest. Hence, we can talk of vortex structures evolving under "stable" dynamics according to the following:

Definition: A vortex structure is said to be *(Lyapunov) stable* if it evolves under signature-preserving flows that conserve topology, geometric signature and vortex coherency.

Stable structures are therefore those able to evolve over a considerable distance (much larger than their typical length-scale), while preserving geometric signature. On the other hand, if they unfold and break-up in a short time (compared with the typical time-scale of the phenomenon), we say that they are (Lyapunov) unstable. The classification and study of flow patterns that show geometric and topological complexity, while maintaining features of (Lyapunov) stability, represent an important aspect of future research.

7 Conclusions

Self-organization of vortical tangles into coherent regions of long-lived structural complex flows seems to be a generic feature of turbulent flows. A detailed analysis of these flows requires the use of a combination of geometric and topological techniques in order to classify and measure structural complexity. In this paper we have discussed applications of various concepts that will help to develop a measure theory for complexity. In particular we have shown how classical geometric quantities (such as global measures of curvature, writhing number and relative linking estimates), topological quantities (such as link and knot type, linking number, asymptotic crossing number) and new concepts based on the degree of 'tropicity' of a flow structure (to characterize tubeness, sheetness and bulkiness of vortex flows) can find useful applications in fluid dynamics. We have introduced the concept of directional tropicity to evaluate relative linking and estimate the degree of knottedness of a vortex pattern and we have extended the classical concept of stability to include those dynamics that preserve geometric signature during evolution. The development of a measure theory to study structural complexity finds useful applications in many physical contexts such as in polymer physics (see, for example, Kantor & Hassold, 1988; Agnes & Rasetti, 1994) and in chemical physics (see, for example, Winfree, 1994), where structured patterns are present. The concepts discussed in this paper, integrated with complementary measures of complexity developed for dynamical systems (Ghrist *et al.*, 1997), random systems (Millett & Sumners (Ed.), 1994; Kholodenko & Rolfsen, 1996), stochastic flows and entropic systems (Badii & Politi, 1997), will be a very useful tool for carrying out efficient computational diagnostics of complex flow patterns.

Acknowledgments

The author would like to express his gratitude to Alan Jones of ISIS for his kind hospitality at the European Joint Research Centre of Ispra, where this work has originated. Financial support from the UK Particle Physics and Astronomy Research Council (PPARC) Grant no. GR/L63143 is also kindly acknowledged.

References

AGNES, C. & RASETTI, M. 1994 Piecewise-linear-category approach to entanglement. *Nuovo Cimento* 3D, 121–136.

ALDINGER, J., KLAPPER, I. & TABOR, M. 1998 Formulae for the calculation and estimation of writhe. *To appear.*

ARNOLD, V.I & KHESIN, B.A. 1998 *Topological Methods in Hydrodynamics.* Applied Mathematical Sciences Series **125**, Springer.

BADII, R. & POLITI, A. 1997 *Complexity.* Cambridge Nonlinear Science Series **6**, Cambridge University Press.

BARENGHI, C.F., SAMUELS, D.C., BAUER, G.H. & DONNELLY, R.J. 1998 Vortex lines in a model of turbulent flow. *Phys. Fluids* **9**, 2631–2643.

BERGER, M.A. 1990 Third order link invariants. *J. Phys.* A: *Math. & Gen.* **23**, 2787–2793.

BERGER, M.A. & FIELD, G.B. 1984 The topological properties of magnetic helicity. *J. Fluid Mech.* **147**, 133–148.

CANTARELLA, J., DE TURCK, D. & GLUCK, H. 1998 Upper bounds for the writhing of knots and the helicity of vector fields. *Preprint.*

FERNANDEZ, V.M., SILVER, D. & ZABUSKY, N.J. 1996 Visiometrics of complex physical processes: diagnosing vortex-dominated flows. *Computers in Phys.* **10**, 463–470.

FRAENKEL, L.E. & BERGER, M.S. 1974 A global theory of steady vortex rings in an ideal fluid. *Acta Mathematica* **132**, 13–51.

FREEDMAN, M.H. & HE, Z.-X. 1991 Divergence-free fields: energy and asymptotic crossing number. *Ann. Math.* **134**, 189–229.

FULLER, F.B. 1972 How the writhing number of a curve depends on the curve. *Rev. Roum. Math. Pures Appl.* **17**, 1329–1334.

GHRIST, R.W., HOLMES, P.J. & SULLIVAN M.C. 1997 *Knots and Links in Three-Dimensional Flows.* Lecture Notes in Mathematics **1654**. Springer.

JIMÉNEZ, J. (ED.) 1991 *The Global Geometry of Turbulence.* NATO ASI Series B: Physics **268**. Kluwer.

JIMÉNEZ, J., WRAY, A.A., SAFFMAN, P.G. & ROGALLO, R.S. 1993 The structure of intense vorticity in homogeneous isotropic turbulence. *J. Fluid Mech.* **255**, 65–90.

KANTOR, Y. & HASSOLD, G.N. 1988 Topological entanglements in the percolation problem. *Phys. Rev. Lett.* **60**, 1457–1460.

KEENER, J.P. 1990 Knotted vortex filaments in an ideal fluid. *J. Fluid Mech.* **211**, 629–651.

KHOLODENKO, A.L. & ROLFSEN, D.P. 1996 Knot complexity and related observables from path integrals for semiflexible polymers. *J. Phys. A: Math. Gen.* **29**, 5677–5691.

KIDA, S. 1981 A vortex filament moving without change of form. *J. Fluid Mech.* **112**, 397–409.

LESIEUR, M. (ED.) 1991 *Images de Mecanique des Fluides Numerique.* CNRS. Institute de Méchanique de Grenoble, Grenoble, France.

MILLETT, K.C. & SUMNERS, D.W. (ED.) 1994 *Random Knotting and Linking.* Series on Knots and Everything **7**, World Scientific, Singapore.

MILNOR, J. 1950 On the total curvature of knots. *Ann. Math.* **52**, 248–257.

MILNOR, J. 1953 On total curvatures of closed space curves. *Math. Scand.* **1**, 289–296.

MOFFATT, H.K. & RICCA, R.L. 1992 Helicity and the Călugăreanu invariant. *Proc. R. Soc. Lond.* A **439**, 411–429.

MOFFATT, H.K. & TSINOBER, A. (ED.) 1990 *Topological Fluid Mechanics.* Cambridge University Press.

MOFFATT, H.K., ZASLAVSKY, G.M., COMTE, P. & TABOR, M. (ED.) 1992 *Topological Aspects of the Dynamics of Fluids and Plasmas.* NATO ASI Series E: Applied Sciences **218**. Kluwer.

POLIFKE, W. & LEVICH, E. 1990 Entangledness of vortex lines in turbulent flows. In *Topological Fluid Mechanics* (ed. H.K. Moffatt & A. Tsinober), pp. 637–646. Cambridge University Press.

RICCA, R.L. 1993 Torus knots and polynomial invariants for a class of soliton equations. *Chaos* **3**, 83–91 [1995 Erratum. *Chaos* **5**, 346].

RICCA, R.L. 1995 Geometric and topological aspects of vortex filament dynamics under LIA. In *Small-Scale Structures in Three-Dimensional Hydro and Magnetohydrodynamics Turbulence* (ed. M. Meneguzzi *et al.*), pp. 99–104. Lecture notes in Physics **462**. Springer.

RICCA, R.L. 1998a Applications of knot theory in fluid mechanics. In *Knot Theory* (ed. V.F.R. Jones *et al.*), pp. 321–346. Banach Center Publications, Institute of Mathematics **42**, Polish Academy of Sciences, Warsaw.

RICCA, R.L. 1998b New developments in topological fluid mechanics: from Kelvin's vortex knots to magnetic knots. In *Ideal Knots* (ed. A. Stasiak *et al.*), pp. 255–273. Series on Knots and Everything **19**, World Scientific, Singapore.

RICCA, R.L. & BERGER, M.A. 1996 Topological ideas and fluid mechanics. *Phys. Today* **49** (12), 24–30.

RICCA, R.L., SAMUELS, D.C. & BARENGHI, C.F. 1999 Evolution of vortex knots. *J. Fluid Mech.* **391**, 29–44.

SAFFMAN, P.G. 1992 *Vortex Dynamics.* Cambridge University Press.

SCHWARZ, K.W. 1988 Three dimensional vortex dynamics in superfluid helium. *Phys. Rev.* B **38**, 2398–2417.

SHE, Z.-S., JACKSON, E. & ORSZAG, S.A. 1990 Intermittent vortex structures in homogeneous isotropic turbulence. *Nature* **344**, 226–228.

TABOR, M. 1992 Stretching and alignment in general flow fields: classical trajectories from Reynolds number zero to infinity. In *Topological Aspects of the Dynamics of Fluids and Plasmas* (ed. H.K. Moffatt ıet al.), pp.83–110. NATO ASI Series E: Applied Sciences **218**. Kluwer.

THOMSON, J.J. 1883 *A Treatise on the Motion of Vortex Rings.* MacMillan & Co., London.

VINCENT, A. & MENEGUZZI, M. 1991 The spatial structure and statistical properties of homogeneous turbulence. *J. Fluid Mech.* **225**, 1–25.

WINFREE, A.T. 1994 Persistent tangles of vortex rings in generic excitable media. *Nature* **371**, 233–236.

ZABUSKY, N.J., SILVER, D. & PELZ, R. 1993 Visiometrics, juxtaposition and modeling. *Phys. Today* **46**, 24–31.

PROJECTIONS OF CODIMENSION TWO EMBEDDINGS

Dennis Roseman
The University of Iowa
Iowa City, IA

December 2, 1999

Abstract

We consider smooth knottings of compact n-dimensional manifolds (not necessarily orientable) in $\mathbf{R}^{n+2}$ (or $\mathbf{S}^{n+2}$). We generalize to higher dimensions the classical notion of a knotting in general position with respect to projection. It is shown that any knotting is equivalent to one which is in general position with respect to projection. In higher dimensions, unlike classical knot theory, a generic projection is not necessarily an immersion. Thus we need to consider maps such that the set of non-immersion points is well behaved. The geometry of projections is examined and we use this information to show that any smooth knotting of an orientable n-manifold in $\mathbf{R}^{n+2}$ is smoothly isotopic to one whose projection into $\mathbf{R}^{n+1}$ is an immersion. As corollary we obtain an elementary geometric proof that an orientable n-dimensional submanifold of $\mathbf{R}^{n+2}$ has trivial normal bundle.

1980 Mathematics Subject Classification: Primary 57R40, 57R42, 57R45, 57R52, 57Q45. Key words and phrases: knots, projections of knots, immersions, normal crossings, jet transversality.

1 Introduction

We consider codimension two smooth embeddings of a compact n-dimensional manifold M^n into $\mathbf{R}^{n+2}$ (or sometimes $\mathbf{S}^{n+2}$). Most of our results hold whether or not M^n is orientable. We will assume that M^n is compact; for the most part this is not an essential hypothesis and is added to keep the proofs short. Here $\mathbf{R}^k$ is the standard k-dimensional Euclidean space, $\mathbf{S}^k$ the standard k-dimensional sphere, and $\mathbf{D}^k$ standard k-dimensional ball, each with the standard differentiable structure. Also, $\mathbf{R}^k_+$ will denote Euclidean half-space, defined to be all points of $\mathbf{R}^k$ with first coordinate greater or equal to zero. We will sometimes simplify the notation M^n to M for an n-dimensional manifold if the dimension is not a critical to the discussion at that point.

We will adopt the following terminology: a smooth codimension two embedding will be called a (smooth) **knotting**, and the corresponding ambient isotopy class will be called a **knot**. The smooth isotopy extension theorem (see, for example, [HR2]) implies that a smooth isotopy gives rise to an ambient isotopy; we will use this without explicit mention.

Historically, knots (knotted circles) were first studied by their projections onto the plane. We generalize this idea to higher dimensions. We define what it means for a knotting to be in general position with respect to projection, then show that any knotting is isotopic to one which is in general position with respect to projection.

One common way of describing higher dimensional knots is the method of "slicing" (see [FX] for consideration of surfaces in $\mathbf{S}^4$, [LM] for 3-manifolds in 5-space). For surfaces in $\mathbf{R}^4$, the advantage of slicing is that the pictures are easier to draw; the disadvantage of slicing is that one considers only a small part of the knot at any one time. Our method of projections offers an alternative method which, at least in some cases, gives a more coherent picture of knots and isotopies. Both methods, slicing and projecting, allow one to compute the knot groups with relative ease. For computations using projections, see [YJ] and [RS2].

We look at some examples and also exploit the geometry of the projection of an orientable manifold and show: If $\psi\colon M^n \to \mathbf{R}^{n+2}$ is a smooth knotting of an orientable manifold, then we may smoothly isotop ψ so that its projection into $\mathbf{R}^{n+1}$ is an immersion.

The results of this paper have applications to the study of isotopies of knots. In [RS4], we define higher dimensional knot moves and show that every isotopy is equivalent to a sequence of knot moves where, in each dimension, there are a finite number of such moves. In [RS5] we describe in detail these knot moves for two-manifolds in four-space and for three-manifolds in five-space.

2 Projections of knots: definitions and examples

In studying knottings of a smooth n-manifold M^n, it is sometimes best to consider knottings into $\mathbf{R}^{n+2}$, sometimes into $\mathbf{S}^{n+2}$. Thus we will wish to consider the idea that a function f is in general position with respect to projection in five basic situations. For the first four of these, we assume that $\partial M = \emptyset$.

(a) f is a smooth embedding of M^n into $\mathbf{R}^{n+2}$
(b) f is a smooth proper embedding of $M^n \times I$ into $\mathbf{R}^{n+2} \times I$
(c) f is a smooth embedding of M^n into $\mathbf{S}^{n+2}$
(d) f is a smooth proper embedding of $M^n \times I$ into $\mathbf{S}^{n+2} \times I$
(e) f is a smooth proper embedding of M^n into $\mathbf{R}^{n+2}_+$

We will primarily focus on (a) and (b). Knottings of type (b) arise from isotopies of knottings of type (a), knottings of type (d) arise from isotopies of knottings of type (c). Situation (e) occurs in certain knotting constructions related to spinning.

In each of these cases we will consider a standard map which will be called a **projection.** For (a) let $f: M^n \to \mathbf{R}^{n+2}$ be a smooth knotting and let $\pi: \mathbf{R}^{n+2} \to \mathbf{R}^{n+1}$ be the map defined by projection in the direction of last coordinate:

$$\pi(x_1, \dots, x_{n+1}, x_{n+2}) = (x_1, \dots, x_{n+1}).$$

For (b), the projection will refer to the map $\pi': \mathbf{R}^{n+2} \times I \to \mathbf{R}^{n+1} \times I$ defined by $\pi' = \pi \times Id_I$. For situation (c) we note that we may assume that $f(M^n)$ misses two disjoint (small) open balls. Thus $f(M^n)$ is contained in the interior of a submanifold W of $\mathbf{S}^{n+2}$ homeomorphic to $\mathbf{S}^{n+1} \times I$. In this case projection will be the map $\pi'': \mathbf{S}^{n+1} \times I \to \mathbf{S}^{n+1}$ which takes (x, t) to x, where we identify W with $\mathbf{S}^{n+1} \times I$. Finally, for (d), we may assume without loss of generality that $M^n \times I$ is a subset of $W \times I$ where W is as above; then $\pi''': W \times I \to \mathbf{S}^{n+1} \times I$ will be defined by $\pi''' = \pi'' \times Id_I$. In situation (e), the projection to consider is $\pi|_{\mathbf{R}^{n+2}_+}: \mathbf{R}^{n+2}_+ \to \mathbf{R}^{n+1}_+$.

In each of these cases we define some terms corresponding to a given knotting and projection. We define them explicitly below only for the situation (a) above. One defines these terms for the other situations by using the other notions of projections, π', π'', or π''' in place of π. If $A \subseteq M$ we define the **projection of** A to be the set $\pi \circ f(A)$ which we will denote A^*. We refer to M^* as the **projection of the knotting.** If $x \in M$ then x^* denotes $\pi \circ f(x)$. The **double point set** of M, denoted by D, will be the closure in M of the set of points $x \in M$ such that $(\pi \circ f)^{-1}$ is not one-to-one at x. The **crossing set of the knotting** is D^*. So, the crossing set is the closure in M^* of the set of all $x^* \in M^*$ such that $(\pi \circ f)^{-1}(x^*)$ contains two or more points. The **branch set of the knotting,** denoted by B, consists of all $x \in M$ such that $\pi \circ f$ is not an immersion at x. (Remark: For knottings general position with respect to projection, soon to be defined, we will have $B \subseteq D$.)

For the familiar knotting of the trefoil knot, the crossing set consists of three points (frequently called "crossings" or "crossing points"), the projection of the knotting is an immersion, thus $B = \emptyset$, and the double point set consists of six points on the circle (see Figure 1).

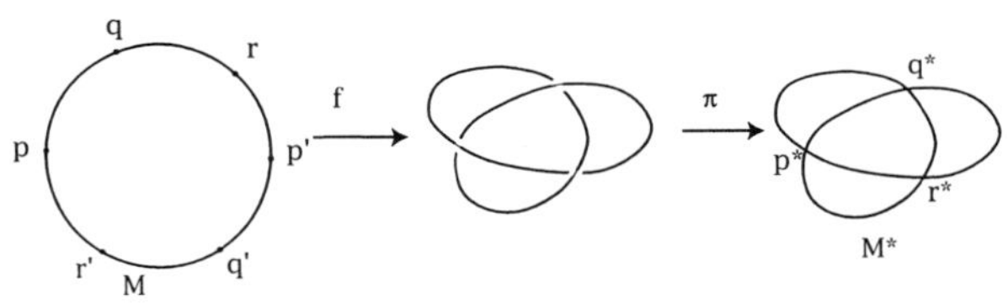

Figure 1: Three crossing points: p^*, q^*, and r^*. On the left the double point set—six points on the circle labeled so that $\pi \circ f(p) = \pi \circ f(p') = p^*$, etc .

Next we consider an example of a knotting $\mathbf{S}^2$ in $\mathbf{R}^4$—the spinning of the trefoil (originally considered in [AR]). If one does the geometry appropriately we will have the projection of the spinning the same as the spinning of the projection, see [RS1]. This construction uses a proper knotted arc $\alpha \subseteq \mathbf{R}^3_+$. In this case each crossing point of α^* gives rise to a circle of crossing points for the spun knot. This knotting of the spun knot has a projection is an immersion. The crossing set consists of three circles (see Figure 2(d)) and a double point set consisting of six circles (see Figure 2(c)).

In general one can take a knotted n-sphere in $\mathbf{R}^{n+2}$ and use the spinning process to obtain an embedding of an $(n+1)$-sphere in $\mathbf{R}^{n+3}$. If we take the spun trefoil and spin

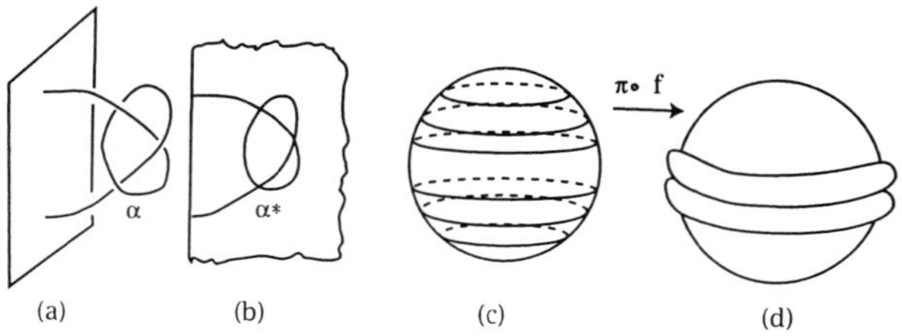

Figure 2: A proper knotted arc α shown in (a), with projection (b). The projection of the spinning of α shown in (d) with crossing set consisting of three disjoint circles; the double point set shown as six circles in (c).

this, we will obtain an embedding of a 3-sphere in 5-space called the spun-spun trefoil. As with the previous example, we may arrange the geometry so that the projection of the spinning is the spinning of the projection. In this case each circle of the crossing set of the spun trefoil will give rise to a torus in the crossing set of the spun-spun trefoil. Thus we can get a projection of the spun-spun trefoil which is an immersion, whose crossing set consists of three (unknotted) tori in $\mathbf{R}^4$, and whose double point set consists of six concentric unknotted tori in $\mathbf{S}^3$. A related (but topologically distinct) embedding is the 2-spun torus knot (for definition see [CP] where the term superspinning is used); here one can arrange the geometry so that the crossing set will be three (unknotted) 2-spheres $\mathbf{R}^4$ and the double point set will be six concentric two-spheres in $\mathbf{S}^3$ [RS3].

In classical knot theory, a knotting is in general position with respect to projection if the projection is an immersion which is at most two-to-one and such that the immersion is self transverse at the crossings. In higher dimensions for generic maps (Definition 4.2) we will have a condition of self transversality but the projection may have more than two points projected to the same point and the projection might not be an immersion.

To illustrate branch points, consider Figure 3(a) and (b). In Figure 3(b), we see a projection of a knotting of $\mathbf{S}^2$ in $\mathbf{R}^4$ where the crossing set is an arc, and the projection has two branch points. The **figure eight curve** is an unknotted circle whose projection has a single self-transverse crossing. One can describe this knotting of $\mathbf{S}^2$ as smoothing of the suspension of the figure eight curve. (The knot corresponding to f is trivial since a suspension of a trivial knot is a trivial knot.) Finally one more example; we could take the above knotting of $\mathbf{S}^2$ in $\mathbf{R}^4$ and spin it, obtaining a knotting of $\mathbf{S}^3$ in $\mathbf{R}^5$. If we arrange our geometry as before so that the spinning of the projection to be the projection of

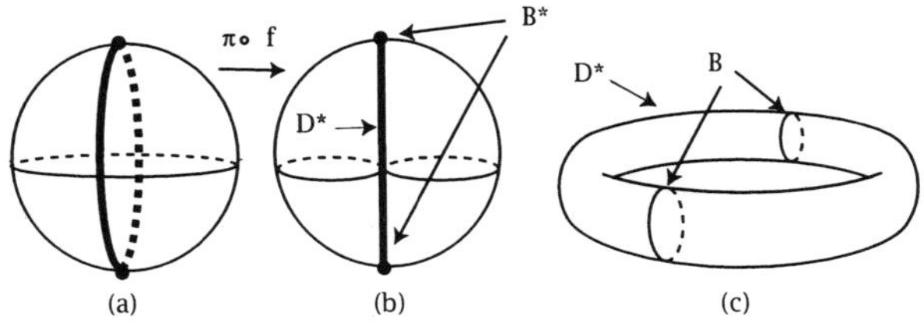

Figure 3: Projection of a simple knotting with two branch points and an arc of crossing points shown in (b); (a) shows the corresponding double point set. Figure (c) shows double point set (as a subset of $\mathbf{S}^3$) of a spinning of the knotting of (b).

the spinning, D will be a standard torus in $\mathbf{S}^3$; B will be two circles in D as indicated in Figure 3(c); and D^* will be an annulus.

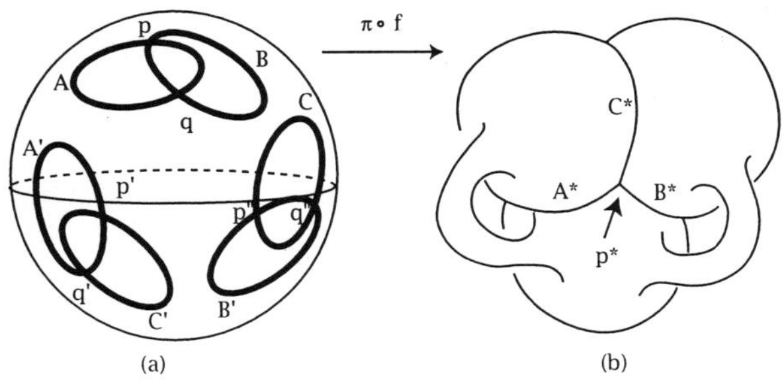

Figure 4: At right in (b), An example of a projection of a sphere with triple points. Double point set shown at left in (a), circles are labeled so that, via $\pi \circ f$, circles A and A' are sent to A^*, p, p' and p'' are sent to p^*, etc.

When studying projections of surfaces in $\mathbf{R}^3$ an important consideration is the set of **triple points** where, in the projection, three points of the surface are identified. This is illustrated in Figure 4 where the knotting is denoted by f. This projection has two triple points, p^* and q^*. We see only p^* in Figure 4(b); q^* is hidden from view. The set D^* will consist of the union of three circles A^*, B^*, and C^* which intersect each other at p^* and q^*. The double point set D is the union of six circles in $\mathbf{S}^2$ as shown in Figure 4(a). The knot described above is trivial since one can readily construct a 3-ball in $\mathbf{R}^4$

with the given sphere as boundary.

3 Realizing Singular Subsets

In the above examples we have shown knottings of $\mathbf{S}^3$ whose double point sets contain an unknotted torus, and another in which the branch set is an unknotted circle. It is natural to ask whether any knotted surface (of any genus) in $\mathbf{S}^3$ could occur as a component of a double point set of a knotting of $\mathbf{S}^3$ in $\mathbf{R}^5$. Also we might ask if a component of the branch set of a knotting could be an arbitrarily chosen knotted circle. The answer is yes to both of these, and in fact the knottings can be chosen to be trivial knots.

Proposition 3.1 *Let F^{n-1} be a smooth $(n-1)$-dimensional submanifold of $\mathbf{S}^n$, with $\partial F^{n-1} = \emptyset$. Then there is a knotting $k\colon \mathbf{S}^n \to \mathbf{R}^{n+2}$ whose projection has double point set D such that F is a component of D. Furthermore, we can find such a knotting k which represents the trivial knot.*

Proof: It is well known that F must be orientable with trivial tubular neighborhood. Let σ be the standard embedding of $\mathbf{S}^n$ into $\mathbf{R}^{n+1}$ and view $\mathbf{R}^{n+1} \subseteq \mathbf{R}^{n+2}$. Let $F' = \sigma(F)$, and consider three tubular neighborhoods. Let N be a tubular neighborhood of F' in $\sigma(\mathbf{S}^n)$, let N' be a tubular neighborhood of F' in $\mathbf{R}^{n+1}$ and let N'' be a tubular neighborhood of F' in $\mathbf{R}^{n+2}$. All these tubular neighborhoods are trivial and we may chose trivializations, ϕ, ϕ' and ϕ'', respectively, so that the following diagram commutes.

$$\begin{array}{ccccc} N & \subseteq & N' & \subseteq & N'' \\ \downarrow{\phi} & & \downarrow{\phi'} & & \downarrow{\phi''} \\ F \times \mathbf{D}^1 & \subseteq & F \times \mathbf{D}^2 & \subseteq & F \times \mathbf{D}^3. \end{array}$$

In the bottom row we use the identity for the first factor and the standard inclusions $\mathbf{D}^1 \subseteq \mathbf{D}^2 \subseteq \mathbf{D}^3$ for the second factor. Furthermore, we may arrange it so that the last coordinate of $\mathbf{D}^3$ corresponds to the last coordinate of $\mathbf{R}^{n+2}$. Thus $\pi|_{N''}$ corresponds to the map of $F \times \mathbf{D}^2$ which is the identity on F and projection on the last coordinate of $\mathbf{D}^3$.

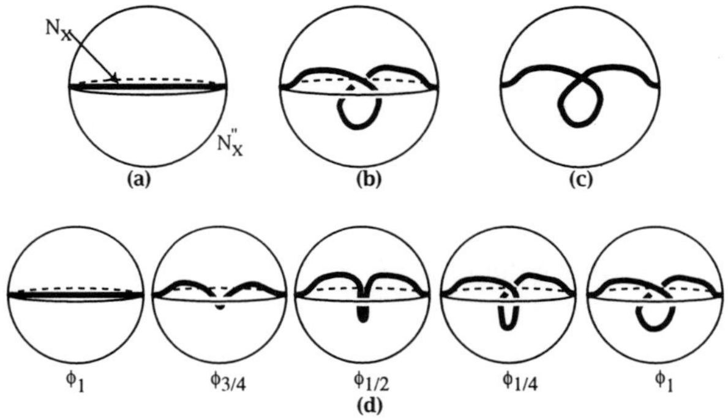

Figure 5: Replacing a standard arc pair shown in (a) with a "pigtail" shown in (b) with projection as in (c); see proof of Theorem 3.1. In (d) is shown an isotopy between embeddings (a) and (b).

For every $x \in F$ let N_x be the fiber of x in N and N''_x the fiber of x in N''. The pair (N''_x, N_x) is equivalent to the standard ball pair $(\mathbf{D}^3, \mathbf{D}^1)$, see Figure 5(a). For each $x \in F'$, remove N_x and replace it by the curve shown in Figure 5(b), a "pigtail", whose projection is the simple loop as shown in Figure 5(c). It is easy to see that the knotting k which we obtain in this way will have crossing set diffeomorphic to F and the double point set of this knotting will be the union of two manifolds, each ambiently isotopic to F. Furthermore, it is clear that k is isotopic to σ. ∎

Proposition 3.2 *Let F^{n-1} be a smooth orientable $(n-1)$-dimensional submanifold of $\mathbf{S}^n$ with $\partial F \neq \emptyset$. Then there is a knotting $k\colon \mathbf{S}^n \to \mathbf{R}^{n+2}$ double point set is homeomorphic to the double of F and whose projection has branch set ∂F. Also, the knot corresponding to k is trivial.*

Proof: A smooth manifold F with boundary has a smooth collar neighborhood of the boundary diffeomorphic to $\partial F \times [0,1]$, and we may find a diffeomorphism

$$\eta\colon \left(F \underset{\partial F = \partial F \times \{0\}}{+} \partial F \times [0,1] \right) \to F.$$

Let $F' = \eta(F)$. We next proceed as in the previous proposition to define N, N', and N'' to be regular neighborhoods of F. As before, for each $x \in F$, (N''_x, N_x) is homeomorphic to the standard ball pair.

To complete our construction we use an isotopy which straightens out the pigtail, as shown in Figure 5(d). This isotopy, ϕ_t, $0 \le t \le 1$, projects to an immersion with a single crossing point except at $t = 1$. We now define the knotting k by replacing N_x with the pigtail for all $x \in F'$. For the rest of F, we view points of $F - Int(F')$ as $(x, t) \in \partial F \times [0, 1]$ and we use the arc which is the image of ϕ_t. That is, as we near the boundary of F, we pull the pigtail so that the kink disappears right at the boundary. This results in a knotting whose branch point set is ∂F and k represents a trivial knot. Also the double point set of this knotting is the double of F. ∎

4 General Position for Knottings

In the previous examples, the projections were of a particularly nice kind—the double point sets, crossing sets, branch sets were (possibly immersed) submanifolds. The following general position theorem assures us that we need consider only such examples. Any knotting is isotopic to one whose projection is of the above-mentioned type. We will use the following definition:

Definition 4.1 *Let $f: M^n \to \mathbf{R}^{n+1}$ be a smooth map with branch set B and double point set D. We will say f is a* **generic map** *if*

(a) *B is a closed $(n-2)$-dimensional submanifold of M^n.*
(b) *B is a submanifold of D and for any $b_0 \in B$ there is a small $(n-1)$-dimensional open subdisk N with $b_0 \in N$, $N \subseteq D$ such that $N - B$ has two components N_0 and N_1, each of which is an $(n-1)$-disk which is embedded by the restriction of $\pi \circ f$ but with $N_0^* = N_1^*$.*
(c) *$(D - B)^*$ is the union of immersed closed $(n-1)$-dimensional submanifolds of M^* with normal crossings.*
(d) *$(\pi \circ f)|_B$ is an immersion of B with normal crossings.*
(e) *B^* meets the crossing set of D^* transversely.*
(f) *The crossing set of B^* is transverse to the crossing set of $(D - B)^*$.*

Definition 4.2 *Let $f: M^n \to \mathbf{R}^{n+2}$ be a knotting, with branch set B and double point set D. We say f is in* **general position with respect to the projection π** *if $\pi \circ f: M \to \mathbf{R}^{n+1}$ is a generic map.*

Remarks:

1. Condition (b) implies that $B \subseteq D$.

2. An immersion of a disk in $\mathbf{R}^3$ with a canonical branch point is sometimes referred to as a "cross cap" [GL-GL]. Condition (b) in higher dimensions could be loosely described as follows: locally, the projection of a neighborhood of the branch set looks like "the cross cap crossed with D^{n-2}."

3. An immersion ϕ has **normal crossings** if at a point x_0^* of self-intersection the images of the tangent planes $\{T_x\}_{x \in \phi^{-1}(x_0)}$, under the differential $d\phi$, are in general position (see [LV], for example, for definitions).

4. If $n = 1$, we are dealing with projections of classical knots and links and our definition becomes the usual one: $\pi \circ f$ is an immersion (since $B = \emptyset$) and the crossing set a set of isolated points, each of which is a transverse intersection of two arcs.

   If $n = 2$, B^* is a set of isolated points, and condition (b) implies that B^* intersects neither $M - B^*$ nor the double point sets of D^*. Then D^* is a union of immersed arcs and circles. The double point sets of D^* are exactly those points of D^* which are triple points of $\pi \circ f$. At such a point x there is a neighborhood N of x and a diffeomorphism between the triple $(N, N \cap M^*, x)$ and $(\mathbf{R}^3, X, \vec{0})$ where X is the union of the three coordinate planes in $\mathbf{R}^3$. Points of D^* which do not correspond to points of B^* or to triple points are then "pure double points" and for such a point x, there is a neighborhood W of x such that $(N, N \cap M^*, x)$ is diffeomorphic to $(\mathbf{R}^3, Y, \vec{0})$ where Y is the union of two coordinate planes.

5. Condition (b) implies that $(\pi \circ f)|_B$ is an immersion of B into $\mathbf{R}^{n+1}$; thus the significance of condition (d) is the portion about normal crossings.

6. If $n \leq 4$, then condition (d) implies that B^* is an embedded submanifold of $\mathbf{R}^{n+1}$. If we consider knottings of 5-dimensional manifolds in $\mathbf{R}^7$ then B^* would be an immersed 3-dimensional submanifold of $\mathbf{R}^6$ and consideration of self intersections of B^* of condition (e) would become meaningful.

7. Condition (c) in the definition of general position could use an example to clarify its significance. Given a smooth knot $c \subseteq \mathbf{R}^3$ we may construct an embedding of a 2- sphere in $\mathbf{R}^4$ by taking the suspension $\Sigma(c)$ of c. Here $\Sigma(c)$ is defined to be the union of two disks D_1 and D_{-1}. For $i = 1$ or -1, D_i is the union of all straight

line segments in $\mathbf{R}^4$ from $(0, 0, 0, i)$ to a point of c. If c is a nontrivial knot then this sphere is not isotopic to a smoothly embedded sphere (because it is locally knotted at the suspension points). But if c is unknotted then we can smooth the suspension points and obtain a smooth sphere (which will be a trivially knotted sphere). If we take the circle c as shown in Figure 6(a) and smooth $\Sigma(c)$, we obtain a sphere whose projection is shown in Figure 6(c). The set B corresponds to the two suspension points and D is the union of four arcs as shown, see Figure 6 (b). The neighborhood of B in D does not satisfy our condition for general position. However, we can isotop $\Sigma(c)$ so that its projection will look Figure 6(e), for which the sets B and D are shown as in Figure 6(d).

A general method for constructing such isotopies is found treated in Theorem 6.1.

The basic result about this general position is the following.

Theorem 4.3 *(General position for knottings.) A knotting $\psi: M^n \to \mathbf{R}^{n+2}$ is isotopic to a map ϕ which is in general position with respect to projection.*

The the proof of Theorem 4.3 will use Lemma 4.4 which basically says that for a given knotting, if one alters the projection slightly in $\mathbf{R}^{n+1}$, one can obtain a knotting in $\mathbf{R}^{n+2}$, equivalent to the original knotting, which projects to the altered subset. Note that we will need the alteration to be small in the C^1 topology.

Lemma 4.4 *Let $f: M^n \to \mathbf{R}^{n+2}$ be a knotting. There is an $\epsilon > 0$ such that if $g': M^n \to \mathbf{R}^{n+1}$ is a smooth function which, in the C^1 topology, is within ϵ of $\pi \circ f$, then there is a knotting $g: M^n \to \mathbf{R}^{n+2}$ such that g is isotopic to f and $\pi \circ g = g'$.*

Proof: It is well known that the set of embeddings is an open subset of the space of continuous maps from one manifold to another. In the C^1 metric, smooth functions are close if corresponding points are close *and* the corresponding differentials are close. (See, for example, [HR2] for definitions and basic background.)

Chose $\epsilon > 0$ so that if $\rho(f, g) < \epsilon$ then g is a knotting where ρ is the C^1 metric. Given $g': M^n \to \mathbf{R}^{n+1}$ a smooth map, with $\rho(g', \pi \circ f) < \epsilon$, $m \in M^n$ write

$$f(m) = (m_1, \dots, m_{n+1}, m_{n+2}) \text{ and } g'(m) = (m'_1, \dots, m'_{n+1}).$$

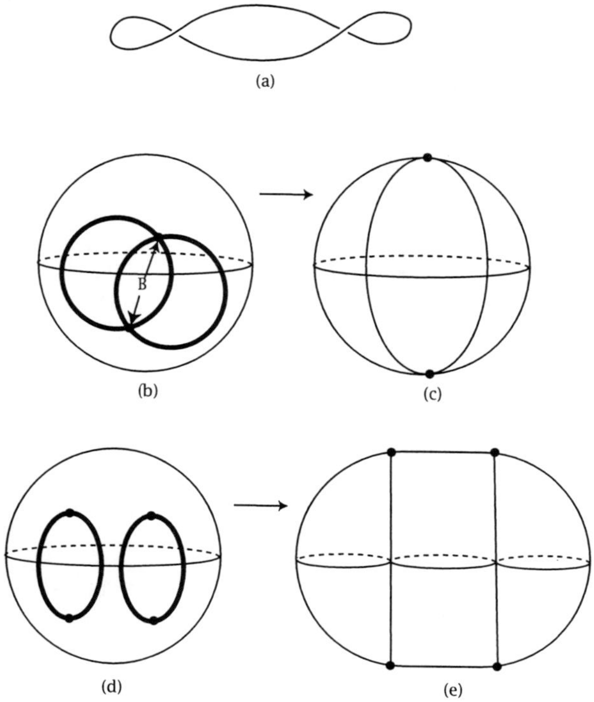

Figure 6: In (c), a non-generic projection; (e) shows generic projection of isotopic knotting.

Now define $g(m) = (m'_1, \ldots, m'_{n+1}, m_{n+2})$. Then $\rho(f,g) = \rho(\pi \circ f, g')$ so $S(f,g) < \epsilon$ and g is a knotting with $\pi \circ g = g'$. ∎

<u>Proof of Theorem 4.3</u>: Let ψ be the given knotting. We will consider the space of r-jets $J^r(M^n, \mathbf{R}^{n+1})$ for $r = 0, 1$, (see, for example [GL-GL] or [LV]). Consider the prolongation of ψ, $j^1\psi$; $j^1\psi$ is a map from M^n into the smooth manifold $J^r(M^n, \mathbf{R}^{n+1})$.

Let S_1 be the set of all σ in $J^1(M^n, \mathbf{R}^{n+1})$ of co-rank 1; then S_1 is a smooth submanifold of $J^1(M^n, \mathbf{R}^{n+1})$ of codimension 2. Now by the jet transversality theorem, the set of maps in $C^\infty(M^n, \mathbf{R}^{n+1})$ whose prolongation in $J^1(M^n, \mathbf{R}^{n+1})$ is transverse to S_1 is residual. Let

ψ_0' be one such map. Let $B_0 = (\psi_0')^{-1}(S_1)$ and let ψ_0 be a knotting of M^n in $\mathbf{R}^{n+2}$ close to ψ given by Lemma 4.4. Since ψ_0 is an embedding, the rank $\pi \circ \psi_0$ is either 0 or 1. The points where it is 0 are immersion points of $\pi \circ \psi_0$; the points where it has rank 1 are the branch points. Thus the branch point set of ψ_0 is B_0 a codimension 2 submanifold of M^n, and we have property (a) in the definition of general position satisfied for any ψ_0 obtained from a ψ_0' where $j^1(\psi_0') \pitchfork S^1$ as above.

Next we consider property (b). For this we will use a result of Morin [MR] which, when applied to our situation, tells us the following: "Let $b_0 \in B_0$; then there is a local system of coordinates $(t_1, \dots, t_{n-1}, x)$ for a disk N about b_0 and a local system of coordinates about $b_0^* \subseteq \mathbf{R}^{n+1}$ so that the map ψ_0 is given by $(t_1, \dots, t_{n-1}, x) \to (t_1, \dots, t_{n-1}, t_1 x, x^2)$." Clearly $\psi_0|_N$ is one-to-one except if $t_1 = 0$ and if $t_1 = 0$, ψ_0 is two-to-one if $x \neq 0$, one-to-one if $x = 0$. Thus in these coordinates the branch set is all points with $t_1 = 0$ and $x = 0$; the points with $t_1 = 0$, $x \neq 0$ are all mapped two-to-one.

Clearly, $(\pi \circ f)|_B$ is an immersion. To satisfy condition (d) we note that the set of maps of B into $\mathbf{R}^{n+1}$ with B normal crossings is a residual set (see, for example, [LV]). Moreover, maps from M into $\mathbf{R}^{n+1}$ which, when restricted to the branch set, give immersions with normal crossings is residual.

We next need to make sure that conditions (c) and (e) are satisfied. For this we may need to take more careful choices of our approximation of $\pi \circ \psi$. The set of maps with normal crossing is residual ([LV]) and so we may find a dense set of maps in $C^\infty(M^n, \mathbf{R}^{n+1})$ transverse to S_1 with normal crossings. Let ψ_1' be such a map close to $\pi \circ \psi$; then we may define a knotting $\psi_1 : M^n \to \mathbf{R}^{n+2}$ as before. If d^* is a point in the crossing set of ψ_1, then, since the crossings are normal, each point of $(\pi \circ \psi_1)^{-1}(d^*)$ will be a normal crossing point of D, the double point set of ψ_1. Our analysis of local coordinates about the branch points of ψ_1 shows that the double point set is an immersion point of D. Another consequence of this is that the branch set of ψ_1 does not intersect the double point set X of D.

However it is possible that B^* and X^* could intersect, and that the crossing set of B^* could meet the crossing set of D^*. If so, we use ordinary transversality to get a close approximation ϕ to ψ_1 so that $\phi' \pitchfork S_1$, ϕ' has normal crossings, $\phi'(B)$ is transverse to $\phi'(X)$, the crossing set of B^* is transverse to the crossing set of D^*. The lifting of ϕ' to a knotting $\phi : M^n \to \mathbf{R}^{n+2}$ will have all the properties of general position. ∎

5 Overpoints and underpoints

From the local models at points for generic maps, it is clear that D and D^* are unions of $(n-1)$-dimensional manifolds and $(\pi \circ \phi)|_D$ is basically a two-to-one map. We wish to have global notations for these manifolds as illustrated in Figure 7.

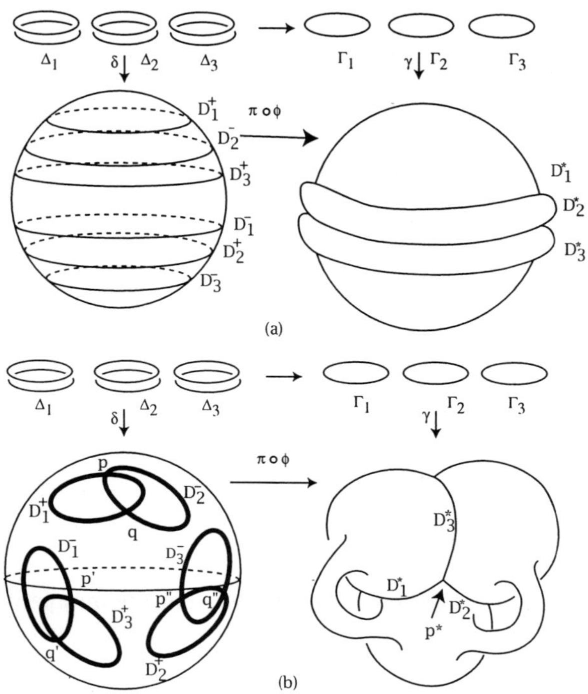

Figure 7: Two projections of knottings showing over and under manifolds.

Let ϕ be a knotting of M in general position with respect to π. Define

$$T = \{x \in M | (\pi \circ \phi)^{-1}(x^*) \text{ contains three points or more } \}.$$

We call T the **triple point set**. Let $P = D - (B \cup T)$; we call P the **set of pure double**

points—P is the set such that $\pi \circ \phi$ is exactly two-to-one. It follows from the definition of general position that P is an open $(n-1)$-manifold, dense in D.

The statement that D^* is the union of immersed $(n-1)$-dimensional manifolds can be expressed as follows. There is a compact $(n-1)$-dimensional manifold Γ (Γ is not necessarily connected, possibly with nonempty boundary) and an immersion $\gamma\colon \Gamma \to \mathbf{R}^{n+1}$ such that $\gamma(\Gamma) = D^*$ and such that $\gamma|_{\gamma^{-1}(P^*)}$ is one-to-one. (This last condition excludes the possibility that $\gamma\colon \Gamma \to D^*$ is a nontrivial covering space.) Points of $\partial\Gamma$ correspond to non-immersion points.

Let Γ_i be the components of Γ, $\gamma_i = \gamma|_{\Gamma_i}$ and $D_i^* = \gamma(\Gamma_i)$. Similarly let Δ be an $(n-1)$-dimensional manifold (Δ will have empty boundary, but may not be connected) where $\delta\colon \Delta \to M$ be an immersion such that $\delta(\Delta) = D$ and $\delta|_{\delta^{-1}(P)}$ is one-to-one. Let $D_i = (\pi \circ \phi)^{-1}(D_i^*)$ and let Δ_i denote $\delta^{-1}(D_i)$. There is a unique map $\alpha\colon \Delta \to \Gamma$ such that the following diagram commutes:

$$\begin{array}{ccc} \Delta & \xrightarrow{\alpha} & \Gamma \\ \downarrow{\scriptstyle \delta} & & \downarrow{\scriptstyle \gamma} \\ D & \longrightarrow & D^*. \end{array}$$

Also, $\alpha|_{\Delta-\delta^{-1}(B)}$ is a two-to-one map onto $\Gamma - \gamma^{-1}(B^*)$.

Figure 7 shows two knottings, For each Γ is the disjoint union of three circles. The knotting of Figure 7 (a) is the same as that of Figure 2, and Figure 7 (b) corresponds to Figure 4.

The definition of α can be made for generic maps of M^n into $\mathbf{R}^{n+1}$. For projections of knottings, the height function gives further structure to Δ.

Let $h\colon \mathbf{R}^{n+1} \to \mathbf{R}^1$ be projection on the last coordinate, we will refer to h as the height function. (Recall our standard projection is along this coordinate.

Suppose $\xi \in \Gamma$ such that $\alpha^{-1}(\xi)$ consists of two points, $\tilde{x}$ and $\tilde{x}'$. Now $x = \delta(\tilde{x})$ and $x' = \delta(\tilde{x}')$ are distinct points in M with $x^* = (x')^*$. We say $\tilde{x}$ and $\tilde{x}'$ are **paired by projection**.

Since ϕ is a knotting, we must have $h(\phi(x)) \neq h(\phi(x'))$. If $h(\phi(x)) > h(\phi(x'))$ we will say that $\tilde{x}$ is an **overpoint**. Otherwise we will say $\tilde{x}$ is an **underpoint**.

The next proposition deals with the simpler case that a projection has no branch points.

Proposition 5.1 *If $\partial\Gamma_i = \emptyset$ then δ_i has two components, Δ_i^+ and Δ_i^-, where Δ_i^- is the set of overpoints of Δ_i^+ and Δ_i^- is a set of underpoints, with $\pi \circ \phi \circ \delta(\Delta_i^+) = \pi \circ \phi \circ \delta(\Delta_i^-) = D_i^*$*

Proof: Let $\alpha_i = \alpha|_{\Delta_i}$, then $\alpha_i \colon \Delta_i \to \Gamma_i$ a two-to-one map (since $B^* \cap D_i^* = \emptyset$). Since Γ_i is connected, it will suffice to show α_i is a covering map and that the set of underpoints and the set of overpoints are components of Δ_i. Let $\xi \in \Gamma_i$ and $\gamma(\xi) = x^*$. Then $x^* \in D_i^*$. Let $\alpha^{-1}(\xi) = \{\tilde{x}^+ \cup \tilde{x}^-\}$ where $\tilde{x}^+$ is the overpoint. There is a small $(n+1)$-dimensional ball N in $\mathbf{R}^{n+1}$ containing x^* such that $M^* \cap N$ looks like k n-disks which intersect each other transversely. The intersections of these disks is a collection of $(n-1)$-disks which we denote by $\{X_j^*\}$ where $1 \leq j \leq \binom{k}{2}$.

Consider a particular choice of $(n-1)$-disk X_s^* where $X_s^* \subseteq D_i^*$. Since $(\pi \circ \phi)^{-1}(X_s^*)$ is two disjoint $(n-1)$-disks, we may find disjoint $(n-1)$-disks, $\widetilde{X_s^+}$ and $\widetilde{X_s^-}$ in Δ_i such that $\pi \circ \phi \circ \delta(\widetilde{X_s^+}) = \pi \circ \phi \circ \delta(\widetilde{X_s^-}) = X_s^*$ with $\tilde{x}^+ \in \widetilde{X_s^+}$ and $\tilde{x}^- \in \widetilde{X_s^-}$. By continuity of $h \circ \phi \circ \delta$ and connectedness of $\widetilde{X_s^+}$, it follows that since $\tilde{x}^+$ is an overpoint, all points of $\widetilde{X_s^+}$ are overpoints. Thus the set of overpoints of Δ_i is open and nonempty. Similarly, the set of underpoints of Δ_i is open and nonempty. ∎

Let $D_i^+ = \delta(\Delta_i^+)$, $D_i^- = \delta(\Delta_i^-)$; Figures 7(a) and (b) show examples of these sets.

The above proposition assumes we have the projection of a knotting and not just an image of a generic map. For example consider the immersion of the projective plane into $\mathbf{R}^3$ known as Boy's surface. This is a generic map, and Γ consists of a single circle as does Δ; the map α is a two-fold covering. Boy's surface is not the projection of a knotting—if it were, we would have an embedded projective plane in $\mathbf{R}^4$ with trivial normal bundle, contradicting a result of Whitney [WH]. Also, there are knottings with $D_i^+ \cap D_i^- \neq \emptyset$ (see Figure 11 of [GL] for this as well as for further information on Boy's surface).

More generally, if there are branch points then $\partial\Gamma_i \neq \emptyset$ and we have:

Proposition 5.2 *If $\partial\Gamma_i \neq \emptyset$ then $\Delta_i - \delta^{-1}(B)$ has two components Δ_i^+ and Δ_i^- where Δ_i^+ is the set of overpoints, Δ_i^- the set of underpoints such that $\pi \circ \phi \circ \delta(\Delta_i^+) = \pi \circ \phi \circ \delta(\Delta_i^-) =$*

$\gamma(\Gamma_i - \partial\Gamma_i)$. Moreover, if $\overline{\Delta_i^+}$ denotes the closure of Δ_i^+ in Δ_i then $\alpha|_{\overline{\Delta_i^+}}$ is a homeomorphism of $\overline{\Delta_i^+}$ onto Γ_i, and Δ_i is the double of the manifold $\overline{\Delta_i^+}$.

Proof: As in the proof of Proposition 5.1, we may argue that $\Delta_i - \delta^{-1}(B)$ has two components: the set of overpoints, Δ_i^+ and the set of underpoints Δ_i^- . Similarly, also, we may show that $\alpha|_{\Delta_i - \delta^{-1}(B)}$ is a covering map from $\Delta_i - \delta^{-1}(B)$ to $\Gamma_i - \gamma^{-1}(B)$. From our knowledge of the local structure of the double point set about the branch set we can see that $\partial(\overline{\Delta_i^+}) = \partial(\overline{\Delta_i^-}) = \Delta_i \cap \delta^{-1}(B)$, and where $\alpha|_{\overline{\Delta_i^+}}$ maps $\partial(\overline{\Delta_i^+})$ homeomorphically onto $\partial\Gamma_i$. ∎

We use the following notation: $B_i = \delta(\Delta_i) \cap B$.

Up to this point, all our results hold whether or not the manifold M^n is orientable. A standard embedding of the projective plane P^2 in $\mathbf{R}^4$ has a projection known as the "cross-cap". Here the crossing set is a single arc and the double point set a single circle whose normal bundle is not a product. The following proposition shows that this can't occur in the orientable case.

Proposition 5.3 *If ϕ is a knotting of an orientable manifold M^n in general position with respect to projection, then the double point set D is the union of immersed orientable $(n-1)$-manifolds with trivial normal bundle.*

Proof: Consider $\mathring{D}_i^+ = \delta(\Delta_i^+)$ (similar arguments will also hold for $\mathring{D}_i^- = \delta(\Delta_i^-)$. For a codimension one immersion of a manifold N into an orientable manifold, N is orientable if and only if the normal bundle of the immersion is trivial. Choose an orientation for M. As in Hirsch [HR2] we may view this as a coherent family of orientations of the tangent planes of M. Suppose x and x' are paired by projection. Let w be a unit vector in TM_x normal to $\mathring{D}_i^+$ at x^*. The image of under the differential of $\pi \circ f$ of w, followed by the image of the ordered basis for $TM_{x'}$ gives an ordered basis for $T\mathbf{R}^{n+1}{}_{x^*}$. If this represents the orientation for $T\mathbf{R}^{n+1}{}_{x^*}$, let $t_x = w$, otherwise let $t_x = -w$. This choice t_x of orientation of the normal fiber to $\mathring{D}_i^+$ at each $x \in \mathring{D}_i^+$ gives a coherent family of orientations. Thus the normal bundle of D is orientable. An orientable 1-dimensional vector bundle over a paracompact space is trivial; thus $\mathring{D}_i^+$ has trivial normal bundle.

For those i for which $\partial\overline{\Delta_i^+} = \emptyset$, $\mathring{D}_i^+ = D_i^+$ there is nothing more to show (similarly for such D_i^-). If $\partial\overline{\Delta_i^+} \neq \varnothing$ then by Proposition 5.2, Δ_i is the double of $\overline{\Delta_i^+}$. As we have seen, Δ_i^+ is, as is Δ_i. Thus in this case also we see $D_i = \delta(\Delta_i)$ is an immersion of an orientable $(n-1)$-manifold and has trivial normal bundle. ∎

Corollary 5.4 *If a knotting $\psi: M^n \to \mathbf{R}^{n+2}$ is in general position with respect to projection, the branch set B is the union of orientable $(n-2)$-dimensional submanifolds of M^n, each embedded with trivial normal bundle.*

Proof: Each component B_i of B corresponds to the boundary of $\overline{\Delta_i^+}$ which, by Proposition 5.3 is orientable, thus B_i must be orientable. Since B_i has trivial normal bundle in D_i, and D_i has trivial normal bundle M, then B_i must have trivial normal bundle in M. ∎

6 Projections and Immersions

Although branch points are generic, there is a technique for eliminating branch points if the manifold M^n is orientable, Theorem 6.1. However, as noted for the cross cap, it may not be possible to do this in the nonorientable case.

Theorem 6.1 *Let $\psi: M^n \to \mathbf{R}^{n+2}$ be a knotting, M^n orientable. Then we may isotop ψ to a new embedding whose projection into $\mathbf{R}^{n+1}$ is an immersion.*

Before we prove this theorem, we note two corollaries.

Corollary 6.2 *If $\psi: M^n \to \mathbf{R}^{n+2}$ is an embedding, M^n orientable, then M^n has a trivial normal bundle in $\mathbf{R}^{n+1}$.*

Proof of Corollary 6.2: Use Theorem 6.1 to isotop ψ to ϕ whose projection is an immersion. Since M^n is orientable, the normal bundle of $\pi \circ \phi$ is orientable and thus trivial. We may now obtain a product framing of the normal bundle of $\phi(M)$ by using a normal vector which projects to a vector normal to $\pi \circ \phi(M)$ and the upward pointing vector in the direction of the projection π. ∎

By weakening the conclusion of Theorem 6.1 from isotopy to regular homotopy we obtain:

Corollary 6.3 *If* $\psi\colon M^n \to \mathbf{R}^{n+2}$ *is smooth knotting,* M^n *orientable then* ψ *is regularly homotopic in* $\mathbf{R}^{n+2}$ *to a map* ψ' *where* ψ' *is an immersion of* M^n *into* $\mathbf{R}^{n+1}$. ∎

We remark that Corollary 6.3 has an alternate proof. One could prove, independently, that M^n has trivial normal bundle (for example, by considering the normal bundle as a one-dimensional complex bundle and using characteristic class arguments), giving an alternate proof of Corollary 6.2. In particular there is a normal vector field to M^n in $\mathbf{R}^{n+2}$. By [HR1] we could regularly homotop ψ to an immersion in $\mathbf{R}^{n+1}$. However, the proofs we present here are elementary and geometric in nature and do not rely on characteristic classes or general immersion theory.

The following example may help to follow the basic construction in proof of Theorem 6.1. Consider an embedding $\psi\colon \mathbf{S}^2 \to \mathbf{R}^4$ whose projection is shown in Figure 3(b). This projection has branch set consisting of two points b and b' and an arc of crossing points D^*, see Figure 8 (a).

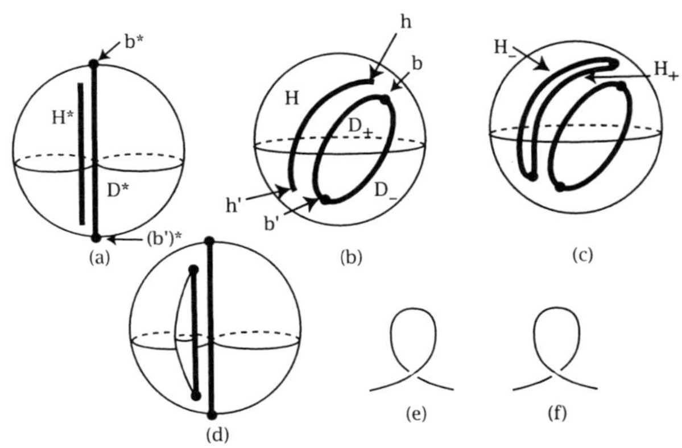

Figure 8: Adding H; (b) shows double point set,(a) shows projection. Altered map shown in (c) as double points, (d) for projection.

The double point set D, shown in Figure 8 (b), is a single circle. This circle is the union of an arc of overpoints, D_+, and an arc of underpoints, D_-. Let $H \subseteq \mathbf{S}^2$ be a parallel copy of D_+, drawn on $\mathbf{S}^2$— parallel in the sense of the product structure of the tubular neighborhood of $D_+ \cup D_-$ on $\mathbf{S}^2$. Figure 8 (b) shows H; its image H^* in the projection shown in Figure 8 (a). Label the endpoints of H by h and h' so that h is near b and h' is near b', as shown. We now wish to add a pigtail along H and construct a new knotting (equivalent to ψ) by the method used in Proposition 3.2. We do this so that the double point set is as shown in Figure 8(c) and the projection looks like that shown in Figure 8(d). The projection of this new knotting has crossing set consisting of two disjoint arcs. The corresponding double point set will then be the union of two disjoint circles in $\mathbf{S}^2$. In Figure 8 (c) shows the double point set: $D \cup H_+ \cup H_-$, where H_+ is the set of overpoints, H_- the underpoints for the new arc.

To define this knotting, we must make two choices. First, if we consider a small neighborhood N^* of H^* in M^*, N^* will be two-sided in $\mathbf{R}^3$ and we have a choice on which side the curl will be put. We will chose the side so that the projection will be as shown in Figure 8 (d). We might describe this as putting the tail on the outside rather the inside. Secondly, in order that H_+ is next to D_+ as shown (rather than the other possibility—where H_- is next to D_+) also involves a choice, namely, the direction of curl of the pigtail. Given the projection of Figure 8 (d) there are two choices we have for the single crossing of the pigtail. The correct choice will correspond to Figure 8(c). This is a matter of choosing between the two possible crossings types of the single self crossing of the loop. Consider the loops as shown in Figure 8(e) and (f). We chose between these two loops so that we will have H_+ next to D_+. Figures 9 and 10 may clarify the distinctions of these choices; Figure 9 uses the point of view of slices, Figure 10 (a) and (b) uses the point of view of projection.

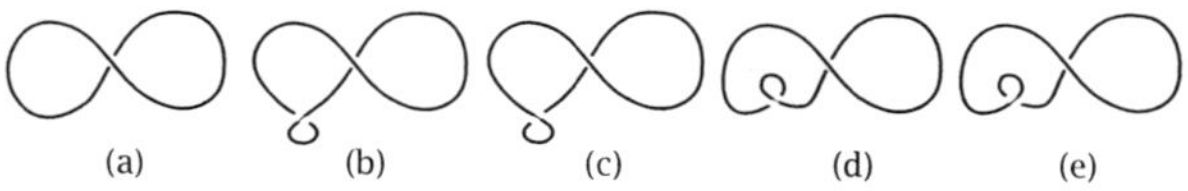

Figure 9: Four choices for adding a loop to curve shown in (a).

In Figure 9 (a) we see a slice of the sphere of Figure 3 (b). In Figures 9 (b),(c),(d) and (e) we show the slice we would obtain with our four possible choices: the loops in Figures 9 (b) and (c) are on the same side of the projection, each is on the opposite side of the choices shown in Figures (d) and (e). The choice we want is the one shown in Figure 9 (b). The choices shown in Figure 9 (c) and (d) will be inappropriate for our construction; the distinction, expressed in terms of projections of these curves, the sum of the signed crossing numbers is 2 whereas the other choices this sum is zero 0. The choice shown in Figure 9 (e) can be made to work but is slightly more complicated to describe.

Alternatively one can see the distinction of our choices by looking at the projection, locally about two of the paired branch points. What we want is that the local picture locally looks like Figure 10 (a) (with height relations at the two branch points consistent) and not the situation shown in Figure 10 (b).

Let $B_0 = \{b\} \cup \{b'\}$ denote the branch set of the original knotting; our new knotting has two additional branch points $\partial H = \{h\} \cup \{h'\}$. A tubular neighborhood N of B_0 in S^2 will consist of two disjoint disks. By placing H close to D_+, we may assume that the disk containing b contains h and the disk containing b' contains h'. In projection, these two disks will look as shown in Figure 10 (a).

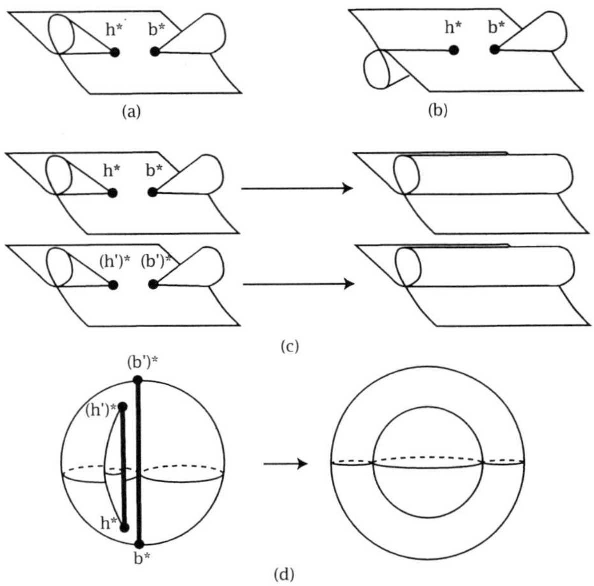

Figure 10: Correct local choice of side shown in (a), incorrect choice in (b). Elimination of branch points shown in projection in (d) using local changes shown in (c).

We may now remove the branch points of ψ' in pairs by an isotopy so that the resulting projection looks like Figure 10(c). (For more detail, see the discussion of knot move of type (d) in [RS5]; this isotopy may be described as two knot moves of type (d), one for each disk of N.) For a global picture see Figure 10 (d) where we show the change produced by our smooth isotopy: from a projection with two disjoint arcs of crossing points to a projection which is an immersion with crossing set a single circle.

We next address three the technical problems which arise in generalizing the above example. First, will need to analyze our definitions so that we can articulate, for the general case, which is the right choice of side on which to put our pigtail. Secondly, the set D_+ might not be an embedded submanifold of M^n, but only immersed. If D_+ has self crossings, our parallel copy H will also have self crossings and we will need to examine this possibility. Finally, in higher dimensions B_0^* might have self intersections. We show that this is not a problem since we define our isotopy as a sequence of locally supported isotopies.

Proof of Theorem 6.1: By Theorem 4.3, we may assume, without loss of generality, that ψ is in general position with respect to projection. We begin by defining H. Referring to the notations of Section 2, the double point set D is the union of immersed manifolds D_i. Let Δ^+ be the union of all $\Delta_i^+ \subseteq \Delta$ such that $\partial\Delta_i^+ \neq \emptyset$, then it follows from Proposition 5.2 that $\delta(\partial\Delta^+) = B$. Now by Proposition 5.3, the immersion $\delta|_{\Delta^+}$ has a trivial normal bundle, denote this by ν_+; ν_+ is an immersion, $\nu_+ \colon \Delta^+ \times \mathbf{D}^1 \to M^n$. Now define $\tilde{H}$ to be the subset of $\Delta^+ \times \mathbf{D}^1$ corresponding to $\Delta^+ \times \{\frac{1}{2}\}$. Let $H = \delta(\tilde{H})$. The height function gives a structure to H at the self crossings of H as is done for the projection link, see [R3]. So at any crossing point the disks which locally are crossing transversely can be given an ordering which is globally well defined since it is obtained from the (globally defined) height function. Figures 12, 13 and 16 discussed below give examples of this situation.

We now construct a new knotting ψ' by spinning the trivial knot about H, using a modification (to the case $\partial H \neq set$ and knot to spin is trivial) of the construction found in [RS3]. Consider an embedding of an arc p in the 3-disk with a single crossing in its projection; p is the pigtail as shown in Figure 5(b). We consider $\psi(H)$. We think of $\psi(H)$ as "a copy of $\tilde{H}$ drawn M." Using the language of [RS3], we spin the knot p along the projection of $\tilde{H}$. Since $\tilde{H}$ has boundary, we need to shrink the loop p near $\partial\tilde{H}$ as we did in the proof of Proposition 3.2.

In more detail, our construction proceeds as follows. First let $\tilde{N} = \Delta^+ \times [1/4, 3/4]$, then $\psi \circ \nu_+|\tilde{N}$ is a trivial tubular neighborhood to $\psi(H)$ in $\psi(M)$. Let $N = \delta(\tilde{N})$. Note that $\psi(N)$ has trivial two-dimensional tubular neighborhood in $\mathbf{R}^{n+2}$; this tubular neighborhood is the Whitney sum of the pullback of the (trivial) normal bundle to N^*

in $\mathbf{R}^{n+1}$ and the restriction of the kernel of the projection map. Let $\tilde{x} \in \tilde{H} - \partial\tilde{H}$, then $\tilde{x} = (\tilde{d}_+, \frac{1}{2})$ where $\tilde{d}_+ \in \Delta^+ - \partial\Delta^+$. Let $d_+ = \delta(\tilde{d}_+)$, $x = \delta(\tilde{x})$. Let d_- denote the point, paired by projection with d_+ and $\tilde{d}_- = \delta^{-1}(d_-)$. Recall Δ has a trivial tubular neighborhood $\nu\colon \Delta\times\mathbf{D}^1 \to M^n$. Let $\tilde{F}_+$ be the subset of $\Delta\times\mathbf{D}^1$ corresponding to $\{\tilde{d}_+\}\times\mathbf{D}^1$ and let $\tilde{F}_-$ correspond to $\{\tilde{d}_-\}\times\mathbf{D}^1$.

Also let $F_+ = \nu(\tilde{F}_+)$; $F_- = \nu(\tilde{F}_-)$. Let us consider d_+^* (recall $d_+^* = d_-^*$). In $\mathbf{R}^{n+1}$, F_+^* and F_-^* meet transversely at d_+^* and determine a plane P which is normal to the crossing set of M^* at d_+^*. Now $x^* \in F_+^*$, see Figure 11 (a) where arrows indicate orientation of the fibers F_+^* and F_-^*.

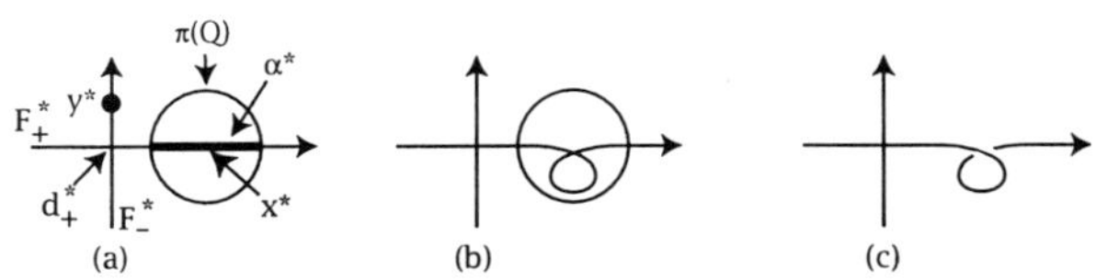

Figure 11: In (a) and (b) subsets of P are shown. Figure (c) projects onto P as shown in (b).

Let $\alpha = \{d_+\} \times [1/4, 3/4]$; $\alpha = \delta(\alpha)$. Then α^* is shown in Figure 11 (a). Consider $\psi(x)$, then let Q be a small normal disk to $\psi(H)$ at $\psi(x)$ in $\mathbf{R}^{n+2}$; Q is a 3-disk which projects into P as shown. Now $\psi(\alpha)$ will look like a straight arc in Q. We will replace α by a pigtail so that the projection will be as in Figure 11 (b). Now consider that F_+^* has two sides in P. Let $\tilde{y}$ be the point in $\Delta \times \mathbf{D}^1$, $\tilde{y} = \{\tilde{d}_-\} \times \{\frac{1}{2}\}$; let $y = \nu(\tilde{y})$. Call the component of $P - F_+^*$ which contains y^* the "inside", call the other component the "outside." Note that in Figure 11 (b) we have put the pigtail on the "outside". We need, furthermore, to specify the direction of the curl of the tail; this is indicated in Figure 11(c).

For a surface in $\mathbf{R}^4$ the local change in the projection is shown in the transition from Figure 12 (a) to Figure 12 (b). In Figure 12 (a) we show an sub-arc D of D_+ and a parallel sub-arc of H is shown. Placing a loop at each point of H gives an embedding whose projection looks like Figure 12 (b). If H has no self crossings, the above local change will define a new embedding of M; at the end of this proof we will be able to smoothly isotop this knotting to one whose projection is an immersion.

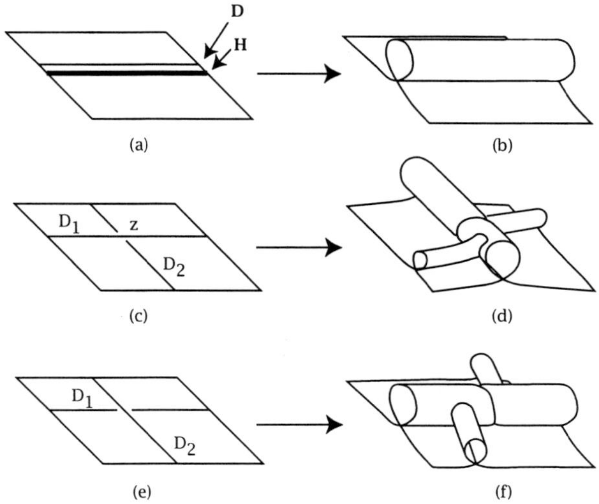

Figure 12: Local changes for a single subdisk of M.

We next consider self crossings of D_+ (and thus of H). Let K denote the self crossing set of H. Suppose there is a point z which is a self-crossing point of D_+, as in figure 12 (c). Here D_1 and D_2 are sub-arcs of D_+. In this case, z will be a triple point of the projection. So there are two other points of M z_1 and z_2 with $z^* = z_1^* = z_2^*$. Label so that z_1 belongs to a sub-disk of M containing the points of M paired by projection to points of D_1, and z_2 belongs to a sub-disk of M containing the points of M paired by projection to points of D_2. If $h(z_2) < h(z_1)$ then we will draw D_1 as passing over D_2 as in Figure 12 (c); otherwise we will draw as D_1 passing under D_2, Figure 12 (e).

When we do our construction, we will make our local changes in sequence. Overpasses are processed first, underpasses last. The order is significant as shown in Figures 12 (d) and (f). Using this order assures that this locally defined construction is globally consistent.

A more complete view of triple points is shown in Figure 13 for a surface in $\mathbf{R}^4$. At a triple point z, one has three arcs of crossing points meeting transversely. In Figure 13 (a) we show the changes in the case that exactly one of these subarcs corresponds

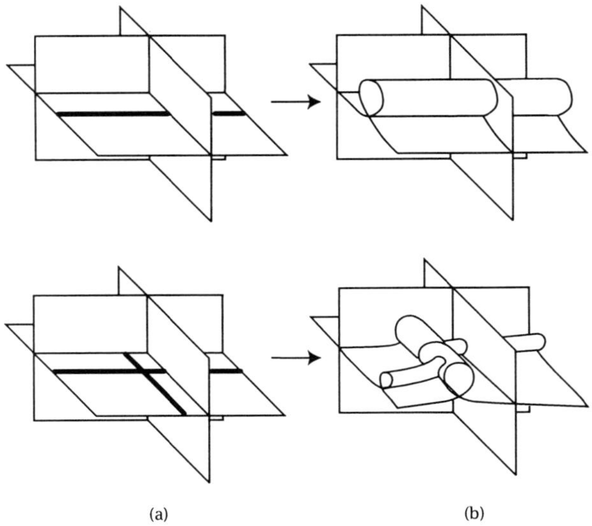

Figure 13: Examples of change of projection at triple point.

to a sub-arc of D_+. In Figure 13 (b) we show the changes in the case that exactly two of these sub-arcs is correspond to sub-arcs of D_+ (compare with Figure 12 (d)).

A simple example where these cases occur is shown in Figure 14 (a) we show a (trivial) knotting of a sphere in $\mathbf{R}^4$ using a sequence cross sections by hyper-planes. The standard projection of this knotting will be in general position with respect to projection. This projection will have four branch points and two triple points. One can see that the crossing set will be the image of two arcs and a single circle. The crossing set is shown in Figure 14 (b).

Figure 15 (a) shows the double point set for the knotting of Figure 14 (a). Then H is a union of two arcs shown in Figure 15 (b). There are is one self crossing of H, and two other crossings of H with the double point set, see Figure 15 (b). After performing our modifications, the double point set will look as in Figure 15 (c). One then cancels the paired branch points as in Figure 10. The resulting knotting (whose projection is an immersion) has double point set as shown in Figure 15 (d).

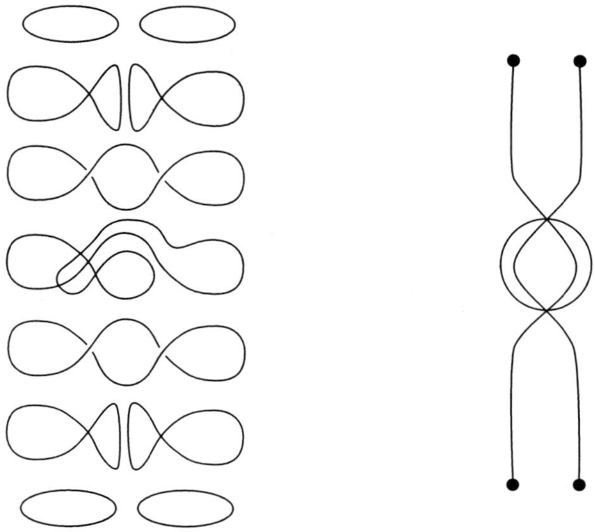

(a) (b)

Figure 14: Example of knotting with self-crossings of D_+; (a) shows slices and (b) shows (only) the crossing set.

Another global picture of this process is shown in Figure 16. In Figure 16 (a) we show the projection of this knotting with two branch points, sliced into three parts and separated to show details. In Figure 16 (b) we show projection of H as it appears in the middle slice. Figure 16 (c) shows H and the double point set of the knotting of (a). Figure 16 (d) shows projection of modified knotting prior to cancellation of branch points.

The case that all three are sub-arcs of D_+ is not illustrated but it is similar to the two sub-arc case. All three 2-disks which meet at z are replaced by embedded disks each of which, looks like Figure 12 (d) or Figure 12 (f).

Now in the general case, when $n = dim(M) > 2$, if $h \in K$ is a self crossing point of multiplicity k, we alter our embedding locally in k steps. Near points of K we use our height relations to give us a multiple fibering and thus we can define a multiple knotting

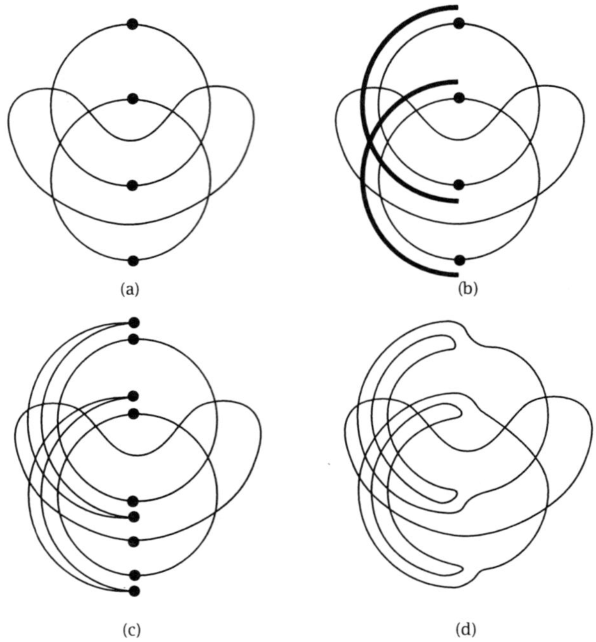

Figure 15: Changes in double point set for knotting of Figure 14 (a); H shown as bold arcs in (b).

using p. It is important to note that, in general, we expect K to be $(n-2)$-dimensional. In the case $n = 2$, as in our examples, K is a set of isolated points and, either of the choices shown in Figure 11 (c) and (e) will give the desired knotting. But if $n = 3$, then K might be 1-dimensional and one cannot arbitrarily assign multiple fibering choices locally to points of K and have a knotting well defined globally. By using the height function to define the local ordering of the multiple fibering, we are assured of a well defined knotting.

Once we have defined H and the new embedding obtained by placing a pigtail along H, we obtain a new knotting ψ' so that, if B_0 denotes the original branch points of ψ, then the branch set of ψ will correspond to $B_0 \cup \partial H$. We use ψ' to eliminate pairs branch

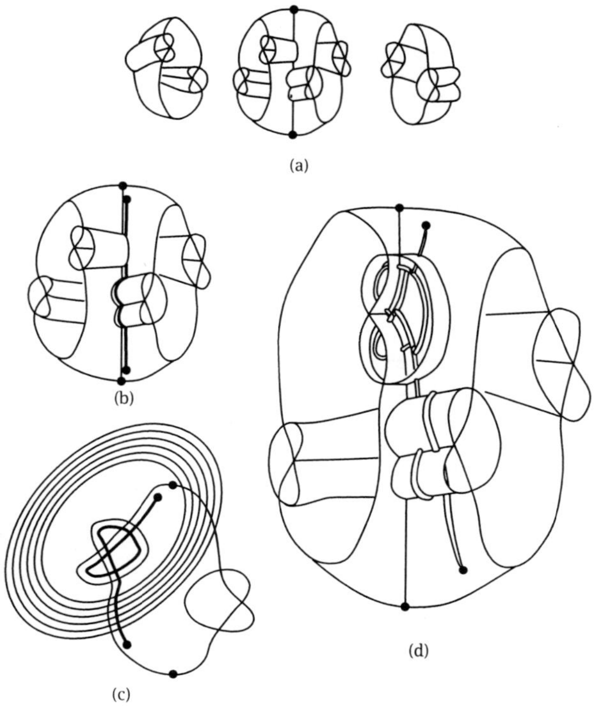

Figure 16: Eliminating branch points, a more complex example.

points. If B_0^* is an embedded manifold (then so is $(\partial H)^*$) we can describe this as follows: in a neighborhood of B_0^* in $\mathbf{R}^{n+1}$, M^* will look like the product of B_0^* and a disk such as one of those pictured in Figure 9 (b), where one branch point of the disk is in $(\partial H)^*$ the other in B_0^*.

We now define an isotopy of ψ' which will "cancel these branch points." Locally, the projection of M about a point of B_0 looks like a subdisk of B_0 crossed with a disk with two branch points. The isotopy which "cancels" the branch points gives rise to a local isotopy of our manifold and by our construction, these local isotopies will fit together to give us the global isotopy of ψ' obtaining a new embedding that has no branch points.

The construction of this portion of isotopy does not depend on the projection of the knotting; thus the possibility that B might have self crossings is not of consequence. ∎

It is important to note here that although the new embedding we obtain is close to ψ in the C^0 topology, it is *not* close in the C^1 topology—putting the pigtail in clearly changes derivatives by a large amount. Of course we cannot get rid of the branch points of ψ *and* be close in the C^1 topology since generic branch points are stable.

In [GL] we also find a quite different proof of Theorem 6.1 for orientable surfaces in $\mathbf{R}^4$.

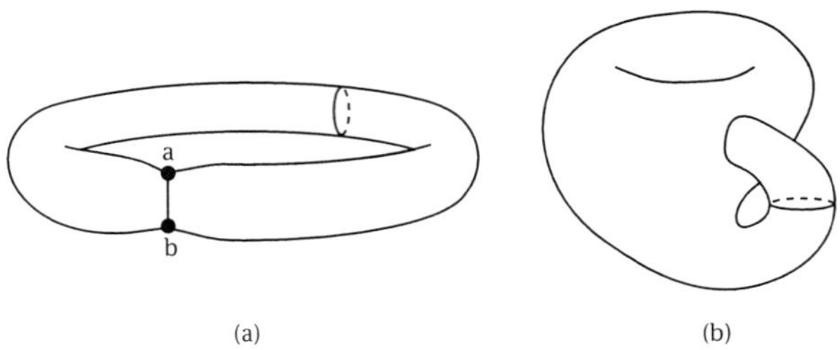

Figure 17: Elimination of branch points for a Klein bottle.

We remark that the hypothesis that M is orientable is not completely necessary; an examination of the proof shows that all we need is for D to have trivial tubular neighborhood in M. For example, consider the projection of the Klein bottle with two branch points as shown in Figure 17(a). Our techniques in this case apply and change the embedding so that the projection is the more familiar immersion as shown in Figure 17(b).

7 Additional remarks

The history of this paper is similar to that of [RS5] and has been discussed there. Based on a preprint of the current paper, Carter and Saito further analyzed the problem of cancelling branch points for non-orientable manifolds, [C-S].

8 Bibliography

[AR] Artin, E., "Zur Isotopie zweidimensionalen Flachen im $\mathbf{R}^4$ ", Abh. Math. Sem. Univ. Hamburg (1925), pp. 174-177.

[CP] Cappell, S., "Superspinning and knot complements", Topology of Manifolds, Georgia (1969), Markham Publ., pp. 358-383.

[CS] Carter, Scott, and Saito, M. , "Canceling Branch Points on Projections of Surfaces in 4-Space," Proc. of the AMS. 116, No 1. (1992), pp. 229-237.

[FX] Fox, R., "A quick trip through knot theory", Topology of 3-Manifolds and Related Topics (Proc. Univ. Georgia Inst.,1961), Prentice-Hall (1962), pp. 120-167.

[GL] Giller, C., "Towards a classical knot theory for surfaces in $\mathbf{R}^4$", Illinois J. Math. 26, No. 4 (1982), pp. 591-631.

[GL-GL] Golubitsky, M. and Guillemin, V., Stable Mappings and Their Singularities, Grad. Texts in Math., No. 14, Springer-Verlag (1973).

[HR1] Hirsch, M., "Immersions of manifolds", Trans. Amer. Math. Soc. 93(1959), pp. 242-276.

[HR2] Hirsch, M., Differential Topology, Graduate Texts in Mathematics No. 33, Springer-Verlag (1976).

[LM] Lamonaco, S., "Five dimensional knot theory", Low Dimensional Topology, Contemp. Math. Vol. 20, A.M.S. (1983), pp. 249-270.

[LV] Levine, H., "Singularities of Differentiable Mappings", Proceedings of Liverpool Singularities Symposium I, Lecture Notes in Math., Vol. 192 (1971), Springer- Verlag, pp. 1-89.

[MR] Morin, B., "Formes cononiques des singularities d'une application differentiable, Comptes Rendus Acad. Sci. Paris", 260(1965), pp. 5662-5665.

[RS1] Roseman, D., "Woven knots are spun knots", Osaka J. Math. 11(1974), pp. 307-312.

[RS2] Roseman, D., "Projections of knots", Fund. Math. 89(1975), pp. 99-110.

[RS3] Roseman, D., "Spinning knots about submanifolds; spinning knots about projections of knots", Topology and its App. 31 (1989) pp. 225-241.

[RS4] Roseman, D., "Elementary moves for higher dimensional knots," Preprint.

[RS5] Roseman, D., "Reidemeister-type moves for surfaces in four dimensional space", Knot Theory, Banach Center Publications, Vol 42, Inst. Math. Polish Acad. Sci. (1998),pp. 347-380.

[WH] Whitney, H., "On topology of differentiable manifolds", Lectures in Topology, University of Michigan Press (1941).

[YJ] Yajima, T., "On the fundamental groups of knotted 2-manifolds in the 4-space", J. Math. Osaka City Univ. 13(1962), pp. 63-71.

A KLEIN BOTTLE WHOSE SINGULAR SET CONSISTS OF THREE DISJOINT SIMPLE CLOSED CURVES

AKIKO SHIMA[†]
Graduate School of Mathematical Sciences
University of Tokyo
Current address:Department of Mathematics Tokai University
1117 Kitakaname Hiratuka Kanagawa 259-1207, Japan

ABSTRACT
Let F be an embedded Klein bottle in $S^4 \setminus \{\infty\}$. If the singular set $\Gamma(F^*)$ of the projection F^* of F into $\mathbf{R}^3$ consists of at most three disjoint simple closed curves, then F bounds a solid Klein bottle in the 4-sphere S^4, i.e., F can be moved to the standard Klein bottle.

Keywords: Knotted surfaces, Klein bottle

Introduction

Aiso [1] classified simply knotted spheres with less than 6 crossing circles. Let T be an embedded torus in $S^4 \setminus \{\infty\}$. In [7] and [8], it was shown that if the singular set $\Gamma(T^*)$ of the projection T^* of T into $\mathbf{R}^3$ consists of at most two disjoint simple closed curves, then T can be moved to the standard position by an ambient isotopy of S^4. In [9], it was shown that if the singular set $\Gamma(T^*)$ consists of three disjoint simple closed curves, then T is ambient isotopic to either the standard torus, the spun torus of the trefoil knot, the twist spun torus of the trefoil knot, or the torus obtained by attaching a handle to the spun 2-sphere of the trefoil knot. The following is the main result in this paper.

Theorem. *Let F be an embedded Klein bottle in $S^4 \setminus \{\infty\}$. If the singular set $\Gamma(F^*)$ of the projection F^* of F into $\mathbf{R}^3$ consists of at most three disjoint simple closed curves, then F bounds a solid Klein bottle in the 4-sphere S^4, i.e., F can be moved to the standard Klein bottle.*

We will work in the PL category. All submanifolds are assumed to be locally flat. Let $\mathbf{R}^n$ be the n-dimensional Euclidean space, S^n the n-dimensional sphere,

[†]The author was partially supported by Research Fellowships of the Japan Society for the Promotion of Science for Young Scientists.

and ∞ its north pole. In this paper we always regard the sphere S^n as one-point compactification of $\mathbf{R}^n$, and ∞ is understood the point at infinity.

Let F be a compact surface. A map f of F into S^3 is in *general position*, if at every point of $f(F)$, $f(F)$ is locally a sheet, or transversal intersection of either 2 or 3 sheets, or a branch point. A branch point is also known as "Whitney's umbrella" or "a pinch point". A point $x \in f(F)$ is called a *double point* if $f^{-1}(x)$ consists of two points. A point $x \in f(F)$ is called a *triple point* if $f^{-1}(x)$ consists of three points.

Let F be an embedded closed surface in $S^4 \setminus \{\infty\}$. In this paper we assume that $p|F$ is in general position, where p is the projection of $\mathbf{R}^4$ onto $\mathbf{R}^3$ defined by $p(x_1, x_2, x_3, x_4) = (x_1, x_2, x_3)$. With every point P or subset F of $S^4 \setminus \{\infty\}$, we associate the point $P^* = p(P)$ or the subset $F^* = p(F)$. Let f be a map in general position. Then the *singular set* $\Gamma(F^*)$ is the set of all double points, triple points and branch points of f. Set $\Gamma(F) = p^{-1}(\Gamma(F^*)) \cap F$.

A regular neighborhood of a trivial knot in S^3 is said to be a *standard solid torus.* All homology groups are with coefficients in $\mathbf{Z}$. We denote by $|A|$ the number of the components of A, and by $\overline{A}$ the closure of A.

The paper is organized as follows. In section 1, we give a diagrammatic representation of a singular set $\Gamma(F)$. In section 2, we show that if the singular set $\Gamma(F^*)$ consists of one simple closed curve, then F bounds a solid Klein bottle. In section 3, we consider conditions satisfied by a singular set $\Gamma(F)$. In section 4, we show that if the singular set $\Gamma(F^*)$ consists of two disjoint simple closed curves, then F bounds a solid Klein bottle. In section 5, we prove the main theorem.

1. Simple closed curves in a Klein bottle

In this section, we give a classification of simple closed curves in a Klein bottle. And we give a diagrammatic representation of a singular set $\Gamma(F)$.

An equivalence relation $\sim$ on $[0,1] \times [0,1]$ is defined by $(0,t) \sim (1, 1-t)$ and $(t, 0) \sim (t, 1)$. We assume that a Klein bottle is identified as $[0,1] \times [0,1]/\sim$.

It is easy to show the following lemma.

Lemma 1.1. *Let c be a simple closed curve in a Klein bottle F. Then (F, c) is homeomorphic to $(F, c_1), (F, c_2), (F, c_3)$ or (F, c_4), where c_1 bounds a disk in F, $c_2 = [0,1] \times \frac{1}{2}/\sim$, $c_3 = \frac{1}{2} \times [0,1]/\sim$, and $c_4 = [0,1] \times \frac{1}{4} \cup [0,1] \times \frac{3}{4}/\sim$.*

Remark 1.2. (1) A regular neighborhood of c_i $(i = 1, 3, 4)$ is an annulus.

(2) Let $c_5 = [0,1] \times 0/\sim$. Then, the two simple closed curves c_2 and c_5 have a Möbius band neighborhood.

Lemma 1.3. *Let F be an embedded Klein bottle in $S^4 \setminus \{\infty\}$, and let c be a simple closed curve in F such that c is transverse to $\Gamma(F)$, and c^* does not intersect triple points and branch points of F^* in S^3. Then*

$$|c \cap \Gamma(F)| = \begin{cases} \textit{even}, & \textit{if } N \textit{ is an annulus} \\ \textit{odd}, & \textit{if } N \textit{ is a Möbius band,} \end{cases}$$

where N is a regular neighborhood of c in F.

Proof. By a slight modification of c in F, we may assume that c^* is a simple closed curve in S^3. By pushing out c^* to the positive normal direction induced from the orientation of $[0,1] \times (0,1)$, we obtain a simple closed curve $\tilde{c}$ in S^3 such that $\tilde{c} \cap \Gamma(F^*) = \emptyset$, and

$$|\tilde{c} \cap F^*| = \begin{cases} |c \cap \Gamma(F)|, & \text{if } N \text{ is an annulus,} \\ |c \cap \Gamma(F)| + 1, & \text{if } N \text{ is a Möbius band,} \end{cases}$$

where N is a regular neighborhood of c in F. It is well known that if d is a simple closed curve in S^3 such that d is transverse to F^* and $d \cap \Gamma(F^*) = \emptyset$, then $d \cap F^*$ consists of even points. This implies Lemma 1.3. □

From Lemma 1.3, we have the following lemma.

Lemma 1.4. *Let F be as above. Suppose that $\Gamma(F^*)$ consists only of double points. Let Γ be the union of the components of $\Gamma(F)$ which do not bound disks in F. Then Γ is one of the following:*

(i) the union of an odd number of simple closed curves isotopic to c_3 in Lemma 1.1, or

(ii) the union of two simple closed curves c_2 and c_5 in Remark 1.2 (2) and simple closed curves isotopic to c_4 in Lemma 1.1.

Remark 1.5. Note that if γ^* is a component of $\Gamma(F^*)$, and if γ^* is a simple closed curve, then $p^{-1}(\gamma^*) \cap F$ consists of two disjoint simple closed curves.

Next, we give a diagrammatic representation of a singular set $\Gamma(F)$. Let F be an embedded closed surface in $S^4 \setminus \{\infty\}$. We assume that $\Gamma(F^*)$ consists only of double points. Given the components $\Sigma_1, \ldots, \Sigma_m$ of $F \setminus \Gamma(F)$, Σ_i is a surface with holes. In particular, if F is a Klein bottle, then by Lemma 1.4 Σ_i is a 2-sphere with holes. We construct the graph G_F as follows. The vertices are in one to one correspondence with the regions $\{\Sigma_i\}$, and the edges are in one to one correspondence with the components of $\Gamma(F)$. For a component c_t of $\Gamma(F)$, there exists a unique pair (Σ_i, Σ_j) such that $\Sigma_i \cap N \neq \emptyset$ and $\Sigma_j \cap N \neq \emptyset$, where N is a regular neighborhood of c_t in F. Then we connect the vertices $v(\Sigma_i)$ and $v(\Sigma_j)$ by the edge $e(c_t)$. In particular, if Σ_i is not a 2-sphere with holes, then we call $v(\Sigma_i)$ the *special vertex* and we denote it by a star.

Suppose that F is a Klein bottle, and $\Gamma(F^*)$ consists of disjoint simple closed curves. By Lemma 1.4, the graph G_F satisfies one of the following two conditions:

(G1) There exists a unique 1-cycle C such that C consists of an odd number of edges and $G_F \setminus C$ is a collection of trees.

(G2) There are two 1-cycles C_1 and C_2 and there is only one path L connecting C_1 and C_2 such that $G_F \setminus (C_1 \cup C_2 \cup L)$ is a collection of trees.

Here $G_F \setminus C$ and $G_F \setminus (C_1 \cup C_2 \cup L)$ are the graphs consisting of all vertices and edges which are not contained in C and $C_1 \cup C_2 \cup L$, respectively. By Remark 1.5,

$\Gamma(F)$ consists of an even number of disjoint simple closed curves on F. Hence, the graph G_F consists of an even number of edges.

Lemma 1.6. *The graphs consisting of at most six edges satisfying one of above conditions are enumerated in Figure 1.*

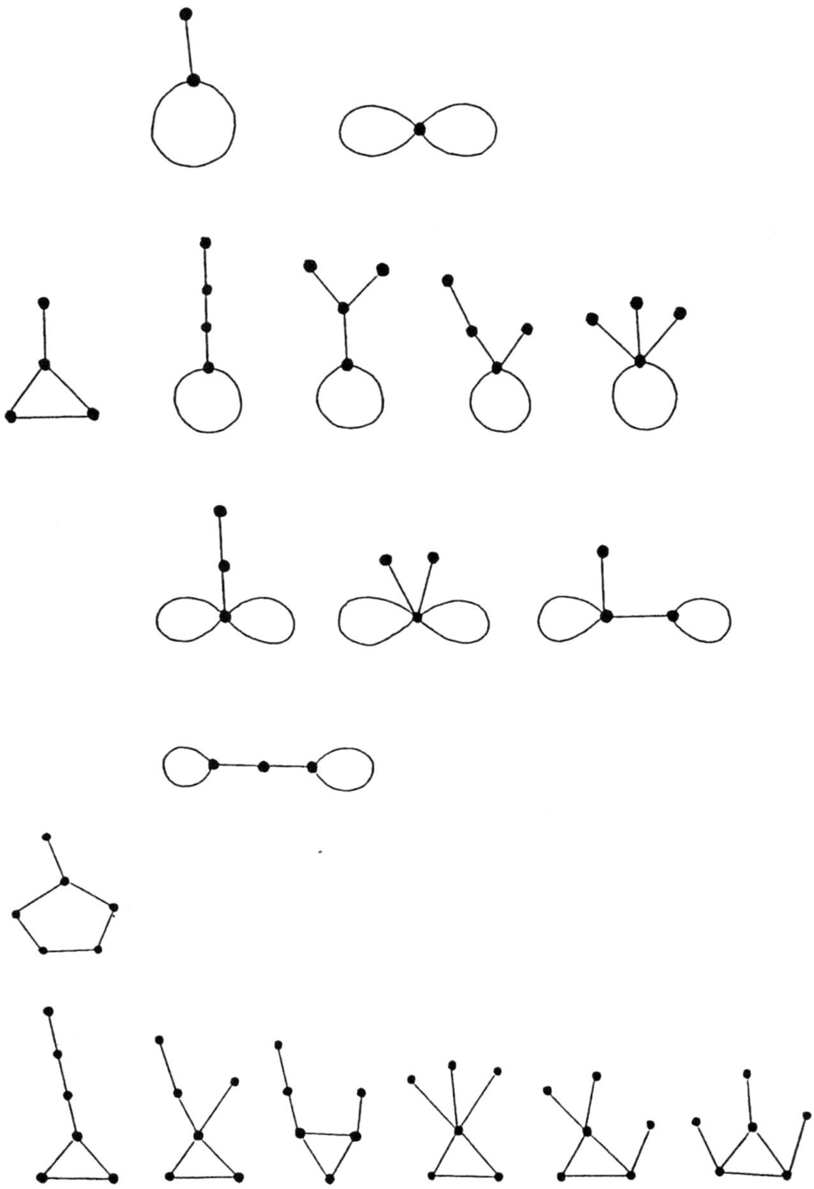

Fig. 1.

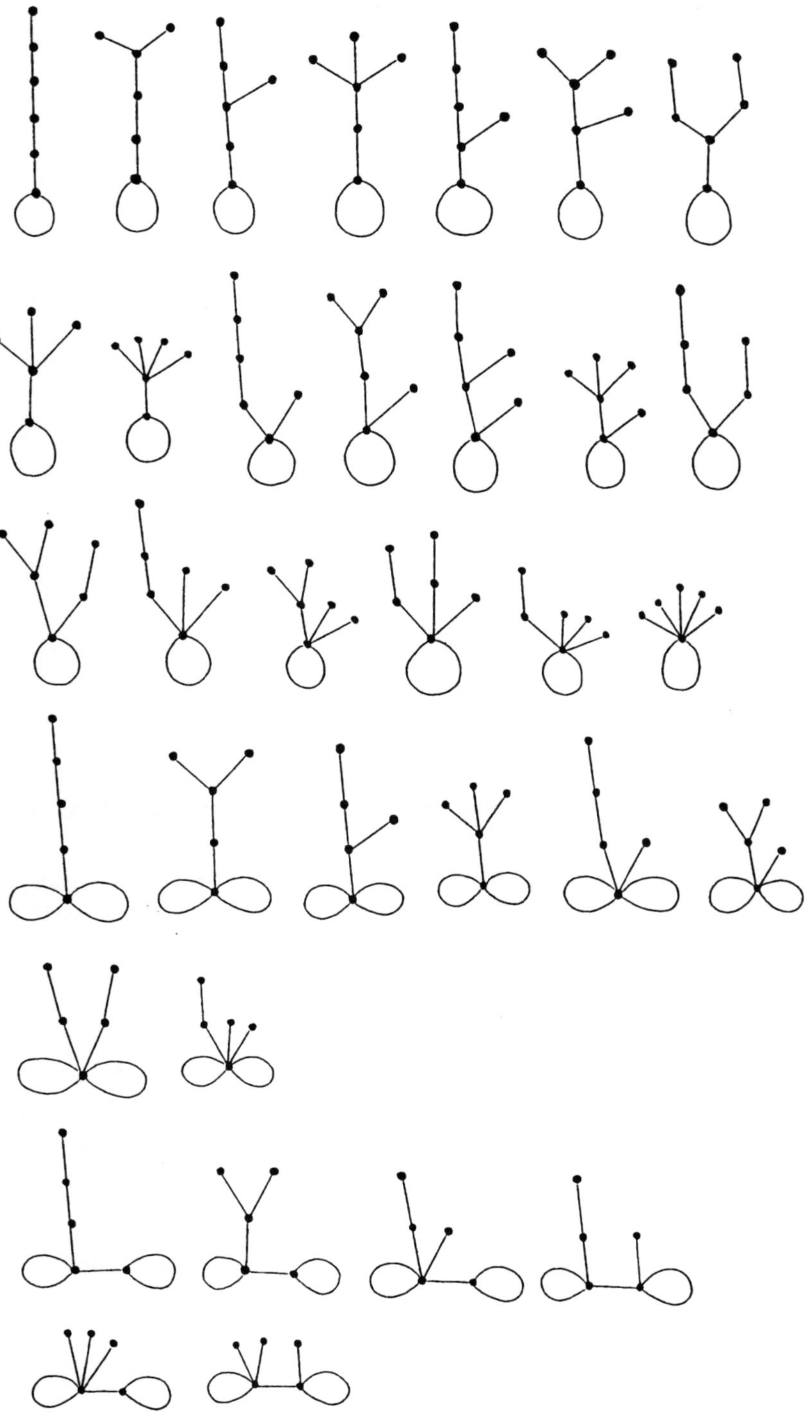

Fig. 1. (continued)

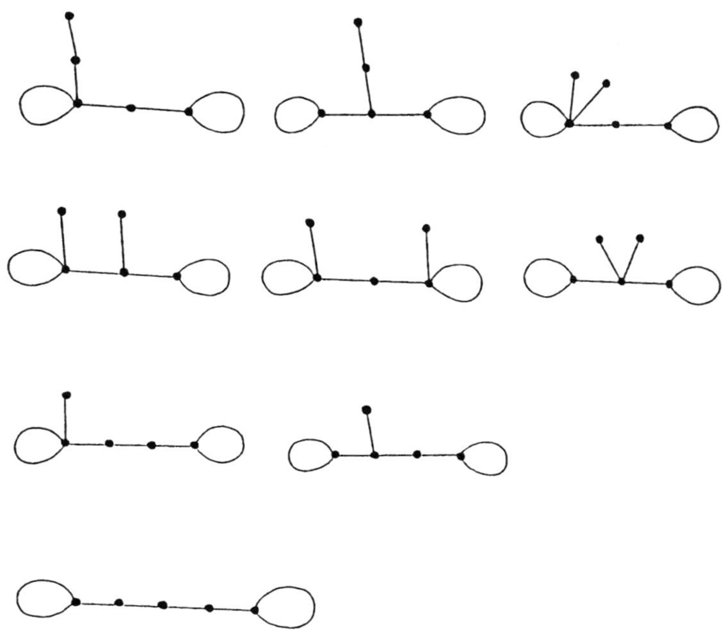

Fig. 1. (continued)

Finally, we consider solid tori in S^3.

Lemma 1.7. *([8, Lemma 2.2] and [9, Lemma 7.4]) Let V_1, V_2 be solid tori in S^3 such that $V_1 \cap V_2 = \partial V_1 \cap \partial V_2$ is an annulus or a simple closed curve. Let c be a simple closed curve in $V_1 \cap V_2$ which is a generator of $\pi_1(V_1 \cap V_2)$.*

(i) If $V_1 \cup V_2 \cup N$ is a solid torus, then c is a generator in $H_1(V_1)$ or $H_1(V_2)$.

(ii) If $V_1 \cup V_2 \cup N$ is not a solid torus, then c is not a generator in $H_1(V_1)$ and $H_1(V_2)$, and V_1, V_2 are standard solid tori in S^3.

Here N is a regular neighborhood of c in S^3 if $V_1 \cap V_2$ is a simple closed curve, and $N = V_1 \cap V_2$ otherwise.

2. A singular set $\Gamma(F^*)$ consists of one simple closed curve

Let F be an embedded Klein bottle in $S^4 \setminus \{\infty\}$. In this section, we show that if the singular set $\Gamma(F^*)$ consists of one simple closed curve, then F bounds a solid Klein bottle.

An immersed Klein bottle in S^3 is *Banchoff's* immersion, if there exists an immersed 1-sphere C in the interior of a disk D such that C is the union of the boundary of two disks intersecting at one point x, and K is obtained from $C \times S^1$ in a solid torus $D \times S^1 \subset S^3$ by cutting a disk $D \times y$ where $y \in S^1$, and then by twisting 180° and by reglueing. Note that the simple closed curve $x \times S^1$ is the singular set of Banchoff's immersion.

Lemma 2.1. *Let f be an immersion of a Klein bottle F into S^3 such that the* *singular set of $f(F)$ consists of one simple closed curve γ^*. If the preimage c, c' of γ^* have Möbius band neighborhoods, then the immersed surface $f(F)$ is isotopic to Banchoff's immersion.*

Proof. Since c, c' have Möbius band neighborhoods, $F \setminus (c \cup c')$ is an open annulus. Given a regular neighborhood N of γ^* in S^3, the surface $B_1 = \overline{f(F) \setminus N}$ is an annulus, and $f(F) \cap N$ consists of two Möbius bands. Note that if b is a component of ∂B_1, then $[b] = \pm 2$ in $H_1(N)$. Let B_2 be an annulus in ∂N with $\partial B_1 = \partial B_2$, then $B_1 \cup B_2$ is a torus in S^3. By the solid torus theorem [6], $B_1 \cup B_2$ bounds a solid torus V in S^3. Then V satisfies that $V \cap N$ is an annulus in ∂V and ∂N, or $V \supset N$.

Case 1. $V \cap N$ is an annulus in ∂V and ∂N.

If $V \cup N$ is a solid torus, then the annulus B_2 is a boundary parallel to an annulus in $\partial(V \cup N)$. Since $[b] = \pm 2$ in $H_1(N)$, the intersection of B_2 and a meridonal disk D of $V \cup N$ consists of proper disjoint two arcs. We see that $(D \cap N) \cap f(F)$ is proper two arcs whose intersection is one point. Hence, $D \cap f(F)$ is an immersed circle with exactly one self-intersection. This implies that $f(F)$ is isotopic to Banchoff's immersion.

If $V \cup N$ is not a solid torus, then $W = \overline{S^3 \setminus V \cup N}$ is a solid torus by the solid torus theorem [6]. Moreover, by Lemma 1.7, V and N are standard solid tori. Hence, $W \cup N = \overline{S^3 \setminus V}$ is a solid torus. This case can be proved in a similar way as above by replacing V by W in the argument.

Case 2. $V \supset N$.

By [8, Lemma 2.1], V is decomposed into two solid tori $N, \overline{V \setminus N}$. Let $W = \overline{V \setminus N}$. This case can proved in a similar way to Case 1 by replacing V by W in the argument. □

We define the *D-surgery* to be the operation shown in Figure 2. (This is a surgery at a disk.) Let $\gamma^* \subset \Gamma(F^*)$ be the intersection curve in Figure 2, c_1 and c_2 be the preimages of γ^* such that c_1 bounds a disk D with $\Gamma(F) \cap \operatorname{Int} D = \emptyset$, then the D-surgery replaces the regular neighborhood of D^* by a pair of disks that do not intersect with F^*, and connects the centers of the disks by an arc the interior of which intersects F^* at one point at the center of D^*.

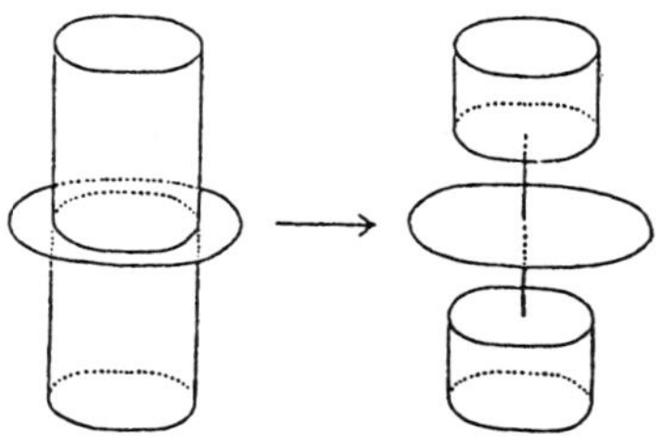

Fig. 2.

We define the *B-move* (branch points moves) to be the operation shown in Figure 3. The symbol $\longleftrightarrow$ means that pictures pointed by an arrow are projections of ambient isotopic surfaces. For canceling branch points, see [4] or [5].

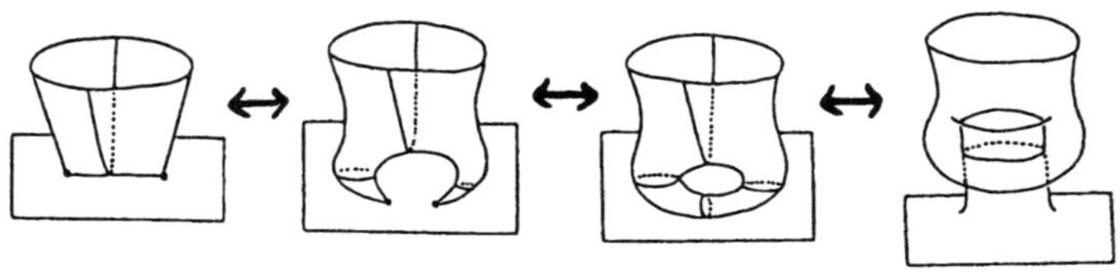

Fig. 3.

Proposition 2.2. *Let F be an embedded Klein bottle in $S^4\backslash\{\infty\}$. If the singular set $\Gamma(F^*)$ consists of one simple closed curve, then F bounnds a solid Klein bottle.*

Proof. By Lemma 1.6, there are two cases of the graph G_F obtained from $\Gamma(F)$.

Case 1. G_F consists of one 1-cycle and one edge.

There is a component of $\Gamma(F)$ bounding a disk D in F. Perform the D-surgery at the disk D^*. Then we obtain an embedded 2-sphere S and an arc α such that $\operatorname{Int}\alpha$ intersects S at exactly one point, and $\partial\alpha \subset S$. We may assume that the arc α is untied. Hence, the projection F^* is obtained from a 2-sphere by attaching a 1-handle along α. Therefore, F bounds a solid Klein bottle.

Case 2. G_F consists of two cycles.

By Lemma 2.1, F^* is isotopic to Banchoff's immersion. The surface F is ambient isotopic to an embedded Klein bottle with the projection of Case 1. This was shown in [3, p 230]. This completes the proof of Proposition 2.2. □

3. Necessary conditions on singular sets

In this section, we consider conditions satisfied by a singular set $\Gamma(F)$, and we consider when the number of components in a singular set $\Gamma(F^*)$ can be reduced without increasing triple points.

Lemma 3.1. *([7, Lemma 3.1]) Let F be an embedded closed surface in $S^4\backslash\{\infty\}$. Let γ^* be a component of $\Gamma(F^*)$ which is a simple closed curve, and c_1, c_2 the preimage of γ^*. If γ^* satisfies one of the following conditions, then F is ambient isotopic to another Klein bottle F' such that $\Gamma(F'^*) = \Gamma(F^*) - \gamma^*$ by an ambient isotopy of S^4.*

(i) There exist disks D_1, D_2 in F with $\partial D_i = c_i$, and $\operatorname{Int} D_i \cap \Gamma(F) = \emptyset$.

(ii) There exist an annulus A in F and a solid torus V in S^3 such that $\partial A = c_1 \cup c_2$, $\partial V = A^*$, $\operatorname{Int} V \cap F^* = \emptyset$,, and γ^* is a generator of $H_1(V) \cong \mathbf{Z}$.

(iii) There exists an annulus A in F with $\partial A = c_1 \cup c_2$, $[c_i] = 1$ in $\pi_1(F)$, and $\operatorname{Int} A \cap \Gamma(F) = \emptyset$.

Lemma 3.2. *([9, Lemma 3.3]) Let F be as above. Let γ_1^*, γ_2^* be components of $\Gamma(F^*)$ which are simple closed curves. Let c_1, c_2 be the preimage of γ_1^*, and d_1, d_2 the preimage of γ_2^*. If γ_1^*, γ_2^* satisfy the following condition (i), F is ambient isotopic to another Klein bottle F' such that $\Gamma(F'^*) = \Gamma(F^*) - \gamma_i^*$ ($i = 1$ or 2) by an ambient isotopy of S^4.*

(i) There exist disks D_1, D_2, D_3 in F with $\partial D_1 = c_1$, $\partial D_2 = c_2$, $\partial D_3 = d_1$, $D_1 \supset D_2 \cup D_3$, $D_2 \cap D_3 = \emptyset$, $\text{Int}\, D_1 \cap \Gamma(F) = c_2 \cup d_1 \cup d_2$, $\text{Int}\, D_2 \cap \Gamma(F) = \emptyset$, and $\text{Int}\, D_3 \cap \Gamma(F) = \emptyset$.

Lemma 3.3. *Let F, γ_i^*, c_i, d_i ($i = 1, 2$) be as above. If γ_1^*, γ_2^* satisfy the following condition (i), F is ambient isotopic to another Klein bottle F' such that $\Gamma(F'^*) = \Gamma(F^*) - \gamma_i^*$ ($i = 1$ or 2) by an ambient isotopy of S^4.*

(i) The surface F is a Klein bottle. There exist a disk D and an annulus A in F such that $\partial D = c_1$, $\partial A = d_1 \cup d_2$, $\text{Int}\, D \cap \Gamma(F) = \emptyset$, $\text{Int}\, A \cap \Gamma(F) = \emptyset$, and c_2, d_1, d_2 are isotopic to c_3 in Lemma 1.1.

Proof. By the solid torus theorem [6], the torus A^* bounds a solid torus V in S^3. Let B be an annulus in F with $\partial B = c_2 \cup d_i$ and $\text{Int}\, B \cap (c_2 \cup d_1 \cup d_2) = \emptyset$, then the surface $(D \cup B)^*$ is a singular disk. By Dehn's lemma, there exists a non-singular disk E with $\text{Int}\, E \cap A^* = \emptyset$ and $\partial E = c_2^*$.

If $E \subset V$, then $F^* \subset V$. Move F by some ambient isotopy without changing the topology of the singular set $\Gamma(F)$. We may assume that V is standard. Then γ_1^* is a longitude of the solid torus $\overline{S^3 \setminus V}$, and γ_1^* can be cancelled by Lemma 3.1 (ii). If $E \cap V = \partial E$, then V is a standard solid torus. In a similar way as above, γ_1^* can be cancelled. □

To represent a singular set $\Gamma(F)$, it is convenient to use colored graphs as follows. Let F be an embedded closed surface in $S^4 \setminus \{\infty\}$. We assume that $\Gamma(F^*)$ consists only of double points. The *graph associated with the singular set* $\Gamma(F)$ is the pair of the graph G_F obtained from $\Gamma(F)$ defined in section 1 and a map h of the set of edges of G_F into $\mathbf{Z}$ satisfying that for any two edges e, e' of G_F, $h(e) = h(e')$ if and only if $c^* = c'^*$, where c, c' are simple closed curves of $\Gamma(F)$ corresponding to edges e, e', respectively.

The map h is regarded as colors on the edges, and the colors are indicated by different types of lines, such as dotted lines and thick lines, in the figures.

Remark 3.4. Suppose that F is a Klein bottle. If the graph associated with $\Gamma(F)$ has two cycles e, e', then $h(e) = h(e')$.

The above lemmas can be described by using graphs associated with singular sets. Let F be an embedded surface in $S^4 \setminus \{\infty\}$ whose singular set consists only of double points. If the graph with associated with $\Gamma(F)$ satisfies one of the following conditions, then the number of components in $\Gamma(F^*)$ can be reduced by Lemma 3.1 and 3.2. See Figure 4.

(a) There exist two edges e, e' with a vertex of valency 1 such that $h(e) = h(e')$.

(b) There exist two edges v_1v_2, v_2v_3 with a vertex v_2 of valency 2 such that

$h(v_1v_2) = h(v_2v_3)$ and v_1v_2, v_2v_3 are not contained any 1-cycle and any path connecting 1-cycles.

(c) There exist four edges $v_1v_2, v_2v_3, v_4v_3, v_3v_5$ such that v_1, v_4 are vertices of valency 1, v_2 is a vertex of valency 2, v_3 is a vertex of valency 3, $h(v_1v_2) = h(v_3v_5)$, and $h(v_2v_3) = h(v_4v_3)$.

(d) There exist four edges $v_1v_2, v_3v_2, v_2v_4, v_4v_5$ such that v_1, v_3 are vertices of valency 1, v_4 is a vertex of valency 2, v_2 is a vertex of valency 3, $h(v_1v_2) = h(v_4v_5)$, and $h(v_3v_2) = h(v_2v_4)$.

(e) The surface F is a Klein bottle. The graph G_F has only one 1-cycle C. There exist four edges $v_1v_2, v_3v_4, v_4v_5, v_6v_7$ such that the edges v_3v_4, v_4v_5, v_6v_7 are contained in C, v_1 is a vertex of valency 1, v_4 is a vertex of valency 2, $h(v_1v_2) = h(v_6v_7)$, and $h(v_3v_4) = h(v_4v_5)$ (it may be that v_1 and v_4 are the same vertices).

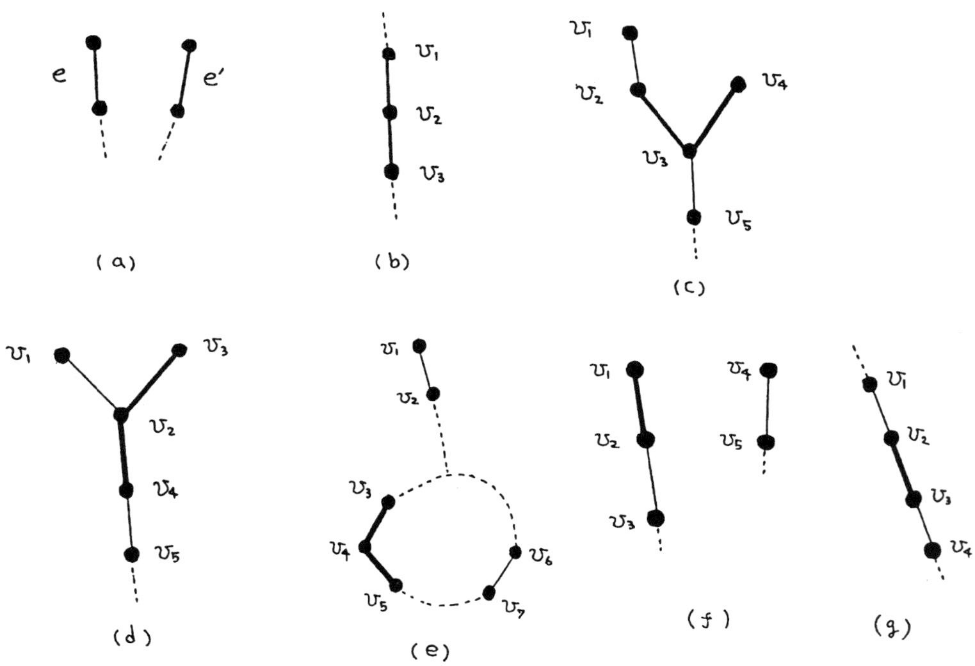

Fig. 4.

Lemma 3.5. *([9, Lemma 3.5]) Let F, γ_i^*, c_i, and d_i $(i = 1, 2)$ be as above. Consider the following conditions.*

(i) There exist disks D_1, D_2, D in F with $\partial D_1 = c_1$, $\partial D_2 = c_2$, $\partial D = d_2$, $\operatorname{Int} D_1 \cap \Gamma(F) = d_1$, $D_1 \cap D_2 = \emptyset$, $D \cap (D_1 \cup D_2) = D_i$ $(i = 1$ or $2)$.

(ii) There exists an annulus A in F with $\partial A = c_1 \cup c_2$, $\operatorname{Int} A \cap \Gamma(F) = d_1$, and $[d_1] \neq 1 \in \pi_1(A)$.

Then there does not exist such an embedded surface in S^4 satisfying (i) or (ii).

Also, the above lemmas can be described by using graphs associated with singular sets. Let F be an embedded surface in $S^4 \setminus \{\infty\}$ whose singular set consists only of double points. The graph (G_F, h) does not satisfy the following conditions by Lemma 3.5. See Figure 4 (f) and (g).

(f) There exist four edges $v_1v_2, v_2v_3, v_4v_5, v_6v_7$ such that v_1, v_4 are vertices of valency 1, v_2 is a vertex of valency 2, $G_F \setminus v_6v_7$ consists of a tree G_1 and a graph G_2, one of G_1 and G_2 contains v_1v_2, the other contains v_4v_5, and $h(v_2v_3) = h(v_4v_5)$, and $h(v_1v_2) = h(v_6v_7)$.

(g) There exist three edges v_1v_2, v_2v_3, v_3v_4 such that v_2, v_3 are vertices of valency 2, and $h(v_1v_2) = h(v_3v_4)$.

We define the D-surgery for the graph associated with a singular set. Let F be an embedded connected closed surface in $S^4 \setminus \{\infty\}$ whose singular set consists only of double points. Let (G_F, h) be the graph associated with the singular set $\Gamma(F)$. Perform the D-surgery at some curve γ^* of $\Gamma(F^*)$, then we obtain an embedded surface $\tilde{F}$ consisting of at most two components. The graph associated with $\Gamma(\tilde{F})$ is obtained as follows. Let c, c' be the preimage of γ^* such that c bounds a disk D in F, $\operatorname{Int} D \cap \Gamma(F) = \emptyset$. First, delete the edges $e_c, e_{c'}$ corresponding to c, c'. If $G_F - (e_c \cup e_{c'})$ consists of two components, and if two edges e, e' with $h(e) = h(e')$ are contained in different components, then contract the edges e, e'. In particular, if one 1-cycle C of G_F is deleted or contracted, then replace the vertex in C by the special vertex. In this way, we obtain the graph $(G_{\tilde{F}}, \tilde{h})$, where $\tilde{h}$ is induced from h. We call it *a graph obtained by the D-surgery*. The Figure 6 is examples of graphs obtained by the D-surgery.

Lemma 3.6. *Let $\tilde{F}, (G_{\tilde{F}}, \tilde{h})$ be as above. Then the graph $(G_{\tilde{F}}, \tilde{h})$ satisfies the following conditions.*

(i) If G_F has only one 1-cycle, then $G_{\tilde{F}}$ is a tree not having the special vertex, or a graph satisfying the condition (G1).

(ii) If G_F has two cycles, then one of $G_{\tilde{F}}$ is a graph satisfying the condition (G2).

Here (G_F, h) is the graph associated with $\Gamma(F)$.

Proof. The surface $\tilde{F}$ is obtained from F by the D-surgery at the disk D^*. Let c be the component of $\Gamma(F)$ bounding the disk D such that $\operatorname{Int} D \cap \Gamma(F) = \emptyset$. Let c' be the component of $\Gamma(F)$ with $c^* = c'^*, c \neq c'$. By Lemma 1.1, there are four cases of c'.

Case 1. c' bounds a disk in F.

The surface $\tilde{F}$ consists of a 2-sphere F_1 and a Klein bottle F_2. Hence, G_{F_1} is a tree not having the special vertex, and G_{F_2} satisfying (G1) or (G2).

Case 2. c' is isotopic to c_2 in Lemma 1.1.

There does not exist this case. Because if c' has a Möbius band neighborhood, then so does c. Hence, c does not bound a disk in F. This is a contradiction.

Case 3. c' is isotopic to c_3 in Lemma 1.1.

The surface $\tilde{F}$ is a 2-sphere, and the graph $G_{\tilde{F}}$ is a tree not having the special vertex.

Case 4. c' is isotopic to c_4 in Lemma 1.1.

There does not exist this case. Because if c' is isotopic to c_4 in Lemma 1.1, then $\tilde{F}$ consists of two projective planes. Hence, $\tilde{F}^*$ is two immersed projective planes whose singular set consists only of double points. By Banchoff's formula [2], $\tilde{F}^*$ has at least two triple points. This is a contradiction. □

4. A singular set $\Gamma(F^*)$ consists of two disjoint simple closed curves

In this section, we show that if the singular set $\Gamma(F^*)$ consists of two disjoint simple closed curves, then F bounds a solid Klein bottle.

First, we define a spun Klein bottle in S^4. Let B^n be a unit n-ball, and $r_\theta : B^3 \to B^3$ the θ-rotation map through the z-axis for $\theta \in [0, 2\pi]$. Let $S^4 = \partial B^5 = \partial B^3 \times B^2 \cup B^3 \times \partial B^2$. An embedded Klein bottle F in S^4 is a *spun Klein bottle* if there exist an integer a and a knot K in B^3 such that

(i) K intersects two points with the z-axis,

(ii) $r_\pi(K) = K$, and

(iii) $F = \{(r_{(a+\frac{1}{2})\theta}(x), \theta) \mid x \in K, \theta \in \partial B^2\} \subset B^3 \times \partial B^2$,

where we identify ∂B^2 and $[0, 2\pi]/0 \sim 2\pi$. We denote the spun Klein bottle F by $Kl^a(K)$.

Remark 4.1. Since the map $\phi : \partial B^3 \times \partial B^2 \to \partial B^3 \times \partial B^2$ defined by $\phi(x, \theta) = (x, \pm 2\theta)$ is ambient isotopic to the identity map, $Kl^a(K)$ is ambient isotopic to $Kl^{a\pm 2}(K)$.

Theroem 4.2. *Let F be an embedded Klein bottle in $S^4 \setminus \{\infty\}$. If the singular set $\Gamma(F^*)$ consists of two disjoint simple closed curves, then F bounds a solid Klein bottle in S^4.*

Proof. By Proposition 2.2, it suffices to prove that the number of components in $\Gamma(F^*)$ can be reduced. Let (G_F, h) be the graph associated with $\Gamma(F)$. If G_F has at least three vertices of valency 1, then by Lemma 3.1 (i) the number of components in $\Gamma(F^*)$ can be reduced. Moreover, if (G_F, h) satisfies the conditions (a) or (b), the number of components in $\Gamma(F^*)$ can be reduced. By Remark 3.4, if G_F has two cycles e, e', then $h(e) = h(e')$. Hence, we may consider graphs associated with $\Gamma(F)$ listed in Figure 5.

Case 1. Since (G_F, h) satisfies the condition (e), the number of components in $\Gamma(F^*)$ can be reduced.

Cases 2 and 3. Since (G_F, h) satisfies the condition (g), there does not exist a Klein bottle with such a singular set.

Cases 4, 5 and 6.

Perform the D-surgery at the curve corresponding to the dotted edge with a vertex of valency 1. Then we obtain an associated graph $(G_{\tilde{F}}, \tilde{h})$ with two components. See Figure 6. One of components of $G_{\tilde{F}}$ is a special vertex. This is a contradiction by Lemma 3.6. Hence, there does not exist a Klein bottle with such a singular set.

Case 7. Let c_1, c_2, d_1, d_2 be the components of $\Gamma(F)$ such that d_i $(i = 1, 2)$ has a Möbius band neighborhood. Given an annulus A in F with $\partial A = c_1 \cup c_2$, the

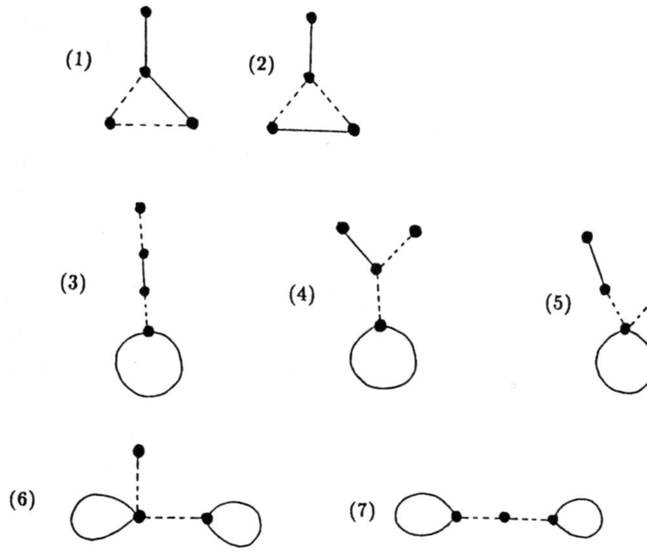

Fig. 5.

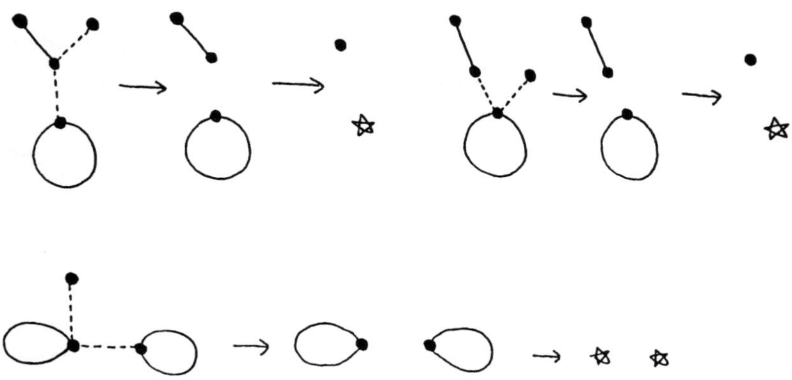

Fig. 6.

immersed Klein bottle $(\overline{F \setminus A})^*$ satisfies the condition in Lemma 2.1. Hence $(\overline{F \setminus A})^*$ is isotopic to Banchoff's immersion. Since A^* is an embedded torus, by the solid torus theorem [6] the torus A^* bounds a solid torus V in S^3.

(7-a) $(\overline{F \setminus A})^* \subset V$.

Move F by some ambient isotopy without changing the topology of the singular set $\Gamma(F)$, we may assume that V is standard. Then F is ambient isotopic to a spun Klein bottle $Kl^a(K)$ for some knot K with $F^* = (Kl^a(K))^*$. By Remark 4.1, we may assume that $a = 0$ or 1, and c_1^* is a $(2, \pm 1)$-torus knot. Hence, c_1^* is a longitude curve of the solid torus $\overline{S^3 \setminus V}$. By Lemma 3.1, c_1^* can be cancelled.

(7-b) $(\overline{F \setminus A})^* \cap V = c_1^*$.

Let N_1 be a regular neighborhood of d_1^* in S^3. Let X be the closure of the component of $S^3 \setminus (\overline{F \setminus A})^*$ such that $[c_1^*] = \pm 2 \in H_1(X \cup N_1)$. Then $X \cup N_1$ is a solid torus. If $V \subset X$, then c_1^* is a generator of $H_1(V)$. Hence, c_1^* can be cancelled by Lemma 3.1. Suppose that $V \cap X = c_1^*$. Let N_2 be a regular neighborhood of c_1^* in S^3. If $V \cup X \cup N_1 \cup N_2$ is a solid torus, then by Lemma 1.7 (i) c_1^* is a generator of $H_1(V)$. Hence, c_1^* can be cancelled by Lemma 3.1. If $V \cup X \cup N_1 \cup N_2$ is not a solid torus, then by Lemma 1.7 (ii) V is a standard solid torus. In this case, we can show by replacing V by the solid torus $\overline{S^3 \setminus V}$ in a similar way to Case (7-a). □

Remark 4.3. Let F be Banchoff's immersion in S^3. Let V be a solid torus such that $\gamma = \partial V \cap F$ is a simple closed curve except the singular set of F. Using Case 7 in Theorem 4.2, we can show that γ is a generator of $H_1(V)$ or $F \subset V$.

5. A singular set $\Gamma(F^*)$ consists of three disjoint simple closed curves

In this section, we prove the main theorem.

Theroem 5.1. *Let F be an embedded Klein bottle in $S^4 \setminus \{\infty\}$. If the singular set $\Gamma(F^*)$ consists of three disjoint simple closed curves, then F bounds a solid Klein bottle in S^4.*

Proof. By Theorem 4.2, it suffices to prove that the number of components in $\Gamma(F^*)$ can be reduced. By Lemma 1.6, G_F is one of graphes in Figure 1. By the condition (a), we may assume that G_F has at most three vertices of valency 1. By Remark 3.4, if G_F has two cycles C_1, C_2, then $h(C_1) = h(C_2)$, and we may assume that G_F has at most two edges with a vertex of valency 1. We can list all the possible associated graphs (G_F, h) as follows. First, list all pairs of edges e, e'. Set $h(e) = h(e')$. In particular, if G_F has three edges with a vertex of valency 1, then choose only one edge from these edges with a vertex of valency 1. And if G_F has two cycles C_1, C_2, choose $e = C_1$ and $e' = C_2$. Next, list all possible choices of two dotted edges in G_F excluding e and e'. Then, the rest two edges are thick edges. See Figure 8. We do not need to consider unnumbered graphs in Figure 8. Because, these graphs satisfy one of the conditions (a),(b),(c),(d),(e),(f),(g) and

(h) Perform the D-surgery at an edge with dotted line. Then we obtain a tree satisfying the condition (g).

(i) The graph has only one 1-cycle consisting of three edges. Perform the D-surgery at an edge with dotted line. Then we obtain a tree and a graph with a 1-cycle consisting of two edges.

(j) The graph has only one cycle. Perform the D-surgery at an edge with dotted line. Then we obtain a tree and a tree with a special vertex.

(k) The graph has only two cycles C_1, C_2. Perform the D-surgery at an edge with dotted line. Then we obtain two graphs with a special vertex.

We give some examples satisfying the condition (h),(i),(j) or (k). See Figure 7.

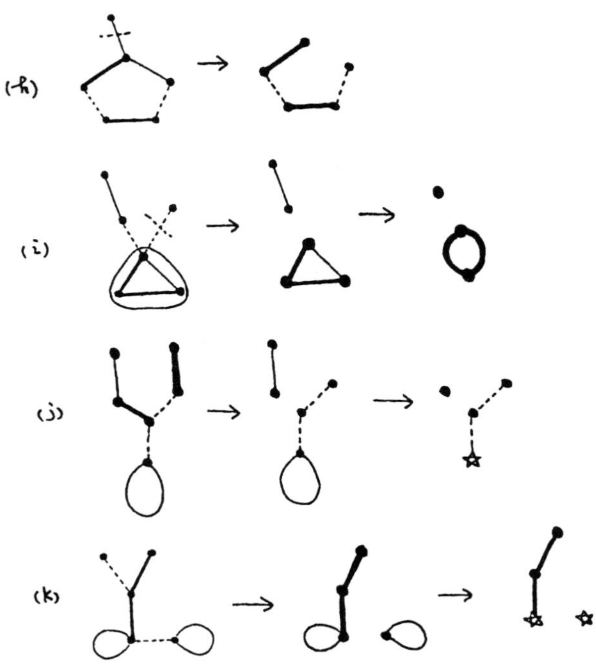

Fig. 7.

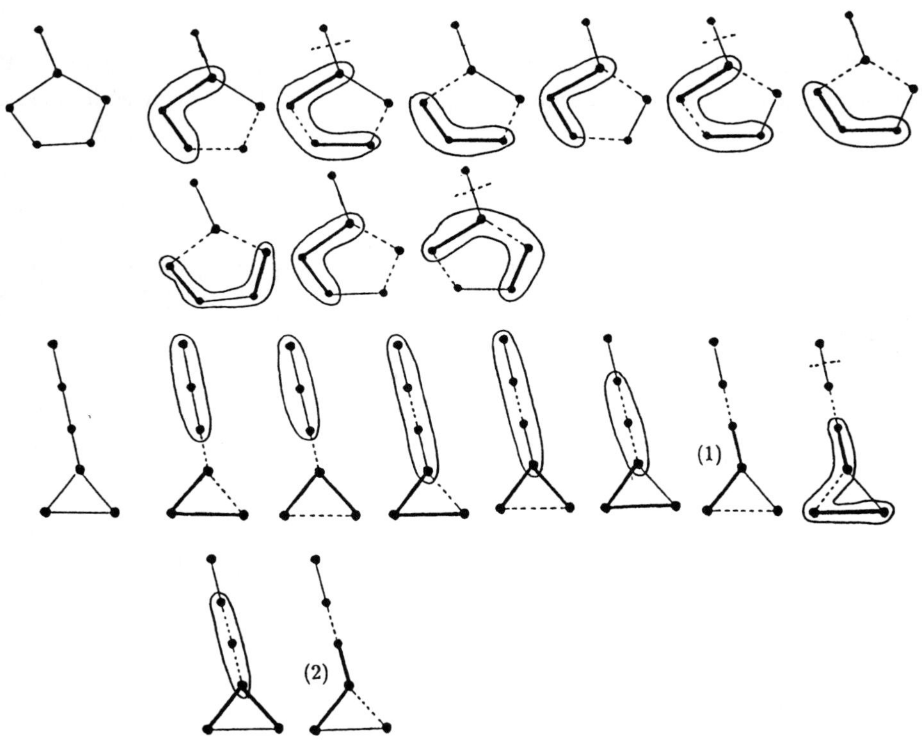

Fig. 8.

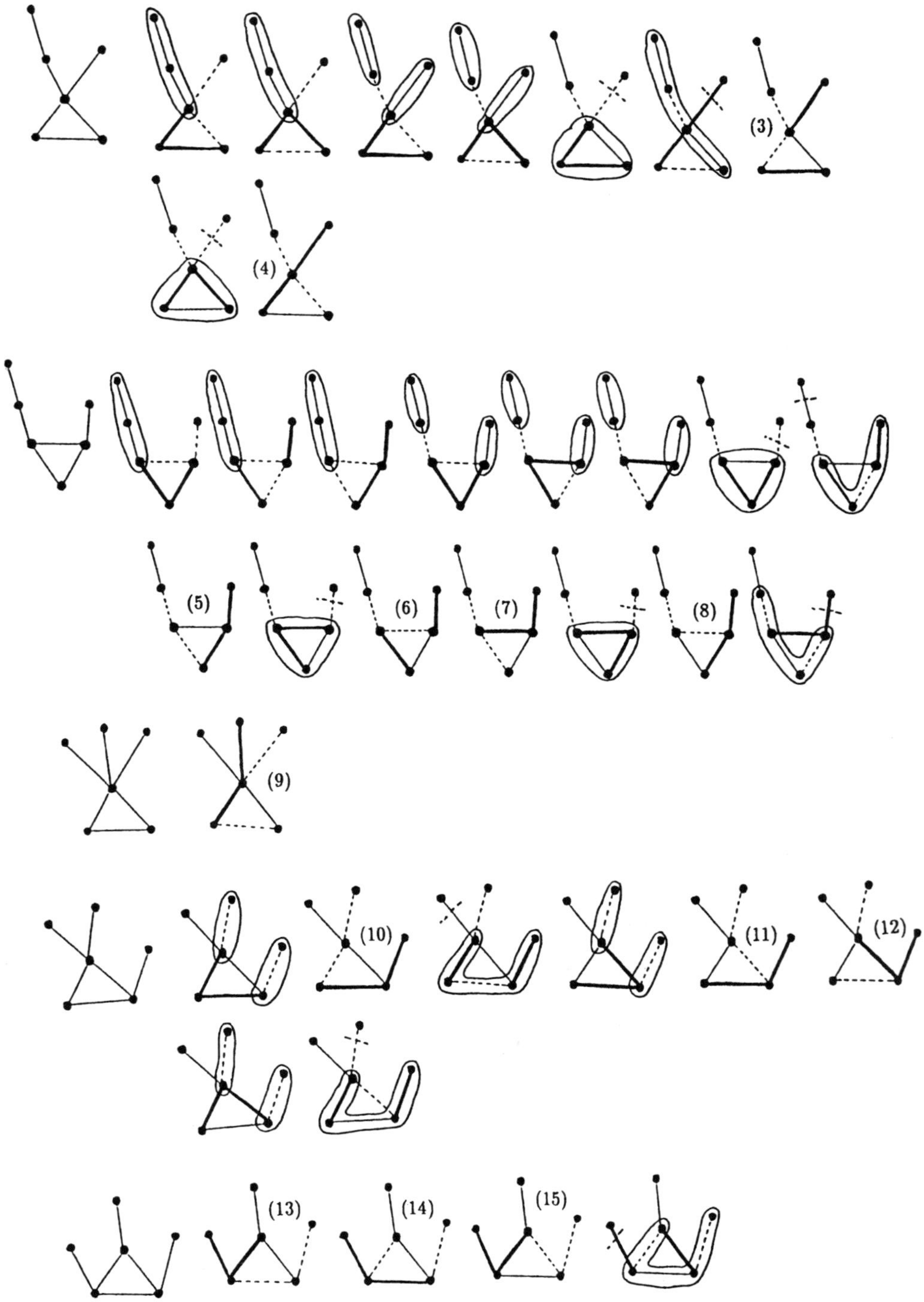

Fig. 8. (continued)

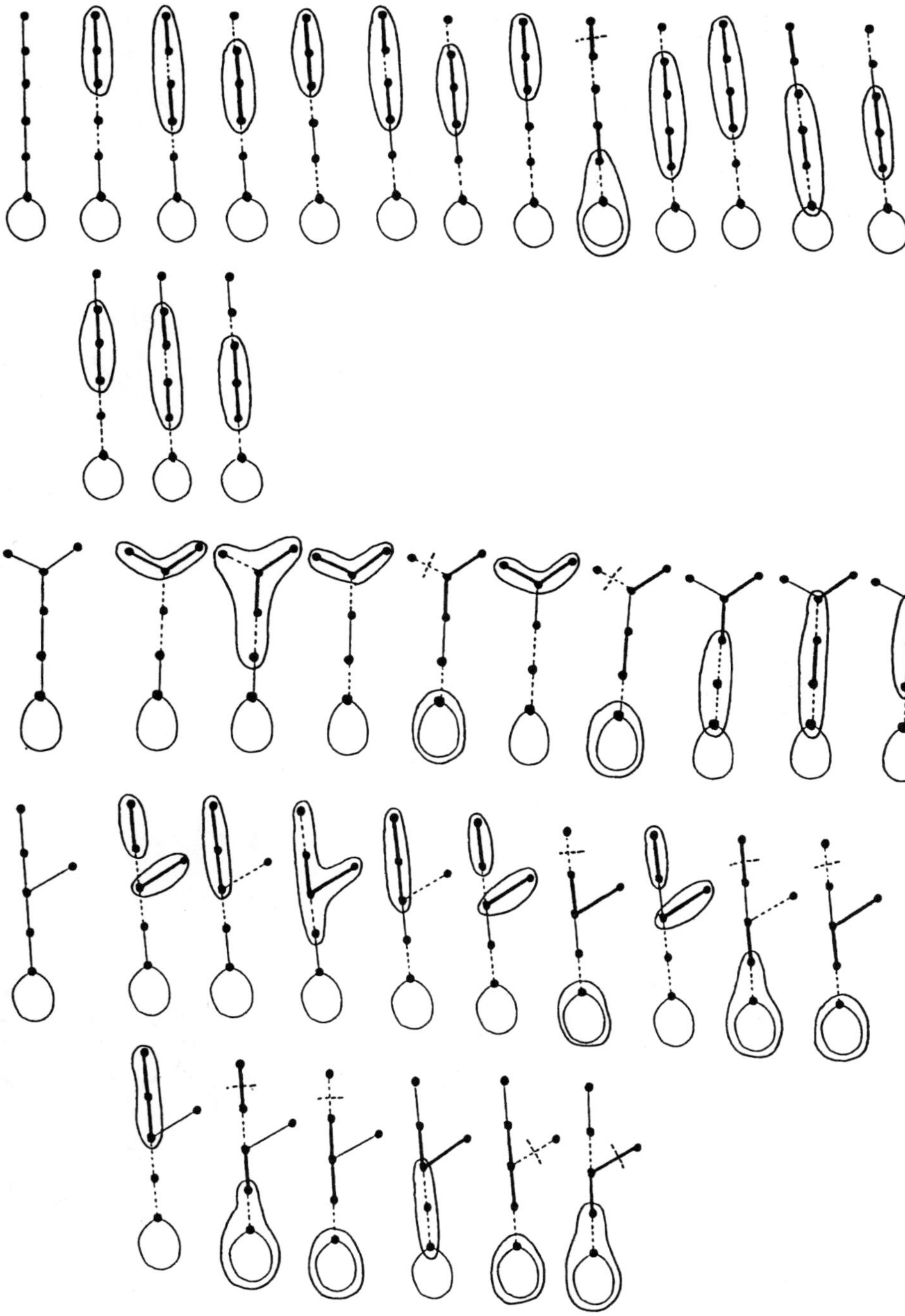

Fig. 8. (continued)

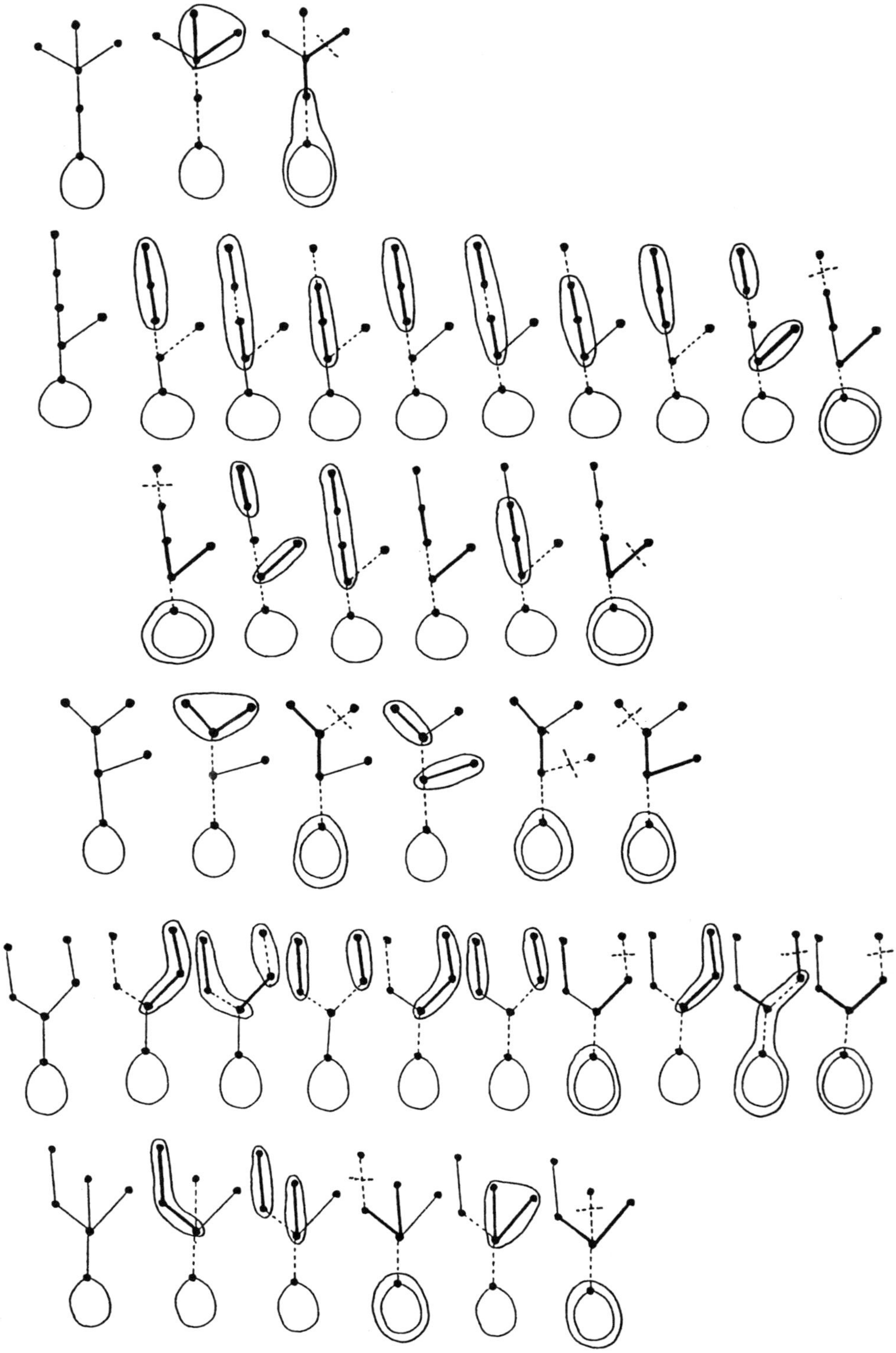

Fig. 8. (continued)

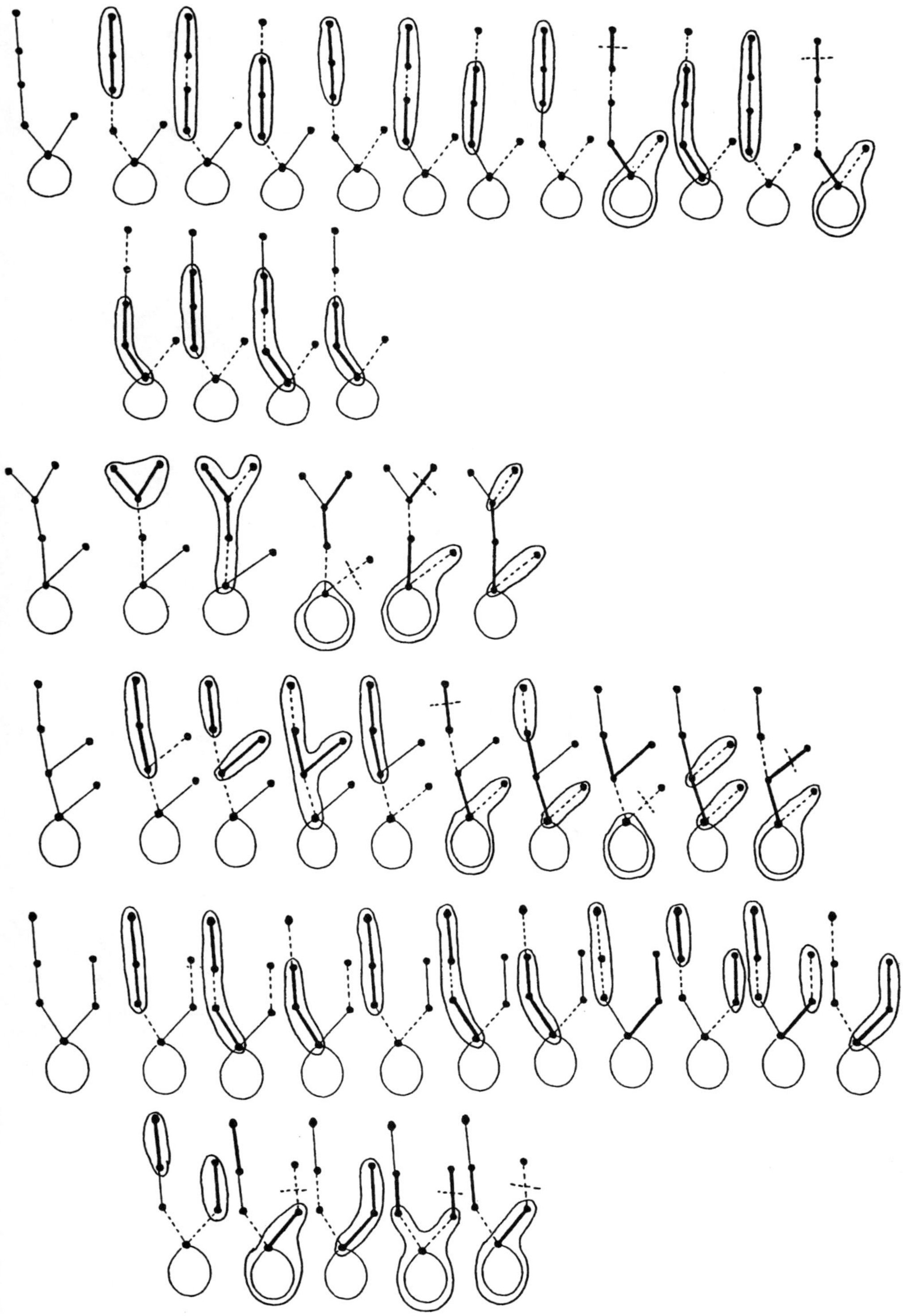

Fig. 8. (continued)

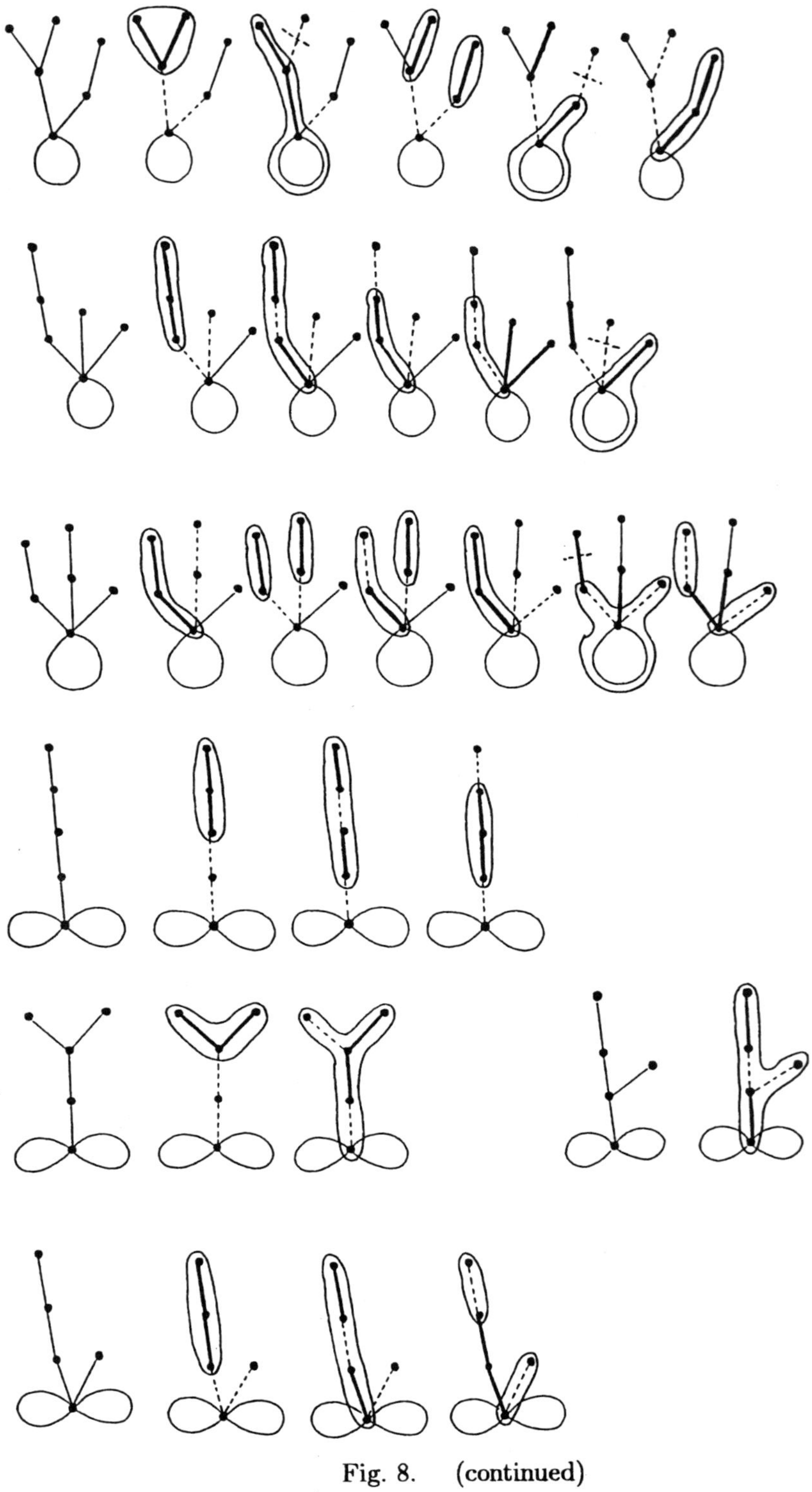

Fig. 8. (continued)

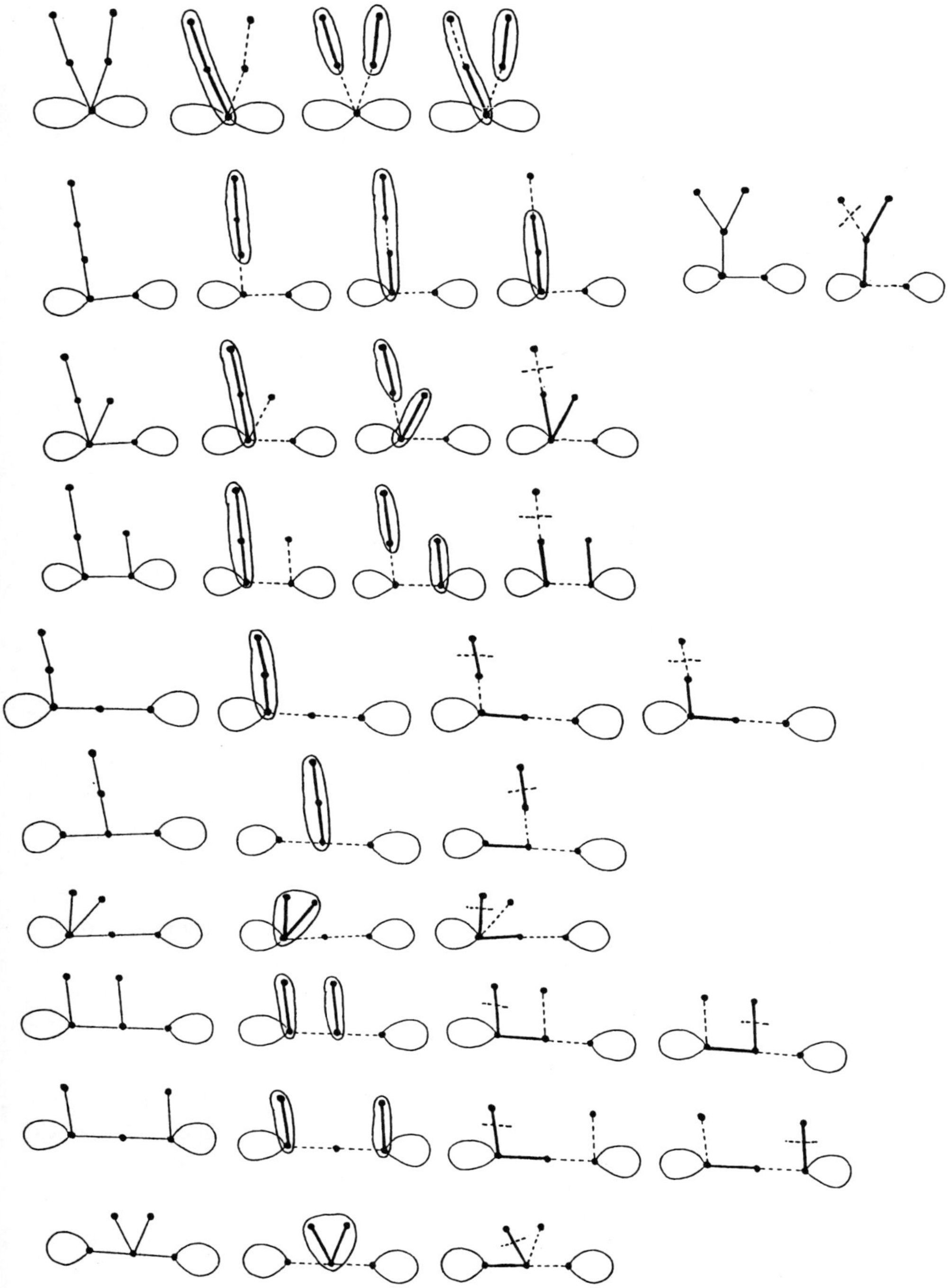

Fig. 8. (continued)

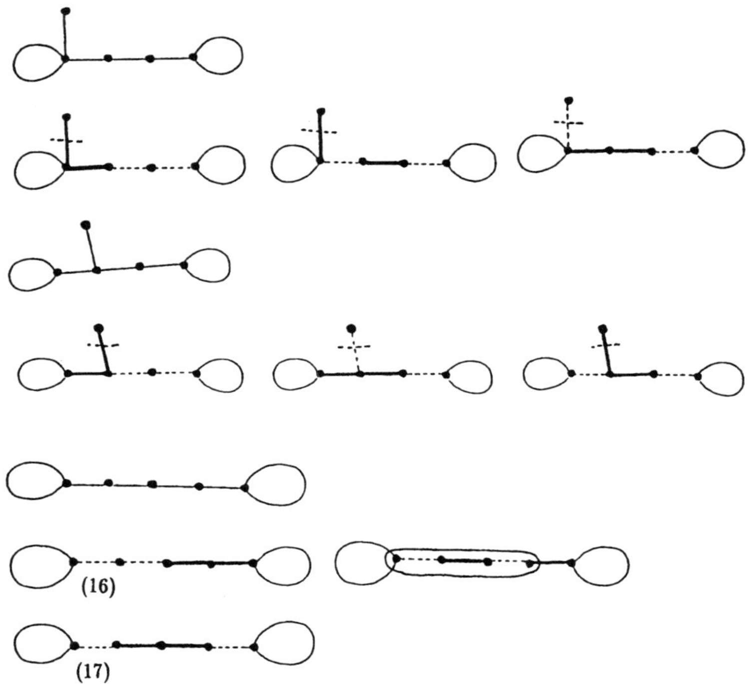

Fig. 8. (continued)

Case (13).

Perform D-surgeries along the disks (see Figure 9 (1)). We obtain an embedded 2-sphere S, an immersed 2-sphere S', and arcs β_1, β_2 in S^3 such that the singular set of S' is one simple closed curve, $S \cap S' = \emptyset$, $\beta_1 \cap S$ consists of two points, $\beta_2 \cap S$ consists of one endpoint of β_2, $\beta_2 \cap S'$ consists of two points, and $\beta_1 \cap S'$ consists of one endpoint of β_1. By the Schöenflies theorem [6, p 34], S bounds a 3-ball. Hence, $S \cup S' \cup \beta_1 \cup \beta_2$ is as shown in Figure 9 (2). Also the arc inside may have a different knotting. These knottings can be untied by an ambient isotopy of S^4. We can check in Figure 9 that the number of double curves in $\Gamma(F^*)$ can be reduced by an ambient isotopy.

Cases (6),(8),(9),(10),(12) and (14).

Perform D-surgeries along the disks (see Figure 10 (1)). We obtain F^* in a similar way to Case (14) (see Figure 10 (2)). We can check in Figure 11 that the number of double curves in $\Gamma(F^*)$ can be reduced by an ambient isotopy.

Case (11).

The graph is the same as graph of Case (10).

Case (15).

The graph is the same as graph of Case (14).

Cases (1),(2),(3),(4),(5) and (7).

Suppose that there exists a Klein bottle with one of these types of singular sets. Perform D-surgeries along some disks. Then we obtain arcs and embedded closed surfaces. But we cannot connect arcs. This is a contradiction.

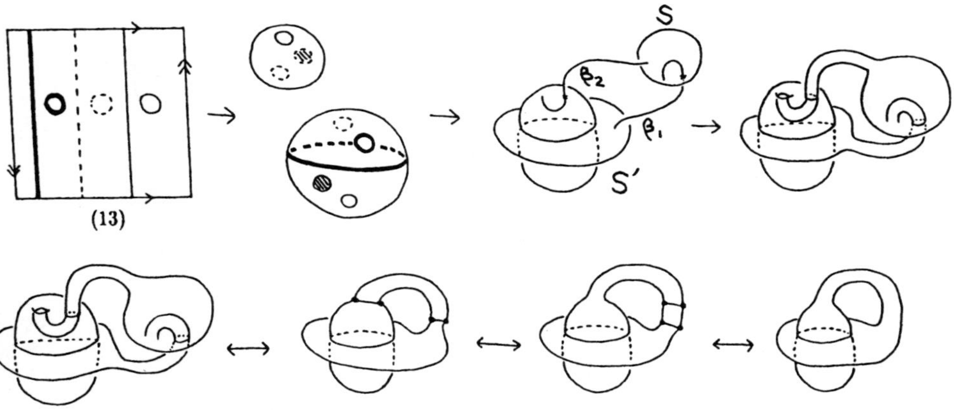

Fig. 9.

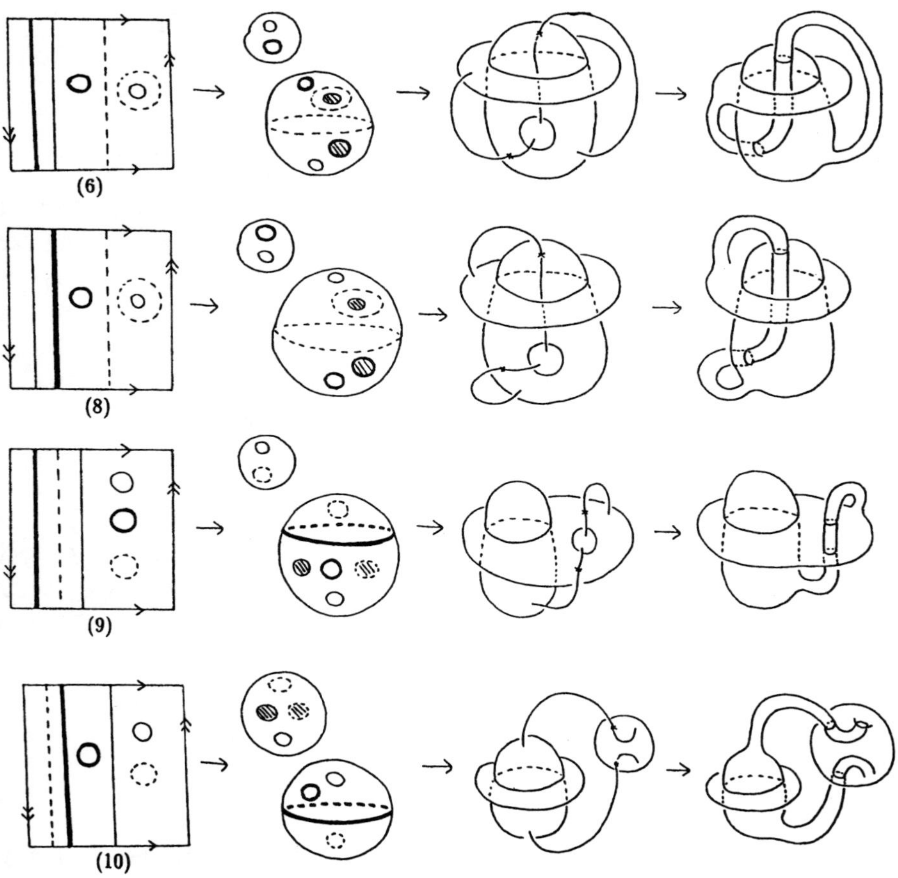

Fig. 10.

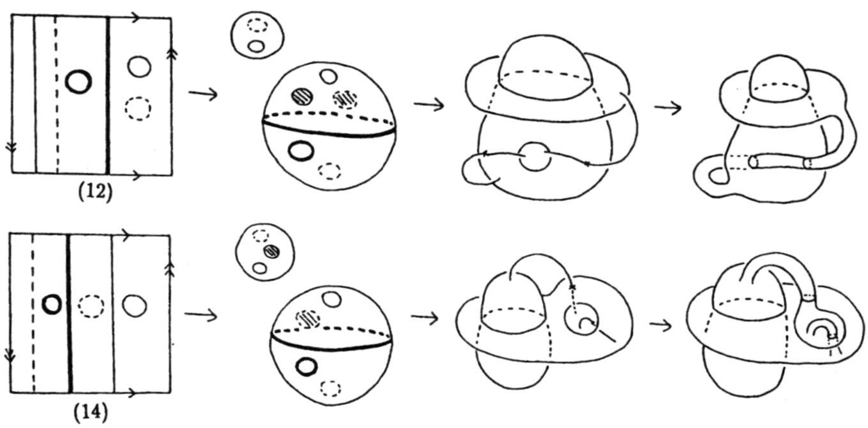

Fig. 10. (continued)

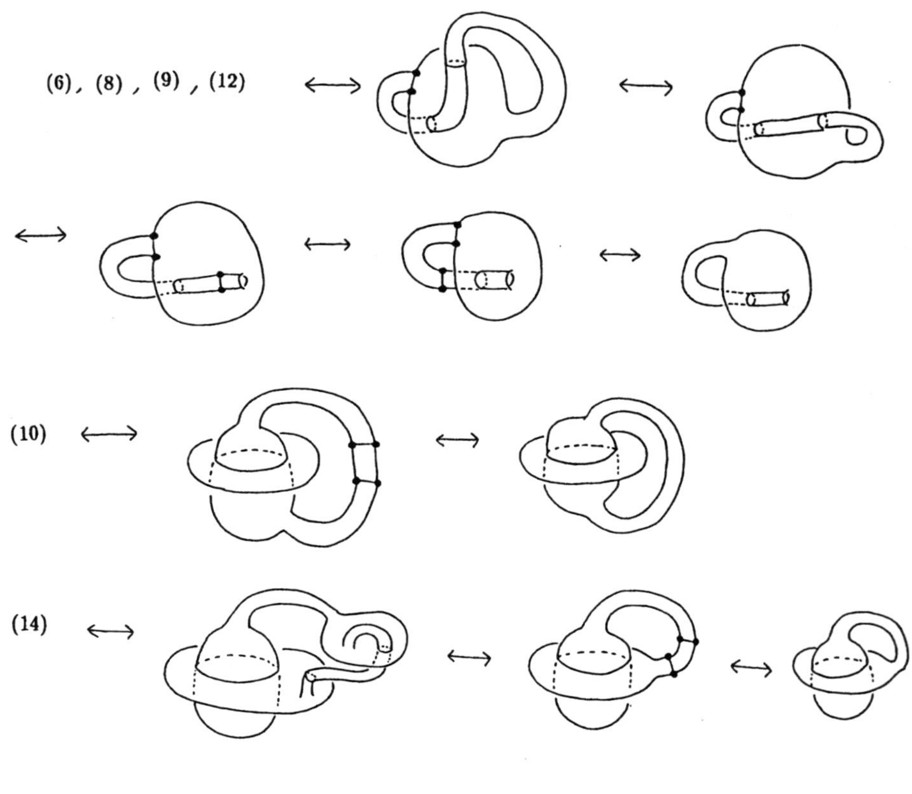

Fig. 11.

Case (16).

Let e_1, e_2, e_3, e_4 be the edges of G_F such that $e_i \cap e_{i+1}$ is one vertex for $i = 1, 2, 3$, $h(e_1) = h(e_2)$, and $h(e_3) = h(e_4)$. Let $c_1, c_2, c_3, c_4, d_1, d_2$ be the components of $\Gamma(F)$

such that c_i is corresponding to e_i. Given two annuli A_1, A_2 in F with $\partial A_1 = c_1 \cup c_2$ and $\partial A_2 = c_3 \cup c_4$, the immersed Klein bottle $(\overline{F \setminus A_1 \cup A_2})^*$ satisfies the condition in Lemma 2.1. Hence, $(\overline{F \setminus A_1 \cup A_2})^*$ is Banchoff's immersion. Since A_i^* $(i = 1, 2)$ is an embedded torus, by the solid torus theorem [6] there exists a solid torus V_i in S^3 with $\partial V_i = A_i^*$. By Remark 4.3, γ_i^* is a generator of $H_1(V_i)$ or $(\overline{F \setminus A_1 \cup A_2})^* \subset V_i$, where $\gamma_1^* = c_1^*$, $\gamma_2^* = c_3^*$. If $(\overline{F \setminus A_1 \cup A_2})^* \subset V_i$, then the immersed surface F^* is the projection of a spun Klein bottle into S^3. We can assume that γ_i^* is a generator of $H_1(V_i)$ in a similar way to Case 7 in Theorem 4.2. By Lemma 3.1 (ii), γ_1^* or γ_2^* can be cancelled.

Case (17). In a similar way to Case (16), we can reduce the number of double curves in $\Gamma(F^*)$. $\square$

References

[1] H. Aiso, *On the classification of simply knotted spheres with less than 6 crossing circles; original title "Crossing circle ga 5 hon ika dearu simply knotted sphere no bunrui ni tsuite" (in Japanese)*, Master thesis, Department of Mathematics University of Tokyo, (1984).

[2] T. Banchoff, *Triple points and singularities of projections of smoothly immersed surfaces*, Proc. Amer. Math. Soc.,**46**, (1974), 402–406.

[3] S. Carter, How Surfaces Intersect in Space, World Scientific, (1995).

[4] S. Carter and M. Saito, *Canceling branch points on projections of surfaces in 4-space*, Proc. Amer. Math. Soc., **116**, (1992), 229–237.

[5] T. Homma and T. Nagase, *On elementary deformation of maps of surfaces into 3-manifolds I*, Yokohama Math. J., **33**, (1985), 103–119.

[6] D. Rolfsen, Knots and Links, Publish or Perish, Berkeley, Calif., (1976).

[7] A. Shima, *An unknotting theorem for tori in S^4*, Rev. Math. Univ. Compute. Madrid, **11**, (1998), 299–309.

[8] A. Shima, *An unknotting theorem for tori in S^4* II, Kobe J. Math., **13**, (1996), 9–25.

[9] A. Shima, *On simply knotted tori in S^4*, J. Math. Sci. Univ. Tokyo, **4**, (1997), 279–339.

SKEIN MODULES AND TQFT

ADAM S. SIKORA

Department of Mathematics, University of Maryland
College Park, MD 20742

ABSTRACT

We prove that the Kauffman bracket skein module of a 3-manifold M at a $4r$-th root of unity divided by the Jones-Wenzl idempotent depends only on ∂M. [1]

Keywords: Kauffman bracket, skein module, TQFT.

Let r be a fixed natural number, $r = 2, 3, \ldots$, and let A be a primitive $4r$-th root of $1 \in \mathbb{C}$. The Kauffman bracket skein module of an oriented 3-manifold M, denoted by $\mathcal{S}(M)$, is $\mathbb{C}$-linear space of formal linear combinations of framed, unoriented links in M, considered up to an ambient isotopy in M and the following skein relations:

Let Δ_n be the value of the n-th Chebyshev polynomial:

$$\Delta_n = d\Delta_{n-1} - \Delta_{n-2},\ \Delta_0 = 1,\ \Delta_1 = d = -(A^2 + A^{-2}),$$

and let f_n be the n-th Jones-Wenzl idempotent:

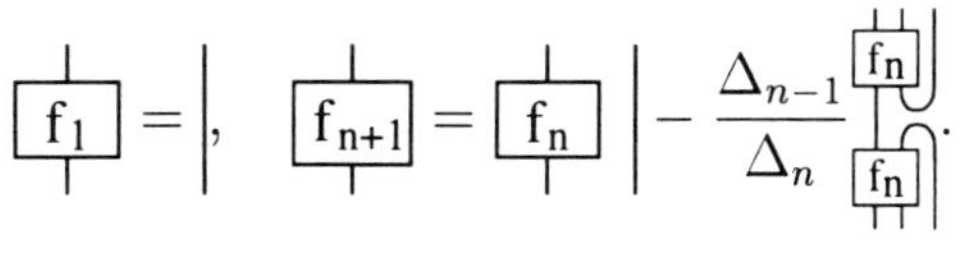

We denote by $K_r(M)$ the quotient of $S(M)$ by the submodule generated by all linear skeins involving f_{r-1}. Obviously, $K_r(S^3) = \mathcal{S}(S^3) \simeq \mathbb{C}$.

[1]This fact was proved by J. Roberts first, and then by myself. Since the result has created much interest and Robert's proof has not been published yet, I decided to publish my proof.

Theorem 1 (Roberts)
If M is a compact 3-manifold then $K_r(M)$ depends on ∂M only, i.e. if M_1, M_2 are compact 3-manifolds such that $\partial M_1 \simeq \partial M_2$ then $K_r(M_1)$ and $K_r(M_2)$ are isomorphic $\mathbb{C}$-linear spaces.

A 3-vertex $\{a, b, c\}$ is a tangle

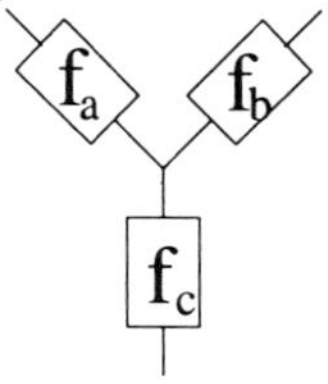

Any 3-vertex satisfies the relations: $a + b + c = 0 \bmod 2$, $a + b \geq c, a + c \geq b, b + c \geq a$. A 3-vertex is admissible if $a + b + c \leq 2r - 4$. J. Roberts proved that all non-admissible 3-vertices vanish in $K_r(M)$, for any 3-manifold M. It is also possible to prove that $K_r(M)$ is isomorphic as a $\mathbb{C}$-linear space to the vector space assigned to ∂M by the topological quantum field theory associated with the Reshetikhin-Turaev quantum SU_2 invariant.

We precede the proof of Theorem 1 with a few standard facts and definitions. For any framed knot K in M, K_ω denotes a linear skein defined as in the example:

K =  $\quad K_\omega = \sum_{n=0}^{r-2} \Delta_n$ 

If L is a framed link then L_ω denotes a linear skein formed by decorating all the components of L with ω.

Lemma 2
Let L and K be a framed link and a framed knot in M respectively. Let L' be a framed link obtained from L by sliding one of its components along K. Then $L \cup K_\omega = L' \cup K_\omega$ in $K_r(M)$.

Lemma 3

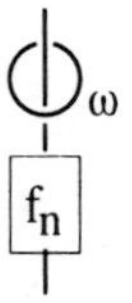

$= \begin{cases} 0 & \text{if } 1 \leq n \leq r-2 \\ \Omega & \text{if } n = 0 \end{cases}$ *in $K_r(M)$, where $\Omega = \sum_{i=0}^{r-2} \Delta_i^2$.*

Moreover, $\Omega \neq 0$.

The Lemmas 2 and 3 are well-known for $M = S^3$. However, the proofs of these lemmas given for example in [L] and [P-S] for $M = S^3$ hold for any 3-manifold M.

Proof of Theorem 1:
For any framed link $L \subset M$ we will denote a regular neighborhood of L in M by $V(L)$ and the surgery along L in M by M_L. $V(L)$ is always a disjoint union of solid tori in M. Let M_1, M_2 be two 3-manifolds such that $\partial M_1 \simeq \partial M_2$. Then there exist framed links $L_1 \subset M_1$, $L_2 \subset M_2$ which satisfy the following conditions:

- There is a homeomorphism $\phi : M_1 \setminus V(L_1) \to M_2 \setminus V(L_2)$
- $(M_1)_{L_1} = M_2,\ (M_2)_{L_2} = M_1.$

We define a homomorphism $f_1 : K_r(M_1) \to K_r(M_2)$ in the following way. Let K be any framed link in M_1. We can isotope K outside of $V(L_1)$ and therefore consider it as an element of $K_r(M_1 \setminus V(L_1))$. Then $\phi(K)$ is a framed link in $M_2 \setminus V(L_2)$ and we define $f_1(K)$ to be $\phi(K) \cup (L_2)_\omega \in K_r(M_2)$. At a first sight it seems that f_1 is ill-defined since the operation of pushing $K \subset M_1$ outside $V(L_1) \subset M_1$ is not unique. The picture presents two (generally inequivalent) results of such operation.

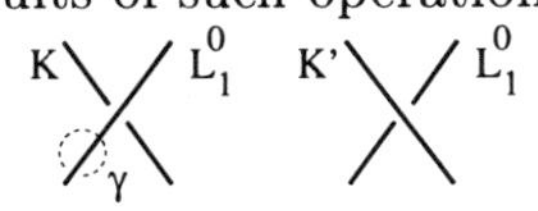

L_1^0 is a component of L_1. Notice that we can get K' by sliding K along the meridian γ going around $V(L_1^0)$. This closed curve will correspond under the homeomorphism ϕ to a closed curve $\phi(\gamma)$ parallel to the boundary of $V(L_2^0)$, for some connected component L_2^0 of L_2. Moreover, the requirement $M_1 = (M_2)_{L_2}$ implies that $\phi(\gamma)$ is parallel to one of the boundary components of L_2^0. (L_2^0 is considered here as an annulus embedded into M_2.) Therefore $\phi(K')$ is $\phi(K)$ slid along L_2^0. Since $f_1(K)$ is a union of L_2 decorated by ω and of $\phi(K)$, Lemma 2 implies that sliding K along L_2^0 will not change the value of $f_1(K) \in K_r(M_2)$. Hence $f_1(K) = f_1(K')$ and f is a well defined homomorphism $K_r(M_1) \to K_r(M_2)$.

Let $f_2 : K_r(M_2) \to K_r(M_1)$ be defined analogously. We want to prove that $f_2 \circ f_1$ and $f_1 \circ f_2$ are of the from $c \cdot Id_{K_r(M_1)}$, $c \cdot Id_{K_r(M_2)}$, respectively, for some non-zero $c \in \mathbb{C}$. If K is a framed link in M_1 then $f_1(K) = K \cup (L_2)_\omega \in K_r(M_2)$. Let us push $(L_2)_\omega$ outside $V(L_2)$ in the direction parallel to the framing of L_2. Now $K \cup (L_2)_\omega$ lies in $M_2 \setminus V(L_2)$. What is $\phi^{-1}(L_2) \in M_1 \setminus V(L_1)$? The condition $M_1 = (M_2)_{L_2}$ implies that each of the components of $\phi^{-1}(L_2)$ is a framed knot surrounding the meridian of one of the tori $\partial V(L_1) \subset \partial(M_1 \setminus V(L_1))$. Therefore $f_2 \circ f_1(K) = K \cup (L_1)_\omega \cup (\phi^{-1}(L_2))_\omega$ where $(L_1)_\omega \cup (\phi^{-1}(L_2))_\omega$ is of the form

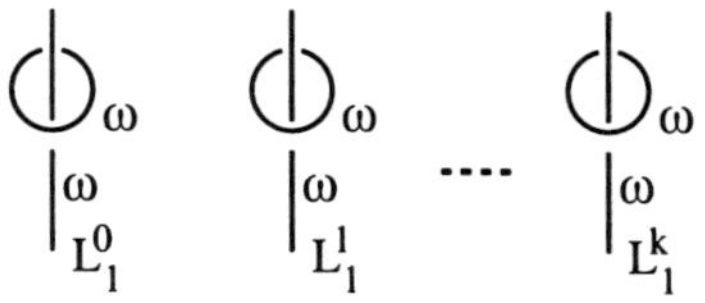

where $L_1^0, L_1^1, ..., L_1^k$ are the connected components of L_1. By Lemma 3 $(L_1)_\omega \cup (L_2)_\omega = \Omega^{k+1} \in \mathbb{C}, \Omega^{k+1} \neq 0$. Therefore $f_2 \circ f_1$ is an automorphism of $K_r(M_1)$. This implies that f_1 is a monomorphism and f_2 is an epimorphism. The same argument for $f_1 \circ f_2$ shows that f_1 is epi and f_2 is mono. Therefore both f_1 and f_2 are isomorphisms. □

References

[L] W. B. R. Lickorish, An Introduction to Knot Theory, Graduate Texts in Mathematics, Springer 1997.

[P-S] V. V. Prasolov, A.B. Sossinsky, Knots, Links, Braids and 3-manifolds, Translations of Mathematical Monographs, AMS, Vol 154.

VIRTUAL KNOT GROUPS

Daniel S. Silver

Susan G. Williams

ABSTRACT. Virtual knot groups are characterized, and their properties are compared with those of classical knot groups. A coloring theory, generalizing the usual notion of Fox n-coloring, is introduced for virtual oriented links.

1. INTRODUCTION. We recall that a **diagram** D for a classical knot $k \subset \mathbf{R}^3$ is a 4-valent plane graph resulting from a regular projection of k. Extra information, conveyed by a *trompe l'oeil* effect at each vertex, which is then called a **diagram crossing**, makes it possible to reconstruct the knot (Figure 1 (a)).

It is well known that two knots are isotopic if and only if a diagram for one can be transformed into a diagram for the other by a finite sequence of **Reidemeister moves** (see first column of Figure 2). As usual, we regard isotopic knots as the same. Hence we may think of a knot as an equivalence class of diagrams.

In 1996 L. Kauffman introduced the concept of a virtual knot [**Ka1**]. He did this by allowing a third type of crossing, a so-called virtual crossing (Figure 1 (b)), and then generalizing the Reidemeister moves (Figure 2).

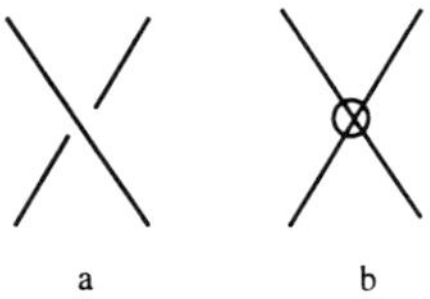

Figure 1: (a) Classical Crossing; (b) Virtual Crossing

Both authors partially supported by National Science Foundation Grant DMS-9704399

1991 *Mathematics Subject Classification.* Primary 57Q45; Secondary 54H20, 20E06, 20F05.

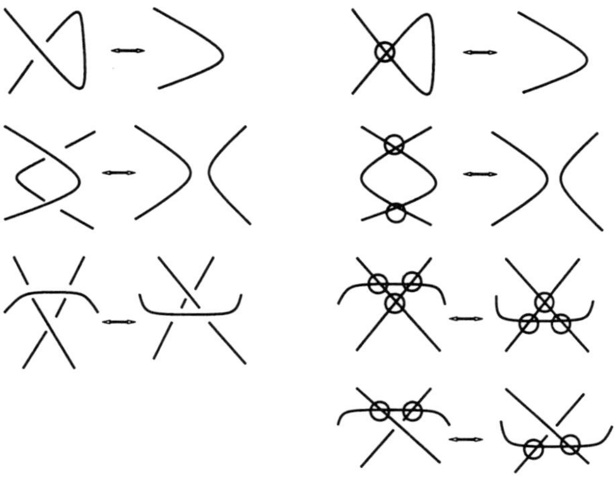

Figure 2: Generalized Reidemeister Moves

A **virtual knot** is now defined to be an equivalence class of diagrams. (Virtual links and oriented virtual knots or links are defined in the obvious way.) A theorem of M. Goussarov, M. Polyak and O. Viro states that if diagrams of a classical knot are equivalent under generalized Reidemeister moves, then they are equivalent under classical Reidemeister moves [**GuPoVi**]. In this sense, the theory of classical knots is embedded in that of virtual knots.

According to [**Ka3**] Kauffman's motivation comes from (1) the study of knots in thickened surfaces of higher genus (classical knot theory being the study of knots in the thickened sphere); (2) the desire to "complete" the set of classical knots in such a way that knots correspond to Gauss codes under a suitable equivalence relation.

We are propelled by a third motivation, one that is group-theoretical. As explained in [**Ka2**], [**Ka3**], a group can be associated to any virtual knot diagram by a generalization of Wirtinger's algorithm: Having first oriented the diagram, one associates a generator to each arc. By an arc we mean a curve in the diagram that begins and ends at classical undercrossings, possibly running through one or more virtual crossings. The relations among the generators are determined by the classical crossings in the usual way (see [**Kaw**], for example). It is clear that reversing the orientation of the diagram does not affect the group. More significantly, equivalent diagrams determine the same group. Hence we can speak of $G(k)$, the **group of a virtual knot** k.

Example 1.1. (Cf. [**Ka2**], [**Ka3**]) The virtual knot k, a modified trefoil, is displayed in Figure 3 together with generators for its group $G(k)$, which has presentation

$$\langle x, y, z \mid xy = zx,\ zx = yz,\ yx = zy\rangle \cong \langle x, y \mid xyx = yxy,\ yxy^{-1} = xyx^{-1}\rangle.$$

It is well known that $G(k)$ is the group of the 2-twist spun trefoil [**Ze**]. Its commutator subgroup is the cyclic group $\mathbf{Z}/3$ of order three.

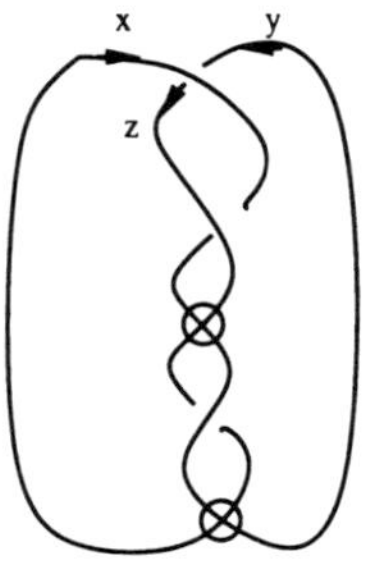

Figure 3: Virtual Knot k

By a theorem of L. Neuwirth [**Ne**], the commutator subgroup of any classical knot group is free whenever it is finitely generated. Hence the group in Example 1.1 is not that of a classical knot. On the other hand, every classical knot group is a virtual knot group. Hence the collection of virtual knot groups is larger than the collection of classical knot groups. We denote the collection of all virtual knot groups by $\mathcal{VK}^1$.

2. CHARACTERIZATION OF VIRTUAL KNOT GROUPS.

We characterize virtual knot groups combinatorially and also topologically. The topological characterization requires the following.

Definition 2.1. Let $B_1^3, \ldots, B_n^3 \subset \mathbf{R}^4$ be a family of mutually disjoint 3-balls, and let $f_1, \ldots, f_n$ be a collection of embeddings $f_i : B^2 \times [0,1] \to \mathbf{R}^4$, $1 \le i \le n$, with mutually disjoint images such that

$$f_i(B^2 \times [0,1]) \cap \partial B_k^3 = \begin{cases} f_i(B^2, 0), & \text{if } k \equiv i \pmod{n}; \\ f_i(B^2, 1), & \text{if } k \equiv i+1 \pmod{n}; \\ \emptyset, & \text{otherwise.} \end{cases}$$

The embedded torus

$$\left(\bigcup_{i=1}^{n} \partial B_i^3\right) \cup \left(\bigcup_{i=1}^{n} f_i(\partial B^2 \times [0,1])\right) - \left(\bigcup_{i=1}^{n} f_i(\text{int } B^2 \times \{0,1\})\right)$$

is called a **ribbon torus** (of n fusions).

Remark. A ribbon torus can be regarded as the result of "piping together" standardly embedded, separated 2-spheres in R^4. The definition of a more familiar notion, that of of a **ribbon 2-knot** (of n fusions), can be recovered from Definition 2.1 by deleting one embedding from the collection $\{f_1, \ldots, f_n\}$. (See [**Kaw**], for example.)

Theorem 2.2. Let G be a group. The following statements are equivalent.

(i) $G \in \mathcal{VK}^1$.

(ii) G has a presentation of the form $\langle x_1, \ldots, x_n \mid x_{i+1} = u_i^{-1} x_i u_i \quad (i \in \mathbf{Z}_n)\rangle$, where $u_1, \ldots, u_n$ are elements of the free group $F(x_1, \ldots, x_n)$.

(iii) G is isomorphic to $\pi_1(\mathbf{R}^4 - T)$, for some ribbon torus T.

Proof. That (i) implies (ii) is an immediate consequence of the Wirtinger algorithm applied to any diagram of a virtual knot with group G.

The equivalence of (ii) and (iii) follows from the theorem of Seifert and Van Kampen. The proof for the analgous statement about ribbon 2-knots can be found in [**Yan**].

We prove that (ii) implies (i). Consider a group presentation as in (ii). By introducing new generators and defining relators, we can assume without any loss of generality that each u_i is one of $x_1^{\pm 1}, \ldots, x_n^{\pm 1}$. For if some u_i is equal to $x_{i_1}^{\epsilon_1} \cdots x_{i_k}^{\epsilon_k}$, where $\epsilon_j = \pm 1$, we can replace the generator x_i by $x_{i,0}, \ldots, x_{i,k-1}$, where $x_{i,0} = x_i$, $x_{i,1} = x_{i_1}^{-\epsilon_1} x_{i,0} x_{i_1}^{\epsilon_1}$, $\ldots, x_{i,k-1} = x_{i_{k-1}}^{-\epsilon_k} x_{i,k-2} x_{i_{k-1}}^{\epsilon_k}$.

We begin constructing a virtual knot diagram, using the first relation $x_2 = x_{j_1}^{-\epsilon_1} x_1 x_{j_1}^{\epsilon_1}$, as in Figure 4.

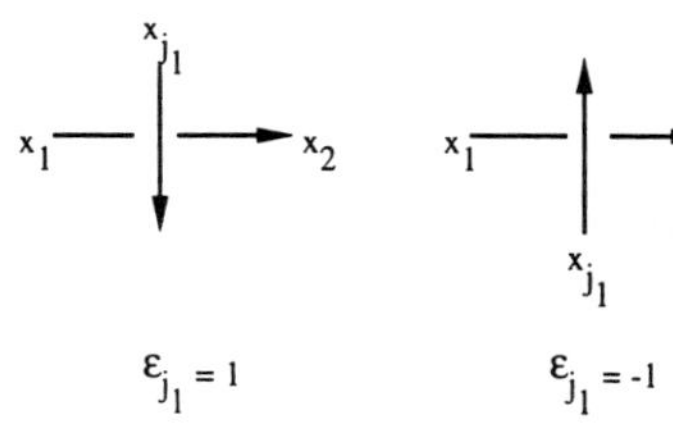

Figure 4: Beginning of Virtual Knot Diagram

We extend the arc labeled x_2 in the same way, using the next relation $x_3 = x_{j_2}^{-\epsilon_2} x_2 x_{j_2}^{\epsilon_2}$. If an arc labeled x_{j_2} has already been drawn, then *we do not draw a second such arc*; we simply continue the existing arc, passing under the arc labeled x_{j_2} in the appropriate direction, and re-emerging as x_3. If an arc labeled x_3 has previously been drawn, then we join the re-emerging arc to it. We introduce virtual crossings as needed.

We repeat the above procedure for each of the arcs labeled $x_3, x_4, \ldots, x_n$. ■

Corollary 2.3. Assume that G has a presentation of the form $\langle x, y \mid r\rangle$, such that (1) $G/[G,G] \cong \mathbf{Z}$; and (2) the normal closure $\langle\langle x \rangle\rangle$ is equal to G. Then $G \in \mathcal{VK}^1$.

Proof. By introducing a generator a with defining relation $a = yx^{-1}$, and then using the relation to eliminate y, we can assume without any loss of generality that G has a presentation of the form $\langle x, a \mid r\rangle$, where $a \in [G,G]$. The hypothesis (2) implies that the exponent sum of a in r is 1 or -1; we can replace r with an equivalent relation $a = w$, where $w \in F(x,a)$ has zero exponent sum in a. Then $a = \prod_{i=1}^m u_i^{-1}x^{\epsilon_i}u_i$, for some $u_1, \ldots, u_m \in F(x,a)$, and $\epsilon_i = \pm 1$, $1 \le i \le m$.

Adapting an argument in [**Yaj**], we introduce generators $y_1, \ldots, y_m$ and defining relations $y_i = u_i^{-1}xu_i$, and obtain

$$G \cong \langle x, a, y_1, \ldots, y_m \mid a = \prod_{i=1}^m u_i^{-1}x^{\epsilon_i}u_i,\ y_1 = u_1^{-1}xu_1,\ \ldots,\ y_m = u_m^{-1}xu_m\rangle$$

$$\cong \langle x, a, y_1, \ldots, y_m \mid a = \prod_{i=1}^m y_i^{\epsilon_i},\ \ y_1 = u_1^{-1}xu_1,\ \ldots,\ y_m = u_m^{-1}xu_m\rangle$$

We rewrite each word $u_1, \ldots, u_m$, replacing each occurrence of $a^{\pm 1}$ by $(\prod_{i=1}^m y_i^{\epsilon_i})^{\pm 1}$. Finally, we use the first relation to eliminate the generator a. The final presentation satisfies condition (ii) of Theorem 2.2. ∎

We recall that a group G is **residually finite** if for any nontrivial element $g \in G$ there exists a finite-index subgroup $N \triangleleft G$ such that $g \notin N$. Every classical knot group is residually finite by a theorem stated in [**Th**] (see [**He**] for a proof). However, virtual knot groups need not have this property, as the following example shows.

Example 2.4. The Baumslag-Solitar group $G = \langle x, y \mid x^{-1}y^2x = y^3\rangle$ is in $\mathcal{VK}^1$, by Corollary 2.3. It is not residually finite [**Ba**].

Another important property of classical knot groups that fails in the virtual category is coherence. A group is **coherent** if every finitely generated subgroup is finitely presented. By a theorem of P. Scott [**Sc**] the fundamental group of any 3-manifold is coherent. However, in [**Si**] we gave examples of ribbon tori T such that the commutator subgroup of $\pi_1(\mathbf{R}^4 - T)$ is finitely generated but not finitely presented. An explicit example visualized by 3-dimensional cross-sections can be found in [**Ph**].

3. VIRTUAL TANGLE GROUPS. Any Wirtinger presentation for the group of a virtual knot k has the same number of generators as relations. It is well known that if k is a classical knot, then any relation is a consequence of the others. In fact, this property holds for a Wirtinger presentation of the group of any spatial graph (see [**St**], for example.) However, this property can fail when k is a virtual knot. For example, the group in Example 1.1 has no presentation with fewer relations than generators [**Le**]. In this section we will examine some consequences of this failure.

We define a **virtual (n-string) tangle** by diagrams in the obvious way. The **group** $G(t)$ of a virtual tangle t is then defined as for virtual knots, and it can be described by a Wirtinger presentation.

It is convenient to regard a virtual knot diagram as contained the 2-sphere $S^2 = \mathbf{R}^2 \cup \infty$ rather than the plane. This is harmless. Indeed, one can check that the final result of isotoping an arc past ∞ can also be achieved in the plane using generalized Reidemeister moves. Instead of simply pulling the arc over or under the other arcs as we would in the classical case, we introduce virtual crossings and apply generalized Reidemeister moves.

Given a diagram in S^2 for a virtual knot k, we can *cut* k, that is, remove the interior of a small disk neighborhood about a point p on the diagram that is away from any crossing, and thereby obtain a diagram of a virtual 1-string tangle t. It is well known that when k is a classical knot, $G(t) \cong G(k)$. One way to see this is to regard k as a spatial graph with a single vertex, the point p. The Wirtinger relation corresponding to p is $x' = x''$. If we view x' and x'' as generators of $G(t)$ corresponding to the input and output arcs of the tangle t, then since $x' = x''$ is is a consequence of the other Wiritinger relations for the group of the spatial graph, it is also a relation in $G(t)$.

Example 3.1. If we cut a diagram for a virtual knot k, then the group of the resulting virtual 1-string tangle t need not be isomorphic to the group of k. Figure 5 (a) displays a diagram for a virtual knot k with infinite cyclic group. If we cut at the indicated point, then the group of the resulting virtual tangle, indicated in Figure 5 (b), is isomorphic to that of the trefoil. The verification is straightforward.

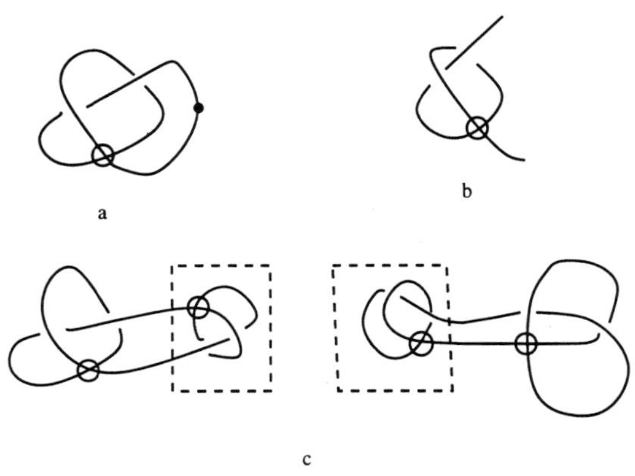

Figure 5: (a) Virtual Knot; (b) Associated 1-String Tangle; (c) Connected Sums

If we attempt to form the connected sum $k\sharp k$ of two copies of k, then the result will depend on the sites where the diagrams are cut and the connection is performed. In Figure 5 (c) we see two connected sums $(k\sharp k)_1$ and $(k\sharp k)_2$: the first has group isomorphic to that of the trefoil, the second has infinite cyclic group. The verification is again routine, and we leave it to the reader.

If we try to deform the diagram for one of the connected sums in Figure 5 (c) into the other by sliding the indicated box along the diagram, a geometric trick for proving that connected sum of oriented classical knots is well defined, then we are forced to apply "forbidden moves" [**GoPoVi**]. These illegal moves appear in Figure 6.

Figure 6: Forbidden Moves

It is interesting to observe that the first forbidden move does not affect the group of the virtual knot. The proof is not difficult. However, using the the generalized Reidemeister moves, the flipping operation and the first forbidden move, we can transform $(k\sharp k)_1$ into $(k\sharp k)_2$. We conclude that the group of the flip of $(k\sharp k)_1$ must be the group of $(k\sharp k)_2$. We have thus arrived at the conclusion of [**GoPoVi**] (see also [**Ka3**]) that the group of a virtual knot and the group of its flip need not be isomorphic. Hence associated to each virtual knot is a *group pair.*

Problem 3.2. Characterize virtual knot group pairs.

4. COLORING VIRTUAL LINK DIAGRAMS. The well-known technique of Fox n-coloring associates a $\mathbf{Z}/n$-module $\mathrm{Col}_{\mathbf{Z}/n}(l)$ to any classical link l. Elements are n-**colorings** of a diagram D of the link: assignments of elements of $\mathbf{Z}/n$ (**colors**) to the arcs of D such that at any crossing, the sum of the colors of the undercrossing arcs is equal to twice the color of the overcrossing. All of this can be done as well for virtual links, with virtual crossings imposing no conditions on the colors. In [**Ka2**] and [**Ka3**] Kauffman notes this fact, and uses it to show that a certain virtual knot k with Jones polynomial equal to 1 is nontrivial. He produces k from a trefoil knot diagram by applying a certain move that does not affect the module of 3-colorings. The conclusion that k is nontrivial then follows from the well-known fact that the trefoil has more 3-colorings than the trivial knot.

In [**SiWi1**] and [**SiWi2**] a generalization of n-coloring for classical oriented links l is introduced. For any positive integer r and any topological group Σ, a space $\mathrm{Col}_{\Sigma,r}(l)$ of

(Σ, r)-colorings is defined. When $r = 2$ the orientation of l is unimportant. When $r = 2$ and Σ is $\mathbf{Z}/n$ with the discrete topology, the space is $\mathrm{Col}_{\mathbf{Z}/n}(l)$.

We will show that the coloring spaces $\mathrm{Col}_{\Sigma,r}(l)$ are also invariants of virtual links. The move that Kauffman applied is the square of a move we denote by T. We show that for any positive integer r, the iterated move T^r leaves $\mathrm{Col}_{\Sigma,r}(l)$ unchanged.

The spaces $\mathrm{Col}_{\Sigma,r}(l)$, for various r, are all contained in a single space $\mathrm{Col}_{\Sigma,\infty}(l)$ that we define first.

Let Σ be any topological group, and consider the group $\Sigma^{\mathbf{Z}}$ of all bi-infinite sequences (α_j) of elements $\alpha_j \in \Sigma$. The **shift map** $\sigma : \Sigma^{\mathbf{Z}} \to \Sigma^{\mathbf{Z}}$ that sends (α_j) to the sequence (α'_j), where $\alpha'_j = \alpha_{j+1}$, is an automorphism.

Definition 4.1. (cf. [**SiWi2**]) Let Σ be any topological group. Assume that D is a diagram of a virtual oriented link.

(i) A $\Sigma^{\mathbf{Z}}$**-coloring** of D is an assignment of elements $C \in \Sigma^{\mathbf{Z}}$ to the arcs of D such that at any classical crossing

$$(\sigma C_k)C_j = (\sigma C_i)C_k, \tag{3.1}$$

where C_k corresponds to the overcrossing, while C_i, C_j correspond to the undercrossings, and C_j is to the left as we travel in the preferred direction along the arc labeled C_k (see Figure 7).

(ii) A $\Sigma^{\mathbf{Z}}$-coloring is **periodic** if there exists a positive integer r such that $\sigma^r(C) = C$, for every assigned label C. In such a case, we say that the coloring has **period** r.

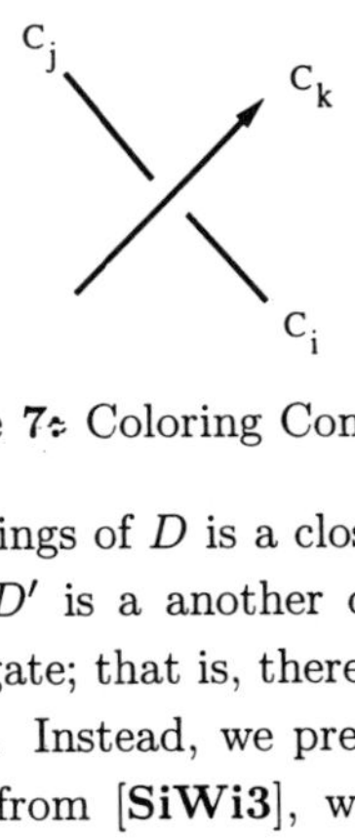

Figure 7: Coloring Convention

The set $\mathrm{Col}_{\Sigma,\infty}(D)$ of $\Sigma^{\mathbf{Z}}$-colorings of D is a closed σ-invariant subspace of $\Sigma^{\mathbf{Z}}$. One can check combinatorially that if D' is a another diagram for l, then $\mathrm{Col}_{\Sigma,\infty}(D)$ and $\mathrm{Col}_{\Sigma,\infty}(D')$ are topologically conjugate; that is, there exists a homeomorphism h between the spaces such that $\sigma \circ h = h \circ \sigma$. Instead, we prefer to sketch the following algebraic argument. The basic ideas come from [**SiWi3**], where the reader can find additional details.

Groups of virtual links are defined via Wirtinger presentations just as they are for virtual knots. Let G be the group of the virtual oriented link l. Consider the homomorphism $\chi : G * \langle x \mid \rangle \to \mathbf{Z}$ that maps x and each generator of G to 1. Give the kernel K

the discrete topology, and denote by $\mathrm{Hom}(K, \Sigma)$ the space of homomorphisms ρ, with the compact-open topology. (Intuitively, homomorphisms are close if they agree on a large subgroup of K.) The map $\sigma_x : \mathrm{Hom}(K, \Sigma) \to \mathrm{Hom}(K, \Sigma)$ given by $\sigma_x\rho(a) = \rho(x^{-1}ax)$, for all $a \in K$, is a homeomorphism.

Proposition 4.2. *If D is a diagram of a virtual oriented link, then $\mathrm{Col}_{\Sigma,\infty}(D)$ and $\mathrm{Hom}(K, \Sigma)$ are topologically conjugate; that is, there exists a homeomorphism h between them such that $\sigma_x \circ h = h \circ \sigma$.*

Proof. It is a consequence of the Reidemeister-Schreier Theorem [**MaKaSo**] that the kernel K has a presentation of the form $\langle a_j, b_j, \ldots \mid r_j, s_j, \ldots \rangle$, where j ranges over $\mathbf{Z}$. Here the number of generator families $a_j, b_j, \ldots$ is finite, as is the number of relator families $r_j, s_j, \ldots$. Also, the relators $r_{j+\nu}, s_{j+\nu}, \ldots$ are obtained from $r_j, s_j, \ldots$ by adding ν to the subscripts of each symbol that appears. The generators of K arise in a natural way. If $x_1, x_2, \ldots$ are the generators of G, then $a_0, b_0, \ldots$ are x_1x^{-1}, $x_2x^{-1}, \ldots$, respectively. For each j, the symbols $a_j, b_j, \ldots$ represent $x^{-j}a_0x^j, x^{-j}b_0x^j, \ldots$.

A homomorphism $\rho : K \to \Sigma$ consists of an assignment of elements $\alpha_j, \beta_j, \ldots$ to the generators $a_j, b_j, \ldots$, such that each relation $r_j, s_j, \ldots$ becomes the identity element upon substitution. Since a typical Wirtinger relation $cx \cdot bx = ax \cdot cx$ is equivalent to $x^{-j-1}cx^{j+1} \cdot x^{-j}bx^j = x^{-j-1}ax^{j+1} \cdot x^{-j}cx^j$ and hence to $c_{j+1}b_j = a_{j+1}c_j$, for any j, such homomorphisms correspond bijectively to the $\Sigma^{\mathbf{Z}}$-colorings of D. The correspondance is shift-commuting and continuous. ∎

In view of Proposition 4.2 we can define $\mathrm{Col}_{\Sigma,\infty}(l)$ to be $\mathrm{Col}_{\Sigma,\infty}(D)$, where D is any diagram from l. For any positive integer r we define the subspace $\mathrm{Col}_{\Sigma,r}(l)$ of (Σ, r)-colorings in the following way.

The normal subgroup N_r of $G * \langle x \mid \ \rangle$ generated by the rth powers of the generators $x, x_1, x_2, \ldots$, is independent of the diagram D of l, since the generators corresponding to any diagram D' obtained from D by a generalized Reidemeister move are conjugates of the generators corresponding to D. We can describe N_r as the normal closure $\langle\langle x^r, (ax)^r, (bx)^r, \ldots \rangle\rangle$, which is equal to $\langle\langle x^r, x^{-r}(ax)^r, x^{-r}(bx)^r, \ldots \rangle\rangle$. The elements $x^{-r}(ax)^r$, $x^{-r}(bx)^r, \ldots$ can be rewritten as $a_ra_{r-1}\cdots a_1$, $b_rb_{r-1}\cdots b_1$, $\ldots$, and it is now a straightforward matter to check that $N_r \cap K$ is the normal closure in K of the elements $a_{r+j}a_{r+j-1}\cdots a_{j+1}$, $b_{r+j}b_{r+j-1}\cdots b_{r+j}, \ldots$, where j ranges over $\mathbf{Z}$. Hence the (discrete) group $K/N_r \cap K$ is an invariant $\hat{K}_r$ of the virtual oriented link l. If l is a classical oriented link, then $\hat{K}_r$ is isomorphic to the fundamental group of the r-fold cyclic cover of S^3 branched over l (see [**SiWi2**]).

Definition 3.3. Let l be a virtual oriented link, Σ a topological group and r a positive integer. A **(Σ, r)-coloring** of l is a homomorphism $\rho : K/N_r \cap K \to \Sigma$. The set

$\mathrm{Hom}(K/N_r \cap K, \Sigma)$ of all such homomorphisms, endowed with the compact-open topology, is dentoted by $\mathrm{Col}_{\Sigma,r}(l)$.

From what we have said, $\mathrm{Col}_{\Sigma,r}(l)$ is an invariant of the virtual oriented link l. We can think of a (Σ, r)-coloring as a $\Sigma^{\mathbf{Z}}$-coloring of a diagram D such that the product in reverse order of any r consecutive coordinates of any assigned label is the identity. Such a coloring is necessarily periodic with period r. (Cf. Definition 4.2 of [**SiWi2**].)

Consider the moves T and $\bar{T}$ described in Figure 8. In [**Ka2**] and [**Ka3**] Kauffman observes that the composition T^2 applied to any virtual knot diagram leaves the module of n-colorings unaffected, for any n. The following result is in the spirit of [**Pr**].

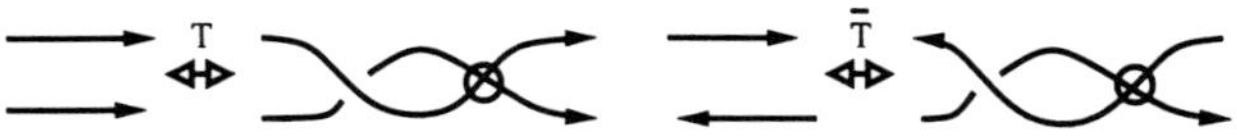

Figure 8: Moves T and $\bar{T}$

Proposition 4.4. Let l, l' be oriented virtual links, Σ a topological group and r a positive integer. If a diagram of l' can be obtained from a diagram of l by a finite sequence of T^r-, $\bar{T}^r$-moves and generalized Reidemeister moves, then $\mathrm{Col}_{\Sigma,r}(l')$ is homeomorphic to $\mathrm{Col}_{\Sigma,r}(l)$.

Proof. It suffices to consider the case that D' is obtained from D by a single T^r-move or a single $\bar{T}^r$-move. We will consider the first, and leave the second case, which is similar, to the reader.

If we have performed a T^r-move at a site of D, then any elements $C, C' \in \Sigma^{\mathbf{Z}}$ assigned to the left-most arcs determine unique labels for each of the new arcs that result from the move. (See Figure 9.) The top and bottom right-most arcs receive the labels C and $(\sigma^r C \cdot \sigma^{r-1} C \cdots \sigma C)^{-1} \cdot \sigma^r C' \cdot (\sigma^{r-1} C \cdot \sigma^{r-2} C \cdots C)$, respectively, which are C and C', in view of Definition 4.3. In this way the (Σ, r)-colorings of D correspond bijectively to those of D', and the result follows. ∎

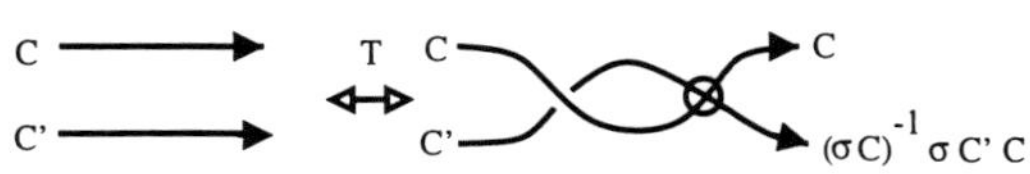

Figure 9: Effect of a T-move on (Σ, r)-colorings

We can also understand the effect of a T^r-move on the group of a virtual oriented link. Let x_0 and x_1 be the Wirtinger generators corresponding to the top and bottom arcs at the site, respectively, before the move is performed. Cut the diagram D along the bottom arc, obtaining a virtual tangle t. The generator x_1 is replaced by two generators x_1', x_1''.

The Wirtinger relations for $G(t)$ corresponding to crossings outside of the site must be rewritten using the appropriate choice of x_1', x_1''. Once this has been done, the T^r-move adds the relation $x_1'' = x_0^{-r} x_1 x_0^r$.

References.

[Ba] G. Baumslag, "Residually finite one-relator groups," Bull. Amer. Math. Soc. **73** (1967), 618 – 620.

[GoPoVi] M. Goussarov, M. Polyak and O. Viro, "Finite type invariants of classical and virtual knots," preprint: math.GT/1981/9810073.

[He] J. Hempel, "Residual finiteness for 3-manifolds," in Combinatorial Group Theory and Topology (Alta, Utah, 1984), *Annals of Math. Studies* **111** Princeton Univ. Press, Princeton, 1987.

[Ka1] L. H. Kauffman, Talks at MSRI Meeting in January 1997, AMS Meeting at University of Maryland, College Park in March 1997.

[Ka2] L. H. Kauffman, "Virtual knot theory," to appear in the European Journal of Combinatorics.

[Ka3] L. H. Kauffman, "An introduction to virtual knot theory," preprint.

[Kaw] A. Kawauchi, A Survey of Knot Theory, Birkhäuser Verlag, Berlin 1996.

[Le] J. Levine, "Some results on higher dimensional knot groups," in LNM 685: Knot Theory (Proc. Sem., Plans-sur-Bex, 1977), 243 –273, Springer-Verlag, Berlin 1978.

[MaKaSo] W. Magnus, A. Karrass and D. Solitar, Combinatorial Group Theory, Dover Publications, Inc., New York, 1976.

[Ne] L. Neuwirth, "Interpolating manifolds for knots in S^3," *Topology* **2** (1964), 359 – 365.

[Ph] D. Phillips, "Visualizing a torus 2-knot with no minimal Seifert manifold," Master's Thesis, Univ. of South Alabama, 1997.

[Pr] J. H. Przytycki, "t_k moves on links," *Contemp. Math.* **78** (1988), 615 – 656.

[Si] D. S. Silver, "Free group automorphisms and knotted tori in S^4," *J. Knot Theory and its Ramifications* **6** (1997), 95 – 103.

[SiWi1] D. S. Silver and S. G. Williams, "Generalized n-colorings of links," in Knot Theory, Banach Center Publications 42, Warsaw 1998, 381–394.

[SiWi2] D. S. Silver and S. G. Williams, "Coloring link diagrams with a continuous palette," *Topology*, to appear.

[SiWi3] D. S. Silver and S. G. Williams, "Augmented group systems and shifts of finite type," *Israel J. Math.* **95** (1996), 231 – 251.

[St] J. Stillwell, Classical Topology and Combinatorial Group Theory, Springer-Verlag, Berlin 1980.

[**Th**] W. Thurston, "Three dimensional manifolds, Kleinian groups and hyperbolic geometry," *Bull. Amer. Math.* **6** (1982), 357 – 381.

[**Yaj**] T. Yajima, "Wirtinger presentations of knot groups," *Proc. Japan Acad.* **46** (1970). 997 – 1000.

[**Yan**] T. Yanagawa, "Knot-groups of higher dimensional ribbon knots," *Kobe Math. Sem. Notes* **8** (1980), 573 – 591.

[**Ze**] E. C. Zeemann, "Twisting spun knots," *Trans. Amer. Math. Soc.* **115** (1965), 471 – 495.

Dept. of Mathematics and Statistics, University of South Alabama, Mobile, AL 36688

e-mail: silver@mathstat.usouthal.edu, williams@mathstat.usouthal.edu

A ZETA-FUNCTION FOR A KNOT USING $SL_2(\mathbb{F}_{p^s})$ REPRESENTATIONS

JEFFREY M. SINK
Telcordia Technologies
6 Corporate Place
Piscataway, NJ 08854

An new invariant of a knot, called the **zeta-function**, is introduced. It is defined by counting the non-diagonalizable representations of the fundamental group into $SL_2(\mathbb{F}_q)$, up to conjugacy, and forming a formal power series, similar to a zeta-function from algebraic geometry. The structure of this invariant in the general case is given. The special case of $(2, n)$ torus knots is studied, and basic topological properties are noted.

Keywords: Zeta-Function, Knot Theory, Casson Invariant, Representations

1. Introduction

The Casson-Walker invariant, $\lambda(M)$, which was developed by Casson, and is described in [1], was originally defined as an invariant only of integer homology spheres, then later generalized to rational homology spheres, by Walker [7]. It is usually described as "counting irreducible representations of $\pi_1(M)$ into $SU_2(\mathbb{C})$, up to conjugacy, with signs". More technically, it looks at the representation spaces of the two handlebodies in a handlebody decomposition, modded out by conjugacy, and sees how they intersect. It uses a intersection pairing on this space to arrive at weights or "signs" for the intersection points. It does not really count representations, but we can say that if $\lambda(M) \neq 0$, then there exists a non-trivial representation.

We attempted to develop a characteristic-p Casson invariant. To accomplish this, we first needed to know which of the properties of the Casson invariant we would like to preserve in going to characteristic-p. We will count representations into $SL_2(\mathbb{F}_q)$, where $\mathbb{F}_q$ is a finite field. We chose this group because the finite field version of SU_2 (where complex conjugation is replaced by conjugacy in a double cover by $\mathbb{F}_{q^2}$) does not have the same nice properties as $SU_2(\mathbb{C})$. And the Lie algebras su_2 and sl_2 are pretty much the same.

Next, we look at the representation spaces, this time represented by varieties and not topological spaces as in the Casson case, there is no nice intersection pairing, since these spaces are not projective varieties. So we instead simply count the representations into $SL_2(\mathbb{F}_q)$. Of course, that will be a very poor invariant, so we then fix a prime p, and count representations into all $SL_2(\mathbb{F}_{p^s})$. As in the Casson invariant, we will count our representations up to conjugacy, but without signs. We will also eliminate the diagonalizable representations. This corresponds to counting only irreducible representations in the original $SU_2(\mathbb{C})$ case. We must take care to notice that the ideas of irreducibility and diagonalization are separate and distinct when we have representations on a vector field over a finite field. Simply put, because the field is not algebraically closed we do not have eigenvalues of all the matrices in the field. In the literature, these are known as "modular representations".

We put these together into a formal power series, and we see connections with algebraic geometry related to the Weil conjectures and zeta-functions. Roughly speaking, if we have a system of equations with coefficients in a finite field $\mathbb{F}_q$, then the numbers, N_s of solutions to the system of over $\mathbb{F}_{q^s}$ form a nice formal series, called the zeta-function:

$$Z(q,T) = \exp\left(\sum_{s=1}^{\infty} \frac{N_s T^s}{s}\right).$$

The first Weil conjecture claims that this is a rational function of T. The conjectures then require stronger assumptions on the system of equation, but give detailed information on the number of equations. We will only need the portion of the Weil conjectures proved by Dwork [2] in 1959. We do not have a system of homogeneous equations, so therefore we do not have projective varieties.

In counting our representation, up to conjugacy, into $SL_2(\mathbb{F}_{p^s})$, we reduce the problem to one of counting the number of solutions to a system of equations. The problem is that the coefficients of the system depend on the field $\mathbb{F}_{p^s}$. Further, we do not want all these solutions, as some of them double count conjugacy classes. We then define the numbers N_s as counting representations, up to conjugacy, into $SL_2(\mathbb{F}_{p^s})$. We form the formal generating series of the zeta-function, and call it the *zeta-function of a knot.* We then prove Theorem 2:

Theorem *The ζ-function of a knot $Z(q,t)$ is the square root of a rational function. The rational function is a ratio of polynomials with coefficients in the integers.*

Let $\lambda(p,T)$ be the modified zeta-function obtained by counting only non-diagonal representations (Definition 2). For a $(n,2)$ torus knot, we further prove Theorem 3:

Theorem *Given $\mathcal{K}$ is a $(n,2)$ torus knot, n an odd integer, and a prime number p, $p \equiv 1$ (4), then $\lambda_{\mathcal{K}}{}^{-2}(p,T)$ is a polynomial. Furthermore, this polynomial is a product of cyclotomic polynomials, and has the property that it is invariant under the transformation $T \mapsto T^{-1}$.*

It is interesting to note that the Alexander polynomials of these knots are also cyclotomic polynomials, but not the same ones.

We also study how the invariant transforms under topological moves, namely Theorem 4:

Theorem *Let $\mathcal{K}_1$ be a $(n,2)$ torus knot, and let $\mathcal{K}_2$ be an $(m,2)$ torus knot $(n,m$*

odd integers). The connected sum of $\mathcal{K}_1$ with $\mathcal{K}_2$, $\mathcal{K}_1 \sharp \mathcal{K}_2$, has the property that $\lambda_{\mathcal{K}_1 \sharp \mathcal{K}_2}{}^{-2}(p,T)$ is a polynomial over the integers. Further, this polynomial is a product of cyclotomic polynomials.

In all cases, these rational functions, when expanded as a formal power series in T, are of the form $Z(p,T) = \sum_n a_n T^n$, where $a_0 = 1$ and all the $a_n \in \mathbb{Z}$. In this way it looks a lot like the Ohtsuki-Murakami invariant (though it is distinct).

2. Algebraic and Topological Background

So as stated, we will be counting the representations of the fundamental group G of the knot group into $SL_2(\mathbb{F}_q)$, counted up to conjugacy. That means that two such representations ϕ, θ are equivalent if the whole representation $\phi(g) = S\theta(g)S^{-1}$ for a fixed $S \in SL_2(\mathbb{F}_q)$ and for all $g \in G$.

In practice how do we do this? Well, first we get a presentation for the knot group. For example, the Wirtinger presentation is easily attained, but very rarely would this be the ideal choice. Next one must develop a set of conjugacy class representatives. Such a list is given below, see Table 1. What we will need these for is as follows: We choose a certain element of the knot group, usually a generator of the group, and we can then say that any representation of the knot group, possibly after acting on it by conjugation, sends this specified element of the knot group to one of the conjugacy class representatives. Then the problem of counting representations (up to conjugacy) comes down to finding all the matrices which the other generators can be sent to such that their images still satisfy the group relations. But for a fixed field, it is easily seen that this gives a system of equations with coefficients over that field, and with variables taking values in that field.

To be more explicit, take any presentation of the knot group, with s generators and t relations. Of course, the knot group has deficiency 1, so we can always find a presentation with $s = t + 1$, but we will not need that here. Let the presentation be written as:

$$G = < X_1, X_2, \ldots, X_s | r_1, r_2, \ldots, r_t > \tag{2.1}$$

Without loss of generality assume that X_1 gets sent to a conjugacy class representative. To be more rigorous we must consider each conjugacy class separately. By that we mean, that the total number of representations (up to conjugacy) of the knot group is simply equal to the sum of the number of representation in which X_1 goes to each of the conjugacy class representatives. Pick any conjugacy class representative from the list. Let X_1 be sent to that conjugacy class representative, which is represented by a matrix with elements in the field. Now to find the number of representations, we first have to send each of the other X_i into $SL_2(\mathbb{F}_q)$, so for each generator X_i, we are picking four elements of the field $\mathbb{F}_q$, namely a_i, b_i, c_i, d_i, so that $a_i d_i - b_i c_i = 1$. This gives us a total of $s - 1$ equations to solve. Then at the same time, one must insure that these matrices satisfy the required relations r_j. For each r_j, it is seen that one must multiply the corresponding matrices together, set it equal to the identity matrix I, and that gives a system of 4 equations for each relation r_j. So in total there are $s-1+4r$ equations to solve, and the degree of the equations involved is related to the length of the words in each of the relations r_j.

So we have reduced the problem of counting representations down to the problem of counting solutions to systems of polynomial equations over finite fields, with coefficients over that same field. This, of course, is a very classical problem, but unfortunately, a lot of the machinery that often comes to help in situations like this does not help us here, since our equations are not homogeneous (because $ad-bc=1$ is certainly not homogeneous). But we still have some tools available, and we will study.

3. Defining the Zeta-Function and the Weil conjectures

A better way of encoding all the information coming from counting representations into all the groups $SL_2(\mathbb{F}_{q^i})$ is seen to come from algebraic geometry. The classical problem in algebraic geometry is to find the number of solutions of a system of equations over a finite field. Well, actually, the *really* classical case is solving the system over the complex, reals, or rationals, but close behind that is the finite field case.

If we have a system of polynomial equations, in several variables, with coefficients in a fixed finite field $\mathbb{F}_q$, then we look at the solutions of this system over $\mathbb{F}_q$, $\mathbb{F}_{q^2}$,$\mathbb{F}_{q^3}$, and so on. Notice that it makes perfect sense to look at solutions over these other fields, since the coefficients are elements of $\mathbb{F}_q \subset \mathbb{F}_{q^s}$, so the equations can be considered as a set of equations over any one of these fields. We define a formal power series $Z(q,T)$ as follows:

$$Z(q,T) = \exp(\sum_{s=1}^{\infty} \frac{N_s T^s}{s}), \text{ where } N_s = \#\{\text{solutions over } \mathbb{F}_{q^s}\} \tag{3.2}$$

In 1949, Weil announced a series of conjectures for the nature of these functions $Z(q,T)$. For our purposes, we will involve ourselves only in the two easiest conjectures.

Theorem 1. *For any system of equations defined over a finite field $\mathbb{F}_q$, and for the $Z(q,T)$ defined in (3.2), then we can say that $Z(q,T) \in \mathbb{Z}[[T]]$ and that $Z(q,T)$ is the ratio of two polynomials, each of which is a polynomial over the integers.*

Proof. The proof involves p-adic analysis and was obtained by Dwork in 1959. The proof will be omitted. See [5] for a complete proof. □

Furthermore, if we make more assumptions on the structure of the system of equations, namely that the system is projective and non-singular, then we can say much more about the structure of the polynomials in the numerator and denominator of the function $Z(q,T)$. We will not go into these cases, since the hypotheses do not seem to apply for our calculations, except to mention that some of the conclusions of these stronger results seem to follow even in our setting, where they are unexpected.

4. Definition of the ζ-Function of a Knot

So how does this apply to us? We will now define an invariant of a knot in exactly the same way.

Definition 1 (The ζ-function of a knot). *Let $\mathcal{K}$ be a knot in 3-space. Fix a prime p. Let N_s be the number of representations of the knot group into $SL_2(\mathbb{F}_{p^s})$, up to conjugacy. Then the ζ-function is simply the formal power series*

$$Z_{\mathcal{K}}(\mathbb{F}_p, T) = \exp(\sum_{s=1}^{\infty} \frac{N_s T^s}{s}). \tag{4.3}$$

So what does this ζ-function have to do with the other ζ-function coming from solving systems of equations? Well, based on our previous arguments, we see that the case where N_s is the numbers of representations is very close to the case where N_s is the number of solutions of a system of equations over $\mathbb{F}_{p^s}$.

In particular, we noted before that the number of representation into a certain conjugacy class can be seen to give rise to a system of equations. So there are $p^s + 4$ conjugacy classes in $SL_2(\mathbb{F}_{p^s})$, and for each one of them we get a system of equations over $\mathbb{F}_{p^s}$. So if N_s is the total number of representations into $SL_2(\mathbb{F}_{p^s})$, then we break it down as being the representations that have the first generator conjugate to the identity, plus the representations that have the first generator conjugate to negative the identity, and so forth. Let N_s^i, $i = 1, \ldots, p^s + 4$ be the number of representations of the knot group into $SL_2(\mathbb{F}_{p^s})$, such that the first generator of the knot group is sent to the i-th conjugacy class representative. So then we can see:

$$N_s = N_s^1 + N_s^2 + N_s^3 + \cdots + N_s^{p^s+1} \tag{4.4}$$

But notice that the ζ-function is multiplicative in the following sense:

$$\begin{aligned} Z(p,T) &= \exp(\sum_{s=1}^{\infty} N_s \frac{T^s}{s}) = \exp(\sum_{s=1}^{\infty} \sum_i N_s^i \frac{T^s}{s}) \\ &= \exp(\sum_i \sum_{s=1}^{\infty} N_s^i \frac{T^s}{s}) = \prod_i \exp(\sum_{s=1}^{\infty} N_s^i \frac{T^s}{s}) = \prod_i Z_i(\mathbb{F}_p, T) \end{aligned} \tag{4.5}$$

Now we only have to calculate the N_s^i for each i, find the ζ-function, then multiply them all together. But as noted before, the N_s^i are calculated as the number of solutions of a system of equations over $\mathbb{F}_{p^s}$. In fact, the calculations are actually a little easier if the conjugacy classes are groups together into 8 types, as given in Table 1. The ζ-function is then the product of the 8 ζ-functions corresponding to each type.

But now we should be thinking that the Weil Conjectures (Dwork's part, Theorem 1) apply. The problem is that the Weil conjectures are based on the problem where the coefficients are sitting in one fixed base field, and we are looking at all possible solutions over covering fields, by considering these coefficients as sitting in these covering fields. But we do not have that in our case, since we have a system of equations for each finite field $\mathbb{F}_{p^s}$, where the coefficients of that system of equations come from the conjugacy class representative, and that has elements in $\mathbb{F}_{p^s}$, not just in the one base field $\mathbb{F}_p$. Further, we see from the table of representatives (Table 1), that the conjugacy classes representatives depend on the choice of an element $\epsilon \in \mathbb{F}_{p^s}$, such that $\sqrt{\epsilon} \notin \mathbb{F}_{p^s}$. So in this sense, the equations we solve to count representations into the SL_2 over that field do indeed depend on the field itself, so strictly the Weil Conjectures do not seem to apply to each of these systems. But still we will make the further claim:

Theorem 2. *The ζ-function $Z(p,T)$ of a knot is the square root of a rational function. The rational function is a ratio of polynomials with coefficients in the integers.*

Before beginning the proof we must establish a few basic facts about these types of functions.

Lemma 1. *If we have a system of equations defined over a finite field $\mathbb{F}_q$, and for the $Z(q,T)$ defined in Definition 3.2, then*

$$Z(q,T) = 1 + a_1 T + a_2 T^2, \ldots \tag{4.6}$$

where the a_i are all positive integers. Note that a_1 is just the number of solutions over $\mathbb{F}_q$.

Proof. This easy proof can be seen in [5], and will be omitted □

Lemma 2. *If we have a system of equations defined over a finite field $\mathbb{F}_q$, and for the $Z(q,T)$ just defined in Definition 3.2, then $Z(q,T)$ is an analytic function on a domain which includes the origin, when we consider the function as a complex function of the complex variable T.*

Proof. Notice that if we are dealing with a system of polynomials in k variable, over a field q^s, then there are *at most* q^{ks} different solutions to the system. So then $N_s \leq q^{ks}$. But then the coefficients a_i given in Lemma 1, are increased by replacing every N_s by q^{ks}, since we get the a_i from the N_s only by multiplying and adding positive numbers. But after we replaced all the N_s by q^{ks} we get:

$$\exp\left(\sum_{s=0}^{\infty} q^{ks} T^s / s\right) = \exp(-\log(1 - q^k T)) = \frac{1}{1 - q^k T} = \sum_{j=0}^{\infty} q^{kj} T^j \tag{4.7}$$

But this function converges for complex T with $|T| < q^{-k}$. And our $Z(T)$ is a power series whose coefficients are less than these in absolute value (actually they are both positive), so $Z(T)$ also converges in this region, by a comparison test. □

Proof. (Of Theorem 2) We will have to do this case by case, based on the set of conjugacy class representatives seen below. See Table 1. We will show that the whole ζ-function is rational, by showing that our ζ-function is really the product or quotient of ζ-functions to which Dwork-Weil applies, thereby showing it is rational itself by showing it is the quotient or product of rational functions.

Recall that the reason that Dwork-Weil does not already apply is that we get a system of equations for each conjugacy class, with the coefficients of the system coming from the elements of the conjugacy class representative used. In particular, we have the situation that the conjugacy class representatives have elements which do not lie in the base field, but depend on the field $\mathbb{F}_{p^s}$.

In the following, let X be the element in the knot group which we will use to determine the conjugacy class of the representation. So we assume that X gets sent to a conjugacy class representative.

(i) If X gets sent to the identity element I, then Dwork-Weil applies, since the elements of this matrix are constants in the base field $\mathbb{F}_p$, no matter what field we are counting representations over.
(ii) If X gets sent to the negative of the identity element, $-I$, then Dwork-Weil applies, since the elements of this conjugacy class representative are constants in the base field $\mathbb{F}_p$, no matter what field we are counting representations over.
(iii) If $X \mapsto \left(\begin{smallmatrix}1 & 1\\ 0 & 1\end{smallmatrix}\right)$, then again Dwork-Weil applies, as above.
(iv) If $X \mapsto \left(\begin{smallmatrix}1 & \epsilon\\ 0 & 1\end{smallmatrix}\right)$ we get our first interesting case to deal with. We have to show that the ζ-function from this case is rational. But the coefficients depend on the element ϵ, which must be chosen for each field we want to solve the equations over.
We handle this by considering ϵ not as a fixed element that must be chosen each time, but rather as a variable, just like all the others in the problem. We denote by $\mathcal{P}_\epsilon(x_1, x_2, \ldots, x_n)$ this system of equations in n variables, with a parameter ϵ. Denote the number of solutions of this system over $\mathbb{F}_{p^s}$ by N_s^ϵ. Also let $\mathcal{P}(\epsilon, x_1, x_2, \ldots, x_n)$ be the same system in $n+1$ variables, where the ϵ is considered as a variable and not as a parameter. Some more useful notation:

$$\begin{aligned} N_s^\epsilon &= \#\{\text{solutions to } \mathcal{P}_\epsilon(x_1, x_2, \ldots, x_n) \text{ over } \mathbb{F}_{p^s}\} \\ N_s &= \#\{\text{solutions to } \mathcal{P}(\epsilon, x_1, x_2, \ldots, x_n) \text{ over } \mathbb{F}_{p^s}\} \\ N_s^{\text{res}} &= \#\{\text{solutions to } \mathcal{P}(\epsilon, x_1, x_2, \ldots, x_n) \text{ over } \mathbb{F}_{p^s}, \sqrt{\epsilon} \in \mathbb{F}_{p^s}\} \\ N_s^{\text{non}} &= \#\{\text{solutions to } \mathcal{P}(\epsilon, x_1, x_2, \ldots, x_n) \text{ over } \mathbb{F}_{p^s}, \sqrt{\epsilon} \notin \mathbb{F}_{p^s}\}. \end{aligned} \tag{4.8}$$

The crucial fact to observe is that for our system, the number N_s^ϵ is the same no matter what value ϵ assumes, as long as ϵ does not have its square root in the field (which we refer to as quadratic non-residue). This fact comes from the algebraic fact that the number of representations of the knot group into a conjugacy class cannot depend on the conjugacy class representative chosen to count them. And all ϵ's in the field generate the same conjugacy class in this way. So the number of solutions of this system does not depend on which

ϵ we choose. So in particular, one can see:

$$N_s^{\text{non}} = \frac{p^s+1}{2} N_s^\epsilon, \ \forall \epsilon, \ \sqrt{\epsilon} \notin \mathbb{F}_{p^s}. \tag{4.9}$$

The factor in front comes from the fact that in a field with q elements, $(q+1)/2$ of the elements have square roots.
Now we have to look at N_s^{non}. Clearly we have $N_s = N_s^{\text{res}} + N_s^{\text{non}}$. Dwork-Weil applies to a ζ-function made from the N_s, since it is now a system in $n+1$ variables with coefficients in the base field $\mathbb{F}_p$. We need to show that N_s^{res} gives a rational ζ-function. Notice that N_s^{res} is the number of solutions of the system obtained by taking $\mathcal{P}(\epsilon, x_1, x_2, \ldots, x_n)$ and adjoining the single new equation $w^2 = \epsilon$, where w is a new variable. This insures that any ϵ will have a square-root in the field, and notice that Dwork-Weil applies to this new system. We do have to divide out for the two different values that the new variable w can assume, namely $w = \pm\sqrt{\epsilon}$. These values are different for all values but $\epsilon = 0$. To handle that case we take the system $\mathcal{P}(\epsilon, x_1, x_2, \ldots, x_n)$ and simply adjoin the equation $\epsilon = 0$. Let $N_s^{w^2=\epsilon}$ be the number of solutions of the first system. Let $N_s^{\epsilon=0}$ be the number of solutions of the second system made. Then one can see:

$$\begin{aligned} N_s^{\text{res}} &= \frac{N_s^{w^2=\epsilon} - N_s^{\epsilon=0}}{2} + N_s^{\epsilon=0} \\ &= \frac{N_s^{w^2=\epsilon} + N_s^{\epsilon=0}}{2}. \end{aligned} \tag{4.10}$$

So we were trying to figure out the ζ-function formed by the N_s^ϵ, for a non-residue ϵ. We have seen:

$$\begin{aligned} N_s^{\text{non}} &= \frac{p^s+1}{2} N_s^\epsilon \\ N_s &= N_s^{\text{res}} + N_s^{\text{non}} \\ N_s^{\text{res}} &= \frac{N_s^{w^2=\epsilon} + N_s^{\epsilon=0}}{2} \end{aligned}$$

Combining these we get:

$$N_s = N_s^{\text{res}} + \frac{p^s+1}{2} N_s^\epsilon$$

$$N_s = \left[\frac{N_s^{w^2=\epsilon} + N_s^{\epsilon=0}}{2}\right] + \frac{p^s+1}{2} N_s^\epsilon \tag{4.11}$$

Note that Dwork-Weil applies to each of the quantities: N_s, $N_s^{w^2=\epsilon}$, $N_s^{\epsilon=0}$, since they each made the parameter into an ordinary variable, possibly with the addition of another equation.

By a similar symmetry argument, we can show the ζ-function corresponding to $\frac{N_s^{w^2=\epsilon}+N_s^{\epsilon=0}}{2}$ is itself a rational function. Basically this argument goes that when ϵ has a square-root in the field, then the conjugacy class we are counting into is either given by the identity matrix, I, or the matrix $\left(\begin{smallmatrix}1&1\\0&1\end{smallmatrix}\right)$. In either case the number of solutions corresponds to the number of representations into these conjugacy classes, and must be invariant. So then the ζ-function corresponding to $N_s^{\text{res}} = \frac{N_s^{w^2=\epsilon}+N_s^{\epsilon=0}}{2}$ is a rational function.
Denote by Z_1 and Z_2 the ζ-functions corresponding to N_s and N_s^{res} respectively.
Multiplying Equation 4.11 by 2 we get:

$$2N_s = N_s^{\text{res}} + (p^s+1)N_s^{\epsilon} \tag{4.12}$$

Forming the ζ-function we see:

$$\begin{aligned}
Z(\mathcal{P}(x_1,x_2,\ldots,x_n),T) &= \exp\left(\sum_{s=1}^{\infty} N_s\frac{T^s}{s}\right) = \exp\left(\sum_{s=1}^{\infty} N_s^{\text{res}} + (p^s+1)N_s^{\epsilon}\frac{T^s}{s}\right)\\
&= \exp\left(\sum_{s=1}^{\infty}[N_s^{\text{res}}+N_s^{\epsilon}]\frac{T^s}{s}\right)\exp\left(\sum_{s=1}^{\infty}p^sN_s^{\epsilon}\frac{T^s}{s}\right)\\
Z(\mathcal{P}(x_1,x_2,\ldots,x_n),T) &= \exp\left(\sum_{s=1}^{\infty}[N_s^{\text{res}}+N_s^{\epsilon}]\frac{T^s}{s}\right)\exp\left(\sum_{s=1}^{\infty}N_s^{\epsilon}\frac{(pT)^s}{s}\right)\\
Z(T) &= Z^{\text{res}}(T)Z^{\epsilon}(T)Z^{\epsilon}(pT)
\end{aligned} \tag{4.13}$$

We know that $Z(T)$ and $Z^{\text{res}}(T)$ are both rational functions, because of the above discussion. So $Z^{\epsilon}(T)Z^{\epsilon}(pT)$ is a rational function. We need to show $Z^{\epsilon}(T)$ itself is a rational function. But this is obtained when we notice that we have already shown that $Z^{\epsilon}(T)$ is a holomorphic function in some neighborhood of the origin in the complex plane. By looking at where its roots and poles are, we can see that $Z^{\epsilon}(T)$ itself is rational. And so we have shown the ζ-function coming from the case where $X \mapsto \left(\begin{smallmatrix}1&\epsilon\\0&1\end{smallmatrix}\right)$ is a rational function.

(v) If $X \mapsto \left(\begin{smallmatrix}-1&1\\0&-1\end{smallmatrix}\right)$, then again Dwork-Weil applies directly, as before.

(vi) $X \mapsto \left(\begin{smallmatrix}-1&\epsilon\\0&-1\end{smallmatrix}\right)$ This case is EXACTLY like the case with $X \mapsto \left(\begin{smallmatrix}1&\epsilon\\0&1\end{smallmatrix}\right)$, and the arguments go through with out any changes whatsoever.

(vii) $X \mapsto \left(\begin{smallmatrix}x&0\\0&x^{-1}\end{smallmatrix}\right)$, for some $x \in \mathbb{F}_{p^s}$, $x \neq \pm 1$, since those cases were handled above.

Well, actually to prove this we will throw the other cases back in and let $x = \pm 1$. In fact, instead of looking at the number of solutions of this system of equations for each invertible x, we will look at x itself as being a variable that can range (as we did with the ϵ above). We will define corresponding notation for this part of the proof. Let $\mathcal{P}_x(x_1,x_2,\ldots,x_n)$ be the system of equations given by trying to count the representations with $X \mapsto \left(\begin{smallmatrix}x&0\\0&x^{-1}\end{smallmatrix}\right)$, with a fixed $x \neq \pm 1$, after first multiplying through sufficient powers of x to get rid of any powers of x^{-1}. Let $\mathcal{P}(x, x_1, x_2, \ldots, x_n)$ be the same system, except leave x as a variable able to range over the same field as all the other x_i. Let the corresponding number of solutions of these equations as N_s^x and N_s. Notice that the numbers we are interested in are the N_s^x, except now since we have eliminated the powers of x^{-1}, we have introduced possible solutions for $x = 0$. We will handle these soon:

$$\begin{aligned} N_s^x &= \#\{\text{solution to } \mathcal{P}_x(x_1, x_2, \ldots, x_n) \text{ over } \mathbb{F}_{p^s}, x \text{ fixed}\} \\ N_s &= \#\{\text{solution to } \mathcal{P}(\epsilon, x_1, x_2, \ldots, x_n) \text{ over } \mathbb{F}_{p^s}\} \\ N_s^{x=\pm 1} &= \#\{\text{ solution to } \mathcal{P}(x, x_1, x_2, \ldots, x_n) \text{ over } \mathbb{F}_{p^s}, x^2 = 1\} \\ N_s^{x=0} &= \#\{\text{solution to } \mathcal{P}(x, x_1, x_2, \ldots, x_n) \text{ over } \mathbb{F}_{p^s}, x = 0\} \end{aligned} \tag{4.14}$$

As mentioned before, we will handle all the possible values of x at the same time. It is clear to see that $N_s = \sum_{x \in \mathbb{F}_q} N_s^x$. Further, notice that $N_s^0 = N_s^{x=0}$, and so the ζ-function corresponding to the N_s^0 is a rational function, since it is defined by the equations over the base field $\mathbb{F}_p$. Further still, since $N_s^{x=\pm 1} = N_s^{x=1} + N_s^{x=-1}$, the ζ-function corresponding to $N_s^{x=\pm 1}$ is just the product of the ζ-function for $N_s^{x=1}$ and $N_s^{x=-1}$. Both of these ζ-functions are rational, since Dwork-Weil applies directly. They were also seen as so in the above cases.

So then, we have the relation: $N_s \sum_{x \in \mathbb{F}_q} N_s^x$, and the quantity we are looking for is $N_s^{\text{total}} = \frac{N_s - N_s^{x=0} - N_s^{x=\pm 1}}{2}$. This number is the total number of representations into conjugacy classes of type $\left(\begin{smallmatrix} x & 0 \\ 0 & x^{-1} \end{smallmatrix}\right)$. The 1/2 factor comes from the fact that the number of solutions $N_s^x = N_s^{x^{-1}}$, and we are really only interested in one of these cases. This comes from the topological fact that we are counting representations, and x and x^{-1} generate the same conjugacy class, so the number of solution must be the same.

But notice that the N_s,$N_s^{x=0}$, and $N_s^{x=\pm 1}$ both generate ζ-functions which are rational functions, since the Dwork-Weil Theorem applies directly. Then it is easy to see that the ζ-function formed by:

$$N_s^{\text{total}} = \frac{N_s - N_s^{x=0} - N_s^{x=\pm 1}}{2} \tag{4.15}$$

is just the square-root of a rational function. Namely, if Z_1, Z_2, Z_3 are the ζ-functions of N_s,$N_s^{x=0}$, and $N_s^{x=\pm 1}$, respectively, then the ζ-function of N_s^{total} is then:

$$Z(T) = \sqrt{\frac{Z_1(T)}{Z_2(T) Z_3(T)}} \tag{4.16}$$

So in particular the product of the ζ-functions of all the N_s^x is the square root of a rational function. It is interesting that we showed that the product is rational, without having to showing that each N_s^x gives a rational function.

It is really only the product we are after, since the ζ-function of the knot $\mathcal{K}$ is just the product of all these functions.

(viii) $X \mapsto \begin{pmatrix} x & y \\ \epsilon y & x \end{pmatrix}$. This case is very similar to the previous two difficult cases. It is accomplished by considering x, y, and ϵ as variables, and doing the same sort of tricks to control for the fact that (x, y) gives the same class as $(x, -y)$, and the case $y = 0$, which was already handled. The proof will be omitted.

□

5. Counting the Representations of the $T_{n,2}$ Torus Knot Group

We will be counting the representations of the fundamental group G of the $(n, 2)$ torus knot in $SL_2(\mathbb{F}_q)$, counted up to conjugacy. That means that two such representations ϕ, θ are equivalent if $\phi(g) = S\theta(g)S^{-1}$ for a fixed $S \in SL_2(\mathbb{F}_q)$ and for all $g \in G$. We get a Wirtinger presentation for this knot group rather easily and a quick change of generators of the group gives the friendly presentation $< X, Y | X^2 = Y^n >$. Now for the group that our representations will be going into, we basically have to find the conjugacy classes in $SL_2(\mathbb{F}_q)$, and these are well known (see [4]), and are listed in Table 1. The element ϵ is an element of $\mathbb{F}_q$ such that $\sqrt{\epsilon} \notin \mathbb{F}_q$.

In this table we have certain redundancies; namely in Case 7, the matrices $\begin{pmatrix} x & 0 \\ 0 & x^{-1} \end{pmatrix}$ and $\begin{pmatrix} x^{-1} & 0 \\ 0 & x \end{pmatrix}$ are in the same conjugacy class. Also, in that same case 7, we do not allow $x = 1$ or $x = -1$, since they are covered in Cases 1 and 2 respectively. Also, in Case 8, we notice that $\begin{pmatrix} x & y \\ \epsilon y & x \end{pmatrix}$ gives the same conjugacy class as $\begin{pmatrix} x & -y \\ -\epsilon y & x \end{pmatrix}$, and further, we do not allow $y = 0$, since then that reduces to Cases 1 or 2.

The methodology goes as follows: the generators of G get sent somewhere into $SL_2(\mathbb{F}_q)$, and we will assume, after conjugation, that Y, the generator which is raise to the n-th power, is sent to one of the representatives of the conjugacy classes. Call this representative B. We then need only find out how many $A \in SL_2(\mathbb{F}_q)$ satisfy $A^n = B^2$.

After some elementary calculations, we derive Table 2 for the number of representations, up to conjugacy, into $SL_2(\mathbb{F}_q)$.

6. Finding the ζ-Functions

What we are going to do now is use the previously discovered information on the counting of the representations to try to evaluate the following formal expression:

$$Z(p, T) = \exp\left(\sum_{s=1}^{\infty} \frac{N_s T^s}{s}\right) \tag{6.17}$$

Where, T is a formal indeterminate, and N_s is the number of representations into $SL_2(\mathbb{F}_{p^s})$. We will keep p fixed, once and for all.

Table 1. Table of Conjugacy Classes in sl_2

Case no.	Representative	No. of Elts in Class	No. Classes
1	$I = \begin{pmatrix} 1 & 0 \\ 0 & 1 \end{pmatrix}$	1	1
2	$-I = \begin{pmatrix} -1 & 0 \\ 0 & -1 \end{pmatrix}$	1	1
3	$\begin{pmatrix} 1 & 1 \\ 0 & 1 \end{pmatrix}$	$\frac{q^2-1}{2}$	1
4	$\begin{pmatrix} 1 & \epsilon \\ 0 & 1 \end{pmatrix}$	$\frac{q^2-1}{2}$	1
5	$\begin{pmatrix} -1 & 1 \\ 0 & -1 \end{pmatrix}$	$\frac{q^2-1}{2}$	1
6	$\begin{pmatrix} -1 & \epsilon \\ 0 & -1 \end{pmatrix}$	$\frac{q^2-1}{2}$	1
7	$\begin{pmatrix} x & 0 \\ 0 & x^{-1} \end{pmatrix}$	$q(q+1)$	$\frac{q-3}{2}$
8	$\begin{pmatrix} x & y \\ \epsilon y & x \end{pmatrix}$	$q(q-1)$	$\frac{q-1}{2}$

Now notice that if N_s^i is the number of representations that come from Case i above then: $N_s = N_s^1 + N_s^2 + N_s^3 + \cdots + N_s^8$:

$$\begin{aligned} Z(p,T) &= \exp(\sum_{s=1}^{\infty} N_s \frac{T^s}{s}) = \exp(\sum_{s=1}^{\infty}\sum_i N_s^i \frac{T^s}{s}) \\ &= \exp(\sum_i \sum_{s=1}^{\infty} N_s^i \frac{T^s}{s}) = \prod_i \exp(\sum_{s=1}^{\infty} N_s^i \frac{T^s}{s}) \\ &= \prod_i Z_i(\mathbb{F}_p, T) \end{aligned} \tag{6.18}$$

Where $Z_i = \exp(\sum_{s=1}^{\infty} N_s^i \frac{T^s}{s})$, so basically we have a situation where we can just find the ζ for individual cases, then multiply the answers together when all done.

6.0.1. *Case 1*

Recall that in this situation we always had 2 representations, no matter what

Table 2. Number of Representations into each Conjugacy Class in $SL_2(\mathbb{F}_q)$

Case no.	Number of Representations
1	2
2	$\begin{cases} q(q+1) & q \equiv 1\ (4) \\ q(q-1) & q \equiv 3\ (4) \end{cases}$
3	2
4	2
5	$\begin{cases} 1 & p \mid n \\ \text{Case 2} & p \nmid n \end{cases}$
6	$\begin{cases} 1 & p \mid n \\ \text{Case 2} & p \nmid n \end{cases}$
7	$\begin{cases} \frac{(n,q-1)-1}{2} q(q+1) + \frac{q-1}{2} - (n,q-1) - 1 & \text{If } q \equiv 1\ (4) \\ \frac{(n,q-1)-1}{2} q(q-1) + q - 3 & \text{If } q \equiv 3\ (4) \end{cases}$
8	$\frac{p^s+1}{2} - 1 - \frac{1}{2}(1 + (-1)^{\frac{p^s-1}{2}})$

$q = p^s$ was. So we can say $N_s^1 = 2$, so

$$\begin{aligned} Z_1 &= \exp(\sum_{s=1}^{\infty} N_s^1 \frac{T^s}{s}) = \exp(\sum_{s=1}^{\infty} 2\frac{T^s}{s}) \\ &= \exp(2\sum_{s=1}^{\infty} \frac{T^s}{s}) = \exp(-2\log(1-T)) = \frac{1}{(1-T)^2} \end{aligned} \tag{6.19}$$

Notice that in this case this Z_1 is exactly the ζ-function (in the sense of the Weil Conjectures and the work of Deligne) of the system of equations:

$$\begin{matrix} a(a+d) = 2 & b(a+d) = 0 \\ c(a+d) = 0 & d(a+d) = 2 \end{matrix} \tag{6.20}$$

The other calculations follow similarly, except Case 7, which looks ugly in Table 2, but also works out.

So then after calculating the ζ-function in each individual case, we multiply them all together to get the ζ-function for the total number of representations of the $(2, n)$ torus knots into $SL_2(\mathbb{F}_q)$, where n is odd and q is not a power of 2. The results are given in Chart 6.21.

$$
\begin{aligned}
Z_1 &= \frac{1}{(1-T)^2}\\
Z_2 &= \frac{1}{(1-p^2T)(1-pT)}\\
Z_3 &= \frac{1}{(1-T)^2}\\
Z_4 &= \frac{1}{(1-T)^2}\\
Z_5 &= \begin{cases} 1 & \text{If } p \not| n \\ \frac{1}{(1-p^2T)(1+(-1)^{\frac{p+1}{2}}pT)} & \text{If } p|n \end{cases}\\
Z_6 &= \begin{cases} 1 & \text{If } p \not| n \\ \frac{1}{(1-p^2T)(1+(-1)^{\frac{p+1}{2}}pT)} & \text{If } p|n \end{cases}\\
Z_7 &= \sqrt{\frac{1-T}{1-pT}}\frac{\left(\prod_{i|\phi(n)}(T^i-1)^{F_2(i)/i}\right)}{\sqrt{\left(\prod_{i|\phi(n)}((pT)^i-1)^{F_1(i)/i}((p^2T)^i-1)^{F_1(i)/i}\right)}}\\
Z_8 &= \frac{(1-(-1)^{\frac{p-1}{2}}T)^{1/2}(1-T)}{\sqrt{(1-pT)}}
\end{aligned}
\tag{6.21}
$$

Where,

$$
F_1(a) = \sum_{d|a}\mu(d)(\gcd(p^{a/d}-1,n)-1) = \sum_{d|a}\mu(d)\gcd(p^{a/d}-1,n) - \sum_{d|a}\mu(d)
$$

$$
F_1(a) = \sum_{d|a}\mu(d)\gcd(p^{a/d}-1,n) - \begin{cases} 1 & a=1 \\ 0 & a>1 \end{cases}, \tag{6.22}
$$

and $\phi(n)$ is the Euler ϕ-function. Note that $a|F_1(a)$ and $a|F_2(a)$ for all a, and $F_1(a), F_2(a)$ are both non-negative integers and

$$
\begin{aligned}
\gcd(p^a-1,n)+1 &= \sum_{d|a}F_2(d)\\
\gcd(p^a-1,n)-1 &= \sum_{d|a}F_1(d)\\
\gcd(p^a-1,n) &= 1+\sum_{d|a}F_1(d).
\end{aligned}
\tag{6.23}
$$

7. Towards a Characteristic-p Casson Invariant

In the original development of the Casson invariant, see [1], it is necessary to restrict the representation spaces to the irreducible representations, because the reducible representations correspond exactly to singularities in the intersection, which would prohibit us from assigning a tangent space at those points.

In our situation it is not as clear what part of the original theory might apply here. In the original spaces, irreducible representations were precisely the diagonalizable representations. In our case these two ideas are not the same. Because we are dealing with representations over a finite field, we have representations which are irreducible, but not diagonalizable. This comes from the fact that the field is not algebraically closed, so in fact does not have all the eigenvalues of the matrices in the field.

It turns out, in the case of the $(n, 2)$ torus knots, that when we restrict ourselves to the non-diagonal representations, our ζ-function goes from being a ratio of integer polynomials, to having its numerator be precisely 1, so that it is the multiplicative inverse of a polynomial over the integers.

Definition 2 (Modified Zeta-function of a Knot). *Let $\mathcal{K} \subseteq S^3$ be a knot, and p be a fixed prime. We define $\lambda_{\mathcal{K}}(p, T)$ as:*

$$\lambda_{\mathcal{K}}(p,T) = \exp\left(\sum_{s=1}^{\infty} \frac{N_s' T^s}{s}\right), \tag{7.24}$$

where N_s' is the number of non-diagonal representations of $\pi_1(S^3 \backslash \mathcal{K})$ into $SL_2(\mathbb{F}_{p^s})$, counted up to conjugacy. This differs from the original ζ-function $Z_{\mathcal{K}}(p, T)$, because diagonal representations are now excluded. Diagonal representations are representations in which all the matrices can be simultaneously diagonalized.

Theorem 3. *Given $\mathcal{K}$ is a $(n, 2)$ torus knot, n an odd integer, and a prime number p, $p \equiv 1 \ (4)$, then ${\lambda_{\mathcal{K}}}^{-2}(p, T)$ is a polynomial. Furthermore, this polynomial is a product of cyclotomic polynomials, and has the property that it is invariant under the transformation $T \mapsto T^{-1}$.*

Proof. The major part of showing this is noting where the different parts of the ζ-function, given in Table 6.21, come from. Since we are counting things up to conjugacy, if the conjugacy class representative which X is get sent to is not diagonal, then that representation is not diagonalizable. We will now go through case by case, and determine which representations are diagonal, and what the removal of these representations does to the ζ-function.

(i) Both representations in Case 1 are diagonal, so the removal of these diagonal representation will completely remove the contribution of Case 1 to the ζ-function.

(ii) In Case 2, $Y \to \left(\begin{smallmatrix} -1 & 0 \\ 0 & -1 \end{smallmatrix}\right)$, and $X^2 = Y^n = \left(\begin{smallmatrix} -1 & 0 \\ 0 & -1 \end{smallmatrix}\right)$ so if X is going to be diagonal, $X = \left(\begin{smallmatrix} \alpha & 0 \\ 0 & \alpha \end{smallmatrix}\right)$, then $\alpha^2 = -1$, and these equations have exactly two solutions exactly when $q \equiv 1 \ (4)$. So now in the case when $q \equiv 3 \ (4)$, nothing changes, but when $q \equiv 1 \ (4)$ we lose two representations. This causes an extra

$(1-T)^2$ to appear in the numerator of the ζ-function. This is easily seen, and can be thought of as the ζ-function for Case 2 divided by the ζ-function for Case 1, since Case 1 always just had 2 representations.

(iii) Case 3, obviously no diagonal representations, since Y itself is not diagonal.

(iv) Case 4, obviously no diagonal representations, since Y itself is not diagonal.

(v) Case 5, obviously no diagonal representations, since Y itself is not diagonal.

(vi) Case 6, obviously no diagonal representations, since Y itself is not diagonal.

(vii) Case 7, the diagonal representations in Case 7 appear above as Case 7.2, in which case ALL of the representations were diagonal. If we follow these representations through the calculations, they become the ratio $\sqrt{\frac{1-T}{1-pT}}$ which sits in front of the ζ-function for Case 7, as well as the numerator for Case 7, $\left(\prod_{i|\phi(n)}(T^i-1)^{F_2(i)/i}\right)$. So by removing these diagonal representations we simply get the denominator of the Case 7 ζ-function

$$\frac{1}{\sqrt{\left(\prod_{i|\phi(n)}((pT)^i-1)^{F_1(i)/i}((p^2T)^i-1)^{F_1(i)/i}\right)}} \tag{7.25}$$

(viii) Case 8, obviously no diagonal representations, since Y itself is not diagonal.

So now we can see what is left over. Looking at the Chart 6.21, we see that Case 1 is gone, Case 2 has acquired a $(1-T)^2$ it the numerator, and the square roots in the numerator of Case 7 are gone. So if we assume $p \equiv 1$ (4), for simplicity, we get Case 3 cancelling out the numerator of Case 2, and Case 4 cancelling out the numerator of Case 8, leaving no numerators in the ζ-function. □

This seems like a weird theorem, since we threw away perfectly good representations to get the result, but it is very consistent with the development of the Casson invariant, and in some sense should be thought of as a characteristic-p Casson invariant.

8. How the ζ-Function Transforms under Topological Moves

We have seen that the ζ-function, $Z(p,T)$ for a knot is the square-root of a rational function. We will now study how the ζ-function transforms under simple topological transformations.

Given two links, one of the simplest ways of creating another link is to take the disjoint union of the two links, i.e., the two links into the same 3-space, but do not have them linked in anyway. Clearly then the number of representations of π_1 of the disjoint union is the product of the representations of each. But because we are working modulo conjugation, things are not so nice. In this case the number of representations does not multiply. We will speak more on this in a second.

We have shown that the ζ-function of a $(n, 2)$, torus knot is the inverse of a polynomial. This a very unique property, with some interesting implications. It will be useful in understanding how this ζ-function changes under another, closely related topological operation.

So supposed we have a knot $\mathcal{K} \subset S^3$, such that the ζ-function of $\mathcal{K}$, is the square root of the multiplicative inverse of a cyclotomic polynomial.

What does that tell us about the structure of the N_i, the coefficients in the definition of the ζ-function? If $\exp(\sum_s N_s T^s/s)$ is a rational function, we can say a bit more about the N_i, by looking at the roots of the numerator and denominator of the rational function. We can say that there exist complex numbers $\alpha_1, \alpha_2, \ldots, \alpha_n$ and $\beta_1, \beta_2, \ldots, \beta_m$, where n, m are the degrees of the numerator and denominator respectively, and the α's are closed under (complex) conjugation, as are the β's, such that for all s:

$$N_s = \alpha_1^s + \alpha_2^s + \cdots + \alpha_n^s - \beta_1^s - \beta_2^s - \cdots - \beta_m^s \tag{8.26}$$

Let us look at one of the α_i, α_1 say, appearing in the formula for N_s. If we represent $\alpha = re^{i\theta}$ then we must have some limit on what values θ may have. For example, θ must be a rational number times 2π. This comes from the fact that these α are algebraic numbers, and must add up to an integer. So as s increases we think of α_1^s as having a periodic part ($e^{i\theta}$) and a non-periodic part (r). So each N_s is a sum of terms each of which is of the form: $r^s f(s)$, where $f(s)$ is some periodic function of s. Let us look now at a simple case where we have a ζ-function whose N_s are all periodic.

If $Z(T) = \exp\left(\sum_{s=1} \frac{f(s)T^s}{s}\right)$, when $f(s)$ is a function into the integers, with period d, then by a straightforward calculation we get:

$$Z(T) = \prod_{a|d} (1 - T^a)^{-F(a)/a} \tag{8.27}$$

In this case, $f(s) = \sum_{a|s} F(a)$, and we see that f is just the Möbius transform of F. Of course, as before, we can invert, using Möbius inversion and solve for $F(a)$. If instead of just $f(s)$, we have $N_s = p^s f(s)$ then clearly we have:

$$\begin{aligned} Z(T) &= \exp\left(\sum_{s=1} \frac{p^s f(s) T^s}{s}\right) \\ &= \exp\left(\sum_{s=1} \frac{f(s)(pT)^s}{s}\right) = \prod_{a|d} (1 - (pT)^a)^{-F(a)/a} \end{aligned} \tag{8.28}$$

And similarly, a ζ-function which has N_s as sum of such terms (various periods θ and "size" r) will simply be the product of the ζ-function of each part.

But it is very difficult to think of a situation where the number of representations from one knot would ADD with the number of representations of the other. For example, if we take the disjoint union of two knots, the resulting link has a fundamental group equal to the product of the fundamental groups, and there for the number of representations will, for the most part, multiply. (There is a little issue of conjugacy, but it ends up almost being multiplication). If we take the connected sum of two knots, the fundamental groups combine as an amalgamated product, and if we fix under conjugation the element we are amalgamating over, the number of representations (up to conjugation) actually does multiply.

Let us look at, then, at the connected sum of a $(n,2)$ torus knot with itself. We have shown that the ζ-function, $\lambda_{\mathcal{K}}(p,T)$, of such a knot is the square-root of the multiplicative inverse of a polynomial. Further, this polynomial is symmetric in that $f(T) = f(T^{-1})$. The proof of this fact really relied on the fact that the numbers N_s in the definition of the ζ-function, $Z(T) = \sum \frac{N_s T^s}{s}$, have the property that $N_s = \sum_{a|s} F_a$, for some integers F_a. More important though, was the fact that these numbers F_a, defined by the Möbius transform of N_s, are all positive integers, and that was seen by noting the fact that $i|j \Rightarrow N_i < N_j$. Of course in the original case of $(2,n)$ torus knots, the N_i are gcd's and such.

So if we look at the ζ-function of the connected sum of a torus knot with itself, the new knot will have data $N'_s = N_s^2, \forall s$, and will share all the same properties. The original N_s was a sum of different parts (corresponding to the different conjugacy classes above), and in each part it was roughly of the form $N_s = p^{as}\lambda(s)$ where a is some integer ($a = 1, 2, \text{or } 3$), and $\lambda(s)$ is a periodic function in s. The product of two such N_s is obviously of the same form, and the result follows that this will also have the same properties, namely that there will exist integers F'_a so that $N'_s = \sum_{a|s} F_a$. It is true that the F_a are non-negative integers because $i|j \Rightarrow N_i < N_j \Rightarrow N_i^2 < N_j^2$, since the N_i are non-negative integers. Thus we have proven:

Theorem 4. *Let $\mathcal{K}_1$ be a $(n,2)$ torus knot, and let $\mathcal{K}_2$ be an $(m,2)$ torus knot (n,m odd integers) and the ζ-function $\lambda_{\mathcal{K}}(p,T)$ is defined as above in Definition 2. The connected sum of $\mathcal{K}_1$ with $\mathcal{K}_2$, $\mathcal{K}_1 \sharp \mathcal{K}_2$, has the property that ${\lambda_{\mathcal{K}_1 \sharp \mathcal{K}_2}}^{-2}(p,T)$ is a polynomial over the integers. Further, this polynomial is a product of cyclotomic polynomials.*

9. Conjectures

In the previous work, we have studied the properties of two different zeta-functions of knots and links in S^3. The first, $Z(p,T)$ uses the data N_s which count representations in $SL_2(\mathbb{F}_{p^s})$, up to conjugacy. The second, $\lambda(p,T)$ is defined similarly, except we further eliminate diagonal representations. We proved in Theorem 2 that for a knot or link, $Z(p,T)$ is the square root of a rational function. We then proved that for the $(n,2)$ torus knots $\lambda^{-2}(p,T)$ is a polynomial.

We make the following obvious conjecture:

Conjecture 1. *If $\mathcal{L} \subset S^3$ is any link, then $\lambda_{\mathcal{L}}^{-2}(p,T)$ is a polynomial.*

Further, is parallel withe Generalized Riemann Hypothesis proven by Deligne we make the folowing conjecture.

Conjecture 2 (Generalized Riemann Hypothesis for Knots). *If $\mathcal{L} \subset S^3$ is any link, then $\lambda_{\mathcal{L}}^{-2}(p,T)$ is the product of cyclotomic polynomials*

Note that by Theorem 4, we have that all $(n,2)$ torus knots, and their connected sums all have this property. It is not too much more difficult to show that if two knots have this property, then their direct sum has this property.

References

[1] S. Akbulut and J. McCarthy, *Casson's invariant for oriented homology 3-spheres: an exposition*, Princeton University Press, (1990).

[2] B. Dwork, *On the rationality of the zeta function of an algebraic variety*, Amer. Jour. Math., *82* (1960) 631-648.

[3] W. Fulton, *Intersection theory*, Springer-Verlag, (1984).

[4] W. Fulton and J. Harris, *Representation theory: a first course*, Springer-Verlag, (1991).

[5] N. Koblitz, *P-adic Numbers, p-adic Analysis, and Zeta-functions*, Springer-Verlag, (1977).

[6] L. Rozansky, *On p-adic properties of the Witten-Reshetikhin-Turaev invariant*, Preprint QA/9806075, preprint, 1998.

[7] K. Walker, *An extension of Casson's invariant to rational homology spheres*, Bull. Amer. Math. Soc., *22* (1990) 261-267.

BRAID COMMUTATORS AND DELTA FINITE-TYPE INVARIANTS

Theodore B. Stanford †
Mathematics Department
United States Naval Academy
572C Holloway Road
Annapolis, MD 21402

stanford@nadn.navy.mil

ABSTRACT

Delta finite-type invariants are defined analogously to finite-type invariants, using delta moves instead of crossing changes. We show that they are closely related to the lower central series of the commutator subgroup of the pure braid group.

0. INTRODUCTION

We consider in this paper delta finite-type invariants of knots and links (ΔFT invariants). In the case of links, these are the same invariants as defined by Mellor [3]. We shall prove some properties of ΔFT invariants which closely resemble properties of finite-type invariants in the usual sense (FT invariants). In the same way that FT invariants are based on sets of crossing changes in knot or link diagrams, ΔFT invariants are based on sets of delta moves in knot or link diagrams. In the same way that FT invariants are closely related to $\gamma_n(P)$, the the lower central series of the pure braid group P, ΔFT invariants are closely related to $\gamma_n(P')$, the lower central series of the commutator subgroup of P. Here $P = P_k$, the pure braid group on k strands.

We will show that two knots have matching ΔFT invariants of order $< n$ if and only if they are equivalent modulo $\gamma_n(P')$, just as two knots have matching FT invariants of order $< n$ if and only if they are equivalent modulo $\gamma_n(P)$ (see [7]). Since $\gamma_n(P') \subset \gamma_{2n}(P)$, it follows that an FT invariant of order $< 2n$ is a ΔFT invariant of order $< n$. It may turn out that all ΔFT invariants of order $< n$ occur this way. However, there is a difference between $\gamma_n(P)$ and $\gamma_n(P')$ which may make the ΔFT invariants more than just a relabeling of the FT invariants: For any k and any n, the quotient group $\gamma_n(P_k)/\gamma_{n+1}(P_k)$ is finitely-generated, whereas even $\gamma_1(P_3')/\gamma_2(P_3')$ is not finitely-generated.

For links of more than one component, there are known advantages to working with ΔFT invariants instead of FT invariants. Murakami and Nakanishi [4] showed that two links are equivalent by a sequence of delta moves if and only if they have the same pairwise linking numbers. Thus, ΔFT invariants detect linking number right at order 0, and so essentially each link homology class gets its own set of invariants of higher orders. Invariants

† Research supported in part by the Naval Academy Research Council

which are not FT over all links may be ΔFT within one homology class. For example, Mellor [3] has shown that Milnor's triple linking number is ΔFT but not FT (it is not even well-defined over all links).

Appreciation: I would like to thank the organizers of the "Knots in Hellas" conference, where this work was presented.

1. DEFINITIONS

The basic terminology here is motivated by the standard ideas from FT invariants. See for example Birman [1], Gusarov [2], Ohyama [5], or Taniyama [8].

Notation 1.1. Let L be a link diagram with n disjoint sets of disjoint disks $S_1, S_2, \dots S_n$. Inside each disk, the diagram is supposed to look like one side of Figure 1.2. For any subset $T \subset \{1, 2, 3, \dots n\}$, denote by L_T the link obtained by applying a delta move in each disk in S_i for all $i \in T$.

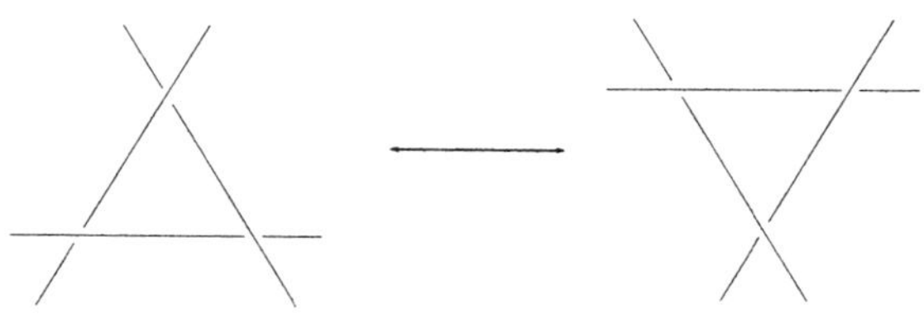

Figure 1.2

Definition 1.3. Two links L and L' are said to be Δ_n*-equivalent* if there exists a diagram as above such that $L_\emptyset = L$ and $L_T = L'$ for all $T \neq \emptyset$. A link Δ_n-equivalent to an unlink is said to be Δ_n*-trivial.*

I do not know if this is an equivalence relation in general. If not, then one can consider the equivalence relation that it generates. It will follow from Theorem 2.1 that the definition does give an equivalence relation when restricted to knots.

Notation 1.4. Let $\mathbb{Z}\mathcal{L}$ be the $\mathbb{Z}$-module freely generated by all link types, and let C_n be the submodule generated by all linear combinations

$$\sum_{T \subset \{1,2,\dots n\}} (-1)^{|T|} L_T \tag{1.5}$$

for all diagrams L as above.

Definition 1.6 Let v be a link invariant taking values in some abelian group A. Extend v linearly to $\mathbb{Z}\mathcal{L}$. Then v is said to be ΔFT *of order* $< n$ if it vanishes on C_n.

Proposition 1.7. *If two links L and L' are Δ_n-equivalent and v is a ΔFT invariant of order $< n$, then $v(L) = v(L')$.*

Proof: Choose a diagram for L as in Definition 1.3. Use this diagram to form the alternating sum as in Notation 1.4. □

Notation 1.8. If $b \in B_k$ is a braid, then we denote by $\lfloor b \rceil$ the usual closure of b, connecting the top ends to the bottom ends to make a knot or link. This extends to a $\mathbb{Z}$-linear map from $\mathbb{Z}B_k$ to $\mathbb{Z}\mathcal{L}$. Set $t_k = \sigma_{k-1}^{-1}\sigma_{k-2}^{-1}\cdots\sigma_1^{-1}$.

If $p \in P_k$, then $\lfloor pt_k \rceil$ is a knot, and any knot may be written this way. Also, $t_k^{-1}pt_k$ is p "shifted" to the right. These are very useful facts.

Definition 1.9. Two links L and L' are said to be $\gamma_n(P')$-equivalent if there exists a positive integer k and braids $p, b \in B_k$ with $L = \lfloor b \rceil$, $L' = \lfloor pb \rceil$, and $p \in \gamma_n(P'_k)$. Here $G' = [G, G]$ and $\gamma_n(G) = [G, \gamma_{n-1}(G)]$, with $\gamma_1(G) = G$ for a group G.

The groups $\gamma_n(P'_k)$ form a subcoherent sequence of pure braid subgroups, and therefore Definition 1.9 gives an equivalence relation. See [7].

Definition 1.10. A knot invariant v is said to be *additive* if $v(K\#K') = v(K) + v(K')$.

2. THEOREMS

Just as with FT invariants, there are several different ways of characterizing the knots whose ΔFT invariants match up to a given finite order.

Theorem 2.1. *For two knots K and K', the following are equivalent:*

A. K and K' are $\gamma_n(P')$-equivalent.

B. K and K' are Δ_n-equivalent.

C. For any ΔFT invariant v of order $< n$, $v(K) = v(K')$.

D. For any additive ΔFT invariant v of order $< n$, $v(K) = v(K')$.

Gusarov [2] showed that the equivalence classes of knots with matching FT invariants of order $< n$ form a group under the operation of connected sum. The same is true for ΔFT invariants.

Theorem 2.2. *For all n, the Δ_n-equivalence classes of knots form a group under connected sum.*

Theorems 2.1 and 2.2 apply only to knots. The next theorem applies to both knots and links.

Theorem 2.3. *For any link L and any positive integer n, there exist an infinite number of prime, nonsplit, alternating links L' such that $v(L) = v(L')$ for all ΔFT invariants of order $< n$.*

3. PROOFS

We showed in [7] that the $P^{(n)}$-equivalence classes of knots (where $P^{(n)}$ is the derived series) form a group under connected sum, so Theorem 2.2 follows from Theorem 2.1 because $P^{(n+1)} \subset \gamma_n(P')$. We showed in [6] how to modify a link L to create an infinite family of prime, nonsplit, alternating links in the same $\gamma_n(P)$-equivalence class as L, and we showed how to make this procedure work for $P^{(n)}$-equivalence of knots in [7]. The same procedure is easily modified to give a proof of Theorem 2.3, given Propositions 1.7 and 3.1.

Proposition 3.1. *If two links are $\gamma_n(P')$-equivalent, then they are Δ_n-equivalent.*

Proof: It suffices to show that a braid in $\gamma_n(P')$ is Δ_n-trivial, in the sense that there exists a braid diagram K such that Notation 1.1 and Definition 1.3 apply. The proof is by induction on n. We first need to see that any braid in P' is Δ_1-trivial. Let $p_{i,j}$ be the standard generator of P_k which links the ith and the jth strands. Any commutator of form $[p_{h,i}, p_{i,j}] = p_{h,i}p_{i,j}p_{h,i}^{-1}p_{i,j}^{-1}$. may be transformed into a trivial braid by a delta move. Take any standard set of relations among the $p_{i,j}$, and add in all relations of the form $[p_{h,i}, p_{i,j}]$. The result is a presentation of P_k abelianized, and therefore any braid in P_k' may be undone by delta moves.

For the induction step, consider a braid $[p, q]$, where $p \in \gamma_n(P')$ and $q \in P'$. Write down the obvious diagram for $[p, q]$, in terms of diagrams for p and q which have n and 1 disjoint sets of disks, respectively, as in Notation 1.1. The first n sets of disks in $[p, q]$ will come from the disks in p together with the disks in p^{-1}. The last set will come from the disks in q and q^{-1}. □

Proposition 3.1 gives us (A) $\Rightarrow$ (B) in Theorem 2.1, Proposition 1.7 gives us (B) $\Rightarrow$ (C), and (C) $\Rightarrow$ (D) is trivial. We need to show then that (D) $\Rightarrow$ (A).

If the module $C_n \subset \mathbb{Z}\mathcal{L}$ is pulled back to the group ring $\mathbb{Z}B_k$ via the closure map, the result is an easily-described ideal:

Proposition 3.2. *For each positive integer k, let $I = I_k \subset \mathbb{Z}B_k$ be the ideal generated by all $p - 1$ with $p \in P_k'$. The module C_n is generated by all $\lfloor x \rceil$ with $x \in I_k^n$ for some k.*

Proof: Let L be a diagram as in Notation 1.1. Using a standard trick from FT invariants (see Gusarov [2]), we may restrict ourselves to the relations (1.5) where each S_i contains a single disk. We would like to apply an Alexander-type theorem as in Birman [1]. Whether or not a delta move can be fit locally into a braid diagram depends on the pattern of orientations of the strands. (If our links aren't oriented, then choose an arbitrary orienation for each component.) Each dot in Figure 3.3 indicates a possible delta move, and all four delta moves shown have the same effect. The moves on the left and right ends of the figure cannot occur locally in a braid because one of the three strands will be forced to go "backwards". However, the figure demonstrates that we can instead always choose moves like the two in the middle, which can occur locally in a braid. We may then write K as a closed braid diagram, and we may write each delta move as $[p_{h,i}, p_{i,j}] - 1$. We then find that the relation determined by L is in fact the closure of an element of I^n. □

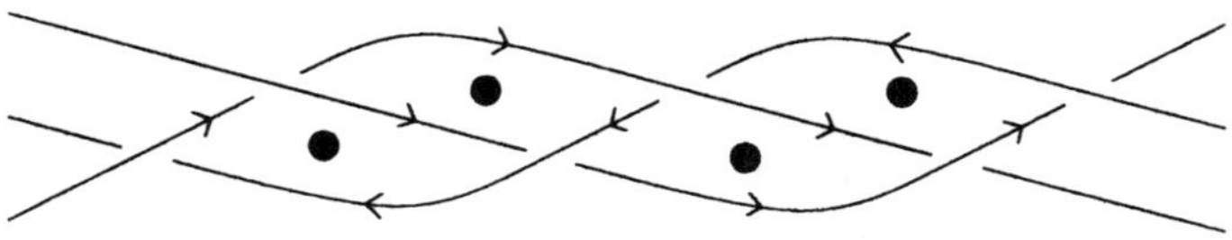

Figure 3.3

Now, restricting ourselves to the case of knots, we may conjugate if necessary and see that C_n is generated by expresssions of the form

$$\lfloor (x_1 - 1)(x_2 - 1) \dots (x_n - 1) y t_k \rceil \tag{3.4}$$

where $x_i \in P'_k$ and $y \in P_k$. The only thing that prevents us from applying exactly the same analysis as in Section 2 of [7] is that y is not necessarily in P'_k. This is fixed as follows.

The expression (3.4) is a $\mathbb{Z}$-sum of 2^n different knots. If we take the connected sum of each of these with the knot $\lfloor y^{-1} t_k \rceil$, then we may write the resulting $\mathbb{Z}$-sum of knots as

$$\lfloor (x_1 - 1)(x_2 - 1) \dots (x_n - 1) y (t_{2k}^{-k} y^{-1} t_{2k}^{k}) t_{2k} \rceil \tag{3.5}$$

If v is an additive invariant, then it vanishes on (3.4) if and only if it vanishes on (3.5). Now we use the same idea as in [7], to slide the y^{-1} around and around until it cancels with the y. That is, we consider relations of the form

$$\lfloor (x_1 - 1)(x_2 - 1) \dots (x_n - 1) z y (t_{2k}^{-m} y^{-1} t_{2k}^{m}) t_{2k} \rceil \tag{3.6}$$

where $0 \leq m \leq k$, $y \in P_k$, and $z \in P'_{2k}$. Relation (3.5) is of this form when $m = k$. When $m = 0$, then y and y^{-1} cancel and we have the form we need. Conjugating by $t_{2k}^{-m} y t_{2k}^{m}$, we obtain

$$\lfloor (x'_1 - 1)(x'_2 - 1) \dots (x'_n - 1) z' y (t_{2k}^{-m+1} y^{-1} t_{2k}^{m-1}) t_{2k} \rceil \tag{3.7}$$

where $x'_i = t_{2k}^{-m} y t_{2k}^{m} x_i t_{2k}^{-m} y^{-1} t_{2k}^{m}$ and $z' = (t_{2k}^{-m} y t_{2k}^{m} z t_{2k}^{-m} y^{-1} t_{2k}^{m})(t_{2k}^{-m} y t_{2k}^{m} y t_{2k}^{-m} y^{-1} t_{2k}^{m} y^{-1})$. This is the same relation as (3.6), and it is written in the same form except that m has been decreased by one. Hence a relation of the form (3.4) may be replaced by a relation of the same form, with $y \in P'$.

REFERENCES

[1] J. S. Birman. *New points of view in knot theory.* Bulletin of the American Mathematical Society 28 (1993) 253–287.

[2] M. Gusarov *On n-equivalence of knots and invariants of finite degree.* "Topology of Manifolds and Varieties", 173–192. Advances in Soviet Mathematics 18, American Mathematical Society, 1994.

[3] B. Mellor. *Finite-type link homotopy invariants II: Milnor's $\bar{\mu}$ invariants.* `math.GT/9812119`.

[4] H. Murakami and Y. Nakanishi. *On a certain move generating link-homology.* Mathematische Annalen 284 (1989), 75–90.

[5] Y. Ohyama. *Vassiliev invariants and similarity of knots.* Proceedings of the AMS 123 (1995), 287–291.

[6] T. Stanford. *Braid commutators and Vassiliev invariants.* Pacific Journal of Mathematics 174 (1996), 269–276.

[7] T. Stanford. *Vassiliev invariants and knots modulo pure braid subgroups.* `math.GT/9805092`.

[8] K. Taniyama. *On similarity of knots.* Gakujutsu Kenkyu, School of Education, Waseda University. Series of Mathematics 41 (1993), 33–36.

QUANTUM-LIKE PROPERTIES OF KNOTS AND LINKS.

ANDRZEJ STASIAK

Laboratoire d'Analyse Ultrastructurale
Université de Lausanne
1015 Lausanne-Dorigny, Switzerland.
e-mail: Andrzej.Stasiak@lau.unil.ch

Knot theory was boosted in its development when in 1860ties Kelvin proposed that knots made out of vortex lines of ether constitute elementary particles of matter. Tait and other associates of Kelvin started to catalogue various types of knots with a hope to create a periodic table of knots which would correspond to periodic tables of chemical elements. However, development of atomic physics, first with classical and then with quantum models of atoms failed to show a connection between knots and atoms. More recently, though, the superstring theory revived the idea that knots can be relevant to understanding elementary particles. Studying ideal geometric configuration of knots we noticed that their writhe is quantized into units of 4/7 and 10/7 whereby the total writhe of a knot is determined by a simple arithmetic sum of four elementary types of crossings which need to be "removed" to convert a standard minimal crossing diagram of a given knot into a diagram of a trivial knot without nugatory crossings. There are parallel and antiparallel crossings of positive and negative sign. Interestingly parallel and antiparallel crossings of the same sign show frequently superposition of states. This resembles the quantum properties of elementary particles and their known superposition principle. A model of physical knotted systems is discussed which can change energy states by adopting discrete shapes with associated discrete energies whereby the writhe difference between any two discrete shapes is n∗constant.

We presented recently a concept of ideal geometric representations of knots[1, 2]. As ideal we defined the shortest possible trajectory of a flexible cylindrical tube that allows us to close a given type of knot. These configurations are ideal from geometric point of view as they allow us to maximize the volume to surface ratio of a cylinder forming a given type of knots[1, 3, 4]. These geometric properties of ideal knots cause that some knotted physical systems tend to adopt ideal or almost ideal representations. So for example knotted bistable chemical systems are expected to relax in a form of ideal knots and links[5]. Knotted magnetic flux lines tend to shorten their length while keeping uniform diameter and maintaining the same volume, this causes that they evolve toward ideal configurations[3]. If superstrings are knotted then their tendency to shorten can also bring them to ideal geometric forms. From these points of view ideal geometric configurations of knots can be of a high physical importance.

1 Quantization of writhe in ideal knots

Writhe is a measure of chirality of closed curves in space. Purely topological interest in writhe is paralleled by such practical aspects as design of modern fusion devices or understanding the behavior of magnetic flux lines in solar corona[6]. Knots minimizing their energy, like the knotted magnetic flux lines[3, 7] or knots relaxed according to simple geometric principles (shortening the length of a string a given knot is made of) adopt characteristic shapes with a specific writhe[1, 8]. Studying the writhe of ideal geometric representations of knots we observed earlier that writhe values within torus and twist families of knots showed two different arithmetic progressions as if each new toroidal or twist turn were introducing always the same characteristic contribution to the writhe. In addition it was observed that the mean writhe values of randomly fluctuating knotted polymers forming a given knot coincides with writhe value of the corresponding ideal knot[1]. Here I provide some explanations of this quantization of writhe and show how the writhe of ideal configurations of knots and links can be predicted using standard tabular diagrams of the corresponding knots and links.

Planar curves as a circle or ellipse are locally and globally achiral and have writhe zero. However when an initially planar curve is twisted or writhed, it acquires writhe and shows intersections or crossings when it is observed from various directions. There are two principal types of crossings; positive (left-handed) and negative (right-handed) that score as +1 and -1 in the calculation of writhe[9] (see Fig. 1). The writhe value of a given projection corresponds to the arithmetic sum contributed by positive and negative crossings seen in this projection. A global 3-D writhe (Wr) is the average of directional writhe values when a 3-D curve is orthogonally projected along all directions equisampling the sphere and can be calculated using a modification of Gauss integral equation:

$$\mathrm{Wr} = \frac{1}{4\pi} \oint_C \oint_C \frac{(\mathbf{dr}_1 \times \mathbf{dr}_2)\mathbf{r}_{12}}{r_{12}^3} \qquad (1)$$

where $\mathbf{r}_1$ and $\mathbf{r}_2$ are vectors whose ends upon integration run over the 3-D curve C [10]. Using this formula we have previously calculated writhe values of axial trajectories of ideal geometric forms of various types of knots obtained in computer simulations[1]. We noticed then that within torus type of knots (see Fig. 1b) for each new right-handed toroidal turn the writhe has grown by ca. 2.86. Twist type of knots (see Fig.1c) showed less steep increase of writhe where each new twist turn contributed only ca. 1.14 to the total writhe.

Why there is such a difference in writhe contributions of toroidal and twist turns? Looking at the regularly wound portion of the knot we can not tell the difference between toroidal and twist turns (see Fig.1a). However, the difference becomes visible when we assign a direction to the knot. In toroidal knots the coils facing each other show the same polarity while in twist knots coils facing each other show opposite polarity. Thus although there are only two standard designations of crossings: positive and negative, there are in fact four types of perceived arrangements of the facing coils: parallel right-handed (Fig. 1b) parallel left-handed as well as antiparallel right-handed (Fig. 1c) and antiparallel left-handed. In ideal geometric configurations of knots right-handed parallel turns as those observed in right-handed torus-type of knots contribute the writhe of 2.86 per full turn (2 crossings) or 1.43 per half turn (1 crossing). There is a higher than 1 contribution to 3-D writhe per each right-handed parallel crossing of standard diagrams since only in planar minimal crossing representations of a given knot the writhe contributed by parallel right-handed half-turn of facing coils amounts to 1. In knots living in 3-D space there are many directions of projections where the right-handed parallel winding of the facing coils give in addition to the right-handed crossings between opposing coils (inter-coil crossings) also right-handed crossings between consecutive segments of the same coil (intra-coil crossings)(Fig. 1d). Thus in the parallel right-handed turns the positive writhe contributed by the right-handed inter-coil crossing adds up with the positive writhe contributed by right-handed self winding of individual coils.

In the case of right-handed winding of antiparallel coils, as in twist type of knots, the situation is very much different. Despite the right-handed winding of the individual coils their antiparallelity results in negative sign of inter-coil crossings (Fig. 1c), therefore negative writhe contributed by crossings of opposing coils is partially cancelled by the intra-coil positive writhe caused by right-handed winding of individual coils (Fig. 1e). Consistent with this we noticed previously that in right-handed antiparallel coils, as those observed in twist type of knots, each complete turn contributes -1.14 to the writhe of ideal knot i.e. -0.57 per crossing. Since the local geometry of windings of opposing coils is insensitive to their relative direction (Fig.1) thus in parallel and antiparallel turns the intra-coil contribution should be the same while the contribution of inter-coil crossings reverses its sign.

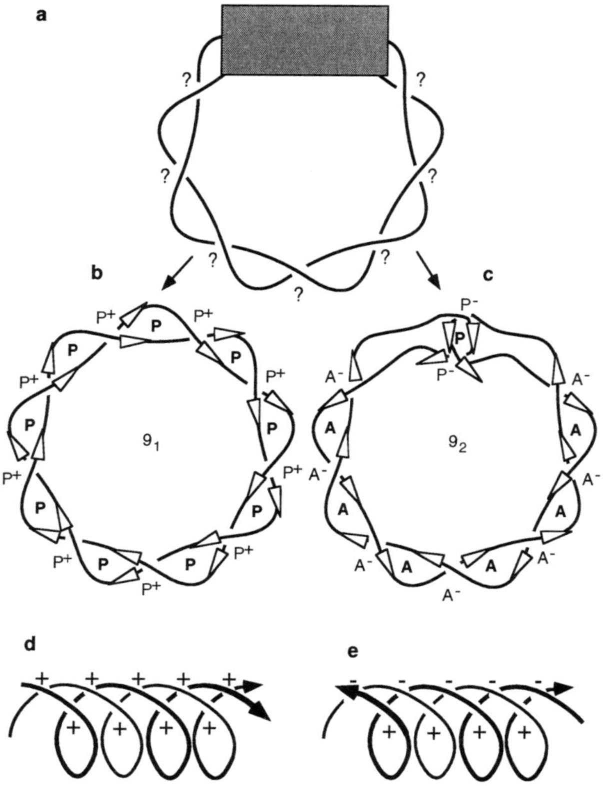

Fig. 1. Difference between torus and twist type of knots. a. Without knowing the complete trajectory of a knot one can not tell torus from twist turns and one is unable to find out if the observed crossings are positive (right-handed) or negative (left-handed). However one can easily find out what is the direction of self-winding or coiling of individual segments (it is right-handed here). b. Torus knot. Upon tracing connectivity of all segments in the knot one can assign directions to the entire knot. In the case of torus knots one observes that directions on the opposing coils are "parallel" when progressing from a crossing **m** to a crossing **n**. In the shown case of right-handed torus knot all crossings have positive signs. The sign of crossings depends on the sense of rotation needed to align direction of the segment closer to an observer with the direction of the segment further from the observer. The rotation has to be smaller than 180° and needs to be applied to the segment closer to the observer. In positive crossings the sense of this rotation is counter clockwise, in negative crossing it is clockwise[9]. c. Twist knot. The directions of the opposing coils on a diagram are "anti-parallel" when progressing from a crossing **m** to **n** and this causes that the crossings have negative sign despite the right-handed direction of winding of individual coils. d. In torus knots inter- and intra-coil crossings have the same sign. e. In twist knots inter- and intra-coil crossings have opposite sign.

Indeed a brief look at above shown values demonstrates that 1.43 = 1 + 0.43 and -0.57 = -1 + 0.43 , where the +1 and -1 arise due to inter-coil crossings and +0.43 is intra-coil contribution per right-handed parallel or antiparallel half-turn in an ideal knot. Of course in the case of left-handed winding of parallel or antiparallel coils the intra- and inter-coil contributions to the writhe change signs and we have writhe contributions of -1.43 per left-handed toroidal crossing and +0.57 per left-handed twist crossing. It then simply follows that the sum of writhe of one positive parallel and one positive anti-parallel half-turn amounts to exactly 2 whereby this sum for the corresponding negative half-turns amounts to -2.

In general, the writhe of one portion of space curve is not additive with the writhe of another portion[11]. However, in ideal knots we observed earlier that the writhe value of ideal configurations of composite knots was a simple sum of writhe values of contributing factor knots in their ideal configurations[8]. Therefore the fact that in twist or torus knots each twist or torus type of turns introduces a constant specific contribution to the writhe led us to think that in ideal knots writhe values of basic building blocks of a knot simply add to the writhe of the entire knot. Indeed, upon obtaining by numerical simulations ideal configurations of knots with up to 9 crossings Pieranski observed that the same values of writhe are frequently attained by several different knots[12] as if they were composed of the same set of the elements contributing to the writhe. Thorough analysis of the observed writhe of ideal configurations of alternating knots reveals that their values can be expressed as **n**∗0.57 where **n** is an integer number including zero. Observation that writhe of alternating ideal knots adopt values corresponding to the multiplicity of the writhe of antiparallel half-turns (Wr = 0.57) and the apparent lack of contribution of parallel half-turns (Wr = 1.43) is due to the fact that toroidal half-turns in knots are always present in pairs, whereby the contributed writhe of ca 2.86 is practically equal to the writhe contribution of 5 antiparallel half-turns of the same sign. Assuming that the observed writhe values contributed by parallel and antiparallel turns reveal simple mathematical relations between them and remembering that the sum of writhe of one positive parallel (Wr(P)) and writhe of one positive antiparallel (Wr(A)) half-turn amounts to exactly 2, we arrive to the following two equations:

$$Wr(P) + Wr(A) = 2 \qquad (2)$$

$$2Wr(P) = 5Wr(A) \qquad (3)$$

Solving these equations we obtain writhe value of 4/7 for positive antiparallel half-turns Wr(A) and 10/7 for positive parallel half-turns Wr(P). Thus intra-coil contribution to writhe amounts to +3/7 and -3/7 for right- and left-handed crossings and there is inter-coil contribution of +1 for positive and -1 for negative crossings. At this point a formal proof that writhe values contributed by individual crossings in ideal configurations adopt very strictly these rational quantum values is still missing, however the available numerical data strongly support the proposed here quantization of writhe[2].

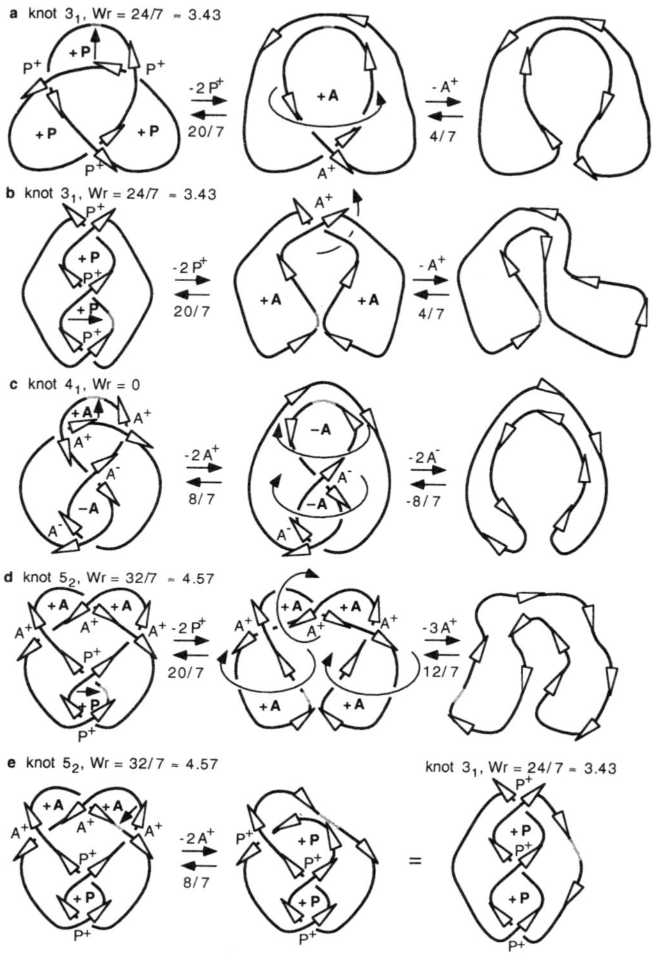

Fig. 2. Simplification of knots by strand passages and untwistings.
Progressive simplification of minimal diagrams of different knots allows us to identify the types of crossings composing a given knot. This is needed to predict the writhe of corresponding ideal knots. To help "tracing" strand passages, the "gate" of passage is drawn in grey and the arrows drawn within the respective double-arc fields indicate the direction of the passage. a, b. Simplification of two different minimal diagrams of trefoils. Notice that in both cases after the first strand passage the remaining crossing changes its type from a positive parallel to a positive antiparallel. c. Simplification of achiral 4_1 knot predicts its writhe value to be zero. d, e. Different simplification paths of minimal diagrams of 5_2 knot give the same writhe score.

2 Predicting the writhe of ideal knots

Assuming the additivity of writhe in ideal knots and knowing the value of the writhe contributed by right- and left-handed parallel and antiparallel turns I tried to predict the total writhe of various ideal knots and compare it with this of their ideal configurations obtained in elaborated computer simulations[1, 8, 12, 13] whereby the value of writhe was calculated using the formula (1). To predict the writhe value of ideal knots it is first necessary to correctly identify the number and handedness of parallel and antiparallel turns a given knot is composed of. To this aim we start with standard (minimal crossing) diagrams of knots, assign direction to them and then perform sequential passages of the facing coils through each other to eliminate pairs of parallel or antiparallel crossings. We can also perform untwisting of individual anti-parallel turns or smoothing of respective individual crossings so that at the end the knot diagram is changed into a diagram of an unknot with no nugatory crossings and thus having the writhe equal to zero (see Fig. 2).

A conceptually similar procedure of progressive simplification of knots in order to detect their chirality was recently described[14] but only integer values were associated to eliminated crossings. We assume here that each eliminated parallel or antiparallel type of crossing changes the writhe by 10/7 and 4/7 respectively, whereby the sign depends on the handedness of eliminated crossings. We then calculate the total change of writhe associated with the simplification of a given knot into unknot. If each type of crossing contributes its specific writhe in an additive way then the total change of writhe during a complete simplification of knot diagrams should be equal to the -Wr of the starting ideal knot. In order to be able to tell which of the observed crossings should be treated as parallel or antiparallel we first identify double-arc fields on the diagrams of knots (fields enclosed by two arcs only). When directions on both arcs point from a crossing **m** to a crossing **n** (or on both arcs from **n** to **m**) then the field and the crossings flanking it are of parallel type (P) (see Fig. 1 and 2). When the direction on one arc is from a crossing **m** to a crossing **n** while on the other arc this direction is opposite, the field and the crossing flanking it are of antiparallel type (A) (Fig. 1 and 2). A passage of segments through each other which eliminates two parallel crossings changes the writhe by 20/7 (2Wr(P)) (see Fig. 2) while a passage which eliminates two antiparallel crossings changes the writhe by 8/7 (2Wr(A)). The signs of the corresponding changes depend on the handedness of the eliminated crossings. As we simplify diagrams of knots we frequently have situations where double arc fields are not more present on the graphs and the remaining crossing can be simply eliminated by 180° untwisting of one domain of the knot in respect to the rest of the knot. Such untwistings eliminate individual twist type of crossings (antiparallel) and thus change the writhe by ±4/7 depending on the handedness of the eliminated crossing.

Figure 2 shows examples of sequential simplification of knots and lists corresponding changes of writhe. Right-handed trefoil knot (Fig. 2a) has initially three double arc fields and each is positive parallel (+**P**). The first simplifying passage of segments therefore eliminates two parallel right-handed crossings ($2P^+$) and thus decreases the writhe by 20/7. After first strand passage a further simplification just requires untwisting of the remaining positive twist crossing. Twist crossings are defined as antiparallel and elimination of one positive antiparallel crossing (A^+) reduces the writhe by 4/7. As a result we obtain a planar diagram without any crossings and thus having the writhe equal to zero. Simplification of standard diagrams of a trefoil into a planar unknot thus accounted for a change of writhe by 24/7 = 3.42857. This value closely coincides with 3.41 which was the computed value of writhe for the simulated configuration of the ideal trefoil[1] and with 3.42793 writhe value obtained in higher accuracy simulations (Ben Laurie, private communication). Of course the score of the simplification process should be independent of the chosen minimal crossing diagram of a knot. Fig. 2b shows another type of minimum crossing diagram of right-handed trefoil knot. This diagram has only two double arc fields, which are of positive parallel type and are thus flanked by positive parallel crossings. Simplifying strand passage within one of these fields decreases thus the corresponding writhe by 20/7 and leaves a diagram with one positive antiparallel crossing having the residual writhe of 4/7. Thus also during the simplification of this different minimal diagram of the trefoil the total change of writhe amounts to 24/7. It is good to mention here that during the simplification of both types of minimal diagrams of trefoils one crossing changes its type from parallel to antiparallel (this point will be discussed later).

Simplification of figure-eight knot (Fig. 2c) eliminates two antiparallel right-handed and two antiparallel left-handed crossings. Therefore the total change of writhe is equal to zero which corresponds to the computed earlier writhe of ideal figure-eight knot[1] and is consistent with globally achiral nature of this knot[15]. In the case of figure-eight knot all crossings maintain their type during simplification involving strand passages and untwisting of antiparallel turns (other simplification procedures can change the type of crossings, see below). Of course the changes of writhe during the simplification toward unknot should be independent of the path of the simplification one chooses. I have tried different simplification paths on a number of alternating knots and links and always found the total value of writhe change to be independent of the taken path of the simplification (Fig. 2d and e). However the order and total composition of the reduced crossings can change with the chosen path of simplification (see next paragraph). Knot 5_2 shown in Fig. 2d,e has in its minimal crossing representation one double arc field with parallel arcs which is flanked by positive parallel crossings and two double arc fields with antiparallel arcs flanked by positive antiparallel crossings. Elimination of the parallel double arc field reduces the writhe by 20/7 and leaves a graph with 3 positive antiparallel half-turns each contributing 4/7 to the writhe. Thus the total change of writhe amounts to 32/7 or 4.57 which closely coincides with 4.56 i.e. the writhe value calculated for ideal configuration of 5_2 knot[2]. Simplification procedure does not need to proceed to the end. One can stop the simplification at any intermediate step and if we know the writhe of the remaining knot we can simply add it to the current simplification score to obtain the writhe of the starting knot. Figure 2e shows that the reduction of two positive antiparallel crossings (2*4/7) from the knot 5_2 leads to formation of a trefoil knot, the writhe of which we determined earlier to be 24/7. Thus we arrive again to the original writhe of ideal 5_2 knot being 32/7 and we see that the writhe score of simplifications starting with elimination of parallel or of antiparallel crossings is the same.

Is the method of writhe prediction also good for much more complicated knots in their ideal configuration? To answer this question one would need first to be sure what is the writhe of more complicated ideal knots. Up to now we do not have an analytical method of calculating the trajectories of ideal knots and only know some approximate solutions obtained by computer simulations, whereby the more complex knot the higher a chance that simulations will not find the actual ideal configuration. Despite these shortcomings I checked for all alternating knots with up to 8 crossings how the predicted writhe value agrees with the simulation data[1, 12]. The results are listed in Table I. It is visible that the agreement is generally very good. However, it is not yet possible to tell if the slight differences between the results arise due to imperfections of the presented here method of writhe prediction or are due to the fact that simulation method can end up in some local minima in the configuration space and miss the actual ideal configuration.

3 Non alternating knots.

Up to now we have considered different cases of alternating knots that always show alternation of under- and overpassing when we pass through consecutive crossings along the minimal diagram of the knot. Is the method of writhe prediction also valid for diagrams of non alternating knots where in the minimal crossing diagram we do not observe a regular alternation of under- and overpassing? To answer this question we took the tabular diagram of a non alternating knot 8_{20}[16, 17]. There are only three non alternating knots 8_{19}, 8_{20} and 8_{21} which have the minimal number of crossings smaller than nine, however the proportion of non alternating knots grows among more complex knots. Since in non alternating knots interweaving of knots trajectories is not perfect, this provides substantial freedom of rearranging the diagram of the knot while maintaining the same minimal number of crossings and thus producing different minimal diagrams which are formally equally suitable for the simplification procedure. However in these different minimal diagrams of non alternating knots some of double arc fields can be replaced by double arc fields of another type (see Fig. 3). Therefore in the case of non alternating knots the simplification score depends on the arbitrarily chosen type of minimal diagram. However if different types of starting diagrams are considered then the mean score of their simplification corresponds to the writhe of the ideal configuration of a given non alternating knot.

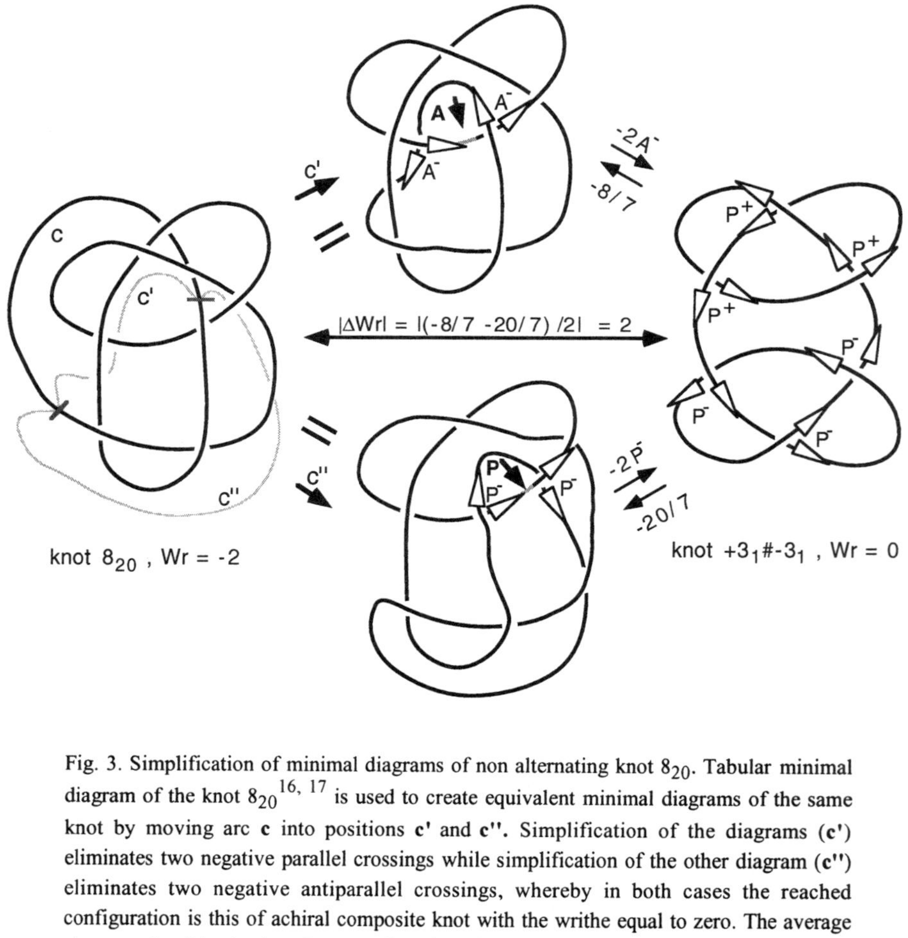

Fig. 3. Simplification of minimal diagrams of non alternating knot 8_{20}. Tabular minimal diagram of the knot 8_{20}[16, 17] is used to create equivalent minimal diagrams of the same knot by moving arc **c** into positions **c'** and **c''**. Simplification of the diagrams (**c'**) eliminates two negative parallel crossings while simplification of the other diagram (**c''**) eliminates two negative antiparallel crossings, whereby in both cases the reached configuration is this of achiral composite knot with the writhe equal to zero. The average change of writhe is therefore equal to 2 and thus ideal configuration of left-handed enantiomer of the knot 8_{20} is expected to have the writhe of -2.

Figure 3 shows how two different minimal diagrams of non alternating knot 8_{20} can be constructed. Simplification of these two diagrams toward a diagram of a composite square knot requires in one case a nullification of two negative antiparallel crossings and in another case a nullification of two negative parallel crossings (Fig.3). The average change of writhe is thus -2 (the mean of -20/7 and of -8/7). Since the square knot is achiral it has the expected and calculated value of writhe equal to zero[1]. Therefore the ideal representation of the knot 8_{20} is expected to have the writhe of +2 or -2 depending on the handedness of the knot. Indeed it was reported earlier the writhe of simulated ideal configuration of right-handed enantiomorph of non alternating knot 8_{20} is equal to +2 [1]. In the case of more complex non alternating knots one would need to analyze all minimal diagrams of different types giving a different score of writhe simplification. This makes that the writhe prediction of ideal non alternating knots is more complicated than the writhe prediction of ideal alternating knots.

4 Superposition states of crossings.

Discussing quantization of writhe values of individual crossings it is good to point attention to another quantum property of the analyzed crossings. During simplification procedure some crossings can change their type. In the case shown on Figure 2a the starting diagram of a trefoil has three positive parallel crossings but after first simplification one crossing changes its type into a positive antiparallel crossing. The crossing that changes its type can not be pinpointed in the original diagram. Initially all crossings are identical and only after arbitrary elimination of other two crossings and scoring them (determining) as positive parallel the remaining crossing becomes determined as positive antiparallel. An alternative way of describing the crossings of a trefoil is thus to treat them as a mixture (superposition) of states where each crossing is partially parallel and antiparallel at the same time, however the superposition collapses during the simplification of the knot. The score of writhe simplification is independent of the actual emergence of crossings from their superposition states as the total sum of the writhe score is maintained.

TYPE	Experimental writhe	Predicted writhe
3_1	3.39 (1)	3.43
4_1	0.00 (2)	0.00
5_1	6.25 (2)	6.29
5_2	4.56	4.57
6_1	1.13	1.14
6_2	2.77	2.86
6_3	0.02	0.00
7_1	9.10	9.14
7_2	5.74	5.71
7_3	7.40	7.43
7_4	5.82	5.71
7_5	7.43	7.43
7_6	3.43	3.43
7_7	0.66	0.57
8_1	2.29	2.29
8_2	5.63	5.71
8_3	0.00	0.00
8_4	1.62	1.71
8_5	5.60	5.71
8_6	3.97	4.00
8_7	2.83	2.86
8_8	1.19	1.14
8_9	0.04	0.00
8_{10}	2.78	2.86
8_{11}	3.88	4.00
8_{12}	0.00	0.00
8_{13}	1.21	1.14
8_{14}	4.03	4.00
8_{15}	7.91	7.71
8_{16}	2.76	2.86
8_{17}	0.01	0.00
8_{18}	0.08	0.00
2_1^2	2.00 (1)	2.00
4_1^2	4.86 (1)	4.86

Table I. Comparison of the computed writhe values obtained for simulated ideal configurations of knots with the writhe values predicted from minimal diagrams of corresponding knots. 1) Data provided by Ben Laurie, 2) Data from Ref. 1, all other writhe values of simulated ideal knots are from Ref. 2.

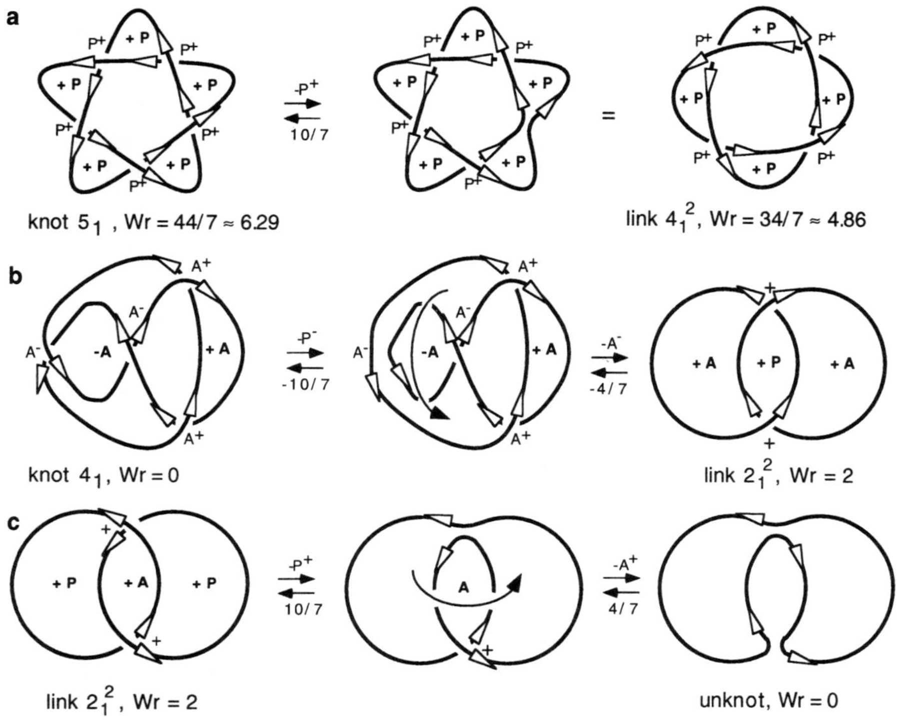

Fig. 4. Smoothing of crossings in knots and links. a. Upon smoothing of one parallel crossing knot 5_1 changes into link 4_1^2. The corresponding change of writhe corresponds to this of one parallel crossing. b. The elimination of one antiparallel crossing of a given sign in knot 4_1 in a conjunction with a simultaneous change of this knot into a link 2_1^2 causes that the effective change of writhe is the same as during the elimination of one parallel crossing of the same sign. c. In a diagram of Hopf link we are unable to find out which crossing is parallel and which is antiparallel as both crossing "belong" at the same time to double arc fields of different types. In reality both crossings are identical and have the properties of perfectly superpositioned parallel and antiparallel crossings as the two catenated rings in ideal configuration of Hopf link are perfectly perpendicular to each other and are therefore exactly in the middle between parallel and antiparallel orientation. However upon randomly chosen smoothing of one of the crossing it becomes clear that the remaining crossing is of antiparallel type and thus the smoothed crossing needed to be of parallel type, as only then the total score of writhe prediction is correct.

5 Gauge factor for knots simplification.

Simplification of knots and links can be also performed using smoothing moves where a given crossing on a diagram is eliminated by "smoothing" it whereby orientation of segments is maintained after new reconnections.(see Fig. 4).During smoothing a knot changes into a link (or contrary) and this introduces an interesting gauge factor which is needed to maintain a writhe score unchanged. During smoothing of parallel crossings a gauge factor does not enter and the change of writhe simply amounts to ±10/7 depending on the sign of the crossing. So for example a knot 5_1 with the writhe of 44/7 is changed by smoothing of one of its crossing into a link 4_1^2 with the writhe of 34/7 (see Fig. 4 and 6). However smoothing of antiparallel crossings requires that the smoothed crossing should be treated and scored as a parallel crossing of the same sign. One can say that smoothing determines the crossing as a parallel one. Figure 4b shows smoothing of an achiral knot 4_1. In the starting diagram all crossings are apparent as antiparallel. Upon first smoothing the knot changes into a link and we see that after a simple untwisting of one of the antiparallel crossings we reach the state of Hopf link. There the remaining two crossings belong at the same time to parallel and antiparallel double arc fields (Fig. 4b) and thus should be treated as an entangled pair of parallel and antiparallel crossings of positive sign (see the next paragraph). The writhe of right-handed enantiomorph of ideal Hopf link was calculated to be equal to two (see Table I). Thus, to maintain the total score of writhe simplification of achiral 4_1 knot being equal to zero it is required that the smoothed antiparallel negative (A^-) crossing should be treated as a parallel negative (P^-) crossing. Therefore when the smoothing is applied, the 4_1 knot reveals itself as composed of one pair of parallel crossings of opposite signs and one pair of antiparallel crossings of opposite signs. Comparing with simplification of 4_1 knot presented in Fig. 2 we see that when the smoothing is applied one pair of antiparallel crossings of opposite sign got changed into one pair of parallel crossings of opposite sign (other examples of replacements of crossing-anticrossing pairs will be discussed later). The gauge factor during the smoothing of antiparallel crossings enters only during the change of the knot into a link or contrary. If a second smoothing just clips off one or more turns leaving them unlinked to the rest of the knot or link then the corresponding writhe score is the same as this of a simple untwisting of an antiparallel turn.

6 Hopf link and EPR paradox.

The method of simplifications of minimal diagrams to determine the writhe of ideal geometric configuration applies also to links. Lets consider first the simplest link with a topological notation 2_1^2 called also Hopf link. Figure 4 shows that the minimal diagram of this link has neighboring double-arc fields of two different types (parallel and antiparallel). Therefore crossings in Hopf link can be considered as parallel and antiparallel at the same time. This would be analogous to a superposition state between the parallel and antiparallel state. In fact in ideal Hopf link both catenated rings are perfectly planar so there is no intrasegmental contribution to writhe. Therefore the writhe per crossing is exactly one (intersegmental contribution only) which is of course halfway in between the writhe values for parallel and antiparallel crossings (10/7 and 4/7 respectively). However when we perform simplification of the Hopf link by smoothing we notice that the superposition state collapses. Upon smoothing of one of the crossings we see that the remaining crossing becomes in the same moment determined as antiparallel crossing (Fig. 4c) and can be then simplified by untwisting (writhe score of 4/7). As discussed already for 4_1 knot, the smoothing process "determines" the smoothed crossing as a parallel one and results in a writhe score of 10/7 (or -10/7 depending on the sign of the crossing). Therefore when we perform smoothing of a Hopf link at a given crossing this crossing emerges from superposition as parallel crossing and the remaining crossing emerges as antiparallel crossing. This gives a total writhe score of 2 which is the same as the writhe of ideal configuration of Hopf link. It is difficult not to mention here the conceptual similarity between undetermined state of individual crossings (until the topological operation is performed) and the well known quantum superposition principle popularized by EPR paradox or Schrödinger cat example where the superposition collapses upon physical measurement. If for example the considered Hopf link is of such dimensions that one crossing is several light years distant from the other crossing then in the very moment when smoothing was performed at one of the crossings the other looses its superposition state and becomes determined as antiparallel crossings. Formally this example is very similar to the behavior of anti-correlated spins in pairs of photons emitted at the same time in diverging directions from a light source. As long as the measurements of spins are not performed, each photon can be considered as being potentially able to adopt with equal probabilities spins parallel or antiparallel to the direction of propagation. However as soon as the spin is measured and determined in one of the photons, the spin of the other photon gets determined and it shows opposing direction to the spin in the first photon. The collapse of the superposition happens at the same time at both photons irrespectively of their physical distance from each other. This particular quantum property of superposition collapse was contested by Einstein, Podolsky and Rosen (EPR paradox) as violating the principle that the influence can not spread with infinite speed[18]. Schrödinger cat example of superposition of states (cat which is dead and alive at the same time) was intended to demonstrate that superposition is difficult to be imagined for classical (not quantum) objects[19]. However, if we take same paper strips, scissors and glue (very classical objects) we can quite well see and contemplate that crossings in a Hopf link made out of two paper strips belong at the same time to double arc fields of parallel and antiparallel type so that we are unable to tell if they are parallel or antiparallel. In fact these two crossings show a superposition of states and have writhe value intermediate between these of parallel and antiparallel crossings. If we smooth randomly one of the crossing (cut and paste) then we see that with probability 1/2 each crossing can emerge from the superposition as parallel or antiparallel whereby we will always have one parallel and one anti parallel crossing. We can also appreciate the fact the collapse of the superposition states happens simultaneously at two crossings irrespective of their actual distance to each other. There is no violation of relativity theory as in fact there is no information exchange between the two crossings (nothing changes in the immediate vicinity of non-manipulated crossings). We thus see that such classical objects like knots and links can help us to understand the formalism of quantum physics. Also the concept of gauge fields or gauge bosons can be very well illustrated with these classical objects as knots and links. Smoothing of one of the antiparallel crossings in 4_1 knot, changes the writhe value associated to this crossing and by changing a knot into a link it immediately introduces the compensating change into an ensemble of other crossings. Amazingly one can not even see that the change is introduced as seemingly nothing changes in the remaining portion of the diagram. Only during further simplification we realize that one of the crossings also changed its type (see Fig 4). Using quantum physics formalism we can draw Feynman type diagrams relating the state of the two concerned crossings (Fig. 5). When one of the crossings is eliminated it exchanges with the rest of the knot a massless particle carrying the quantum of writhe whereby the time life of the particle is so short that it can not be even observed as such and we only see its effect. In this respect the instantaneous exchange of quantum of writhe corresponds to the exchange of virtual gauge bosons known from the quantum physics[20].

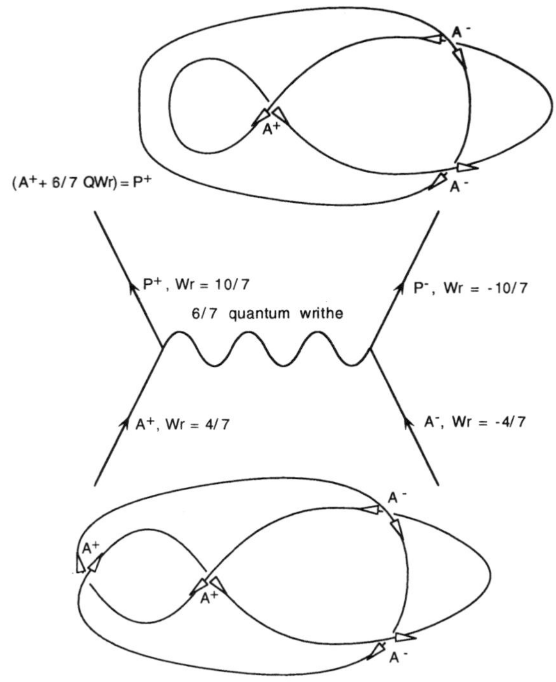

Fig. 5. Feynman diagram of writhe changes during smoothing of an antiparallel crossing. The eliminated antiparallel crossing behaves as if it takes with itself an additional quantum of writhe being equal to 6/7. This quantum of writhe is depleted from the rest of the knot.

7 Linking number and the writhe scoring method.

In continuing to demonstrate that the method of writhe scoring for each type of crossings allows to predict writhe of ideal knots and links we consider now the link 4_1^2 in its right handed-form and where the direction assigned to both components is parallel (Fig. 6a). In its standard diagram this link has 4 double arc fields which are all parallel right-handed. Upon the first simplification passage which reduces the writhe by 2.86 (20/7) the catenane reaches Hopf-link stage with the writhe of 2. Thus right-handed parallel 4_1^2 catenanes are expected to have the writhe of 4.86 (34/7). Indeed it is the value of writhe obtained for the simulated ideal shape of this link (Ben Laurie, personal communication). We can also apply the smoothing operation to the link 4_1^2. Upon first smoothing reducing one right-handed parallel crossing we reach the right-handed trefoil stage (Fig.6b). Since the writhe of trefoil was determined already to be 3.43 (Fig.2) and the reduction of one parallel right-handed crossing changes the writhe by 1.43 thus we can conclude again that writhe of the link 4_1^2 amounts to 4.86. When we change the direction of one of the component of the original diagram of the link 4_1^2 we obtain antiparallel orientation of the two components and double arcs field are now of antiparallel type and are flanked by left-handed crossings (Fig. 6c). The first simplification passage changes the writhe by 1.14 and we are left with negative Hopf link having the writhe of -2. Thus the total writhe of the antiparallel 4_1^2 link is expected to be -3.14. The value of writhe calculated for ideal representation of antiparallel 4_1^2 link is -3.15 (Ben Laurie, personal communication). Can we perhaps go further and predict the writhe resulting from intra-component crossings in 4_1^2 link? To do this we have to remember that the writhe of individual components depends only on their internal crossings and is independent of inter-ring crossings. As indicated earlier each right-handed winding associated with one crossing introduces into knots an intra-coil writhe of 3/7. In case of links this contribution of writhe would be redistributed into the writhe of individual components. When the right-handed link 4_1^2 is simplified as shown in Fig. 4a, the first strand passage eliminates two crossings and changes intra-component writhe by 6/7. After the first strand passage the Hopf link state is reached, where each of the components is already planar and thus has the writhe zero. Therefore we can conclude that writhe of one component plus writhe of the other component of the link 4_1^2 should amount to 6/7 or 0.857. In fact simulations of ideal configurations of the link 4_1^2 [21] revealed that sum of writhe values of the two rings forming the link amounts to 0.855 (Ben Laurie, personal communication). It is good to point out here that writhe values of individual components in a link does not depend on the change of their direction. In contrast to it the total writhe for the whole link changes when direction of one of the components is reversed and linking of the two components changes the sign. Link 4_1^2 that winds in right-handed direction has a linking of +2 for parallel and -2 for antiparallel orientation. The formula: $Wr(a+b) = Wr(a) + Wr(b) + 2Lk(ab)$ [22] connects the total writhe of the whole link composed of elements a and b with the sum of writhe of the individual components and the linking number between the two components. It provides independent verification of our method of writhe scoring. Since $Wr(a) + Wr(b) = 6/7$ and the linking number of the link 4_1^2 is plus or minus 2, therefore the total writhe of the link should be $6/7 + 4 = 4.86$ for parallel orientation of the two rings and $6/7 - 4 = -3.14$ for the antiparallel orientation. Indeed, as shown earlier, our method of writhe estimation allowed us to obtain the same corresponding values of the total writhe for parallel and antiparallel 4_1^2 links. To obtain these values we did not need to consider the linking number of two components and the writhe of the components. Therefore one can conclude that the writhe prediction method works very well for all tried knots and links. Available data support also the proposal that the writhe of all knot and links in their ideal geometric configuration assumes quantized values.

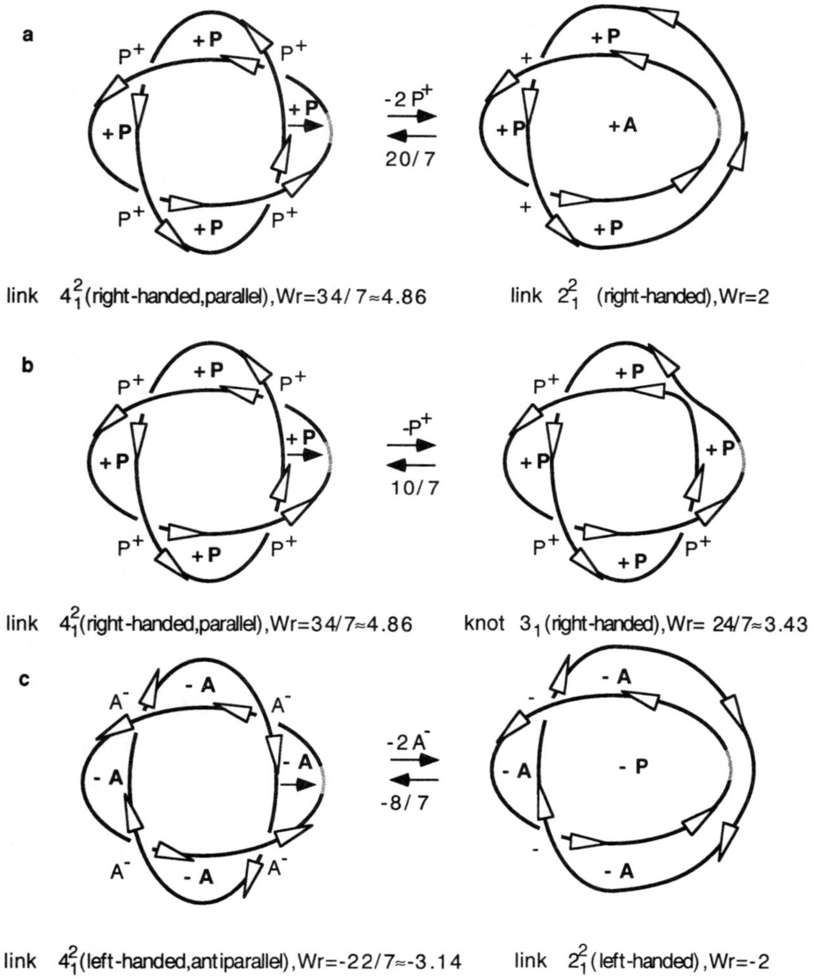

Fig. 6. Simplification of links. a, b. Simplification by passage of segments (a) and by smoothing (b) gives the same writhe score. c. Writhe of the links depends on the relative orientation of two composing curves. The difference of the writhe upon change of the relative orientation of two curves is twice bigger than the difference in the linking number resulting from the change of relative orientation of the two curves.

8 Knotted magnetic flux lines.

In our discussion of ideal knots and links we now make a quantum leap and consider magnetic flux lines closed into knots. From theoretical work of Moffatt we know that knotted magnetic flux lines are expected to be quite stable and that they will tend to shorten their length and thus will stabilize adopting shapes close to ideal geometric configurations of the corresponding knots[3, 7]. From experimental observations of magnetic flux lines in solar corona we know that from time to time they undergo reconnection[6] which is topologically analogous to the smoothing of knots and links. During reconnections of magnetic flux lines in solar corona great amounts of energy are being released.

The discussed above elements provide a good framework for the following model. Lets imagine a very complicated knot formed by a microscopic magnetic flux line. It will tend toward ideal configuration (shortest possible path) of this knot and should remain in this form until local fluctuations will allow the reconnection between the touching segments of the knots. The tendency of magnetic flux lines to get shorter will select these reconnection events which lead to a simplification of the knot and permit shortening of the flux lines. After temporal stabilization at ideal configuration of newly formed knot or link a new reconnection event can happen which will allow a further shortening of the flux line. This process can be repeated until the lowest energy state is achieved; unknot or a simple ideal knot where the orientation of segments in ideal configuration is not suited for further reconnection events[23, 24]. Shortening of magnetic flux lines upon reconnections can be an extremely rapid process that is associated with emission of the energy released during the shortening. The magnetic flux line in the lowest energy state or in one of excited states can also absorb some energy that could then cause lengthening of the line and introduce wild vibration into a knot. During this phase some reconnections between segments of the flux line could increase the knot complexity and thus stabilize it for some time in a higher energy state. As the high energy knot considered here is a very complex one there are many possible pathways of its simplification (Fig. 7). Some reconnection events could allow simplification in one stroke and other in many different steps. However every simplification step resulting from the recombination changes the writhe in steps of 4/7 or 10/7 and in various combinations of these values. If we draw a schematic cascade of possible changes of writhe we obtain a spectrum where every level is separated from any other level by exact natural multiplicity of writhe value of 2/7 (Fig. 7). The spectrum does not need to be a regular monotone one as some of the levels may not be realizable as knots or links. Of course the writhe spectrum is not the energy spectrum and decrease of energy does not need to cause the decrease of writhe so for example the writhe can grow during simplification of complicated achiral knots. Numerical simulations would be needed to find out if progressive shortening of magnetic flux lines forming different knots indeed causes strict energy quantization of accessible knot types. The pictured here spontaneous simplification of knotted magnetic flux lines can be relevant to such macro objects like magnetic flux lines in solar corona and to such micro objects as magnetic flux lines in atomic nucleus. In early 70ties Jehle discussed that closed quantized magnetic flux line provide an interesting model for electrons[25] whereby knotted magnetic flux lines were also proposed to constitute neutrinos and quarks[26]. Present observation of quantized writhe of lowest energy states of different knots might stimulate new interest in these early theoretical studies. It is important to stress here that in the presented here models the quantization of writhe does not arise from ad hoc postulate of permitted energy levels (as in Bohr model) but is intrinsic to the physical objects considered here.

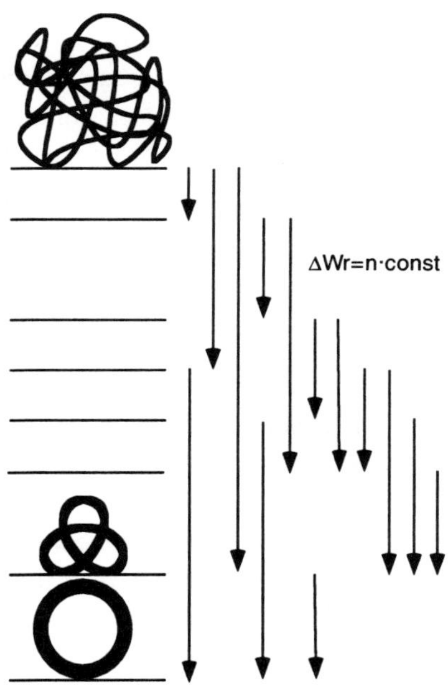

Fig. 7. Quantum states during relaxation of a knot. A knotted magnetic flux line can relax toward a ground state via reconnections occurring when the knot is very close to its ideal configuration[6]. The residence time in a form close to ideal configuration can be very long as compared to the time needed to find new ideal configuration after the reconnection. During relaxation and shortening of the magnetic flux line the energy of the system goes down and the initial high energy can be progressively lost by emitting photons with energies corresponding to the energy difference between two different stable conformation of knots made out of the same magnetic flux line. In the shown example the starting knot has all crossings with a positive contribution to writhe. Therefore its relaxation is always associated with the reduction of writhe.

9 Topology and quantization of writhe.

Until now in topology only the linking number was known to show quantization properties. Irrespectively of the actual shape of a given type of link its linking number is always a topological invariant and adopts only integer values. Writhe changes with the actual shape of the knot or link and for any given knot it can in principle adopt any value from $-\infty$ to $+\infty$. However ideal shapes of knots and links are in fact topological invariants of knots and links and thus their writhe becomes also a topological invariant. This does not explain yet why the writhe is quantized for ideal configuration of knots. Knots in their ideal configurations lose all unnecessary length and thus each knot becomes composed of integer number of torus or twist half-turns. So for example trefoil knot in its ideal configuration has exactly two toroidal half-turns and one twist half-turn. If there would be less than two toroidal half-turn or less than one twist half-turn the knot could not be closed. If there would be a fraction more than the minimal required number of twist and torus half-turns then the configuration would not be ideal. In minimal diagrams of knots and links an integer number of twist and torus half-turns is projected on a plane. Therefore minimal diagrams are very well suited to reveal what is the number of torus and twist half-turns a given knot is composed of. Quantization of writhe is not limited to ideal knots and links only. Also random knots as for example formed by knotted cyclic polymers in solution show quantization of writhe but to observe it one needs to consider time averaged writhe of a given knotted polymer undergoing a stochastic thermal motion. We showed earlier that time averaged writhe of randomly fluctuating knotted polymers forming a given knot is very close to the writhe of ideal knots[1]. Average writhe of a big sample of random knots of a given type built on a cubic lattice is also the same as the writhe of the corresponding ideal knot[27]. Therefore observed here quantization of writhe of knots and links might be a very general phenomenon related to different type of knotted physical objects.

10 Conclusions.

Some formal similarities between quantum description of elementary particles and the description of knots and links are discussed here. It remains to be seen if this similarity is only formal or if indeed understanding of knots and links will help us to understand elementary particles and quantum physics.

Acknowledgements.

I thank Jacques Dubochet for his constant friendly support and encouragement, Piotr Pieranski, Corinne Cerf and Akos Dobay for discussions. This work was supported by Swiss National Foundation grant 31-42158.94 to Jacques Dubochet and AS.

References

1. V. Katritch, J. Bednar, D. Michoud, R. G. Scharein, J. Dubochet and A. Stasiak, 1996, Geometry and physics of knots, *Nature* **384**, 142-145.
2. A. Stasiak, J. Dubochet, V. Katritch and P. Pieranski, 1998, Ideal knots and their relation to the physics of real knots, in *Ideal Knots* (eds. A. Stasiak, S. Katritch and L. H. Kauffman) 1-19 (World Scientific: Singapore).
3. H. K. Moffatt, 1990, The energy spectrum of knots and links, *Nature* **347**, 367-369.
4. A. Y. Grosberg, A. Feigel and Y. Rabin, 1996, Flory-type theory of a knotted ring polymer, *Phys. Rev. A* **54**, 6618-6622.
5. A. Malevanets and R. Kapral, 1996, Links, knots, and knotted labyrinths in bistable systems, *Phys. Rev. Lett.* **77**, 767-770.
6. R. L. Ricca and M. A. Berger, 1996, Topological ideas and fluid mechanics, *Physics Today* 28-34.
7. A. Y. K. Chui and H. K. Moffatt, 1995, The energy and helicity of knotted magnetic flux tubes, *Proc. Roy. Soc. Lond. A* **451**, 609-629.
8. V. Katritch, W. K. Olson, P. Pieranski, J. Dubochet and A. Stasiak, 1997, Properties of ideal composite knots, *Nature* **388**, 148-151.
9. A. D. Bates and A. Maxwell, 1993,*DNA Topology* 1-120 (IRL Press: Oxford).
10.A. V. Vologodskii, 1992,*Topology and physics of circular DNA* 1-179 (CRC Press: Boca Raton).
11. F. B. Fuller, 1978, Decomposition of the linking number of a closed ribbon: a problem from molecular biology, *Proc. Natl. Acad. Sci. USA* **75**, 3557-3561.
12. P. Pieranski, 1998, In search of ideal knots, in *Ideal Knots* (eds. A. Stasiak, W. Katritch and L. H. Kauffman) 20-41 (World Scientific: Singapore).
13. B. Laurie, 1998, Annealing ideal knots and links: methods and pitfalls, in *Ideal Knots* (eds. A. Stasiak, V. Katritch and L. H. Kauffman) 42-51 (World Scientific: Singapore).
14. C. Cerf, 1997, Nullification writhe and chirality of alternating links, *Journal of Knot Theory and Its Ramification* **6**, 621-632.
15. C. Liang and K. Mislow, 1994, On amphicheiral knots, *J. Math. Chem.* **15**, 1-34.

16. D. Rolfsen, 1976,*Knots and links* (Publish or Perish Press.: Berkeley, CA.).
17. C. C. Adams, 1994,*The Knot Book* (W.H. Freeman and Company: New York).
18. A. Einstein, B. Podolsky and N. Rosen, 1935, Can quantum-mechanical description of physical reality be considered complete?, *Phys. Rev.* **47**, 777-780.
19. K. Mainzer, 1996,*Symmetries of Nature : a handbook for philosophy of nature and science* (Walter de Gruyter: Berlin).
20. F. Wilczek, 1998, The standard model transcended, *Nature* **394**, 13-15.
21. B. Laurie, V. Katritch, J. M. Sogo, J. Dubochet and A. Stasiak, 1998, Geometry and physics of catenanes applied to study DNA replication, *Biophysical Journal* **74**, 2815-2822.
22. C. Liang, C. Cerf and K. Mislow, 1996, Specification of chirality for links and knots, *J. Math. Chem.* **19**, 241-263.
23. H. Aref and I. Zawadzki, 1991, Linking of vortex rings, *Nature* **354**, 50-53.
24. M. V. Melander and F. Hussain, 1994, Topological vortex dynamics in axisymmetric viscous flows, *J. Fluid Mech.* **260**, 57-80.
25. H. Jehle, 1971, Relationship of flux quantization to charge quantization and the electromagnetic coupling constant, *Physical Review D* **3**, 306-345.
26. H. Jehle, 1972, Flux quantization and particle physics, *Physical Review D* **6**, 441-457.
27. E. J. Janse van Rensburg, E. Orlandini, D. W. Sumners, M. C. Tesi and S. G. Whittington, 1997, The writhe of knots in the cubic lattice, *Journal of Knot Theory and its Ramification* **5**, 31-44.

Mutant links distinguished by degree 3 Gauß sums

A. Stoimenow

e-mail: stoimeno@informatik.hu-berlin.de,
WWW: http://www.informatik.hu-berlin.de/~stoimeno

Current version: September 30, 1999 First version: December 2, 1997

Abstract

We give examples showing that the Fiedler solid torus degree 3 Gauß sum invariants can be used to detect mutation of links.
Keywords: mutation, Gauß sums *AMS subject classification:* 57M25

1 Introduction

The notion of a mutant was introduced by Conway [Co]. Mutants have been later intensively studied and received much attention especially because of the difficulties they provide to distinguish them [P, P2, LL, LM, MC, MT, MR, MS, CD]. Beside the problem of detecting orientation they are the largest class of links left indistinguishable by the knot polynomials [H, J, Ka]. Although the knot polynomials and the related quantum theory are not completely useless for detecting mutation, the applicable methods result in calculations, which are (almost) infeasible already for the simplest examples, see [MR, MC, MS].

A conceptually new approach to finding invariants was initiated by Fiedler for braids [Fi2] and later by Viro-Polyak [PV] for invariants of finite degree [BN, BL, Va, Vo]. Recently, Fiedler [Fi] generalized the approach to invariants of degree 3 for knots in the solid torus [Go] and more generally in orientable S^1- and line bundles over surfaces. These invariants can be used to study two component links $K \cup T$ in $\mathbb{R}^3$ with an unknotted component T by the fact, that for isotopic links $K_1 \cup T_1 \sim K_2 \cup T_2$ in $\mathbb{R}^3$ it holds $K_1 \sim K_2$ in the solid torus $S^3 \setminus T_1 \simeq S^3 \setminus T_2$.

Now if $K_1 \sim K_2$ are isotopic knots in $\mathbb{R}^3$, then $K_i \cup T_i$, $i = 1,2$ are isotopic links, where T_i is the meridian of K_i (K_i may be replaced by some satellite around itself). This rendered it possible to examine knots in $\mathbb{R}^3$ by the new invariants in this way. Unfortunately, it turned out [St], that so the Fiedler invariants do not give more than the usual degree 3 Vassiliev invariant for knots in $\mathbb{R}^3$, explaining the preceeding series of disappointing experimental results [FS], in particular the failure to distinguish mutants. Further experiments with mutants suggested that the situation might not be better, when we place the trivial component somewhere else than to be the meridian of the knotted component (or of its companion). This opened the possibility, that the Fiedler invariants do not detect mutation at all.

Here we present examples of mutations in which the rotated tangle contains parts of both components. Computation with the program of [FS] showed, that in this case the mutants could be distinguished.

It is worth remarking that this method to distinguish mutants will not work with the various solid torus link polynomial invariants coming from Markov traces on Hecke algebras of type B [HK, La, GL] as they also satisfy skein relations, and so the same arguments which apply for the HOMFLY polynomial on mutants (see [Li]) similarly apply for its analogues in the solid torus as well. For the polynomials of generalized Hecke algebras of [La2] the skein arguments do not apply in full generality, but concrete examples of distinguished mutants are not known (at least to me) due to lack of practical computations. In any case, however, the arguments apply, when the geometric linkning number in the mutated tangle of the strands belonging to T and to K is ± 2 and the strand of T has no self-crossings, as in our examples below.

But anyway, there is never an absoulte certainty that computers really do what we tell them or we really tell them what we want, so we encourage an independent verification of our result.

The spirit of this paper is intended to be similar to this of [MR]. We would, however, like to convince the reader, that the new invariants are simpler than quantum invariants and appear as a promising alternative.

The concept of these new invariants was outlined in [FS]. However, no one of the formulas given there serves in our case, so, to make the result verifiable, we need to make new definitions. But then, also to encourage more experiments, we felt it would be better to give a complete list of all invariants.

2 Gauß diagrams and Gauß sums

We start by briefly recalling the definition of our invariants. See [FS] for details.

Consider a knot $K : S^1 \hookrightarrow \mathbb{R}^3$ (S^1 and $\mathbb{R}^3$ oriented). Decompose $\mathbb{R}^3 = \mathbb{R}^2 \oplus \mathbb{R}$ so that the projection (henceforth called *knot diagram*) of K into $\mathbb{R}^2$ is generic. To this projection we can assign a *Gauß diagram* (GD), a circle with oriented chords, by connecting points in S^1 mapped to a crossing and orienting the chord from the preimage of the undercrossing to the preimage of the overcrossing. See [PV].

Figure 1 shows the knot 6_2 in its standard projection and the corresponding Gauß diagram.

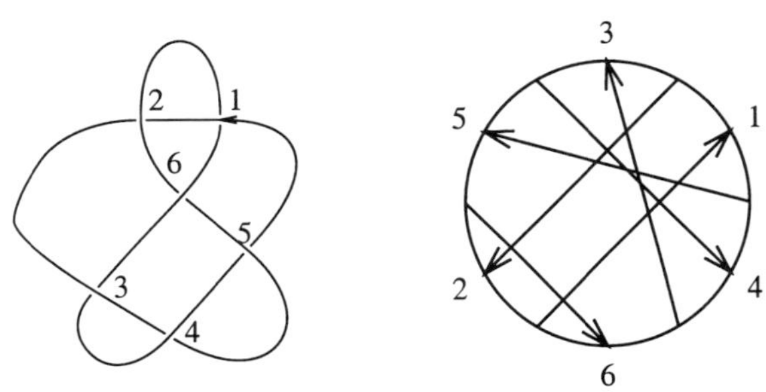

Figure 1: The knot 6_2 and its Gauß diagram.

A *Gauß sum* of degree k is a term assigned to a knot diagram, which is of the following form

$$\sum_{\substack{\text{ordered choices of } k \text{ crossings of the} \\ \text{knot diagram, whose arrows in the GD} \\ \text{form a given subdiagram}}} \text{function}\big(\text{ data, assigned to the crossings }\big).$$

Each summand we will call *weight* and the function *weight function*. We will denote the summation by the subdiagram itself, which we will also call *configuration*.

Now we need to specify the data assigned to the crossings.

Definition 2.1 *The* winding index *of a plane curve $C \subset \mathbb{R}^2 = \mathbb{C}$ around a point $p \notin C$ is*

$$w(C,p) := \frac{1}{2\pi i} \oint_C \frac{1}{z-p} dz.$$

Pictorially it measures how many times the curve "walks" around p, counting reverse walk negatively.

Definition 2.2 *The* Whitney index $n(C)$ *of a plane curve* C *is the degree of the map*

$$\frac{C'}{||C'||} : S^1 \longrightarrow S^1 .$$

The Whitney index of a knot diagram is the Whitney index of its underlying plane curve.

Definition 2.3 *The* writhe $w(D)$ *of a knot diagram* D *is the sum of the writhes of all crossings (see figure 2).*

Figure 2: The writhe of a crossing.

Example 2.1 The standard projection of 6_2 on figure 1 has Whitney index 1 and writhe -2.

Definition 2.4 *A* smoothing *of a crossing is the procedure*

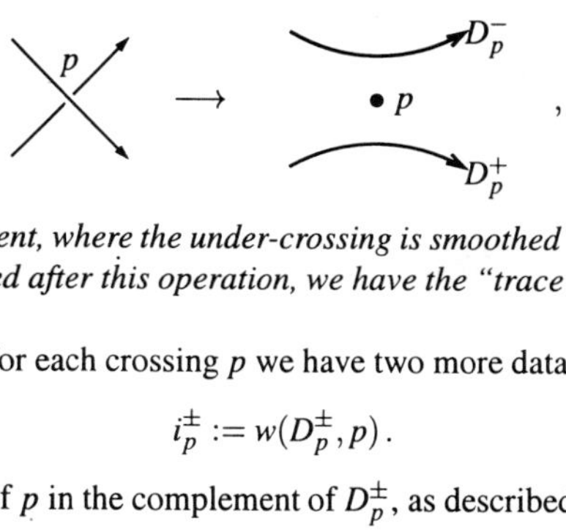

where D_p^+ *denotes the component, where the under-crossing is smoothed to the over-crossing. Note, that beside the link diagram resulted after this operation, we have the "trace" of* p *in its complement.*

Apart from its writhe w_p, for each crossing p we have two more data:

$$i_p^{\pm} := w(D_p^{\pm}, p) .$$

Here by p we mean the trace of p in the complement of $D_p^{\pm}$, as described above. Set

$$i_p := i_p^+ + i_p^- , \qquad \delta_p := i_p^+ - i_p^- .$$

Now consider a two component link $K \cup T$ in S^3 where T is the trivial knot (unknot). Let K, T, S^3 be oriented. Deform $K \cup T$ in $S^3 = \mathbb{R}^3 \cup \{\infty\}$ so that $\infty \in T$. This isotopy is unique up to isotopy. Such a link we can represent choosing an appropriate projection $\mathbb{R}^3 \to \mathbb{R}^2$ as knot with a point in its complement, on which T projects, assuming the orientation of T to be from the sheet of paper to the reader's eye.

Definition 2.5 *The* type *of a crossing* p *is* $w(D_p^+, T)$ mod 2.

All invariants are regular isotopy invariants. A regular isotopy is an ambient isotopy preserving the writhe w and Whitney index n of K.

We distinguish two cases according to the parity of linking number $lk(K,T) = w(K,T)$.

For the definition we need the following atomary terms. Here near any arrow in the configuration its name and the type of the crossing is indicated by '0', '1' or by '*', if both types are allowed. A chord denotes an arrow whose orientation does not matter. The default weight function is $w_p w_q w_r$, which is not written.

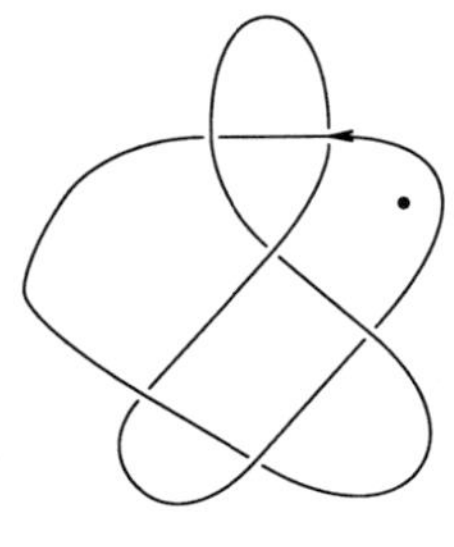

Figure 3: The knot 6_2 with its meridian depicted in our favourable way.

A_1 = 

$w_p \cdot w_q \cdot (i_p^+ + i_p^- - i_q^+ - i_q^-)$

A_2 = 

$w_p \cdot w_q \cdot (i_p^+ + i_p^- - i_q^+ - i_q^-)$

A_3 = 

$w_p \cdot w_q \cdot (i_p^+ + i_p^- - i_q^+ - i_q^-)$

A_4 = 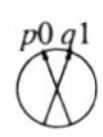

$w_p \cdot w_q \cdot (i_p^+ + i_p^- - i_q^+ - i_q^-)$

A_{13} = 

A_{14} = 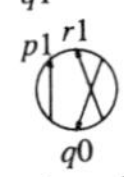

A_{15} = 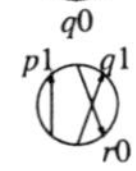

A_{16} = 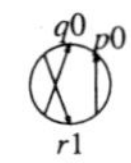

$A_5 = \frac{1}{3} \cdot$ 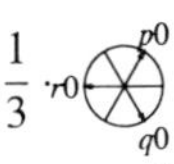

$A_6 = \frac{1}{3} \cdot$ 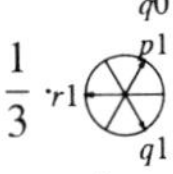

A_7 = 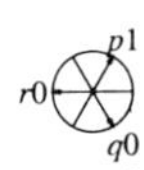

A_8 = 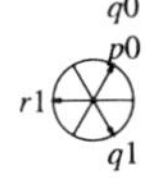

A_{17} = 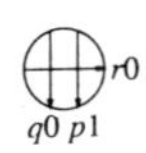

A_{18} = 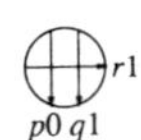

A_{19} = 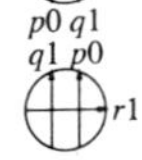

A_{20} = 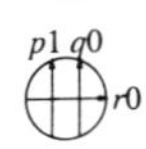

A_9 = 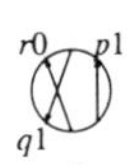

A_{10} = 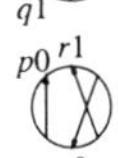

A_{11} = 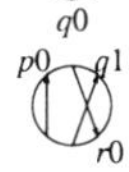

A_{12} = 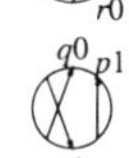

A_{21} = 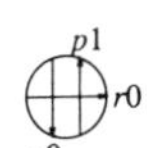

A_{22} = 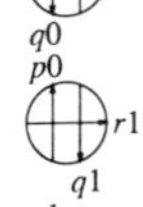

A_{23} = 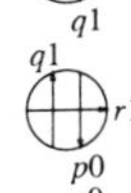

A_{24} = 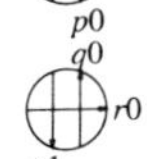

$A_{25} =$ 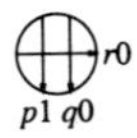

$A_{26} =$ 

$A_{27} =$ 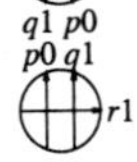

$A_{28} =$ 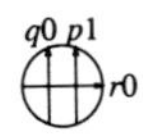

$A_{29} =$ 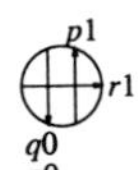

$A_{30} =$ 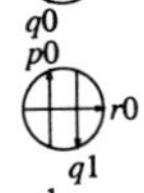

$A_{31} =$ 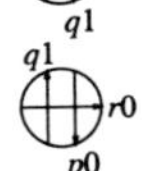

$A_{32} =$ 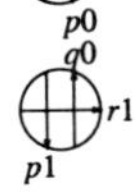

$A_{33} = \frac{1}{2} \cdot$ 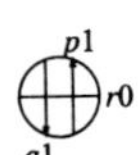

$A_{34} = \frac{1}{2} \cdot$ 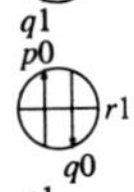

$A_{35} = \frac{1}{2} \cdot$ 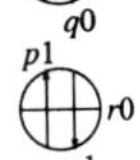

$A_{36} = \frac{1}{2} \cdot$ 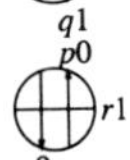

$A_{37} = \frac{1}{2} \cdot$ 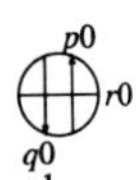

$A_{38} = \frac{1}{2} \cdot$ 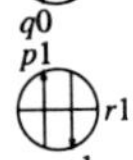

$A_{39} = \frac{1}{2} \cdot$ 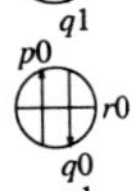

$A_{40} = \frac{1}{2} \cdot$ 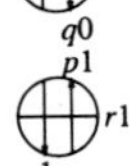

$A_{41} =$ 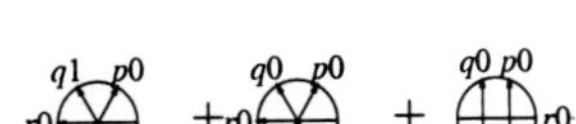

$+\frac{1}{4} \cdot$  $(w_p + w_q)$

$A_{42} =$ 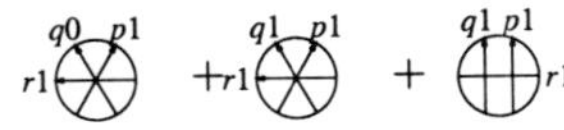

$+\frac{1}{4} \cdot$  $(w_p + w_q)$

$A_{43} =$  w_p

$A_{44} =$  w_p

$A_{45} =$ 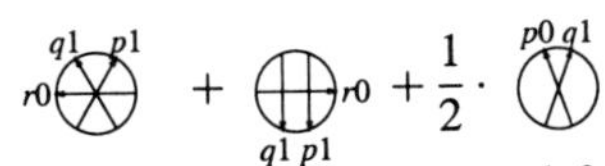 w_p

$A_{46} =$ 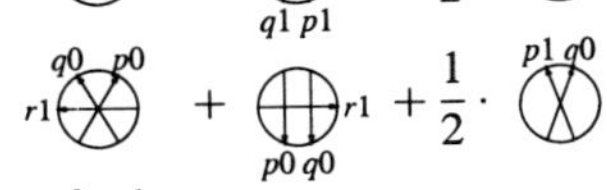 w_p

$A_{47} =$ 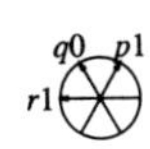

$A_{48} =$ 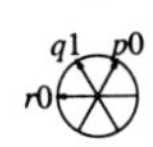

$A_{49} = \frac{1}{4} \cdot$ 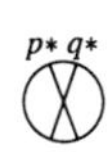 $w_p \cdot w_q \cdot (i_p^+ + i_p^- + i_q^+ + i_q^-)$

$A_{50} = \frac{1}{2} \cdot$ 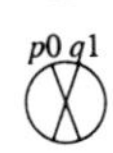 $w_p \cdot w_q \cdot (i_p^+ + i_p^- + i_q^+ + i_q^-)$

$A_{51} =$ 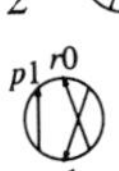

$A_{52} =$ 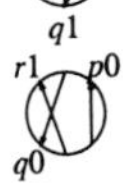

$A_{53} =$ 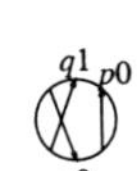

$A_{54} =$ 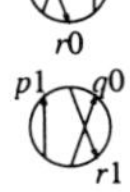

$A_{55} =$ 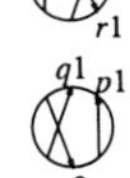

$A_{56} =$ 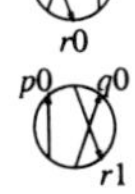

A_{57} = 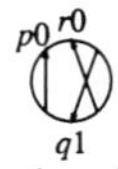

A_{58} = 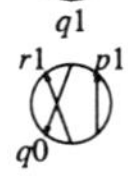

A_{59} = 

A_{60} = 

A_{61} = 

A_{62} = 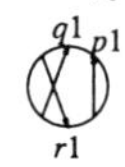

A_{63} = 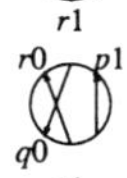

A_{64} = 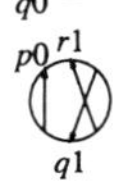

A_{65} = 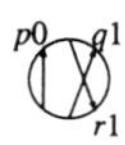

A_{66} = 

A_{67} = 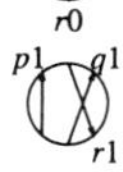

A_{68} = 

A_{69} = 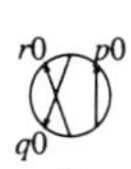

A_{70} = 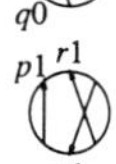

A_{71} = 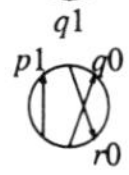

A_{72} = 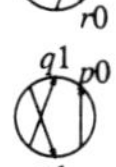

A_{73} = 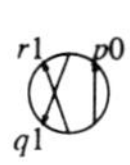

A_{74} = 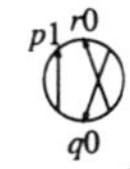

A_{75} = 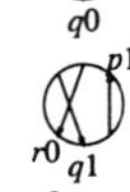

A_{76} = 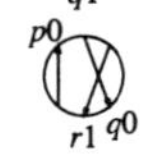

A_{77} = 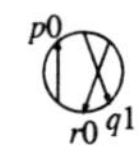

A_{78} = 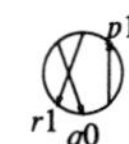

A_{79} = 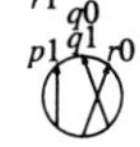

A_{80} = 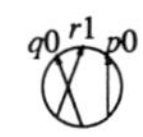

A_{81} = 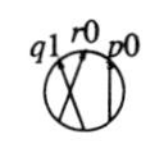

A_{82} = 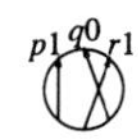

A_{83} = 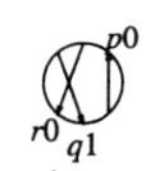

A_{84} = 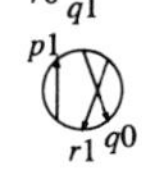

A_{85} = 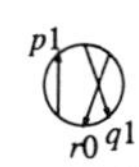

A_{86} = 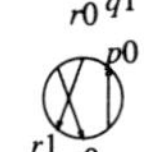

A_{87} = 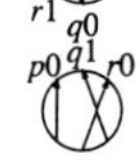

A_{88} = 

A_{89} = 

A_{90} = 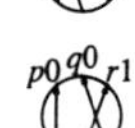

2.1 Even linking number invariants

For the case $w(K,T) \equiv 0 \pmod 2$ we obtained 39 invariants. They are given below.

$$
\begin{aligned}
In_1 &= A_{56} - A_{57} - A_{87} + A_{90}, \\
In_2 &= A_{55} - A_{58} - A_{59} + A_{62} - A_{88} + A_{89}, \\
In_3 &= A_{52} - A_{53} - A_{83} + A_{86}, \\
In_4 &= A_{51} - A_{54} - A_{67} + A_{70} - A_{84} + A_{85}, \\
In_5 &= A_{14} - A_{15} - A_{67} + A_{70} - A_{79} + A_{82}, \\
In_6 &= A_{13} - A_{16} - A_{80} + A_{81}, \\
In_7 &= -A_9 + A_{12} - A_{59} + A_{62} - A_{75} + A_{78}, \\
In_8 &= -A_{10} + A_{11} - A_{76} + A_{77}, \\
In_9 &= -A_1 + A_6 + A_7 + A_9 + A_{14} - A_{17} + 2 \cdot A_{21} + A_{23} + A_{25} + A_{29} + A_{51} \\
&\quad + A_{58} + A_{59} + A_{63} + A_{70} + A_{74}, \\
In_{10} &= A_1 - 2 \cdot A_6 - 2 \cdot A_7 - A_8 + A_{10} + A_{13} + A_{17} - 2 \cdot A_{21} - A_{23} - A_{25} - A_{29} + A_{52} \\
&\quad + A_{57} + A_{64} + A_{73}, \\
In_{11} &= A_1 - 2 \cdot A_6 - 2 \cdot A_7 - A_8 + A_{10} + A_{16} + A_{17} - 2 \cdot A_{21} - A_{23} - A_{25} - A_{29} + A_{53} \\
&\quad + A_{57} + A_{64} + A_{72}, \\
In_{12} &= -A_1 + A_6 + A_7 + A_9 + A_{15} - A_{17} + 2 \cdot A_{21} + A_{23} + A_{25} + A_{29} + A_{54} \\
&\quad + A_{58} + A_{59} + A_{63} + A_{67} + A_{71}, \\
In_{13} &= -A_5 + A_{60} + A_{69}, \\
In_{14} &= -A_5 + A_{60} + A_{68}, \\
In_{15} &= -A_9 + A_{12} + A_{55} - A_{58} - A_{59} + A_{62} - A_{63} + A_{66}, \\
In_{16} &= -A_{10} + A_{11} + A_{56} - A_{57} - A_{64} + A_{65}, \\
In_{17} &= -A_{60} + A_{61}, \\
In_{18} &= -A_7 + 2 \cdot A_{17} + A_{18} - A_{19} - A_{21} - A_{24} + A_{48}, \\
In_{19} &= 2 \cdot A_6 + 2 \cdot A_7 + A_8 + A_{19} + A_{21} + A_{23} + A_{24} + 2 \cdot A_{25} + A_{26} + A_{29} + 2 \cdot A_{43} + A_{47}, \\
In_{20} &= -A_1 + A_{18} - A_{19} + A_{21} - A_{24} - A_{43} + A_{46}, \\
In_{21} &= -2 \cdot A_1 - A_3 + 2 \cdot A_6 + 2 \cdot A_7 - 2 \cdot A_{17} - A_{18} + 2 \cdot A_{19} + 2 \cdot A_{21} + A_{23} + 4 \cdot A_{25} + A_{26} \\
&\quad + A_{29} - 2 \cdot A_{33} + 2 \cdot A_{43} + 2 \cdot A_{45}, \\
In_{22} &= -4 \cdot A_1 - A_3 + 2 \cdot A_6 + 2 \cdot A_7 - 2 \cdot A_{17} + A_{18} + 2 \cdot A_{21} - 2 \cdot A_{22} + 3 \cdot A_{23} + 4 \cdot A_{25} + A_{26} \\
&\quad + A_{29} - 2 \cdot A_{33} + 2 \cdot A_{43} + 2 \cdot A_{44}, \\
In_{23} &= 4 \cdot A_1 + A_3 - A_6 + A_8 + 2 \cdot A_{17} + A_{19} - 2 \cdot A_{21} + A_{22} - 2 \cdot A_{23} + 2 \cdot A_{24} - 2 \cdot A_{25} + A_{26} \\
&\quad - A_{29} + 2 \cdot A_{33} + A_{42}, \\
In_{24} &= A_5 - A_7 + 2 \cdot A_{17} + A_{18} - A_{19} - A_{21} - A_{24} + A_{41}, \\
In_{25} &= 2 \cdot A_1 + A_3 + 2 \cdot A_6 + 2 \cdot A_8 + 2 \cdot A_{17} + A_{18} - A_{23} - 2 \cdot A_{25} - A_{26} + A_{29} + 2 \cdot A_{33} + 2 \cdot A_{40}, \\
In_{26} &= 2 \cdot A_5 + A_{39}, \\
In_{27} &= 4 \cdot A_1 + A_3 + 2 \cdot A_6 + 2 \cdot A_8 + 2 \cdot A_{17} - A_{18} - 2 \cdot A_{21} + 2 \cdot A_{22} - A_{23} + 2 \cdot A_{24} - 2 \cdot A_{25} + A_{[illegible]} \\
&\quad - A_{29} + 2 \cdot A_{33} + 2 \cdot A_{38}, \\
In_{28} &= 2 \cdot A_5 + A_{37}, \\
In_{29} &= -2 \cdot A_1 - A_3 + 2 \cdot A_6 + 4 \cdot A_7 + 2 \cdot A_8 - 2 \cdot A_{17} - A_{18} + 4 \cdot A_{21} + A_{23} + 2 \cdot A_{25} + A_{26} \\
&\quad + 3 \cdot A_{29} + 2 \cdot A_{36},
\end{aligned}
$$

$$
\begin{aligned}
In_{30} &= -3\cdot A_1 - A_3 - 2\cdot A_{17} + A_{21} - A_{22} + A_{23} - A_{24} + 2\cdot A_{25} - A_{33} + A_{35},\\
In_{31} &= -2\cdot A_1 + A_3 + 2\cdot A_6 + 4\cdot A_7 + 2\cdot A_8 - 2\cdot A_{17} + A_{18} + 4\cdot A_{21} + 3\cdot A_{23} + 2\cdot A_{25} - A_{26}\\
&\quad + A_{29} + 2\cdot A_{34},\\
In_{32} &= -A_{21} + A_{24} - A_{29} + A_{32},\\
In_{33} &= A_1 + A_{17} - A_{21} - A_{25} + A_{31},\\
In_{34} &= A_1 + A_{17} - A_{21} + A_{22} - A_{23} - A_{25} + A_{30},\\
In_{35} &= 2\cdot A_1 - A_{18} + A_{19} - A_{21} + A_{22} - A_{23} + A_{24} - A_{25} + A_{28},\\
In_{36} &= -2\cdot A_1 + A_{18} - A_{19} + A_{21} - A_{22} + A_{23} - A_{24} - A_{26} + A_{27},\\
In_{37} &= -A_{17} - A_{18} + A_{19} + A_{20},\\
In_{38} &= 2\cdot A_1 + A_3 + A_4,\\
In_{39} &= -A_1 + A_2
\end{aligned}
$$

2.2 Odd linking number invariants

For the case $w(K,T) \equiv 1 \pmod 2$ we obtained 36 invariants. They are given in terms of the expressions

$$
\begin{aligned}
In_1 &= A_{56} - A_{57} - A_{60} + A_{61} - A_{87} + A_{90},\\
In_2 &= A_{55} - A_{58} - A_{59} + A_{62} - A_{88} + A_{89},\\
In_3 &= A_{52} - A_{53} - A_{68} + A_{69} - A_{83} + A_{86},\\
In_4 &= A_{51} - A_{54} - A_{67} + A_{70} - A_{84} + A_{85},\\
In_5 &= -A_{67} + A_{70} - A_{79} + A_{82},\\
In_6 &= -A_{68} + A_{69} - A_{80} + A_{81},\\
In_7 &= -A_9 + A_{12} - A_{59} + A_{62} - A_{75} + A_{78},\\
In_8 &= -A_{10} + A_{11} - A_{60} + A_{61} - A_{76} + A_{77},\\
In_9 &= 3\cdot A_1 + A_3 - 2\cdot A_5 - 6\cdot A_6 - 4\cdot A_7 - 4\cdot A_{13} + 4\cdot A_{14} + A_{17} + A_{19} - 6\cdot A_{21} + 4\cdot A_{22} - A_{23}\\
&\quad + A_{24} - 2\cdot A_{25} - 5\cdot A_{29} + 3\cdot A_{30} - A_{49} + 8\cdot A_{51} + 8\cdot A_{70} + 8\cdot A_{74},\\
In_{10} &= 3\cdot A_1 + A_3 - 6\cdot A_5 - 2\cdot A_6 - 4\cdot A_8 + 4\cdot A_{13} - 4\cdot A_{14} + A_{17} + A_{19} + 2\cdot A_{21} - 4\cdot A_{22} - A_{23}\\
&\quad + A_{24} - 2\cdot A_{25} + 3\cdot A_{29} - 5\cdot A_{30} + A_{49} + 8\cdot A_{52} + 8\cdot A_{69} + 8\cdot A_{73},\\
In_{11} &= 3\cdot A_1 + A_3 - 6\cdot A_5 - 2\cdot A_6 - 4\cdot A_8 + 4\cdot A_{13} - 4\cdot A_{14} + A_{17} + A_{19} + 2\cdot A_{21} - 4\cdot A_{22} - A_{23}\\
&\quad + A_{24} - 2\cdot A_{25} + 3\cdot A_{29} - 5\cdot A_{30} + A_{49} + 8\cdot A_{53} + 8\cdot A_{68} + 8\cdot A_{72},\\
In_{12} &= 3\cdot A_1 + A_3 - 2\cdot A_5 - 6\cdot A_6 - 4\cdot A_7 - 4\cdot A_{13} + 4\cdot A_{14} + A_{17} + A_{19} - 6\cdot A_{21} + 4\cdot A_{22} - A_{23}\\
&\quad + A_{24} - 2\cdot A_{25} - 5\cdot A_{29} + 3\cdot A_{30} - A_{49} + 8\cdot A_{54} + 8\cdot A_{67} + 8\cdot A_{71},\\
In_{13} &= -3\cdot A_1 - A_3 + 2\cdot A_5 - 2\cdot A_6 - 4\cdot A_7 + 8\cdot A_{12} + 4\cdot A_{13} + 4\cdot A_{14} - A_{17} - A_{19} - 2\cdot A_{21}\\
&\quad + 4\cdot A_{22} + A_{23} - A_{24} + 2\cdot A_{25} - 3\cdot A_{29} + 5\cdot A_{30} + A_{49} + 8\cdot A_{55} + 8\cdot A_{62} + 8\cdot A_{66},\\
In_{14} &= -3\cdot A_1 - A_3 - 2\cdot A_5 + 2\cdot A_6 - 4\cdot A_8 + 8\cdot A_{11} + 4\cdot A_{13} + 4\cdot A_{14} - A_{17} - A_{19} + 6\cdot A_{21}\\
&\quad - 4\cdot A_{22} + A_{23} - A_{24} + 2\cdot A_{25} + 5\cdot A_{29} - 3\cdot A_{30} - A_{49} + 8\cdot A_{56} + 8\cdot A_{61} + 8\cdot A_{65},\\
In_{15} &= -3\cdot A_1 - A_3 - 2\cdot A_5 + 2\cdot A_6 - 4\cdot A_8 + 8\cdot A_{10} + 4\cdot A_{13} + 4\cdot A_{14} - A_{17} - A_{19} + 6\cdot A_{21}\\
&\quad - 4\cdot A_{22} + A_{23} - A_{24} + 2\cdot A_{25} + 5\cdot A_{29} - 3\cdot A_{30} - A_{49} + 8\cdot A_{57} + 8\cdot A_{60} + 8\cdot A_{64},\\
In_{16} &= -3\cdot A_1 - A_3 + 2\cdot A_5 - 2\cdot A_6 - 4\cdot A_7 + 8\cdot A_9 + 4\cdot A_{13} + 4\cdot A_{14} - A_{17} - A_{19} - 2\cdot A_{21}\\
&\quad + 4\cdot A_{22} + A_{23} - A_{24} + 2\cdot A_{25} - 3\cdot A_{29} + 5\cdot A_{30} + A_{49} + 8\cdot A_{58} + 8\cdot A_{59} + 8\cdot A_{63},\\
In_{17} &= A_1 + A_3 - 4\cdot A_5 - A_{17} - A_{19} + 2\cdot A_{21} + A_{23} - A_{24} - 2\cdot A_{25} + A_{29} - 3\cdot A_{30} - 4\cdot A_{43}
\end{aligned}
$$

$$
\begin{aligned}
& -2\cdot A_{47}+2\cdot A_{48}, \\
In_{18} &= -A_1+A_{18}-A_{19}+A_{21}-A_{24}-A_{43}+A_{46}, \\
In_{19} &= -A_1-A_3+2\cdot A_5+2\cdot A_6+A_{17}+A_{19}-A_{23}-A_{24}+2\cdot A_{25}+A_{29}+A_{30} \\
& +2\cdot A_{43}+2\cdot A_{45}, \\
In_{20} &= -3\cdot A_1-A_3+2\cdot A_5+2\cdot A_6+A_{17}+2\cdot A_{18}-A_{19}-2\cdot A_{22}+A_{23}-A_{24}+2\cdot A_{25} \\
& +A_{29}+A_{30}+2\cdot A_{43}+2\cdot A_{44}, \\
In_{21} &= -A_1+A_3-2\cdot A_6+2\cdot A_7+A_{17}+2\cdot A_{18}-A_{19}+2\cdot A_{21}-2\cdot A_{22}+A_{23}+A_{24} \\
& +2\cdot A_{25}-A_{29}-A_{30}+2\cdot A_{42}, \\
In_{22} &= -A_1+A_3-2\cdot A_5+2\cdot A_8+A_{17}+2\cdot A_{18}-A_{19}+3\cdot A_{23}-A_{24}+2\cdot A_{25}-A_{29} \\
& -A_{30}+2\cdot A_{41}, \\
In_{23} &= 3\cdot A_1+A_3+4\cdot A_6+2\cdot A_7+2\cdot A_8+A_{17}+A_{19}+2\cdot A_{21}-A_{23}+A_{24}-2\cdot A_{25}+3\cdot A_{29} \\
& -A_{30}+4\cdot A_{40}, \\
In_{24} &= 3\cdot A_1+A_3+4\cdot A_5+2\cdot A_7+2\cdot A_8+A_{17}+A_{19}-2\cdot A_{21}+4\cdot A_{22}-A_{23}+A_{24}-2\cdot A_{25} \\
& -A_{29}+3\cdot A_{30}+4\cdot A_{39}, \\
In_{25} &= -A_1+A_3+4\cdot A_6+2\cdot A_7+2\cdot A_8+A_{17}-3\cdot A_{19}+2\cdot A_{21}+3\cdot A_{23}+A_{24}+2\cdot A_{25} \\
& -A_{29}-A_{30}+4\cdot A_{38}, \\
In_{26} &= -A_1+A_3+4\cdot A_5+2\cdot A_7+2\cdot A_8-3\cdot A_{17}+A_{19}+2\cdot A_{21}+3\cdot A_{23}+A_{24}+2\cdot A_{25} \\
& -A_{29}-A_{30}+4\cdot A_{37}, \\
In_{27} &= -5\cdot A_1-3\cdot A_3+4\cdot A_5+2\cdot A_7+2\cdot A_8+A_{17}-3\cdot A_{19}+2\cdot A_{21}-A_{23}-3\cdot A_{24}+2\cdot A_{25} \\
& +3\cdot A_{29}+3\cdot A_{30}+4\cdot A_{36}, \\
In_{28} &= -5\cdot A_1-3\cdot A_3+4\cdot A_6+2\cdot A_7+2\cdot A_8-3\cdot A_{17}+A_{19}+2\cdot A_{21}-A_{23}-3\cdot A_{24}+2\cdot A_{25} \\
& +3\cdot A_{29}+3\cdot A_{30}+4\cdot A_{35}, \\
In_{29} &= 3\cdot A_1+A_3+4\cdot A_5+2\cdot A_7+2\cdot A_8+A_{17}+A_{19}-2\cdot A_{21}+4\cdot A_{22}-A_{23}+A_{24}-2\cdot A_{25} \\
& -A_{29}+3\cdot A_{30}+4\cdot A_{34}, \\
In_{30} &= 3\cdot A_1+A_3+4\cdot A_6+2\cdot A_7+2\cdot A_8+A_{17}+A_{19}+2\cdot A_{21}-A_{23}+A_{24}-2\cdot A_{25}+3\cdot A_{29} \\
& -A_{30}+4\cdot A_{33}, \\
In_{31} &= -A_{21}+A_{24}-A_{29}+A_{32}, \\
In_{32} &= -A_{22}+A_{23}-A_{30}+A_{31}, \\
In_{33} &= 2\cdot A_1-A_{18}+A_{19}-A_{21}+A_{22}-A_{23}+A_{24}-A_{25}+A_{28}, \\
In_{34} &= -A_{25}+A_{27}, \\
In_{35} &= 2\cdot A_1-A_{18}+A_{19}-A_{21}+A_{22}-A_{23}+A_{24}-A_{25}+A_{26}, \\
In_{36} &= -A_{17}-A_{18}+A_{19}+A_{20}, \\
In_{37} &= -A_{13}+A_{16}, \\
In_{38} &= -A_{14}+A_{15}, \\
In_{39} &= 2\cdot A_1+A_3+A_4, \\
In_{40} &= -A_1+A_2
\end{aligned}
$$

From these terms all In_i with $i<9$ or $i>16$ and In_i+In_{i+1} for $i=9,11,13,15$ are invariants.

3 The examples

Figure 4 shows 2 link mutants of an unknot and a positive trefoil. The mutation "interchanges" the two components.

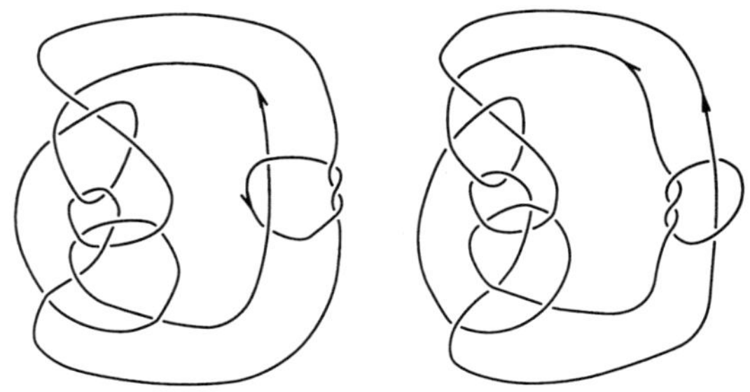

Figure 4: The link mutants with $lk = 1$.

The same links are transformed into our favourable diagram and shown on figure 5.

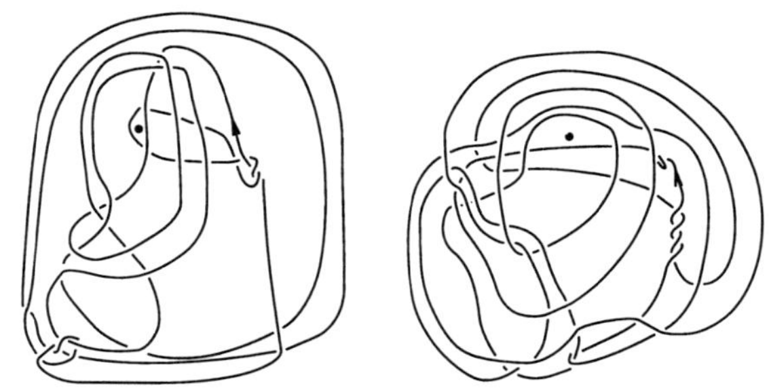

Figure 5: The link mutants of figure 4 depicted in our favourable way.

They have the same linking number $lk = w(K,T) = 1$ (as should be), and are adjusted K to have the same n and w. We verified that the degree 3 Vassiliev invariant vt_3 of [FS] on K is in both cases 4 (as for the positive trefoil). The computer calculation took just about 2 minutes and gave the following result:

```
file: k-mut1
lk(K,T)=1
w=0
n=3
invt     k

  1      0
  2      0
  3      0
  4      0
  5      0
  6      0
  7      0
  8      0
  9      -18
 10      -34
 11      -34
 12      -18
```

```
file: k-mut2a
lk(K,T)=1
w=0
n=3
invt     k

  1      0
  2      0
  3      0
  4      0
  5      0
  6      0
  7      0
  8      0
  9      -1
 10      -19
 11      -19
 12      -1
```

13	10	13	-7
14	10	14	-5
15	10	15	-5
16	10	16	-7
17	-2	17	-2
18	0	18	0
19	2	19	2
20	2	20	2
21	2	21	2
22	2	22	2
23	2	23	2
24	2	24	2
25	2	25	2
26	2	26	2
27	2	27	2
28	2	28	2
29	2	29	2
30	2	30	2
31	0	31	0
32	0	32	0
33	0	33	0
34	0	34	0
35	0	35	0
36	0	36	0
37	0	37	0
38	0	38	0
39	0	39	0
40	0	40	0

If the mutants were the same, both diagrams would have to be isotopic and would have the same invariants. However, $In_{13} + In_{14}$ is once 20 and another time -12.

Figure 6 shows 2 link mutants of an unknot and a positive trefoil with $lk = 0$. The mutation "interchanges" again the two components.

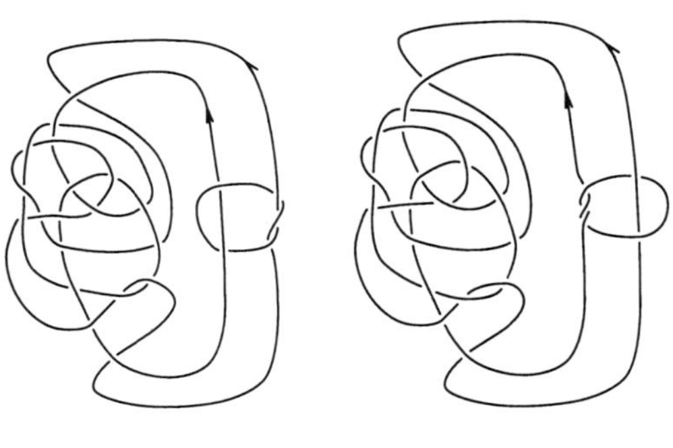

Figure 6: The link mutants with $lk = 0$.

The same links are transformed into our favourable diagram and shown on figure 7, and are adjusted K to have the same n and w. We verified again that vt_3 on K is in both cases 4.

The computer calculation gave the following result:

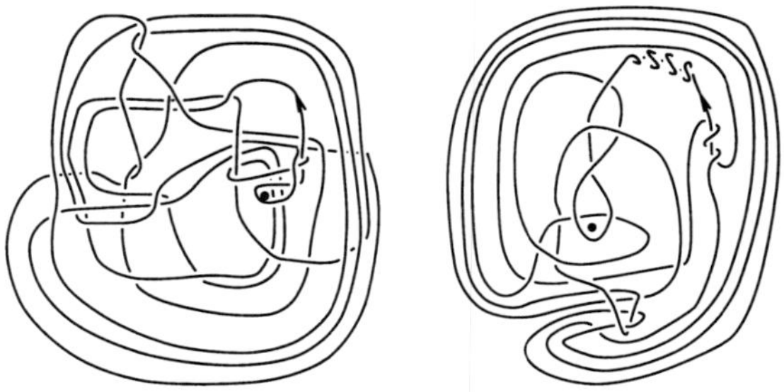

Figure 7: The link mutants of figure 6 depicted in our favourable way.

file: k-mut3
mod=2
w=7
n=2

invt	k lk=0
1	0
2	0
3	0
4	0
5	0
6	0
7	0
8	0
9	214
10	-168
11	-168
12	214
13	-43
14	-43
15	0
16	0
17	0
18	0
19	624
20	0
21	624
22	624
23	-188
24	-432
25	-188
26	-216
27	-188
28	-216
29	624
30	0
31	624
32	0

file: k-mut4
mod=2
w=7
n=2

invt	k lk=0
1	0
2	0
3	0
4	0
5	0
6	0
7	0
8	0
9	2
10	-4
11	-4
12	2
13	5
14	5
15	0
16	0
17	0
18	0
19	4
20	0
21	4
22	4
23	0
24	0
25	0
26	0
27	0
28	0
29	4
30	0
31	4
32	0

33	0	33	0
34	0	34	0
35	0	35	0
36	0	36	0
37	0	37	0
38	0	38	0
39	0	39	0

Here the difference is even clearer.

Acknowledgement. I would wish to thank to T. Fiedler and S. Lambropoulou for their helpful remarks.

4 References

[BN] D. Bar-Natan, *On the Vassiliev knot invariants*, Topology **34** (1995) 423–472.

[BL] J. S. Birman and X-S. Lin, *Knot polynomials and Vassiliev's invariants*, Invent. Math. **111** (1993) 225–270.

[CD] S. Chmutov, S. Duzhin, *Kontsevich integral*, Preprint of Max-Planck-Institut für Mathematik (Bonn, August 1997); available via anonymous file transfer from `pier.botik.ru`, subdirectory `pub/local/zmr`, file name `ki.ps.gz`.

[Co] J. H. Conway, *On enumeration of knots and links*, in "Computational Problems in abstract algebra" (J. Leech, ed.), 329–358. Pergamon Press, 1969.

[Fi] T. Fiedler, *New knot invariants*, UPS Toulouse preprint, to appear.

[Fi2] —— " ——, *A small state sum for knots*, Topology **32 (2)** (1993), 281–294.

[FS] —— " —— and A. Stoimenow, *New knot and link invariants*, submitted to these Proceedings.

[GL] M. Geck and S. Lambropoulou, *Markov traces and knot invariants related to Iwahori–Hecke algebras of type B*, J. reine angew. Math. **482** (1997), 191–213.

[Go] V. Goryunov, *Finite Order Invariants of Framed Knots in a solid torus and in Arnold's J^+-Theory of plane curves*, "Geometry and Physics", Lecture Notes in Pure & Appl. Math. **184**, M. Dekker, New York, 1996, 549–556.

[HK] J. Hoste and M. Kidwell, *Dichromatic link invariants*, Trans. Amer. Math. Soc. **321 (1)** (1990), 197–229.

[H] P. Freyd, J. Hoste, W. B. R. Lickorish, K. Millett, A. Ocneanu and D. Yetter, *A new polynomial invariant of knots and links*, Bull. Amer. Math. Soc. **12** (1985), 239–246.

[J] V. F. R. Jones, *A polynomial invariant of knots and links via von Neumann algebras*, Bull. Amer. Math. Soc. **12** (1985), 103–111.

[Ka] L. H. Kauffman, *An invariant of regular isotopy*, Trans. Amer. Math. Soc. **318** (1990), 417–471.

[La] S. Lambropoulou, *Solid torus links and Hecke algebras of B-type*, Proceedings of the Conference on Quantum Topology (D. N. Yetter, ed.), World Scientific, 1994, 225–245.

[La2] —— " —— , *Knot theory related to generalized and cyclotomic Hecke algebras of type B*, Mathematica Gottingensis Schriftenreihe, Heft **25** (1997).

[Li] W. B. R. Lickorish, *Linear skein theory and link polynomials*, Top. Appl. **27** (1987), 265–274.

[LL] —— " —— and A. S. Lipson, *Polynomials of 2-cable-like links*, Proc. Amer. Math. Soc. **100** (1987), 355–361.

[LM] —— " —— and K. C. Millett, *A polynomial invariant for oriented links*, Topology **26 (1)** (1987), 107–141.

[MS] H. R. Morton and H. B. Short, *The 2-variable polynomial of cable knots*, Math. Proc. Camb. Philos. Soc. **101** (1987), 267–278.

[MC] —— " —— and P. R. Cromwell, *Distinguishing mutants by knot polynomials*, Jour. of Knot Theory and its Ramifications **5(2)** (1996), 225–238.

[MT] —— " —— and P. Traczyk, *The Jones polynomial of satellite links around mutants*, In 'Braids', ed. Joan S. Birman and Anatoly Libgober, Contemporary Mathematics **78**, Amer. Math. Soc. (1988), 587–592.

[MR] —— " —— and H. J. Ryder, *Mutants and* $SU(3)_q$ *invariants*, preprint 1997.

[PV] M. Polyak and O. Viro, *Gauss diagram formulas for Vassiliev invariants*, Int. Math. Res. Notes **11** (1994) 445–454.

[P] J. Przytycki, *Search for different links with the same Jones' type polynomials: Ideas from graph theory and statistical mechanics*, Panoramas of Mathematics, Banach Center Publications, Vol. **34**, Warszawa 1995.

[P2] —— " ——, *Equivalence of cables of mutants of knots*, Canad. J. Math., **41 3(2)**, 1989, 250–273.

[St] A. Stoimenow, *Gauss sum invariants, Vassiliev invariants and braiding sequences*, to appear in J. Of Knot Theory and Its Ram.

[Va] V. A. Vassiliev, *Cohomology of knot spaces*, Theory of Singularities and its Applications (Providence) (V. I. Arnold, ed.), Amer. Math. Soc., Providence, 1990.

[Vo] P. Vogel, *Algebraic structures on modules of diagrams*, Université Paris VII preprint, June 1996.

E_6 TURAEV-VIRO-OCNEANU INVARIANT OF LENS SPACE $L(p,1)$

KOUTAROU SUZUKI *
NTT Laboratories
1-1 Hikarinooka, Yokosuka, Kanagawa, 239-0847 Japan
koutarou@isl.ntt.co.jp

ABSTRACT

The Turaev-Viro-Ocneanu invariant of three manifolds is a generalization of the Turaev-Viro invariant of three manifolds, and is derived from the paragroup of the subfactor. We investigate the E_6 Turaev-Viro-Ocneanu invariant derived from the E_6 subfactor. We give a formula for the invariant of the lens space $L(p,1)$. The value of the E_6 Turaev-Viro-Ocneanu invariant can be a complex number, in contrast with the fact that the value of the Turaev-Viro invariant must be a real number because of the relation "Turaev-Viro invariant"=|"Reshetikhin-Turaev invariant"$|^2$. Thus the E_6 Turaev-Viro-Ocneanu invariant can distinguish the lens space $L(p,1)$ from the same one with reversed orientation $L(p,p-1)$ for $p = 12k+3, 12k+4, 12k+8, 12k+9$ while the Turaev-Viro invariant can not.

Keywords: Turaev-Viro-Ocneanu invariant, quantum $6j$-symbol, paragroup, E_6 subfactor, lens space.

1. Introduction

The Turaev-Viro-Ocneanu invariant of three manifolds [1] is a generalization of the Turaev-Viro invariant of three manifolds [2] defined by using the state sum on a triangulation of a three manifold, and is derived from the subfactor theory. In the case of the Turaev-Viro-Ocneanu invariant, the dimension of an intertwiner space can be more than one. Accordingly we give colors not only to each edge but also to each face of a simplex of a simplicial decomposition of a three manifold. The Turaev-Viro-Ocneanu invariant is defined by using the quantum $6j$-symbols derived from the paragroup which appear in the classification of hyperfinite II_1-subfactors by the A_n, D_{2n}, E_6, E_8 Coxeter graph [3]. In the case of the A_{k+1} Coxeter graph, we have the $SU(2)$ level k Turaev-Viro invariant. In the case of the E_6, E_8 Coxeter graph, on the other hand, there are few computations of the invariants.

By using the E_6 quantum $6j$-symbols derived by M.Izumi [4] from the E_6 subfactor, we investigate the E_6 Turaev-Viro-Ocneanu invariant.

We compute the invariant for the lens space $L(p,1)$. (M.Wakui calculated it for $p = 1,2$ [5]. Using a different method, K.Habiro made calculations for $p = 1,2,3$, and M.Okamoto and C.Sato for $p = 4,5$ [6].)

*This work was done when the author was in Graduate School of Mathematical Sciences, University of Tokyo.

Theorem 1 *The Turaev-Viro-Ocneanu invariant associated with the E_6 quantum $6j$-symbol for the lens space $L(p,1)$ is given by*

$$Z(L(p,1)) = \frac{1}{12}(5 + 2(-1)^p + i^p + 2e^{-\frac{5}{6}\pi i p} + (1 + (-1)^p)e^{-\frac{1}{3}\pi i p}).$$

By Theorem 1 we see that invariants $Z(L(p,1))$ have period 12, and these values are as follows.

p	$12k+1$	$12k+2$	$12k+3$	$12k+4$	$12k+5$	$12k+6$
$Z(L(p,1))$	$(3-\sqrt{3})/12$	$1/2$	$(1-i)/4$	$(3+\sqrt{3}i)/6$	$(3+\sqrt{3})/12$	$1/2$

$12k+7$	$12k+8$	$12k+9$	$12k+10$	$12k+11$	$12k+12$
$(3+\sqrt{3})/12$	$(3-\sqrt{3}i)/6$	$(1+i)/4$	$1/2$	$(3-\sqrt{3})/12$	1

From this table we see that the values of the invariant $Z(L(p,1))$ for $p = 12k+3, 12k+4, 12k+8, 12k+9$ are complex numbers, in contrast with the fact that the value of the Turaev-Viro invariant must be a real number because of the relation "Turaev-Viro invariant"=|"Reshetikhin-Turaev invariant"$|^2$. If we reverse the orientaion of the three manifold, the value of the Turaev-Viro-Ocneanu invariant becomes the complex conjugate of the value of the original one. This means that the E_6 Turaev-Viro-Ocneanu invariant can distinguish the lens space $L(p,1)$ from the same one with reversed orientation $L(p,p-1)$ for $p = 12k+3, 12k+4, 12k+8, 12k+9$ while the Turaev-Viro invariant can not.

Corollary 2 *The Turaev-Viro-Ocneanu invariant associated with the E_6 quantum $6j$-symbol distinguishes $L(p,1)$ from $L(p,p-1)$ for $p = 12k+3, 12k+4, 12k+8, 12k+9$.*

In section 2 we recall the definition of the Turaev-Viro-Ocneanu invariant following Evans and Kawahigashi [3]. In section 3 we construct a table of the E_6 quantum $6j$-symbols by using Izumi's result [4]. In section 4 we give a proof of Theorem 1.

2. Turaev-Viro-Ocneanu Invariant

In this section we recall the definition of the Turaev-Viro-Ocneanu invariant following Evans and Kawahigashi [3]. Let T be a triangulation of a closed three manifold M. We fix an order of the vertices of triangulation T, and assign the orientation to each edge according to the order of vertices.

We define the coloring of triangulation T. A coloring λ of edges is a map from the set of oriented edges $|v_0v_1|$ $(v_0 < v_1)$ of triangulation T to basis $\{X_i\}$ of fusion rule algebra $\mathcal{A}$

$$\lambda : \{oriented\ edges\ of\ T\} \longrightarrow \{X_i\}.$$

For a coloring λ of edges, we assign an element of orthonormal basis $\{s_i\}$ of intertwiner space $\mathcal{H}^{\lambda(|v_0v_2|)}_{\lambda(|v_0v_1|)\lambda(|v_1v_2|)}$ to each face $|v_0v_1v_2|$ $(v_0 < v_1 < v_2)$ of triangulation T. We call this assignment a coloring ϕ of faces

$$\phi(|v_0v_1v_2|) = s_i \in \mathcal{H}^{\lambda(|v_0v_2|)}_{\lambda(|v_0v_1|)\lambda(|v_1v_2|)}.$$

The coloring of edges and faces yields colored triangulation (T, λ, ϕ) and colored simplex (σ, λ, ϕ) as

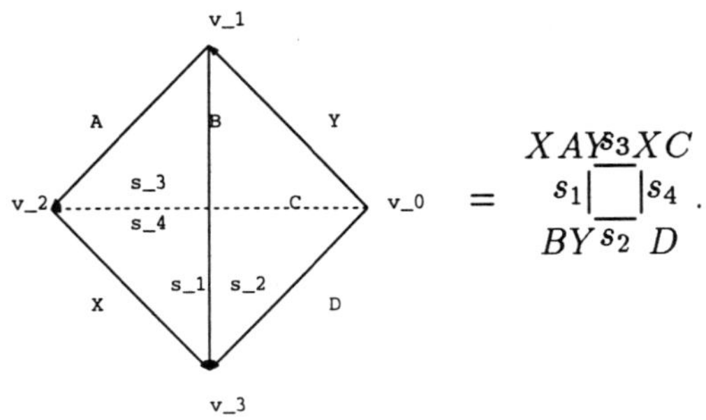

We denote the colored simplex by the colored rectangle as depicted above; the four edges of the rectangle correspond to the four faces of the simplex and the four vertices of the rectangle to the four oriented paths on the simplex from vertex v_0 to vertex v_3.

For each colored simplex (σ, λ, ϕ) of colored triangulation (T, λ, ϕ), we have the quantum $6j$-symbol $Z(\sigma, \lambda, \phi) \in \mathbf{C}$ as explained in section 3. Using this we define the Turaev-Viro-Ocneanu invariant.

Definition 3 (Turaev-Viro [2],Ocneanu [1],Evans-Kawahigashi [3]) *For triangulation T of a closed three manifold M and an order of its vertices, we define*

$$Z(M,T) = \omega^{-V} \sum_{\lambda} [\prod_{E:edge\ of\ T} \mu(\lambda(E))] \sum_{\phi} [\prod_{\sigma:simplex\ of\ T} Z(\sigma, \lambda, \phi)]$$

where λ/ϕ runs over all colorings of edges/faces, and $V = \#\{vertices\ of\ T\}$, $\omega = \sum_i \mu(X_i)^2$.

Now we consider two local moves of the triangulation depicted in Figure 1, and call these moves Pachner move I and II. Using these moves, we have the following theorem by Pachner [7].

Theorem 4 (Pachner [7]) *Let M be a compact three manifold possibly with boundary ∂M. Let T_1 and T_2 be two arbitrary triangulations of M that coincide with each other on boundary ∂M i.e. $T_1|_{\partial M} = T_2|_{\partial M}$. These two triangulations T_1 and T_2 can be transformed into each other by a finite sequence of Pachner moves I and II which does not change the triangulation on boundary ∂M.*

By the tetrahedral symmetry of quantum $6j$-symbols, $Z(M,T)$ does not depend on the order of the vertices of triangulation T. By the pentagon identity and the unitarity of quantum $6j$-symbols, $Z(M,T)$ is invariant under Pachner move I and II of a triangulation. Therefore, by Theorem 4, $Z(M,T)$ does not depend on triangulation T, and is an invariant of closed three manifolds. This yields the following theorem.

Theorem 5 (Turaev-Viro [2],Ocneanu [1],Evans-Kawahigashi [3]) *$Z(M,T)$ does not depend on triangulation T and the order of its vertices. Thus it is an invariant of closed three manifolds.*

We call this the Turaev-Viro-Ocneanu invariant and denote it by $Z(M)$.

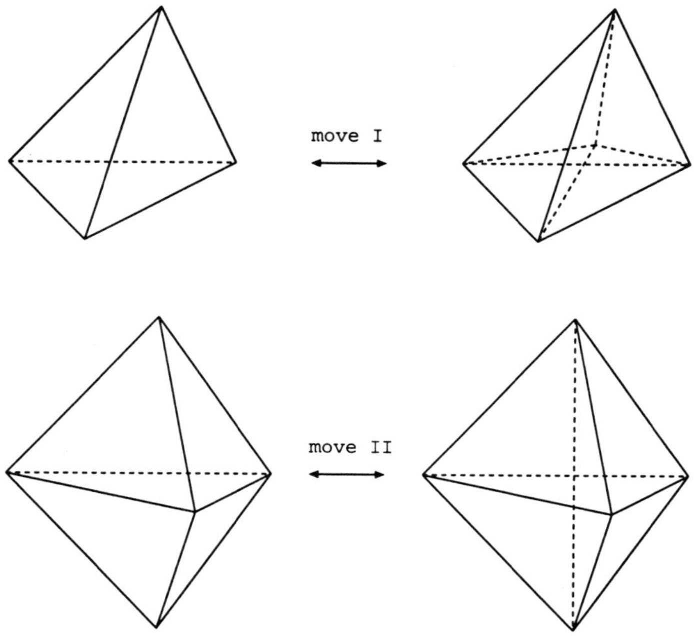

Figure 1: Pachner move I,II

3. E_6 Quantum $6j$-Symbol

In this section we make a table of the E_6 quantum $6j$-symbols by using Izumi's result [4]. M.Izumi gave the following description of the E_6 quantum $6j$-symbols of the E_6 subfactor [4].

The E_6 fusion rule algebra $\mathcal{A}$ is an algebra over $\mathbf{C}$ with basis $\{id, \rho, \alpha\}$ that satisfies the following multiplication rules

$$id\ a = a\ id = a \ \ (a = id, \alpha, \rho), \quad \alpha^2 = id, \quad \rho\alpha = \alpha\rho = \rho, \quad \rho^2 = id + \alpha + 2\rho.$$

The quantum dimension $\mu \in \mathbf{R}_{>0}$ of these elements is given by

$$\mu(\rho) = d = 1 + \sqrt{3}, \ \mu(\alpha) = \mu(id) = 1$$

and we set

$$\omega = \mu^2(id) + \mu^2(\alpha) + \mu^2(\rho) = 6 + 2\sqrt{3}.$$

Let $\mathcal{O}_4$ be a C^*-algebra over $\mathbf{C}$ generated by S_1, S_2, S_3, S_4 and unit element 1 which satisfy the following relations

$$S_i^* S_j = \delta_{i,j} 1 \ (i, j = 1, 2, 3, 4), \ \ S_1 S_1^* + S_2 S_2^* + S_3 S_3^* + S_4 S_4^* = 1.$$

E_6 fusion rule algebra $\mathcal{A}$ acts on C^*-algebra $\mathcal{O}_4$ as

$$\alpha(S_1) = S_2, \ \ \alpha(S_2) = S_1, \ \ \alpha(S_3) = S_3, \ \ \alpha(S_4) = -S_4,$$

$$
\begin{aligned}
\rho(S_1) &= \frac{S_1+S_2}{d} + \frac{e^{\pi i/4}S_3S_3 + e^{-\pi i/4}S_4S_4}{\sqrt{d}}, \\
\rho(S_2) &= \left(\frac{S_1-S_2}{d} + \frac{e^{\pi i/4}S_4S_3 + e^{-\pi i/4}S_3S_4}{\sqrt{d}}\right)U, \\
\rho(S_3) &= c_1\left(\frac{S_1+S_2}{\sqrt{2}}S_3^* + \frac{S_1-S_2}{\sqrt{2}}S_4^*\right) \\
&\quad +c_2(S_3(S_1S_1^*+S_2S_2^*)+S_4(S_1S_1^*-S_2S_2^*)) \\
&\quad +c_3(S_3S_3S_3^*+S_4S_3S_4^*)+c_4(S_3S_4S_4^*+S_4S_4S_3^*), \\
\rho(S_4) &= c_1\left(\frac{S_1+S_2}{\sqrt{2}}S_3^* - \frac{S_1-S_2}{\sqrt{2}}S_4^*\right) \\
&\quad +ic_2(S_3(S_1S_1^*+S_2S_2^*)-S_4(S_1S_1^*-S_2S_2^*)) \\
&\quad +ic_3(S_3S_3S_3^*-S_4S_3S_4^*)+ic_4(S_3S_4S_4^*-S_4S_4S_3^*).
\end{aligned}
$$

$$\text{where}\quad U = S_1S_1^* - S_2S_2^* + S_3S_4^* + S_4S_3^*,$$

$$d = 1+\sqrt{3},\ c_1 = \frac{1}{\sqrt{d}}e^{-\frac{5\pi}{6}i},\ c_2 = \frac{1}{\sqrt{2}}e^{\frac{7\pi}{12}i},\ c_3 = -\frac{1}{d},\ c_4 = \frac{1}{\sqrt{2}}e^{-\frac{\pi}{4}i}.$$

Notice that element U satisfies the following relations

$$U^* = U,\quad UU = 1,$$

$$\alpha(U) = -U,\quad \rho(U) = S_1S_2^* + S_2S_1^* + i(S_3US_4^* - S_4US_3^*).$$

An orthonormal basis of intertwiner spaces $\mathcal{H}_{a,b}^c$ of E_6 fusion rule algebra $\mathcal{A}$ is given by

$$\mathcal{H}_{id,a}^a = \mathbf{C}1,\ \mathcal{H}_{id,a}^a = \mathbf{C}1\ (a = id, \alpha, \rho),$$

$$\mathcal{H}_{\alpha,\alpha}^{id} = \mathbf{C}1,\ \mathcal{H}_{\alpha,\rho}^{\rho} = \mathbf{C}1,\ \mathcal{H}_{\rho,\alpha}^{\rho} = \mathbf{C}U,$$

$$\mathcal{H}_{\rho,\rho}^{id} = \mathbf{C}S_1,\ \mathcal{H}_{\rho,\rho}^{\alpha} = \mathbf{C}S_2,\ \mathcal{H}_{\rho,\rho}^{\rho} = \mathbf{C}S_3 \oplus \mathbf{C}S_4.$$

Next we consider the following coloring of a triangulation of a three manifold

$$\lambda : \{\text{edges of triangulation}\} \longrightarrow \{id, \alpha, \rho\} : \text{coloring of edges},$$

$$\phi : \{\text{faces of triangulation}\} \longrightarrow \{1, U, S_1, S_2, S_3, S_4\} : \text{coloring of faces}.$$

By these colorings λ, ϕ we have a colored triangulation and a colored simplex.

M.Izumi [4] gave the following description of the E_6 quantum $6j$-symbols Z and the pre-normalized E_6 quantum $6j$-symbols W for the colored simplex

$$Z\left(\begin{array}{ccc} XAY & s_3 & XC \\ s_1 & \square & s_4 \\ BY & s_2 & D \end{array}\right) = \frac{1}{\sqrt{\mu(X)\mu(Y)}}W\left(\begin{array}{ccc} XAY & s_3 & XC \\ s_1 & \square & s_4 \\ BY & s_2 & D \end{array}\right),\ W\left(\begin{array}{ccc} XAY & s_3 & XC \\ s_1 & \square & s_4 \\ BY & s_2 & D \end{array}\right) = s_2^*s_1^*X(s_3)s_4 \in \mathbf{C}1 \cong \mathbf{C}.$$

(We denote the colored simplex by the colored rectangle as explained above.) Using this description we have the following table. Non-zero pre-normalized E_6 quantum $6j$-symbols W are as follows (we omit symbol $W(\cdot)$).

$$c_1 = \frac{1}{\sqrt{d}}e^{-\frac{5\pi}{6}i}, \quad c_2 = \frac{1}{\sqrt{2}}e^{\frac{7\pi}{12}i}, \quad c_3 = -\frac{1}{d}, \quad c_4 = \frac{1}{\sqrt{2}}e^{-\frac{\pi}{4}i}$$

$\begin{array}{c} \rho\rho\rho \underline{S_l}\ \rho\rho \\ S_i|\square|S_j \\ \rho\rho \overline{S_m}\ \rho \end{array} =$

(i,l,j,m)	
(3,3,3,3)	c_3
(3,3,4,4)	c_4
(3,4,3,3)	ic_4
(3,4,4,4)	ic_3
(4,3,3,4)	c_4
(4,3,4,3)	c_3
(4,4,3,4)	$-ic_3$
(4,4,4,3)	$-ic_4$
otherwise	0

$\begin{array}{c} \rho\rho\rho \underline{S_i}\ \rho\rho \\ S_j|\square|S_2 \\ \rho\rho \overline{S_2}\ \alpha \end{array} =$

	j=3	j=4
i=3	c_2	$-c_2$
i=4	ic_2	ic_2

$\begin{array}{c} \rho\rho\rho \underline{S_i}\ \rho\rho \\ S_j|\square|S_1 \\ \rho\rho \overline{S_1}\ id \end{array} =$

	j=3	j=4
i=3	c_2	c_2
i=4	ic_2	$-ic_2$

$\begin{array}{c} \rho\rho\rho \underline{S_2}\ \rho\alpha \\ S_i|\square|U \\ \rho\rho \overline{S_j}\ \rho \end{array} =$

	j=3	j=4
i=3	0	$\frac{1}{\sqrt{d}}e^{-\frac{\pi}{4}i}$
i=4	$\frac{1}{\sqrt{d}}e^{\frac{\pi}{4}i}$	0

$\begin{array}{c} \rho\rho\rho \underline{S_1}\ \rho id \\ S_i|\square|1 \\ \rho\rho \overline{S_j}\ \rho \end{array} =$

	j=3	j=4
i=3	$\frac{1}{\sqrt{d}}e^{\frac{\pi}{4}i}$	0
i=4	0	$\frac{1}{\sqrt{d}}e^{-\frac{\pi}{4}i}$

$\begin{array}{c} \rho\rho\rho \underline{S_i}\ \rho\rho \\ S_2|\square|S_j \\ \alpha\rho\ \overline{1}\ \rho \end{array} =$

	j=3	j=4
i=3	$\frac{c_1}{\sqrt{2}}$	$-\frac{c_1}{\sqrt{2}}$
i=4	$\frac{c_1}{\sqrt{2}}$	$\frac{c_1}{\sqrt{2}}$

$\begin{array}{c} \rho\rho\rho \underline{S_i}\ \rho\rho \\ S_1|\square|S_j \\ id\rho\ \overline{1}\ \rho \end{array} =$

	j=3	j=4
i=3	$\frac{c_1}{\sqrt{2}}$	$\frac{c_1}{\sqrt{2}}$
i=4	$\frac{c_1}{\sqrt{2}}$	$-\frac{c_1}{\sqrt{2}}$

$\begin{array}{c} \rho\rho\rho \underline{S_2}\ \rho\alpha \\ S_2|\square|U \\ \alpha\rho\ \overline{1}\ \rho \end{array} = -\frac{1}{d}$ $\quad \begin{array}{c} \rho\rho\rho \underline{S_2}\ \rho\alpha \\ S_1|\square|U \\ id\rho\ \overline{1}\ \rho \end{array} = \frac{1}{d}$ $\quad \begin{array}{c} \rho\rho\rho \underline{S_1}\ \rho id \\ S_2|\square|1 \\ \alpha\rho\ \overline{1}\ \rho \end{array} = \frac{1}{d}$ $\quad \begin{array}{c} \rho\rho\rho \underline{S_1}\ \rho id \\ S_1|\square|1 \\ id\rho\ \overline{1}\ \rho \end{array} = \frac{1}{d}$

$\begin{array}{c} \rho\rho\alpha \underline{U}\ \rho\rho \\ S_i|\square|S_j \\ \rho\alpha\ \overline{U}\ \rho \end{array} =$

	j=3	j=4
i=3	0	i
i=4	$-i$	0

$\begin{array}{c} \rho\rho\alpha \underline{U}\ \rho\rho \\ S_1|\square|S_2 \\ id\alpha\ \overline{1}\ \alpha \end{array} = 1$ $\quad \begin{array}{c} \rho\rho\alpha \underline{U}\ \rho\rho \\ S_2|\square|S_1 \\ \alpha\alpha\ \overline{1}\ id \end{array} = 1$

$\begin{array}{c} \rho\alpha\rho \underline{1}\ \rho\rho \\ U|\square|S_i \\ \rho\rho\ \overline{S_j}\ \rho \end{array} =$

	j=3	j=4
i=3	0	1
i=4	1	0

$\begin{array}{c} \rho\alpha\rho \underline{1}\ \rho\rho \\ U|\square|S_2 \\ \rho\rho\ \overline{S_2}\ \alpha \end{array} = -1$ $\quad \begin{array}{c} \rho\alpha\rho \underline{1}\ \rho\rho \\ U|\square|S_1 \\ \rho\rho\ \overline{S_1}\ id \end{array} = 1$

$\begin{array}{ccc} \rho\rho id & 1 & \rho\rho \\ S_i & \square & S_j \\ \rho id & 1 & \rho \end{array} =$

	j=3	j=4
i=3	1	0
i=4	0	1

$\begin{array}{ccc} \rho\rho id & 1 & \rho\rho \\ S_2 & \square & S_2 \\ \alpha id & 1 & \alpha \end{array} = 1 \qquad \begin{array}{ccc} \rho\rho id & 1 & \rho\rho \\ S_1 & \square & S_1 \\ idid & 1 & id \end{array} = 1$

$\begin{array}{ccc} \rho id\rho & 1 & \rho\rho \\ 1 & \square & S_i \\ \rho\rho & S_j & \rho \end{array} =$

	j=3	j=4
i=3	1	0
i=4	0	1

$\begin{array}{ccc} \rho id\rho & 1 & \rho\rho \\ 1 & \square & S_2 \\ \rho\rho & S_2 & \alpha \end{array} = 1 \qquad \begin{array}{ccc} \rho id\rho & 1 & \rho\rho \\ 1 & \square & S_1 \\ \rho\rho & S_1 & id \end{array} = 1$

$\begin{array}{ccc} \rho\alpha\alpha & 1 & \rho id \\ U & \square & 1 \\ \rho\alpha & U & \rho \end{array} = 1 \qquad \begin{array}{ccc} \rho\alpha id & 1 & \rho\alpha \\ U & \square & U \\ \rho id & 1 & \rho \end{array} = 1 \qquad \begin{array}{ccc} \rho id\alpha & 1 & \rho\alpha \\ 1 & \square & U \\ \rho\alpha & U & \rho \end{array} = 1 \qquad \begin{array}{ccc} \rho idid & 1 & \rho id \\ 1 & \square & 1 \\ \rho id & 1 & \rho \end{array} = 1$

$\begin{array}{ccc} \alpha\rho\rho & S_i & \alpha\rho \\ 1 & \square & 1 \\ \rho\rho & S_j & \rho \end{array} =$

	j=3	j=4
i=3	1	0
i=4	0	−1

$\begin{array}{ccc} \alpha\rho\rho & S_2 & \alpha\alpha \\ 1 & \square & 1 \\ \rho\rho & S_1 & id \end{array} = 1 \qquad \begin{array}{ccc} \alpha\rho\rho & S_1 & \alpha id \\ 1 & \square & 1 \\ \rho\rho & S_2 & \alpha \end{array} = 1$

$\begin{array}{ccc} \alpha\rho\alpha & U & \alpha\rho \\ 1 & \square & 1 \\ \rho\alpha & U & \rho \end{array} = -1 \qquad \begin{array}{ccc} \alpha\alpha\rho & 1 & \alpha\rho \\ 1 & \square & 1 \\ id\rho & 1 & \rho \end{array} = 1 \qquad \begin{array}{ccc} \alpha\alpha\alpha & 1 & \alpha id \\ 1 & \square & 1 \\ id\alpha & 1 & \alpha \end{array} = 1$

$\begin{array}{ccc} \alpha\rho id & 1 & \alpha\rho \\ 1 & \square & 1 \\ \rho id & 1 & \rho \end{array} = 1 \qquad \begin{array}{ccc} \alpha\alpha id & 1 & \alpha\alpha \\ 1 & \square & 1 \\ idid & 1 & id \end{array} = 1 \qquad \begin{array}{ccc} \alpha idid & 1 & \alpha id \\ 1 & \square & 1 \\ \alpha id & 1 & \alpha \end{array} = 1 \qquad \begin{array}{ccc} \alpha id\rho & 1 & \alpha\rho \\ 1 & \square & 1 \\ \alpha\rho & 1 & \rho \end{array} = 1 \qquad \begin{array}{ccc} \alpha id\alpha & 1 & \alpha\alpha \\ 1 & \square & 1 \\ \alpha\alpha & 1 & id \end{array} = 1$

$\begin{array}{ccc} id\rho\rho & S_i & id\rho \\ 1 & \square & 1 \\ \rho\rho & S_j & \rho \end{array} =$

	j=3	j=4
i=3	1	0
i=4	0	1

$\begin{array}{ccc} id\rho\rho & S_2 & id\alpha \\ 1 & \square & 1 \\ \rho\rho & S_2 & \alpha \end{array} = 1 \qquad \begin{array}{ccc} id\rho\rho & S_1 & idid \\ 1 & \square & 1 \\ \rho\rho & S_1 & id \end{array} = 1$

$\begin{array}{ccc} id\rho\alpha & U & id\rho \\ 1 & \square & 1 \\ \rho\alpha & U & \rho \end{array} = 1 \qquad \begin{array}{ccc} id\alpha\rho & 1 & id\rho \\ 1 & \square & 1 \\ \alpha\rho & 1 & \rho \end{array} = 1 \qquad \begin{array}{ccc} id\alpha\alpha & 1 & idid \\ 1 & \square & 1 \\ \alpha\alpha & 1 & id \end{array} = 1$

$\begin{array}{ccc} id\rho id & 1 & id\rho \\ 1 & \square & 1 \\ \rho id & 1 & \rho \end{array} = 1 \qquad \begin{array}{ccc} id\alpha id & 1 & id\alpha \\ 1 & \square & 1 \\ \alpha id & 1 & \alpha \end{array} = 1 \qquad \begin{array}{ccc} ididid & 1 & idid \\ 1 & \square & 1 \\ idid & 1 & id \end{array} = 1 \qquad \begin{array}{ccc} idid\rho & 1 & id\rho \\ 1 & \square & 1 \\ id\rho & 1 & \rho \end{array} = 1 \qquad \begin{array}{ccc} idid\alpha & 1 & id\alpha \\ 1 & \square & 1 \\ id\alpha & 1 & \alpha \end{array} = 1$

4. Proof of Theorem

In this section we give a proof of Theorem 1. To compute the invariant of the lens space $L(p,1)$, we take the following simplicial decomposition of the lens space $L(p,p-1)$ with p simplices [8]

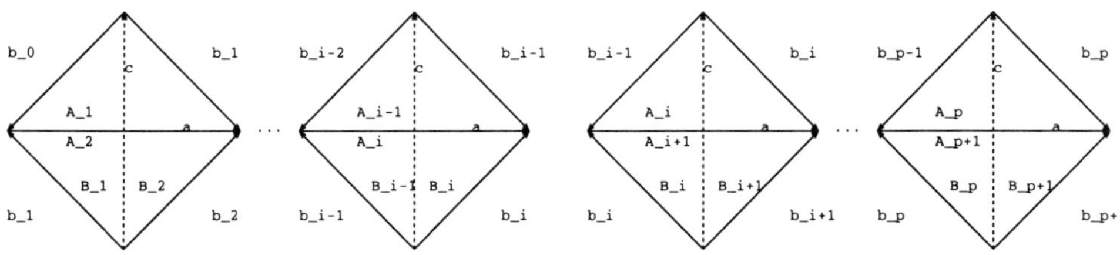

$$\text{where } b_0 = b_p, b_1 = b_{p+1}, A_1 = A_{p+1}, B_1 = B_{p+1}.$$

These colored simplices correspond to the following rectangles

$$\begin{array}{ccc} ab_0c & B_1 & ab_1 \\ A_1 & \boxed{} & A_2 \\ b_1c & B_2 & b_2 \end{array} \quad \cdots \quad \begin{array}{ccc} ab_{i-2}c & B_{i-1} & ab_{i-1} \\ A_{i-1} & \boxed{} & A_i \\ b_{i-1}c & B_i & b_i \end{array} \quad \begin{array}{ccc} ab_{i-1}c & B_i & ab_i \\ A_i & \boxed{} & A_{i+1} \\ b_ic & B_{i+1} & b_{i+1} \end{array} \quad \cdots \quad \begin{array}{ccc} ab_{p-1}c & B_p & ab_p \\ A_p & \boxed{} & A_{p+1} \\ b_pc & B_{p+1} & b_{p+1} \end{array}$$

$$\text{where } b_0 = b_p, b_1 = b_{p+1}, A_1 = A_{p+1}, B_1 = B_{p+1}.$$

Notice that the normalization factors cancel with the factor of coloring of edges as

$$\prod_{i=1}^{p} \left(\frac{1}{\sqrt{\mu(b_i)}} \right)^2 \times \mu(a)\mu(c) \prod_{i=1}^{p} \mu(b_i) = \mu(a)\mu(c),$$

and that this simplicial decomposition has two vertices. Thus we have the following formula for the lens space $L(p,p-1)$

$$Z(L(p,p-1)) = \omega^{-2} \sum_{coloring} \mu(a)\mu(c) \times \prod_{i=1}^{p} W\left(\begin{array}{ccc} ab_{i-1}c & B_i & ab_i \\ A_i & \boxed{} & A_{i+1} \\ b_ic & B_{i+1} & b_{i+1} \end{array} \right)$$

where we take the sum over all colorings satisfying the following conditions

$$ab_{i-1} = b_i, b_{i-1}c = b_i \quad (i = 1, \cdots, p).$$

Noticing that this summation is nothing but the multiplication of matrices, we can reduce the computation of this formula to the computation of the trace of the p-th power of some matrices. Thus we define matrices A_{ac} by

$$\left(\begin{array}{cc} awc & D \quad az \\ C & \\ zc & \end{array} , \begin{array}{cc} axc & B \quad ay \\ A & \\ yc & \end{array} \right)\text{-entry of } A_{ac} = \delta_{y,w} W\left(\begin{array}{ccc} axc & B & ay \\ A & \boxed{} & C \\ yc & D & z \end{array} \right).$$

By using these matrices, we have

$$Z(L(p,p-1)) = \omega^{-2}\sum_{a,c}\mu(a)\mu(c)Tr(A_{ac}^p) = \omega^{-2}\sum_{a,c}\mu(a)\mu(c)\sum_{i=1}^{rkA_{ac}}(\lambda_{ac}^i)^p$$

where $\{\lambda_{ac}^i\}_i$ are the eigenvalues of matrix A_{ac}. By using the table of the E_6 quantum $6j$-symbols, we have the following eigenvalues

$$\{\lambda_{idid}^i\} = \{1,1,1\}, \{\lambda_{id\alpha}^i\} = \{1\}, \{\lambda_{id\rho}^i\} = \{1,1\}, \{\lambda_{\alpha id}^i\} = \{1\}, \{\lambda_{\rho id}^i\} = \{1,1\},$$
$$\{\lambda_{\alpha\alpha}^i\} = \{1,-1,-1\}, \{\lambda_{\alpha\rho}^i\} = \{1,-1\}, \{\lambda_{\rho\alpha}^i\} = \{1,-1\},$$
$$\{\lambda_{\rho\rho}^i\} = \{1,1,-1,-i,e^{5/6\pi i},e^{5/6\pi i},e^{1/3\pi i},-e^{1/3\pi i}\}.$$

Using these eigenvalues, we can compute the invariant $Z(L(p,p-1))$ of the lens space $L(p,p-1)$. The invariant $Z(L(p,1))$ of the lens space $L(p,1)$ is given by the complex conjugate of the invariant $Z(L(p,p-1))$ of the lens space $L(p,p-1)$.

Acknowledgments

The author would like to thank M.Asaeda, K.Habiro, M.Kosuda, S.Nakabo, M.Okamoto, C.Sato, and M.Wakui. He also would like to express his appreciation for Prof. H.Murakami and Prof. T.Kohno.

This work was done when the author was in Graduate School of Mathematical Sciences, University of Tokyo. The author was partially supported by Research Fellowships of the Japan Society for the Promotion of Science for Young Scientists.

References

[1] A.Ocneanu, *Quantized group, string algebras and Galois theory for algebras*, Operator algebras and applications Vol.2 (Warwick 1987) (ed. D.E.Evans and M.Takesaki) London Math. Soc. Lecture Note Series Vol.136,119-172 (1988).

[2] V.G.Turaev, O.Ya.Viro, *State sum invariants of 3-manifolds and quantum 6j-symbols*, Topology **31** (1992) 865-902.

[3] D.E.Evans, Y.Kawahigashi, *Quantum symmetries on operator algebras*, Oxford University Press (1998).

[4] M.Izumi, *Subalgebras of Infinite C^*-Algebras with Finite Watatani Indices I: Cuntz Algebras*, Commun. Math. Phys. **155** (1993) 157-182.

[5] M.Wakui, *private communications*, (1998).

[6] M.Okamoto, C.Sato, *Turaev-Viro-Ocneanu invariant and its calculations from the view point of TQFT*, RIMS Kokyuroku 1053 (japanese) (1998).

[7] U.Pachner, *P.L. homeomorphic manifolds are equivalent by elementary shellings*, Europ. J. Comb. **12** (1991) 129-145.

[8] D.Rolfsen, *Knots and Links*, Publish or Perish Inc..

Periodic knots with delta-unknotting number one

Dedicated to Professor Fuichi UCHIDA on his 60th birthday

Yoshiaki UCHIDA[1]
Department of Mathematical Sciences
Yamagata University
Yamagata Yamagata 990-8560, JAPAN
yuchida@sci.kj.yamagata-u.ac.jp

ABSTRACT
Delta-unknotting operation is a local move, as shown in Figure 2. Murakami-Nakanishi [MN] showed this move to be an unknotting operation. We will show that for many periodic knots the Δ-unknotting number is greater than one.

Keywords: periodic knot, unknotting operation, unknotting number, Δ-unknotting number

1. Introduction

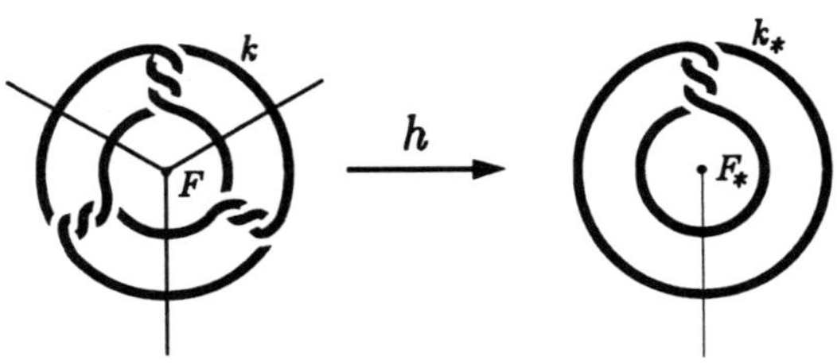

Figure 1

Let k be a knot in $\mathbb{S}^3 = (\mathbb{R}^3 \cup \{\infty\})$. We say that this knot has a period p, if there is a periodic map f of $(\mathbb{R}^3, k)$ which satisfies (1) f is $2\pi/p$-rotation about a line F in $\mathbb{R}^3$, (2) $F \cap k = \phi$. Let $h : \mathbb{R}^3 \to \mathbb{R}^3/f$ be the projection. Put $F_* = h(F)$

[1]This work was mainly done while the author was visiting UNIVERSIDAD COMPLUTENSE DE MADRID with financial support from the Japanese Ministry of Education in the form of an Overseas Research Fellowship. The author would like to express his gratitude to Professor Montesinos for his hospitality.

and $k_* = h(k)$. We define the linking number $d = lk(k,l)$ to be the linking number of the knot $l = F \cup \{\infty\}$. See Figure 1. The Δ-unknotting operation is a local move on a knot diagram, denoted in Figure 2.

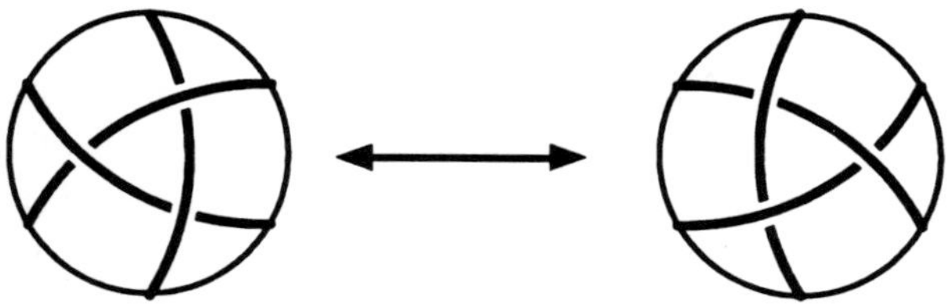

Figure 2

We define the Δ-unknotting number of a knot k, as the minimum number of Δ-unknotting operations necessary to change a diagram of k into a diagram for the unknot, where the minimum is taken over all diagram for k. We will denote the Δ-unknotting number of k by $u_\Delta(k)$.

2. Periodic knots and Δ-unknotting number

In [NNU], we showed that for a torus knot $T(a,b)$ of type (a,b), a, $b > 0$, $(a,b) = 1$, its Δ-unknotting number is $(a^2-1)(b^2-1)/24$. And in [U], we gave a conjecture that if period p is sufficiently large, then the Δ-unknotting number is greater than one. In this paper we will give a partial answer.

THEOREM.. *Let k be a knot with period p where p is prime and $d = lk(k,l)$. If $d^2 - 1 \not\equiv \pm 24 \pmod p$ then $u_\Delta(k) \neq 1$.*

Note. There exist periodic knots with period 2, 3, 4 whose Δ-unknotting number is one. For example, see Figure 3. If k has period 5, then its Δ-unknotting number is not one. Because k is a period 5 knot, then $(d,5) = 1$ for $d = lk(k,l)$. So, $d^2 - 1 \not\equiv \pm 24 \pmod p$.

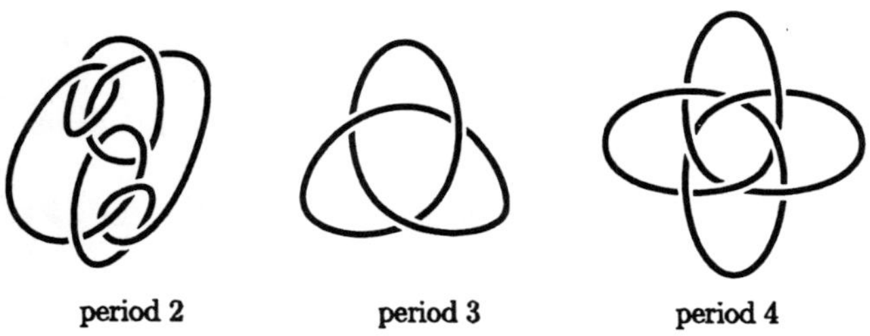

Figure 3

PROOF OF THEOREM: Let k_Δ be a knot obtained from k by one Δ-unknotting operation, $\nabla(k)$ the Conway polynomial of k, and $a_i(k)$ i-th coefficient of $\nabla(k)$. In [Ok], Okada showed that $|a_2(k_\Delta) - a_2(k)| = 1$. It follows that $u_\Delta(k) \geq |a_2(k)|$. For more details of the Conway polynomial, please see [Kau]. From Okada's Theorem, if $|a_2(k)| \neq 1$, then $u_\Delta(k) \neq 1$. Now, we will show that if $d^2 - 1 \not\equiv \pm 24 \pmod p$, then $|a_2(k)| \neq 1$. According to [Kaw p.122], $\nabla(k) \equiv \nabla(T(p,d))\nabla(k_*)^p \pmod p$, where $k_* = h(k)$. (In [Kaw p.122], it is incorrectly stated that $\nabla(k) \equiv \nabla(T(p,d))^{p-1}\nabla(k_*)^p \pmod p$. This equation does not hold for, for example, a torus knot $T(p,d)$. Because $T(p,d)$ has period p and $k_* = h(T(p,d))$ is a trivial knot. Then $\nabla(T(p,d)) \equiv \nabla(T(p,d))^{p-1} \pmod p$. It is incorrect. This equation must be $\nabla(k) \equiv \nabla(T(p,d))\nabla(k_*)^p \pmod p$. Because for Alexander polynomial, the following equation is well-known. $\Delta(k) \equiv \rho_d(t)^{p-1}\Delta(k_*,t)^p \pmod p$ where $\rho_d(t) = 1 + t + \cdots + t^{d-1}$ for a period p knot, where p is prime. For a torus knot $T(p,d)$, $\Delta(T(p,d)) = (t^{pd}-1)(t-1)/(t^p-1)(t^d-1) = \rho_d(t^p)/\rho_d(t) \equiv (\rho_d(t))^{p-1} \pmod p$, then $\rho_d(t)^{p-1}$ is corresponding to a torus knot factor $\nabla(T(p,d))$, and $\Delta(k_*)^p$ is corresponding to $\nabla(k_*)^p$.)

The following equations are well-known.

$$a_0(k) = \begin{cases} 1, & \text{if } k \text{ ia a knot} \\ 0, & \text{if } k \text{ is a link} \end{cases}$$
$$a_1(k) = \begin{cases} lk(k), & \text{if } k \text{ is a 2-compoment link } k_1 \cup k_2 \\ 0, & \text{otherwise} \end{cases}$$

And the second coefficient of Conway polynomial of a torus knot $T(p,d)$ of type (p, d) is $(p^2-1)(d^2-1)/24$. We get the following equations;

$$\nabla(T(p,d)) = 1 + \frac{1}{24}(p^2-1)(d^2-1)z^2 + \cdots + a_{2m}z^{2m}$$
$$\nabla(k_*) = 1 + a_2{}'z^2 + \cdots + a_{2n}{}'z^{2n} \qquad (k_* \text{ knot}, a_i, a_i{}' \in \mathbb{Z}).$$

Now, we get

$$\begin{aligned}\nabla(T(p,d))\nabla(k_*)^p &= \left(1 + \frac{1}{24}(p^2-1)(d^2-1)z^2 + \cdots + a_{2m}z^{2m}\right) \\ &\quad \times \left(1 + pa_2{}'z^2 + \cdots + (a_{2n}{}')^p z^{2np}\right) \\ &= 1 + \left(\frac{1}{24}(p^2-1)(d^2-1) + pa_2{}'\right)z^2 \\ &\quad + \cdots + a_{2m}(a_{2n}{}')^p z^{2m+2np}.\end{aligned}$$

Then $a_2(k) \equiv (p^2-1)(d^2-1)/24 + pa_2{}' \equiv (p^2-1)(d^2-1)/24 \equiv -(d^2-1)/24 \pmod p$. Now, by Okada's theorem [Ok] we have, if $d^2 - 1 \not\equiv \pm 24 \pmod p$ then $u_\Delta(k) \neq 1$.

Note that there exists a periodic knot k, such that $a_2(k) = 1$ and $u_\Delta(k) \neq 1$. Let k be the knot obtained as the closure of the braid shown in Figure 4, where the box m means m-half twists. In the case of $m = -34$, $a_2(k) = 1$ and k has period 7. And performing the unknotting operation 7 times, we can get a torus knot of type (7,5). It is known that its unknotting number is 12. Then the unknotting number of the knot k is greater than or equal to $12 - 7 = 5$. And $u_\Delta(k) \geq u(k)/2$. So, $u_\Delta(k) \geq 3$.

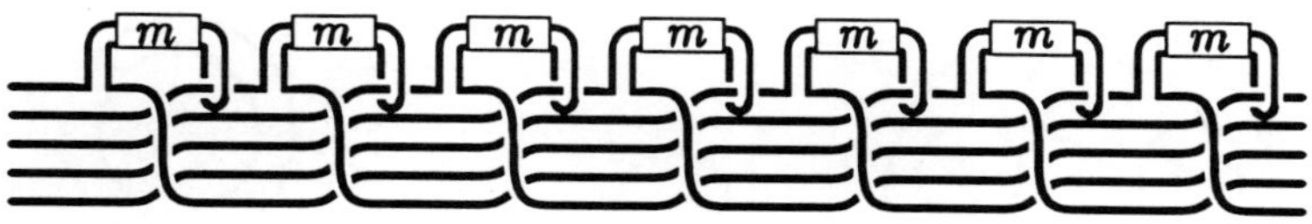

Figure 4

Conjecture. Let k be a periodic knot with period p. If $p \geq 5$, then $u_\Delta(k) \geq 2$.

3. Periodic knots and unknotting number

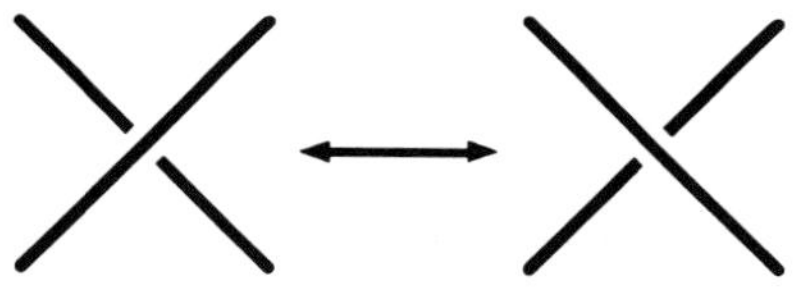

Figure 5

It is well-known that any knot diagram can be changed into a diagram for the unknot by repeated applications of the move denoted in Figure 5 at appropriate crossings. The unknotting number $u(k)$ for a knot is the minimum number of such moves necessary to change a diagram of k into a diagram for the unknot, where the minimum is taken over all diagrams for k.

For any natural number p, there exists a periodic knot with period p, such that its unknotting number is equal to one. In Figure 6 (i), we replace the loop by an arc denoted in Figure 6 (ii). Then we can get a knot k_p. It is easy to see that the knot k_p has period p. In this figure, $p = 5$. If we exchange one of the crossings indicated in Figure 6, we can get a trivial knot. Now, we show that k_p is not trivial. Then this knot's unknotting number is one. Let $h : \mathbb{S}^3(= \mathbb{R}^3 \cup \{\infty\}) \to \mathbb{S}^3$ be the projection defined in section 1. We consider h is a covering map from $\mathbb{S}^3$ to $\mathbb{S}^3$ branched along a trivial knot $l_* = h(l)$. Let assume that k_p is trivial. Then $n(\geq 2)$-fold cyclic branched covering space branched along k_p is also the 3-sphere, denoted this covering by $f : (\mathbb{S}^3, \tilde{k}_p) \to (\mathbb{S}^3, k_p)$, where $\tilde{k}_p$ is the preimage of k_p. Then $h \circ f$ is an abelian covering of $\mathbb{S}^3$ branched along two-compoment link $l_* \cup k_* (= h(k_p))$. By [S. Theorem 1], $l_* \cup k_*$ must be a torus link $T(2,2)$. But $l_* \cup k_*$ is not a torus link $T(2,2)$, because its Jones polynomial is $-t^{-11/2}+3t^{-9/2}-4t^{-7/2}+4t^{-5/2}-3t^{-3/2}+t^{-1/2}-3t^{3/2}+3t^{5/2}-4t^{7/2}+3t^{9/2}-t^{11/2}$, and for $T(2,2)$, it is $-t^{1/2}-t^{5/2}$, by Kodama program [Ko]. Then k_p is not trivial. Note that for this knot, $d=1$.

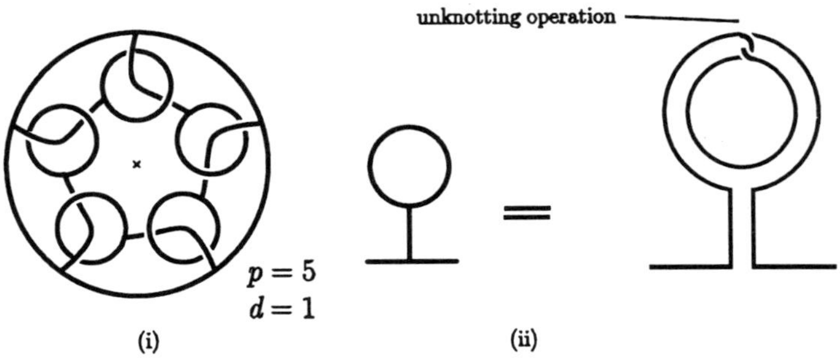

Figure 6

Conjecture. Let k be an unknotting number one periodic knot with period p and $d = lk(k,l)$, where l is the axis of the rotation. If $p=3$, then $|d| \leq 2$ and if $p \geq 4$ then $|d|=1$.

The unknotting number of the knots in Figure 7 is one.

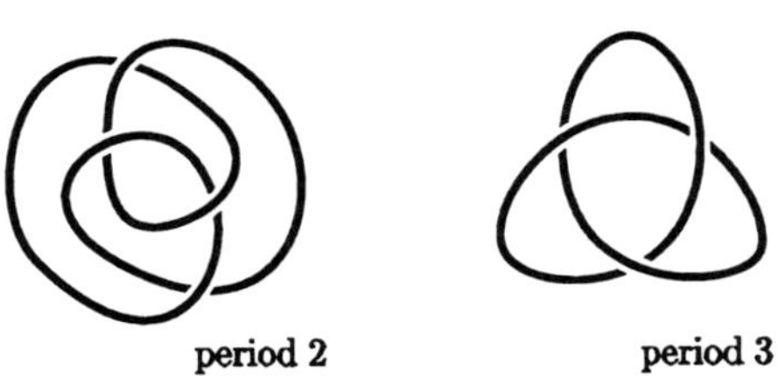

Figure 7

References

[Kau] L. Kauffman, On knots, Ann. of Math. Studies, (1987) **115**, Princeton Univ. Press.

[Kaw] A. Kawauchi (ed.), *A Survey of Knot Theory*, Birkhäuser Basel (1996).

[Ko] K. Kodama, Kodama Program ftp://ftp.math.kobe-u.ac.jp/pub/knot/

[MN] H. Murakami, Y. Nakanishi, *On a certain move generating link-homotopy*, Math. Ann., (1989) **284**, 75–89.

[NNU] K. Nakamura, Y. Nakanishi, Y. Uchida *Delta-unknotting number for knot.* J. Knot Theory and Its Ramifications, (1998) **7**, 639–650.

[Ok] M. Okada, *Delta-unknotting operation and the second coefficient of the Conway polynomial*, J. Math. Soc. Japan, (1990) **42**, 713–717.

[S] M. Sakuma, *On regular coverings of links*, Math. Ann., (1982) **260**, 303–315.

[U] Y. Uchida *Two-bridge knots with generalized unknotting number one Proceedings of Knot96*, World Sci. Pub. (1997) pp.109-113

THE KONTSEVICH INTEGRAL AND ALGEBRAIC STRUCTURES ON THE SPACE OF DIAGRAMS

SIMON WILLERTON

Institut de Recherche Mathématique Avancée, Université Louis Pasteur et CNRS, 7 rue René Descartes, 67084 Strasbourg, France

ABSTRACT

This paper is part expository and part presentation of calculational results. The target space of the Kontsevich integral for knots is a space of diagrams; this space has various algebraic structures which are described here. These are utilized with Le's theorem on the behaviour of the Kontsevich integral under cabling and with the Melvin-Morton Theorem, to obtain, in the Kontsevich integral for torus knots, both an explicit expression up to degree five and the general coefficients of the wheel diagrams.
Keywords: Vassiliev invariants, torus knots, Kontsevich integral.

INTRODUCTION

The Kontsevich integral, Z, is a map from the set of oriented knots to a space, $\mathcal{A}$, of "diagrams". For example it will be seen in Section 8 that the value taken on a negative trefoil — the $(2, -3)$-torus knot — begins

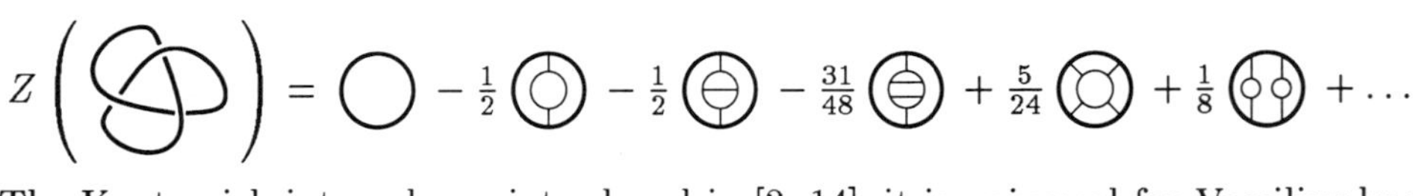

The Kontsevich integral was introduced in [2, 14]; it is universal for Vassiliev knot invariants in the sense that every finite-type invariant factors through the Kontsevich integral, and it is universal in a certain sense for knot invariants coming from quantum groups. Other approaches to defining this invariant have been introduced but none give in easily to attempts at actual calculation.

The space of diagrams has a rich algebraic structure, the following will be described in this paper: it has two products making it a graded Hopf algebra in two ways, it has a set of simultaneously diagonalizable "Adams operations" and the space of primitive elements in $\mathcal{A}$ has the structure of a module over Vogel's algebra Λ. How all of these structures interact and how they relate to topological properties of knots is not yet properly understood. However, they can be utilized to describe the Kontsevich integral for torus knots and to identify the Alexander polynomial in terms of wheel diagrams. These last two things are taken in this paper as motivations for the algebraic structures described herein.

The paper is organized as follows. The first section contains a brief sketch of the original definition of the Kontsevich integral. The second section introduces the target space $\mathcal{A}$ and gives an explicit basis up to degree five — the specific choice

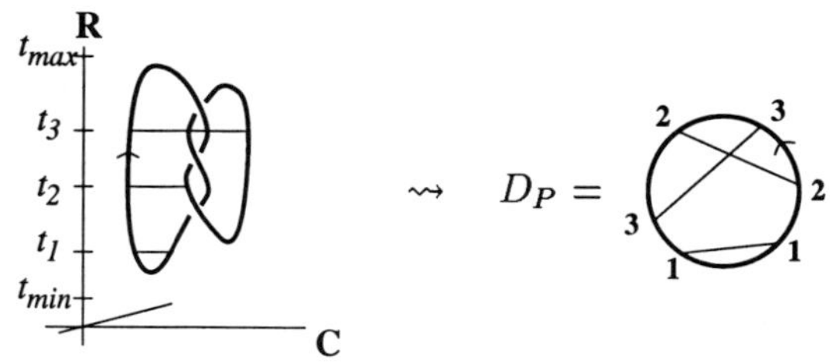

FIGURE 1. A Morse knot together with a set of pairings P for $m = 3$, and the corresponding diagram D_P.

of basis is made to interact nicely with the algebraic structure described later. The Hopf algebra structure of $\mathcal{A}$ coming from the connect sum product is presented in Section 3, together with some consequences of the Hopf algebra structure theorems, such as identifying $\mathcal{A}$ as being freely generated by the space spanned by connected diagrams. Section 4 shows how the theory naturally extends to framed knots and how the target space is extended to ${}^f\!\mathcal{A}$ by the addition of an extra generator to encode the framing. In Section 5 the Adams operations are introduced, as are symmetrized diagrams which are eigenvectors for these operations; also wheel diagrams are defined and used with the disjoint union product to give a description of the element Ω. The module structure of the primitive elements of $\mathcal{A}$ over Vogel's algebra Λ is the subject of the sixth section. Cabling knots and Le's theorem on the behaviour of the Kontesevich integral are contained in the seventh section. Section 8 presents the result of calculating the Kontsevich integral of torus knots by using the machinery so far introduced, and also gives a criterion for Vassiliev invariants of torus knots. The final section shows how the Melvin-Morton Theorem allows the identification of the wheels coefficients from the Alexander polynomial and, in particular, does this explicitly for torus knots. Appendix A lists the change of basis data from round diagrams to symmetrized diagrams used in Section 8 and Appendix B compares the results here with those of Alvarez and Labastida.

This paper is an extended version of the talk that I gave at Knots in Hellas '98, and this supersedes the earlier preprint [27].

1. THE KONTSEVICH INTEGRAL.

Here I will try to give a flavour of the original definition of the Kontsevich integral — it still remains somewhat mysterious — one of the goals of this paper is to show that it can be calculated for certain knots without performing any sorts of integral. The target space $\mathcal{A}$ will not be defined until the next section, so here it should be thought of as "linear combinations of diagrams".

Suppose that K is a Morse knot, that is a knot embedded in $\mathbb{R}^3 = \mathbb{R}_t \times \mathbb{C}_z$ such that the projection to the $\mathbb{R}$ co-ordinate is a Morse function. Then define

$$\overline{z}(K) := \sum_{m=0}^{\infty} \frac{1}{(2\pi i)^m} \int_{\substack{t_{\min}<t_1<\cdots<t_m<t_{\max} \\ t_j \text{ non-critical}}} \sum_{\substack{\text{pairings} \\ P=(\{z_j,z_j'\})}} (-1)^{\#\downarrow P}[D_P] \bigwedge_{j=1}^{m} \frac{dz_j - dz_j'}{z_j - z_j'},$$

where a set of pairings P is a set of unordered pairs $\{z_j, z_j'\}$ so that $(t_j, z_j), (t_j, z_j') \in \mathbb{R} \times \mathbb{C}$ are distinct points on the knot on the same level, where $\#{\downarrow}P$ is the number of points in the pairing such that the orientation is pointing downwards and where D_P is the diagram obtained by drawing the preimage of the knot — a circle — and drawing chords connecting the points corresponding to the pairs. See Figure 1.

Considered as an element of $\mathcal{A}$, this gives an invariant of Morse knots, but is not invariant if one introduces a "hump", i.e. a minima-maxima pair with respect to the $\mathbb{R}$-axis, which increases the number of critical points. It is necessary to introduce the following correction term: define $\Omega := \overline{z}\,($$)^{-1} \in \mathcal{A}$, where this is a Morse embedding of the unknot with four critical points. Ω is an important element which will also be returned to later in the paper. Now the Kontsevich integral for a Morse knot K with $c(K)$ critical points is defined to be

$$Z(K) := \overline{z}(K).\Omega^{c(K)/2-1} \in \mathcal{A}.$$

For a pleasant survey going in to some of the finer points of this definition, the reader is directed [8].

It is important to note here that there are two normalizations of the Kontsevich integral in common use: the other can be defined by $Z_q(K) := \overline{z}(K).\Omega^{c(K)/2}$. It is clear from this that the two normalizations differ by a factor of Ω and that $Z(\text{unknot}) =$ , while $Z_q(\text{unknot}) = \Omega$, and so Ω is often referred to as "the Kontsevich integral of the unknot". Both normalizations have their advantages: Z is multiplicative under connect sum, while Z_q is better behaved under cabling operations. In this work I will use exclusively the multiplicative normalization.

There are a couple of other ways of defining such a knot invariant (see [6]): one is via quasi-Hopf algebras and Drinfeld associators, this is essentially a discretized version of the above; the other way is the so-called Bott-Taubes approach using configuration space integrals, this is believed to give the same answer as the Kontsevich integral above.

2. The space of diagrams $\mathcal{A}$.

The target space $\mathcal{A}$ of the Kontsevich integral will be defined here, and a basis given up to degree five.

First it is necessary to define the notion of a *diagram* (which I will also call a round diagram) — this was called a Chinese character diagram in [2]. A diagram is a trivalent connected graph such that each vertex is cyclically oriented and such that there is a distinguished oriented cycle called the *Wilson loop*. A diagram is graded by half the number of vertices that it has. By convention a diagram is drawn such that the Wilson loop is an anticlockwise oriented circle, and such that each vertex is also oriented anticlockwise. See Figure 2. The complement of Wilson loop is termed the *internal graph*.

Define the rational vector space $\mathcal{A}_n$ to be the vector space spanned by the diagrams of degree n and subject to the so-called STU and 1T relations:

$\mathcal{A}_n := \langle\text{diagrams of degree } n\rangle_{\mathbb{Q}} \,/$

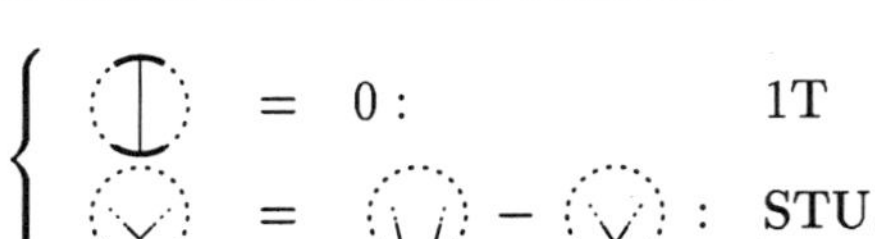

FIGURE 2. A degree three diagram. Note that the four-valent vertex is not a vertex of the diagram, but just a result of the planar drawing.

Note that the 1T relation means that a diagram vanishes if it has a chord which is not intersected by any other part of the internal graph. It is also worth noting that the antisymmetry (AS) relation is a direct consequence of the STU relation: antisymmetry says that reversing the orientation of one vertex in a diagram is the same as multiplying the diagram by minus one; pictorially,

Look now at the spaces of low order, and find explicit bases. Degree zero is very easy:

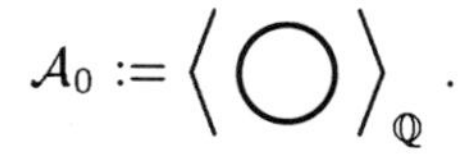

Degree one is almost as easy:

It starts getting interesting at degree two:

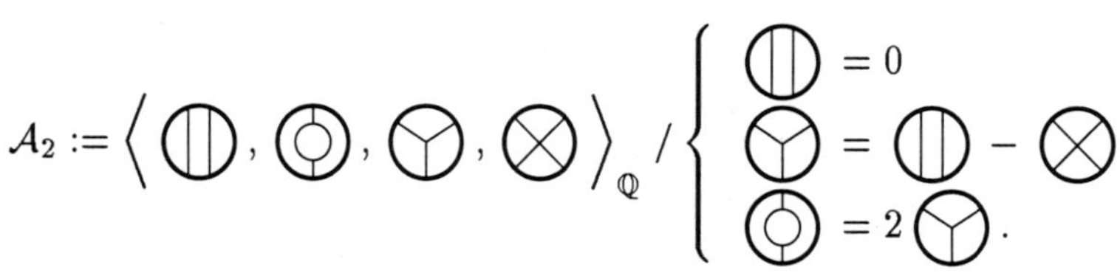

That has four generators and three relations, so pick the following generator:

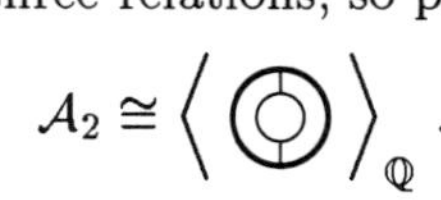

Degree three is more awkward:

It transpires that $\mathcal{A}_3$ is also one dimensional and the generator can be chosen to be as follows:

For later use, we will just write down bases for $\mathcal{A}_4$ and $\mathcal{A}_5$:

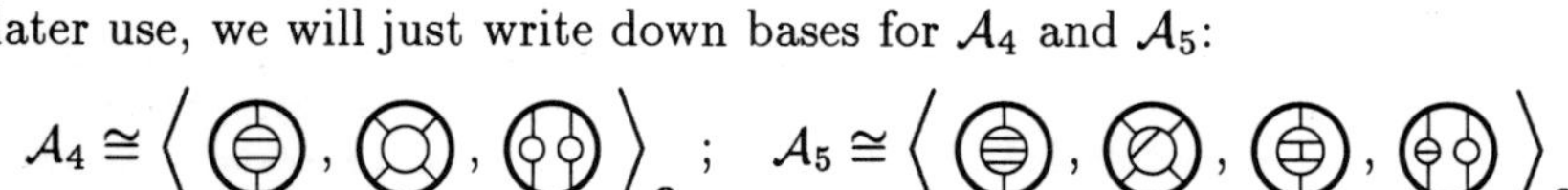

n	0	1	2	3	4	5	6	7	8	9	10	11	12
$\dim \mathcal{A}_n$	1	0	1	1	3	4	9	14	27	44	80	132	232

TABLE 1. The dimensions of the spaces $\mathcal{A}_n$ up to $n = 12$ as given in [2, 12].

A rationale for this choice of basis should become apparent during the of course of the paper; it has primarily been chosen to simplify the calculation of the results in Appendix A. The dimension of $\mathcal{A}_n$ is only known for $n \leq 12$ and these dimensions are tabulated in Table 1. See [10] for some results on asymptotic lower bounds on the dimensions of these spaces.

Define $\mathcal{A}$ to be the projective limit of the direct sum $\bigoplus_{n=0}^{\infty} \mathcal{A}_n$, so an element of $\mathcal{A}$ is a possibly infinite linear combination of diagrams which is finite in each degree. This space has lots of algebraic structure which will be described in the sections below.

3. Hopf algebraic structure of $\mathcal{A}$.

In this section I will introduce a Hopf algebra structure for $\mathcal{A}$ coming from the connect sum product, and show consequences of the Hopf algebra structure theorems, notably that the Kontsevich integral of a knot can be expressed as an exponential of a sum of connected diagrams.

3.1. **Product and coproduct on $\mathcal{A}$.** Firstly, $\mathcal{A}$ has a natural product $\mathcal{A} \otimes \mathcal{A} \to \mathcal{A}$. This is defined on diagrams in terms of a connect sum operation.
E.g.

The resulting diagram is, of course, dependent on where the cuts are made in the two diagrams; however, the result is well-defined modulo the STU relation (see [2]). This makes $\mathcal{A}$ into a graded, commutative algebra.[1] The unit is the diagram which has no internal graph.

This product is related to the connect sum operation, #, on knots via the Kontsevich integral, as the following theorem demonstrates.

Theorem 1 (Kontsevich). *For K and L knots: $Z(K \# L) = Z(K).Z(L)$.*

Secondly, $\mathcal{A}$ has a coproduct $\Delta \colon \mathcal{A} \to \mathcal{A} \otimes \mathcal{A}$. This is defined on a diagram by taking all of the ways that the internal graph can be decomposed into an ordered pair of complementary sets of components.
E.g.

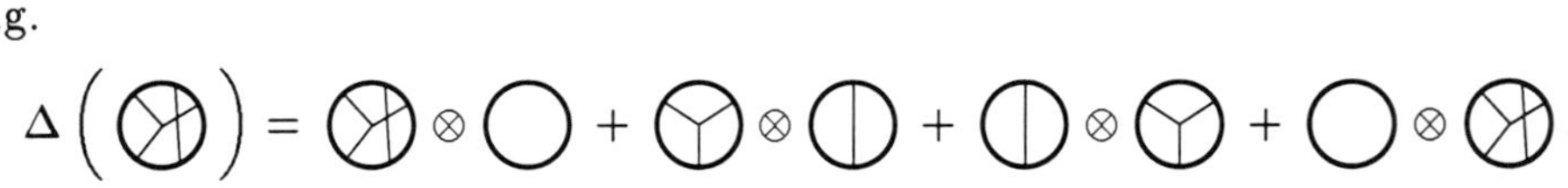

[1]This means commutative in the genuine sense and not the commutative-graded sense. Philosophically this is because $\mathcal{A}$ all lives in homological degree zero. The structure theorems of [20] deal with commutative-graded algebras, but this is not a problem as, if necessary, the grading of $\mathcal{A}$ can be doubled so that everything lives in even degree.

The counit is the function $\mathcal{A} \to \mathbb{Q}$ which maps the diagram with no internal graph to 1 and which vanishes on all other diagrams. This coproduct is graded and cocommutative, and is also compatible with the product in the sense that it is a map of algebras: $\Delta(a.b) = \Delta(a).\Delta(b)$. This makes $\mathcal{A}$ into a connected, graded Hopf algebra *à la* Milnor and Moore[2] [20].

3.2. **Consequences of Hopf algebra structure theorems.** Now we can utilize the nice structure theorems of commutative, cocommutative Hopf algebras.

There are two special subsets of a Hopf algebra. The first is the vector space of *primitive elements*; this is defined to be the set of elements which satisfy $\Delta(x) = x \otimes 1 + 1 \otimes x$. This space naturally forms a Lie algebra under the commutator bracket, but in the commutative case of $\mathcal{A}$ the bracket is trivial. The space of primitive elements of $\mathcal{A}$ certainly contains those diagrams whose internal graph is connected — the so-called *connected* diagrams.

Connected diagrams: , . Non-connected diagrams: 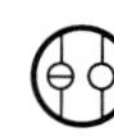, 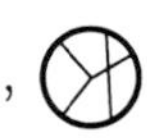.

In fact by the following theorem it suffices to know these diagrams.

Theorem 2 (Bar-Natan). *The vector space of primitive elements of $\mathcal{A}$ is (projectively) spanned by connected diagrams.*

One of the key structure theorems of Milnor and Moore is that a commutative and cocommutative Hopf algebra is (naturally isomorphic to) the polynomial algebra over its primitive elements. Translating this into the current context we obtain:

Theorem 3. *The algebra $\mathcal{A}$ is the completed polynomial algebra on the connected diagrams.*

In particular this means that if one has a basis for the space of connected diagrams then this can be extended to a basis of $\mathcal{A}$ by taking the monomials in the basis of connected diagrams. Now look back to the bases given in Section 2 and see that this is precisely what I have done. The basis elements are all connected except one in degree four and one in degree five which are products of lower degree connected basis elements.

The second important subset of a Hopf algebra is the set of *group-like elements*; these are the elements which satisfy $\Delta(g) = g \otimes g$. This set forms a group under the product; the inverse operation is just the antipode map. This set is important in the current context because of the following theorem.

Theorem 4 (Kontsevich). *The Kontsevich integral maps knots to group-like elements.*

A theorem of Quillen [22] asserts that, under reasonable hypotheses, the usual exponential map, $\exp(x) := \sum_{i=0}^{\infty} x^i/i!$, defines a bijection of sets from the primitive elements to the group-like elements, with the inverse being given by logarithm. Combining with Theorem 4 this gives

Theorem 5. *The Kontsevich integral of a knot is expressible as the exponential of a sum of connected diagrams.*

---

[2]This implies that it is uniquely equipped with an antipode, but this is not really of concern here.

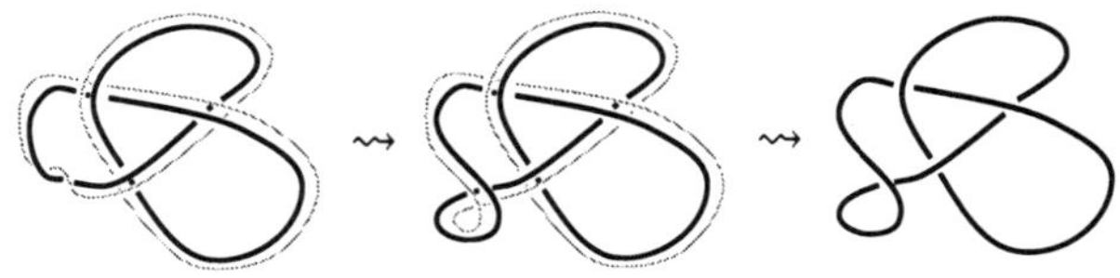

FIGURE 3. This framed trefoil can be drawn in the blackboard framing by adding a kink.

For instance the example from the introduction can be rewritten as

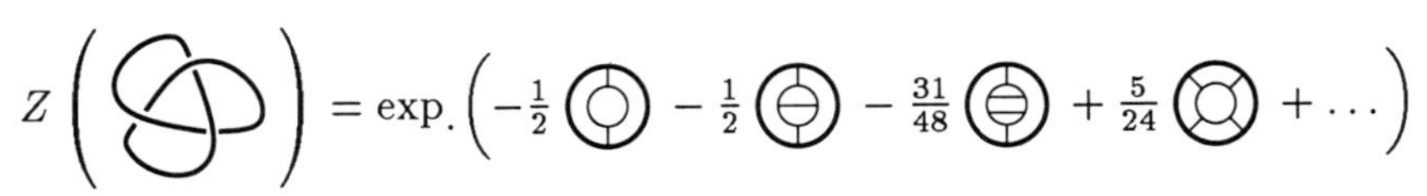

Here $\exp_{.}$ means that the connect sum product is used in the definition of the exponential map; this is to prevent confusion when the disjoint union product, $\amalg$, is introduced later.

The Baker-Campbell-Hausdorff formula describes how the product on the group-like elements is pulled back to the primitive elements; in the commutative case this reduces to just addition on the space of primitive elements. Together with Theorem 1, this means that not only is $\ln_{.}\circ Z$ well defined on knots, but that it is additive under connect sum of knots: $\ln_{.}\circ Z(K\#L) = \ln_{.}\circ Z(K) + \ln_{.}\circ Z(L)$.

4. FRAMED VERSION OF THE THEORY.

The Kontsevich integral can be naturally defined for framed knots. The target space of this has one extra generator which lives in degree one — this keeps count of the framing.

4.1. **Framed knots.** A framed knot is a knot equipped with a homotopy class of sections of the normal bundle. This can be thought of as a parallel strand running very close to the knot. A knot diagram is said to be blackboard framed if the framing curve lies parallel to the knot in the plane of the paper. Any framed knot can be drawn in the blackboard framing, possibly by adding kinks, see Figure 3.

For a given knot, the framings are classified by the framing number (also called the writhe) which is the linking number of the knot and its framing curve: for a framed knot K, this will be denoted by $F(K)$.

4.2. **The Hopf algebra $^f\!\mathcal{A}$.** The Kontsevich integral for framed knots will be defined below to take values in a Hopf algebra of diagrams defined similarly to $\mathcal{A}$, but without the 1T relation, namely:

$$^f\!\mathcal{A} := \langle\text{diagrams}\rangle_{\mathbb{Q}}/\text{STU}.$$

Thus $\mathcal{A} \cong \mathcal{A}/1\text{T}$. In fact, $\mathcal{A}$ can be identified as a sub-Hopf algebra of $^f\!\mathcal{A}$ by an inclusion $\varphi\colon \mathcal{A} \hookrightarrow {}^f\!\mathcal{A}$ [26]. Furthermore $^f\!\mathcal{A}$ has just one extra primitive generator — this is the diagram in degree one: ⦶. So there is the natural isomorphism $^f\!\mathcal{A} \cong \mathcal{A} \oplus$ ⦶$.{}^f\!\mathcal{A}$.

4.3. **The framed Kontsevich integral.** The Kontsevich integral for oriented framed knots, taking values in ${}^f\!\mathcal{A}$, can be defined[3] by

$$ {}^fZ(K) := Z(K).\exp_.(F(K)\bigcirc\!\!\!\!\!|\;/2)\,. $$

This looks a little *ad hoc*, but it is equivalent to other definitions, see [18]. Note that this extra generator of ${}^f\!\mathcal{A}$ is encodes precisely the framing data and on the level of logarithms

$$ \ln_.\circ{}^fZ(K) = \ln_.\circ Z(K) + F(K)\bigcirc\!\!\!\!\!|\;/2. $$

5. Adams operations and symmetrized diagrams.

5.1. **Adams operations.** On the algebra ${}^f\!\mathcal{A}$ there is a set of operations, $\{\psi^m\}_{m\in\mathbb{Z}}$, called the Adams operations. For m a positive integer, ψ^m is defined on a diagram by lifting the vertices on the Wilson loop to the pth connected cover of the Wilson loop.
E.g.

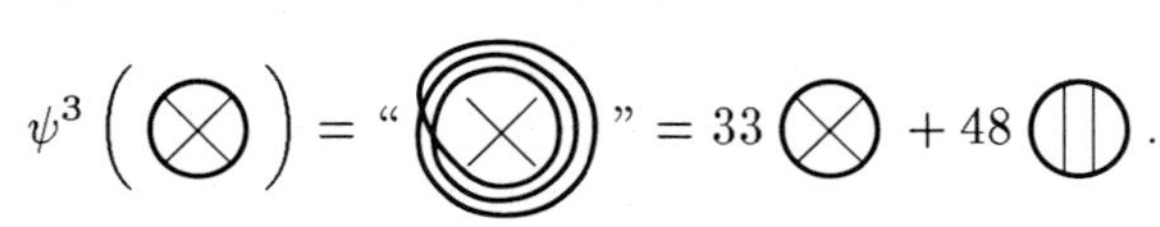

For m a negative integer, one takes ψ^{-m} of the diagram with the orientation of the Wilson loop reversed — note it is unknown whether reversing the orientation of the Wilson loop induces the identity map on ${}^f\!\mathcal{A}$ or not.

The Adams operations satisfy the relation $\psi^m \circ \psi^p = \psi^{mp}$, but they are not algebra maps, so in general $\psi^m(a.b) \neq \psi^m(a)\psi^m(b)$. However the $\{\psi^m\}$ are co-algebra maps and in fact ${}^f\!\mathcal{A}$ can be equipped with a different multiplication for which the $\{\psi^m\}$ are Hopf algebra maps — see below.

5.2. **Symmetrized diagrams and the disjoint union product.** In certain situations, such as doing calculations involving the Adams operations or for expressing the element Ω, it is convenient to use a basis of $\mathcal{A}$ which is an eigenbasis for the Adams operations; such a basis can be given in terms of *symmetrized diagrams* (in [2] they are called Chinese characters).

A symmetrized diagram is a graph with univalent vertices, and cyclicly ordered, trivalent vertices, such that each connected component of the graph has at least one univalent vertex. The univalent vertices are also called legs. The element of ${}^f\!\mathcal{A}$ that such a graph represents is obtained by averaging all possible ways of attaching the univalent vertices to a Wilson loop.
E.g.

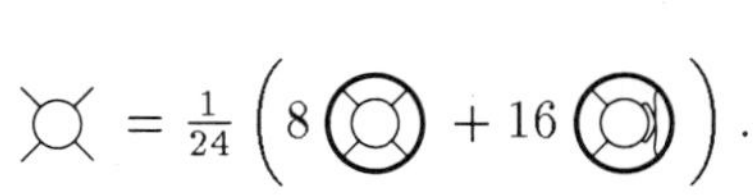

It is not difficult to show that these are eigenvectors for the Adams operations: a symmetrized diagram with u univalent vertices, is an eigenvector for the Adams operation ψ^p with eigenvalue p^u. Also it is not difficult to show that symmetrized diagrams span ${}^f\!\mathcal{A}$: one can use the easy fact that a diagram can be written as

[3] Note that this will not quite give a universal Vassiliev invariant of framed knots, a term involving the parity of the framing is required — see [11].

the sum of the symmetrized diagram obtained by removing the Wilson loop and symmetrized diagrams with fewer legs. If one defines the following IHX relation:

 : IHX,

then one can in fact take the vector space generated by symmetrized diagrams modulo the IHX and AS relations, and the above map to ${}^{f}\mathcal{A}$ induces an isomorphism. Note that the connected symmetrized diagrams span the primitive subspace of ${}^{f}\mathcal{A}$.

The disjoint union of symmetrized diagrams, $\amalg$, gives ${}^{f}\mathcal{A}$ a *second product* which is also compatible with the co-product; thus it gives ${}^{f}\mathcal{A}$ a second Hopf algebra structure. With respect to this second structure the Adams operations are Hopf-algebra maps. There seems not to be any topological interpretation of this disjoint union product; however, it can be related to the connect sum product by the (complicated) *wheeling* map [5].

5.3. **Wheels.** Certain symmetrized diagrams — the *wheels* — play a particularly important rôle in the theory, both in the Alexander polynomial (see Section 9) and the explicit formula for the element Ω described below.

If one considers the connected symmetrized diagrams of degree $n \geq 2$ (i.e. with a total of $2n$ vertices) then those with more than n univalent vertices are trivial in ${}^{f}\mathcal{A}$. If n is odd then those with n univalent vertices also vanish; but if n is even then define the non-trivial diagram w_n to be the *wheel with n legs*.
E.g.

$w_2 :=$ 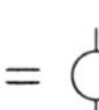 ; $\qquad w_4 :=$ 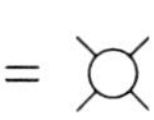 .

Each w_n spans the "top" eigenspace of the degree n primitive diagrams for the Adams operations.

These wheels can be used together with the disjoint union product to give a closed formula for the element $\Omega \in \mathcal{A}$ defined in Section 1. Firstly define the modified Bernoulli numbers, $\{b_{2n}\}_{n=1}^{\infty}$ via

$$\sum_{n=1}^{\infty} b_{2n} x^{2n} = \tfrac{1}{2} \ln \frac{\sinh(x/2)}{x/2}.$$

The following was conjectured in [4] and proved in [5]:

Theorem 6 (Bar-Natan, Le, Thurston). $\Omega = \exp_{\amalg}\left(\sum_{n=1}^{\infty} b_{2n} w_{2n}\right)$.

For instance, writing 1 for the empty symmetrized diagram of degree zero,

$\Omega = 1 + \frac{1}{48}$ 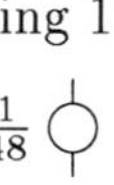 $- \frac{1}{5760}$ 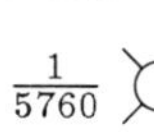 $+ \frac{1}{2.48^2}$ 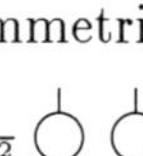 $+ \{\text{terms of degree } \geq 6\}$.

This will be needed in the calculation in Section 8.

6. Primitive elements as a module over Vogel's algebra Λ.

The space of primitive elements of $\mathcal{A}$ admits a module structure, introduced in [25], which turns out to be very useful for performing low order calculations by hand; it is also very useful in the context of weight systems coming from Lie algebras, and it was used by Vogel to prove that not all weight systems come from simple Lie super-algebras — this will not be addressed further here, but the reader is directed to [19, 12, 13, 25].

It is important to note that I am working over the rationals, so I denote by Λ what is denoted $\Lambda \otimes \mathbb{Q}$ in [25].

Define a Vogel diagram to be a graph with cyclicly oriented trivalent vertices and three univalent vertices labelled bijectively with the set $\{1, 2, 3\}$. A Vogel diagram has degree (#trivalent vertices − 1)/2. Consider the space of linear combinations of Vogel diagrams modulo the IHX and AS relations, then Λ is defined to be the subspace consisting of all elements u such that[4] if σ is a permutation of the label set then $\sigma(u) = \text{signature}(\sigma)u$.

Given a Vogel diagram and a connected diagram in $\mathcal{A}$, one can form a new connected diagram by putting the Vogel diagram in the place of a trivalent vertex, taking note of orientations. For example

The constraint on the elements of Λ ensures precisely that this descends to a well-defined action of Λ on the primitive elements of $\mathcal{A}$, for instance it does not depend on which trivalent vertex is replaced. Similarly Λ acts on itself, giving it a graded algebra structure.

Up to degree six, the algebra Λ is freely generated by the following elements:

$\frac{1}{2}x_1 := t :=$  ; $x_3 :=$  ; $x_5 :=$  .

It was conjectured by Vogel that Λ is freely generated by $\{x_{2i-1}\}_{i=1}^{\infty}$ but Kneissler [12] proved a relation between them in degree ten: it is unknown whether these elements generate Λ.

Kneissler also showed that in degrees less than six, the primitive space of $\mathcal{A}$ is freely generated over Λ by the diagrams  and 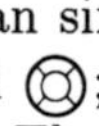; although more than just the round wheel diagrams are needed in degree six. The reader can check that this is precisely the form that my basis of primitive elements takes.

I do not know how to sensibly extend this to a Λ action on the whole of $\mathcal{A}$, nor how to interpret the action topologically.

7. Cabling knots and the theorem of Le

Cabling is a natural operation on framed knots, with respect to which the Kontsevich integral is particularly well behaved, as is demonstrated by the theorem of Le described below.

For m and p coprime integers and K a framed knot, the (m,p)-cable, $K^{(m,p)}$, of K can be defined by taking a blackboard framed knot diagram of K, drawing m parallels of the knot in the plane of the paper and inserting p copies of the tangle T_m shown in Figure 4(i), to obtain a blackboard framed knot diagram representing $K^{(m,p)}$ — see Figure 4(ii).

The algebraic structure of ${}^f\!\mathcal{A}$ described in the previous sections, together with the element Ω, permits the following description of how the framed Kontsevich integral behaves under cabling:

[4] If one is working over a ring in which 2 is not invertible then an extra condition is required on the elements of Λ, see [25].

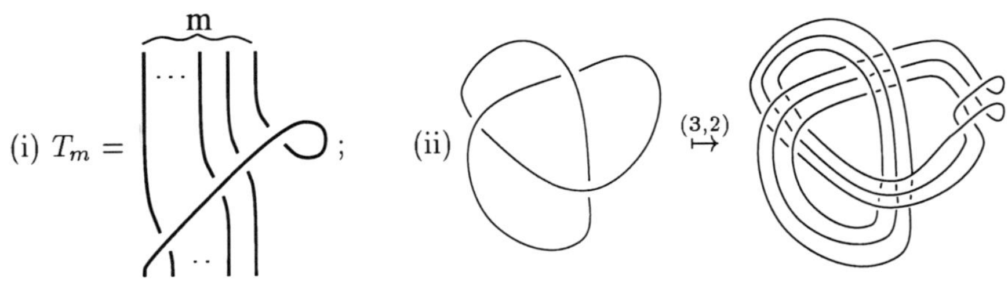

FIGURE 4. (i) The tangle used in the definition of the (m,p)-cable. (ii) The $(3,2)$-cable of a blackboard framed trefoil.

Theorem 7 (Le [17]). *For m and p coprime integers, and K a framed knot,*

$$ {}^fZ(K^{(m,p)}).\Omega = \psi^m\left({}^fZ(K).\Omega.\exp_{.}\left(\tfrac{p}{m}\tfrac{1}{2}\Phi\right)\right). $$

Note that if any of the weight system coming from a semi-simple Lie algebras is applied to both sides, then the consequent equality follows from an analysis of the theorem in [21] which Morton attributes to Strickland [24] and Rosso and Jones [23], and it was because of this that I conjectured this theorem in an earlier version of this note before discovering that Le had already proved it.

This theorem will be used in the next section to calculate the Kontsevich integral up to degree five for torus knots.

8. THE KONTSEVICH INTEGRAL AND VASSILIEV INVARIANTS OF TORUS KNOTS.

If m and p are coprime integers then the (m,p)-torus knot, denoted $T(m,p)$, is the knot type that has a representative which is embedded on the standard torus, wrapping m times around longitudinally and p times around meridinally. For instance, the positive trefoil is the $(2,3)$-trefoil. The (m,p)-torus knot can equivalently be described as the (m,p)-cable of the zero-framed unknot. This latter description allows the calculation of the Kontsevich integral of the (m,p)-torus knot via the previous theorem, as the framed Kontsevich integral of the zero-framed unknot is just the diagram in degree zero.

To calculate $Z(T(m,p))$ proceed as follows: convert the expression for Ω given in Section 5 to round diagrams using the change of basis data in Appendix A; form the connect sum product with $\exp_{.}\left(\frac{p}{m}\frac{1}{2}\Phi\right)$; convert back to symmetrized diagrams; apply ψ^m, recalling that symmetrized diagrams are eigenvectors; convert back to round diagrams and form the connect sum product with Ω^{-1}; finally change to the zero framing — this will require multiplying by $\exp_{.}(-mp\,\Phi/2)$.

This calculation is easily performed by hand up to degree four and the reader is strongly encouraged to do this. The following degree five result was obtained with a little bit of help from `maple`:

$Z\left(T\left(m,p\right)\right) = \exp_{.}\Big((m^2-1)(p^2-1)\Big(-\frac{1}{48}$  $+ \frac{mp}{288}$  $- \frac{9m^2p^2-m^2-p^2-1}{11520}$ 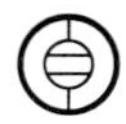

$+ \frac{(m^2+1)(p^2+1)}{5760}$ 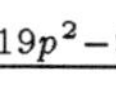 $+ \frac{mp(71m^2p^2-19m^2-19p^2-9)}{345600}$ 

$+ \frac{mp(m^2+1)(p^2+1)}{345600}$  $- \frac{mp(2m^2p^2+m^2+p^2)}{17280}$  $+ \dots\Big)\Big)$

Note that the coefficients of the round wheels,  and , are respectively $-(m^2-1)(p^2-1)/48$ and $(m^4-1)(p^4-1)/5760$; this will be generalized in the next section. First however, the methodology of this section leads to the next theorem which I had not been able to prove using more naïve methods — for example the methods discussed in [28].

Theorem 8. *If v is a rational, type n, Vassiliev knot invariant then, as a function of m and p, $v(T(m,p))$ is a polynomial, symmetric in m and p and of degree at most n in each of m and p.*

Proof. The symmetry in m and p follows from the fact that the (m,p)-torus knot and the (p,m)-torus knots are of the same knot-type (see [7]). Because the Kontsevich integral is universal for Vassiliev invariants it suffices to prove that in the Kontsevich integral the coefficient of a basis symmetrized diagram has the properties described in the theorem. Fix a basis of symmetrized diagrams of degree n. Pick any element of this basis. By the linearity of change of basis from round diagrams to symmetrized diagrams, the coefficient of this element in $\Omega.\exp_{.}(\frac{p}{m}\frac{1}{2}$ $)$ is a polynomial in $(\frac{p}{m})$ of degree at most n. The action of ψ^m will just be to multiply by some power of m. By the symmetry in m and p the result must be a symmetric polynomial of degree at most n in m and p. □

9. The Alexander polynomial, wheel coefficients and torus knots.

From work on the Melvin-Morton Theorem it is known that the Alexander polynomial of a knot corresponds to the coefficients of the wheels in its Kontsevich integral. This is used to identify the coefficients of the wheels for the torus knots.

9.1. Identifying the Alexander polynomial in the Kontsevich integral.

For a knot K, let $A_K(t)$ be the Alexander polynomial normalized so that $A_K(t) = A_K(-t)$ and $A_K(1) = 1$. Define the weight system $W_{\mathrm{AC}}\colon \mathcal{A} \to \mathbb{Q}[[h]]$ by first defining it on connected symmetrized diagrams as

$$W_{\mathrm{AC}}(D) = \begin{cases} -2h^{2n} & \text{if } D = w_{2n}, \\ 0 & \text{if } D \text{ has less than } \deg D \text{ legs,} \end{cases}$$

then extend it linearly to the space of primitive elements and multiplicatively, with respect to the connect sum product, to the whole of $\mathcal{A}$ (or equivalently [15], extend W_{AC} multiplicatively with respect to the II product).

The AC in W_{AC} stands for Alexander-Conway, the reason being given by the next theorem which follows from work on the Melvin-Morton Theorem.

Theorem 9 ([3, 16]). *If K is a knot then, as formal power series in h,*

$$A_K(e^h) = W_{AC} \circ Z(K).$$

This can be used to identify the coefficients of the wheels in the Kontsevich integral in the following manner. Take logarithms of both sides, as W_{AC} is multiplicative with respect to the disjoint union product, one finds

$$\ln A_K(e^h) = W_{\mathrm{AC}} \circ \ln_{\amalg} \circ Z(K).$$

From the definition of W_{AC} and the fact that $\ln_{\amalg} \circ Z(K)$ can be written as the sum of connected symmetrized diagrams, one deduces:

Theorem 10. *With respect to a polynomial basis of symmetrized diagrams, the coefficient of w_{2n} in $Z(K)$ is equal to $-\frac{1}{2}$ times the coefficient of h^{2n} in $\ln A_K(e^h)$.*

Note that if a basis of connected round diagrams is taken which in degree $2n$ consists of the round wheel together with diagrams with fewer univalent vertices, then the coefficient of the round wheel with n legs is the same as the coefficient of w_{2n} in the above theorem. This can now be applied to the case of torus knots.

9.2. **Wheels coefficients for torus knots.** For the torus knot $T(m,p)$ it is known that the Alexander polynomial is given by

$$A_{T(m,p)}(t) = \frac{(t^{mp/2} - t^{-mp/2})(t^{1/2} - t^{-1/2})}{(t^{m/2} - t^{-m/2})(t^{p/2} - t^{-p/2})}.$$

Substituting $t = e^h$, rewriting the bracketted terms as sinhs, dividing the numerator and the denominator of the right hand side by $h^2 mp$, taking logarithms of both sides and recalling the definition of the modified Bernoulli numbers $\{b_{2n}\}$ from Section 5, one obtains

$$\ln\left(A_{T(m,p)}(e^h)\right) = \sum b_{2n}(m^{2n} - 1)(p^{2n} - 1)2h^{2n}.$$

It then follows from the theorem above that the coefficient of w_{2n}, the wheel with $2n$ legs, in $Z(T(m,p))$ is precisely $-(m^{2n} - 1)(p^{2n} - 1)b_{2n}$. This can be seen to be in accord with the calculation of Section 8.

Acknowledgements.

Much of this work was done whilst I was supported by a Royal Society Research Fellowship at the University of Melbourne.

I would like to thank Sergei Chmutov and Andrew Kricker for useful conversations, Jacob Mostovoy and Jens Lieberum for comments on earlier versions of this paper, and Thang Le for sharing his unpublished work with me.

References

1. M. Alvarez and J. M. F. Labastida, *Vassiliev invariants for torus knots*, J. Knot Theory Ramifications **5** (1996), 779–803.
2. D. Bar-Natan, *On the Vassiliev knot invariants*, Topology **34** (1995), no. 2, 423–472.
3. D. Bar-Natan and S. Garoufalidis, *On the Melvin-Morton-Rozansky conjecture*, Invent. Math. **125** (1996), 103–133.
4. D. Bar-Natan, S. Garoufalidis, L. Rozansky, and D. P. Thurston, *Wheels, wheeling, and the Kontsevich integral of the unknot*, preprint, see q-alg/9703025, March 1997.
5. D. Bar-Natan, T. Q. T. Le, and D. Thurston, in preparation.

6. D. Bar-Natan and A. Stoimenow, *The fundamental theorem of Vassiliev invariants*, Proc. of the Aarhus Conf. Geometry and physics (New York) (J. E. Andersen, J. Dupont, H. Pedersen, and A. Swann, eds.), lecture notes in pure and applied mathematics 184, Marcel Dekker, 1997, pp. 101–134.
7. G. Burde and H. Zieschang, *Knots*, Studies in Mathematics, no. 5, de Gruyter, 1985.
8. S. V. Chmutov and S. V. Duzhin, *The Kontsevich integral*, Program System Institute (Pereslavl-Zalessky, Russia) preprint, 1997.
9. S. V. Chmutov and A. N. Varchenko, *Remarks on the Vassiliev knot invariants coming from* sl_2, Topology **36** (1997), 153–178.
10. O. T. Dasbach, *On the combinatorial structure of primitive Vassiliev invariants III — a lower bound*, Düsseldorf University preprint, see `math/9806086`.
11. C. Kassel and V. Turaev, *Chord diagram invariants of tangles and graphs*, Duke Math. J. **92** (1998), 497–552.
12. J. A. Kneissler, *The number of primitive Vassiliev invariants up to degree twelve*, University of Bonn preprint, see `q-alg/9706022`.
13. ______, *Relations in the algebra* λ, University of Bonn preprint, October 1997.
14. M. Kontsevich, *Vassiliev's knot invariants*, Adv. Sov. Math. **16** (1993), no. 2, 137–150.
15. A. Kricker, *Alexander-Conway limits of many Vassiliev weight systems*, J. Knot Theory Ramifications **6** (1997), 687–714.
16. A. Kricker, B. Spence, and I. Aitchison, *Cabling the Vassiliev invariants*, J. Knot Theory Ramifications **6** (1997), 327–358.
17. T. Q. T. Le, unpublished.
18. T. Q. T. Le and J. Murakami, *The universal Vassiliev-Kontsevich invariant for framed oriented links*, Compositio Math. **102** (1996), 41–64.
19. J. Lieberum, *On Vassiliev invariants not coming from semisimple Lie algebras*, University of Strasbourg preprint, see `q-alg/9706005`.
20. J. Milnor and J. Moore, *On the structure of Hopf algebras*, Ann. of Math. (2) **81** (1965), 211–264.
21. H. R. Morton, *The coloured Jones function and Alexander polynomial for torus knots*, Math. Proc. Cambridge Philos. Soc. **117** (1995), 129–135.
22. D. Quillen, *Rational homotopy theory*, Ann. of Math. (2) **90** (1969), 205–295.
23. M. Rosso and V. F. R. Jones, *On the invariants of torus knots derived from quantum groups*, J. Knot Theory Ramifications **2** (1992), 97–112.
24. P. M. Strickland, *On the quantum enveloping algebra invariants of cables*, University of Liverpool preprint, 1990.
25. P. Vogel, *Algebraic structures on modules of diagrams*, University of Paris VII preprint, 1995.
26. S. Willerton, *Vassiliev knot invariants and the Hopf algebra of chord diagrams*, Math. Proc. Cambridge Philos. Soc. **119** (1996), 55–65.
27. ______, *On universal Vassiliev invariants, cabling, and torus knots*, University of Melbourne preprint, 1998.
28. ______, *Vassiliev invariants as polynomials*, Knot Theory (V. F. R. Jones, J. Kania-Bartoszynska, J. H. Przytycki, P. Traczyk, and V. G. Turaev, eds.), Banach Center Publications, vol. 42, Polish Academy of Sciences, 1998, pp. 457–463.

Appendix A. Change of basis data.

Here is the change of basis data, up to degree five, between round diagrams and symmetrized diagrams, used in the calculation of Section 8. It was worked out by hand by expanding each symmetrized diagram and then using the STU relations to rewrite in terms of the round diagram basis. The Λ-module structure certainly simplified the calculations. The calculations took more than a week and the degree six result seems to be beyond the limit of my stamina. I don't know how much could be gained by automating the procedure, but I think there is still much structure to be exploited, for instance there is the following.

Note how sparse these matrices are. The presence of identity matrices along the diagonal is explained by the fact mentioned above that a round diagram can be written as the symmetrized diagram obtained by removing the Wilson loop plus diagrams with fewer legs. Le [17] conjectured that the connect sum of l round diagrams can be written as a linear combination of symmetrized diagrams with at least l legs: this would explain some of the zeroes in the bottom left corners. The bottom row could be calculated if it was known, as was also suggested by Le, that $\exp_{\cdot}(⦶/2) = \Omega \amalg \exp_{\amalg}(|/2)$. Also some of the zeroes seem to be reflecting Λ-module structure, but I can not yet make that more precise.

Degree zero:

 $= 1.$

Degree one:

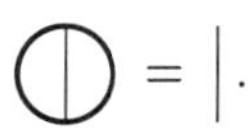

Degree two:

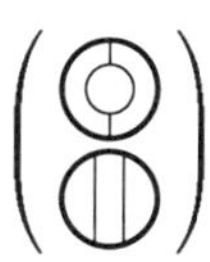 $= \begin{pmatrix} 1 & 0 \\ \frac{1}{6} & 1 \end{pmatrix}$ 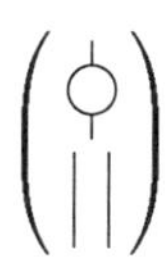 .

Degree three:

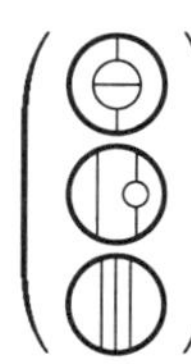 $= \begin{pmatrix} 1 & 0 & 0 \\ \frac{1}{3} & 1 & 0 \\ 0 & \frac{1}{2} & 1 \end{pmatrix}$ 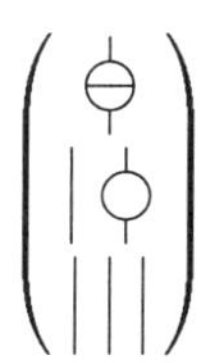 .

Degree four:

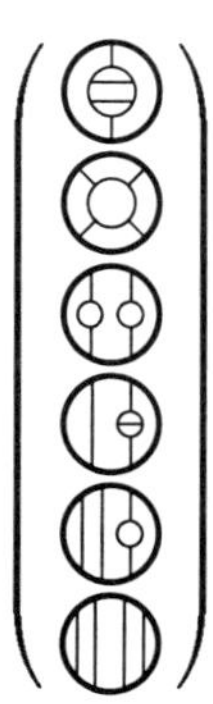 $= \begin{pmatrix} 1 & 0 & 0 & 0 & 0 & 0 \\ \frac{1}{3} & 1 & 0 & 0 & 0 & 0 \\ \frac{2}{3} & 0 & 1 & 0 & 0 & 0 \\ \frac{1}{3} & 0 & 0 & 1 & 0 & 0 \\ 0 & 0 & \frac{1}{6} & \frac{2}{3} & 1 & 0 \\ 0 & -\frac{1}{15} & \frac{1}{12} & 0 & 1 & 1 \end{pmatrix}$ 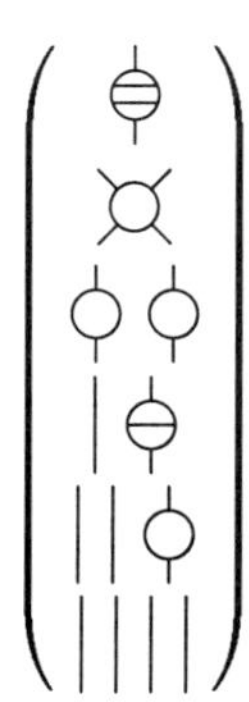 .

Degree five:

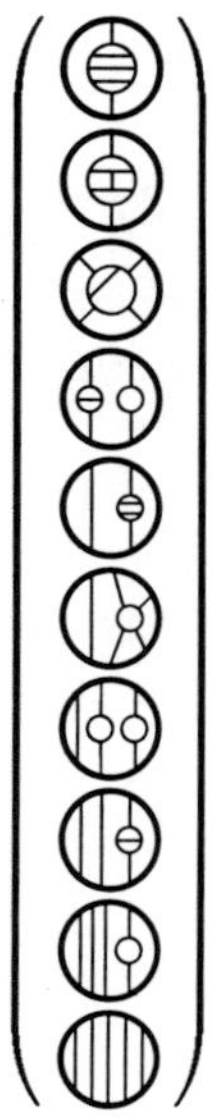

$$= \begin{pmatrix} 1 & 0 & 0 & 0 & 0 & 0 & 0 & 0 & 0 & 0 \\ 0 & 1 & 0 & 0 & 0 & 0 & 0 & 0 & 0 & 0 \\ \frac{1}{3} & 0 & 1 & 0 & 0 & 0 & 0 & 0 & 0 & 0 \\ \frac{2}{3} & 0 & 0 & 1 & 0 & 0 & 0 & 0 & 0 & 0 \\ \frac{1}{3} & 0 & 0 & 0 & 1 & 0 & 0 & 0 & 0 & 0 \\ \frac{2}{15} & -\frac{1}{30} & \frac{2}{3} & 0 & \frac{1}{3} & 1 & 0 & 0 & 0 & 0 \\ 0 & 0 & 0 & \frac{2}{3} & \frac{2}{3} & 0 & 1 & 0 & 0 & 0 \\ 0 & 0 & 0 & \frac{1}{6} & \frac{2}{3} & 0 & 0 & 1 & 0 & 0 \\ 0 & 0 & -\frac{2}{15} & \frac{1}{6} & 0 & 0 & \frac{1}{2} & 1 & 1 & 0 \\ 0 & 0 & 0 & 0 & 0 & -\frac{1}{3} & \frac{5}{12} & 0 & \frac{5}{3} & 1 \end{pmatrix}$$

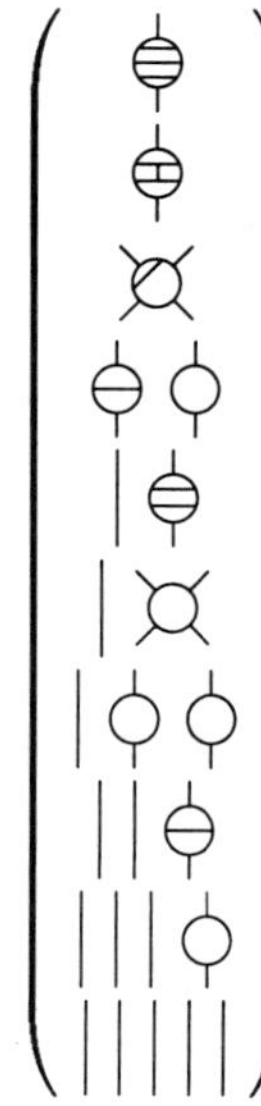

.

Appendix B. Comparison with Alvarez and Labastida's results

In [1], Alvarez and Labastida calculate Vassiliev invariants up to degree six for the torus knots by using the values on the torus knots of the HOMFLY, Kauffman and Akutsu-Wadati polynomials, together with a Chern-Simons approach to the universal Vassiliev invariant. Their method will not give all Vassiliev invariants of torus knots because the invariants coming from Lie algebras do not span the space of Vassiliev invariants [25] (although they do so up to at least degree twelve [12]). However, up to degree five, their results suggest that $\ln Z(T(m,p))$ might be the following:

$(m^2-1)(p^2-1)\Big(\frac{1}{6}$  $+ \frac{mp}{18}$  $+ \frac{9m^2p^2-m^2-p^2-1}{360}$ 

$+ \frac{(m^2+1)(p^2+1)}{360}$ 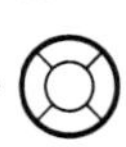 $+ \frac{mp(69m^2p^2-21m^2-21p^2-11)}{5400}$ 

$+ \frac{mp(11m^2p^2+m^2+p^2-9)}{5400}$  $+ \frac{mp(m^2+1)(p^2+1)}{900}$  $\Big)$.

Most of the basis element for $\mathcal{A}$ which they use differ only by a scalar from the basis elements used in the calculations above; the only non-trivial change of basis relation is[5]

 $= \frac{3}{2}$  $- \frac{1}{6}$  $- \frac{1}{12}$  .

[5] One of the signs given in this expression in [9] is incorrect.

Rewriting, in the basis of Alvarez and Labastida, the expression derived in Section 4, the following expression for $\ln_{.} Z(T(m,p))$ up to degree five is obtained:

$(m^2-1)(p^2-1)\Big(-\frac{1}{24}$  $+\frac{mp}{144}$  $-\frac{9m^2p^2-m^2-p^2-1}{5760}$ 

$+\frac{(m^2+1)(p^2+1)}{5760}$  $+\frac{mp(69m^2p^2-21m^2-21p^2-11)}{172800}$ 

$-\frac{mp(11m^2p^2+m^2+p^2-9)}{172800}$  $-\frac{mp(m^2+1)(p^2+1)}{28800}$ 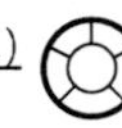 $\Big).$

Note that this expression differs from that of Alvarez and Labastida in the following two respects: for a diagram of degree n, there is a factor of 2^n difference — this is probably explained taking a different norm on the Lie algebras; and diagrams with an odd number of external vertices differ in sign — this is probably explained by an orientation convention.

PROBLEMS

Edited by H.R.Morton

Participants were asked to contribute problems from areas which were covered at the meeting. The rule of thumb suggested for proposing a problem was that it should have been thought about previously by the proposers themselves.

The resulting collection of problems, edited slightly and grouped roughly by topic, appears below. Other problems can also be found in the contributed articles.

PROBLEM 1 (H. Murakami). *Expansion of the Jones polynomial around* $t = -1$.

Let $V(L)$ be the Jones polynomial of a link L normalized so that $V = 1$ on the trivial knot. A link is called algebraically split if the linking number of each pair of components vanishes.

For an m-component algebraically split link L we denote by $L^{(k_1,k_2,\ldots,k_m)}$ the parallel link obtained by replacing the ith component with k_i parallels using the zero-framing.

M. Sakuma [4] proved that $V(L^{(k_1,k_2,\ldots,k_m)})$ is divisible by $(t+1)^{\frac{1}{2}\sum k_i}$ when all k_i are even.

The following generalisation to include some odd k_i is proved in [1].

Theorem 1.1 (H. Murakami). *Write* $k_i = 2n_i + 1$ *when* k_i *is odd and* $k_i = 2n_i$ *when* k_i *is even. Then* $V(L^{(k_1,k_2,\ldots,k_m)})$ *is divisible by* $(t+1)^{\sum n_i}$.

I conjecture that it is divisible by $(t+1)$ at least one more time. That is:

Problem 1.1. *Let* L *be an* m*-component zero-framed algebraically split link, and write* $k_i = 2n_i+1$ *when* k_i *is odd and* $k_i = 2n_i$ *when* k_i *is even. Then* $V(L^{(k_1,k_2,\ldots,k_m)})$ *is divisible by* $(t+1)^{1+\sum n_i}$, *provided that* $k_i > 2$ *for some* i.

If this is true then we can improve the result in [1]. In particular we can express the coefficient of $\xi - 1$ of the quantum $SU(2)$-invariant of a $\mathbb{Z}/p\mathbb{Z}$-homology three-sphere associated with non-trivial first cohomology class in terms of the orders of the first integral homology of the manifold and that of its double cover (in [1], it is proved that the quantum invariant is in $\mathbb{Z}[1/2,\xi]$ or in $\sqrt{-1}\,\mathbb{Z}[1/2,\xi]$ with ξ the primitive pth root of unity).

Remark 1.1. The conjecture is true when all k_i are even. It has also been verified for the Borromean rings with all $k_i \leq 4$.

I have tried in vain to prove the conjecture by using T. Ohtsuki's universal invariant [2]. More careful use of the technique used in [3, §6] might work.

Remark 1.2. After submitting the problem, I was informed by Kazuo Habiro that he can prove the following theorem.

Theorem 1.2 (K. Habiro). *If a link* ℓ *is the boundary of an* s*-component surface (possibly containing non-orientable components), then* $V(\ell)$ *is divisible by* $(t+1)^{s-1}$.

The following stronger divisibility result for boundary links can then be proved. For an m-component zero-framed *boundary link* B, $V(B^{(k_1,k_2,\ldots,k_m)})$ is divisible by $(t+1)^{(\sum k_i)-1}$. In particular, $V(K^k)$ is divisible by $(t+1)^{k-1}$ for any knot K, as suggested by H. Morton.

References

1. H. Murakami, *Quantum $SU(2)$-invariants for three-manifolds associated with non-trivial cohomology classes modulo two*, preprint, Waseda University and Mittag-Leffler Institute, 1999.
2. T. Ohtsuki, *Colored ribbon Hopf algebras and universal invariants of framed links*, J. Knot Theory Ramifications **2** (1993), 211–232.
3. ———, *A polynomial invariant of integral homology 3-spheres*, Math. Proc. Cambridge Philos. Soc. **117** (1995), no. 1, 83–112.
4. M. Sakuma, *An evaluation of the Jones polynomial of a parallel link*, Math. Proc. Cambridge Philos. Soc. **104** (1988), 105–113.

PROBLEM 2 (A. Stoimenow). *Q polynomial and Vassiliev invariants*

The Q polynomial contains the Vassiliev knot invariant of degree 2: in [K], Kanenobu proved the relation $Q'(-2) = V''(1)$ for the evaluation at -2 of the first derivative of the Q polynomial [BLM, Ho].

Problem 2.1. *Are the evaluations at -2 of the higher derivatives of Q also Vassiliev invariants? Is the n-th one of degree $2n$?*

Remark 2.1. Kanenobu gives a formula for $Q(x)$ for a 2-bridge link in terms of the Jones polynomial $V(t)$ of the link, with $x = -(t+t^{-1})$, [K2, theorem 1]. Explicitly the formula says that

$$x(Q(x)-1) = 2(V(t)V(t^{-1})-1).$$

By differentiating Kanenobu's formula at $t=1$ the derivatives of Q evaluated at -2 are polynomials in the derivatives of V evaluated at 1 on rational links and hence are Vassiliev invariants on them.

References

BL. J. S. Birman and X-S. Lin, *Knot polynomials and Vassiliev's invariants*, Invent. Math. **111** (1993) 225–270.

BLM. R. D. Brandt, W. B. R. Lickorish and K. Millett, *A polynomial invariant for unoriented knots and links*, Inv. Math. **74** (1986), 563–573.

Ho. C. F. Ho, *A polynomial invariant for knots and links – preliminary report*, Abstracts Amer. Math. Soc. **6** (1985), 300.

K. T. Kanenobu, *An evaluation of the first derivative of the Q polynomial of a link*, Kobe J. Math., **5** (1988), 179–184.

K2. ———, *Relations between the Jones and Q polynomials of 2-bridge and 3-braid links*, Math. Ann. **285** (1989), 115–124.

PROBLEM 3 (Gyo Taek Jin).

The crookedness of a knot K embedded in $\mathbb{R}^3$ with respect to a unit vector $\vec{v}$ is the number of connected components of the preimage of the set of local maximum values of the orthogonal projection $K \to \mathbb{R}\vec{v}$, denoted by $b_{\vec{v}}(K)$. The superbridge index of K, denoted by $s[K]$, is defined to be "min max $b_{\vec{v}}(K)$" where the maximum is taken over all unit vectors and the minimum taken over all equivalent embeddings of K. This invariant was introduced by Kuiper [6] who discovered that the torus knot of type (p,q) with $2 \leq p < q$ has superbridge index $\min\{2p,q\}$. The bridge index $b[K]$ can be defined in a similar way by "$\min\min b_{\vec{v}}(K)$."

For any nontrivial knots, the following three conditions hold [7, 2, 6, 5], where $\beta[K]$ is the braid index of K.

$$b[K_1 \sharp K_2] = b[K_1] + b[K_2] - 1 \tag{0.1}$$

$$b[K] < s[K] \leq 2\beta[K] \tag{0.2}$$

$$s[K_1 \sharp K_2] \leq \max\{2\beta[K_1] + \beta[K_2], \beta[K_1] + 2\beta[K_2]\} - 1 \tag{0.3}$$

Problem 3.1. *For any nontrivial knots K_1 and K_2, prove the inequalities*

$$s[K_1 \sharp K_2] \leq \max\{s[K_1] + b[K_2], b[K_1] + s[K_2]\} - 1 \tag{0.4}$$

$$s[K_1 \sharp K_2] \geq \max\{s[K_1] + b[K_2], b[K_1] + s[K_2]\} - 1 \tag{0.5}$$

Remark 3.1. If K_i is a torus knot of type (p_i, q_i) with $2 \leq p_i < q_i/2$, for $i = 1, 2$, the conditions (0.3) and (0.4) are identical. In general, (0.4) is sharper than (0.3) and is satisfied by all torus knots [5].

The following is a weaker version of (0.4).

Problem 3.2. *For any nontrivial knots K_1 and K_2, prove the inequality*

$$s[K_1 \sharp K_2] \leq s[K_1] + s[K_2] - 2 \tag{0.6}$$

Let $p[K]$ denote the minimal number of edges required to form a polygonal embedding of K. For any nontrivial knots, the following conditions hold [1, 4].

$$2s[K] \leq p[K] \tag{0.7}$$

$$p[K_1 \sharp K_2] \leq p[K_1] + p[K_2] - 3 \tag{0.8}$$

Problem 3.3. *For any nontrivial knots K_1 and K_2, prove the inequality*

$$p[K_1 \sharp K_2] \leq p[K_1] + p[K_2] - 4 \tag{0.9}$$

If K_i is a torus knot of type (p_i, q_i) with $2 \leq p_i < q_i < 2p_i$, then (0.9) holds [4]. The inequalities (0.7) and (0.9) were used to show that (0.6) is satisfied for such torus knots.

The equality in (0.4) and (0.5) holds if K_i is a torus knot of type $(p_i, p_i + 1)$, for $p_i \geq 2$ and $i = 1, 2$, in particular, for the granny knot and the square knots. It also holds for the composite knot of a trefoil knot and a figure eight knot [4, 5]. The inequality (0.5) can be weakened as follows.

Problem 3.4. *For any nontrivial knots K_1 and K_2, prove the inequality*

$$s[K_1 \sharp K_2] \geq \max\{s[K_1], s[K_2]\} + 1 \tag{0.10}$$

Remark 3.2. If (0.10) is true, then there are infinitely many $(2n-1)$-superbridge knots for each $n \geq 3$ [3]. By Kuiper's result we know that there are infinitely many $2n$-superbridge torus knots for each $n \geq 2$.

References

1. C.C. Adams, B.M. Brennan, D.L. Greilsheimer and A.K. Woo, *Stick numbers and composition of knots and links*, J. Knot Theory Ramifications 6 (1997), 149–161.
2. H. Doll, *A generalized bridge number for links in 3-manifolds*, Math. Ann. 294 (1992), 701–717.
3. C.B. Jeon and G.T. Jin, *There are only finitely many 3-superbridge knots*, to appear in J. Knot Theory Ramifications (special issue of *Knots in Hellas 1998, Delphi*).
4. G.T. Jin, *Polygon indices and superbridge indices of torus knots and links*, J. Knot Theory Ramifications 6 (1997), 281–289.
5. G.T. Jin, *Superbridge index of composite knots*, preprint.
6. N. Kuiper, *A new knot invariant*, Math. Ann. 278 (1987), 193–209.
7. H. Schubert, *Über eine numerische Knoteninvariante*, Math. Zeit. 61 (1954), 245–288.

PROBLEM 4 (N. Askitas). *Unknotting operations.*

Grids.

A *grid local move* of type $(d_1, r_1) \times (d_2, r_2)$ is that local move which makes the substitution of the tangles in Figure 1 below.

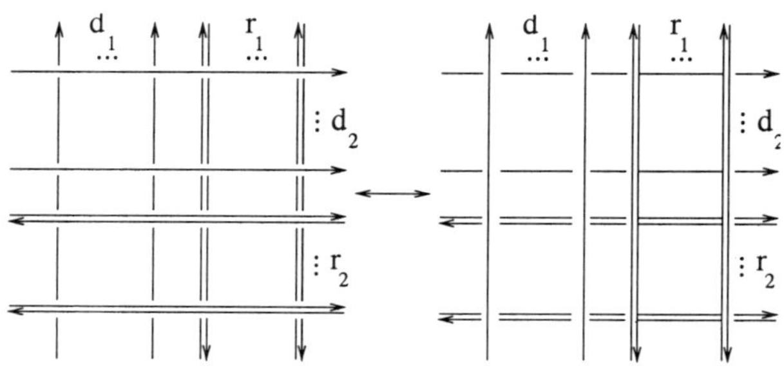

Figure 1.

Problem 4.1. *Which grid local moves are unknotting operations?*

I conjecture that they all are (modulo possibly Arf Invariant issues). In [A1] it is shown that grids of type $(2,0) \times (n,0)$ are unknotting operations for any integer $n \geq 0$. In [A2] it is proven that grids of type $(3,0) \times (n,0)$ and $(4,0) \times (n,0)$ are also unknotting operations for any $n \geq 0$ it is then not hard to see that any grid local move of type $(p,0) \times (q,0)$ with $\gcd(p,q) = 1,2,3,4$ is an unknotting operation (modulo the fact that a grid of type $(4,0) \times (2n,0)$ preserves the Arf Invariant). In [A3] I was able to prove that grids of type $p \times q$ with $\gcd\{p,q\} = 6$ also unknot modulo the fact that when $pq \equiv 0 \bmod 8$ the Arf invariant is preserved.

Twisting.

Let a $p\,\Xi\,(n)$-local move be that move which gives n $\pm$-full twists on p similarly oriented strings (so that a $2\,\Xi\,(1)$ move is nothing but the ordinary crossing change).

Problem 4.2. *For which values of p is a $p\,\Xi\,(1)$ move an unknotting operation?*

Remark 4.1. For $p = 2$ it is the ordinary crossing change. For $p = 3$ it is a result of Y.Ohyama [O] and for $p = 4$ it is proved by this author in [A1].

Additivity.

Let λ be an unknotting operation. It defines a metric on the space $\mathcal{K}$ of all knots by defining $d_\lambda(K, J)$ for any two knots K, J to be equal to the minimum number of λ moves necessary to pass from K to J. The λ-unknotting number $u_\lambda(K)$ of a knot K is then nothing but its distance $d_\lambda(K, O)$ to the unknot O. In general if $K\#J$ is the connected sum of knots we have

$$u_\lambda(K\#J) \leq u_\lambda(K) + u_\lambda(J)$$

It is an old conjecture that for the ordinary crossing change the unknotting number is additive. In [A4] I gave examples which imply the non-additivity of the unknotting number of a host of unknotting operations (basically all the ones I knew). The one operation I failed to show it for is the Δ-move.

Problem 4.3. *Is u_Δ additive, i.e. is*

$$u_\Delta(K\#J) = u_\Delta(K) + u_\Delta(J)?$$

References

A1. N. A. Askitas, *Multi-# unknotting operations: A new family of local moves on a knot diagram and related invariants of knots.* J. Knot Theory Ramifications, 7 (1998), 857-871.

A2. N. A. Askitas, *Grids as unknotting operations.* Submitted to the Proceedings of the Delphi Conference.

A3. N. A. Askitas, *A grid of type $p \times q$ with* gcd $\{p, q\} = 6$ *unknots.* Preprint.

A4. N. A. Askitas, *λ-unknotting number one knots need not be prime.* Talk delivered at the Knots In Hellas 98 conference in Delphi, Greece.

O. Y. Ohyama, *Twisting and unknotting operations.* Revista Mat. Univ. Comp. Madrid, 7 (1994), 289-305.

PROBLEM 5 (A. Stoimenow). *Vassiliev invariants and some conjectures on braid representations.*

The origin of the problems I'd like to propose is the following

Problem 5.1. *Prove that*

$$\dim\{\ \textit{Vassiliev invariants of degree} \leq k\ \} \leq C^k$$

for some constant $C > 1$.

As the usual approach of counting chord diagrams [CD] does not look very promising, in [St3] I started a different method which gave solutions in cases where Vassiliev invariants are restricted to some knot classes, e. g. arborescent knots. For many of these classes it might well be that they cover all knots, and this is the origin of the diagramss I propose here. Hence, any one of the conjectures in the following implies a positive solution of the problem and they are all of the form "some class covers all knots" or equivalently "any knot has the property used to describe this class". The reasons why these conjectures solve the problem are explained in [St, St2].

I start with a conjecture related to a conjecture of Przytycki [Pr] (quoted, *inter alia*, in Rob Kirby's problem book [Kr]). A diagram is called *matched* if its crossings

can be paired up so as to look like

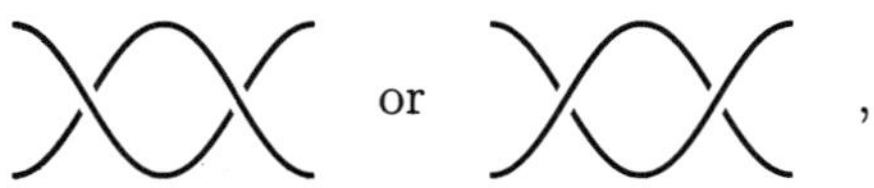

where orientation of the strands is arbitrary.

So, in contrast to Przytycki, I allow parallel orientation of the strands. A matched diagram according to Przytycki's definition I shall call reversely matched. Thus a diagram is *reversely matched* if it is matched and the orientation of the strands in every pair is reverse.

The method works for knots with matched diagrams: in [St] it was proved that if one restricts Vassiliev invariants to the class $\mathcal{M}$ of knots with matched diagrams then

$$\dim\{\text{ Vassiliev invariants of degree} \leq k \text{ on } \mathcal{M}\} \leq C^k .$$

But possibly Przytycki's conjecture [Pr], that there are knots with no reversely matched diagram, is true, so the following should be treated with some caution.

Problem 5.2. *Prove that any knot has a matched diagram.*

To get possibly more realistic problems, we turn to braids.

Trying a different approach, it was possible in [St] to construct large classes of knots via braids, for which the exponential bound holds (in the same sense as for knots with matched diagrams). A problem related to the first such class is like a braid version of problem 5.2:

Problem 5.3. *Prove that any knot is the closure of a braid*

$$\gamma = \sigma_1 \cdot \dots \cdot \sigma_{n-1} \cdot \prod_{i=1}^{r} \sigma_{k_i}^{p_i} \in B_n,$$

where $n, r, k_i \geq 1$ and all p_i even.

Several weaker versions of problem 5.3 suffice:

- replace σ_{k_i} by β_{k_i,j_i} where $\beta_{k,j}$ is the positive band (in Rudolph's [Ru] sense) interchanging strands k and j, under the restriction that for all i we have $|k_i - j_i| < C$ for some global constant C, or more generally
- replace $\sigma_{k_i}^{p_i}$ by any pure braid with crossing number (that is, the minimal length of words in the Artin generators representing it) $< C$ for some global constant C.

Problem 5.4. *Prove that any knot is the closure of a braid*

$$\gamma = \prod_{i=1}^{r} \sigma_{p_i}^{r_i} \in B_n,$$

for some $n, r \in \mathbf{N}$ with $p_i > 0, r_i \in \mathbf{Z}$, such that even (index) generators appear with exactly one odd power (that is, for each even $1 \leq p < n$ there is exactly one i with $p_i = p$ and r_i odd).

Or more generally the following weaker property suffices: $\exists \{c_i\}_{i\in\mathbf{N}}$ with $c_{i+1} - c_i \in [1, C]$ for some global constant C with C and c_i independent of γ, s. t. $\forall k$:

σ_{c_k} appears with exactly one odd power in (5.4) (that is, for each $k \in \mathbf{N}\ \exists!i$ with $p_i = c_k$ and r_i odd).

Problem 5.5. *Any knot is the closure of a braid*

$$\gamma = \prod_{i=1}^{r} \sigma_{p_i}^{r_i},$$

such that $|\{\, 1 \leq i < r \,:\, p_{i+1} < p_i + 2 \,\}| < C$ *for some global constant* $C \in \mathbf{N}$.

These are the braids presentable by (stacking up) at most C "blocks" (of arbitrary but equal strand number) involving a product of commuting generators as in figure 2.

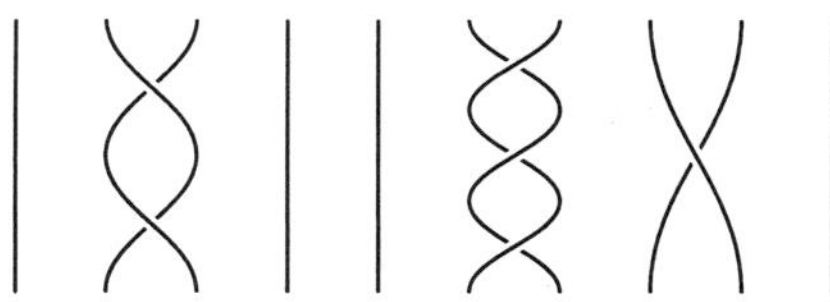

Figure 2. The block corresponding to $\sigma_2^{-2}\sigma_6^3\sigma_8^{-1}$

In any of the previous problems 'any knot' may be replaced by 'any alternating knot' or 'any closed positive braid knot'.

References

CD. S. V. Chmutov and S. V. Duzhin, *An upper bound for the number of Vassiliev knot invariants*, Jour. of Knot Theory and its Ramifications, **3(2)** (1994), 141–151.

Kr. R. Kirby, *Problems of low-dimensional topology*, book available on `http://math.berkeley.edu/~kirby`.

Pr. J. Przytycki, *t_k moves on links*, in "Braids", Santa Cruz, 1986 (J. S. Birman and A. L. Libgober, eds.), Contemp. Math. **78**, 615–656.

Ru. L. Rudolph, *Braided surfaces and Seifert ribbons for closed braids*, Comment. Math. Helv. **58** (1983), 1–37.

St. A. Stoimenow, *The braid index and the growth of Vassiliev invariants*, preprint.

St2. ———, *On finiteness of Vassiliev invariants and a proof of the Lin-Wang conjecture via braiding polynomials*, preprint.

St3. ———, *Gauß sum invariants, Vassiliev invariants and braiding sequences*, preprint.

PROBLEM 6 (J. Przytycki). *Skein polynomials based on the cubic relation.*

The following problems are about the existence and uniqueness of invariants of unoriented links which satisfy a cubic braid relation (as does the Kauffman 2-variable polynomial).

Problem 6.1. (1) *Show that there is a polynomial invariant of unoriented links,* $P_1(L) \in Z[x,t]$ *which satisfies:*

(i) Initial conditions: $P_1(T_n) = t^n$, *where* T_n *is a trivial link of* n *components.*

(ii) Skein relation: $P_1(L_0) + xP_1(L_1) - xP_1(L_2) - P_1(L_3) = 0$ *where* L_0, L_1, L_2, L_3 *is a standard, unoriented skein quadruple* (L_{i+1} *is obtained from* L_i *by a right-handed half twist on two arcs involved in* L_i).

(2) *Show that there is a polynomial invariant of unoriented framed links,* $P_2(L) \in Z[A^{\pm 1}, t]$ *which satisfies:*
   (i) Initial conditions*:* $P_2(T_n) = t^n$*,*
   (ii) Framing relation*:* $P_2(L^{(1)}) = -A^3 P_2(L)$ *where* $L^{(1)}$ *is obtained from a framed link* L *by a positive twist on its framing.*
   (iii) Skein relation*:*

$$P_2(L_0) + A(A^2 + A^{-2})P_2(L_1) + (A^2 + A^{-2})P_2(L_2) + AP_2(L_3) = 0.$$

(3) *Show that there is a polynomial invariant of unoriented framed links,* $P_3(L) \in Z[a^{\pm 1}, x, y]$ *which satisfies:*
   (i) Initial conditions*:*

$$P_3(T_n) = \left(\frac{-a^3(x + y + xy + x^2) + a^7(x + y + 1)^2 - a^{-1}}{x + y + xy + y^2} \right)^n,$$

   (ii) Framing relation*:* $P_3(L^{(1)}) = aP_3(L)$*,*
   (iii) Skein relation*:*

$$P_3(L_0) + axP_3(L_1) + a^2 yP_3(L_2) - a^3(x + y + 1)P_3(L_3) = 0.$$

(4) *In all cases (1)-(3), show that the polynomial is uniquely defined.*

Remark 6.1. Uniqueness for $P_1(L)$ and $x = 0$ is equivalent to the Montesinos-Nakanishi conjecture on 3-moves on links [1].

Remark 6.2. If P_2 exists then it distinguishes the knot 9_{42} from its mirror image $\bar{9}_{42}$. The knots 9_{42} and $\bar{9}_{42}$ share the same Homfly and Kauffman polynomials.

References

1. H.R. Morton, Problems, in *Braids*, Ed. J.S.Birman, A.Libgober, AMS Contemporary Math., 78(1988), 557-574.

PROBLEM 7 (H. Morton). *Conjugacy of positive permutation braids.*

Let $\pi \in S_n$ be a permutation. The *positive permutation braid* $\beta_\pi \in B_n$ is the positive n-braid determined by the following properties:

- It is a positive braid whose ith string joins the point i at the top to the point $\pi(i)$ at the bottom.
- Every pair i, j of strings crosses at most once.

Equivalently, positive permutation braids are exactly the 'fragments' of the half-twist braid Δ_n, see [2].

Problem 7.1. *Recognise 'geometrically' when two positive permutation braids are conjugate in* B_n.

Remark 7.1. Two braids β and γ are conjugate if and only if the oriented links consisting of the closure of the braid together with the braid axis are equivalent. Consequently link invariants of these two links will help to sort the braids into conjugacy classes.

In this problem I have in mind a more algorithmic criterion.

Remark 7.2. If β_π and $\beta_{\pi'}$ are conjugate then certainly π and π' must be conjugate in S_n. By the extension of Garside's conjugacy algorithm in [2] it will be possible to pass from β_π to $\beta_{\pi'}$ by a sequence of other positive permutation braids, each conjugate to the next by a positive permutation braid.

It is clear that the number of crossings, $c(\beta)$, in the braid is a conjugacy invariant. Indeed the closure of each permutation braid is fibred, and the fibre surface has Euler characteristic $\chi = n - c(\beta)$.

Remark 7.3. It would be interesting to refine the calculations by decomposing the braids into 'fragments' of the braid δ_n considered in [1].

REFERENCES

1. J.S.Birman, K.H.Ko and S.J.Lee, *A new approach to the word problem in the braid groups.* Preprint 1997, Columbia University. To appear in Adv. Math.
2. E.A.Elrifai and H.R.Morton, *Algorithms for positive braids.* Quart. J. Math. Oxford (2), 45 (1994), 479-497.

PROBLEM 8 (A. Stoimenow). *Genus and the Homfly polynomial.*

In [Be] Bennequin showed that if a knot is presented as the closure of an n-string braid β then the Euler characteristic of any orientable spanning surface F for the knot satisfies the inequality $w(\beta) - n \leq -\chi(F)$, where $w(\beta)$ is the writhe of β. The Bennequin number $b(K)$ of a knot K, defined as $b(K) = \max(w(\beta) - n + 1)$, over all braid presentations of K, then satisfies $b(K) \leq 2g(F)$ and hence $b(K) \leq 2g(K)$ where $g(K)$ is the genus of K.

When the Homfly polynomial $P_K(v, z)$ of a knot K is written as a Laurent polynomial in v the lowest degree $\min\deg_v P$ in v satisfies the inequality $b(K) \leq \min\deg_v P$. In [Mo] Morton conjectured that $\min\deg_v P \leq 2g(K)$.

In the accompanying table are the simplest possible candidates for a counterexample to this conjecture.

Problem 8.1. *Decide whether any of the following thirteen knots (recorded in Dowker-Thistlethwaite notation* [DT]*) has a Seifert surface of genus* $\max\deg\Delta$.

Such a knot would be a counterexample to an inequality conjectured by Morton [Mo]: $2g \geq \min\deg_v P$. Unfortunately, for any of them surfaces of the Seifert algorithm of genus $\max\deg\Delta$ do not exist. The knots below are the only examples in Thistlethwaite's tables [HTW] with ≤ 16 crossings, for which the conjectured inequality of Morton is false when replacing $2g$ by any of its lower bounds 2, $\max\deg\nabla$, $|\sigma|$ or the bound of (writhe - number of Seifert circles + 1) applied to all the diagrams of the knot that Thistlethwaite's tool `knotfind` generates from the diagram included in the table. (This last number is in turn a lower bound for the Bennequin number $b(K)$, using Yamada's construction of a closed braid presentation from the given diagram).

The four numbers $\max\deg\nabla = 2\max\deg\Delta$, $\min\deg_v P$, $\max\deg_z P$ and (writhe – number of Seifert circles + 1) are recorded after the code for the knot at the end of each entry.

16_{494921} :	4	8	16	2	18	−22	−24	6	14	28	−12	−10	32	30	20	26	4 6 6 4
16_{575572} :	4	10	12	−16	2	8	−22	−6	−26	−30	−24	−14	−20	−32	−18	−28	4 6 6 4
16_{575703} :	4	10	12	−16	2	8	−24	−6	−28	−30	−32	−26	−14	−22	−20	−18	4 6 6 4
16_{585206} :	4	10	12	−18	2	8	−22	−26	−6	−32	−30	−14	−28	−16	−24	−20	4 6 6 4
16_{585619} :	4	10	12	−18	2	8	−24	−26	−6	−30	−32	−28	−16	−14	−22	−20	4 6 6 4
16_{592788} :	4	10	12	−20	2	8	−24	−26	−28	−6	−32	−30	−18	−16	−14	−22	4 6 6 4
16_{614650} :	4	10	−14	16	−30	2	−18	−24	22	−12	−28	−6	−32	−8	−20	−26	4 6 6 4
16_{788939} :	4	12	14	−16	−22	2	10	−20	−26	−6	−32	−28	−30	−18	−24	−8	6 8 8 6
16_{851600} :	4	12	18	16	22	2	26	10	6	−32	8	28	30	14	24	−20	6 8 8 6
16_{883841} :	4	12	−20	−16	−22	2	18	−28	−8	32	−10	−6	−30	−14	−24	−26	6 8 8 6
$16_{1007218}$:	4	14	26	−18	−24	−22	2	12	−30	28	−32	−10	16	20	6	−8	2 4 6 2
$16_{1116868}$:	6	8	14	2	−18	−28	4	24	−12	30	26	32	16	20	−10	22	6 8 8 6
$16_{1118397}$:	6	8	14	2	22	−18	4	24	26	−30	−28	32	16	−12	−20	10	6 8 8 6

This list is also available on [St2] or in diagram form, on request.

Problem 8.2. *A similar appealing inequality to the one conjectured by Morton is* $\max \deg_z P \geq 2g(K)$*. Is this always true? Possible counterexamples are the four knots in figure 9 of* [St].

Remark 8.1 (Editor). For comparison, it is known that $2\hat{g}(K) \geq \max \deg_z P$ where $\hat{g}(K)$ is the smallest genus of any spanning surface which can be obtained by applying Seifert's algorithm to a diagram of K. While this is the same as $g(K)$ for knots with alternating or positive diagrams, it can be substantially greater in other cases. Stoimenow adds the comment that the only prime examples which he knows in which $2\hat{g}(K) > \max \deg_z P$, a necessary condition for a counterexample to problem 8.2, are the four knots quoted above.

It is also known that $\min \deg_v P \leq \max \deg_z P$, so there is no immediate obstruction to the two conjectures. In all the cases above, these two are equal, so the genus must be given exactly by them if both conjectures are to be satisfied.

References

Be. D. Bennequin, *Entrelacements et équations de Pfaff*, Astérisque, **107-108** (1983), 87–161.

DT. C. H. Dowker and M. B. Thistlethwaite, *Classification of knot projections*, Topol. Appl. **16** (1983), 19–31.

HTW. J. Hoste, M. Thistlethwaite and J. Weeks, *The first 1,701,936 knots*, Math. Intell. **20 (4)** (1998), 33–48.

Mo. H. Morton, *Problems*, in "Braids", Santa Cruz, 1986 (J. S. Birman and A. L. Libgober, eds.), Contemp. Math. **78**, 557–574.

St. A. Stoimenow, *Knots of genus two*, preprint.

St2. ———, web document `http://www.informatik.hu-berlin.de/~stoimeno/mexs.dat`.

PROBLEM 9 (M. Thomé). *Brunnian links.*

An n-component link is called *Brunnian* if every proper sublink is trivial.

Problem 9.1. *Construct for each n a Brunnian link for which one component L_1 lies on the boundary of the standard genus $(n-1)$ handlebody lying in the complement of the remaining $n-1$ components in such a way that when any of these components* *is removed then L_1 bounds a disc on the boundary of the handlebody of genus $n-2$ formed by filling in the corresponding hole.*

Remark 9.1 (Editor). The Hopf link and the Borromean rings can be made to look like this, and the proposer gives an example for $n = 4$. There may well be an inductive construction in which a neighbourhood of L_1 and a band going round the extra hole are combined to produce a curve in the larger handlebody having the right properties. It is not clear whether there is any way to do this so that every component may be chosen as L_1.

PROBLEM 10 (J. Przytycki, M. Veve). *Billiard knots.*

A *billiard room*, M, is a Riemannian 3-manifold with a PL-smooth boundary. A *billiard knot* K is a simple closed trajectory of a point (thought of as a billiard ball) along geodesic lines in M subject to perfectly elastic reflection at smooth points of ∂M. For a given M let $\mathcal{K}(M)$ be the family of knots supported by M. For a given knot K let $\mathcal{M}(K)$ be the family of manifolds supporting K.

There are two fundamental problems in the theory of billiard knots:

Problem 10.1. (a) *Analyze* billiard knots, $\mathcal{K}(M)$, *in any Riemannian manifold* M.

(b) *Describe the manifolds, $\mathcal{M}(K)$, which support a given knot K.*

We define a transitive relation, $\leq$, on manifolds by the condition that $M_1 \leq M_2$ if $\mathcal{K}(M_1) \subset \mathcal{K}(M_2)$. We also define a transitive relation on knots by $K_1 \leq K_2$ if every manifold which supports K_2 also supports K_1.

In further problems we limit ourself to the set, $\mathcal{N}$, of compact connected submanifolds of R^3.

Problem 10.2. *Is the relation $\leq$ on $\mathcal{N}$ reflexive, i.e. is it true that if $M_1 \leq M_2$ and $M_1 \geq M_2$ then $M_1 = M_2$?*

The expected answer is negative.

Problem 10.3. *Is D^3 the minimal element with respect to the relation $\leq$? Is it minimal among convex manifolds? Is D^3 smaller than any finite polyhedron? Any finite convex polyhedron?*
Characterize the manifolds $\mathcal{M}(T_1)$ which support the trivial knot T_1.

Problem 10.4. (a) *Is there a manifold $M \in \mathcal{N}$ which supports every knot?*

(b) [2] *Is there a finite polyhedron which supports every knot?*

(c) [2] *Is there a polytope (finite convex polyhedron) which supports every knot?*

(d) *Is every knot supported by one of the Platonic solids?*

(e) [3] *Is every knot supported by an elliptical prism.*

Remark 10.1. If we relax the condition of finiteness in (b) we can build a connected simply-connected infinite polyhedron supporting every knot.

Remark 10.2. We know [4, 1, 2] that an infinite number of knots are supported by a cube and by a cylinder but they all satisfy some symmetry properties (or ribbon condition).

Problem 10.5. [5] *For any polytope P, show that $\mathcal{K}(P)$ is infinite.*

Problem 10.6. *Let $b = s_{i_1}s_{i_2}...s_{i_k}$ be an n-braid with $n > 1$ odd, and $b' = s_{j_1}^{-1}s_{j_2}^{-1}...s_{j_k}^{-1}$ where $j_t = n - i_t$. Show that the closure of the n-braid bb' is not a billiard knot in a cube.*

Remark 10.3. Notice that this knot is strongly + amphicheiral.

References

1. M.G.V.Bogle, J.E.Hearst, V.F.R.Jones, L.Stoilov, *Lissajous knots*, Journal of Knot Theory and Its Ramifications, 3(2), 1994, 121-140.
2. V.F.R.Jones, J.H.Przytycki, *Lissajous knots and billiard knots*, Banach Center Publications, Vol. 42, *Knot Theory*, 1998, 145-163.
3. C.Lamm, *There are infinitely many Lissajous knots*, Manuscripta Mathematica, 93, 1997, 29-37.
4. C.Lamm, D.Obermeyer, *A note on billiard knots in a cylinder*, Journal of Knot Theory and Its Ramifications, 8(3), 1999, 353-366.
5. J.H.Przytycki, Symmetric knots and billiard knots, in *Ideal Knots*, Vol. 19 in Series on Knots and Everything, Ed. A.Stasiak, V.Katrich, L.Kauffman, World Scientific, 1999, 374-414.

PROBLEM 11 (J. Scott Carter). *Gauss words.*

An *(unsigned) Gauss word* is a sequence taken from letters $a_1, \ldots a_n$ in which each letter appears exactly twice. For example *abcabc*, *abab*, and *abcbac* are unsigned Gauss words. Gauss words are associated to generically immersed curves in surfaces. The association is obtained by starting at a point of the curve and recording the crossing sequence. Gauss words are applicable to virtual knots. The *genus* of a Gauss word is the minimal genus (oriented) surface in which the Gauss word is the word associated to an immersed curve.

Problem 11.1. *Develop a linear-time algorithm that is analogous to the algorithm of Read and Rosentiehl* [3] *that will determine the genus of a Gauss word.*

Remark 11.1. Note there is an obvious exponential time algorithm which assigns opposing exponents to the pairs of letters in the words, computes the genus, and finds the one of minimal genus. In the examples above, *abcabc* is a genus 0 word; the words *abab* and *abcbac* are genus 1.

Problem 11.2. *Develop a census of Gauss words. I have in mind a Monte Carlo simulation in which a random (signed) Gauss word of a fixed number of crossings is constructed and the genus is determined. What percentage of Gauss words with crossing number n fit on a surface of genus g?*

Problem 11.3. *Mix Gauss codes and Morse codes (pun intended) to study the difference between virtual knots and knots that are embedded in thickened surfaces.*

Here Morse codes would encrypt maximum and minimum points of a curve where a height function is chosen on the pair consisting of a surface and an immersed curve. Morse codes have the potential to distinguish curves that are related by Dehn twists.

References

1. J. Scott Carter, *Classifying immersed curves*, Proc. Amer. Math. Soc. 111 (1991), 281–287.
2. C. F. Gauss, Werke VIII, 271-286.
3. R. C. Read and P. Rosenstiehl, *On the Gauss crossing problem*. Combinatorics, (Proc. Fifth Hungarian Colloq., Keszthely, 1976), Vol. II, pp. 843–876, Colloq. Math. Soc. Janos Bolyai, 18, North-Holland, Amsterdam-New York, 1978.

PROBLEM 12 (Adam S. Sikora). *Action of cyclic groups on three-dimensional manifolds.*

It can be shown that if a cyclic group $\mathbb{Z}_p$, p prime, acts on a closed orientable surface F with fixed points then the number of such fixed points is determined by the induced $\mathbb{Z}_p$-action on $H_1(F, \mathbb{F}_p)$. Here, $\mathbb{F}_p$ denotes the field of p elements. In order to state this result precisely, it is useful to introduce the following piece of terminology. If N is an $\mathbb{F}_p[\mathbb{Z}_p]$-module i.e. an F_p-linear space with a $\mathbb{Z}_p$-action, then we will define the H^1-rank of N to be

$$\dim_{\mathbb{F}_p} H^1(\mathbb{Z}_p, N).$$

Theorem 12.1. *If $\mathbb{Z}_p$, p prime, acts on a closed, orientable surface F, by orientation preserving homeomorphisms and if $F^{\mathbb{Z}_p} \neq 0$, then this action has $2 +$ rk $H^1(F, \mathbb{F}_p)$ fixed points.*

The proof of this result is an easy application of the localization theorem for (Tate) equivariant cohomology.

Surprisingly, no analogous results for group actions on 3-manifolds are known. Let $\mathbb{Z}_p$, $p \neq 2$ prime, act on a closed, orientable manifold M. It can be shown show that the fixed point set of such an action is a disjoint union of circles. Denote their number by c and assume that that $c \neq 0$. One can show using equivariant cohomology that $c \leq 1 +$ rk $H_1(M, \mathbb{F}_p)$. We propose the following:

Problem 12.1. *Show that* $c = 1 +$ *rk* $H_1(M, \mathbb{F}_p)$ *mod 2*

Remark 12.1. We have proved the above result in the case when the $\mathbb{Z}_p$-action on $H_1(M, \mathbb{F}_p)$ is either trivial, or free, or it decomposes into a sum of free and trivial actions [1]. The proof relies on a detailed analysis of the differentials in the Leray-Serre spectral sequence convergent to $M \times_G EG$, where $G = \mathbb{Z}_p$.

In particular, if the $\mathbb{Z}_p$ action on $H_1(M, \mathbb{F}_p) = \mathbb{F}_p^k$ is trivial then

$$c \leq k + 1 \quad \text{and} \quad c = k + 1 \bmod 2.$$

Moreover, one can show that given any pair (c, k), $c \geq 1$, satisfying the above two conditions there exists a manifold M with $H_1(M, \mathbb{F}_p) = \mathbb{F}_p^k$, and a $\mathbb{Z}_p$-action on M with $M^{\mathbb{Z}_p} = \underbrace{S^1 \cup \ldots \cup S^1}_{k}$. Therefore the result in the problem cannot be improved.

Remark 12.2. The result is false for $p = 2$.

References

1. A. S. Sikora, Cyclic group actions on manifolds and Poincare duality on spectral sequences, in preparation.

ORGANISING COMMITTEE

C.McA. GORDON, University of Texas at Austin,
V.F.R. JONES, University of California at Berkeley,
S. LAMBROPOULOU, University of Göttingen,
S. NEGREPONTIS, University of Athens,
J.H. PRZYTYCKI, George-Washington University.

LIST OF PARTICIPANTS

D. ANDERMANN, Goettingen Univ., Germany. dorothee.andermann@sap.com
E. APPLEBOIM, Technion Univ., Israel. eaea@techunix.technion.ac.il
L. ARMAS-SANABRIA, UNAM, Mexico. larmas@math.unam.mx
A. AROUCHE, King Saud Univ., Saudi Arabia. aarouche@ksu.edu.sa
N. ASKITAS, MPI Bonn, Germany. askitas@mpim-bonn.mpg.de
D. BAR-NATAN, Hebrew Univ., Israel. drorbn@math.huji.ac.il
G. BARAD, Univ. of Bucharest, Rumania. gefry@chow.mat.jhu.edu
A. BELIAKOVA, Univ. of Berne, Switzerland. beliak@math-stat.unibe.ch
M. BELOLIPETSKY, Novosibirsk State Univ., Russia. mbel@math.nsc.ru
S. BENVENUTI, Univ. of Pisa, Italy. benvenut@paley.dm.unipi.it
A.B. BERGER, Univ. of Berne, Switzerland. bergerab@math-stat.unibe.ch
J. BIRMAN, Columbia Univ., USA. jb@cpw.math.columbia.edu
S. BOYER, Univ. de Quebec a Montreal, Canada. boyer.steven@uqam.ca
T.E. BRENDLE, Columbia Univ., USA. tbrendle@math.columbia.edu
D. BUCK, Univ. of Texas, USA. dbuck@fireant.ma.utexas.edu
G. BUCK, Saint Anselm College, USA. gbuck@anselm.edu
J.A. CALVO, Williams College, USA. jacalvo@math.ucsb.edu
S. CARTER, Univ. of South Alabama, USA. carter@knots.mathstat.usouthal.edu
C. CERF, Univ. libre de Bruxelles, Belgium. ccerf@ulb.ac.be
N. CHBILI, Monastir Univ., Tunisia. chbili@esstt.rnu.tn
T. COCHRAN, RICE Univ., USA. cochran@math.rice.edu
O. COLLIN, Harvard Univ., USA. collin@math.harvard.edu
D. COOPER, UCSB, USA. cooper@math.ucsb.edu
P. CONTASTEFANO, .Dept. of Physics, Univ. of Rome, Italy.
I. DARCY, Univ. of Texas at Dallas, USA. darcy@utdallas.edu
O. DASBACH, Univ. California, Riverside, USA. kasten@math.ucr.edu
O. DEARRICOTT, State Univ. Stony Brook, USA. dearrico@math.sunysb.edu
F. DELOUP, Univ. de Strasbourg I, France. deloup@irma.u-strasbg.fr
R. DIAZ, Univ. Complutense de Madrid, Spain. radiaz@eucmos.sim.ucm.es
A. DERUELLE, Univ. de Provence a Marseille, France. deruelle@gyptis.univ-mrs.fr
A. DOBAY, Univ. Lausanne, Switzerland. Akos.Dobay@lau.unil.ch

M. EISERMANN, Bonn Univ., Germany. uzs7j0@math.uni-bonn.de
J. ELLEGAARD ANDERSEN, Aarhus Univ., Denmark. andersen@mi.aau.dk
M. EUDAVE-MUNOZ, Inst. de Math. UNAM, Mexico. mario@math.unam.mx
V. FARMAKI, Univ. of Athens, Greece. vgeorgil@atlas.uoa.gr
T. FIEDLER, Univ. Paul Sabatier, Toulouse, France. fiedler@picard.ups-tlse.fr
E. FOMINYKH, Chelyabinsk State Univ., Russia. fominykh@csu.ac.ru
S. GAROUFALIDIS, Harvard Univ. USA. stavros@math.brandeis.edu
R. GELCA, Univ. of Michigan, USA. rgelca@math.lsa.umich.edu
B. GEMEIN, Univ. Düsseldorf, Germany. gemein@rz.uni-duesseldorf.de
C. GILLE, Univ. of Nantes, France. gille@math.univ-nantes.fr
C.McA. GORDON, Univ. Texas at Austin, USA. gordon@math.utexas.edu
N. HABEGGER, Univ. of Nantes, France. habegger@math.univ-nantes.fr
C. HAYASHI, Gakushuin Univ., Japan. 19952761@gakushuin.ac.jp
C. HOG-ANGELONI, Frankfurt Univ., Germany. cyn@math.uni-frankfurt.de
P. HOIDN, Univ. de Lausanne, Switzerland. phoebe.hoidn@lau.unil.ch
M. HUTCHINGS, Stanford Univ., USA. hutching@math.stanford.edu
T. IKEDA, Univ. of Tokyo, Japan. caa64440@pop06.odn.ne.jp
S. JABLAN, Belgrade Univ., Yugoslavia (Serbia). jablans@mi.sanu.ac.yu
G.T. JIN, Korea Adv. Inst. Sci. Technol., S. Korea. trefoil@math.kaist.ac.kr
V.F.R. JONES, Univ. Berkeley, USA. vfr@math.berkeley.edu
E. KALFAGIANNI, Michigan State Univ., USA. ekal@math.rutgers.edu
N. KAMADA, Osaka City Univ., Japan. naoko@sci.osaka-cu.ac.jp
S. KAMADA, Osaka City Univ., Japan. kamada@sci.osaka-cu.ac.jp
T. KANENOBU, Osaka City Univ., Japan. kanenobu@sci.osaka-cu.ac.jp
J. KANIA-BARTOSZYNSKA, Boise State Univ., USA. kania@diamond.idbsu.edu
Y. KASAHARA, Kochi Univ. Technology, Japan. kasahara@ele.kochi-tech.ac.jp
A. KATAVOLOS, Univ. of Athens, Greece. akatavol@eudoxos.dm.uoa.gr
L. KAUFFMAN, Univ. of Illinois at Chicago, USA. kauffman@uic.edu
A. KAWAUCHI, Osaka City Univ., Japan. kawauchi@sci.osaka-cu.ac.jp
T. KAWAMURA, Univ. of Tokyo, Japan. kawamura@ms.u-tokyo.ac.jp
E. KEYMAN, Univ. of Osmangazi, Turkey. ekeyman@bim.fen.ogu.edu.tr
J. KNEISSLER, Philips Res. Labs Aachen, Germany. jkneiss@pfa.research.philips.com
K.H. KO, Korea Adv. Inst. Sci. and Technology, S. Korea. knot@knot.kaist.ac.kr
K. KOBAYASHI, Tokyo Woman's Christian Univ., Japan. kazuaki@twcu.ac.jp
U. KOSCHORKE, Siegen Univ., Germany. koschorke@mathematik.uni-siegen.de
C. KOUROUNIOTIS, Univ. of Crete, Greece. christos.kourouniotis@kcl.ac.uk
S. KOLD HANSEN, Univ. of Aarhus, Denmark. kold@mi.aau.dk
K. KYPARISSIDOU, Univ. Göttingen, Germany. Kypariss@math.uni-goettingen.de
S..LAMBROPOULOU, Univ. Göttingen, Germany. sofia@uni-math.gwdg.de
C. LAMM, Bonn Univ., Germany. lamm@math.uni-bonn.de
S. LANDO, Indep. Univ. Moscow, Russia. lando@lando.mccme.rssi.ru

W. LI, Oklahoma State Univ., USA. wli@littlewood.math.okstate.edu
T. LE, SUNY Buffalo, USA. letu@math.buffalo.edu
S. LOMONACO, CSEE Dept., Univ. of Maryland, USA. lomonaco@umbc.edu
M. LUSTIG, Univ. Aix-Marseille III, France. Martin.lustig@math.u-3mrs.fr
M. MACKAAY, Univ. of Algarve, Portugal. mmackaay@mozart.si.ualg.pt
P.G. MANCHON EUIT Industrial U.P.M, Spain. pmanchon@ma.upm.es
G. MASBAUM, CNRS, Paris VII, France. masbaum@math.jussieu.fr
D. MATEI, Northeastern Univ., USA. dmatei@mail1.dac.neu.edu
Y. MATHIEU, Univ. Provence á Marseille, France. mathy@gyptis.univ-mrs.fr
D. MATIGNON, Univ. de Provence á Marseille, France. matignon@math.utexas.edu
A. MEDNYKH, Novosibirsk Univ., Russia. mednykh@math.nsc.ru
J. MIGHTON, Univ. of Toronto, Canada. mighton@math.toronto.can
Y. MILADI, Univ. Quebec á Montreal, Canada. yassir@math.uqam.ca
D. MILLER, William Paterson Univ., USA. damiller@pilot.njin.net
K. MILLETT, Univ. of California, USA. millett@math.ucsb.edu
H. MIYAZAWA, Tsuda College, Japan. aida@tsuda.ac.jp
H. MORTON, Liverpool Univ., UK. su14@liverpool.ac.uk
J. MOSTOVOY, UNAM, Mexico. jacob@matcuer.unam.mx
Y. MORIAH, Technion, Israel. ymoriah@techunix.technion.ac.il
K. MOTEGI, Nihon Univ., Japan. motegi@math.chs.nihon-u.ac.jp
T. MOTOHASHI, Waseda Univ., Japan. tomoe@mn.waseda.ac.jp
K. MURASUGI, Univ. of Toronto, Canada. murasugi@math.toronto.edu
H. MURAKAMI, Waseda Univ., Japan. hitoshi@uguisu.co.jp
Y. NAKAGAWA, Yamaguchi Pref. Univ., Japan. Nakagawa@ws1.yawaguchi-pu.ac.jp
Y. NAKANISHI, Kobe Univ., Japan. nakanisi@math.kobe-u.ac.jp
S. NEGREPONTIS, Univ. of Athens, Greece. snegrep@eudoxos.dm.uoa.gr
F. NEUMANN, Univ. Goettingen, Germany. neumann@uni-math.gwdg.de
J. O'HARA, Tokyo Metropolitan Univ., Japan. ohara@comp.metro-u.ac.jp
M. OKAMOTO, Waseda Univ., Japan. miyuki3@mn.waseda.ac.jp
M. OZAWA, Waseda Univ., Japan. oto@yyy.or.jp
E. PEREZ, Univ. Santiago de Compostela, Spain. esther@fpaxp1.usc.es
A. PILZ, Siegen Univ., Germany. pilz@mathematik.uni-siegen.de
M. POLYAK, Tel-Aviv Univ., Israel. polyak@math.tau.ac.il
R. POPESCU, Columbia Univ., USA. radu@shire.math.columbia.edu
J. PRZYTYCKI, George-Washington Univ., USA. przytyck@math.gwu.edu
T. PRZYTYCKA, John Hopkins Univ., USA. przytyck@grserv.med.jhmi.edu
D. RADFORD, Univ. of Illinois at Chicago, USA. U21523@UICVM.UIC.EDU
M. RAMPICHINI, Univ. of Milano, Italy. rampichini@vmimat.mat.unimi.it
R. RANDELL, Univ. of Iowa, USA. richard-randell@uiowa.edu
R. RICCA, Univ. College London, UK. renzo.ricca@jrc.it
J. ROBERTS, Edinburgh Univ., UK. justin@maths.ed.ac.uk

W. ROSICKI, Gdansk Univ., Poland. wrosicki@ksinet.univ.gda.pl
B. RUINI, Univ. of Modena, Italy. ruini@unimo.it
N. SAYARI, UQAM, Canada. sayari@math.uqam.ca
M. SCHARLEMANN, UC Santa Barbara, USA. mgscharl@math.ucsb.edu
A. SHIMA, Univ. of Tokyo, Japan. shima@ms.u-tokyo.ac.jp
K. SHIMOKAWA, Univ. of Tokyo, Japan. BZM03125@niftyserve.or.jp
H. SHORT, Univ. de Provence a Marseille, France. hamish@claudia.univ-mrs.fr
A. SIKORA, Univ. of Maryland at College Park, USA. asikora@math.umd.edu
D. SILVER, Univ. of South Alabama, USA. silver@mathstat.usouthal.edu
J. SIMON, Univ. of Iowa, USA. jsimon@math.uiowa.edu
J. SINK, Univ. of Michigan jsink@umich.edu
T. STANFORD, US Naval Academy, USA. stanford@nadn.navy.mil
E. STARR, Vassar College, USA. starr@vassar.edu
A. STASIAK, Lausanne Univ., Switzerland. Andrzej.Stasiak@lau.unil.ch
I. STASSEN, Univ. of Berne, Switzerland. stassen@math-stat.unibe.ch
A. STOIMENOW, MPI Bonn, Germany. alex@mpim-bonn.mpg.de
M. SOKOLOV, George-Washington Univ., USA. sokolov@math.gwu.edu
De W. SUMNERS, Florida State Univ., USA. sumners@zeno.math.fsu.edu
K. SUZUKI, Univ. of Tokyo, Japan. koutarou@ms.u-tokyo.ac.jp
S. SUZUKI, Waseda Univ., Tokyo, Japan. sssuzuki@mn.waseda.ac.jp
K. TANIYAMA, Tokyo Woman's Christian Univ., Japan. taniyama@twcu.ac.jp
M. TEICHER, Bar-Ilan Univ., Israel. teicher@macs.biu.ac.il
M. TERAGAITO, Hiroshima Univ., Japan. mteraga@sed.hiroshima-u.ac.jp
M THOME, Bibliotheque Publique d' Information, France. grupedumardi@europost.org
P. TRACZYK, Warsaw Univ., Poland. traczyk@mimuw.edu.pl
K. TRUSHKIN, Moscow State Univ., Russia. Kostya@trushkin.dnttm.rssi.ru
T. TSUKAMOTO, George-Washington Univ., USA. tsuka@tkg.att.ne.jp
Y. UCHIDA, Yamagata Univ., Japan. yuchida@sci.kj.yamagata-u.ac.jp
V. VERSHININ, Novosibirsk Univ., Russia. versh@math.nsc.ru
M. VEVE, George-Washington Univ., USA. mikev@hunterlab.com
P. VOGEL, Univ. Paris VII, France. vogel@math.jussieu.fr
B. WAJNRYB, Technion, Israel. wajnryb@techunix.technion.ac.il
R. WEIDMANN, Ruhr-Univ., Bochum, Germany. richard.weidmann@gmx.de
B. WIEST, Univ. British Columbia, Canada. bertw@pims.math.ca
S. WILLERTON, IRMA, Strasbourg, France. willert@math.u-strasbg.fr
S. WILLIAMS, Univ. South Alabama, USA. williams@knots.mathstat.usouthal.edu
N. WRINKLE, Columbia Univ., USA. wrinkle@math.columbia.edu
Y. YAMADA, Univ. of Electro-Communication, Japan. yyyamada@matha.e-one.uec.ac.jp
D. YETTER, Kansas State Univ., USA. dyetter@math.ksu.edu
J. ZABLOW, CUNY, New York, USA. jzablow@gc.cuny.edu
M. ZINNO, Columbia Univ., USA. Matzinno@math.columbia.edu

adidas

GREECE